Advanced Signal Analysis and its Applications to Mathematical Physics

Advanced Signal Analysis and its Applications to Mathematical Physics

HARISH PARTHASARATHY
Division of Electronics and Communication
Netaji Subhash Institute of Technology
New Delhi

I.K. International Publishing House Pvt. Ltd.
NEW DELHI • BANGALORE

Published by

I.K. International Publishing House Pvt. Ltd.
S-25, Green Park Extension, Uphaar Cinema Market
New Delhi – 110 016 (India)
E-mail: info@ikinternational.com
website: www.ikbooks.com

ISBN 978-93-80026-54-1

Reprint 2013

Published by Krishan Makhijani for I.K. International Publishing House Pvt. Ltd., S-25, Green Park Extension, Uphaar Cinema Market, New Delhi – 110 016 and Printed by Rekha Printers Pvt. Ltd., Okhla Industrial Area, Phase II, New Delhi – 110 020.

Preface

The mathematical tools used in signal analysis involve differential and difference equations, integral equations, matrix algebra and calculus, complex analysis and probability theory and random processes. This book applies these tools to problems in various branches of physics like fluid dynamics, electromagnetism and quantum theory. The book will be of use to research workers in signal processing as well as to research workers in physics and applied mathematics. Partial differential equations have been introduced here as an additional tool in signal analysis since they are used to describe quantum, electromagnetic and fluid dynamical phenomena not to forget Einstein's equations of gravitation. The book will be of use to signal processing experts who are interested in developing tools for the analysis of signals arising in real systems.

I am grateful to Prof. Raj Senani for his constant encouragement and support for providing the lab facilities. I am also grateful to Tarun Kumar Rawat for helping me with the typesetting work.

Harish Parthasarathy

Contents

Chapter 1

Probability Theory

1. Martingale inequalities:

Let X_n be a discrete time integrable submartingale on the finite set $\{0, 1, ..., N\}$. Let

$$T = inf\{n : X_n \geq \lambda\}$$

if this set is nonempty and if empty, set $T = N$. T is evidently a stop time and hence by the optional stopping theorem, we have

$$\mathbb{E}(X_N) \geq \mathbb{E}(X_T) = \mathbb{E}(X_T.\chi_{sup_n X_n \geq \lambda}) + \mathbb{E}(X_T.\chi_{sup_n X_n < \lambda})$$

Now on the set $\{sup_n X_n \geq \lambda\}$, we have $X_T \geq \lambda$ and on the set $\{sup_n X_n < \lambda\}$, we have $T = N$. So this inequality becomes

$$\mathbb{E}(X_N) \geq \lambda.P(sup_n X_n \geq \lambda) + \mathbb{E}(X_N.\chi_{sup_n X_n < \lambda})$$

from which we conclude that

$$\mathbb{E}(X_N \chi_{sup_n X_n \geq \lambda}) \geq \lambda.P(sup_n X_n \geq \lambda)$$

In particular, if X_n is a martingale, then replacing X_n by $|X_n|$ (which is a submartingale, see remark below), it follows that

$$\mathbb{E}(X_N) \geq \lambda.P(sup_n X_n \geq \lambda)$$

Remark:Suppose X_n is a martingale. $\phi(x) = |x|$ is a convex function over $\mathbb{R}$. Hence by Jensen's inequality,

$$\mathbb{E}(|X_{n+1}||\mathcal{F}_n) \geq |\mathbb{E}(X_{n+1}|\mathcal{F}_n)| = |X_n|$$

proving that $|X_n|$ is a submartingale to which the above inequality can be applied. Suppose X_n is a positive submartingale. Then also we get from the above result,

$$\mathbb{E}(X_N) \geq \lambda.P(sup_n X_n \geq \lambda)$$

Suppose X is a martingale or a positive submartingale indexed by the finite set $\{0, 1, ..., N\}$. If $p \geq 1$, then $|X_n|^p$ is a submartingale. Indeed, we first note that $\phi(x) = |x|^p$ is convex on $\mathbb{R}$ and hence if X_n is a martingale,

$$\mathbb{E}(|X_{n+1}|^p|\mathcal{F}_n) \geq |\mathbb{E}(X_{n+1}|\mathcal{F}_n)|^p = |X_n|^p$$

on the other hand, suppose X_n is a positive submartingale, then by convexity of $\phi(x) = |x|^p$ on $\mathbb{R}$, we have

$$\mathbb{E}(|X_{n+1}|^p|\mathcal{F}_n) \geq |\mathbb{E}(X_{n+1}|\mathcal{F}_n)|^p = \mathbb{E}(X_{n+1}|\mathcal{F}_n))^p$$
$$\geq X_n^p = |X_n|^p$$

It follows from the above inequality that

$$\mathbb{E}|X_N|^p \geq \lambda.P(sup_n|X_n| \geq \lambda)$$

Now let X be martingale or a positive submartingale. Set

$$X^* = sup_n|X_n|$$

We have just proved that

$$\lambda.P(X^* \geq \lambda) \leq \mathbb{E}(|X_N|\chi_{X^* \geq \lambda})$$

Thus, for a fixed $k > 0$, we have

$$\mathbb{E}(min(X^*, k)^p) = \mathbb{E}(\int_0^{min(X^*,k)} p\lambda^{p-1}d\lambda) = \mathbb{E}(\int_0^k p\lambda^{p-1}\chi_{X^* \geq \lambda})$$

$$\int_0^k p\lambda^{p-1}P(X^* \geq \lambda)d\lambda \leq \int_0^k p\lambda^{p-2}\mathbb{E}(|X_N|\chi_{X^* \geq \lambda})$$

$$= p\mathbb{E}(|X_N| \int_0^{min(X^*,k)} \lambda^{p-2}d\lambda) = \frac{p}{p-1}\mathbb{E}(|X_N|(min(X^*, k)^{p-1})$$

By Holder's inequality, we have for non-negative Y,

$$\mathbb{E}(|X_N|Y^{p-1}) \leq \mathbb{E}(|X_N|^p)^{1/p}\mathbb{E}(Y^p)^{(p-1)/p}$$

and hence, we get from the above,

$$\mathbb{E}(min(X^*, k)^p) \leq (\mathbb{E}|X_N|^p)^{1/p}\mathbb{E}(min(X^*, k)^p)^{1-1/p}$$

giving

$$\mathbb{E}(min(X^*, k)^p) \leq (p/(p-1))^p\mathbb{E}|X_N|^p$$

Letting $k \to \infty$ gives us the fundamental inequality

$$\mathbb{E}(X^{*p}) \leq (p/(p-1))^p\mathbb{E}|X_N|^p$$

Remark:The above inequalities admit immediate generalizations to continuous martingales.

2. Stochastic differential equations; Existence and uniqueness of solutions;

Let $(\Omega, \mathcal{F}_t, P)$ be a filtered probability space and Z a continuous semi-martingale. Let f, g be defined as follows: We denote the Wiener space $C(\mathbb{R})$ (continuous functions on the real line) by $\mathbf{W}$. Let $w(s,\omega)$ be the coordinate process on $\mathbf{W}$, ie, $w(t,\omega) = \omega(t)$. Define $\mathcal{B}_t = \sigma(w(s), s \leq t)$. $f : \mathbb{R}_+ \times \mathbf{W} \to \mathbb{R}$ is predictable is assumed to be predictable with respect to the filtration $\mathcal{B}_t$. Assume that g is also of the same kind. Assume further that the following Lipshitz condition is satisfied:

$$|f(t,w) - f(t,w')| + |g(t,w) - g(t,w')| \leq K.sup_{s \leq t}|w(s) - w'(s)|$$

If $y \in \mathbb{R}$ is any given number and we define the process $\bar{y}(s) = y \forall s \geq 0$, then we further assume that

$$\mathbb{E}(\sup_{s \leq t}(\int_0^s f(u, \bar{y}du)^2)) < \infty$$

Using these conditions, we wish to show that there is a unique process X such that

$$X_t = x + \int_0^t f(x, X_.)dZ_s$$

and that X is an adapted process. Let U be a well behaved process, so that the stochastic integral $\int_0^t f(s,U)dZ_s$ is defined. Let V be another well behaved process. Define

$$\Phi_t(U,V) = \mathbb{E}(sup_{s \leq t}|U_s - V_s|^2)$$

Define

$$(SU)_t = x + \int_0^t f(s,U)dZ_s, (SV)_t = x + \int_0^t f(s,V)dZ_s$$

Then we have

$$\Phi_t(SU,SV) = \mathbb{E}\{sup_{s \leq t}(\int_0^t (f(s,U) - f(s,V))dM_s + \int_0^t (f(s,U) - f(s,V))dA_s)^2\}$$

where

$$Z_t = M_t + A_t$$

is the Doob-Meyer decomposition of the semimartingale Z into the sum of a martingale and a predictable process. Note that $A_t = A_{1t} - A_{2t}$ where A_1, A_2 are predictable increasing processes. Hence, A is a process of finite variation. Using the inequality $(a+b)^2 \leq 2(a^2 + b^2)$ and $sup(a_t + b_t) \leq supa_t + supb_t$, it follows that

$$\Phi_t(SU,SV) \leq 2\mathbb{E}sup_{s \leq t}(\int_0^t (f(s,U) - f(s,V))dM_s)^2 + 2\mathbb{E}sup_{s \leq t}(\int_0^t |f(s,U) - f(s,V)||dA_s|)^2$$

We now use Doob's inequality in the following form: Let X be a continuous martingale, then for $p \geq 1$

$$\mathbb{E}sup_{s \leq t}|X_s|^p \leq (p/(p-1))^p \mathbb{E}|X_t|^p$$

Specializing to $p = 2$ gives

$$\mathbb{E}sup_{s \leq t}|X_s|^p \leq 4\mathbb{E}|X_t|^2$$

We take as our martingale

$$X_t = \int_0^t (f(s,U) - f(s,V))dM_s$$

to get

$$\mathbb{E}_{sup s\leq t}(\int_0^s (f(s,U) - f(s,V))dM_s)^2 \leq 4\mathbb{E}(\int_0^t (f(s,U) - f(s,V))dM_s)^2$$

Also

$$(\int_0^t (f(s,U) - f(s,V))|dA_s|)^2 \leq (\int_0^t (f(s,U) - f(s,V))^2|dA_s|) \int_0^t |dA_s|$$

by the Cauchy-Schwarz inequality. Applying this to the above gives

$$\Phi_t(SU,SV) \leq 8\mathbb{E}(\int_0^t (f(s,U) - f(s,V))dM_s)^2 + 2\mathbb{E}(\int_0^t |dA_s|(\int_0^t (f(s,U) - f(s,V))^2|dA_s|)$$

Now,

$$\mathbb{E}(\int_0^t (f(s,U) - f(s,V))dM_s)^2 = \mathbb{E}\int_0^t (f(s,U) - f(s,V))^2 d < M,M >_s$$

If we assume that $|dA_t| \leq dt$, ie the measure defined by $|dA|$ on the real line is dominated by the Lebesgue measure, then combining the above facts results in the inequality

$$\Phi_t(SU,SV) \leq 8\mathbb{E}[\int_0^t (f(s,U) - f(s,V))^2 d < M,M >] + 2t\mathbb{E}[\int_0^t (f(s,U) - f(s,V))^2|dA_s|]$$

Making use of the Lipshitz condition satisfied by f, we finally end up with

$$\Phi_t(SU,SV) \leq 2K^2(4+t)\int_0^t \Phi_s(U,V)ds$$

This is the fundamental inequality using which existence and uniqueness of the solution to stochastic differential equations driven by semimartingale's is obtained.

3. Markov processes and transition kernels:

Definition: $(\Omega, \mathcal{F}, P)$ is a given probability space and X is a process defined on this space assuming values in the measurable space (E, ϵ). The process is said to be Markov if for all $A \in \epsilon$ and $t > s$. Associated with each Markov process X is a transition kernel $P_t : E \times \epsilon \to [0,1]$ so that for every $A \in \epsilon$, the map $x \to P_t(x, A)$ is measurable and for each $x \in E$, the map $A \to P_t(x, A)$ is a probability measure. Sometimes we allow $P_t(x, E) < 1$ for some values of x. Physically, this corresponds to the situation that the particle starting from x dies or disappears from the state space. The transition probability kernel of a Markov process satisfies the Chapman-Kolmogorov equation:

$$P_{t+s}(x, A) = \int_E P_t(x, dy)P_s(y, A)$$

Suppose $\pi(x, dy)$ is a transition probability measure. Let π^n denote the composition of π with itself n times, ie,

$$\pi^n(x, A) = \int_{y_1,...,y_{n-1}\in E} \pi(x, dy_1)\pi(y_1, dy_2)...\pi(y_{n-1}, A), A \in \epsilon$$

Let $a > 0$ and define

$$P_t(x, A) = \sum_{n=0}^{\infty} exp(-at)((at)^n/n!)\pi^n(x, A)$$

Then P_t is a transition probability function satisfying the Chapman-Kolmogorov equation. Indeed we have

$$\int P_t(x, dy)P_s(y, A) = \sum_{n,m=0}^{\infty} exp(-a(t+s))((at)^n/n!)((as)^m/m!)\pi^{n+m}(x, A)$$

$$= \sum_{n,m=0}^{\infty} exp(-a(t+s))((at)^n(as)^m/n!m!)\pi^{n+m}(x, A)$$

$$= \sum_{n=0}^{\infty} exp(-a(t+s))(\sum_{n=0}^{k}(at)^n(as)^{k-n}/n!(k-n)!)\pi^k(x, A)$$

$$= \sum_{n=0}^{\infty} exp(-a(t+s))((a(t+s))^k/k!)\pi^k(x, A) = P_{t+s}(x, A)$$

where use of the binomial theorem has been made.

Motion with uniform velocity: Suppose the position of a particle moving along the real line is given by $x(t) = x + vt$ where x is the starting point. Then $\{x(t)\}$ is a Markov process with transition probability function

$$P_t(x, A) = \chi_A(x + vt)$$

We can verify that

$$P_t f(x) = f(x + vt)$$

so that

$$P_s P_t f(x) = f(x + v(t+s)) = P_{t+s} f(x)$$

Martingale down crossings inequality: This is the fundamental inequality based on which all the martingale convergence theorems are based. Let T be a subset of $\mathbb{R}$ and let $f : T \to \mathbb{R}$. Let $F = \{t_1, t_2, ..., t_d\} \subset T$ so that $t_1 < t_2 < ... < t_d$. Let $a < b$ be two real numbers. Define

$$s_1 = min(t_i : f(t_i) > b), s_2 = min(t_i > s_1 : f(t_i) < a), s_3 = min(t_i > s_2 : f(t_i) > b)$$

etc if ϕ is the null set then we set $min(\phi) = t_d$. Define

$$D(f, F, [a, b]) = sup(n : s_{2n} < t_d)$$

It is clear from a graph that $D(f, F, [a, b])$ is the number of downcrossings by the function f restricted to F of the interval $[a, b]$. Define

$$D(f, T, [a, b]) = sup(D(f, F, [a, b]) : F \subset T, F finite)$$

Let X be a submartingale and T a countable subset of the positive real line. First assume that T is a finite set. The $s_k's$ defined above are stopping times with respect to $\mathcal{F}_{t_k}$. Let $A_k = \{s_k < t_d\}$. Then,

$A_k \in \mathcal{F}_{t_k}$ and obviously $A_{k+1} \subset A_k$. On A_{2n-1} we have $X_{s_{2n-1}} > b$ and on A_{2n}, we have $X_{s_{2n}} < a$. Thus, application of the optional stopping theorem gives

$$0 \le \int_{A_{2n-1}} (X_{s_{2n-1}} - b)dP \le \int_{A_{2n-1}} (X_{s_{2n}} - b)dP$$

$$\le (a-b)P(A_{2n}) + \int_{A_{2n-1}-A_{2n}} (X_{s_{2n}} - b)dP$$

Now on A_{2n}^c, we have $s_{2n} = t_d$ and hence we get

$$(b-a)P(A_{2n}) \le \int_{A_{2n-1}-A_{2n}} (X_{t_d} - b)dP$$

4. Let $\mathcal{F}_t$ be a filtration and S, T $\mathcal{F}_t$-stop times. Then (a) $min(S,T)$ is a stop time. Indeed,

$$\{min(S,T) \le t\} = \{S \le t\} \cup \{T \le t\} \in \mathcal{F}_t$$

since $\mathcal{F}_t$ is a σ-field and S,T are stop times implies $\{S \le t\}, \{T \le t\} \in \mathcal{F}_t$. (b) $max(S,T)$ is a stop time. Indeed,

$$\{max(S,T) \le t\} = \{S \le t\} \cap \{T \le t\} \in \mathcal{F}_t$$

(c) The sets $\{S = T\}, \{S \le T\}, \{S < T\}$ are all in $\mathcal{F}_S \cap \mathcal{F}_T$. Indeed,

$$\{S = T\} \cap \{S \le t\}$$

First we show that $\{S \le T\} \in \mathcal{F}_S \cap \mathcal{F}_T$. To prove this, it suffices to prove that $\{S \le T\} \cap \{S \le t\} \in \mathcal{F}_t$ and $\{S \le T\} \cap \{T \le t\} \in \mathcal{F}_t$ for all t. The former would imply that $\{S \le T\} \in \mathcal{F}_S$ and the latter that $\{S \le T\} \in \mathcal{F}_T$. Suppose we have proved that $\{S \le T\} \in \mathcal{F}_T$. Then since $\mathcal{F}_T$ is a σ-field, it would follow that $\{T < S\} \in \mathcal{F}_T$. Now $S + 1/n$ is also a stop time and hence it would follow that $\{T < S + 1/n\} \in \mathcal{F}_T$. Taking intersections over $n = 1, 2, ...$, it would follow that $\{T \le S\} \in \mathcal{F}_T$ and on interchanging S and T, we would get that $\{S \le T\} \in \mathcal{F}_S$. Thus to establish the claim, it suffices to show that $\{S \le T\} \in \mathcal{F}_T$, or equivalently that $\{S \le T\} \cap \{T \le t\} \in \mathcal{F}_t$. Let E denote the set of all rationals in $(0, t]$. Then we claim that $\{S < T\} \cap \{T \le t\}$ is the same as $\bigcup_{q \in E} \{S \le t - q, t - q \le T \le t\}$. Indeed suppose $T \le t$ and $S < T$. Then there exists a $q \in E$ such that $S \le t - q \le T \le t$. The converse is also obvious. Now $\{S \le t - q\} \in \mathcal{F}_{t-q} \subset \mathcal{F}_t$ for all $q > 0$ and also $\{t - q \le T \le t\} = \{T \le t\} \cap \{T \le t - q\}^c \in \mathcal{F}_t$. The proof now follows by forming unions over all $q \in E$ and using the fact that $\mathcal{F}_t$ is a σ-field. Since now $\{S < T\} \in \mathcal{F}_T$, it follows by the standard $1/n$ argument that $\{S \le T\} \in \mathcal{F}_T$ and the proof is complete. To prove that $\{S = T\}$ is an element of $\mathcal{F}_T$, we use the fact that $\{S \le T\} \in \mathcal{F}_T$ and $\{S < T\} \in \mathcal{F}_T$. Take the difference of the two sets and note that

$$\{S = T\} = \{S \le T\} - \{S < T\} = \{S \le T\} \cap \{S < T\}^c$$

5. If $S \le T$ are stopping times, then $\mathcal{F}_S \subset \mathcal{F}_T$. Let $E \in \mathcal{F}_S$. Then

$$E \cap \{T \le t\} = E \cap \{T \le t\} \cap \{S \le T\} = E \cap \{T \le t\} \cap \{S \le t\} \in \mathcal{F}_t$$

proving the claim.

6. Definition of the Bessel process: Let $d = 1, 2, ...$ and let $B_i(t), i = 1, 2, ..., d$ be d standard independent Brownian motion processes. Define the process

$$X(t) = (\sum_{i=1}^{d} B_i(t)^2)^{1/2} = f(\mathbf{B}(t))$$

where

$$f(\mathbf{x}) = (\sum_{i=1}^{d} x_i^2)^{1/2}$$

Then,

$$f_{,i} = x_i/f, f_{,ij} = \delta_{ij}/f - x_i x_j/f^3$$
$$\nabla^2 f = \sum_i f_{,ii} = (d-1)/f$$

so that by Ito's formula,

$$dX(t) = df(\mathbf{B}(t)) = \sum_{i=1}^{d} B_i(t) dB_i(t)/X(t) + (d-1)dt/2X(t)$$

Define a process

$$Z(t) = \int_0^t \sum B_i(s) dB_i(s)/X(s)$$

Then, Z is a martingale and

$$(dZ)^2 = \sum B_i^2 dt/X^2 = dt$$

so that by Levy's characterization, Z is a Brownian motion process. We can write the above sde as

$$dX(t) = dZ(t) + (d-1)dt/X(t)$$

X is called a Bessel process of order $d-1$. For general $\delta \geq 0$, we define a process $X = X_\delta$ so that

$$X(t) = x + Z(t) + \delta \int_0^t ds/X(s)$$

where Z is a standard Brownian motion starting at the origin. Define the process

$$Y(t) = X^2(t)/2$$

Then,

$$dY = XdX + (dX)^2/2 = (\delta + 1/2)dt + X.dZ$$

We set $\epsilon = \delta + 1/2$. Then assuming that $Y(0) = x$, we get

$$Y(t) = x + \epsilon.t + \int_0^t \sqrt{Y(s)} dZ(s)$$

Suppose Z' is another process independent of Z and U satisfies

$$U(t) = x' + \epsilon'.t + \int_0^t \sqrt{U(s)}dZ'(s)$$

Then,

$$Y(t) + U(t) = x + x' + (\epsilon + \epsilon')t + \int_0^t (\sqrt{Y(s)}dZ(s) + \sqrt{U(s)}dZ'(s))$$

Now define a process $W()$ so that $W(0) = 0$ and

$$dW(t) = [\sqrt{Y(t)}dZ(t) + \sqrt{U(t)}dZ'(t)]/\sqrt{Y(t) + U(t)}$$

Then, W is a martingale and

$$(dW)^2 = dt$$

implying that W is a standard Brownian motion. We have with $V(t) = Y(t) + U(t)$,

$$dV(t) = x + x' + (\epsilon + \epsilon')t + \int_0^t \sqrt{V(s)}dW(s)$$

Suppose the probability distribution function of the process Y starting at x is denoted by $Q_{x,\epsilon}(.)$. Then the probability distribution of the process U starting at x' is $Q_{x',\epsilon'}(.)$ and the probability distribution function of the process V starting at $x + x'$ is given by $Q_{x+x',\epsilon+\epsilon'}(.)$. It follows that

$$Q_{x+x',\epsilon+\epsilon'} = Q_{x,\epsilon} * Q_{x',\epsilon'}$$

7. Definition of the Poisson process and of the spatial Poisson random field. Suppose $t_1, t_2, \ldots$ are random variables with the property that $t_{i+1} - t_i, i = 1, 2, \ldots$ are independent exponentially distributed random variables having mean $1/\lambda$. Let $\theta(t)$ denote the unit step function. Then define

$$N(t) = \sum_{i=1}^{\infty} \theta(t - t_i)$$

We have $dN(t) = 1$ if $t \in \{t_i\}$ and $dN(t) = 0$ if $t \notin \{t_i\}$. For any well behaved function $f(.)$, we have

$$X = \int_0^{\infty} f(t)dN(t) = \sum_{i=1}^{\infty} f(t_i)$$

and

$$\mathbb{E}X = \lambda \int_0^{\infty} f(t)dt$$

$$\mathbb{E}X^2 = \int f(t)^2 \lambda dt + (\lambda \int f(t)dt)^2$$

since

$$dN(t).dN(t) = dN(t)$$

and

$$\mathbb{E}dN(t) = \lambda.dt$$

8. Suppose $X(t)$ is a stochastic process with finite dimensional density functions $f_k(x_1, ..., x_k, t_1, ..., t_k)$. We wish to evaluate the moments of $Z = \int_0^T g(t)X(t)dt$ in terms of these density functions. Observe that

$$\mathbb{E}Z^m = \int g(t_1)...g(t_m)\mathbb{E}(X(t_1)...X(t_m))dt_1...dt_m$$

$$= \int g(t_1)...g(t_m)\int x_1...x_m f_m(x_1, ..., x_m, t_1, ..., t_m)dt_1...dt_m$$

$$= \int x_1...x_m \phi(x_1, ..., x_m)dx_1...dx_m$$

where

$$\phi(x_1, ..., x_m) = \int g(t_1)...g(t_m)f_m(x_1, ..., x_m, t_1, ..., t_m)dt_1...dt_m$$

9. Defintion of a strong Markov process. A filtered probability space $(\Omega, \mathcal{F}, \{\mathcal{F}_t\}, P)$ is given and on this spacc, a proccss $\{X_t : t \geq 0\}$ is given. This process is assumed to be adapted to the filtration, ie, for each $t \geq 0$, the random variable X_t is measurable with respect to $\mathcal{F}_t$. The process is said to be Markov if for every $t > s$ and every bounded Borel function $g : \mathbb{R} \to \mathbb{R}$, we have

$$\mathbb{E}(g(X_t)|\mathcal{F}_s) = \mathbb{E}(g(X_t)|X_s)a.s$$

An equivalent definition is that if $P(A|\mathcal{F}_t)$ denotes the conditional probability of $A \in \mathcal{F}$ given $\mathcal{F}_t$, then for every Borel set B, we have

$$P(X_t \in B|\mathcal{F}_s) = P(X_t \in B|X_s)a.s, t > s$$

For a strong Markov process, the time variables s, t are replaced respectively by stoptimes S, T satisfying $S \leq T$.

10. Convergence of random variables: If $(\Omega, \mathcal{F}, P)$ is a probability space and $X_n, n = 1, 2, ..., X$ a sequence of random variables on this space, we say that $\{X_n\}$ converges in probability to X if for every $\delta > 0$, one has

$$lim_{n\to\infty}P(|X_n - X| > \delta) = 0$$

If the random variables X_n, X assume values in a metric space (Z, d), then we say that $\{X_n\}$ converges in probability to X if for every $\delta > 0$, one has

$$lim_{n\to\infty}P(d(X_n, X) > \delta) = 0$$

or more precisely,

$$lim_{n\to\infty}P(\{\omega : d(X_n(\omega), X(\omega)) > \delta\}) = 0$$

We say that X_n converges in distribution to X if for every Borel set $B \subset Z$ such that $P(X \in \partial B) = 0$, we have

$$lim_{n\to\infty} P(X_n \in B) = P(X \in B)$$

If $Z = \mathbb{R}$ so that $d(x, y) = |x-y|$ and $F(x) = P(X \leq x)$, then the condition for convergence in distribution is equivalent to the condition that if x is any point of continuity of $F(.)$, then,

$$lim_{n\to\infty} P(X_n \leq x) = F(x)$$

The condition for convergence in distribution is equivalent to the condition that if $f : Z \to \mathbb{R}$ is any bounded Borel function, then

$$lim_{n\to\infty} \mathbb{E}[f(X_n)] = \mathbb{E}[f(X)]$$

If X_n converges in probability to X, then it also converges in distribution to X. Indeed, let $\delta > 0$ and let f be bounded and continuous. Then,

$$|\mathbb{E}f(X_n) - \mathbb{E}f(X)| \leq \mathbb{E}|f(X_n) - F(X)| = \mathbb{E}[|f(X_n) - f(X)|\chi_{d(X_n,X)\leq\delta}] + \mathbb{E}[|f(X_n) - f(X)|\chi_{d(X_n,X)>\delta}]$$

Now, let $sup|f(x)| = M$. Then,

$$\mathbb{E}[|f(X_n) - f(X)|\chi_{d(X_n,X)>\delta}] \leq M.P(d(X_n, X) > \delta) \to 0$$

For each $x \in Z$, choose a $\delta_x > 0$ such that $d(x, y) < \delta_x$ implies $|f(x) - f(y)| < \epsilon$. Here, ϵ is any given positive real number. We have $[-M, M] \subset \bigcup_{x\in Z} B(f(x), \epsilon)$ and hence there exists a finite set $\{x_1, ..., x_m\}$ such that

$$[-M, M] \subset \bigcup_{i=1}^{m} B(f(x_i), \epsilon)$$

Define

$$O_i = f^{-1}(B(f(x_i), \epsilon))$$

Then O_i is open, contains x_i and

$$Z = \bigcup_{i=1}^{m} O_i$$

Moreover, if $u, v \in O_i$, then

$$d(f(u), f(v)) \leq d(f(u), f(x_i)) + d(f(v), f(x_i)) < 2\epsilon$$

Also, $x \in O_i$ and $d(x, y) < \delta$ implies $d(x, x_i) < \delta, d(x, y) < \delta$ implies $d(x_i, y) < 2\delta, d(f(x), f(x_i)) < \delta$ implies $d(f(x_i), f(y)) < \epsilon, d(f(x), f(x_i)) < \epsilon$ implies $d(f(x), f(y)) < 2\epsilon$ provided that we choose $\delta = \frac{1}{2} min_j \delta_{x_j}$. Define $V_1 = O_1, V_i = O_i - \bigcup_{j=1}^{i-1} O_j$. Then, $V_j's$ are disjoint, m in number and $x \in V_i, d(x, y) < \delta$ imply $d(f(x), f(y)) < 2\epsilon$. It follows that

$$\mathbb{E}[|f(X_n) - f(X)|\chi_{d(X_n,X)<\delta}] = \sum_i \mathbb{E}[|f(X_n) - f(X)|\chi_{X\in V_i}\chi_{d(X_n,X)<\delta}]$$

$$\leq 2\epsilon \sum P(X \in V_i) = 2\epsilon$$

Letting $\epsilon \to 0$ completes the proof. A sequence of random variables $X_n, n = 1, 2, ...$ converges to a r.v X (all defined on the same space) in L^p norm provided

$$\mathbb{E}|X_n - X|^p \to 0, n \to \infty$$

A sequence $X_n, n = 1, 2, ...$ converges almost surely to X if the set

$$E = \{\omega \in \Omega : X_n(\omega) \to X(\omega), n \to \infty\}$$

has probability one. L^p-convergence for $p \geq 1$ implies convergence in probability, an elementary consequence of the Chebyshev inequality.

11. Explain the notion of weak convergence of probability measures on a probability space. Give different equivalent versions of weak convergence. Hint: Let $\{P_n : nm = 1, 2, ...\}, P$ be a sequence of probability measures on the measurable space $(\Omega, \mathcal{F})$. Say that $\{P_n\}$ converges weakly to P if for every bounded random variable X defined on this space, we have

$$lim_{n\to\infty} \int XdP_n = \int XdP$$

12. Central limit theorem: If $X_n, n = 1, 2, ...$ is a sequence of independent identically distributed integrable random variables and we set $\mathbb{E}X_n = \mu, V(X_n) = \sigma^2$, then show that $(S_n - n\mu)/\sigma\sqrt{n}$ converges in distribution to an $N(0, 1)$ random variable.

13. If X_n is a sequence of identically distributed independent random variables with zero mean and unit variance, and we define $S_n = X_1 + X_2 + ... + X_n, n \geq 1$, and for every $t \geq 0$, we define

$$B_n(t) = n^{-1/2} S_{[nt]}$$

then show that the processes $B_n(.)$ converge in distribution to the Brownian motion process.

14. Let $\mathcal{F}_n, n = 0, 1, 2, ...$ be a discrete time filtration on a probability space and let $A_n, n = 0, 1, 2, ...$ be predictable with respect to this filtration, ie, A_n is $\mathcal{F}_{n-1}$ measurable. Then, let $M_n, n = 0, 1, ...$ be a martingale with respect to this filtration. Define $Y_0 = 0$ and for $n \geq 1$, define

$$Y_n = \sum_{k=1}^{n} A_k(M_k - M_{k-1}), n \geq 1$$

Then, Y_n is $\mathcal{F}_n$-measurable and since for $n \geq 1$,

$$Y_n = Y_{n-1} + A_n(M_n - M_{n-1})$$

if follows that

$$\mathbb{E}[(Y_n - Y_{n-1})|\mathcal{F}_{n-1}] = A_n\mathbb{E}[(M_n - M_{n-1})|\mathcal{F}_{n-1}] = 0$$

it follows that Y is also a martingale. Y is called the martingale transform of A by M. An increasing process A_n is called natural if for every bounded martingale $M_n, n = 0, 1, 2, ...$, we have

$$\mathbb{E}(M_n A_n) = \mathbb{E}\sum_{k=1}^{n} M_{k-1}(A_k - A_{k-1})$$

We can write this equation as

$$\mathbb{E}(M_nA_n) = \mathbb{E}[\sum_{k=1}^{n} M_{k-1}A_k - \sum_{k=1}^{n} M_{k-1}A_{k-1}]$$

$$= \mathbb{E}[\sum_{k=1}^{n} M_{k-1}A_k - \sum_{k=0}^{n-1} M_kA_k]$$

or equivalently,

$$0 = \mathbb{E}[\sum_{k=0}^{n} M_kA_k] - \mathbb{E}[\sum_{k=1}^{n} M_{k-1}A_k] = \mathbb{E}\sum_{k=1}^{n}(M_k - M_{k-1})A_k]$$

since by hypothesis, $M_0 = 0$. This means that the process A_n is natural iff its martingale transform Y by any martingale M satisfies $\mathbb{E}Y_n = 0$ for all n. It may be shown (see Karatzas and Shreve) that an increasing process is natural iff it is predictable.

15. An alternate version of the Girsanov theorem. Let W be standard BM and let X be an adapted process. Let P be the probability measure of the space (relative to which W is BM). Define for $t \geq 0$ and $A \in \mathcal{F}_t$,

$$\tilde{P}_t(A) = \mathbb{E}[Z_t(X)\chi_A]$$

where

$$Z_t(X) = exp(\int_0^t X_s dW_s - \frac{1}{2}\int_0^t X_s^2 ds)$$

Then

$$dZ_t(X) = Z_t(X)dW_t$$

and hence

$$Z_t(X) = 1 + \int_0^t Z_s(X)dW_s$$

It follows that $Z(X)$ is a P martingale. Hence, if $t > s$ and $A \in \mathcal{F}_s$, we have

$$\tilde{P}_t(A) = \mathbb{E}[Z_t(X)\chi_A] = \mathbb{E}[Z_s(X)\chi_A] = \tilde{P}_s(A)$$

This shows that the family of probability measures $\{P_t\}$ is consistent. Define the process

$$S_t = W_t - X_t$$

We then claim that S is BM relative to the measure $\tilde{P}$. This can be proved by straight forward arguments.

16.Martingale problem of Strook and Varadhan: Suppose L is a second order partial differential operator of the form

$$Lf = -\sum_i b_i(x)\frac{\partial f}{\partial x_i} + \frac{1}{2}\sum_{i,j} a_{ij}(x)\frac{\partial^2 f}{\partial x_i \partial x_j}$$

Let X_t be a stochastic process such that for all well behaved functions f,

$$M^f(t) = f(X_t) - \int_0^t Lf(X_s)ds$$

is a martingale. Then, X_t is a diffusion process. Note that if X is a diffusion with drift $b(x)$ and diffusion $\sigma_{ij}(x)$, then

$$df(X_t) = Lf(X_t)dt + \sum_{i,j} \frac{\partial f(X_t)}{\partial x_i}\sigma_{ij}(X_t)dB_j(t)$$

where B is the driving Brownian for X_t. Then, it follows that

$$f(X_t) - \int_0^t Lf(X_s)ds$$

is a martingale. Other martingales can also be constructed from a diffusion process. Suppose $X(t)$ satisfies the sde

$$dX(t) = f(X(t))dt + g(X(t))dB(t)$$

consider the process

$$Z(t) = exp(a(X(t) - \int_0^t (f(X(s))ds - \int_0^t a^2g^2(X(s))/2)ds))$$

write

$$U(t) = a(X(t) - \int_0^t f(X(s)) - (a^2/2)\int_0^t g^2(X(s))ds$$

Then,

$$Z(t) = exp(U(t))$$

and we have

$$dZ(t) = Z(t)(dU(t) + (dU(t))^2/2)$$

Now,

$$dU(t) = a.(dX(t) - f(X(t))dt) - (a^2/2)g^2(X(t))dt = ag(X(t))dB(t) - (a^2/2)g^2(X(t))dt$$

$$(dU(t))^2 = a^2g^2(X(t))dt$$

so that

$$dU(t) + (dU(t))^2/2 = ag(X(t))dB(t)$$

and this proves that

$$dZ(t) = aZ(t)g(X(t))dB(t)$$

implying that $Z(.)$ is a martingale. This can be generalized even further. Let h be any smooth function. Then,

$$dh(X(t)) = h'(X(t))dX(t) + \frac{1}{2}h''(X(t))(dX(t))^2$$

$$= h'(X(t))(f(X(t))dt + g(X(t))dB(t)) + \frac{1}{2}h''(X(t))g^2(X(t))dt$$

So define

$$p(x) = h'(x)f(x) + \frac{1}{2}h''(x)g^2(x)$$

and the process

$$Z_h(t) = exp(h(X(t)) - \int_0^t p(X(s))ds)$$

Then,

$$dZ_h(t) = Z_h(t)h'(X(t))g(X(t))dB(t)$$

proving that Z_h is a martingale.

17. Dirichlet problem:Assume that D is an open bounded subset of $\mathbb{R}^d$ $g : D \to \mathbb{R}$ and $f : \partial D \to \mathbb{R}$ are well behaved. Starting from some point $x \in D$ assume that τ_D is the exit time of a d-dimensional Brownian motion from the boundary of D. Define

$$u(x) = \mathbb{E}(f(B(\tau_D)|B_0 = x) + \mathbb{E}(\int_0^{\tau_D} g(B(t))dt|B(0) = x)$$

Then we have

$$u(x) = f(x), x \in \partial D$$

since $\tau_D = 0$ when $B(.)$ starts at any point on ∂D. Note how τ_D is defined:

$$\tau_D = inf\{t \geq 0 : B(t) \in \partial D\}$$

Further, by the Markov property for $B(.)$, we have that

$$u(x) = lim_{h\to 0}\mathbb{E}(\mathbb{E}(f(B(\tau_D))|B_h)|B_0 = x) + g(x)h + \mathbb{E}(\mathbb{E}(\int_h^{\tau_D} g(B(t))dt|B(h))|B(0) = x)$$

or equivalently,

$$u(x) = \mathbb{E}(u(B(h))|B(0) = x) + g(x)h + o(h)$$

Application of Ito's formula then gives

$$u(x) = u(x) + \frac{h}{2}\Delta u(x) + hg(x) + o(h)$$

so that

$$\frac{1}{2}\Delta u(x) + g(x) = 0, x \in D$$

This illustrates how Brownian motion can be used to solve the Dirichlet problem with source. Supose D is the ball of radius r in $\mathbb{R}^d$, ie,

$$D = \{x \in \mathbb{R}^d : \| x \| < r\}$$

Define

$$u(x) = (r^2 - \| x \|^2)/d, x \in D, f = 0, g = 1$$

Then,

$$\Delta u = -1 = -g$$

and

$$u(x) = 0 = f(x), x \in \partial D$$

It follows from the above discussion that

$$u(x) = \mathbb{E}(\tau_D | B(0) = x)$$

the expected value of the exit time of the Brownian motion from the boundary of D.

18. Let $W(t)$ be standard one dimensional BM and suppose that $v(t,x)$ satisfies the pde

$$-\frac{\partial v(t,x)}{\partial t} + k(x)v(t,x) = \frac{1}{2}\Delta v(t,x) + g(t,x), 0 \le t \le T, x \in \mathbb{R}$$

with the boundary condition

$$v(T,x) = f(x)$$

Let $0 \le t \le T$ and define the process

$$X(s) = v(t+s, W(s))exp(-\int_0^s k(W(u))du)$$

We have by the Ito rule

$$d\mathbb{E}X(s) = \mathbb{E}dX(s) = \mathbb{E}[\frac{\partial v(t+s, w(s))}{\partial t}ds - k(W(s))v(t+s, W(s))ds + \frac{1}{2}\Delta v(t+s, W(s))ds]$$

$$exp(-\int_0^s k(W(u))du)$$

$$= \mathbb{E}[g(t+s, W(s)).exp(-\int_0^s k(W(u))du]ds$$

We are assuming the initial condition $W(0) = x$. Integrating both sides with respect to s from $s = 0$ to $s = T - t$ and noting that $v(T, W(T-t)) = f(W(T-t))$, we get

$$\mathbb{E}_x[f(W(T-t))exp(-\int_0^{T-t} k(W(u))du)] - v(t,x) =$$

$$\mathbb{E}_x[\int_0^{T-t} g(t+s, W(s))exp(-\int_0^s k(W(u))du)ds]$$

This is the generalized Feynman-Kac formula for the generalized heat equation, ie, heat equation with potential and source.

19.Doob-Meyer decomposition for continuous sub-martingales. Given any continuous submartingale X, ie a process X_t on a filtered probability space satisfying $\mathbb{E}(X_t|\mathcal{F}_s) \ge X_s$ almost surely, we can write $X = M + V$, where X is a continuous martingale and V is a predictable increasing process. The construction is based on analogy with the discrete time scenario. Suppose X_n is a martingale with respect to the filtration $\mathcal{F}_n$. We define

$$A_n = \sum_{k=1}^{n} \mathbb{E}((X_k|\mathcal{F}_{k-1}) - X_{k-1})$$

Then A is increasing and predictable. Let $M_n = X_n - A_n$. Then

$$M_n - M_{n-1} = X_n - \mathbb{E}(X_n|\mathcal{F}_{n-1})$$

so that

$$\mathbb{E}((M_n - M_{n-1}|\mathcal{F}_{n-1}) = 0$$

confirming that M is a martingale. The continuous time version of this decomposition is achieved by using limiting processes.

20.Kolmogorov-Centsov theorem:

21. Problem: The only $\mathcal{B}(\mathbb{R}^{[0,\infty)})$ measurable set that is contained in $C[0,\infty)$ is the empty set. Indeed any element E of $\mathcal{B}(\mathbb{R}^{[0,\infty)})$ is expressible as

$$E = \{\omega \in \mathbb{R}^{[0,\infty)} : (\omega(t_1), \omega(t_2), ...) \in A\}$$

where A is an element of $\mathcal{B}(\mathbb{R} \times \mathbb{R} \times ...)$ Any element ω of E depends only on the values of ω at a countable set of times t_j and hence assuming that $t_1 < t_2 < ...$, we can choose any $i = 1, 2, ...$ and make ω discontinuous in the range (t_i, t_{i+1}) keeping the values of ω at the t_j's the same. This point will also belong to E but since this ω is discontinuous, it follows that $E \notin C[0,\infty)$.

22.Suppose $B(t)$ is standard Brownian motion. We obtain by formal application of Ito's differential rule

$$d|B(t)| = sgn(B(t))dB(t) + \delta(B(t))dt$$

so that defining

$$L(t) = \int_0^t \delta(B(s))ds$$

and

$$W(t) = \int_0^t sgn(B(s))dB(s)$$

we find that W is also a Brownian motion process, L_t is non-decreasing and

$$|B(t)| = W(t) + L(t)$$

Moreover, $L(t)$ is flat off $B(t) \neq 0$. This can be proved by an intuitive argument. $L(t)$ by its definition, increases only when $B(t)$ is zero for otherwise, $\delta(B(t)) = 0$. It follows that

$$\int \chi_{B(t)\neq 0} dL(t) = 0$$

If $X(t)$ is any Brownian motion process and $M(t) = max(X(s) : s \leq t)$, then M is nondecreasing and non-negative and further we have the decomposition

$$M(t) - X(t) = -X(t) + M(t)$$

is a decomposition of $Y(t) = M(t) - X(t)$ into the sum of a Brownian motion process and a non-decreasing process. Further $M(t)$ is flat off $M(t) > X(t)$ since if $M(t) > X(t)$, then there is a an $s < t$ such that $M(t) = X(s)$ and then over a small duration $[t, t+dt]$, we have $M(t+dt) = M(t) = X(s)$ which proves the claim. Application of the Skorhohod theorem (Revuz and Yor, Continuous martingales and Brownian motion, or Karatzas and Shreve, Brownian motion and stochastic calculus) then shows that the pair $(M(t) - X(t), M(t)), t \geq 0$ has the same law as the process $(|B(t)|, L(t)), t \geq 0$.

23. Notion of shift operator. The space on which stochastic processes are defined is $C[0, \infty)$. We define a shift operator $\theta_s : C[0, \infty) \to C[0, \infty)$ so that $\theta_s(w)(t) = w(s + t)$. Suppose P_x is a measure on $C[0, \infty)$ with the initial condition x. Formulate using the shift operator, the condition for the measure P_x to be shift invariant, ie if $w(t), t \geq 0$ is the process corresponding to P, then the law of $w(t), t \geq 0$ given $w(0) = x$ be the same as the law of the process $w(t + s), t \geq 0$ given $w(s) = x$.

Ans:Let B be any Borel subset of $C[0, \infty)$. Then the condition should read

$$P((w(t), t \geq 0) \in B | w(0) = x) = P((w(s + t), t \geq 0) \in B | w(s) = x)$$

which is the same as

$$P_{0,x}(B) = P_{s,x}(\theta_s^{-1}(B))$$

The shift operator θ_s has been used with much success in the classic Continuous martingales and Brownian motion by Revuz and Yor.

24. Discussion with Dr.Shambu Sharma: Let ϵ be a small parameter. Consider the system

$$x_1' = x_2 + \epsilon.f_1(x_1, x_2), x_2' = -x_1 + \epsilon.f_2(x_1, x_2)$$

Then letting

$$E = (x_1^2 + x_2^2)/2$$

we get

$$E' = \epsilon.(x_1 f_1 + x_2 f_2)$$

so that the change in energy around one orbit of the unperturbed system is given by

$$\Delta E = \int_0^T E' dt = \epsilon. \int_0^T (x_1 f_1 + x_2 f_2) dt + O(\epsilon^2)$$

or with neglect of $O(\epsilon^2)$,

$$\Delta E = \epsilon. \int_\Gamma (f_2 dx_1 - f_1 dx_2)$$

where Γ is the closed trajectory of the unperturbed orbit. Let S denote the disc of radius R. Consider the solution $x_1(t) = R.cos(t), x_2(t) = -R.sin(t)$ of the unperturbed system. The change in the energy of the system around one orbit with neglect of $O(\epsilon^2)$ terms is given by applying Green's theorem to the above line integral to get

$$\Delta E = -\epsilon. \int_S (\frac{\partial f_1}{\partial x_1} + \frac{\partial f_2}{\partial x_2}) dx_1 dx_2$$

25.(Revuz and Yor, Continuous martingales and Brownian motion): Notion of a Gaussian random field. Let $\mathcal{H}$ be a Hilbert space and $(\Omega, \mathcal{F}, P)$ a probability space. For each $h \in \mathcal{H}$, let $X(h)$ be a random variable in $L^2(\Omega, \mathcal{F}, P)$ such that the map $h \to X(h)$ is linear and

$$\mathbb{E}(X(h)^2) = \| h \|^2$$

Further, assume that the random variables $\{X(h) : h \in \mathcal{H}\}$ are all jointly Gaussian. Then $X(.)$ is called a Gaussian random field on the space $\mathcal{H}$. Note that if $g, h \in \mathcal{H}$, then

$$\mathbb{E}X(g+h)^2 = \| g+h \|^2 = \| g \|^2 + \| h \|^2 + 2 < g, h >$$

On the other hand,

$$\mathbb{E}X(g+h)^2 = \mathbb{E}(X(g) + X(h))^2 = \mathbb{E}X(g)^2 + \mathbb{E}X(h)^2 + 2\mathbb{E}(X(g)X(h))$$
$$= \| g \|^2 + \| h \|^2 + 2\mathbb{E}(X(g)X(h))$$

and a comparison of the two equations gives us

$$\mathbb{E}(X(g)X(h)) = < g, h >$$

Another formulation of Gaussian random fields is through integration with respect to a Gaussian random measure. Suppose $(A, \mathcal{A}, \mu)$ is a measure space and $(\Omega, \mathcal{F}, P)$ is a probability space. For each $E \in \mathcal{A}$, assume that we are given a random variable $X(E) \in L^2(\Omega, \mathcal{F}, P)$ such that (i) $\{X(E) : E \in \mathcal{A}\}$ is jointly Gaussian, (ii) if $E_n, n = 1, 2, ...$ is a pairwise disjoint countable family in $\mathcal{A}$, then $X(\bigcup_{n=1}^{\infty} E_n) = \sum_{n=1}^{\infty} X(E_n)$ and if $E, F \in \mathcal{A}$ are arbitrary sets, then $\mathbb{E}(X(E).X(F)) = \mu(EF)$. Then for $f \in L^2(A, \mu)$, we can define a stochastic integral $X(f) = \int_A f(x)X(dx)$. This integral is first defined for simple functions f, ie, for f having the form

$$f = \sum_{i=1}^{n} c_i \chi_{A_i}, A_i \in \mathcal{A}$$

so that

$$X(f) = \sum_{i=1}^{n} c_i X(A_i)$$

A simple calculation shows that if g is another simple function, ie,

$$g = \sum_{i=1}^{m} d_i \chi_{B_i}, B_i \in \mathcal{A}$$

then

$$\mathbb{E}(X(f)X(g)) = \sum_{i,j} c_i d_j \mu(A_i B_j) = < f, g >$$

Now use the L^2 approximation theorem, namely that any $f \in L^2(\mu)$ can be expressed as an L^2-limit of simple functions, ie, there exists a sequence $f_n, n = 1, 2, ...$ of simple functions in $L^2(\mu)$ such that $\| f - f_n \| \to 0$ as $n \to \infty$. Then we have

$$\| X(f_n) - X(f_m) \| = \| X(f_n - f_m) \| = \| f_n - f_m \| \to 0, n, m \to \infty$$

Hence there exists a random variable Z in $L^2(\Omega, \mathcal{F}, P)$ such that $\| X(f_n) - Z \| \to 0$ as $n \to \infty$. It is also clear that upto almost sure equivalence, the random variable Z is independent of the approximating sequence $\{f_n\}$ of simple functions. It depends only on the function f and the Gaussian random measure $X(.)$. We can thus denote Z by $X(f)$. Let $f, g \in L^2(\mu)$ and choose sequences f_n, g_n of simple functions converging in L^2 to f, g respectively. Then

$$< X(f), X(g) >= lim < X(f_n), X(g_n) >= lim < f_n, g_n >=< f, g >$$

This confirms the fact that $f \to X(f)$ is a Gaussian random field.

26. Discrete stochastic processes: We assume that the notion of random variable has been dealt with on a given fixed probability space $(\Omega, \mathcal{F}, P)$. For each $n = 0, 1, 2, ...$, we are given a random variable $X_n(\omega)$ on this space then we obtain a discrete time stochastic process $\{X_n\}$. The finite dimensional joint probability distribution functions of this process are given by

$$F(\xi_1, ..., \xi_k; n_1, ..., n_k) = P(\bigcap_{i=1}^{k} \{X_{n_i} \le \xi_i\})$$

The corresponding joint density, when it exists, is given by

$$f(\xi_1, ..., \xi_k; n_1, ..., n_k) = \frac{\partial^k}{\partial \xi_1 ... \partial \xi_k} P(\bigcap_{i=1}^{k} \{X_{n_i} \le \xi_i\})$$

This joint density can be used, for example, to evaluate the statistics of the random process, for instance, if $g : \mathbb{R}^k \to \mathbb{R}$ is a bounded Borel function, then

$$\mathbb{E}g(X_{n_1}, ..., X_{n_k}) = \int_{\mathbb{R}^k} g(\xi_1, ..., \xi_k) f(\xi_1, ..., \xi_k; n_1, ..., n_k) d\xi_1 ... d\xi_k$$

27. Construction of the Wiener measure on $C[0, \infty)$. This measure is constructed along the following lines. We start with a sequence $\{\xi_j : j = 1, 2, ...\}$ of iid random variables having zero mean and variance σ^2 and then define $S_0 = 0, S_k = \sum_{j=1}^{k} \xi_j, k \ge 0$. Define

$$Y_t = S_{[t]} + (t - [t])\xi_{[t]+1}$$

where $[t]$ denotes integer part of t. Define the process $X^{(n)}$ by

$$X_t^{(n)} = \frac{1}{\sigma\sqrt{n}} Y_{nt}$$

It is easy to see that Y_t is a continuous process and hence $X^{(n)}, n = 1, 2, ...$ are continuous processes. We then prove that for every finite set $t_1 < t_2 < ... < t_k$, $(X_{t_1}^{(n)}, X_{t_2}^{(n)}, ..., X_{t_k}^{(n)})$ converges in distribution to $(B_{t_1}, ..., B_{t_k})$, where B is a standard Brownian motion process. This proof is based on characteristic functions and is an elementary application of the central limit theorem. Let P_n denote the measure induced on the Borel subsets of $C[0, \infty)$ by the process $X^{(n)}$ and let P denote the Wiener measure, ie, the measure on the Borel subsets of $C[0, \infty)$ induced by a standard Brownian motion process. Then, the Donsker invariance principle asserts that P_n converges weakly to P. The proof is based on Prohorov's theorem which in short reads tightness of a sequence of measures implies relative compactness. A sequence of probability measures Q_n on a metric space is said to be tight if for each $\epsilon > 0$, there exists a compact set K such that $Q_n(K^c) < \epsilon$. A sequence Q_n of probability measures on a metric space is said to be weakly convergent to a probability measure Q if for every bounded continuous function f on the space, $\int f_n(\omega)dQ_n(\omega) \to \int f(\omega)dQ(\omega)$. A sequence Q_n of probability measures is relatively compact if it has a subsequence that is weakly convergent.

28. Absorbed and reflected Brownian motion: Let $B(.)$ be standard BM and define $M(t) = max_{0 \le s \le t} B(s)$. Then, the reflection principle gives for $x > y$,

$$Pr(M(t) > x, B(t) < y) = Pr(B(t) > 2x - y) = 1 - \Phi((2x - y)/\sqrt{t})$$

In particular, letting y increase to x gives us

$$Pr(M(t) > x) = P(M(t) > x, B(t) \le x) + P(B(t) > x) = 2(1 - \Phi(x/\sqrt{t}))$$

where

$$\Phi(z) = \frac{1}{\sqrt{2\pi}} \int_{-\infty}^{z} exp(-u^2/2)du$$

Let $T_x = min(t : B(t) > x)$. Then, we have

$$\{T_x < t\} = \{M(t) > x\}$$

so that from the reflection principle, we get

$$P(T_x < t, B(t) < y) = 1 - \Phi((2x - y)/\sqrt{t})$$

In the above formulas, we are assuming that the BM starts at 0. Suppose now that the BM starts at some x 0. Let $T_0 = min(t \ge 0 : B(t) = 0\}$. Then,

$$P(T_0 < t) = P(min_{0<s<t} B(s) < 0 | B(0) = x) = P(min_{0<s<t} B(s) < -x | B(0) = 0)$$
$$= P(max_{0<s<t} - B(s) > x | B(0) = 0) = P(M(t) > x) = 2(1 - \Phi(x/\sqrt{t}))$$

For $x > 0$ and $y > 0$, we have

$$P(B(t) < y, T_0 > t | B(0) = x) = P(B(t) < y, min_{0<s<t} B(s) > 0 | B(0) = x)$$
$$= P(B(t) < y - x, min_{0<s<t} B(s) > -x | B(0) = 0)$$
$$= P(B(t) > x - y, max_{0<s<t} B(s) < x) = P(B(t) > x - y, M(t) < x)$$

This probability is non-zero only when $x-y<x$, ie, $y>0$ as assumed. Now,

$$P(B(t)>x-y, M(t)<x) = P(B(t)>x-y) - P(B(t)>x-y, M(t)>x)$$

$$= P(B(t)>x-y) - P(M(t)>x) + P(B(t)<x-y, M(t)>x)$$

It follows that

$$P(B(t)\in dy, T_0>t) = \frac{\partial}{\partial y}[P(B(t)>x-y) + P(B(t)<x-y, M(t)>x)]$$

This evaluates to

$$\frac{1}{\sqrt{t}}(\phi((x+y)/\sqrt{t}) - \phi((x-y)/\sqrt{t}))$$

where $\phi(u) = \frac{1}{\sqrt{2\pi}}exp(-u^2/2)$ is the standard normal density. We have thus calculated the transition probability of absorbed Brownian motion, the absorption being at the origin.

29. Martingales associated with the Brownian motion process: Let $B(t), t\geq 0$ be a standard Brownian motion process. A process $X(t)$ that is continuous, square integrable and a martingale with respect to the Brownian filtration must necessarily be expressible as an Ito stochastic integral, ie,

$$X(t,\omega) = \int_0^T f(s,\omega)dB(s,\omega)$$

where $f(t,\omega)$ is adapted, and square integrable over compact intervals. Note that if $B(.)$ is a Brownian motion with respect to a filtration $\{\mathcal{F}_t : t\geq 0\}$ (not necessarily the filtration generated by the process itself) and if X_t is a continuous square integrable martingale with respect to the filtration $\{\mathcal{F}_t\}$, then it necessarily has the representation

$$X_t = \int_0^t f(s)dB(s) + Z_t$$

where f is adapted with respect to the filtration and the process Z is such that if Y is any adapted process and $I_t(Y) = \int_0^t Y_s dB(s)$, then $< I(Y), Z >= 0$.

30. Large deviations principle:

1. Let $X_1, X_2, ...$ be a sequence of independent $N(0,1)$ variables. Define $U_n = S_n/n$ where $S_n = X_1 + ... + X_n$. It is clear that $U_n = N(0,1/n)$ and hence for any $\delta>0$, it follows that

$$P(|U_n|>\delta) = P(|U_n|\sqrt{n} > \delta\sqrt{n}) = 2(1-\Phi(\delta\sqrt{n}))$$

where Φ is the standard normal distribution, ie,

$$\Phi(x) = (2\pi)^{-1/2}\int_{-\infty}^{x} exp(-u^2/2)du$$

We claim that

$$lim_{n\to\infty} n^{-1}.log(P(|U_n|>\delta) = -\delta^2/2$$

or equivalently, for large n,

$$P(|U_n| > \delta) \approx exp(-n\delta^2/2)$$

2. Σ is a given finite alphabet say $\{a_1, ..., a_d\}$ and $Y_1, ..., Y_n$ is a sequence of iid random variables assuming values in this alphabet according to the law μ, ie, $P(Y_j = a_i) = \mu(a_i), i = 1, 2, ..., d$. For any given sequence $y_1, ..., y_n \in \Sigma$, we write $\mathbf{y} = (y_1, ..., y_n)$ and let $L_n(\mathbf{y})$ denote the empirical distribution of the sequence $\mathbf{y}$, ie,

$$L_n(\mathbf{y})(a_i) = \frac{1}{n}\sum_{j=1}^{n} \chi_{\{a_i\}}(y_j)$$

In other words, $L_n(\mathbf{y})(a_i)$ is the proportion of times the symbol a_i appears in the sequence $\mathbf{y}$. Suppose ν is a probability distribution on Σ such that for every $i = 1, 2, ..., d$, $n\nu(a_i)$ is an integer. Then obviously there exists an n length vector $\mathbf{y}$ in Σ^n such that $L_n(\mathbf{y}) = \nu$. The number of such vectors $\mathbf{y}$ is clearly

$$|T_n(\nu)| = \frac{n!}{\Pi_{i=1}^{d}(n\nu(a_i))!}$$

Let $|\mathcal{L}_n|$ denote the number of all empirical probability distributions on Σ corresponding to n length sequences. Any probability distribution μ of this form can be visualized as a d-length vector $(\mu(a_1), ..., \mu(a_d))$ and since $\mu(a_i) \in \{0/n, 1/n, ..., n/n\}$ for all i, it follows that

$$|\mathcal{L}_n| \le (n+1)^d$$

Suppose ν is any probability distribution on Σ. We then claim that we can find an empirical probability distribution ν' corresponding to an n-length sequence such that $|\nu(a_i) - \nu'(a_i)| \le 1/n, i = 1, 2, ..., n$. Choose integers $i_1, i_2, ..., i_n$ such that $i_j/n \le \nu(a_j) < (i_j + 1)/n, j = 1, 2, ..., d$, where $i_j + 1$ is replaced by n if $i_j = n$. Set $\nu'(a_1) = i_1/n$. Then, $0 \le \nu(a_1) - \nu'(a_1) < 1/n$. Suppose $\nu(a_2) = i_2/n$. Then set $\nu'(a_2) = i_2/n$. Otherwise, if $\nu(a_2) > i_2/n$, then set$\nu'(a_2) = (i_2 + 1)/n$. Then $0 \le \nu(a_1) - \nu'(a_1) < 1/n$ and $-1/n < \nu(a_2) - \nu'(a_2) \le 0$. It follows that $-1/n < \nu(a_1) + \nu(a_2) - (\nu'(a_1) + \nu'(a_2)) < 1/n$. If the quantity within is in $[0, 1/n)$ then choose $\nu'(a_3) = (i_3+1)/n$ provided $\nu(a_3) > i_3/n$ while if $\nu(a_3) = i_3/n$, then set $\nu'(a_3) = (i_3+1)/n$. This would imply that $-1/n < \nu(a_3) - \nu'(a_3) \le 0$ and if the quantity within is in $(-1/n, 0]$, then choose $\nu'(a_3) = i_3/n$. This would imply that $0 \le \nu(a_3) - \nu'(a_3) \le 1/n$. In either case, we get $-1/n < \sum_{i=1}^{3}(\nu(a_i) - \nu'(a_i)) < 1/n$. Continuing this way, we get $\nu'(a_i), i = 1, 2, ..., d-1$ with the property $|\nu(a_i) - \nu'(a_i)| < 1/n$ for $i = 1, 2, ..., d-1$ and $|\sum_{i=1}^{d-1}(\nu(a_i) - \nu'(a_i))| < 1/n$. Note that $\nu'(a_i)$ are all non-negative real numbers. If $\nu(a_d) = 0$, then we have $\sum_{i=1}^{d-1}\nu(a_i) = 1$ and the above inequality becomes $1 - 1/n < \sum_{i=1}^{d-1}\nu'(a_i) < 1 + 1/n$. It follows that $\sum_{i=1}^{d-1}\nu'(a_i) = 1$ since the $\nu'(a_i)'s$ all are in the set $i/n, i = 0, 1, 2, ...$ If $\nu(a_d) > 0$, we get

$$1 - \nu(a_d) - 1/n < \sum_{i=1}^{d-1}\nu'(a_i) < 1 - \nu(a_d) + 1/n$$

Define

$$\nu'(a_d) = 1 - \sum_{i=1}^{d-1}\nu'(a_i)$$

Then, the above implies

$$\nu(a_d) - 1/n < \nu'(a_d) < \nu(a_d) + 1/n$$

This implies that

$$-1/n < \nu'(a_d) < \nu(a_d) + 1/n$$

and hence $\nu'(a_d) \geq 0$. This shows that ν' is an empirical probability distribution. The proof is complete. $\mathcal{L}_n$ is the set of all empirical probability distributions on Σ corresponding to sequences of length n. We have just shown that for each probability distribution ν on Σ, there exists a $\nu' \in \mathcal{L}_n$ such that $|\nu(a_i) - \nu'(a_i)| \leq 1/n$ for $i = 1, 2, ..., d$. It follows that

$$d(\nu, \nu') = \frac{1}{2}\sum_{i=1}^{d} |\nu(a_i) - \nu'(a_i)| \leq d/2n$$

In other words, we have proved that for any probability distribution ν on Σ, we have

$$d(\nu, \mathcal{L}_n) \leq d/2n$$

Recall that $d = |\Sigma|$, the number of elements in Σ.

Remark:given two probability distributions μ, ν on the finite alphabet Σ, the distance between the two is defined by

$$d(\mu, \nu) = sup_{A \subset \Sigma}|\mu(A) - \nu(A)|$$

Suppose A_0 is a subset of Σ at which the above supremum is attained. Then

$$d(\mu, \nu) = |\mu(A_0) - \nu(A_0)| = |\mu(A_0^c) - \nu(A_0^c)|$$

and hence,

$$d(\mu, \nu) \leq \sum_{a \in A_0} |\mu(a) - \nu(a)| = \sum_{a \in A_0^c} |\mu(a) - \nu(a)|$$

implying that

$$d(\mu, nu) \leq \frac{1}{2}\sum_{a \in \Sigma} |\mu(a) - \nu(a)|$$

On the other hand, suppose

$$B = \{a \in \Sigma : \mu(a) \geq \nu(a)\}$$

Then

$$\sum_{a \in \Sigma} |\mu(a) - \nu(a)| = \sum_{a \in B} (\mu(a) - \nu(a)) + \sum_{a \in B^c} (\nu(a) - \mu(a))$$

$$= 2\sum_{a \in B} (\mu(a) - \nu(a))$$

implying that

$$\frac{1}{2}|\sum_{a \in \Sigma} |\mu(a) - \nu(a)| \leq d(\mu, \nu)$$

This shows that

$$d(\mu,\nu) = \frac{1}{2}\sum_{a\in\Sigma}|\mu(a)-\nu(a)|$$

Let Σ be a finite alphabet as before of size d and let ν be an empirical probability distribution on Σ corresponding to a sequence of length n. In other words, ν is a probability distribution on Σ and $\nu(a_i) \in \{0/n, 1/n, ..., n/n\}$ for all $i = 1, 2, ..., d$. $T_n(\nu)$ is the set of all n-length sequences having empirical distribution ν. Let $\mathbf{y} \in T_n(\nu)$ and let μ be any probability distribution on Σ. Suppose the random vector $\mathbf{Y}$ of length n consists of iid random variables, each having distribution μ, ie,

$$P(Y_i = a_{j_i}, i = 1, 2, ..., n) = \Pi_{i=1}^n \mu(a_{j_i})$$

Then,

$$P(\mathbf{Y} = \mathbf{y}) = \Pi_{i=1}^d \mu(a_i)^{n\nu(a_i)} = exp(n\sum_{i=1}^d \nu(a_i)log(\mu(a_i)))$$

For any probability distributions θ_1, θ_2 on Σ, we define

$$H(\theta_1) = -\sum_{i=1}^d \theta(a_i)log(\theta(a_i)), H(\theta_2|\theta_1) = \sum_{i=1}^d \theta_2(a_i)log(\theta_2(a_i)/\theta_1(a_i))$$

$H(\theta_1)$ is, in the language of Shannon, the entropy of θ_1 and $H(\theta_2|\theta_1)$ is the relative entropy of θ_2 given θ_1. We then have

$$-\sum \nu(a_i)log(\mu(a_i)) = -\sum_i \nu(a_i)log(\nu(a_i)) + \sum_i \nu(a_i)log(\nu(a_i)/\mu(a_i)) = H(\nu) + H(\nu|\mu)$$

and hence,

$$P(\mathbf{Y} = \mathbf{y}) = exp(-n(H(\nu) + H(\nu|\mu))$$

Now assume that $\nu \in \mathcal{L}_n$ and that the random vector $\mathbf{Y}$ consists of iid components, each having the distribution ν. Then, $\nu' \in \mathcal{L}_n$, we have

$$P(L_n(\mathbf{Y}) = \nu') = |T_n(\nu')|\Pi_{i=1}^d \nu(a_i)^{n\nu'(a_i)}$$

This is justified as follows. We have

$$P(L_n(\mathbf{Y}) = \nu') = \sum_{\mathbf{y}\in T_n(\nu')} P(\mathbf{Y} = \mathbf{y})$$

For any $\mathbf{y} \in T_n(\nu')$, we have

$$P(\mathbf{Y} = \mathbf{y}) = \nu(y_1)...\nu(y_n) = \nu(a_1)^{n\nu'(a_1)}...\nu(a_d)^{n\nu(a_d)}$$

since a_k occurs $n.\nu'(a_k)$ times in $\mathbf{y}$ for each $k = 1, 2, ..., d$. Thus,

$$P(L_n(\mathbf{Y}) = \nu)/P(L_n(\mathbf{Y}) = \nu') = [|T_n(\nu)|/|T_n(\nu')|][\Pi_i \nu(a_i)^{n\nu(a_i)}/\Pi_i \nu(a_i)^{n\nu'(a_i)}]$$

Now observe that

$$T_n(\nu) = n!/\Pi_i(n\nu(a_i))!$$

so that

$$P(L_n(\mathbf{Y}) = \nu')/P(L_n(\mathbf{Y}) = \nu) = \Pi(n\nu'(a_i))!\nu(a_i)^{n\nu(a_i)}/(n\nu(a_i))!\nu(a_i)^{n\nu'(a_i)}$$

Now let m, l be positive integers with $m \geq l$. Then,

$$m!/l! = m(m-1)...(m-l+1) \geq l.l...l(m - ltimes)$$

or

$$m!/l! \geq l^{m-l}$$

Suppose $m < l$. Then

$$m!/l! = 1/[l(l-1)...(l-m+1)] \geq 1/(l.l....l) = 1/l^{l-m} = l^{m-l}$$

Thus, for all positive integers l, m, we have

$$m!/l! \geq l^{m-l}$$

and hence

$$P(L_n(\mathbf{Y} = \nu')/P(L_n(\mathbf{Y}) = \nu) \geq 1$$

It follows that

$$1 = \sum_{\nu' \in \mathcal{L}_n} P(L_n(\mathbf{Y}) = \nu') \leq |\mathcal{L}_n| P(L_n(\mathbf{Y}) = \nu) = |\mathcal{L}_n||T_n(\nu)|exp(-nH(\nu))$$

This furnishes us with an important inequality on $|T_n(\nu)|$.

31.An application of the LMS algorithm to frequency estimation. The desired signal is given by

$$V(t) = A.exp(j(\omega t + \phi)) + n(t) = exp(j\omega).A.exp(j(\omega(t-1) + \phi)) + n(t)$$

This equation suggests that ω can be estimated by applying the LMS algorithm to the equation

$$\hat{V}(t) \approx exp(j\omega)V(t-1)$$

Since the frequency appears in the phase of the signal, we may match the phases instead. Let $\hat{\omega}(t)$ be the estimate of the frequency at time t. Then define the weight

$$w(t) = exp(j\hat{\omega}(t))$$

and write the above approximate equation as

$$\hat{V}(t) = w(t)V(t-1)$$

from which we deduce that

$$Arg(\hat{V}(t)) = \hat{\omega}(t) + Arg(V(t-1))$$

The LMS algorithm for $\hat{\omega}(t)$ gives

$$d\hat{\omega}(t) = -dt.\mu\frac{d}{d\hat{\omega}(t)}(Arg(V(t)) - Arg(\hat{V}(t)))^2 = -dt.\mu\frac{d}{d\hat{\omega}(t)}(Arg(V(t)) - \hat{\omega}(t) - Arg(V(t-1)))^2$$

$$= 2\mu.dt.(Arg(V(t)) - \hat{\omega}(t) - Arg(V(t-1)))$$

The discretized version of this algorithm for estimating the frequency ω adaptively reads

$$\hat{\omega}[n+1] - \hat{\omega}[n] = 2\mu.\Delta(Arg(V[n]) - \hat{\omega}[n] - Arg(V[n-1]))$$

where Δ is the time step size of discretization. We now derive the stochastic differential equation from which convergence properties of the frequency estimate can be obtained. Note that

$$log(V(t)) = log(A.exp(j(\omega t + \phi)) + n(t)) = \omega t + \phi + log(1 + A^{-1}exp(-j(\omega t + \phi))n(t))$$

$$\approx \omega t + \phi + A^{-1}exp(-j(\omega t + \phi))n(t)$$

Thus,

$$Arg(V(t)) = Im(logV(t)) \approx \omega t + \phi - n(t).sin(\omega t + \phi)/A$$

Thus,

$$d\hat{\omega}(t) = 2\mu.dt(\omega t + \phi - n(t).sin(\omega t + \phi)/A - \hat{\omega}(t) - \omega(t-1) - \phi + n(t-1).sin(\omega(t-1) + \phi)/A)$$

32. Suppose X is the observation. Under hypothesis H_0, X has got the density $p_0(X)$ while under hypothesis H_1, it has the density $p_1(X)$. Define $\Lambda(X) = p_1(X)/p_0(X)$. The likelihood ratio test (LRT) is given by select H_1 if $\Lambda(X) \geq \lambda$ and select H_0 if $\Lambda(X) < \lambda$ where λ is a given threshold. The probability of error is

$$Pr(\epsilon) = P_0\int_{\Lambda(x)\geq\lambda} p_0(x)dx + P_1\int_{\Lambda(x)<\lambda} p_1(x)dx$$

where P_0, P_1 are respectively the apriori probabilities of H_0 and H_1. The Neymnan-Pearson test involves choosing the decision region so that $Pr(H_0|H_1)$ is minimized for a given $Pr(H_1|H_0)$, say α. Let Z_1 be the region in which if X falls, we decide H_1. Then using Lagrange multipliers, we must select Z_1 so that

$$(1 - \int_{Z_1} p_1(x)dx) + \mu(\int_{Z_1} p_0(x)dx - \alpha)$$

is minimized, or equivalently

$$\int_{Z_1}(p_1(x) - \mu p_0(x))dx$$

is maximized. It is clear that this happens when and only when

$$Z_1 = \{x : p_1(x)/p_0(x) \geq \mu\}$$

That is when the test is the likelihood ratio test. The threshold μ is obtained from the condition

$$\int_{Z_1} p_0(x)dx = \alpha$$

33.

1. Notion of experiment, elementary outcomes, examples:throw of a die, throw of a coin etc.

2.Σ field of events on a probability space, properties of fields and σ fields.

3.Σ field generated by a class of subsets of a set.

4. Independence of events.

5. The Borel-Cantelli lemmas.

6. Definition of measurable functions and random variables.

7. Probability distribution functions, absolute continuity of one measure with respect to another, probability density functions.

8. Examples of discrete probability distribution functions:

(a) $\{p_1, ..., p_n, ...\}$ are probabilties on a countable set E with $\sum p_i = 1$. Then the distribution function is given by

$$F(x) = \sum_{x_i \leq x} p_i$$

where we take

$$E = \{x_1, x_2, ...\}, p_i = Pr(x_i)$$

(b) Bernoulli distribution as a special case of (a): The set here is $E = \{0, 1\}$ and $Pi(0) = p, Pr(1) = q = 1 - p$. The probability distribution function is $F(x) = 0, x < 0, F(x) = p, 0 \leq x < 1, F(x) = 1, x \geq 1$. If $f : E \to \mathbb{R}$ is a random variable, then its expectation is given by

$$\mathbb{E}f = pf(0) + (1-p)f(1) = \mu$$

The variance of the random variable f is given by

$$V(f) = (f(0) - \mu)^2 p + (f(1) - \mu)^2 (1-p)$$

(c) Binomial distribution and its derivation from independent Bernoulli trials:

(d) Poisson distribution as a limit of the Binomial distribution.

9. Examples of continuous probability distribution functions:

[a] Uniform, [b] Normal, [c] Exponential, [d] χ-square, [e] Beta distribution.

9. Mean and variance of random variables, covariance matrix of random vectors.

10. Characteristic function and moment generating function of random variables.

11. Conditional expectation, conditional probability density of one random vector given another.

12. Random walk process

13. Central limit theorem.

14. Bayesian hypothesis testing.

15. Conditional expectation as a minimum mean square estimate.

16. Introduction to random processes in discrete and continuous time:

[a] Wide sense stationary process, [b] Strict sense stationary process, [c] Markov chains and processes, [d] Martingale process.

Estimation of a probability density by the moment matching technique: Let $X_1, ..., X_n$ be random variables all samples from the same probability distribution $p(X|\theta)$. Here, θ is an unknown parameter or a parameter vector. We consider functions $\phi_i, i = 1, 2, ..., p$, where p is the number of parameters in the vector θ. Then since $\frac{1}{N}\sum_{k=1}^{N}\phi_i(X_k)$ is an unbiased estimator of $\mathbb{E}\phi_i(X)$, we may hope to obtain a good estimate of θ by solving the equations

$$\frac{1}{N}\sum_{k=1}^{N}\phi_i(X_k) = \int \phi_i(X)p(X|\theta)dX, i = 1, 2, ..., p$$

Suppose $\phi_1, ..., \phi_N$ are

34.Large deviations, Cramer's theorem.

Let $X_1, ..., X_n, ...$ be iid random variables and define

$$Z_n = S_n/n, S_n = X_1 + X_2 + ... + X_n$$

Let μ_n be the probability distribution of Z_n. We have

$$\mu_n([x, \infty)) = \mathbb{E}\chi_{Z_n \geq x} \leq \mathbb{E}exp(n\lambda(Z_n - x))$$

fore all $\lambda \geq 0$. Let

$$M(\lambda) = \mathbb{E}exp(\lambda X_1)$$

Then, we get from the above,

$$\mu_n([x, \infty)) \leq exp(-n\lambda x)\mathbb{E}exp(\lambda X_1) = exp(-n\lambda x)M(\lambda)^n$$

Define

$$\Lambda(\lambda) = logM(\lambda)$$

Λ is a convex function. This follows from Holder's inequality. Indeed, let $\theta \in [0, 1]$. Then,

$$M(\theta\lambda_1 + (1-\theta)\lambda_2) = \mathbb{E}exp((\theta\lambda_1 + (1-\theta)\lambda_2)X_1)$$

$$= \mathbb{E}[(exp\lambda_1 X_1))^\theta (exp(\lambda_2 X_1)^{1-\theta}]$$

$$\leq (\mathbb{E}exp(\lambda_1 X_1))^\theta (\mathbb{E}exp(\lambda_2 X_1))^{1-\theta}) = M(\lambda_1)^\theta (M(\lambda_2)^{1-\theta}$$

and on taking logarithms, we get

$$\Lambda(\theta\lambda_1 + (1-\theta)\lambda_2) \leq \theta.\Lambda(\lambda_1) + (1-\theta)\Lambda(\lambda_2)$$

proving convexity of Λ. Define

$$\Lambda^*(x) = sup_{\lambda\in\mathbb{R}}(\lambda.x - \Lambda(\lambda))$$

Since $M(0) = 1$, it follows that $\Lambda(0) = 0$ and hence $F(\lambda, x) = \lambda x - \Lambda(\lambda)$ equals zero when $\lambda = 0$. It follows that

$$\Lambda^*(x) \geq 0, x \in \mathbb{R}$$

Since $x \to log(x)$ is concave on $(0, \infty)$, it follows from Jensen's inequality that

$$log\mathbb{E}exp(\lambda X_1) \geq \mathbb{E}log(exp(\lambda X_1) = \lambda\mathbb{E}X_1$$

or equivalently,

$$F(\lambda, \mathbb{E}X_1) = \lambda\mathbb{E}X_1 - \Lambda(\lambda) \leq 0$$

It follows from this and the fact that $F(0, x) = 0$ that

$$\Lambda^*(\mathbb{E}X_1) = 0$$

We next wish to prove that Λ^* is lower semicontinuous. Let $x_n \to x$. Then,

$$liminf\Lambda^*(x_n) = liminf\, sup_\lambda(\lambda x_n - \Lambda(\lambda)) \geq sup_\lambda liminf(\lambda x_n - \Lambda(\lambda))$$

$$= sup_\lambda(\lambda.x - \Lambda(\lambda)) = \Lambda^*(x)$$

Note that

$$\lambda x_n - \Lambda(\lambda) \geq inf_{m\geq k}(\lambda x_m - \Lambda(\lambda)), n \geq k$$

and hence

$$\Lambda^*(x_n) = sup_\theta(\theta x_n - \Lambda(\theta)) \geq inf_{m\geq k}(\lambda.x_m - \Lambda(\lambda)), n \geq k$$

so that

$$\liminf \Lambda^*(x_n) \geq lim_{k\to\infty} inf_{m\geq k}(\lambda.x_m - \Lambda(\lambda)) = \lambda x - \Lambda(\lambda)$$

It follows on taking supremum over all λ on the right side that

$$liminf\Lambda^*(x_n) \geq \Lambda^*(x)$$

proving lower semicontinuity of Λ^*. We next wish to show that $inf_{x\in\mathbb{R}}\Lambda^*(x) = 0$. Let $D_\Lambda = \{\lambda : \Lambda(\lambda) < \infty\}$. Suppose $D_\Lambda = \{0\}$. Then, $\Lambda(\lambda) = \infty$ for all $\lambda \neq 0$ and hence $F(\lambda, x) = \lambda x - \Lambda(\lambda) = -\infty$ for all $\lambda \neq 0$. Also $F(0, x) = 0$ and hence $\Lambda^*(x) = sup_\lambda F(\lambda, x) = F(0, x) = 0$ for all x proving the claim. Suppose now that $\mathbb{E}X_1$ is finite. Then we have shown using Jensen's inequality that $\Lambda^*(\mathbb{E}X_1) = 0$ and since $\Lambda^*(x) \geq 0$ for all x, it follows that $inf_x\Lambda^*(x) = 0$ proving the claim for this case. Now define for $\epsilon \in \mathbb{R}$,

$$f_\epsilon(x) = \epsilon^{-1}(exp((\eta + \epsilon)x) - exp(\eta x))$$

For every Let $\delta > 0$ and $\epsilon \in (-\delta, \delta)$. Then

$$|f_\epsilon(x)| \leq exp(\eta x)(exp(\delta|x|) - 1)/\delta$$

This is based on the fact that if $a > 0$, then $exp(\xi a) - 1)/\xi$ increases with ξ for $\xi > 0$. Indeed,

$$(exp(\xi a) - 1)/\xi = \sum_{m=1}^{\infty} \xi^m a^m / m!$$

increases with ξ for $\xi > 0$. From this inequality and the dominated convergence theorem, it follows that for all real η in $int(D_\Lambda)$ (so that $\mathbb{E}exp(\eta X_1) < \infty$), we have

$$M'(\eta) = \frac{d}{d\eta}\mathbb{E}exp(\eta X_1) = \mathbb{E}(X_1.exp(\eta X_1))$$

or

$$\Lambda'(\eta) = \mathbb{E}(X_1.exp(\eta.X_1))/M(\eta)$$

Note: If $\eta int(\in D_\Lambda)$, then for sufficiently small δ, $\eta \pm \delta \in int(D_\Lambda)$ and hence $\mathbb{E}[exp(\eta X_1).exp(\delta|X_1|)] < \infty$ and this in turn guarantees that dominated convergence can be applied. Finally, suppose that $\Lambda'(\eta) = y$ and define $g(\lambda) = \lambda y - \Lambda(\lambda)$. g is a concave function since by Holder's inequality, as seen above, $\Lambda(.)$ is convex. Also $g'(\eta) = 0$ and hence, by the basic result in calculus on maxima of functions,

$$g(\eta) = sup_\lambda g(\lambda)$$

or equivalently,

$$\eta.x - \Lambda(\eta) = \Lambda^*(x)$$

We now have for all $x \geq \mathbb{E}X_1$ and $\lambda < 0$,

$$\lambda x - \Lambda(\lambda) \leq \lambda \mathbb{E}X_1 - \Lambda(\lambda) \leq \Lambda^*(\mathbb{E}X_1) = 0$$

so that

$$\Lambda^*(x) = sup_{\lambda \geq 0}(\lambda.x - \Lambda(\lambda)), x \geq \mathbb{E}X_1$$

It follows that for $x \geq \mathbb{E}X_1$,

$$\mu_n[x, \infty) \leq inf_{\lambda \geq 0} exp(-n\lambda x) M(\lambda)^n = inf_{\lambda \geq 0} exp(-n(\lambda x - \Lambda(\lambda))$$

$$\leq exp(-n.sup_{\lambda \geq 0}(\lambda.x - \Lambda(\lambda))) = exp(-n\Lambda^*(x))$$

Suppose $x \leq \mathbb{E}X_1$. Then define $Y_1 = -X_1$ and we have $-x \geq \mathbb{E}Y_1$. Hence, by the same argument,

$$\mu_n(-\infty, x] = P(Z_n \leq x) = P(-Z_n \geq -x) \leq exp(-n.K^*(-x))$$

where

$$K^*(x) = sup_\lambda(\lambda x - K(\lambda)), K(\lambda) = log\mathbb{E}exp(\lambda Y_1) = log\mathbb{E}exp(-\lambda X_1) = logM(-\lambda) = \Lambda(-\lambda)$$

In other words,

$$K^*(-x) = sup_\lambda(-\lambda x - \Lambda(-\lambda)) = sup_\lambda(\lambda x - \Lambda(\lambda)) = \Lambda^*(x)$$

Thus, we get for all $x \le \mathbb{E}X_1$,

$$\mu_n(-\infty, x] \le exp(-n\Lambda^*(x))$$

Assume that $\mathbb{E}X_1$ is finite. Let F be non-empty and closed. Assume that

$$I_F = inf_{x\in F}\Lambda^*(x) > 0$$

Since $\Lambda^*(\mathbb{E}X_1) = 0$, it follows that $\mathbb{E}X_1 \in F^c$ and hence there exist $x_- < x_+$ such that $\mathbb{E}X_1 \in (x_-, x_+)$, $x_-, x_+ \in F$ and $(x_-, x_+) \in F^c$. Now,

$$(x_-, x_+) \subset F^c$$

implies

$$F \subset (-\infty, x_-] \cup [x_+, \infty)$$

and hence,

$$\mu_n(F) \le \mu_n(-\infty, x_-] + \mu_n[x_+, \infty) \le exp(-n\Lambda^*(x_-)) + exp(-n\Lambda^*(x_+)) \le 2.exp(-nI_F)$$

which is the desired large deviations upper bound. Note that this method works only if both x_- and x_+ are finite. Suppose x_+ is finite but $x_- = -\infty$. Then $x_+ \in F$ and we have $\Lambda^*(x_+) \ge I_F$. Also

$$\mu_n((-\infty, x_-]) = 0$$

and the same inequality is true. The same holds if x_- is finite and $x_+ = \infty$. If however, both x_- and x_+ are infinite, then $F^c = \mathbb{R}$ and F is the empty set which is contrary to assumption.

35.(Discussion with Dr.Shambunath Sharma) :Definition of a stopping time and an optional time: $(\Omega, \mathcal{F}, P)$ is a probability space and $\{\mathcal{F}_t : t \ge 0\}$ is a filtration. A random variable $T : \Omega \to [0, \infty)$ is called a stopping time if for all $t \ge 0$, the event $\{T \le t\} \in \mathcal{F}_t$. It is called an optional time if for all $t > 0$, $\{T < t\} \in \mathcal{F}_t$. Suppose T is a stop time. Then, for each $n = 1, 2, ..., \{T \le t - 1/n\} \in \mathcal{F}_{t-1/n} \subset \mathcal{F}_t$. It follows that

$$\{T < t\} = \bigcup_{n\ge1}\{T \le t - 1/n\} \in \mathcal{F}_t$$

and hence that T is an optional time. Conversely suppose T is an optional time. Then, $\{T < t\} \in \mathcal{F}_t$ for all t and hence for every $n = 1, 2, ...$ and $m \ge n$, $\{T < t + 1/m\} \in \mathcal{F}_{t+1/m} \subset \mathcal{F}_{t+1/n}$. It follows that

$$\{T \le t\} = \bigcap_{m=n}^{\infty} \{T < t + 1/m\} \in \mathcal{F}_{t+1/n}, n = 1, 2, ...$$

Then,

$$\{T \le t\} \in \bigcap_{n=1}^{\infty} \mathcal{F}_{t+1/n} = \mathcal{F}_{t+}$$

If the filtration is right continuous, then $\mathcal{F}_{t+} = \mathcal{F}_t$ and it follows then that T is also a stopping time. If T, S are stop times, so is $T + S$. For $T(\omega) + S(\omega) > t$ implies that either $S(\omega) = 0, T(\omega) > t$, $S(\omega) > t, T(\omega) = 0$, or $S(\omega) > 0, T(\omega) \ge t$, or $0 < T(\omega) < t, T(\omega) + S(\omega) > t$. Further

$$\{0 < T < t, T + S > t\} = \bigcup_{q\in Q\cap(0,t)} \{r < T < t, S > t - r\} \in \mathcal{F}_t$$

36. (Continuation of discussion with Shambunath Sharma): If T is a stop time, then T is $\mathcal{F}_T$-measurable. Note that by definition, $E \in \mathcal{F}_T$ iff $E \cap \{T \leq t\} \in \mathcal{F}_t$ for all $t \geq 0$. To show that T is $\mathcal{F}_T$-measurable, it suffices to show that for all $s \geq 0$, we have $\{T \leq s\} \in \mathcal{F}_T$ which is in turn equivalent to showing that for all $s, t \geq 0$, $\{T \leq s\} \cap \{T \leq t\} \in \mathcal{F}_t$, or equivalently that $\{T \leq min(s,t)\} \in \mathcal{F}_t$. But

$$\{T \leq min(s,t)\} \in \mathcal{F}_{min(s,t)} \subset \mathcal{F}_t$$

where the definition of a stopping time and non-decreasing property of a filtration has been used.

37. Gauss-Markov signals: Suppose $x(t,\omega), t \geq 0$ is a random process defined on a probability space $(\Omega, \mathcal{F}, P)$ that satisfies the first order differential equation

$$\frac{dx(t)}{dt} = -ax(t) + w(t), t \geq 0$$

where $w(.)$ is white Gaussian noise with zero mean and unit spectral density, and $a > 0$. By white we mean that w has autocorrelation function $R_{ww}(\tau) = \mathbb{E}(w(t+\tau)w(t)) = \delta(\tau)$. Assume further that $x(0)$ is a random variable independent of the process $\{w(t), t \geq 0\}$. The solution to the above differential equation is given by

$$x(t) = exp(-at)x(0) + \int_0^t exp(-a(t-s))w(s)ds$$

We have for $t > s$,

$$x(t) = exp(-a(t-s))x(s) + \int_s^t exp(-a(t-u))w(u)du$$

Let $\{\mathcal{F}_t : t \geq 0\}$ be the filtration generated by the process $w(.)$, ie, $\mathcal{F}_t = \sigma(w(u) : u \leq t)$. Then it is clear from the above expression that

$$P(x(t) \in E|\mathcal{F}_s) = P(\int_s^t exp(-a(t-u))w(u)du \in E - x), x = x(s)$$

Now define the random variable

$$Y = \int_s^t exp(-a(t-u))w(u)du$$

Then, Y is zero mean normal with variance

$$\int_s^t exp(-2a(t-u))du = \int_0^{t-s} exp(-2au)du = (2a)^{-1}(1 - exp(-2a(t-s)))$$

It follows that $x(.)$ is a Markov process with probability law of $x(t)$ given $x(s)$ being normal with mean $x(s)$ and variance $(2a)^{-1}(1 - exp(-2a(t-s))$. This is the most elementary example of a Gauss-Markov process. $x(.)$ is also called an Ornstein-Uhlenbeck process.

38. Let S, T be stoptimes relative to a filtration $\mathcal{F}_t, t \geq 0$. We claim that

$$\mathcal{F}_T \cap \mathcal{F}_S = \mathcal{F}_{T \wedge S}$$

First observe that if S, T are stoptimes such that $S \leq T$, then $\mathcal{F}_S \subset \mathcal{F}_T$. For let $E \in \mathcal{F}_S$. Then since $\{S \leq T\} = \Omega$ is the entire sample space, it follows that for all $t \geq 0$, we have

$$E \cap \{T \leq t\} = E \cap \{T \leq t\} \cap \{S \leq T\} = E \cap \{T \leq t\} \cap \{S \leq t\} \cap \{S \leq T\}$$

$$= E \cap \{T \leq t\} \cap \{S \leq t\} = E \cap \{S \leq t\} \cap \{T \leq t\} \in \mathcal{F}_t$$

since

$$E \cap \{S \leq t\}, \{T \leq t\} \in \mathcal{F}_t$$

Now, suppose S, T are arbitrary stop times. Then $S \wedge T \leq T, S$ and hence by the result just proved,

$$\mathcal{F}_{S \wedge T} \subset \mathcal{F}_S \cap \mathcal{F}_T$$

Conversely suppose $E \in \mathcal{F}_S \cap \mathcal{F}_T$. Then

$$E \cap \{S \wedge T \leq t\} = E \cap (\{S \leq t\} \cup \{T \leq t\}) = (E \cap \{S \leq t\}) \cup (E \cap \{T \leq t\}) \in \mathcal{F}_t$$

This completes the proof.

39. Suppose T, S are stoptimes. Then we claim that if $E \in \mathcal{F}_S$,

$$E \cap \{S \leq T\} \in \mathcal{F}_T$$

To prove this, we must show that for all $t \geq 0$,

$$E \cap \{S \leq T\} \cap \{T \leq t\} \in \mathcal{F}_t$$

Now, the right hand side is the same as

$$E \cap \{S \leq t\} \cap \{S \leq T\} \cap \{T \leq t\}$$

and since

$$E \cap \{S \leq t\} \in \mathcal{F}_t$$

it suffices to prove that

$$\{S \leq T\} \cap \{T \leq t\} \in \mathcal{F}_t$$

The set on the right is the same as

$$\{T \leq t\} - \{S > T, T \leq t\}$$

and hence, it suffices to prove that

$$\{S > T, T \leq t\} \in \mathcal{F}_t$$

The set on the right is the same as

$$\bigcup_{r \in Q} \{S > r > T, T \leq t\} = \bigcup_{r \in Q, r \leq t} \{T < r < S\} \cup \bigcup_{r \in Q, r > t} \{T \leq t, S > r\}$$

Now,

$$\{T < r\} \in \mathcal{F}_r \subset \mathcal{F}_t, r \leq t,$$

$$\{S > r\} = \{S \leq r\}^c \in \mathcal{F}_r \subset \mathcal{F}_t, r \leq t,$$

$$\bigcup_{r \in Q, r > t} \{S > r\} = \{S > t\} = \{S \leq t\}^c \in \mathcal{F}_t$$

and the proof is complete.

40. According to the previous problem, if T, S are stoptimes, then

$$\{S \leq T\} \in \mathcal{F}_T$$

It follows on taking complements that

$$\{T < S\} \in \mathcal{F}_T$$

We want to show also that

$$\{T \leq S\} \in \mathcal{F}_T$$

Consider the stoptime $R = S \wedge T$. Then by the result proved, we have

$$\{T \leq R\} \in \mathcal{F}_R \subset \mathcal{F}_T$$

But $\{T \leq R\} = \{T \leq S \wedge T\} = \{T \leq S\}$. Hence

$$\{T \leq S\} \in \mathcal{F}_T$$

41. We have thus proved that $\{S \leq T\}, \{T \leq S\}, \{S < T\}, \{T < S\}$ are all $\mathcal{F}_T$-measurable sets. Interchanging T and S, it follows that the same sets are also $\mathcal{F}_S$-measurable. It follows that these sets are all $\mathcal{F}_{T \wedge S}$ measurable. Now let Z be any L^1 random variable. We want to show that

$$\mathbb{E}(Z|\mathcal{F}_T) = \mathbb{E}(Z|\mathcal{F}_{S \wedge T}), on\{T \leq S\}$$

where T, S are any two stoptimes. Since $\mathcal{F}_{S \wedge T} \subset \mathcal{F}_T$, it suffices to show that if $E \in \mathcal{F}_T$ is arbitrary, then

$$\int_{E \cap \{T \leq S\} \mathbb{E}(Z|\mathcal{F}_{S \wedge T})dP = \int_{E \cap \{T \leq S}} \mathbb{E}(Z|\mathcal{F}_T)dP$$

Now we have already seen that since $E \in \mathcal{F}_T$, it follows that $E \cap \{T \leq S\} \in \mathcal{F}_S$ and since $\{T \leq S\} \in \mathcal{F}_T$ (as shown above), it follows that

$$E \cap \{T \leq S\} \in \mathcal{F}_T \cap \mathcal{F}_S$$

Thus we get

$$\int_{E \cap \{T \leq S\}} Z dP = \int_{E \cap \{T \leq S\}} \mathbb{E}(Z|\mathcal{F}_{S \wedge T})dP$$

and since $E, \{T \leq S\} \in \mathcal{F}_T$, we get

$$\int_{E \cap \{T \leq S\}} Z dP = \int_{E \cap \{T \leq S\}} \mathbb{E}(Z|\mathcal{F}_T)dP$$

and the proof is complete.

42. Progressive measurability of a random process. $(\Omega, \mathcal{F}, P)$ is a probability space. $\mathcal{B}[0,\infty)\otimes\mathcal{F}$ is the σ field on $\mathbb{R}_+\times\Omega$. A set $E\in\mathcal{B}[0,\infty)\otimes\mathcal{F}$ is said to be a progressively measurable set if the process $(t,\omega)\to X_t(\omega)=\chi_E(t,\omega)$ is a progressively measurable stochastic process. This is equivalent to requiring that for all $t\geq 0$,

$$\{(s,\omega): s\leq t, (s,\omega)\in E\}\in\mathcal{B}[0,t]\otimes\mathcal{F}_t$$

All progressively measurable sets form a σ-field called the progressively measurable σ-field. For example if E, F are progressively measurable sets, then

$$\{(s,\omega): s\leq t, (s,\omega)\in E\cup F\}=\{(s,\omega): s\leq t, (s,\omega)\in E\}\cup\{(s,\omega): s\leq t, (s,\omega)\in F\}\in\mathcal{B}[0,t]\otimes\mathcal{F}_t$$

and hence $E\cup F$ is also a progressively measurable set. The same argument carries through for countable unions. Suppose X is a progressively measurable stochastic process. This means that for all $t\geq 0$ and every $B\in\mathcal{B}(\mathbb{R})$, we have

$$\{(s,\omega): s\leq t, X_s(\omega)\in B\}\in\mathcal{B}[0,t]\otimes\mathcal{F}_t$$

Then we claim that $(t,\omega)\to X_t(\omega)$ is measurable with respect to the progressively measurable σ-field. To do so, we must show that if B is any Borel subset of $\mathbb{R}$, then $E=\{(t,\omega): X_t(\omega)\in B\}$ is a progressively measurable set. But this is a consequence of the definition of progressive measurablity of a stochastic process. Indeed,

$$\{(s,\omega): s\leq t, (s,\omega)\in E\}=\{(s,\omega): s\leq t, X_s(\omega)\in B\}\in\mathcal{B}[0,t]\otimes\mathcal{F}_t$$

Suppose X is a progressively measurable process and let $A\in\mathcal{B}(\mathbb{R}_+)$ and $B\in\mathcal{B}(\mathbb{R})$. Then $A\times B\in\mathcal{B}(\mathbb{R}_+)\otimes\mathcal{B}(\mathbb{R})$. We want to show that the process $(t,\omega)\to Y(t,\omega)=\chi_A(t)\chi_B(X_t(\omega))$ is progressively measurable. Then,

$$\{(s,\omega): s\leq t, Y(s,\omega)=1\}=[(A\cap[0,t])\times\Omega]\cap\{(s,\omega): s\leq t, X_s(\omega)\in B\}$$

which is clearly an element of $\mathcal{B}[0,t]\otimes\mathcal{F}_t$. Now suppose $f:\mathbb{R}_+\times\mathbb{R}\to\mathbb{R}$ is a measurable function and let X be progressively measurable. We wish to show that the process $(t,\omega)\to f(t,X_t(\omega))$ is progressively measurable. Since f can be approximated by linear combinations of indicators of sets of the form $A\times B$ with $A\in\mathcal{B}(\mathbb{R}_+)$ and $B\in\mathcal{B}(\mathbb{R})$ and the result is true for $f=\chi_{A\times B}$, we are through ??

[43] Let X be a positive integer valued random variable and Y another positive integer valued random variable. Assume that $P(X=n,Y=m)=p(n,m), n,m=0,1,2,...,$ so that $\sum_{n,m}p(n,m)=1$. Given $X=nandY=m$, let $Z(i,j), i=1,2,...,m, j=1,2,...,n$ be a random variable with distribution

$$P(Z(i,j)>\tau_j, j=1,2,...,n|X=n,Y=m)=G_n(\tau_1,...,\tau_n), i=1,2,...,m$$

Conditioned on $X=n, Y=m$, define the event

$$E_i=\{Z(i,1)>\tau_1,...,Z(i,n)>\tau_n\}^c, i=1,2,...,m$$

Then, conditioned on $X = n$ and $Y = m$, the probability that r of the $E_i's$ will occur is given by

$$Q(n,m,r) = \binom{m}{r}(1 - G_n(\tau_1, ..., \tau_n))^r G_n(\tau_1, ..., \tau_n)^{m-r}$$

and the unconditioned probability that exactly r of the $E_i's$ occur is given by the formula

$$\sum_{n,m} Q(n,m,r)p(n)q(m)$$

[44] Heat equation with random diffusion coefficient.

$$u_{tt}(t,x) = D.u_{xx}(t,x), 0 < x < L, t \geq 0$$

Solution

$$u(t,x) = \sum_n c_n(t)sin(n\pi x/L)$$

$$c_n''(t) = -D(n\pi/L)^2 c_n(t)$$

$$c_n(t) = A_n.cos(n\pi\sqrt{D}t/L) + B_n.sin(n\pi\sqrt{D}t/L)$$

$$u(0,x) = \sum A_n.sin(n\pi x/L), A_n = (2/L)\int_0^L u(0,x)sin(n\pi xL)dx$$

$$u_t(0,x) = \sum B_n(n\pi\sqrt{D}/L)sin(n\pi x/L), B_n = (2/n\pi\sqrt{D})\int_0^L u_t(0,x)sin(n\pi x/L)dx$$

Assuming that

$$Pr(D \in d\xi) = dF(\xi)$$

determine the correlations

$$< u(t_1,x_1).u(t_2,x_2) >$$

and more generally the kth order moments

$$< \Pi_{i=1}^k u(t_i,x_i) >$$

[45] Discussion with Dr.Shambunath Sharma: Definition and properties of the Poisson process. Let $X_i, i = 1,2,...$ be independent identical and exponentially distributed random variables. Thus,

$$Pr(X_i \in dx) = \lambda.exp(-\lambda.x)dx, x \geq 0$$

Define

$$S_0 = 0, S_n = X_1 + X_2 + ... + X_n, n \geq 1$$

Also define

$$N(t) = max(t \geq 0 : S_n \leq t)$$

We can visualize S_n as the time of arrival of the nth passenger in the queue. $X_{n+1} = S_{n+1} - S_n$, is the difference in the times of arrival of the nth and $(n+1)^{th}$ passenger. We clearly have

$$\{N(t) = n\} = \{S_n \leq t < S_{n+1}\}$$

Also,

$$\{N(t) \geq n\} = \{S_n \leq t\}$$

We have

$$\mathbb{E}exp(-sS_n) = [\mathbb{E}exp(-sX_1)]^n = \frac{\lambda^n}{(s+\lambda}{)^n})$$

so that the density of S_n is given by

$$f_{S_n}(x) = [\lambda^n x^{n-1}/(n-1)!]exp(-\lambda x)U(x)$$

We want to show that the process $\{N(t) : t \geq 0\}$ has independent increments. Let $t > s \geq 0$. Then, we shall prove that

$$Pr(N(t) - N(s) \geq n | N(u), u \leq s) = P(N(t) - N(s) \geq n), n = 0, 1, 2, ...$$

We have

$$P(N(t) - N(s) \geq n | N(u), u \leq s, N(s) = r) = P(N(t) \geq n + r | N(u), u \leq s, N(s) = r)$$

Let

$$0 = \tau_0 < \tau_1 < ... < \tau_p = s$$

Consider the event

$$E = \{N(\tau_k) \geq m_k, k = 0, 1, 2, ..., p\}$$

Since N is a non-decreasing process, we may without loss of generality assume that $m_0 \leq m_1 \leq ... \leq m_p$. We then claim that $E \cap \{N(s) = r\} = A \cap \{N(s) = r\}$ for some $A \in \sigma(X_1, ..., X_r\}$. To show this, first note that

$$E \cap \{N(s) = r\} = \{S_{m_k} \leq \tau_k, k = 0, 1, ..., p, S_r \leq s < S_{r+1}\}$$

If $m_k > r$, then $S_r \leq s < S_{r+1}$ would imply that $S_{m_k} > s$ which would imply that the above set is empty. So we may assume that $m_k \leq r$ and then it becomes clear that

$$\{S_{m_k} \leq \tau_k, k = 0, 1, ..., p\} \in \sigma(S_0, ..., S_r\} = \sigma\{X_1, ..., X_r\}$$

This proves the claim. It thus follows that

$$P(N(t) - N(s) \geq n | N(\tau_k) \geq m_k, k = 0, 1, ..., p, N(s) = r) = P(N(t) \geq n + r | A, N(s) = r)$$

for some

$$A \in \sigma(X_1, ..., X_r)$$

The right side can be expressed as

$$p = P(N(t) \geq n+r|A, N(s)=r) = P(S_{n+r} \leq t|A, N(s)=r) = P(S_{n+r} - S_r \leq t - S_r|A, N(s)=r)$$

$$= P(S_{n+r} - S_r \leq t - S_r, N(s)=r|A)/P(N(s)=r|A)$$

$$P(S_{n+r} - S_r \leq t - S_r, N(s)=r|A, S_r = x) = P(S_{n+r} - S_r \leq t - x, s - x < X_{r+1})$$

assuming that $x \leq s$. It is zero if $x > s$. This evaluates to

$$\int_{s-x}^{t-x} F^{(n-1)}(t-x-u)dF(u)$$

Thus,

$$P(S_{n+r} - S_r \leq t - S_r, N(s)=r|A)$$

$$= \int F^{(n-1)}(t-x-u)dF(u)P(S_r \in dx|A)$$

where u ranges over $[s-x, t-x]$ and x ranges over $[0, s]$. Without loss of generality, we assume that on A, $\{S_r \leq y\}$ where $y \leq s$ holds. Then for $x \leq y$, we have

$$P(S_r \in dx|A) = P(S_r \in dx)/P(A) = dF^{(r)}(x)/P(A)$$

so that the above probability becomes

$$P(A)^{-1} \int F^{(n-1)}(t-x-u)dF(u)dF^{(r)}(x)$$

The integral is such that u ranges over $[s-x, t-x]$ and x ranges over $[0, y]$. This is the same as

$$P(A)^{-1}P(S_r \leq y, s \leq S_{r+1} \leq t, S_{n+r} \leq t)$$

On the other hand,

$$P(N(s)=r|A) = P(N(s)=r, A)/P(A)$$

and

$$P(N(s)=r|A) = P(S_r \leq s < S_{r+1}, A)$$

$$= \int_0^y P(X_{r+1} > s-x)P(S_r \in dx|A) = P(A)^{-1}\int_0^y (1-F(s-x))dF^{(r)}(x)$$

Thus,

$$p = (\int_0^y (1-F(s-x))dF^{(r)}(x))^{-1} \int F^{(n-1)}(t-x-u)dF(u)dF^{(r)}(x)$$

where in the second integral, the limits are $x \in [0, y]$ and $u \in [s-x, t-x]$. We aim to show that p is independent of r, y. It may not be easy to show this from the above formulae. Note also that

$$\int_0^y (1-F(s-x))dF^{(r)}(x) = P(s \leq S_{r+1}, S_r \leq y)$$

Thus,

$$p = \frac{P(S_r \le y, s \le S_{r+1}, S_{n+r} \le t)}{P(s \le S_{r+1}, S_r \le y)}$$

$$= P(S_{n+r} < t | S_r < y, S_{r+1} > s)$$

In order to prove that the process $N(.)$ has independent increments, it suffices therefore to show that this p is independent of r, s. Now,

$$P(S_{n+r} \in dt, S_r \in dy, S_{r+1} > s) = \int_{u \in [s,t]} P(S_{n+r} - S_{r+1} \in dt - u) P(S_{r+1} \in du | S_r = y) P(S_r \in dy)$$

$$= ((n-2)!)^{-1}.dtdy \int_s^t \lambda^{n-1}(t-u)^{n-2} exp(-\lambda(t-u))\lambda.exp(-\lambda(u-y))du.\lambda^r y^{r-1} exp(-\lambda y)/(r-1)!$$

$$= dtdy[(n-2)!(r-1)!]^{-1}\lambda^{n-1}\lambda^r y^{r-1} exp(-\lambda t)dtdy \int_s^t (t-u)^{n-2} du$$

This quantity can be expressed in separable form as $\phi(t-s, n)\psi(r, y)dtdy$. Thus,

$$p = (\int_s^t \phi(v-s, n)dv \int_0^y \psi(r, w)dw)/) \int_s^\infty \phi(v-s, n)dv \int_0^y \psi(r, w)dw)$$

$$= \int_s^t \phi(v-s, n)dv / \int_s^\infty \phi(v-s, n)dv$$

This is independent of y, r thereby proving that $N(.)$ has independent increments. We also have an explicit formula for p:

$$p = P(N(t) - N(s) \ge n)$$

$$= \int_{s<u<v<t} (v-u)^{n-2} exp(-\lambda(v-u))dudv / \int_{s<u<v<\infty} (v-u)^{n-2} exp(-\lambda(v-u))dudv$$

We do not evaluate this here.

Note:The density of S_n is given by

$$f_n(x) = \lambda^n x^{n-1} exp(-\lambda x)U(x)/(n-1)!$$

An easier proof that $N(.)$ has independent increments. We have formally,

$$N(t) = \sum_{i=0}^{\infty} \theta(t - S_i)$$

where $X_0 = 0$ and $\theta(t)$ is the unit step function, ie, $\theta(t) = 1$ if $t \ge 0$ and $\theta(t) = 0$ if $t < 0$. Then,

$$\frac{dN(t)}{dt} = \sum_{i=0}^{\infty} \delta(t - S_i)$$

[46]. Discussion with Dr.Shambu Sharma: Let $X_n, n = 0, 1, 2, ..., N, N+1$ be a martingale and let c be any real number. Let $X^* = max(X_n : 0 \leq n \leq N)$. If $X^* > c$, then let $T = min(n \geq 0 : X_n > c)$ else define $T = N$. Then T is a stoptime. Indeed, let $n = 0, 1, 2, ..., N-1$. Then, the event $T = n$ is the same as the event $X_k \leq c, k = 0, 1, ..., n-1, X_n > c$ and this is $\mathcal{F}_n$ measurable. Further, $T = N$ is the union of the events $X^* \leq c$ and $X_n \leq c, n = 0, 1, ..., N-1, X_N > c$ and this is clearly in $\mathcal{F}_N$. Thus by the optional stopping theorem,

$$\mathbb{E}X_N \geq \mathbb{E}X_T = \mathbb{E}X_T\chi_{X^*\leq c} + \mathbb{E}X_T\chi_{X^*>c} = \mathbb{E}X_N\chi_{X^*\leq c} + \mathbb{E}X_T\chi_{X^*>c}$$
$$\geq \mathbb{E}X_N\chi_{X^*\leq c} + c.P(X^* > c)$$

and this gives

$$c.P(X^* > c) \leq \mathbb{E}X_N\chi_{X^*>c}$$

In particular,

$$c.P(X^* > c) \leq \mathbb{E}(X_N^+)$$

This is the famous maximal inequality for martingales (See the book by Revuz and Yor, Continuous martingales and Brownian motion).

47. Doob's inequality: Let $X_n, n = 0, 1, 2, ..., N$ be a submartingale. Then let $X^* = max(X_n : 0 \leq n \leq N)$. Then, for $p > 1$, we have

$$\mathbb{E}(X^{*p}) \leq (p/p-1)^p\mathbb{E}(|X_N|^p)$$

Proof:

$$\mathbb{E}X^{*p} = \mathbb{E}\int_0^{X^*} p\lambda^{p-1}d\lambda = \mathbb{E}\int_0^\infty p\lambda^{p-1}\chi_{X^*>\lambda}d\lambda$$
$$= \int_0^\infty p\lambda^{p-1}P(X^* > \lambda)d\lambda$$
$$\leq \int_0^\infty p\lambda^{p-2}\mathbb{E}(X_N\chi_{X^*>\lambda})d\lambda$$
$$= \mathbb{E}(X_N.\int_0^\infty p\lambda^{p-2}\chi_{X^*>\lambda}d\lambda)$$
$$= \mathbb{E}(X_N((p/p-1)X^{*(p-1)}) = (p/p-1)\mathbb{E}(X_NX^{*(p-1)})$$
$$\leq (p/p-1)(\mathbb{E}|X_N|^p)^{1/p}(\mathbb{E}X^{*p})^{(p-1)/p}$$

on using Holder's inequality with $q = p/p-1$. This gives

$$\mathbb{E}X^{*p} \leq (p/p-1)^p\mathbb{E}|X_N|^p$$

completing the proof.

48. Down-crossings inequality:Let $X_n, n = 0, 1, ..., N+1$ be a sub-martingale. Let $a < b$ be real numbers. Define

$$s_1 = min(n \geq 0 : X_n > b), s_2 = min(n > s_1 : X_n < a), ...$$

In general, we have

$$s_{2k} = min(n > s_{2k-1} : X_n < a), s_{2k+1} = min(n > s_{2k} : X_n > b)$$

if at some stage s_k becomes $N+1$ or undefined, then we set $s_k = N+1$. For example, if $X_n \leq b$ for all $n = 0, 1, ..., N$, then we set $s_1 = N$. It is easy to see that each s_k is a stop time (Prove this). Define

$$A_k = \{s_k \leq N\}$$

On the set A_{2k}, we obviously have $X_{s_{2k}} < a$ and on the set A_{2k+1}, we have $X_{s_{2k+1}} > b$. It follows from the optional sampling theorem and the fact that $A_{2k+1} \subset A_{2k}$ that

$$0 \leq \int_{A_{2k+1}} (X_{s_{2k+1}} - b)dP \leq \int_{A_{2k+1}} (X_{N+1} - b)dP$$

$$= \int_{A_{2k}} (X_{N+1} - b)dP + \int_{A_{2k}-A_{2k+1}} (X_{N+1} - b)dP$$

$$= \int_{A_{2k}} (X_{s_{2k}} - b)dP + \int_{A_{2k}-A_{2k+1}} (X_{N+1} - b)dP$$

$$\leq (a-b)P(A_{2k}) + \int_{A_{2k}-A_{2k+1}} (X_{N+1} - b)dP$$

so that

$$(b-a)P(A_{2k}) \leq \int_{A_{2k}-A_{2k+1}} (X_{N+1} - b)dP$$

Let D denote the number of down crossings of the process in the time interval $[0, N]$. Then obviously, Then, $P(A_{2k}) = P(D \geq k)$ and hence we get

$$\sum_k P(D \geq k) \leq \mathbb{E}(X_{N+1} - b)^+$$

However notice that

$$\sum_k P(D \geq k) = \sum_{k \leq n} P(D = n) = \sum_n (n+1)P(D = n) \geq \sum_n nP(D = n) = \mathbb{E}D$$

We thus obtain the fundamental martingale downcrossing inequality

$$\mathbb{E}D \leq \mathbb{E}(X_{N+1} - b)^+$$

Suppose in particular that the time index range of the sub-martingale is the entire set of non-negative integers and suppose further that $sup_n \mathbb{E}|X_n| < \infty$. Then, it follows that if D is the number of down crossings of the martingale of the set $[a, b]$ over the entire time range $0, 1, 2, ...$, we have that

$$\mathbb{E}(D) \leq b + K < \infty$$

where

$$K = sup_n \mathbb{E}|X_n|$$

Let now X_n be a sub-martingale satisfying the above condition, ie, $sup_n \mathbb{E}|X_n| < \infty$. We are going to show that $lim_{n\to\infty} X_n$ exists almost surely. For suppose this limit does not exist. Then $P(liminf X_n < limsup X_n) > 0$ and hence there exist finite real numbers $\alpha < \beta$ such that the event

$$E = \{liminf X_n < \alpha < \beta < limsup X_n\}$$

has positive probability. It is clear that on the set E, the number D of downcrossings of the interval $[\alpha, \beta]$ is infinite and hence $\mathbb{E}(D) = \infty$ a contradiction to the downcrossing theorem. This proves the claim.

49. Let $T_n, n = 1, 2, ...$ be a sequence of $\{\mathcal{F}_t\}$ stopping times. Then $sup_n T_n$ is a stopping time. Indeed, for $t \geq 0$, we have

$$\{sup_n T_n \leq t\} = \bigcap_n \{T_n \leq t\} \in \mathcal{F}_t$$

proving the claim.

If $\{\mathcal{F}_t\}$ is a right continuous filtration, then $inf T_n$ is a stopping time. Indeed, the event $\{\inf T_n < t + \epsilon\}$ for $\epsilon > 0$ is the same as the event $\bigcup_n \{T_n < t + \epsilon\}$ which belongs to $\mathcal{F}_{t+\delta}$ for all $\delta > \epsilon$. It follows that

$$\{inf T_n \leq t\} = \bigcap_{n\geq 1} \{inf T_n < t + 1/n\} \in \mathcal{F}_{t+\delta}, \delta > 0$$

and letting $\delta \downarrow 0$ gives us

$$\{inf T_n \leq t\} \in \mathcal{F}_{t+}$$

The desired conclusion follows from this.

50. Let X be a left continuous process and define

$$X_t^n = X_0 \chi_{\{0\}}(t) + \sum_{k=0}^{\infty} X_{k/n} \chi_{(k/n,(k+1)/n]}(t)$$

Then we claim that X is the pointwise limit of the processes X^n. Indeed, we have $X_0^n = X_0$ and hence $lim X_0^n = X_0$. Let $t > 0$ and let $\delta > 0$ be given. Choose n large enough so that $1/n < \delta$. Then there exists some $k = 0, 1, ...$ such that $k/n < t \leq (k+1)/n$ and hence $X_t^n = X_{k/n}$. Thus

$$|X_t^n - X_t| = |X_{k/n} - X_t|$$

Suppose δ has been chosen such that $|X_{t-u} - X_t| < \epsilon$ for all $u < \delta$ (this is possible since X is left continuous). Then it follows that since $|t - k/n| < \delta$, we have

$$|X_{k/n}^n - X_t| < \epsilon$$

and hence,

$$|X_t^n - X_t| < \epsilon, n > 1/\delta$$

proving that $lim X_t^n = X_t$. It follows that the σ field on $\Omega \times \mathbb{R}_+$ generated by left continuous processes is contained in the σ field generated by the elementary processes. Now suppose that $\mathcal{F}_t$ is a right continuous filtration. Let $\mathcal{E}$ be the space of elementary processes and let τ be the σ field on $\Omega \times \mathbb{R}_+$ generated by these processes and let τ' be the σ field generated by adapted continuous processes. Let τ'' be the σ field generated by left continuous adapted processes. We have seen that $\tau'' \subset \tau$. Obviously, we also have $\tau' \subset \tau''$. We now show that $\tau \subset \tau'$ so that $\tau = \tau' = \tau''$. Indeed, $\chi_{(u,v]}$ is the limit of continuous functions f_n with compact support contained in $(u, v+1/n]$. If H is $\mathcal{F}_u$ measurable, the process Hf_n is continuous and adapted so that $\tau \subset \tau'$. We have proved that if the filtration is right continuous, then the σ fields on $\Omega \times \mathbb{R}$ generated by adapted left continuous processes and adapted continuous processes are the same. We denote this σ field by $\mathcal{P}$. More precisely, this should be denoted by $\mathcal{P}(\mathcal{F}_t)$ to emphasize the fact that it depends on the filtration. A process Z on Ω is said to be predictable if $(\omega, t) \to Z_t(\omega)$ is $\mathcal{P}$-measurable.

Predictable processes are progressively measurable. Let Z be a predictable process and define for $t \geq 0$ and a Borel set $B \subset \mathbb{R}$, the set

$$E = \{(\omega, s) : Z_s(\omega) \in B, s \leq t\}$$

We wish to show that

$$E \in \mathcal{F}_t \times \mathcal{B}[0, t]$$

It is sufficient to prove the result for continuous adapted processes. Let X_t be continuous and adapted. Then define for any given $t > 0$, the process

$$X_s^n = X_{kt/n}, kt/n < s \leq (k+1)t/n, k = 0, 1, ..., n-1$$

Since X is continuous, it follows that

$$lim_n X_s^n = X_s, s \in [0, t]$$

and hence to prove that the process X is progressively measurable, it suffices to prove that the map $(\omega, s) \to X_s^n(\omega)$ from $\Omega \times [0, t]$ into $\mathbb{R}$ is measurable with respect to $\mathcal{F}_t \otimes B[0, t]$. But observe that if B is any Borel subset of $\mathbb{R}$, we have

$$\{(\omega, s) : X_s^n(\omega) \in B, s \leq t\} = \bigcup_{k=0}^{n-1} X_{kt/n}^{-1}(B) \times (kt/n, (k+1)t/n]$$

and this is an element of $\mathcal{F}_t \otimes B[0, t]$ and we are through. Note that this argument shows that it is sufficient that X be left continuous.

Let S, T be stopping times relative to a right continuous filtration $(\mathcal{F}_t)$ and let $S \leq T$. Define

$$(S, T] = \{(\omega, t) : S(\omega) < t \leq T(\omega)\}$$

Then we claim that this set is in the predictable σ-field. To prove this, it suffices to show that the process $\chi_{(S,T]}(\omega, t)$ is left continuous and adapted. Note that

$$\chi_{(S,T]}(\omega, t) = \chi_{(-\infty,t)}(S(\omega)).\chi_{[t,\infty)}(T(\omega))$$

Now the set of all ω for which this is one is clearly

$$S^{-1}((-\infty, t)) \cap T^{-1}([t, \infty))$$

and this is evidently an element of $\mathcal{F}_t$ since S, T are stopping times. Thus the process is adapted. The process is also left continuous since for a given t, suppose the above evaluates to one at (ω, t). Then $S(\omega) < t$ and $T(\omega) \geq t$. Let t_n increase to t. Then for sufficiently large n, $S(\omega) < t_n$ and $T(\omega) \geq t_n$ so the process evaluates to one at (ω, t_n) too and this proves left continuity of the process. We further wish to show that the set of all subsets of $Omega \times \mathbb{R}_+$ sets of the form $(S, T]$ generates the predictable σ-field. To see this, we need only show that any process of the form $H(\omega).\chi_{(u,v]}(t)$ where H is $\mathcal{F}_u$ measurable can be written as a limit of linear combinations of processes of the form $\chi_{(S,T]}$. Since H can be approximated by indicators of $\mathcal{F}_u$ sets, it suffices to prove this taking $H(\omega) = \chi_E(\omega)$ with $E \in \mathcal{F}_u$. Choose a stop time T so that $\{T = u\} = E$ and $\{T = \infty\} = E^c$. Then, $\{T \geq u\} = E$ and

$$\chi_E(\omega) = \chi_{\infty > T \geq u}(\omega) = \chi_{[u,\infty)}(T(\omega))$$

and this completes the construction.

51. Discussion with Dr. Shambunath Sharma: Continuous time stochastic processes: Let T be an interval in $\mathbb{R}$ and $(\Omega, \mathcal{F}, P)$ a probability space. Then a continuous time stochastic process on this probability space with T as the time index set and Ω as the chance parameter set is a mapping $(t, \omega) \to X_t(\omega) \in \mathbb{R}$ from $T \times \Omega$ into $\mathbb{R}$ such that for each $t \in T$, the mapping $\omega \to X_t(\omega)$ from Ω into $\mathbb{R}$ is measurable. If further the map $(t, \omega) \to X_t(\omega)$ is measurable with respect to the σ-field $(T \cap \mathcal{B}(\mathbb{R})) \otimes \mathcal{F}$, then we say that the stochastic process X is a measurable process. Suppose that for each $t \in T$, we have a σ-field $\mathcal{F}_t \subset \mathcal{F}$ such that $s < t$ implies $\mathcal{F}_s \subset \mathcal{F}_t$, then we call $\{\mathcal{F}_t : t \in T\}$ a filtration. Suppose that the stochastic process X is such that for each $t \in T$, $X_t : \Omega \to \mathbb{R}$ is $\mathcal{F}_t$ measurable, then we say that X is adapted to the filtration $\{\mathcal{F}_t\}$. Suppose that for each $t \in T$, the map $(s, \omega) \to X_s(\omega)$ from $(T \cap (-\infty, t]) \otimes \Omega$ into $\mathbb{R}$ is measurable with respect to the σ-field $[(T \cap (-\infty, t]) \cap \mathcal{B}(\mathbb{R})] \otimes \otimes \mathcal{F}_t$ on the domain space, then we say that X is a progressively measurable stochastic process. For simplicity of notation, we shall take $T = [0, \infty) = \mathbb{R}_+$. Then, if a stochastic process $(t, \omega) \to X_t(\omega)$ from $[0, \infty) \otimes \Omega$ into $\mathbb{R}$ is a measurable map we say that X is a measurable stochastic process. If the mapping $(s, \omega) \to X_s(\omega)$ from $[0, t] \otimes \Omega$ into $\mathbb{R}$ is measurable with respect to the σ-field $([0, t] \cap \mathcal{B}(\mathbb{R})) \otimes \mathcal{F}_t$ for all $t \in \mathbb{R}_+$, we then say that the process X is progressively measurable. A filtration $\mathcal{F}_t, t \geq 0$ is said to be right continuous if for all $t \geq 0$, we have

$$\mathcal{F}_{t+} = \bigcap_{n=1}^{\infty} \mathcal{F}_{t+1/n} = \mathcal{F}_t$$

Suppose $X_t(\omega)$ is a rightcontinuous adapted stochastic process. Define the process Y_t so that

$$Y_t^{(n)} = X_{(k+1)/2^n}, k/2^n \leq t < (k+1)/2^n, k = 0, 1, 2, ... n = 1, 2, ...$$

Then $Y^{(n)}$ converges pointwise to X as $n \to \infty$. Let $t \geq 0$ be arbitrary, fix n and let r be the unique integer such that $r/2^n \leq t < (r+1)/2^n$. Then,

$$\{(s, \omega) : s \leq t, Y_s^{(n)}(\omega) \in B\} =$$

$$[\bigcup_{k=0}^{r-1} ([k/2^n, (k+1)/2^n) \times X^{-1}_{(k+1)/2^n}(B))] \cup ([r/2^n, t] \otimes X^{-1}_{(r+1)/2^n}(B))$$

The first bracket is an element of $\mathcal{B}[0,t] \otimes \mathcal{F}_t$ since $r/2^n \leq t$. The second bracket is an element of $\mathcal{B}[0,t] \otimes \mathcal{F}_{(r+1)/2^n}$. It follows by taking $limn \to \infty$ that the map $(s,\omega) \to X_s(\omega)$ from $[0,t] \times \Omega$ into $X_s(\omega)$ is $\mathcal{B}[0,t] \otimes \mathcal{F}_{t+}$ measurable. This argument does not yield progressive measurability of X. If however the filtration is right continuous, then X becomes progressively measurable.

Problem:Let $X_t(\omega), t \geq 0$ be a right continuous stochastic process on a probability space and let $\mathcal{F}_t = \sigma(X_s : s \leq t)$. Then, show that $\mathcal{F}_{t+} = \mathcal{F}_t$.

52. Let $B(t), t \geq 0$ be Brownian motion and $t > 0$. We wish to prove that the random variables

$$X_n = \sum_{k=0}^{2^n-1} (B((k+1)t/2^n) - B(kt/2^n))^2, n = 1, 2, ...$$

converges almost surely to t. We compute

$$\mathbb{E}X_n = \sum_{k=0}^{2^n-1} t/2^n = t,$$

$$\mathbb{E}X_n^2 = \sum_k \mathbb{E}(B((k+1)t/2^n) - B(kt/2^n))^4 + 2\sum_{k<j} \mathbb{E}\{(B((k+1)t/2^n) - B(kt/2^n))(B((j+1)t/2^n) - B(jt/2^n))\}$$

$$= (3.2^n)(t/2^n)^2 + (2^n(2^n-1))(t/2^n)^2 = 3t^2/2^n + (2^n-1)t^2/2^n$$

so that

$$\mathbb{E}(X_n - t)^2 = 3t^2/2^n + -t^2/2^n = 2t^2/2^n$$

It follows that if $\epsilon > 0$, then

$$P(|X_n - t| > \epsilon) \leq 2t^2/(\epsilon.2^n)$$

and hence

$$\sum_n P(|X_n - t| > \epsilon) < \infty$$

so that

$$P(lim_{n\to\infty} X_n = t) = 1$$

ie X_n converges almost surely to t.

53. Holder continuity: A function $f : \mathbb{R} \to \mathbb{R}$ is locally holder continuous of order α if for all $L > 0$, we have

$$sup(|f(t) - f(s)|/|t-s|^\alpha : |s|, |t| \leq L, t \leq s) < \infty$$

Denoting the left side by $C(L)$, we have in this case

$$|f(t) - f(s)|^\alpha \leq C(L)|t-s|^\alpha, t, s \in [-L, L]$$

Let $X_t, t \in [0,t]$ be a random process such that for some positive γ, c, ϵ we have

$$\mathbb{E}(|X_t - X_s|^\gamma) \leq c|t-s|^{1+\epsilon}, t, s \in [0,1]$$

Then there is a modification Y of X such that

$$\mathbb{E}[sup_{s\neq t}(|Y_t - Y_s|/|t-s|^\alpha)^\gamma] < \infty$$

Let D_m be the set $\{i/2^m : i = 0, 1, ..., 2^m\}$ for $m = 0, 1, 2, ...$ Then $\bigcup D_m$ is the set of dyadic numbers in $[0, 1]$. Let Δ_m be the set of all ordered pairs (s, t) in D_m with the property $|t - s| = 1/2^m$. Thus, Δ_m is the set of all ordered pairs $(k/2^m, (k+1)/2^m), ((k+1)/2^m, k/2^m)$ with $0 \leq k \leq 2^m - 1$. Let

$$K_m = sup(|X_t - X_s| : (s,t) \in \Delta_m)$$

Then,

$$K_m^\gamma \leq \sum_{(s,t)\in\Delta_m} |X_t - X_s|^\gamma$$

By hypothesis,

$$\mathbb{E}|X_t - X_s|^\gamma \leq c|t-s|^\gamma = c/2^{m(1+\epsilon)}, (s,t) \in \Delta_m$$

so that

$$\mathbb{E}(K_m^\gamma) \leq (c/2^{m(1+\epsilon)})2^m = J2^{-m\epsilon}$$

Let $s \in D$. Then there exists an integer N such that $2^N s = k$ is an integer. Thus, $s = k/2^N$. We set $s_n = s, n \geq N$. Now, we can choose a sequence $s_n, n < N$ such that $s_n \leq s_{n+1}$ for $n \leq N-1$ and $s_n \in D_n$ for all n. Further, given s, t in D, we may choose N to be the same for both. We have $s_N = k/2^N$. Suppose $k = b_0 + b_1.2 + ... + b_N 2^N$ where $b_i \in \{0, 1\}$. Then define

$$s_{N-1} = (b_1.2 + ... + b_N 2^N)/2^N = (b_1 + b_2.2 + ... + b_N 2^{N-1})/2^{N-1} \in D_{N-1}$$

and so on we can recursively define s_k for all k. Clearly $s_k \leq s_{k+1}$. We also have $|s_N - s_{N-1}| \leq 1/2^N$ and more generally, for all m, $|s_{i+1} - s_i| \leq 1/2^{i+1}$ with a similar result for t_i. For any m, suppose $|s-t| \leq 1/2^m$. Then if $m \geq N$, we have $s_m = s = s_N \in D_N \subset D_m$ and likewise $t_m = t = t_N \in D_N \subset D_m$ and $|s_m - t_m| \leq 1/2^m$ implying that $(s_m, t_m) \in \Delta_m$. Suppose $m < N$, say $m = N-1$ example. Writing $s = (b_0 + b_1.2 + ... + b_N.2^N)/2^N$ and $t = (c_0 + c_1.2 + ... + c_N.2^N)/2^N$, we have

$$s_{N-1} = s - b_0/2^N, t_{N-1} = t - c_0/2^N$$

so that

$$|s_{N-1} - t_{N_1}| \leq |s-t| + 1/2^N \leq 1/2^{N-1}$$

In other words, for all m, we have that $|s-t| \leq 1/2^m$ implies either $s_m = t_m$ or else $(s_m, t_m) \in \Delta_m$. We now have that for all m,

$$X_s - X_t = \sum_{i\geq m}(X_{s_{i+1}} - X_{s_i}) + X_{s_m} - X_{t_m} + \sum_{i\geq m}(X_{t_i} - X_{t_{i+1}})$$

and hence

$$|X_s - X_t| \leq 2\sum_{i>m+1} K_i + K_m \leq 2\sum_{i\geq m} K_i$$

Define

$$M = sup(|X_t - X_s|/|t-s|^\alpha : s,t \in D, s \neq t)$$

We can write

$$\{(s,t) : s,t \in D, s \neq t\} = \bigcup_{m=0}^{\infty} \{(s,t) \in D : |s-t| \leq 2^{-m}\}$$

We want to show that

$$M \leq sup_{m\in\mathbb{N}}(2^{(m+1)\alpha} sup(|X_t - X_s| : |t-s| \leq 2^{-m}, s,t \in D, s \neq t)$$

Indeed, let $s,t \in D, s \neq t$. Then there exists an m such that $s,t \in D_m$, say $s = k_1/2^m$ and $t = k_2/2^m$ with $k_2 > k_1$. Then,

$$|X_t - X_s| \leq \sum_{k=k_1}^{k_2-1} |X_{(k+1)/2^m} - X_{k/2^m}| \leq (k_2 - k_1) sup(|X_a - X_b| : a,b \in D_m, |a-b| \leq 2^{-m})$$

$$\leq 2^m.sup(|X_a - X_b| : a,b \in D_m, |a-b| \leq 2^{-m})$$

Also

$$|t-s| = |k_2 - k_1|/2^m \geq 1/2^m$$

Combining the two gives us

$$|X_t - X_s|/|t-s|^\alpha \leq 2^{(m+1)\alpha}.sup(|X_a - X_b| : a,b \in D, a \neq b, |a-b| \leq 2^{-m})$$

and this proves the claim. The desired result can be deduced from this.

54. Some definitions: Assume that X is a right continuous process adapted to the filtration $(\mathcal{F}_t)$. Suppose that there exist stoptimes $T_n, n \geq 1$ such that (T_n) is increasing, $limT_n = \infty$ and for every n, $X_{T_n \wedge t}\chi_{T_n > 0}$ is a uniformly integrable $\mathcal{F}_t$-martingale. Then X is called a local martingale. We can drop the condition of uniform integrablility. For suppose T_n is as in the definition and $X_{T_n \wedge t}\chi_{T_n>0}$ is a martingale for all n. Then define $S_n = T_n \wedge n$. S_n is also an increasing sequence of stoptimes converging to infinity. We then claim that $X_{S_n \wedge t}\chi_{S_n>0}$ is a uniformly integrable martingale. Note that $\chi_{S_n>0} = \chi_{T_n>0}$. Since $X_{T_n \wedge t}\chi_{T_n>0}$ is a martingale and since n is a bounded stopping time and $n > S_n$, we have with $Y_n(t) = X_{T_n \wedge t}$, the equation

$$\mathbb{E}[Y_n(n)|\mathcal{F}_{S_n \wedge t}] = Y_n(S_n \wedge t) = X(S_n \wedge t)$$

by the optional stopping theorem. It follows that $X(S_n \wedge t)$ is uniformly integrable for each n. Thus the condition of uniform integrability can be dropped.

Remark:Let $(\Omega, \mathcal{F}, P)$ be a probability space and let X be an integrable random variable. The family of random variables $\mathbb{E}(X|\mathcal{B})$ as $\mathcal{B}$ ranges over all sub-σ-fields of $\mathcal{F}$ is uniformly integrable. For let

$$\Gamma(\mathcal{B}) = \{|\mathbb{E}(X|\mathcal{B})| > c\}$$

Then,

$$\int_{\Gamma(\mathcal{B})} |\mathbb{E}(X|mathcal B)|dP \leq_{\Gamma(\mathcal{B})} \mathbb{E}(|X||\mathcal{B})dP$$

$$= \int_{\Gamma(\mathcal{B})} |X| dP$$

Now,

$$P(\Gamma(\mathcal{B})) \le c^{-1}\mathbb{E}|X|$$

and by choosing c sufficiently large we obtain the desired uniform integrability of $\{\mathbb{E}(X|\mathcal{B})\}$.

55. Shift operator: Let $(\Omega, \mathcal{F}, P)$ be a probability space and let $X_t, t \in \mathbb{R}$ be a stochastic process defined on this probability space. This process induces another probability space as follows. Let $\mathcal{E}$ be the set of all real valued functions on $\mathbb{R}$, ie

$$\mathcal{E} = \mathbb{R}^{\mathbb{R}}$$

For any given $\omega \in \Omega$, the function $t \to X_t(\omega)$ is an element of $\mathcal{E}$. Let $\mathcal{B}$ denote the σ field on $\mathcal{E}$ generated by cylinder sets, ie, by sets of the form

$$C_{t_1,...,t_k} = \{f : \mathbb{R} \to \mathbb{R} : (f(t_1), ..., f(t_k)) \in B\}, B \in \mathcal{B}(\mathbb{R}^k)$$

For each finite subset $\{t_1, ..., t_k\}$ of $\mathbb{R}$ and $B \in \mathcal{B}(\mathbb{R}^k)$, define

$$F_{t_1,...,t_k}(B) = P((X(t_1), ..., X(t_k)) \in B)$$

Then, $F_{t_1,...,t_k} : t_i \in \mathbb{R}, i = 1, 2, ..., k, k = 1, 2, ...$ is a consistent family of finite dimensional probability distribution functions and hence by the Kolmogorov consistency theorem, there exists a probability measure P on $(\mathcal{E}, \mathcal{B})$ whose finite dimensional distributions coincide with the given $F_{t_1,...,t_k}$. In other words, we have for any $B \in \mathcal{B}(\mathbb{R}^k)$,

$$F_{t_1,...,t_k}(B) = P(B \times \Pi_{t\in\mathbb{R}, t\neq t_1,...,t_k}\mathbb{R})$$

For $t \geq 0$ and $\omega \in \mathcal{E}$, define $\theta_t(\omega)(s) = \omega(t+s)$, ie, if $\omega : \mathbb{R} \to \mathbb{R}$ is a given function, then $\theta_t\omega : \mathbb{R} \to \mathbb{R}$ is the function defined by $\theta_t\omega(s) = \omega(t+s)$, or in other words, $\{\omega(s) : s \in \mathbb{R}\}$ maps under θ_t to $\{\omega(t+s) : s \in \mathbb{R}\}$. Alternately, θ_t can be viewed as a mapping from $\Omega \to \Omega$ for which

$$X_s(\theta_t(\omega)) = X_{s+t}(\omega)$$

Suppose X is a stationary process. Then we have for any $B \in \mathcal{B}(\mathbb{R}^k)$,

$$P((X_{s+t_1}), ..., X_{s+t_k}) \in B | X_s = x) = P((X_{t_1}, ..., X_{t_k}) \in B | X_0 = x)$$

and in terms of the shift operator, this can be expressed as

$$P((X_{t_1} o\theta_s, ..., X_{t_k} o\theta_s) \in B | X_0 o\theta_s = x) = P((X_{t_1}, ..., X_{t_k}) \in B | X_0 = x)$$

Thus, if g is any Borel function on $\mathbb{R}^{\mathbb{R}}$, for a stationary process, we have

$$\mathbb{E}g(Xo\theta_s) = \mathbb{E}g(X)$$

for all s. Suppose that X is a process adapted to a filtration $\{\mathcal{F}_t\}$. We wish to show that if $E \in \mathcal{F}_t$, then $\theta_s^{-1}(E) \in \mathcal{F}_{s+t}$. Indeed, we may assume that

$$E = \{(X_{t_1}, ..., X_{t_k}) \in B\}$$

for some $t_1 < t_2 < ... < t_k \leq t$ and $B \in \mathcal{B}(\mathbb{R}^k)$, and then, we have $\omega \in \theta_s^{-1}(E)$ iff $\theta_s(\omega) \in E$ iff $(X_{t_1}(\theta_s(\omega)), ..., X_{t_k}(\theta_s(\omega))) \in B$ iff $(X_{s+t_1}(\omega), ..., X_{s+t_k}(\omega)) \in B$. This gives

$$\theta_s^{-1}(E) = \{(X_{s+t_1}, ..., X_{s+t_k}) \in B\} \in \mathcal{F}_{s+t}$$

The same result is also valid for stop times T in place of s. To prove this, we must show that if T is a stop-time, then $\theta_T^{-1}(E) \in \mathcal{F}_{T+t}$ whenever $E \in \mathcal{F}_t$. For simplicity, take $E = \{X_t \in B\}$ where B is some Borel subset of $\mathbb{R}$. Then $\omega \in \theta_T^1(E)$ iff $\theta_T(\omega) \in E$ iff $X_t(\theta_T(\omega)) \in B$ iff $X_{T+t}(\omega) \in B$. But X_{T+t} is $\mathcal{F}_{T+t}$-measurable provided X is adapted and right continuous and hence we are through. Note that if T is a random-time and X a measurable process, then X_T is a random variable. We can assume without loss of generality that

$$X(t, \omega) = \sum_{k=0}^{n-1} X_k(\omega)\chi_{[t_k, t_{k+1})}(t)$$

where $0 = t_0 < t_1 < ... < t_n$ and X_k is $\mathcal{F}_{t_k}$-measurable. Then,

$$X(T(\omega), \omega) = \sum_{k=0}^{n-1} X_k(\omega)\chi_{t_k \leq T < t_{k+1}}(\omega)$$

Suppose we take χ_E for X_{t_k} so that $E \in \mathcal{F}_{t_k}$. Then,

$$X_k \chi_{t_k \leq T < t_{k+1}} = \chi_{E \cap \{t_k \leq T < t_{k+1}}$$

We have

$$E \cap \{t_k \leq T < t_{k+1}\} \cap \{T \leq t\} = E \cap \{t_k \leq T < t_{k+1} \wedge t\}$$

Case 1:$t_k \leq t$. Then, $E \in \mathcal{F}_t$ and $\{t_k \leq T < t_{k+1} \wedge t\} \in \mathcal{F}_t$. It follows that the above set is in $\mathcal{F}_t$ and hence

$$E \cap \{t_k \leq T < t_{k+1}\} \in \mathcal{F}_T$$

Case 2: $t_k > t$. Then, $E \cap \{t_k \leq T < t_{k+1}\}$ is the empty set and is therefore in $\mathcal{F}_T$. Thus, in all cases, we find that if X is adapted and right continuous, then X_T is $\mathcal{F}_T$-measurable.

We shall now prove that if X is a measurable process and T and random time, then X_T is a random variable. This is elementary. The map $\omega \to (T(\omega), \omega)$ is measurable and the map $(t, \omega) \to X_t(\omega)$ is also measurable, so the composition $\omega \to X(T(\omega), \omega)$ is measurable and we are through.

56. Let X be a continuous process on a probability space with values in a metric space and adapted to a filtration $\{\mathcal{F}_t\}$ and let A be a closed subset of the metric space. Define

$$\tau_A = inf(t \geq 0 : d(X_t, A) = 0\}$$

We then claim that τ_A is a stop time. Indeed, we shall show that

$$\{\tau_A \leq t\} = \{inf_{s \leq t} d(X_s, A) = 0\}$$

For suppose $inf_{s\leq t}d(X_s, A) = 0$. Then there is a sequence $s_n \in [0, t]$ such that $limd(X_{s_n}, A) = 0$. Since $[0, t]$ is compact, by the Bolzano-Weierstrass theorem, there is a subsequence s_{n_k} such that $lims_{n_k} = s \in [0, t]$. Continuity of X then implies that

$$d(X_s, A) = limd(X_{s_n}, A) = 0$$

and hence $\tau_A \leq t$. Conversely, suppose $\tau_A \leq t$. Then there is a sequence s_n of reals such that s_n decreases to τ_A and $d(X_{s_n}, A) = 0$. It follows by taking limits and using continuity of X that

$$d(X_{\tau_A}, A) = limd(X_{s_n}, A) = 0$$

and hence that $inf_{s\leq t}d(X_s, A) = 0$.

Remark: Suppose A is any subset of a metric space (X, d). For any $x \in X$, define

$$d(x, A) = inf(d(x, y) : y \in A)$$

Then $x \to d(x, A)$ is a continuous function from $X \to \mathbb{R}$. Indeed, we have

$$d(x, u) \leq d(x, y) + d(y, u)$$

so that taking inf over all $u \in A$ on both sides gives

$$d(x, A) \leq d(x, y) + d(y, A)$$

This can be rearranged as

$$d(x, A) - d(y, A) \leq d(x, y)$$

Interchanging x and y and using the symmetry of d gives us

$$d(y, A) - d(x, A) \leq d(x, y)$$

Combining the two results, we get

$$|d(x, A) - d(y, A)| \leq d(x, y)$$

which proves uniform continuity of the function $x \to d(x, A)$.

57. Doob-Meyer decomposition: Let $X_n, n = 0, 1, ..., N$ be a submartingale with respect to a filtration $\mathcal{F}_n, n = 0, 1, ..., N$, so that for every n, X_n is $\mathcal{F}_n$-measurable and $\mathbb{E}(X_n|\mathcal{F}_{n-1}) \geq X_{n-1}$ almost surely for all n. Define

$$Z_n = X_n - \mathbb{E}(X_n|\mathcal{F}_{n-1}), n = 0, 1, 2, ...$$

Then from the submartingale property,

$$Z_n \leq X_n - X_{n-1}$$

and hence if we define

$$Y_n = X_n - X_{n-1} - Z_n$$

then $Y_n \geq 0$ and hence with

$$A_n = \sum_{k=0}^{n} Y_k$$

it follows that $\{A_n : n \geq 0\}$ is an increasing process. We have

$$A_n = X_n - M_n, M_n = \sum_{k=0}^{n} Z_k$$

and since

$$\mathbb{E}(Z_n|\mathcal{F}_{n-1}) = 0,$$

it follows that $M_n, n \geq 0$ is a martingale. Further observe that

$$A_n = X_n - \sum_{k=0}^{n}(X_k - \mathbb{E}(X_k|\mathcal{F}_{k-1}))$$

is $\mathcal{F}_{n-1}$ measurable since

$$A_n = \mathbb{E}(X_n|\mathcal{F}_{n-1}) - \sum_{k=0}^{n-1}(X_k - \mathbb{E}(X_k|\mathcal{F}_{k-1}))$$

We thus obtain the Doob-Meyer decomposition of a submartingale X as $X = M + A$, the sum of a martingale and a predictable increasing process. Using this we can prove the optional sampling theorem:If $X_n, n = 0, 1, ..., N$ is a submartingale and S, T are stopping times with $0 \leq S \leq T \leq N$, then for any $E \in \mathcal{F}_S$, we have

$$\mathbb{E}(X_T\chi_E) \geq \mathbb{E}(X_S\chi_E)$$

Indeed, in view of the Doob-Meyer decomposition, it suffices to prove equality in the above when X is a martingale. It suffices in turn to prove that if T is a stop time so that $T \leq N$, then for any $E \in \mathcal{F}_T$, we have

$$\mathbb{E}(X_T\chi_E) = \mathbb{E}(X_N\chi_E)$$

However since $E \cap \{T = k\} \in \mathcal{F}_k$ for all k, we get using the martingale property of X that

$$\mathbb{E}(X_N\chi_{E\cap\{T=k\}}) = \mathbb{E}(X_k\chi_{E\cap\{T=k\}}) = \mathbb{E}(X_T\chi_{E\cap\{T=k\}})$$

and hence summing over $k = 0, 1, ..., N$, we get

$$\mathbb{E}(X_N\chi_E) = \mathbb{E}(X_T\chi_E)$$

completing the proof.

58. Some important notions in Probability theory

1.Example of a stationary stochastic process: Let $x(t) = A.sin(\omega t + \phi), t \in \mathbb{R}$ where ϕ is uniformly distributed over $[0, 2\pi)$. We claim that this process is strict sense stationary. Indeed, let $\tau, t_i \in \mathbb{R}, i = 1, 2, ..., n$. Then

$$x(t_i + \tau) = A.sin(\omega t_i + (\phi + \omega\tau)mod2\pi)$$

and since ϕ is uniformly distributed over $[0, 2\pi)$, it follows that $\phi+\omega\tau(mod2\pi)$ is also uniformly distributed over $[0, 2\pi)$. Calling this variable by ϕ', we get

$$x(t_i + \tau) = A.sin(\omega t_i + \phi'), i = 1, 2, ..., n$$

implying that the random vectors $(x(t_i + \tau) : i = 1, 2, ...)$ and $(x(t_i) : i = 1, 2, ..., n)$ have the same probability distributions.

Additive functionals of a Markov process: Let $X_t, t \geq 0$ be a Markov process adapted to a filtration $\mathcal{F}_t, t \geq 0$. Let A_t be a right continuous non-decreasing process adapted to the filtration $\mathcal{F}_t^0$ generated by the Markov process X and such that

$$A_{t+s} = A_t + A_s o\theta_t$$

This means that suppose we write

$$A_t = A_t(X_u, u \leq t)$$

then

$$A_{t+s}(X_u, u \leq t+s) = A_t(X_u, u \leq t) + A_s(X_{u+t}, u \leq s)$$

For example, let

$$A_t = \int_0^t f(X_u)du$$

where f is a well behaved non-negative Borel function on $\mathbb{R}$. Then,

$$A_{t+s} = \int_0^{t+s} f(X_u)du = \int_0^t f(X_u)du + \int_t^{s+t} f(X_u)du = A_t + \int_0^s f(X_{t+u})du = A_t + A_s o\theta_t$$

Suppose f is a well behaved non-negative Borel function and A and additive functional. We then show that

$$B_t = \int_0^t f(X_u)dA_u$$

is also an additive linear functional. Indeed,

$$B_{t+s} = B_t + \int_t^{t+s} f(X_u)dA_u = B_t + \int_0^s f(X_{t+u})dA_{t+u}$$

$$= B_t + \int_0^s f(X_{t+u})d(A_t + A_u o\theta_t) = B_t + \int_0^s f(X_u o\theta_t)d(A_u o\theta_t) = B_t + B_s o\theta_t$$

59. Define the Brownian motion process and derive the Hermite polynomial martingales from this process using the exponential martingale.

Suppose $B(t), t \geq 0$ is Brownian motion. Then for every real x,

$$M_x(t) = exp(xB(t) - x^2t/2), t \geq 0$$

is a martingale as follows from the Ito differential rule:

$$dM_x(t) = M_x(t).dB(t)$$

or equivalently,

$$M_x(t) = \int_0^t M_x(s)dB(s)$$

Now we know that $exp(tx - t^2/2)$ is the generating function of the Hermite polynomials, ie,

$$exp(tx - t^2/2) = \sum_{n=0}^{\infty} H_n(x)t^n/n!$$

It follows that

$$M_x(t) = exp(x\sqrt{t}.B(t)/\sqrt{t} - x^2t/2) = \sum_n H_n(B(t)/\sqrt{t})t^{n/2}x^n/n!$$

and hence

$$W_n(t) = H_n(B(t)/\sqrt{t})t^{n/2}, n = 0, 1, 2, ...$$

are all Brownian martingales.

60. Let $X(t), t \geq 0$ be a Brownian motion process and let $f(t,s), t, s \geq 0$ be a nonrandom function. Then evaluate the characteristic function of the random variable

$$\xi = \int_{0 \leq s \leq t \leq T} f(t,s)dX(t)dX(s)$$

Hint:Suppose X is a Gaussian random vector of size $n \times 1$ with mean zero and covariance matrix R. Let A be any symmetric matrix. Then,

$$\mathbb{E}exp(t.X^TAX) = \int exp(t.x^TAx)exp(-x^TR^{-1}x/2)dx/(2\pi)^{n/2}|R|^{1/2}$$

$$= (2\pi)^{-n/2}|R|^{-1/2}\int exp(-x^T(R^{-1} - 2tA)x/2)dx$$

and this can be evaluated by applying an orthogonal transformation on the vector x. The method is as follows. Let C be any positive definite matrix with spectral representation $C = UDU^T$, where U is an orthogonal matrix, $D = diag[d_1, ..., d_n], d_i > 0$. Then

$$\int exp(-x^TCx/2)dx = \int exp(-x^TUDU^Tx/2)dx = \int exp(-y^TDy/2)dy = \Pi_i \int exp(-d_iy_i^2/2)dy_i$$

$$= (2\pi)^{n/2}(\Pi d_i)^{-1/2} = (2\pi)^{n/2}|C|^{-1/2}$$

61. Brownian motion applied to circuit theory: The dynamics of the circuit is described by a system of linear stochastic differential equations and the aim is to compute the correlations of the state variables using the Ito integral. Suppose $\mathbf{x}(t)$ is the state and that it obeys the sde

$$d\mathbf{x}(t) = \mathbf{A}\mathbf{x}(t)dt + \mathbf{G}d\mathbf{B}(t)$$

where $\mathbf{B}$ is Brownian motion. The solution is given by the Ito integral

$$\mathbf{x}(t) = \int_0^t exp((t-s)\mathbf{A})\mathbf{G}d\mathbf{B}(s)$$

and its covariance is given by

$$Q(t) = \mathbb{E}(\mathbf{x}(t)\mathbf{x}(t)^T) = \int_0^t exp(t\mathbf{A})\mathbf{G}\mathbf{G}^T exp(t\mathbf{A}^T)dt$$

The integral converges provided that all the eigenvalues of $\mathbf{A}$ have negative real part.

62.Quadratic variations: Let X be a stochastic process and let $\Delta : 0 = t_0 < t_1 < t_2 < ...$ be a subdivision of $\mathbb{R}_+ = [0, \infty)$. For any $t \in [0, \infty)$

$$T_{\Delta,t}(X) = \sum_{i=0}^{k-1}(X_{t_{i+1}} - X_{t_i})^2 + (X_t - X_{t_k})^@$$

where k is the largest integer m for which $t_m \leq t$. The process X is said to be of finite quadratic variation if there exists a process $< X, X >$ for which for each t, $T_{\Delta,t}(X)$ converges in probability to $< X, X >_t$ as $|\Delta| \to 0$. We want to show that if M is a continuous and bounded martingale, then it has finite quadratic variation $< M, M >$ and $< M, M >$ is the unique continuous increasing adapted process vanishing at zero such that $M^2- < M, M >$ is a martingale. The idea of the proof is based on the identity that if $t > s$, then

$$\begin{aligned}\mathbb{E}[(M_t - M_s)^2|\mathcal{F}_s] &= \mathbb{E}(M_t^2 + M_s^2 - 2M_tM_s|\mathcal{F}_s)\\ &= \mathbb{E}[M_t^2|\mathcal{F}_s] - M_s^2\end{aligned}$$

More generally, if $t_i < s < t_{i+1}$, then

$$(M_{t_{i+1}} - M_{t_i})^2 = (M_{t_{i+1}} - M_s)^2 + (M_s - M_{t_i})^2 + 2(M_{t_{i+1}} - M_s)(M_s - M_{t_i})$$

so that using the martingale property,

$$\mathbb{E}[(M_{t_{i+1}} - M_{t_i})^2|\mathcal{F}_s] = \mathbb{E}[(M_{t_{i+1}} - M_s)^2|\mathcal{F}_s] + (M_s - M_{t_i})^2$$

It follows that if $s < t$ and $t_i \leq s < t_{i+1} \leq t_k \leq t < t_{k+1}$, then

$$T_{\Delta,s}(M) = \sum_{j=0}^{i-1}(M_{t_{j+1}} - M_{t_j})^2 + (M_s - M_{t_i})^2$$

$$T_{\Delta,t}(M) = \sum_{j=0}^{k-1}(M_{t_{j+1}} - M_{t_j})^2 + (M_t - M_{t_k})^2$$

so that

$$T_{\Delta,t}(M) - T_{\Delta,s}(M) = \sum_{j=i}^{k-1}(M_{t_{j+1}} - M_{t_j})^2 + (M_t - M_{t_k})^2 - (M_s - M_{t_i})^2$$

For j=i+1,..., we have

$$\mathbb{E}[(M_{t_{j+1}} - M_{t_j})^2|\mathcal{F}_{t_j}] = \mathbb{E}[(M^2_{t_{j+1}} - M^2_{t_j}|\mathcal{F}_{t_j}]$$

implying that for the same j,

$$\mathbb{E}[(M_{t_{j+1}} - M_{t_j})^2|\mathcal{F}_s] = \mathbb{E}[M^2_{t_{j+1}} - M^2_{t_j}|\mathcal{F}_s]$$

so that

$$\mathbb{E}[T_{\Delta,t}(M) - T_{\Delta,s}(M)|\mathcal{F}_s]$$

$$= \mathbb{E}[M^2_{t_k} - M^2_{t_{i+1}}|\mathcal{F}_s] + \mathbb{E}[(M_{t_{i+1}} - M_{t_i})^2|\mathcal{F}_s]$$
$$+\mathbb{E}[(M_t - M_{t_k})^2|\mathcal{F}_s] - (M_s - M_{t_i})^2$$

$$= \mathbb{E}[M^2_{t_k} - M^2_{t_{i+1}} + M^2_{t_{i+1}} + M^2_{t_i} - 2M_{t_i}M_{t_{i+1}} + M_t^2 + M^2_{t_k} - 2M_{t_k}M_t - M_s^2 - M^2_{t_i} + 2M_sM_{t_i}|\mathcal{F}_s]$$

$$= \mathbb{E}[(M_t^2 - M_s^2|\mathcal{F}_s] = \mathbb{E}[(M_t - M_s)^2|\mathcal{F}_s]$$

where we have made use of the following identities.

$$\mathbb{E}[M_{t_i}M_{t_{i+1}}|\mathcal{F}_s] = M_{t_i}M_s = \mathbb{E}(M_{t_i}M_s|\mathcal{F}_s),$$

$$\mathbb{E}[M_{t_k}M_t|\mathcal{F}_s] = \mathbb{E}[M^2_{t_k}|\mathcal{F}_s]$$

The aim now is to prove that if a is fixed and $\Delta_n, n = 1, 2, ...$ a sequence of partitions such that $|\Delta_n| \to 0$, then $T_{\Delta_n,a}$ converges in L^2. Assume that this has been established. The process $T_{\Delta_n,t}(M), 0 \le t \le a$ converges uniformly to a process $< M, M >$ a.s and in L^2 as follows by applying Doob's inequality to the martingale $M - T_\Delta(M)$. Then on passing to the limit in the equation

$$\mathbb{E}[(M_t^2 - M_s^2 - T_{\Delta_n,t}(M) + T_{\Delta_n,s}(M)|\mathcal{F}_s] = 0$$

it follows that $M^2 - < M, M >$ is a martingale.

63. Construction of the stochastic integral with respect to a martingale: Let $\mathbb{H}^2$ denote the collection of all martingales M for which $sup_t\mathbb{E}M_t^2 < \infty$. Then $\mathbb{H}^2$ is a Hilbert space with the norm

$$\| M \| = \mathbb{E}(M^2_\infty)$$

In fact, $sup_t\mathbb{E}M_t^2 < \infty$ implies from the martingale convergence theorem that $M_\infty = lim_{t\to\infty}M_t$ exists. Further let $M^* = sup_t|M_t|$. Then Doob's inequality combined with $sup_t\mathbb{E}M_t^2 < \infty$ implies that $\mathbb{E}M^{*2} < \infty$. We have $|M_t| \le M^*$ for all t and for $t > s$, we have

$$\mathbb{E}(M_t|\mathcal{F}_s) = M_s$$

By the dominated convergence theorem, it follows on taking $limt \to \infty$ that

$$\mathbb{E}(M_\infty|\mathcal{F}_s) = M_s$$

for all s. Hence, $M_\infty = 0$ implies $M_s = 0$ for all s. This guarantees that $\| \, . \, \|$ as defined above is a valid Hilbert space norm. Let H^2 denote the subset of all martingales in $\mathbb{H}^2$ that are continuous. We want to show that H^2 is a closed subspace of $\mathbb{H}^2$. Let M^n be a sequence in H^2 converging in $\mathbb{H}^2$ to M. Doob's inequality gives

$$\mathbb{E}sup_t(M_t^n - M_t)^2 \leq 4\mathbb{E}(M_\infty^n - M_\infty^n)^2 = 4 \parallel M - M^n \parallel^2$$

and hence $sup_t(M_t^n - M_t)^2$ converges in L^2 to zero. It follows from a standard result in measure theory that there is a subsequence $sup_t(M_t^{n_k} - M_t)^2$ that converges almost surely to zero. Since $t \to M_t^{n_k}$ are continuous processes, it follows that $t \to M_t$ is also a continuous process and hence $M \in H^2$ completing the proof. The stochastic integral can be defined using this Hilbert space. Specifically, consider two martingales M, N. Then, $< M, N >$ is a process of finite variation. Suppose K is a predictable process. Then, we can define the process $K. < M, N >$ so that at t, it evaluates to $\int_0^t K_s d < M, N >_s$. The map $N \to K. < M, N >$ is linear and hence by Riesz representation theorem, there exists a martingale $K.M$ such that $K. < M, N >=< K.M, N >$. The details can be filled in as in the book "Continuous martingales and Brownian motion" by Revuz and Yor.

Boltzmann's kinetic transport equation: $f(t, \mathbf{r}, \mathbf{v})$ is the particle distribution function in phase space. Thus, if $n(t, \mathbf{r}) = \int f(t, \mathbf{r}, \mathbf{v}) d^3v$ is the number density of particles and if E is a subset of $\mathbb{R}^3$, then $n(t, E) = \int_{\mathbb{R}^3 \times E} f(t, \mathbf{r}, \mathbf{v}) d^3r d^3v$ is the number of particles at time t having velocities in the range E. The particle distribution function obeys the kinetic transport equation

$$\frac{\partial f}{\partial t} + \mathbf{v}.\frac{\partial f}{\partial \mathbf{r}} + (\mathbf{F}/m).\nabla_v f = C(f)$$

where $\mathbf{F}$ is the force on each particle and $C(f)$ is the Boltzmann collision integral. Suppose $w(\mathbf{p}, \mathbf{p}_1, \mathbf{p}', \mathbf{p}_1')$ $d^3p'd^3p_1'$ is the probability that the collsion between a particle having momentum $\mathbf{p}$ and another having momentum $\mathbf{p}_1$ will lead to the same pair of particles having momenta in the range $d^3\mathbf{p}'$ and $d^3\mathbf{p}_1'$ respectively, then by symmetry, we have

$$w(\mathbf{p}, \mathbf{p}_1, \mathbf{p}', \mathbf{p}_1') = w(\mathbf{p}', \mathbf{p}_1', \mathbf{p}, \mathbf{p}_1)$$

We then obviously have the equation

$$C(f) = \int (f(t, \mathbf{r}, \mathbf{p}')f(t, \mathbf{r}, \mathbf{p}_1') - f(t, \mathbf{r}, \mathbf{p}).f(t, \mathbf{r}, \mathbf{p}_1))w(\mathbf{p}', \mathbf{p}_1', \mathbf{p}, \mathbf{p}_1)d^3p'd^3p_1'$$

Note that during a binary collision, we are assuming that the momentum and energy are conserved, ie, the collision is elastic. The natural question to be asked here if one can arrive at the Fokker-Planck equation from the kinetic transport equation by appropriate approximations.

Remark: $f(t, \mathbf{r}, \mathbf{p}).f(t, \mathbf{r}, \mathbf{p}_1)d^3pd^3p_1$ equals the number of particle pairs per unit volume at the position $\mathbf{r}$ and at time t, that have momenta in the range (d^3p, d^3p_1). This multiplied by $w(\mathbf{p}', \mathbf{p}_1', \mathbf{p}, \mathbf{p}_1)$ $d^3p'd^3p_1'$ gives the number of such pairs that make a transition after collision to pairs having momenta in the range (d^3p', d^3p_1') per unit time. The other term has a similar interpretation. Assume that during a binary collision, an amount of momentum **q** is transferred from the first particle to the second. Then the corresponding transition probability rate can be expressed as $w(\mathbf{p}+\mathbf{q}/2, \mathbf{p}_1-\mathbf{q}/2, \mathbf{q})$. The equilibrium distribution function f_0, in the absence of external forces, satisfies the equation

$$(\mathbf{p}/m).\frac{\partial f_0}{\partial \mathbf{r}} = \int w(p', p_1', p, p_1)(f_0' f_{01}' - f_0 f_{01}) d^3p' d^3p_1'$$

If the temperature is constant throughout the system, then f_0 is independent of **r** and the condition for f_0 to be an equilibrium distribution is that the collision integral should be zero, ie

$$\int w(p', p_1', p, p_1)(f_0' f_{01}' - f_0 f_{01}) d^3p' d^3p_1' = 0$$

Writing

$$f_0(\mathbf{p}) = C.exp(-p^2/2mkT)$$

we find that

$$f_0' f_{01}' = f_0 f_{01}$$

is implied by the law of conservation of energy, ie, by

$$p^2/2m + p_1^2/2m = p'^2/2m + p_1'2/2m$$

The conservation of energy constraint is built into the transition probability function, ie, for p, p_1, p', p_1' that do not satisfy the energy conservation condition, $w(p', p_1', p, p_1) = 0$. Now let us consider a small perturbation of the distribution function from statistical equilibrium, ie,

$$f(t, \mathbf{r}, \mathbf{p}) = f_0(\mathbf{p}) + \delta f(t, \mathbf{r}, \mathbf{p})$$

Then linearizing the Boltzmann equation about the equilibrium, we get

$$\frac{\partial \delta f}{\partial t} + (p/m).\nabla \delta f = \int w(p', p_1', p, p_1)(f_0' \delta f_1' + delta f_0' f_{01}' - f_0 \delta f_1 - \delta f f_{01}) d^3p' d^3p_1' = C(\delta f)$$

say. This is the linearized Boltzmann equation. If small forces F are present, then it gets modified to

$$\frac{\partial \delta f}{\partial t} + (p/m).\nabla \delta f + (F(r,p)/m).\nabla_p f_0 = C(\delta f)$$

Suppose we do not linearize. We shall approximately evaluate the collsion integral:

$$C(f) = \int (f(t,r,p') f_1(t,r,p_1') - f(t,r,p) f(t,r,p_1)) w(p', p_1', p, p_1) d^3p' d^3p_1'$$

$$= \int (f(t,r,p+q/2) f(t,r,p_1-q/2) - f(t,r,p) f(t,r,p_1)) w(p+q/2, p_1-q/2, q) d^3q d^3p_1$$

64. Topics for a course in probability theory

Probability spaces, properties of the probability measure, random variables, joint distribution of random variables, properties of the joint probability distribution function, integration of random variables with respect to a probability measure, expectation, variance, covariance of random variables, random processes, Kolmogorov consistency theorem, stationary and wide sense stationary random processes, transition probability functions and Markov processes, Chapman-Kolmogorov equations, Brownian motion and Poisson process, compound Poisson processes, joint characteristic function of a finite set of random variables, Bochner's theorem, central limit theorem, martingale property, examples of Brownian martingales.

1. Brownian motion: A random process $B(t,\omega), t \geq 0$ and $\omega \in \Omega$ satisfying (1) if $0 = t_0 < t_1 < ... < t_n$, then $B(t_i,\omega) - B(t_{i-1},\omega)$ are independent Gaussian random variables with zero mean and variance $t_i - t_{i-1}, i = 1, 2, ..., n$. A random variable $X(\omega)$ has zero mean if $\int_\Omega X(\omega)dP(\omega) = 0$ and it has variance

$$\sigma^2 = \int_\Omega X^2(\omega)dP(\omega)$$

Stochastic differential equations driven by Brownian motion.

2. Definition of Gaussian process, example of Brownian motion as a prototype Gaussian random process, central limit theorem for random walk processes.

$X_1, X_2, ...$ are independent random variables with $Pr(X_i = s) = 1/2, Pr(X_i = -s) = 1/2$. Define

$$x(nT) = X_1 + ... + X_n$$

Then,

$$Pr(x(nT) = ms) = \binom{n}{m}2^{-n}$$

Thus,

$$\mathbb{E}x(nT) = 0, \mathbb{E}x(nT)^2 = ns^2$$

Write $t = nT$. Then,

$$\mathbb{E}x(t) = 0, \mathbb{E}x^2(t) = s^2t/T$$

so that if $s, T \to 0$ such that $s^2/T = \alpha$, then we get

$$\mathbb{E}x^2(t) \to \alpha t$$

Application of the central limit theorem shows that the process $\{x(t)\}$ in the limit of $T \to 0$ converges in distribution to $\{\sqrt{\alpha}.B(t)\}$ where $B(.)$ is a Brownian motion process. We may also consider the asymmetric random walk process in which $P(X_i = s) = p, P(X_i = -s) = q = 1 - p$. Then

$$\mathbb{E}x(nT) = (p - q)ns, Var(x(nT)) = n(s^2 - (p - q)^2s^2) = 4npqs^2$$

so that in the limit $s, T \to 0, s^2/T = \alpha, nT = t, p, q \to 1/2$ and $(p - q)/s \to \beta$, we get

$$\mathbb{E}x(t) = \alpha\beta t, Var(x(t)) = \alpha t$$

which means that $x(.)$ in the limit becomes Brownian motion with constant drift.

3.Differentiation and integration of random processes, autocorrelation function of a random process after differentiation and integration.

4.Transmission of random processes through linear filters:Computation of the correlation function and higher moments after transmission.

$$y(t) = \int h(t,u)x(u)du, z(t) = \int g(t,u)w(u)du$$

Then

$$R_{yz}(t,s) = \mathbb{E}(y(t)z(s)) = \int h(t,u)g(s,v)R_{xw}(u,v)dudv$$

Let

$$y_i(t) = \int h_i(t,u)x_i(u)du, i = 1,2,...,m$$

Then,

$$\mathbb{E}(y_1(t_1)...y_m(t_m)) = \int h_1(t_1,u_1)h_2(t_2,u_2)...h_m(t_m,u_m)\mathbb{E}(x_1(u_1)...x_m(u_m))du_1...du_m$$

5.If x, y are jointly normal random variables with means μ_x, μ_y, variances σ_x^2, σ_y^2 and covariance σ_{xy}, then show that

$$\mathbb{E}(x|y) = \mu_x + \frac{\sigma_{xy}}{\sigma_y^2}(y - \mu_y)$$

and

$$Var(x|y) = \sigma_x^2 - \sigma_{xy}^2/\sigma_y^2$$

Hint:First note that $\mathbb{E}(x|y)$ is of the form $ay + b$ where a, b are constants. Then by the orthogonality principle, we have $x - \mathbb{E}(x|y)$ is orthogonal to $1, y$. Thus,

$$\mu_x = \mathbb{E}(x) = a\mu_y + b, \mathbb{E}(xy) = \mu_x\mu_y + \sigma_{xy} = \mathbb{E}((ay+b)y) = a(\sigma_y^2 + \mu_y^2) + b\mu_y$$

6.Let $X_1, X_2, ...$ be a sequence of independent zero mean random variables. Define $S_n = X_1 + ... + X_n, n \geq 1$. Then the sequence $\{S_n\}$ converges in mean square iff $\mathbb{E}(S_{m+n} - S_m)^2 \to 0, m \to \infty$ for all n. On the other hand, we note that

$$\mathbb{E}(S_{m+n} - S_m)^2 = Var(S_{m+n} - S_m) = \sum_{k=m+1}^{n} \sigma_k^2$$

where $\sigma_k^2 = Var(X_k)$ and hence it follows that S_n converges in mean square iff $\sum_k \sigma_k^2 < \infty$.

7.Let $c_1, c_2, \ldots$ be square integrable random variables with zero mean and $\omega_1, \omega_2, \ldots$ distinct numbers in $[0, 2\pi)$. Define the discrete time stochastic process

$$x[n] = \sum_i c_i.exp(j\omega_i n), n \in \mathbb{Z}$$

Then

$$\mathbb{E}(x[n+m]\bar{x}[m]) = \sum_{i,j} \mathbb{E}(c_i \bar{c}_j) exp(j(\omega_i - \omega_j)m) exp(j\omega_i n)$$

This is independent of m iff

$$\mathbb{E}(c_i \bar{c}_j) = 0, i \neq j$$

ie, iff the random variables $\{c_i\}$ are uncorrelated.

8.Let $x(t)$ be a WSS random process in continuous time with power spectral density $S_x(\omega)$, ie $R_x(\tau) = \mathbb{E}(x(t+\tau)x(t))$ and $S_x(\omega) = \int_{-\infty}^{\infty} R_x(\tau) exp(-j\omega\tau) d\tau$. Define $y[n] = x(nT)$. Then

$$R_y[m] = \mathbb{E}(y[n+m]y[m]) = R_x(nT)$$

so that the power spectral density of $y[n]$ equals

$$S_y(\omega) = \sum_m R_y[m] exp(-j\omega m) = \sum_m exp(-j\omega m)(2\pi)^{-1} \int_{-\infty}^{\infty} S_x(u) exp(jumT) du$$

$$= (2\pi)^{-1} \int S_x(u) du \sum_m exp(j(uT - \omega)m) = \int S_x(u) du \sum \delta(uT - \omega - 2m\pi) = T^{-1} \sum_m S_x((\omega + 2m\pi)/T)$$

65. Thesis Problem of Dr.Prabhat Chowdhury.

A list of subjects is given and numbered $1, 2, \ldots, N$. The number of publications in an institute in the year i in the journal j is given by $X(i,j)$ where $i = 1, 2, \ldots, K$. We thus have a $K \times N$ matrix $((X(i,j))_{1 \leq i \leq K, 1 \leq j \leq N}$. The mean number of publications over the years in the journal j is given by

$$\mu(j) = \frac{1}{K} \sum_{i=1}^{K} X(i,j), j = 1, 2, \ldots, N$$

The mean number of publications in all the journals in the year i is given by

$$\nu(i) = \frac{1}{N} \sum_{j=1}^{N} X(i,j), i = 1, 2, \ldots, K$$

$\mu(i)$ is thus the column mean and $\nu(i)$ the row mean. The dispersion/variance of the number of publications in the jth journal over the years is given by

$$V(j) = \frac{1}{K} \sum_{i=1}^{K} (X(i,j) - \mu(j))^2$$

and the covariance between the number of publications in the jth and k^{th} journals over the years is

$$Cov(j,k) = \frac{1}{K}\sum_{i=1}^{K}(X(i,j) - \mu(j))(X(i,k) - \mu(k))$$

Suppose we are interested in constructing, based on the available data, a linear time series model giving the number of publications in each year in each journal. As a zeroth order approximation, we assume that the number of publications in the different journals are uncorrelated. We then construct a first order time series model as

$$\hat{X}(i,j) = \sum_{m=1}^{p} a(m,j)X(i-m,j)$$

$\hat{X}(i,j)$ can be regarded as the first order forecasted estimate of the number of publications in the jth journal in the ith year based on the number of publications in the same journal over the past p years. The coefficients $a(m,j)$ of the linear predictor may be obtained by minimizing the error energy:

$$E(a(m,j) : m = 1,2,...,p) = \sum_{i=1}^{K}(X(i,j) - \hat{X}(i,j))^2$$

The optimal equations that minimize this error energy are given by

$$\frac{\partial}{\partial a(m,j)}E(.) = 0$$

and this yields the normal equations

$$\sum_{i=1}^{K} X(i,j)X(i-m,j) = \sum_{n=1}^{p} a(n,j)\sum_{i=1}^{K} X(i-n,j)X(i-m,j)$$

We define the jth journal correlation functions

$$R_j(m,n) = \sum_{i=1}^{K} X(i-n,j)X(i-m,j)$$

and then the above normal equations can be cast as

$$\sum_{n=1}^{p} R_j(m,n)a(n,j) = R_j(m,0), m = 1,2,...,p$$

Using this predictor, we can predict the number of papers in any given journal in any of the future years by successively iterating this difference equation. We can more generally assume that there is a correlation between the publications in different journals. In other words, we are considering a vector valued time series whose value at the ith year is given by the $N \times 1$ vector

$$\mathbf{X}(i) = (X(i,1), X(i,2), ..., X(i,N))^T$$

Then the predictor is constructed as

$$\hat{\mathbf{X}}(i)=\sum_{j=1}^{p}\mathbf{A}(j)\mathbf{X}(i-j), j=1,2,...,K$$

where $\mathbf{A}(j)$ is a $N\times N$ matrix for each j. The predictor matrix coefficients $\{\mathbf{A}(j)\}$ are obtained by minimizing

$$\sum_{i=1}^{N}\|\mathbf{X}(i)-\sum_{j=1}^{p}\mathbf{A}(j)\mathbf{X}(i-j)\|^2$$

Carrying out the minimization using the variational method leads to the normal equations

$$\sum_{i=1}^{N}\mathbf{X}(i)\mathbf{X}(i-j)^T=\sum_{m=1}^{p}\mathbf{A}(m)\sum_{i=1}^{N}\mathbf{X}(i-m)\mathbf{X}(i-j)^T$$

or in term of the correlation matrices

$$\mathbf{R}(j,m)=\sum_{i=1}^{N}\mathbf{X}(i-m)\mathbf{X}(i-j)^T$$

we get

$$\mathbf{R}(j,0)=\sum_{m=1}^{p}\mathbf{A}(m)\mathbf{R}(j,m), j=1,2,...,p$$

66. Probability theory(Problems for the M.Tech course):

1.Let $(X,\mathcal{F}_X)$ and $(Y,\mathcal{F}_Y)$ be measurable spaces and let $\mathcal{F}_X\times\mathcal{F}_Y$ be the σ field on $X\times Y$ generated by measurable rectangles, ie, sets of the form $E\times F$ with $E\in\mathcal{F}_X$ and $F\in\mathcal{F}_Y$. For $B\in\mathcal{F}_X\times\mathcal{F}_Y$, prove that

$$B_x=\{y\in Y:(x,y)\in B\}\in\mathcal{F}_X\forall x\in X$$

Solution: Let $\mathcal{C}$ be the set of all $B\in\mathcal{F}_X\times\mathcal{F}_Y$ for which $B_x\in\mathcal{F}_X$. Suppose $B_n\in\mathcal{C}, n=1,2,....$ Then, $B_{nx}\in\mathcal{F}_X, n=1,2,...$ for each $x\in X$ by hypothesis. We now have

$$(\bigcup B_n)_x=\{y\in Y:(x,y)\in\bigcup_n B_n\}=\{y\in Y:(x,y)\in B_n for some n\}=\bigcup B_{nx}\in\mathcal{F}_X$$

It follows therefore that $\bigcup B_n\in\mathcal{C}$. In other words, $\mathcal{C}$ is closed under countable unions. We now claim that $\mathcal{C}$ is also closed under complementation. Indeed, let $B\in\mathcal{C}$ and let $x\in X$ and $y\in Y$. Then, $y\in(B^c)_x$ iff $(x,y)\in B^c$ iff $(x,y)\notin B$ iff $y\notin B_x$ iff $y\in(B_x)^c$. This shows that $(B^c)_x=(B_x)^c$ and hence if $B\in\mathcal{C}$, then $B_x\in\mathcal{F}_X$ which implies $(B_x)^c\in\mathcal{F}_X$ and this in turn implies that $(B^c)_x\in\mathcal{F}_X$ so that $B^c\in\mathcal{C}$. This proves that $\mathcal{C}$ is closed under complementation. Thus, $\mathcal{C}$ is a σ field. We claim that $\mathcal{C}$ contains all the measurable rectangles. Indeed, let $E\in\mathcal{F}_X$ and $F\in\mathcal{F}_Y$. Then if $x\in E$, we have $(E\times F)_x=F\in\mathcal{F}_Y$ while if $x\notin E$, then $(E\times F)_x=\phi\in\mathcal{F}_Y$. This proves that $E\times F\in\mathcal{C}$. We have thus proved that $\mathcal{C}$ is a

σ-field containing all the measurable rectangles. It follows that $\mathcal{C}$ contains the σ field $\mathcal{F}_X \times \mathcal{F}_Y$ generated by the measurable rectangles and hence $\mathcal{C} = \mathcal{F}_X \times \mathcal{F}_Y$. This completes the proof.

2.Let $(X, \mathcal{F}_X)$ and $(Y, \mathcal{F}_Y)$ be measurable spaces and let $(X \times Y, \mathcal{F}_X \times \mathcal{F}_Y)$ be the product measurable space. Let $f : X \times Y \to \mathbb{R}$ be a measurable map. Then for each $x \in X$, define the map $f_x : Y \to \mathbb{R}$ so that $f_x(y) = f(x, y)$. Show that f_x is measurable.

Solution:Define the map $T_x : Y \to X \times Y$ so that $T_x(y) = (x, y)$. We claim that T_x is measurable. Indeed, let $E \in \mathcal{F}_X \times \mathcal{F}_Y$. Then

$$T_x^{-1}(E) = \{y \in Y : (x, y) \in E\} = E_x$$

and by the previous problem, $E_x \in \mathcal{F}_Y$. This shows that T_x is measurable for every $x \in X$. Now $f_x = foT_x$ and since the composition of two measurable maps is measurable, the measurability of f_x follows.

3.Let $N(t), t \geq 0$ be a Poisson process with mean arrival rate λ. Let $S_n = X_1 + ... + X_n$ be the arrival time of the nth customer. The time of arrival of the first customer after time t is given by $S_{N(t)+1}$. The waiting time for the arrival of the first customer after time t is given by $\delta(t) = S_{N(t)+1} - t$. Find the probability density function of $\delta(t)$. Note that passengers arrive in accord with the Poisson process, and then if a bus arrives at time t, the time that the bus will have to wait for the first passenger is $\delta(t)$.

4.Definition and properties of the integral:$(\Omega, \mathcal{F}, \mu)$ is a measure space and $f : \Omega \to [0, \infty)$ is a measurable function. Let $\mathcal{P} = \{A_i : i = 1, 2, ..., n\}$ be a measurable partition of Ω, ie, for each i, $A_i \in \mathcal{F}$, $A_iA_j = \phi, i \neq j$ and $\Omega = \bigcup_i A_i$. Let $m_i = inf_{\omega \in A_i} f(\omega)$ and let $\int f d\mu$ denote the supremum of $\sum_i m_i\mu(A_i)$ over all partitions $\mathcal{P}$. Then $\int f d\mu$ is called the integral of f with respect to μ. Note that $\int f d\mu$ can be ∞. If $f : \Omega \to \mathbb{R}$ is any measurable function, then we set $f_+(\omega) = max(f(\omega), 0)$ and $f_-(\omega) = max(-f(\omega), 0) = -min(f(\omega), 0)$. Then we have that f_+, f_- are both non-negative measurable functions, and $f = f_+ - f_-$, $|f| = f_+ + f_-$. Suppose either $\int f_+ d\mu$ or $\int f_- d\mu$ is finite. Then, we define

$$\int f d\mu = \int f_+ d\mu - \int f_- d\mu$$

and call $\int f d\mu$ as the integral of f with respect to μ. If both $\int f_+ d\mu$ and $\int f_- d\mu$ are finite, then $\int f d\mu$ is finite and we say that f is μ-integrable. In this case, we have

$$\int |f| d\mu = \int f_+ d\mu + \int f_- d\mu < \infty$$

Suppose $f = \sum_{k=1}^{m} x_k \chi_{A_k}$ where $\{A_1, ..., A_m\}$ is a measurable partition of Ω and $x_k \in \mathbb{R}$. Then we claim that $\int f d\mu = \sum_k x_k \mu(A_k)$. Indeed choosing the partition $\mathcal{P}$ as $\{A_k\}$, we get $m_k = x_k$ and hence $\int f d\mu \geq \sum m_k \mu(A_k) = \sum x_k \mu(A_k)$. On the other hand, let $\{B_i\}$ be any finite measurable partition of Ω. Let $m_i = inf_{\omega \in B_i} f(\omega)$. Then if $A_iB_j = \phi$, it follows that

$$m_j \mu(A_iB_j) = 0 = x_i \mu(A_iB_j)$$

on the one hand, and on the other, if $A_iB_j \neq \phi$, then

$$m_j\mu(A_iB_j) \leq inf_{\omega\in A_iB_j}f(\omega)\mu(A_iB_j) = x_i\mu(A_iB_j)$$

so that we get

$$\sum_j m_j\mu(B_j) = \sum_{i,j} m_j\mu(A_iB_j) \leq \sum_{i,j} x_i\mu(A_iB_j) = \sum_i x_i\mu(A_i)$$

implying that

$$\int fd\mu \leq \sum_i x_i\mu(A_i)$$

proving the claim. We next claim that if $0 \leq f \leq g$, then $\int fd\mu \leq \int gd\mu$. Indeed, let $\{A_i\}$ be a (measurable) partition of Ω and set $m_i = inf_{\omega\in A_i}f(\omega)$ and $m_i' = inf_{\omega\in A_i}g(\omega)$. Then we have $m_i \leq m_i'$ and hence

$$\sum_i m_i\mu(A_i) \leq \sum_i m_i'\mu(A_i)$$

Taking supremum over all partitions result into

$$\int fd\mu \leq \int gd\mu$$

and the proof is complete. We next want to establish the monotone convergence theorem, namely that if $f_n \geq 0$ and f_n increases to f, then $\int f_nd\mu$ increases to $\int fd\mu$. Since $f_n \leq f$, it follows from the result just established that $\int f_nd\mu \leq \int fd\mu$. Let $\{A_i : i = 1, 2, ..., n\}$ be a measurable partition of Ω and let $v_i = inf_{\omega\in A_i}f(\omega)$. Then it suffices to establish that $lim \int f_nd\mu \geq \sum_i v_i\mu(A_i)$. Let $\epsilon > 0$ and define

$$A_{in} = \{\omega \in A_i : f_n(\omega) > v_i - \epsilon\}$$

Since $f_n \uparrow f$, and $f(\omega) \geq v_i$ for all $\omega \in A_i$, it follows that A_{in} increases to A_i as n tends to ∞. We assume that $v_i > \epsilon$ for all i and then we have

$$\int f_nd\mu \geq \sum(v_i - \epsilon)\mu(A_{in}) \rightarrow \sum_i(v_i - \epsilon)\mu(A_i), n \rightarrow \infty$$

Letting ϵ decrease to zero gives

$$lim \int f_nd\mu \geq \sum_i v_i\mu(A_i)$$

and taking supremum over all partitions $\{A_i\}$ results in

$$lin \int f_nd\mu \geq \int fd\mu$$

and the proof is complete. This proof works provided that $v_i > 0$ for all i. If $v_i = 0$ for some i, then let J be precisely the set of all i for which $v_i = 0$. Then we have assuming that $0 < \epsilon < v_j$ for all $j \in J^c$ that

$$\int f_nd\mu \geq \sum_{i\in J^c}(v_i - \epsilon)\mu(A_{in}) \rightarrow \sum_{i\in J^c}(v_i - \epsilon)\mu(A_i)$$

and letting ϵ decrease to zero gives

$$lim \int f_n d\mu \geq \sum_{i \in J^c} v_i \mu(A_i) = \sum_i v_i \mu(A_i)$$

completing the proof of the monotone convergence theorem. The monotone convergence theorem is valid even if convergence of f_n to f holds for almost all μ instead of all μ. We now prove the dominated convergence theorem. Let f_n be a sequence of measurable functions such that $|f_n| \leq g$ where g is measurable and integrable, ie, $\int g d\mu < \infty$. Then $g \pm f_n$ are non-negative measurable functions and application of Fatou's lemma gives us

$$\int liminf(g + f_n)d\mu \leq liminf \int (g + f_n)d\mu,$$

$$\int liminf(g - f_n)d\mu \leq liminf \int (g - f_n)d\mu$$

from which it follows that

$$\int (liminf f_n)d\mu \leq liminf \int f_n d\mu,$$

$$\int limsup f_n \geq limsup \int f_n d\mu$$

so that

$$\int liminf f_n d\mu \leq liminf \int f_n d\mu \leq limsup \int f_n d\mu \leq \int limsup f_n d\mu$$

and if $lim f_n = f$ exists, then we deduce from this that

$$\int lim f_n = lim \int f_n d\mu$$

This is precisely the dominated convergence theorem. Note the proof of Fatou's lemma. Suppose f_n are non-negative functions. Define $g_n = inf_{k \geq n} g_k$. Then $\{g_n\}$ is a non-negative increasing sequence so by monotone convergence, we have

$$lim \int g_n d\mu = \int lim g_n d\mu = \int liminf f_n d\mu$$

However since $g_n \leq f_k$ for all $k \geq n$, we have

$$\int g_n d\mu \leq \int f_k d\mu, k \geq n$$

so that

$$\int g_n d\mu \leq inf_{k \geq n} \int f_k d\mu$$

and hence

$$lim \int g_n d\mu \leq liminf \int f_k d\mu$$

Combining the two gives us Fatou's lemma:

$$\int liminf\, f_n d\mu \leq liminf \int f_n d\mu$$

Discuss Fourier series expansion for mean square periodic signals. Let $x(t), t \geq 0$ be mean square periodic with period T, ie, $x(t+T) = x(t)$ for all $t \in \mathbb{R}$, the equality holding in the L^2 sense. This means that $R(t+T, s) = R(t, s)$ for all $t, s \in \mathbb{R}$. Define

$$c_n = \frac{1}{T} \int_0^T x(t) exp(-in\omega t) dt, \omega = 2\pi/T$$

Then consider

$$S_N(t) = \sum_{n=-N}^{N} c_n exp(in\omega t)$$

We have

$$\mathbb{E}|S_N(t) - x(t)|^2 = \mathbb{E}[(S_N(t) - x(t))S_N^*(t)] + \mathbb{E}[(S_N(t) - x(t))x^*(t)]$$

We observe that

$$\mathbb{E}[(S_N(t) - x(t))x^*(s)] = \sum_{n=-N}^{N} \mathbb{E}(c_n x^*(s)) exp(in\omega t) - R(t, s)$$

$$= \sum_{n=-N}^{N} \frac{1}{T} (\int_0^T \mathbb{E}(x(u)x^*(s)) exp(-in\omega u) du) exp(in\omega t) - R(t, s)$$

$$= \sum_{n=-N}^{N} \frac{1}{T} (\int_0^T R(u, s) exp(-in\omega u) du) exp(in\omega t) - R(t, s)$$

$$\rightarrow 0, N \rightarrow \infty$$

in view of the periodicity of $R(t, s)$ in the first argument. It follows from this expression that

$$\mathbb{E}[(S_N(t) - x(t))c_n^*] = \frac{1}{T} \int_0^T \mathbb{E}[(S_N(t) - x(t))x^*(u)] exp(in\omega u) du = 0$$

$$\rightarrow 0, N \rightarrow \infty$$

for all n.

67. Some additional topics in Probability theory

1.Let $X_1, X_2, ...$ be a sequence of iid exponential random variables with mean $1/\alpha$. Define $M_n = max(X_1, ..., X_n)$ and $b_n = log(n)/\alpha$. We have

$$Pr(M_n - b_n \leq x) = Pr(M_n \leq x + b_n) = Pr(X_1 \leq b_n + x)^n$$

$$= (1 - exp(-\alpha(b_n + x)))^n = (1 - n^{-1}exp(-\alpha x))^n \to exp(-exp(-\alpha x)), n \to \infty$$

This is an example of weak convergence. The random variables $\{M_n - b_n\}$ converge weakly to a random variable X having distribution $exp(-exp(-\alpha x))$.

2. Another example of weak convergence: Let X_n be a binomially distributed random variable with $Pr(X_n = r) = \binom{n}{r}(\lambda/n)^r(1 - \lambda/n)^{n-r}$ for $r = 0, 1, ..., n$. Then X_n converges weakly to a Poisson distribution with mean λ.

3.Suppose $(\Omega, \mathcal{F})$ is a measurable space and f a non-negative measurable function on this space. Suppose $\mu(f > 0) = \infty$. Let $\int^* f d\mu$ denote the infimum of $\sum_i sup_{\omega \in A_i} f(\omega)\mu(A_i)$ over all measurable partitions $\{A_i\}$ of Ω. Show that $\int f^* d\mu = \infty$. Let A_i be any measurable partition of Ω. Since $\{f > 0\}$ is non-empty, it follows that $A_i \cap \{f > 0\}$ is non-empty for some i. It follows that $sup_{\omega \in A_i \cap \{f>0\}} f(\omega) = \alpha_i > 0$. Let J be the set of all i such that $A_i \cap \{f > 0\} \neq \phi$. For each $i \in J$, we have $\alpha_i = sup_{\omega \in A_i \cap \{f>0\}} f(\omega) > 0$. Now

$$\sum_i (sup_{\omega \in A_i} f(\omega)).\mu(A_i) \geq \sum_{i \in J} \alpha_i \mu(A_i \cap \{f > 0\})$$

$$\geq \alpha \sum_{i \in J} \mu(A_i \cap \{f > 0\}) = \alpha.\mu(f > 0) = \infty$$

where

$$\alpha = min(\alpha_i : i \in J)$$

The proof of the claim is complete.

4.Transience and persistence of a Markov chain. Let $X_n, n = 0, 1, 2, ...$ be a Markov chain with state space $\{1, 2, ...\}$ and one step transition probabilities $Pr(X_{n+1} = j | X_n = i) = p_{ij}$. The r-step transition probabilities are

$$Pr(X_{n+r} = j | X_n = i) = \sum_{i_1, ..., i_{r-1}} p_{ii_1} p_{i_1 i_2} ... p_{i_{r-1} j}$$

If the one step transition probabilities are represented by a matrix

$$\mathbf{P} = ((p_{ij}))$$

then the matrix of the r-step transition probabilities is represented by the matrix $\mathbf{P}^r$. Let $f_{ij}^{(n)}$ denote the probability that the chain will visit the state j for the first time at epoch n given that it started in the state i. Thus,

$$f_{ij}^{(n)} = Pr(X_1 \neq j, X_2 \neq j, ..., X_{n-1} \neq j, X_n = j | X_0 = i)$$

Then, we have

$$f_{ij}^{(n)} = \sum_{i_1 \neq j, i_2 \neq j, ..., i_{r-1} \neq j} p_{ii_1} p_{i_1 i_2} ... p_{i_{r-1} j}$$

The probability that the chain will eventually visit the state j starting at the state i is given by

$$f_{ij} = \sum_{n=1}^{\infty} f_{ij}^{(n)}$$

We have

$$p_{ij}^{(n)} = \sum_{k=1}^{n} f_{ij}^{(k)} p_{jj}^{(n-k)}, n \geq 1$$

This equation states that the visit of the chain to the state j at time n starting at the state i, can occur iff the chain visits the state j for the first time at some time $k = 1, 2, ..., n$ followed by a transition from state j to state j in the remaining time $n - k$. We get from the above equation by summing over n,

$$\sum_{n=1}^{\infty} p_{ij}^{(n)} = \sum_{1 \leq k \leq n < \infty} f_{ij}^{(k)} p_{jj}^{(n-k)}$$

$$= f_{ij} . \sum_{n=0}^{\infty} p_{jj}^{(n)}$$

or equivalently, since $p_{jj}^{(0)} = 1$, it follows that

$$f_{jj} == \frac{\sum_{n=1}^{\infty} p_{jj}^{(n)}}{1 + \sum_{n=1}^{\infty} p_{jj}^{(n)}}$$

Hence $f_{jj} = 1$ implies that with probability one, the chain will eventually return to the state j starting from the state j and this in turn implies that $\sum_{n=1}^{\infty} p_{jj}^{(n)} = \infty$ and it also implies the converse, namely that $\sum_{n=1}^{\infty} p_{jj}^{(n)} = \infty$ implies $f_{jj} = 1$, ie the process eventually will return to the state j starting at the state j. If $f_{jj} = 1$, we say that the state j is persistent while if $f_{jj} < 1$, we say that the state j is transient. Thus, the state j is persistent iff $\sum_n p_{jj}^{(n)} = \infty$. Now consider the probability that the chain starting in the state i will visit the state j at times $n_1 < n_2 < ... < n_k$ and will not visit the state j at the intermediate times. The probability for this to happen is

$$f_{ij}^{(n_1)} f_{jj}^{(n_2 - n_1)} ... f_{jj}^{(n_k - n_{k-1})}$$

Summing over all $n_1, n_2, ..., n_k$ gives the probability that the chain will, starting at the state i, visit the state j at least k times. This probability evaluates to

$$f_{ij} f_{jj}^k$$

The probability of the chain starting in the state i to visit the state j infinitely often is obtained by taking the limit $k \to \infty$ in the above expression. If the state j is transient, then $f_{jj} < 1$ and we get

$$Pr(X_n = ji.o | X_0 = i) = 0$$

If the state j is persistent, then $f_{jj} = 1$ and we get

$$Pr(X_n = ji.o | X_0 = i) = f_{ij}$$

Now we define another important notion regarding Markov chains. This is called irreduciblility. The Markov chain $\{X_n\}$ is said to be irreducible if given any two states i, j, there exists some integer $n \geq 0$

such that $p_{ij}^{(n)} > 0$. In other words, the chain is said to be irreducible if given any pair of states, there is nonzero probability of the chain starting in the first state to reach the second state in finite time. In other words, any two states communicate. Now suppose the chain is irreducible. Let i, j be any two states. Then there exist positive integers r, s such that $p_{ij}^{(r)}, p_{ji}^{(s)} > 0$. The following identity is obvious

$$p_{ii}^{(n+r+s)} \geq p_{ij}^{(r)} p_{jj}^{(n)} p_{ji}^{(s)}$$

Suppose the state i is transient. Then $\sum_n p_{ii}^{(n)} < \infty$ and it follows from the above equation and $p_{ij}^{(r)}, p_{ji}^{(s)} > 0$ that $\sum_n p_{jj}^{(n)} < \infty$, ie, the state j is also transient. Likewise, persistence of state i implies persistence of state j. In other words, for an irreducible Markov chain, either all the states are transient or all the states are persistent. We now choose i, j, r as above and note that

$$p_{ij}^{(n+r)} \geq p_{ij}^{(r)} p_{jj}^{(n)}$$

for all positive integers n. We are assuming irreducibility of the chain. If the chain is persistent, then $\sum_n p_{jj}^{(n)} = \infty$ and by the above, it follows that $\sum_n p_{ij}^{(n)} = \infty$. On the other hand, we have with s as above,

$$p_{ii}^{(n+s)} \geq p_{ij}^{(n)} p_{ji}^{(s)}$$

Summing over n, we get the result that if the chain is transient so that $\sum_n p_{ii}^{(n)} < \infty$, then $\sum_n p_{ij}^{(n)} < \infty$. In other words, we have proved the following fundamental result: If a Markov chain is irreducible, then either all the states are transient or else all the states are persistent. In the former case, we have for any pair of states i, j, $Pr(X_n = ji.o|X_0 = i) = 0$ and in the latter case, $Pr(X_n = ji.o|X_0 = i) = f_{ij} > 0$. In fact, we can show that in the persistent case for an irreducible Markov chain, we must have $f_{ij} = 1$ and $\sum_{n\geq 1} p_{ij}^{(n)} = \infty$ for all pairs i, j. Indeed, suppose the chain is persistent. Then, we have $Pr(X_n = ji.o|X_0 = j) = 1$ and hence

$$p_{ij}^{(m)} = Pr(X_m = j|X_0 = i) = Pr(X_m = j, X_n = ii.o|X_0 = i)$$

$$\leq \sum_{n\geq m} Pr(X_m = j, X_{m+1} \neq i, ..., X_{n-1} \neq i, X_n = i|X_0 = i)$$

$$= \sum_{n\geq m} p_{ij}^{(m)} . f_{ij}^{(n)} \leq p_{ij}^{(m)} f_{ij}$$

Since the chain is irreducible, we can choose an m such that $p_{ij}^{(m)} > 0$ and then we get from the above, $f_{ij} = 1$. Now,

$$\sum_{n\geq 1} p_{ij}^{(n)} = f_{ij}(1 + \sum_{n\geq 1} p_{jj}^{(n)})$$

and since $f_{ij} = 1$ and $\sum_{n\geq 1} p_{jj}^{(n)} = \infty$, it must follow that $\sum_{n\geq 1} p_{ij}^{(n)} = \infty$. In this case, we have $Pr(X_n = ji.o|X_0 = i) = 1$ for all pairs i, j. Recall that the probability that the chain starting in the state i visits the state j at least k times equals $f_{ij} f_{jj}^k$ and if the state j is persistent, then $f_{jj} = 1$ in which case, we have the probability that starting in the state i, the chain will visit the state j infinitely often equals f_{ij}. In the irreducible case, we have shown that this probability is unity.

68. Patrick Billingsley, Probability and measure:The book starts by introducing the probability space $([0,1],\mathcal{B}[0,1],\mu)$ where μ is the Lebesgue measure on $[0,1]$, ie, uniform probability distribution. This space has been shown (Borel's normal number theorem) to be equivalent to that generated by an infinite sequence of tosses of a fair coin. If a fair coin is tossed an infinite number of times, then we generate a sequence $(X_1, X_2, ...)$ of iid random variables with $Pr(X_i = 0) = Pr(X_i = 1) = 1/2$. Then consider the random variable

$$X = \sum_{i=1}^{\infty} \frac{X_i}{2^i}$$

This random variable is uniformly distributed over $[0,1]$. Moreover, if $X_1, ..., X_n$ are assumed to be given with respective values $x_1, ..., x_n$, then X assumes values in the range $[a, a+1/2^n]$ with uniform probability distribution where $a = \sum_{i=1}^{n} x_i/2^i$. This fact has been used to construct the Lebesgue measure on $[0,1]$. The book then introduces probability theory in its rigorous measure theoretic setting. λ and π systems are discussed and various theorems concerning these systems and σ-fields generated by classes of sets are discussed. Using the uniform distribution on $[0,1]$, discrete random variables are introduced. For example, if a random variable X assumes values x_i with probability $p_i, i = 1,2,...,n, \sum_{i=1}^{n} p_i = 1$, then such a random variable can be easily realized by starting with the probability space $[0,1]$ with its Lebesgue measure, ie, uniform probability measure and dividing this interval into disjoint intervals $I_1, ..., I_n$ respectively of length $p_1, p_2, ..., p_n$ and defining $X(\omega) = x_i$ if $\omega \in I_i, i = 1,2,...,n$. A slight generalization of this method can be used to realize an infinite sequence of independent discrete random variables. For example, if X, Y are two random variables with $Pr(X = x_i, Y = y_j) = p_{ij}, i = 1,2,...,n, j = 1,2,...,m$, then we divide the interval $[0,1]$ into disjoint intervals I_i of length $p_i = \sum_j p_{ij}, i = 1,2,...,n$ and then divide each interval I_i into disjoint intervals I_{ij} of length $p_{ij}, j = 1,2,...,m$ and finally define $X(\omega) = x_i, Y(\omega) = y_j$ for $\omega \in I_{ij}$.

69. Stroock and Varadhan, Multidimensional diffusion processes.

The first chapter discusses weak convergence of probability measures and the Wiener space of continuous functions on the real line. The Wiener measure gives unit probability to this space and the corresponding construction of the Brownian motion process leads to continuous sample paths. The tool for the proof of the Kolmogorov-Centsov theorem is the Garsia-Rudemich-Rumsey inequality which gives an upper bound on quantities of the form $sup|\phi(t) - \phi(s)|$. The space of continuous functions $C[0,\infty)$ is a subset of the space of all functions $\mathbb{R}^{[0,\infty)}$. The latter space is endowed with the standard σ-field

$$\mathcal{B} = \sigma(\mathbb{R}^{[0,\infty)})$$

This σ-field is generated by the cylindrical sets, ie, sets of the form

$$\{\omega : [0,\infty) \to \mathbb{R} : \omega(t_i) \in B_i, i = 1,2,...,N\}$$

where N is any finite positive integer, $\{t_i\}$ is a subset of $\mathbb{R}$ and B_i are Borel subsets of $\mathbb{R}$. A probability measure P on this space need not give unit outer measure to $C[0,\infty)$. Conditions under which this happens are derived in the text. The Kolomogorov-Centsov theorem gives a condition on the process for it to have a continuous modification. This theorem enables one to construct the Brownian motion process as a stochastic process with continuous sample-paths starting from the transition probability functions. More generally, suppose we are given a transition probability distribution function$P(s,x;t,B)$, where

$s \le t$ and $x \in \mathbb{R}, B \in \mathcal{B}(\mathbb{R})$. Then it is explained how to construct a Markov process starting from this and the Kolmogorov-Centsov theorem gives a condition on the transition probability distribution and the initial distribution that guarantees the existence of a continuous modification of the Markov process so constructed. Elementary proofs of the facts that if τ, σ are stoptimes, then $min(\tau, \sigma), max(\tau, \sigma)$ and $\tau + \sigma$ are also stoptimes are given. A proof of Kolmogorov's inequality for martingales is given, namely, that if $X_k, k = 0, 1, ..., n$ is a martingale, then an inequality like Chebysheve's inequality for the event $\{max_{0\le k\le n}|X_k| > \epsilon\}$ is derived.

70. Karatzas and Shreve:Brownian motion and stochastic calculus. This book gives a nice account of Ito's stochastic calculus as applied to Brownian motion and stochastic differential equations. The book starts by defining a stochastic process as a parametric family of random variables defined on a probability space with the parameter signifying time. The notion of a filtration as an increasing family of sub sigma algebras of the original sigma algebra is described. The notion of a right continuous filtration is given. Then comes the important notion of a stoptime. This concept has been explained with the help of several examples. If $\mathcal{F}_t, t \ge 0$ is a filtration, then a stoptime T is a random variable for which the event $T \le t$ is $\mathcal{F}_t$-measurable no matter what t is. This means that if T is the time at which an event occurs, then the event that the given event will occur before time t can be determined completely from knowledge of the process values upto time t. The authors give the following illustrative example: Let $X_t, t \ge 0$ be a continuous stochastic process and T the first time at which the process hits the level a. Then, the event $T \le t$ occurs iff the process hits the level a before time t and this means that $X_s \ge a$ for some $s \le t$. This event is completely determined by the process values $X_s, s \le t$. The author describes three kinds of methods for comparing two stochastic processes. The first is equality in distribution. Two processes are said to be equal in distribution if all their finite dimensional joint probability distributions coincide, ie, if $X_t, Y_t, t \ge 0$ are two stochastic processes then the two are distribution equal iff for all $n = 1, 2, ...,$ $t_1 < t_2 < ... < t_n$ and $x_1, ..., x_n \in \mathbb{R}$, one has the equality

$$P(X_{t_i} \le x_i, i = 1, 2, ..., n) = P(Y_{t_i} \le x_i, i = 1, 2, ..., n)$$

Two processes that are equal in distribution are also said to have the same law. Note that the two processes can be defined on different probability spaces, so that events of the form $X_t = Y_t$ may not make any sense. When two processes coincide in law, then from the statistical angle, they are indistinguishable, ie, probabilities of all events based on the two processes will have the same value. For example, the probability that the process spends its time inside a Borel subset B over the time interval $[a, b]$ will have the same value for both the processes, provided that we assume that the sample paths of the processes are sufficiently regular. The other kind of equivalence of processes is equal in probability. Two processes $X_t, Y_t, t \ge 0$ are said to be equal in probability if for all $t \ge 0$, one has

$$P(X_t = Y_t) = 1, t \ge 0$$

Equality in probability guarantees equality in distribution. When two processes are equal in probability, we say sometimes that each process is a modification of the other. This can be proved easily using elementary properties of the probability measure, predominantly, the fact that any finite (and in fact countable) intersection of unit probability sets has unit probability. Finally, the authors introduce indistinguishability of processes. The two processes are said to be indistinguishable if for each $T \ge 0$, the set $\{X_t = Y_t, 0 \le t \le T\}$ contains an event of unit probability. Note that if the probability space is complete, then this would be the same as saying that this set is an event of probability one. The authors prove the famous theorem of Kolmogorov and Centsov giving sufficient conditions for a process to have a continuous modification. Then the authors prove the Kolmogorov existence theorem for stochastic processes, namely, that given a consistent family of probability distribution functions, there exists a probability space and a stochastic process defined on this probability space whose finite dimensional probability distribution functions coincide with the given family. The authors next introduce the Brownian motion process in terms of its finite dimensional distribution functions and the constraint that the sample paths of the process are continuous.

Chapter 2

Signal Analysis

1.General problems in signal theory and communication:

Write down the expression for an amplitude modulated signal when the carrier frequency is ω_c, the message signal is $m(t)$ and the modulation index is μ. Determine the Fourier transform of this signal. Assuming that the message signal is wide sense stationary, calculate the time averaged autocorrelation function of the amplitude modulated signal.

Ans:

$$x(t) = A_c(1 + \mu.m(t)).cos(\omega_c t)$$

$$X(j\omega) = \pi.A_x(\delta(\omega - \omega_c) + \delta(\omega + \omega_c)) + \frac{\mu}{2}(M(j(\omega - \omega_c)) + M(j(\omega + \omega_c)))$$

For the second part,

$$< R_{xx}(t + \tau, t) >= \frac{A_c^2}{2} cos(\omega_c \tau) + A_c^2 \mu^2 R_{mm}(\tau) cos(\omega_c \tau)$$

Note that

$$< R_{xx}(t + \tau, t) >= lim_{T \to \infty} T^{-1} \int_0^T R_{xx}(t + \tau, t) dt$$

2.Estimating the parameters of a nonlinear system using the LMS algorithm. Stochastic differential equation approach for. The true system has input-output equation

$$y(t) = g(x(t), \phi) + w(t)$$

where ϕ is a parameter to be estimated and $w(t) = \sigma.dB(t)/dt$ with $B(.)$ as standard Brownian motion. Let $\theta(t)$ be the parameter estimate at time t based on the LMS algorithm. Then,

$$\theta(t + dt) - \theta(t) = -\mu.\frac{\partial}{\partial\theta(t)}(y(t) - g(x(t), \theta(t))^2$$

$$= 2\mu.\frac{\partial g(x(t),\theta(t))}{\partial\theta(t)}(y(t) - g(x(t),\theta(t)))$$

Substituting for $y(t)$ gives us the following stochastic differential equation:

$$d\theta(t) = 2\mu.\frac{\partial g(x(t),\theta(t))}{\partial\theta(t)}((g(x(t),\phi) - g(x(t),\theta(t)))dt + \sigma.dB(t)$$

3. If white Gaussian noise is input to a linear filter, then obtain an expression for the autocorrelation function of the output.

4. Explain how you would measure the difference between two random variables and two random processes using mean square theory.

Ans: If X, Y are two random variables, their difference is measured by $C(X,Y) = \mathbb{E}(X - Y)^2$. If $X(t), t \in \mathbb{R}$ and $Y(t), t \in \mathbb{R}$ are two random processes, their difference can be measured by the power in the difference between the two processes, ie,

$$C(X(.),Y(.)) = lim_{T\to\infty}\frac{1}{2T}\int_{-T}^{T}\mathbb{E}(X(t) - Y(t))^2 dt$$

5. Let $A(t), t \geq 0$ be a right continuous nondecreasing process on a probability space adapted to a right continuous filtration $\mathcal{F}_t, t \geq 0$. By right continuous, we mean that $lim_{\epsilon\to 0}A(t+\epsilon) = A(t)$ for all t. Define

$$C(s) = inf(t : A(t) > s)$$

We claim that $C(s)$ is an $\mathcal{F}_t$-stop time. To see this, we look at the event $E = \{C(s) \leq u\}$. Now observe that this even occurs iff $inf(t : A(t) > s) \leq u$. This event is the same as $A(u+1/n) > s$ for all $n = 1, 2, ...$ It follows that

$$E = \bigcap_{n=1}^{\infty}\{A(u+1/n) > s\}$$

This can be proved by mere inspection. It follows that

$$E = \bigcap_{n=m}^{\infty}(A(u+1/n) > s), m = 1, 2, ...$$

and hence for every $m = 1, 2, ...,$ we have that

$$E \in \mathcal{F}_{u+1/m}$$

It follows that

$$E \in \bigcap_{m=1}^{\infty}\mathcal{F}_{u+1/m} = \mathcal{F}_u$$

This proves $C(s)$ is a stoptime. We next wish to show that $A(t)$ is an $\mathcal{F}_{C(s)}$ stoptime for all t. To this end, we must show that $\{A(t) \leq u\} \in \mathcal{F}_{C(u)}$, or equivalently,

$$\{A(t) \leq u\} \cap \{C(u) \leq s\} \in \mathcal{F}_s$$

To prove this, we shall first show that $A(t) = inf(s : C(s) > t)$. First note that $C(.)$ is right continuous and nondecreasing. To prove right continuity, we use the definition that

$$C(s+\epsilon) = inf(t : A(t) > s+\epsilon), C(s) = inf(t : A(t) > s)$$

This implies that

$$C(s+\epsilon) > C(s)$$

Physically, the first equation implies that it takes at least a time $C(s+\epsilon)$ for $A(t)$ to cross the level $s+\epsilon$, while the second equation implies that it takes at least a time $C(s)$ for $A(t)$ to cross the level s. Hence since $A(.)$ is non-decreasing, it follows that $C(s+\epsilon) \geq C(s)$ from which it follows that $C(s+) \geq C(s)$. On the other hand, let $\delta > 0$ be arbitrary. Then by definition of infimum, $A(C(s+\epsilon)-\delta) \leq s+\epsilon$. Letting $\epsilon \to 0$ gives us $A(C(s+)-\delta) \leq s$. It follows that $C(s+)-\delta < C(s)+1/n$ for all $n = 1, 2, ...$ and letting $n \to \infty$ and then $\delta \to 0$ gives us $C(s+) \leq C(s)$. This proves that $C(s+) = C(s)$, ie, $C(.)$ is a right continuous process. We shall now prove that $A(t) = inf(s : C(s) > t)$. If $C(s) > t$, then by definition, $A(t) \leq s$. It follows that

$$A(t) \leq inf\{s : C(s) > t\}$$

On the other hand, suppose $A(t) < s$. Then by definition of $C(s)$, we have that $C(s) > t$. It follows that

$$\{s : A(t) < s\} \subset \{s : C(s) > t\}$$

It follows that

$$A(t) = inf\{s : A(t) < s\} \geq inf\{s : C(s) > t\}$$

which proves the claim. Now we shall show that

$$\{A(t) \leq u\} \cap \{C(u) \leq s\} \in \mathcal{F}_s$$

Observe that $A(t) < u + 1/n$ implies that $C(u+1/n) > t$. This means that

$$\{A(t) \leq u\} \subset \bigcap_{n=1}^{\infty} \{C(u+1/n) > t\}$$

On the other hand $C(u+1/n) > t$ for all n implies that $A(t) \leq u+1/n$ and hence taking intersections, we get

$$\bigcap_{n=1}^{\infty} \{C(u+1/n) > t\} \subset \{A(t) \leq u\}$$

We have thus proved that

$$\{A(t) \leq u\} = \bigcap_{n=1}^{\infty} \{C(u+1/n) > t\}$$

Thus,

$$\{A(t) \leq u\} \cap \{C(u) \leq s\} = \bigcap_n \{C(u) \leq s\} \cap \{C(u+1/n) > t\}$$

If $t \leq s$ then since $C(x)$ is a stop time for all x, it follows that the right side is an element of $\mathcal{F}_s$. If $t > s$, then the right side is empty because $\bigcap_n \{C(u+1/n) > t\} \subset \bigcap_n \{C(u) \geq t\}$ by right continuity of $C(.)$. This completes the proof that $A(t)$ is an $\mathcal{F}_{C(s)}$ stoptime.

Ref:Daniel Revuz and Marc Yor; Continuous Martingales and Brownian Motion, Springer.

6. Least mean square algorithm applied to Frequency modulation. Assume that the FM wave is given by

$$x(t,\beta) = A.cos(\omega_c t + .\beta.sin(\omega_c t)) + w(t) = f(t,\beta) + w(t)$$

$w(t) = \sigma.dB(t)/dt$ with $B(.)$ as Brownian motion. The modulation index β is assumed to be an unknown parameter to be estimated. The estimate at time t is given by

$$\hat{x}(t,\beta(t)) = A.cos(\omega_c t + \beta(t).sin(\omega_c t))$$

The continuous time version of the LMS algorithm has the form

$$\beta(t+dt) = \beta(t) - \mu.dt.\frac{d}{d\beta(t)}(x(t,\beta) - \hat{x}(t,\beta(t)))^2$$

This gives

$$d\beta(t) = 2\mu.dt.(x(t,\beta) - \hat{x}(t,\beta(t))).\frac{\partial\hat{x}(t,\beta(t))}{\partial\beta(t)}$$

$$= 2\mu(f(t,\beta) - \hat{x}(t,\beta(t)) + \sigma)\frac{\partial\hat{x}(t,\beta(t))}{\partial\beta(t)} + 2\mu.\sigma.\frac{\partial\hat{x}(t,\beta(t))}{\partial\beta(t)}dB(t)$$

The convergence analysis proceeds by expanding $\hat{x}(t,\beta(t))$ around β and deriving differential equations for $\mathbb{E}(\beta(t) - \beta)^m, m = 1,2,....$ For $m = 1,2$, this research work has been carried out by Dr.Tarun, colleague of the author.

7. Determine the impulse response of a second order filter defined by the transfer function

$$H(z) = \frac{1}{(1 - r.exp(i\alpha)z^{-1})(1 - r.exp(-i\alpha)z^{-1})}$$

Use the method of partial fractions.

8. Finite register effects in a second order system. The difference equation governing the system dynamics is

$$y[n] + ay[n-1] + by[n-2] = x[n]$$

Let $\epsilon_1[n]$ be the quantization noise induced by the first multiplier a and $\epsilon_2[n]$ the quantization noise induced by the second multiplier b. Then if $f[n]$ is the ouput noise, we have

$$f[n] + af[n-1] + bf[n-2] = -(\epsilon_1[n] + \epsilon_2[n])$$

Assuming $\epsilon_i, i = 1,2$ to be independent white noise processes, with mean zero and variance σ_e^2, it follow that the output noise power is given by

$$\mathbb{E}f[n]^2 = 2\sigma_e^2 \sum_n h[n]^2 = \frac{\sigma_e^2}{\pi} \int_{-\pi}^{\pi} |H(j\omega)|^2 d\omega$$

where

$$H(j\omega) = \frac{1}{1 + az^{-1} + bz^{-2}}$$

Reference:Oppenheim and Schaffer, Digital Signal Processing.

9. Consider a two dimensional discrete space system defined by the difference equation

$$y[n,m] = a[1]y[n-1,m] + a_2 y[n,m-1] + a_3 y[n-1,m-1] + x[n,m]$$

Show that the two dimensional transfer function of this system is given by

$$H(\omega_1,\omega_2) = \frac{1}{1 - a[1]exp(-j\omega_1) - a[2]exp(-j\omega_2) - a[3]exp(-j(\omega_1+\omega_2))}$$

Show that if each multiplier introduces a white quantization noise $\epsilon_i[n,m], i = 1,2,3$ each one have zero mean and covariance $\sigma^2\delta[n,m]$, then the quantization noise power in the output is given by

$$\sigma_f^2 = (3\sigma^2/(2\pi)^2) \int_{-\pi}^{\pi}\int_{-\pi}^{\pi} |H(\omega_1,\omega_2|^2 d\omega_1 d\omega_2$$

Show that we can write

$$\sigma_f^2 = (3\sigma^2) \sum_{n,m=-\infty}^{\infty} |h[n,m]|^2$$

Show that $h[n,m]$ is given by the coefficient of $exp(-j(\omega_1 n + \omega_2 m))$ in the expansion

$$\sum_{r=0}^{\infty} (a[1]exp(-j\omega_1) + a[2]exp(-j\omega_2) + a[3]exp(-j(\omega_1+\omega_2)))^r$$

Using the trinomial theorem, determine a series expansion for $h[n,m]$.

10. Least mean phase algorithm. The input is a complex vector signal $\mathbf{x}(t)$ and the output is given by a linear combination of these signals plus noise, ie,

$$d(t) = \mathbf{w}^T\mathbf{x}(t) + n(t)$$

where $\mathbf{w}$ is a fixed complex weight and $n(t) = \sigma.dB(t)/dt$ with $B(.)$ as standard Brownian motion. The estimated signal at time t is given by

$$\hat{d}(t) = \mathbf{w}(t)^T\mathbf{x}(t)$$

where $\mathbf{w}(t)$ is a complex time varying weight. We are interested only in obtaining the correct phase information about the output. So we match the phase of $d(t)$ with that of $\hat{d}(t)$ by the least mean square method. The continuous time version of the resulting algorithm, which we call the least mean phase algorithm reads

$$d\mathbf{w}(t) = -dt.\nabla_{\mathbf{w}(t)}(Arg(d(t)) - Arg(\hat{d}(t)))^2$$

Now observe that

$$Arg(\hat{d}(t)) = Arg(w(t)^T x(t)) = Im.log(w(t)^T x(t))$$

so that

$$\nabla_{w(t)} Arg(\hat{d}(t)) = Im(x(t)/w(t)^x(t)) = Im(x(t)/\hat{d}(t))$$

We can verify this formula directly too. Write

$$w(t) = (w_i(t)), w_i(t) = w_{i1}(t) + j.w_{i2}(t),$$
$$x(t) = (x_i(t)), x_i(t) = x_{i1}(t) + jx_{i2}(t)$$

so that

$$\hat{d}(t) = w(t)^T x(t) = \sum_i (w_{i1}x_{i1} - w_{i2}x_{i2}) + j\sum(w_{i1}x_{i2} + w_{i2}x_{i1})$$

so that

$$Arg(\hat{d}(t)) = tan^{-1}[(\sum_i (w_{i1}x_{i2} + w_{i2}x_{i1})/(\sum_i w_{i1}x_{i1} - x_{i2}x_{i2})] = tan^{-1}(\hat{d}_2(t)/\hat{d}_1(t))$$

$$\frac{\partial}{\partial w_{i1}} Arg(\hat{d}(t)) =$$

11. Phase LMS: Let $\mathbf{z}(t)$ be a complex vector valued signal and $d(t)$ a complex linear combination of $\mathbf{y}(t)$ plus noise. Thus, $d(t) = \mathbf{w}_0^T \mathbf{z}(t) + n(t)$. The estimate of the desired signal at time t is given by $\hat{d}(t) = \mathbf{w}(t)^T \mathbf{z}(t)$. We are bothered only about phase information and so we wish to minimize $(Arg(d(t)) - Arg(\hat{d}(t)))^2$. The adaptive algorithm for doing this minimization reads

$$\mathbf{w}(t+dt) = \mathbf{w}(t) - \mu.dt.\nabla_{\mathbf{w}(t)}(Arg(d(t)) - Arg(\hat{d}(t)))^2$$

Now observe that

$$Argz = Im(logz)$$

Suppose w, z are complex numbers. We let $w = a + ib$ and $z = x + iy$. Then, $wz = (ax - by) + i(ay + bx)$. It follows that

$$Arg(wz) = tan^{-1}[(ay + bx)/(ax - by)]$$

and hence

$$\frac{\partial}{\partial a} Arg(wz) = ((ax - by)y - (ay + bx)x)/((ax - by)^2 + (ay + bx)^2)$$

$$\frac{\partial}{\partial b} Arg(wz) = ((ax - by)x + (ay + bx)y)/((ax - by)^2 + (ay + x)^2)$$

Suppose on the other hand, $z_1, ..., z_n, w_1, ..., w_n$ are complex numbers and we wish to find

$$\frac{\partial}{\partial a_i} Arg(\sum_j w_j z_j), \frac{\partial}{\partial b_i}(\sum_j w_j z_j)$$

where

$$w_j = a_j + ib_j$$

Writing

$$z_j = x_j + iy_j$$

gives

$$\sum_j w_j z_j = \sum_j [(a_j x_j - b_j y_j) + i(a_j y_j + b_j x_j)]$$

so that

$$Arg(w^T z) = Arg(\sum_j w_j z_j) = tan^{-1}[(\sum_j (a_j y_j + b_j x_j))/(\sum_j (a_j x_j - b_j y_j))]$$

Thus,

$$\frac{\partial}{\partial a_i} Arg(\sum_j w_j z_j) = ((\sum_j (a_j x_j - b_j y_j))y_i - (\sum_j (a_j y_j + b_j x_j))x_i)/|\sum_j w_j z_j|^2$$

$$= [Re(w^T z)y_i - Im(w^T z)x_i]/|w^T z|^2$$

$$\frac{\partial}{\partial b_i} Arg(w^T z) =$$

$$= [Re(w^T z)x_i + Im(w^T z)y_i]/|w^T z|^2$$

so that

$$(\frac{\partial}{\partial a_i} + j.\frac{\partial}{\partial b_j})Arg(w^T z)$$

$$= [y_i w^T z + j x_i w^T z]/|w^T z|^2 = j\bar{z}_i.w^T z/|w^T z|^2$$

Note that the LMS update equations are

$$da_i(t) = -\mu.dt \frac{\partial}{\partial a_i(t)}(Arg(d(t)) - Arg(\hat{d}(t)))^2,$$

$$db_i(t) = -\mu.dt \frac{\partial}{\partial b_i(t)}(Arg(d(t)) - Arg(\hat{d}(t)))^2$$

or equivalently,

$$da_i(t) = 2\mu.dt \frac{\partial Arg(\hat{d}(t))}{\partial a_i(t)}(Arg(d(t) - Arg(\hat{d}(t))),$$

$$db_i(t) = 2\mu.dt.\frac{\partial Arg(\hat{d}(t))}{\partial b_i(t)}(Arg(d(t)) - Arg(\hat{d}(t)))$$

This gives us the LMS update equation on using $dw_i = da_i + jdb_i$, or $dw = da + jdb$,

$$dw(t) = 2\mu.dt(Arg(d(t)) - Arg(\hat{d}(t)))(j\bar{z}(t)w(t)^T z(t))/|w(t)^T z(t)|^2$$

or

$$dw = 2\mu.dt(Arg(d(t)) - Arg(\hat{d}(t)))(j\bar{z}(t)z(t)^T w(t))/|w(t)^T z(t)|^2$$

Now let

$$d(t) = w_0^T z(t) + \sigma.dB(t)/dt = w_i^T z(t) + n(t)$$

so that we have approximately,

$$logd(t) = log(w_0^T z) + log(1 + n(t)/w_0^T z) = log(w_0^T z) + n(t)/(w_0^T z)$$

Taking imaginary parts, we get

$$Arg(d(t)) \approx Arg(w_0^T z(t)) + n(t)Im((w_0^T z(t))^{-1})$$

and the phase LMS update equation becomes

$$dw(t) = 2\mu.dt(Arg(w_0^T z(t)) + \sigma.Im((w_0^T z(t))^{-1}).dB(t)/dt - Arg(w(t)^T d(t)))j\bar{z}(t)z(t)^T w(t)/|w(t)^T z(t)|^2$$

which can be cast in stochastic differential form

$$dw(t) = 2\mu.(dt.(Arg(w_0^T z(t)) - Arg(w(t)^T z(t))) + \sigma.Im((w_0^T z(t))^{-1}).dB(t))j\bar{z}(t)z(t)^T w(t)/|w(t)^T z(t)|^2$$

12.Define the notion of impulse respose of a linear discrete time time invariant system. Distinguish between the forms of the impulse responses of linear time varying and linear time invariant systems. Give examples illustrating this distinction.

13.Hilbert transform for discrete time signals has the transfer function $H(\omega) = -j.sgn(\omega)$ where $|\omega| \leq \pi$. Determine the impulse response of such a system. Determine the action of such a system on a sinusoidal signal $sin(\omega n + \phi)$.

14.Explain the overlap add method for computing the convolution between a finite duration sequence and an infinite duration sequence.

15. Explain how you would solve the wave equation in one dimension by a two dimensional discretization procedure. If the wave field is perturbed by white Gaussian noise so that it acquires the form

$$\psi_{tt}(t,x) - c^2\psi_{xx}(t,x) = F(t,x)$$

where $F(.,.)$ is white Gaussian noise, then by the discretization of the spatial variables, transform this equation into a system of linear stochastic differential equations. Hint:The discretized system has the form

$$\psi_{tt}(t,n) - \frac{c^2}{\Delta^2}(\psi(t,n) - 2\psi(t,n-1) + \psi(t,n-2)) = F(t,n), n = 1,2,...,N$$

16.Let $\mathcal{H}$ be a Hilbert space with onb $\{e_1,...,e_n\}$. For $r \leq n$, let $\mathcal{H}^{\otimes_a n}$ denote the r-fold antisymmetric tensor product of $\mathcal{H}$ with itself. Let $\mathcal{H}^{\otimes r}$ denote the r-fold tensor product of $\mathcal{H}$ with itself. For any permutation $\sigma \in S_n$, define the operator

$$U_\sigma : \mathcal{H}^{\otimes n} \rightarrow \mathcal{H}^{\otimes n}$$

by the equation

$$U_\sigma(x_1 \otimes x_2 \otimes .... \otimes x_n) = x_{\sigma 1} \otimes x_{\sigma 2} \otimes ... \otimes x_{\sigma n}$$

Define the operator

$$E : \mathcal{H}^{\otimes n} \to \mathcal{H}^{\otimes_a n}$$

by

$$E = \frac{1}{n!} \sum_{\sigma \in S_n} sgn(\sigma) U_\sigma$$

Show that E is the orthogonal projection of $\mathcal{H}^{\otimes n}$ onto $\mathcal{H}^{\otimes_a n}$. Hint:First show that $E^* = E$, then show that $E^2 = E$, then complete the proof by showing that the range of E is the entire $\mathcal{H}^{\otimes_a n}$. Observe that

$$< U_\sigma(x_1 \otimes ... \otimes x_n), (y_1 \otimes ... \otimes y_n) >=< x_{\sigma 1} \otimes ... \otimes x_{\sigma n}, y_1 \otimes ... \otimes y_n >$$

$$= Pi_{i=1}^n < x_{\sigma i}, y_i >= \Pi_{i=1}^n < x_i, y_{\sigma^{-1} i} >$$

$$=< x_1 \otimes ... \otimes x_n, U_{\sigma^{-1}}(y_1 \otimes ... \otimes y_n) >$$

proving that

$$U_\sigma^* = U_{\sigma^{-1}} = U_\sigma^{-1}$$

It follows that

$$E^* = \frac{1}{n!} \sum_{\sigma \in S^n} sgn(\sigma) U_{\sigma^{-1}} = E$$

Also

$$E^2 = \frac{1}{n!^2} \sum_{\sigma, \rho \in S_n} sgn(\sigma\rho) U_{\sigma\rho} = E$$

The last equation needs to be justified !. Replace the summation over the permutations σ and ρ by summation over the permutations σ and $\tau = \sigma\rho$ and then obtain the desired result using

$$\sum_{\sigma \in S_n} 1 = n!$$

17.Baire's category theorem: Let (X, d) be a complete metric space and let A be a subset of X. If $int(Cl(A))$ is empty, we then say that A is meager. Suppose $X = \bigcup_{k=1}^\infty A_k$ where A_k is closed for every k. Then, for some k, we must have that $Int(A_k)$ is non-empty, ie, A_k is non-meager. For suppose $Int(A_k)$ is empty for every k. We are assuming that X is non-empty. Then $A_1 \neq X$, for if $A_1 = X$, then $Int(A_1) = X$ which is non-empty. So A_1^c is a non-empty open subset of X. Let $B(x_1, \epsilon_1) \subset A_1$ for some $x_1 \in A_1^c$ and some $\epsilon_1 > 0$. This is possible since A_1^c is non-empty and open. Then $B(x_1, \epsilon_1/2)$ cannot be contained in A_2 since by hypothesis, A_2 is non-empty. It follows that $B(x_1, \epsilon_1/2) \cap A_2^c$ is non-empty and open. Hence, there exists $x_2 \in X$ and $\epsilon_2 > 0$ such that

$$B(x_2, \epsilon_2) \subset B(x_1, \epsilon_1/2) \cap A_2^c$$

We may assume that $\epsilon_2 < \epsilon_1/2$. Continuing this way, we obtain a sequence $x_1, x_2, ...$ in X and a sequence $\epsilon_1, \epsilon_2, ...$ of real numbers such that $\epsilon_{i+1} < \epsilon_i/2$ and

$$B(x_{i+1}, \epsilon_{i+1}) \subset B(x_i, \epsilon_i/2) \cap A_{i+1}^c$$

Now let $n > m$. Then,

$$B(x_n, \epsilon_n) \subset B(x_{n-1}, \epsilon_{n-1}/2) \subset ... \subset B(x_m, \epsilon_m/2)$$

so that

$$d(x_m, x_n) \leq \epsilon_m/2 \leq \epsilon_1/2^m$$

and hence $\{x_n\}$ is Cauchy in X. Since X is complete, it follows that $x_0 = lim x_n$ exists. We now have for $n > m$,

$$x_n \in B(x_n, \epsilon_n) \subset B(x_m, \epsilon_m/2)$$

and hence, taking $lim n \to \infty$, it follows that

$$x_0 \in B[x_m, \epsilon_m/2] \in B(x_m, \epsilon_m) \subset A_m^c$$

for all m. It follows that $x_0 \in \bigcap_m A_m^c = \phi$, a contradiction. This completes the proof Baire's category theorem.

18.Let $(\Omega, \mathcal{F})$ be a measurable space and let ξ be as spectral measure on $\mathcal{F}$ with values in $P(\mathcal{H})$, the collection of all orthogonal projection operators in a Hilbert space $\mathcal{H}$. Let f be a complex measurable function on the space $(\Omega, \mathcal{F})$ and let D_f denote the set of all $u \in \mathcal{H}$ with the property $\int |f(\omega)|^2 < u, \xi(d\omega)u >< \infty$. Then it is easy to see that D_f is dense in $\mathcal{H}$. Indeed, let E_n denote the set $\{\omega : |f(\omega)| \leq n\}$ and define $u_n = \xi(E_n)u$. Then we have

$$\int |f(\omega)|^2 < u_n, \xi(d\omega)u_n >= \int_{E_n} |f(\omega)|^2 < u, \xi(d\omega)u >\leq n. < u, \xi(E_n)u >\leq n \parallel u \parallel^2$$

Thus, $u_n \in D_f$ and since E_n increases to Ω, it follows that $u_n = \xi(E_n)u$ converges to u.

19.Determine the impulse response using partial fractions of the transfer function

$$H(z) = \frac{1}{(1 - r.exp(-j\theta)z^{-1})(1 - r.exp(j\theta)z^{-1}}$$

20.Discussion with Shambu Sharma. Let v be a vector field with components $(v^i(x))$. Define the Lie derivative along v by

$$L_v(f) = (v, \nabla f) = \sum_i v^i \frac{\partial f}{\partial x^i}$$

Show that if v, w are two vector fields, then $L_v L_w - L_w L_v$ is a first order differential operator and determine the vector field corresponding to this operator.

Hint:

$$L_v(f) = v^i \frac{\partial f}{\partial x^i} = v^i f_{,i}$$

where summation over the repeated index i is implied. Thus,

$$L_w L_v(f) = w^j(v^i f_{,i})_{,j} = w^j v^i_{,j} f_{,i} + w^j v^i f_{,ij}$$

Interchanging w and v gives us

$$L_v L_w(f) = v^j w^i_{,j} f_{,i} + v^j w^i f_{,ij}$$

Taking the difference, we get

$$(L_v L_w - L_w L_v)(f) = (v^j w^i_{,j} - w^j v^i_{,j}) f_{,i}$$

so that

$$L_v L_w - L_w L_v = L_u$$

where

$$u^i = v^j w^i_{,j} - w^j v^i_{,j}$$

summation over the repeated index j being implied.

21.Modification of the LMS algorithm for estimating the amplitudes and phases of a harmonic signal. Assume that the harmonic signal has the form

$$x(t) = \sum_{i=1}^{n} A_i exp(j\omega_i t) + w(t)$$

where $w(t)$ is white noise, ie, a constant σ times the derivative of Brownian motion. The estimated signal is given by

$$\hat{x}(t) = \sum_{i=1}^{n} B_i.exp(j\theta_i t)$$

We are assuming A_i, B_i to be real numbers as also ω_i, θ_i. In fact, B_i, θ_i will be real valued functions of time. The objective is to apply the LMS algorithm to the square of the difference between the amplitudes and phases of the desired and estimated signals. That is, we define the parameter estimate evolutions by

$$dB_i(t) = -\mu.dt.\frac{\partial}{\partial B_i(t)}\{(|\hat{x}(t)|^2 - |x(t)|^2)^2 + \lambda.(Arg(\hat{x}(t)) - Arg(x(t)))^2\},$$

$$d\theta_i(t) = -\mu.dt.\frac{\partial}{\partial \theta_i(t)}\{(|\hat{x}(t)|^2 - |x(t)|^2)^2 + \lambda.(Arg(\hat{x}(t)) - Arg(x(t)))^2\}$$

According to these equations, we are splitting out the total error between the true signal and the estimated signal into an amplitude error component and a phase error component. If we wish to give more weightage to the amplitude information of the signal, then we make λ very small while if we wish to given more weightage to the phase information of the signal, then we make λ very large. For our sinusoidal processes, we have

$$|x(t)|^2 = \sum A_i^2 + 2\sum_{i<j} A_i A_j.cos((\omega_i - \omega_j)t),$$

$$|\hat{x}(t)|^2 = \sum B_i(t)^2 + 2\sum_{i<j} B_i(t)B_j(t).cos(\theta_i(t)t - \theta_j(t)t)$$

Then,

$$\frac{\partial}{\partial B_i}(|\hat{x}(t)|^2 - |x(t)|^2)^2) = 2(|\hat{x}(t)|^2 - |x(t)|^2)\frac{\partial}{\partial B_i}|\hat{x}(t)|^2$$

where

$$\frac{\partial}{\partial B_i}|\hat{x}(t)|^2 = 2B_i + 2\sum_{j\neq i} B_j cos((\omega_j - \theta_j)t)$$

Likewise,

$$Arg(x(t)) = tan^{-1}\frac{\sum_i A_i sin(\omega_i t)}{\sum_i A_i.cos(\omega_i t)}$$

$$Arg(\hat{x}(t)) = tan^{-1}\frac{\sum_i B_i sin(\theta_i t)}{\sum_i B_i cos(\theta_i t)}$$

$$\frac{\partial}{\partial B_i} = (\sum_j B_j cos(\theta_j t)sin(\theta_i t) - \sum_j B_j sin(\theta_j t)cos(\theta_i t))/|\hat{x}(t)|^2$$

$$= \sum_j B_j sin((\theta_i - \theta_j)t)/|\hat{x}(t)|^2$$

22. Metric space:Definition and properties of a metric (symmetry, positivity, triangle inequality), Examples of metric spaces: A finite set with the Hamming (zero one) metric. Distance between a set and a point, distance between two sets.

Completeness:Definition of a Cauchy sequence: $d(x_n, x_m) \to 0, as n, m \to \infty$. If every Cauchy sequence has a limit, we say that the metric space is complete.

Positive definite operators in a Hilbert space: $< Tx, x >\geq 0$ for all $x \in \mathcal{H}$. Then, we write $T \geq 0$. This implies that $T^* = T$, ie, T is self-adjoint. If T, S are self adjoint and $T - S \geq 0$, we write $T \geq S$.

Norm of an operator:

$$\| T \| = sup\{\| Tx \| / \| x \|, x \neq 0\}$$

23. Adjoint of a linear operator in a Hilbert space: $T : \mathcal{H} \to \mathcal{H}$ is a linear operator. Assume T is bounded, ie, $\| T \| = sup(\| Tx \| : \| x \| \leq 1\} < \infty$. Then let

$$f_T(x, y) = < Tx, y >$$

For each y, $x \to f_T(x, y)$ is a bounded linear functional on the Hilbert space and hence by the Riesz representation theorem, there is a unique vector $T^*(y)$ such that

$$< Tx, y > = < x, T^*(y) >, \forall x \in \mathcal{H}$$

The map T^* is linear and bounded and is uniquely determined by T. We have

$$| < x, T^*y > | \leq \| T \| \| x \| \| y \|$$

so that

$$\| T^* \| \leq \| T \|$$

Since $T^{**} = T$, it follows that

$$\| T^* \| \leq \| T^{**} \| = \| T \|$$

so that

$$\| T^* \| = \| T \|$$

24. Convergence of a sequence of vectors in a normed space: Let $(V, \| . \|)$ be a normed linear space and $\{x_n : n = 1, 2, ...\}$ a sequence of vectors in this space. This sequence is said to converge to a vector x in the space if $\| x_n - x \| \to 0$ as $n \to \infty$. We then write $lim x_n = x$. Let $y_n = x_1 + ... + x_n$. Show that if $\sum \| x_n \| < \infty$ and the normed space is complete (Banach), then $lim \sum_{k=1}^{n} x_k$ exists. Hint: Define $S_n = x_1 + ... + x_n$. Then

$$\| S_{n+k} - S_n \| \leq \sum_{m=n+1}^{k} \| x_m \|$$

which when combined with $\sum \| x_n \| < \infty$ shows that $\{S_n\}$ is Cauchy.

25. Riesz representation theorem: Let $\mathcal{H}$ be a Hilbert space and let $f : \mathcal{H} \to \mathbb{C}$ be a non-zero bounded linear functional on $\mathcal{H}$. Then

$$W = \mathcal{N}(f) = \{x \in \mathcal{H} : f(x) = 0\}$$

is a closed subspace of $\mathcal{H}$. More over $W \neq \mathcal{H}$, for otherwise $f = 0$. It follows that $W^{\perp} \neq 0$. Let $0 \neq z \in W^{\perp}$ and choose any $x \in \mathcal{H}$. Consider

$$y = x - f(x)z/f(z)$$

Note that $f(z) \neq 0$, for otherwise, $z \in W \cap W^{\perp} = \{0\}$. Then evidently $f(y) = f(x) - f(x)f(z)/f(z) = 0$, ie, $y \in W$. It follows that $< y, z >= 0$, ie,

$$< x, z > -(f(x)/f(z)) < z, z >= 0$$

or equivalently,

$$f(x) = < x, \bar{f}(z)z / \| z \|^2 >$$

and this proves the Riesz representation theorem that there exists $u \in \mathcal{H}$ such that $f(x) = < x, u >$ for all $x \in \mathcal{H}$.

26. Applications of Complex analysis: Phasor theory for circuit analysis, phasor theory for electromgnetic waves, phasor theory applied to waveguides and antennas, application of complex analysis to fluid dynamics (velocity potential)

1.Phasor theory: If a linear circuit consists of voltage and current sources all at the same frequency ω, then these sources can be written as $v_a(t) = V_a.cos(\omega t+\phi_a), a = 1, 2, ..., N$ where $\{V_a\}, \phi_a$ are real. The phasor notation for these voltages is $\mathbf{V}_a = V_a exp(i\phi_a)$. A resistance R is retained as it is, a capacitance C is replaced by the complex impedance $1/i\omega C$ since if $v(t) = V.cos(\omega t+\phi)$ is the voltage across it, then the current through it is given by $i(t) = C.v'(t) = -\omega.C.V.sin(\omega t + \phi) = \omega.C.V.cos(\omega t + \pi/2)$. The voltage across the capacitor has the phasor representation $V.exp(i\phi)$, while the current through the capacitor has the phasor representation $i\omega C.V.exp(i\phi)$ thereby justifying the representation of the capacitance. Likewise, an inductance L has the impedance $i\omega L$. This follows from the fact that if $i(t) = I.cos(\omega t + \phi)$ is the current through it, then the voltage across it is given by $v(t) = L.i'(t) = -\omega LI.sin(\omega t + \phi)$. The current is then represented by the phasor $I.exp(i\phi)$ and the voltage by the phasor $i\omega L.I$. In this way, a linear circuit with passive time invariant elements excited by sinusoidal sources can be dealt with by purely algebraic methods based on complex algebra.

2. Phasor theory for electromagnetic wave propagation: In Maxwell's equations, the partial derivatives of the electric and magnetic fields with respect to time are replaced by multiplication with $i\omega$ assuming that the exciting charge density and current density have sinusoidal variation. Then the electric field and magnetic field at each point in space can be replaced by phasors and one can derive the Helmholtz equation for these fields:

$$(\nabla^2 + k^2)\psi(\mathbf{x}) = 0$$

In case the fields are to be determined in a region where there is charge and current, the vector and scalar potential satisfy the Helmholtz equations with source terms:

$$(\nabla^2 + k^2)\psi(\mathbf{x}) = s(\mathbf{x})$$

the solution to which is obtained as

$$\psi(\mathbf{x}) = \int G(\mathbf{x}, \mathbf{x}')s(\mathbf{x}')s^3x'$$

where

$$G(\mathbf{x}, \mathbf{x}') = -exp(-ik|\mathbf{x} - \mathbf{x}'|)/4\pi|\mathbf{x} - \mathbf{x}'|$$

In this fashion, an electromagnetic field with sinusoidally varying sources can be completely reduced to phasor analysis wherein only spatial variables are involved, the time variable disappears.

27. General quiz in signals and systems:

[1] Define a signal in terms of pattern of variations over a set. Show that if the set is equipped with the structure of a measure space, then one can successfully define signal energy and correlations but not power.

Note: Let $(X, \mathcal{F}, \mu)$ be a measure space. If $f : X \to \mathbb{R}$ is a non-negative measurable function, then we can define the integral $\int_X f(x)d\mu(x)$. Explain this part. Specifically, show that any nonnegative measurable function is a pointwise limit of simple nonnegative measurable functions and if f is simple, its integral can be readily defined: $f(x) = \sum_{i=1}^n c_i\chi_{E_i}(x)$ where $E_i \in \mathcal{F}, i = 1, 2, ..., n$, then

$$\int fd\mu = \sum_{i=1}^n c_i\mu(E_i)$$

Then show that if $0 \leq f \leq g$ are simple measurable functions, then $\int f d\mu \leq \int g d\mu$. Hence, deduce that the integral $\int f d\mu$ of any nonnegative measurable function can be defined as $lim \int f_n d\mu$ where f_n are nonnegative measurable and simple and increase pointwise to f. Show that the definition of the integral is independent of the approximating sequence. The energy of any measurable signal f may now be defined as $E(f) = \int |f(x)|^2 d\mu(x)$. If f, g are two measurable signals, their cross correlation can be defined as

$$R(f,g) = \int f(x)g(x)d\mu(x)$$

Show that the space $\mathcal{H}$ of all measurable signals f defined on X for which $\int |f(x)|^2 d\mu(x) < \infty$ is a Hilbert space provided that we define the inner product between $f, g \in \mathcal{H}$ as

$$< f, g >= \int f(x)g(x)d\mu(x)$$

Prove the Cauchy-Schwarz inequality

$$| < f, g > |^2 \leq < f, f >< g, g >, f, g \in \mathcal{H}$$

[2] Error correcting codes: Show that if $x = (x_1, ..., x_n)$ and $y = (y_1, ..., y_n)$ are two n-length binary sequences with the the distance between the two being defined by

$$d(x, y) - \sum_{i=1}^{n} |x_i - y_i|$$

and if w is another n length binary sequence with $d(w, 0) \leq t$ interpreted as noise, then if $d(x, y) > 2t$, we can design an error correcting scheme.

Hint: Let x, y be two sequences with $d(x, y) > 2t$. The noise corrupted sequence is $z = x + w$ when x is the transmitted signal and is $z = y + w$ when y is the transmitted signal. If $d(z, x) \leq t$ then decide that x was transmitted while if $d(z, y) \leq t$ decide that y was transmitted. Then this scheme is a t error correcting scheme. Note that the situation $d(z, x), d(z, y) \leq t$ cannot occur since that would imply

$$d(x, y) \leq d(z, x) + d(z, y) \leq 2t$$

which is false.

[3] Matched filter over a finite duration: Let $x(t) = s(t) + w(t), t \in [0, T]$ where $s(.)$ is the noiseless signal and $w(.)$ white Gaussian noise. Let $h(t), t \geq 0$ denote the impulse response of a causal linear time invariant filter. Show that the output at time T can be expressed as

$$y(t) = \int_0^T h(\tau)x(T - \tau)d\tau = y_s(T) + y_w(T)$$

where

$$y_s(T) = \int_0^T h(\tau)s(T - \tau)d\tau, y_w(T) = \int_0^T h(\tau)w(T - \tau)d\tau$$

The signal power at the filter output at time T is given by $P_s = |y_s(T)|^2$ and the noise power at the output at time T by

$$P_w = \mathbb{E}[y_w(T)^2] = \int_0^T h(\tau)^2 d\tau$$

The signal to noise ratio (SNR) at the output at time T is thus given by

$$SNR_o = \frac{P_s}{P_w} = \frac{(\int_0^T h(\tau)s(T-\tau)d\tau)^2}{\int_0^T h(\tau)^2 d\tau}$$

and application of the Cauchy-Schwarz inequality gives

$$SNR_o =\leq \int_0^T s(T-\tau)^2 d\tau = \int_0^T s(\tau)^2 d\tau$$

with equality iff $h(t) = K.s(T-t), t \in [0,T]$. This is the matched filter. In words, the condition for the SNR at the output to be a maximum at some time $T > 0$ is that the filter impulse response be matched to the input signal.

[4] Singularity functions occurring in quantum electrodynamics: Let $x^0 = t, x^1 = x, x^2 = y, x^3 = z$, so that the contravariant position four vector is given by $(x^\mu) = (t, x, y, z)$. The covariant position four vector is given by $(x_\mu) = (t, -x, -y, -z)$. The wave operator (denoted as $\Box$) is defined by

$$\Box = \partial^\mu \partial_\mu = \partial_t^2 - \nabla^2$$

We define the retarded Greens function $D_R(x)$ by the equation

$$\Box D_R(x) = \delta^4(x), D_R(x) = 0, x^0 < 0$$

Then it is easy to see that

$$D_R(x) = C.\theta(x)\delta(x^2)$$

where $\theta(x) = 1$ if $x^0 > 0$ and $\theta(x) = 0$ if $x^0 < 0$. To prove this, we first note that

$$\delta^4(x) = (2\pi)^{-4} \int exp(ik.x) d^4k$$

the integral being over the entire $\mathbb{R}^4$. We write

$$D_R(x) = \int A(k) exp(ik.x) d^4k$$

Substituting this into the pde gives

$$-\int k^2 A(k) exp(ik.x) d^4k = (2\pi)^{-4} \int exp(ik.x) d^4k$$

This gives

$$A(k) = -(2\pi)^{-4}/k^2 = -(2\pi)^{-4}/(k^{02} - |\mathbf{k}|^2)$$

We choose while integrating with respect to k^0, we choose the contour to be the upper half semicircle along with the real line so that both the poles $k^0 = \pm|\mathbf{k}|$ are enclosed when $x^0 > 0$. Then, for $x^0 > 0$ the integral over the upper half semicircle vanishes since if $x^0 > 0$ and k^0 has a positive imaginary part, then ik^0x^0 has a negative real part. It follows that for $x^0 > 0$, we have

$$\int A(k)exp(ik.x)dk^0 = (-(2\pi i)(2\pi)^{-4}/2|\mathbf{k}|)exp(i(|\mathbf{k}|x^0 - \mathbf{k.x})) + ((2\pi i)(2\pi)^{-4}/2|\mathbf{k}|)exp(-i(|\mathbf{k}|x^0 + \mathbf{k.x}))$$

It follows that $D_R(x)$ is proportional to

$$\int |\mathbf{k}|^{-1}.sin(|\mathbf{k}|x^0).exp(-i\mathbf{k.x})d^3k$$

for $x^0 > 0$. for $x^0 < 0$ we choose the lower semicircle so that both the poles are excluded giving zero as required. We want to show that $D_R(x)$ is proportional to $\theta(x)\delta(x^2)$. Note that

$$\delta(x^2) = \delta(x^{02} - |\mathbf{x}|^2) == \frac{1}{2|\mathbf{x}|}(\delta(x^0 - |\mathbf{x}|) + \delta(x^0 + |\mathbf{x}|))$$

so that

$$\theta(x)\delta(x^2) = \frac{1}{2r}\delta(t - r), r = |\mathbf{x}|, t > 0$$

and zero for $t < 0$. To show this, we note first that from above, for $t > 0$,

$$I = \int A(k)exp(ik.x)d^4k = 2\pi \int sin(zt)exp(-izrcos(\theta))z.sin(\theta)dzd\theta$$

where $0 < \theta < \pi$ and $0 < z < \infty$. Evaluating the θ integral first gives except for a proportionality factor,

$$I = -ir^{-1}\int_0^\infty sin(zt)(exp(-izr) - exp(izr))dz = r^{-1}\int_0^\infty sin(tz)sin(rz)dz$$

With neglect of a proportionality factor this is

$$r^{-1}\int_{-\infty}^\infty (exp(itz) - exp(-itz))(exp(-irz) - exp(irz))dz$$

$$= r^{-1}\int_{-\infty}^\infty (exp(iz(t-r)) - exp(iz(t+r)) - exp(-iz(t+r) + exp(-iz(t-r)))dz$$

$$= r^{-1}\delta(t - r)$$

since $t > 0$ implies $\delta(t + r) = 0$. This completes the proof.

Now note that the retarded Green's function is given by

$$D_R(x) = \theta(x)\delta(x^2) = r^{-1}\delta(t - r)$$

after ignoring a constant of proportionality. We have seen that if the contour Γ_R for k^0-integral encloses the poles when $x^0 > 0$ and does'nt enclose any pole when $x^0 < 0$, then

$$D_R(x) = \int_+ exp(ik.x)d^4k/k^2$$

where the subscript + in the integral sign means that the k^0 integral is the straight line passing below the two poles. Similarly, define

$$D_A(x) = \int_{-} exp(ik.x)d^4k/k^2$$

where the subscript $-$ for the integral indicates that the integral is along a line passing just above the poles. Then a simple computation gives that

$$D_A(x) = r^{-1}\delta(t + r)$$

for $t < 0$ and zero for $t > 0$. Let now Γ denote the contour which is a rectangle with infinite length along the k^0 axis and negligible breadth so that both the poles are enclosed by this contour. Then, after ignoring a proportionality constant, we have

$$D(x) = D_R(x) - D_A(x) = \int_\Gamma exp(ik.x)d^4k/k^2 = r^{-1}(\delta(t - r) - \delta(t + r))$$

We shall show that after ignoring a proportionality constant,

$$D(x) = \int exp(ik.x)\delta(k^2)\epsilon(k)d^4k$$

the integral being an ordinary real four dimensional integral. In fact, we have

$$\delta(k^2) = \frac{1}{2|\mathbf{k}|}(\delta(k^0 - |\mathbf{k}|) + \delta(k^0 + |\mathbf{k}|))$$

Define

$$\epsilon(x) = sgn(x^0)$$

Then,

$$\delta(k^2)\epsilon(k) = \frac{1}{2|\mathbf{k}|}(\delta(k^0 - |\mathbf{k}|) - \delta(k^0 + |\mathbf{k}|))$$

It follows that

$$\int exp(ik.x)\delta(k^2)\epsilon(k)d^4k = \int |\mathbf{k}|^{-1}(exp(i(|\mathbf{k}|x^0 - \mathbf{k.x})) - exp(-i(|\mathbf{k}|x^0 - \mathbf{k.x})))d^3k$$

which agrees with

$$\int_\Gamma exp(ik.x)d^4k/k^2$$

obtained using residue calculus to the two poles $\pm|\mathbf{k}|$ enclosed by the contour. We observe that

$$\Box D(x) = \Box \int_\Gamma exp(ik.x)d^4k/k^2 = \int_\Gamma exp(ik.x)d^4k = 0$$

since the integrand in the last expression on the left is an analytic function of k^0, or more specifically, it has no poles enclosed by Γ. Another way to infer this result is to use the other expression,

$$\Box D(x) = \Box \int exp(ik.x)\delta(k^2)\epsilon(k)d^4k = -\int exp(ik.x)k^2\delta(k^2)\epsilon(k)d^4k = 0$$

since

$$\xi.\delta(\xi) = 0$$

[5] $A^{\mu}(x)$ is the four vector potential. It has the expression

$$A_{\mu}(x) = \int d^3k|\mathbf{k}|^{-1/2}(a_{k\mu}exp(ik.x) + a^*_{k\mu}exp(-ik.x))d^3k$$

We introduce the commutation relations

$$[a_{k,\mu}, a_{k',\lambda}] = 0, [a^*_{k,\mu}, a^*_{k',\lambda}] = 0, [a_{k,\mu}, a_{k',\lambda}] = \delta_{\mu\lambda}\delta(\mathbf{k} - \mathbf{k}')$$

Then,

$$[A_{\mu}(x), A_{\lambda}(x')] = \int d^3kd^3k'|\mathbf{k}|^{-1/2}|\mathbf{k}'|^{-1/2}\delta_{\mu\lambda}\delta(\mathbf{k} - \mathbf{k}')(exp(i(k.x - k'.x')) - exp(-i(k.x - k'.x')))d^3kd^3k'$$

$$= \delta_{\mu\lambda}\int d^3k|\mathbf{k}|^{-1}(exp(ik.(x - x')) - exp(-ik.(x - x')))$$

Note that here $k^0 = |\mathbf{k}|$. On the other hand, recall that

$$D(x) = \int exp(ik.x)\delta(k^2)\epsilon(k)d^4k = \int |\mathbf{k}|^{-1}(exp(ik.x) - exp(-ik.x))d^3k = D_R(x) - D_A(x)$$

except for a constant of proportionality. It follows that

$$[A_{\mu}(x), A_{\lambda}(x')] = i\delta_{\mu\lambda}D(x - x')$$

The constant of proportionality i can be inferred as follows: $A_{\mu}(x)$ is a Hermitian operator and hence assuming

$$[A_{\mu}(x), A_{\lambda}(x')] = c\delta_{\mu\lambda}D(x - x')$$

it follows on taking Hermitian conjugate that

$$[A_{\lambda}(x'), A_{\mu}(x)] = \bar{c}\delta_{\mu\lambda}\bar{D}(x - x')$$

But recall that D_R, D_A are real and hence D is also real. Thus we get

$$\bar{c} = -c$$

and hence c is pure imaginary.

6.Vacuum expectations:

28. Fredholm integral equations. Let $a < b$ be finite real numbers and let $k : [a, b] \times [a, b] \to \mathbb{R}$ be continuous. Then since $[a, b] \times [a, b]$ is a compact subset of $\mathbb{R}^2$, it follows that k is bounded, say

$|k(t,s)| \leq c < \infty$ for all $t, s \in [a,b]$. Let $v : [a,b] \to \mathbb{R}$ be continuous and μ a real number. We ask the question:When does the integral equation

$$x(t) - \mu \int_a^b k(t,s)x(s)ds = v(t), t \in [a,b]$$

have a unique solution for $x \in C[a,b]$? Define the operator $T : C[a,b] \to C[a,b]$ by

$$T(x)(t) = v(t) + \mu \int_a^b k(t,s)x(s)ds$$

Then if $x, y \in C[a,b]$, we have

$$|T(x)(t) - T(y)(t)| \leq |\mu| \int_a^b |k(t,s)||x(s) - y(s)|ds \leq c\mu(b-a)d(x,y)$$

or

$$d(T(x), T(y)) \leq \alpha.d(x,y), \alpha = c|\mu|(b-a)$$

where d is the standard L^∞-metric on $C[a,b]$, ie,

$$d(x,y) = sup_{t\in[a,b]}|x(t) - y(t)|$$

It follows from Banach's fixed point theorem that if $\alpha < 1$, T has a unique fixed point or equivalently, the above integral equation has a unique solution. Note that the whole scheme works only because $C[a,b]$ is a Banach space, ie, a complete normed linear space. By the proof of Banach's fixed point theorem, the solution to $T(x) = x$ is given by $x = lim_{n\to\infty}T^n(y)$ where $y \in C[a,b]$ is arbitrary. Equivalently, we choose any $y \in C[a,b]$, and define $x_0 = y$,

$$x_{n+1}(t) = T(x_n)(t) = v(t) + \mu \int_a^b k(t,s)x_n(s)ds, n \geq 0$$

and then

$$x(t) = lim_{n\to\infty}x_n(t)$$

is the unique solution to the integral equation.

Statement and proof of Banach's fixed point theorem: Let (X,d) be a complete metric space and $T : X \to X$ a map such that $d(T(x), T(y)) \leq \alpha.d(x,y)$ for all $x, y \in X$ where $0 < \alpha < 1$. Then, T has a unique fixed point. More generally, let $T : X \to X$ be a map such that for some $N = 1, 2, ...,$ we have $d(T^N(x), T^N(y)) \leq \alpha.d(x,y), \forall x, y \in X$ for some $\alpha \in (0,1)$. Then T has a unique fixed point. For let $y \in X$ be arbitrary. Then we have seen that $x = lim_n T^{Nn}(y)$ is a fixed point of T^N. Now, since $T^{Nn}(x) = x$ for all $n = 1, 2, ...,$ it follows that

$$Tx = TT^{Nn}(x) = T^{Nn}(Tx)$$

and taking limit $n \to \infty$ gives us

$$Tx = x$$

Thus, x is also a fixed point of T. Suppose $Ty = y$ for some $y \in X$. Then, $T^N y = y$, ie y is a fixed point of T^N. But by the uniqueness part of the Banach fixed point theorem, we then have that $y = x$. This proves that T has a unique fixed point.

Volterra integral equation:

Let $X = C[a, b]$ and let

$$D = \{(t, s) : s \leq t, t, s \in [a, b]\}$$

Let $k : D \to \mathbb{R}$ be continuous and let $v \in C[a, b]$. Consider the integral equation

$$x(t) - \mu \int_a^t k(t, s)x(s)ds = v(t), t \in [a, b]$$

We want to know when this integral equation has a unique solution. Define the operator $T : C[a, b] \to C[a, b]$ by

$$T(x)(t) = x(t) - \mu \int_a^t k(t, s)x(s)ds$$

Then, for $x, y \in C[a, b]$, we have

$$|T(x)(t) - T(y)(t)| \leq |\mu|c \int_a^t |x(s) - y(s)|ds$$

where

$$c = sup\{|k(t, s)| : (t, s) \in D\}$$

Thus,

$$d(T(x), T(y)) \leq c|\mu|(b - a)d(x, y)$$

Again,

$$|T^2(x)(t) - T^2(y)(t)| \leq |\mu|c \int_a^t |T(x)(s) - T(y)(s)|ds \leq |\mu|^2c^2 \int_a^t (t - s)|x(s) - y(s)|ds$$

$$\leq |\mu|^2(c^2(b - a)^2/2)d(x, y)$$

etc. In general, we get by iteration,

$$d(T^n(x), T^n(y)) \leq (|\mu|^n c^n (b - a)^n/n!)d(x, y)$$

It follows that for sufficiently large N, we have

$$d(T^N(x), T^N(y)) \leq \alpha.d(x, y), x, y \in C[a, b]$$

for some $\alpha \in (0, 1)$. From the above theorem, we get the result that T has a unique fixed point in $C[a, b]$, no matter what the values of μ, c are. Thus, the Volterra integral equation has a unique solution.

29. Strain tensor, concept of(Landau and Lifshitz, elasticity): Let the point $\mathbf{x} = (x_1, .x_2, x_3)$ inside the elastic body be displaced to the point $\mathbf{x} + \mathbf{u}(\mathbf{x})$, in terms of components, $x_i + u_i$. The point $x_i + dx_i$

then gets displaced to $x_i + dx_i + u_i + du_i$. The length of the original element, ie, prior to stretching is given by $dl^2 = dx_i dx_i$, with summation over the repeated index i being implied. The length of this element after the stretching is given by $dl'^2 = (dx_i + du_i)(dx_i + du_i)$ or equivalently,

$$dl'^2 = dx_i dx_i + 2dx_i du_i + du_i du_i = dx_i dx_i + 2dx_i dx_j u_{i,j} + u_{i,j} u_{i,k} dx_j dx_k = S_{ij} dx_i dx_j$$

where

$$S_{ij} = \delta_{ij} + (u_{i,j} + u_{j,i}) + u_{i,j} u_{i,k}$$

If we assume that the displacement is small, then the quadratic term in the symmetric tensor S_{ij} can be omitted giving

$$S_{ij} \approx \delta_{ij} + (u_{i,j} + u_{j,i})$$

The quantity

$$u_{ij} = u_{i,j} + u_{j,i}$$

is the strain tensor for small displacements. For large displacements, it is given by the accurate formula

$$u_{ij} = u_{i,j} + u_{j,i} + u_{k,i} u_{k,j}$$

The stress tensor is denoted by σ_{ij}. It is the ith component of the elastic force inside the body acting on a unit area at that point whose normal is along the jth direction. The energy stored in the body by the stretching is given by the surface integral $\int \sigma_{ij} u_i dS_j$ which by Gauss' theorem can be written as $\int (\sigma_{ij} u_i)_{,j} dV$. This has two components the first is $\int \sigma_{ij,j} u_i dV$ and the second is $\int \sigma_{ij} u_{i,j} dV$. The first component has no component due to strain and can be termed as the bulk energy. $\sigma_{ij,j}$ is thus the body force per unit volume and the second is the strain component. The strain energy per unit volume is thus

$$\sigma_{ij} u_{i,j} = \frac{1}{2} \sigma_{ij} (u_{i,j} + u_{j,i})$$

assuming that the stress tensor σ_{ij} is symmetric. The net torque due to the stresses of the body about the origin is given by

$$\begin{aligned}
\tau_{ij} &= \int (x_i \sigma_{jk} - x_j \sigma_{ik}) dS_k = \int (x_i \sigma_{jk} - x_j \sigma_{ik})_{,k} dV \\
&= \int (\delta_{ik} \sigma_{jk} - \delta_{jk} \sigma_{ik}) dV + \int (x_i \sigma_{jk,k} - x_j \sigma_{ik,k}) dV \\
&= \int (\sigma_{ji} - \sigma_{ij}) dV + \int (x_i \sigma_{jk,k} - x_j \sigma_{ik,k}) dV
\end{aligned}$$

If the torque is solely due to body forces, then the first term on the right must be zero, ie, $\sigma_{ij} = \sigma_{ji}$, ie, the stress tensor must be symmetric.

30. Relationship between problems of mechanics and signal theory.

Let $(q(t), p(t))$ be the state of a mechanical system at time t. This satisfies a system of first order differential equations called the Hamilton equations of motion:

$$q_i'(t) = \frac{\partial H(q(t), p(t))}{\partial p_i}, p_i'(t) = -\frac{\partial H(q(t), p(t))}{\partial q_i}, i = 1, 2, ..., n$$

Linearize this system about a quiescent trajectory. Then we end up with a first order state variable system:

$$\frac{d\mathbf{x}(t)}{dt} = \mathbf{A}(t)\mathbf{x}(t)$$

where

$$\mathbf{x}(t) = [\delta q(t)^T, \delta p(t)^T]^T,$$

$$a_{ij}(t) = \frac{\partial^2 H(q(t), p(t))}{\partial p_i \partial q_j}, 1 \leq i, j \leq n,$$

$$a_{i,n+j}(t) = \frac{\partial^2 H(q(t), p(t))}{\partial p_i \partial q_j}, 1 \leq i, j \leq n,$$

$$a_{n+i,j}(t) = -\frac{\partial^2 H(q(t), p(t))}{\partial q_i \partial q_j}, 1 \leq i, j \leq n,$$

$$a_{n+i,n+j}(t) = -\frac{\partial^2 H(q(t), p(t))}{\partial q_i \partial p_j}, 1 \leq i, j \leq n$$

State variable equations with time varying matrix coefficients can be solved using the Dyson series, or equivalently the chronological product:

$$\mathbf{x}(t) = \Phi(t)\mathbf{x}(0),$$

$$\Phi(t) = I + \sum_{k=1}^{\infty} \int_{0<t_k<t_{k-1}<\ldots<t_2<t_1<t} A(t_1)A(t_2)\ldots A(t_k)dt_k dt_{k-1}\ldots dt_2 dt_1$$

Note that $\Phi(t)$ satisfies

$$\Phi'(t) = A(t)\Phi(t), t \geq 0, \Phi(0) = I$$

31. Chebyshev polynomials as used in antenna arrays

32. Define the three notions of convergence of a sequence of linear operators on a normed linear space.

33. Convolution and its generalization: Let two measurable functions x, h on $\mathbb{R}$ be given. Assume that h is integrable while x is bounded, ie, $\int_{-\infty}^{\infty} |h(t)|dt < \infty$ and $sup_{t\in\mathbb{R}}|x(t)| = M < \infty$. Then for each $t \in \mathbb{R}$, the signal $z_t : \mathbb{R} \to \mathbb{R}$ defined by $z_t(\tau) = h(\tau)x(t-\tau)$ is integrable. In fact, we have

$$\int_{-\infty}^{\infty} |z_t(\tau)|d\tau \leq M \int_{-\infty}^{\infty} |h(\tau)|d\tau < \infty$$

It follows that the signal $y : \mathbb{R} \to \mathbb{R}$ defined by

$$y(t) = \int_{-\infty}^{\infty} z_t(\tau)d\tau = \int_{-\infty}^{\infty} h(\tau)x(t-\tau)d\tau$$

is well defined and bounded by

$$K = M \int_{-\infty}^{\infty} |h(\tau)|d\tau$$

The signal $y(.)$ however need not be integrable. If however x is also integrable (which is a stronger condition than being bounded), then $y(.)$ will be integrable. In fact, we have by Fubini's theorem

$$\int |y(t)|dt \leq \int |h(t-\tau)||x(\tau)|d\tau dt = \int_{-\infty}^{\infty} |h(t)|dt \int_{-\infty}^{\infty} |x(\tau)|d\tau$$

In this case (ie, the case in which both x, h are integrable), the Fourier transforms of all the signals x, h, y are well defined and we have

$$Y(j\omega) = H(j\omega).X(j\omega)$$

where

$$Z(j\omega) = \int_{-\infty}^{\infty} z(t)exp(-i\omega t)dt$$

34. Problem: Construct examples of signals x, y such that $x \in L^1(\mathbb{R}) - L^2(\mathbb{R})$ while $y \in L^2(\mathbb{R}) - L^1(\mathbb{R})$.

Ans: Let $x(t) = t^{-1/2}.exp(-t), t > 0$ and $x(t) = 0$ if $t \leq 0$. Thios signal is integrable but not square integrable. Let $y(t) = t^{-1}$ if $t \geq 1$ and $y(t) = 0$ if $t < 1$. This signal is square integrable but not integrable.

35. Problem: Let $(X, \| . \|)$ be a normed linear space. Choose vectors $x, y_1, ..., y_n \in X$ and define a function $f : \mathbb{C}^n \to \mathbb{R}$ by the equation

$$f(\alpha_1, ..., \alpha_n) = \| x - \sum_{i=1}^{n} \alpha_i y_i \|$$

Show that f is convex, ie, for any $\lambda \in [0, 1]$, we have

$$f(\lambda.\alpha + (1-\lambda).\beta) \geq \lambda.f(\alpha) + (1-\lambda)f(\beta), \alpha, \beta \in \mathbb{C}^n$$

Soln:

$$\begin{aligned} f(\lambda.\alpha + (1-\lambda).\beta) &= \| x - \sum_{i=1}^{n} (\lambda.\alpha_i + (1-\lambda)\beta_i) y_i \| \\ &= \| \lambda(x - \sum \alpha_i y_i) + (1-\lambda)(x - \sum \beta_i y_i) \| \\ &\leq \lambda. \| x - \sum \alpha_i y_i \| + (1-\lambda) \| x - \sum \beta_i y_i \| \\ &= \lambda.f(\alpha) + (1-\lambda)f(\beta) \end{aligned}$$

proving the claim.

36. Problem: Let $(X, \| . \|)$ be a normed linear space. Let

$$B = \{x \in X : \| x \| \leq 1\}$$

the closed unit ball in X. Show that X is finite dimensional iff B is compact as a metric space.

Soln: Suppose X is infinite dimensional. Choose any vector $e_1 \in X$ such that $\| e_1 \|= 1$. Then $sp(e_1) \neq X$ implies by the Riesz theorem that there is an $e_2 \in X$ such that $\| e_2 \|= 1$ with $d(e_2, sp(e_1)) > 1/2$. Again since $sp(e_1, e_2) \neq X$, by Riesz theorem, there is an $e_3 \in X$ with $\| e_3 \|= 1$ such that $d(e_3, sp(e_1, e_2)) > 1/2$. Continuing this process, it follows that there is an infinite sequence $\{e_1, e_2, ...\} \subset X$ such that $\| e_i \|= 1$ for all i such that $d(e_{i+1}, sp(e_1, ..., e_i)) > 1/2$ for every $i = 1, 2, ...$ It follows in particular that for any $i \neq j$, $\| e_i - e_j \|> 1/2$ and hence no subsequence of $\{e_i : i = 1, 2, ...\}$ can converge in X. It follows that B is not compact. Conversely, suppose X is finite dimensional. Let $x_i, i = 1, 2, ...$ be any infinite sequence in B. Choose a basis $\{e_1, ..., e_N\}$ for X. We can write $x_i = \sum_{j=1}^{N} a(i, j)x_j$ for some complex scalars $a(i, j)$. It follows from the equivalence of all norms on a finite dimensional normed linear space that there exist constants $c, c' > 0$ such that

$$c.\sum_{j=1}^{N} |a(i,j)| \leq \| x_i \| \leq c' \sum_{j=1}^{N} |a(i,j)|$$

for all $i = 1, 2, ...$ In particular, $\{(a(i, 1), a(i, 2), ..., a(i, N)) : i = 1, 2, ...\}$ is a bounded sequence in $\mathbb{R}^N$ and hence by the Bolzano Weierstrass property has a convergent subsequence, say $(a(n_i, 1), a(n_i, 2), ..., a(n_i, N)), i = 1, 2, ...$ Let $(a(1), ..., a(N))$ be the limit of this subsequence. Then, define

$$x = \sum_{i=1}^{N} a(i)x_i$$

We have

$$\| x - x_{n_m} \| \leq \sum_{i=1}^{N} |a(i) - a(n_m, i)| \to 0, m \to \infty, m \to \infty$$

Since $\| x_i \| \leq 1$, it follows then that $\| x \| \leq 1$ and $x_{n_m} \to x$. This proves that B is compact. The proof is complete.

37. Problem: Let (X, d) be a compact metric space and $f : X \to \mathbb{R}$ a continuous function. Show that f attains its supremum at some point $x_0 \in X$.

Ans: Let $M = sup(f(x) : x \in X)$. Then by definition of supremum, there is a sequence $x_n \in X, n = 1, 2, ...$ such that $f(x_n) \to M$. Since X is compact, there is a convergent subsequence $(x_{n_m} : m = 1, 2, ...)$ of (x_n), say $x_{n_m} \to x_0, m \to \infty$. Then by continuity of f, we get

$$M = lim f(x_n) = lim f(x_{n_m}) = f(lim x_{n_m}) = f(x_0)$$

38. Problem:Let X be a normed linear space and let $x \in X$. Let Y be a subspace of X and define $B = \{y \in Y :\| y \| \leq 2 \| x \|\}$. Let $\delta = inf(\| x - y \|: y \in Y)$. Since $0 \in Y$, it follows that $\| x \| \geq \delta$. If $y \notin B$, then $\| y \|> 2 \| x \|$ and then

$$\| x - y \| \geq \| y \| - \| x \| > \| x \| \geq \delta$$

It follows that

$$\delta = d(x, B) = inf(\| x - y \|: y \in B)$$

Now suppose Y is finite dimensional. Then, B is compact and hence by the preceding result, there exist a point $x_0 \in B$ such that $\| x - x_0 \| = \delta$. In other words, the best approximation of any vector in X exists in Y provided Y is a finite dimensional subspace of X.

39. Delta signal as a distribution: Let X be a space of functions. A distribution on X is simply a linear functional $F : X \to \mathbb{C}$. For example, suppose $X = C(\mathbb{R})$, the space of continuous functions on $\mathbb{R}$ and we define $T(f) = f(0)$. Then T is a distribution called the δ distribution. Suppose $X = \mathcal{BC}$ is the space of bounded continuous functions on $\mathbb{R}$ and ϕ is an integrable function on $\mathbb{R}$. Then, define $T : X \to \mathbb{R}$ by

$$T(f) = \int_{\mathbb{R}} f(t)\phi(t)dt$$

This is well defined since letting $M = sup|f(t)|$, we have

$$|T(f)| \leq M \int_{\mathbb{R}} |\phi(t)|dt$$

We introduce the sup norm on $\mathcal{BC}$ so that

$$\| f \| = sup|f(t)|$$

Then,

$$|T(f)| \leq C \| f \|$$

where

$$C = \int |\phi(t)|dt$$

T is thus a bounded linear functional on $\mathcal{BC}$. For the δ functional on $\mathcal{BC}$, we have

$$|T(f)| = |f(0)| \leq \| f \|$$

so that

$$\| T \| \leq 1$$

40.

1.Definition of a field as a set with two commutative binary operations, addition and multiplication such that under addition, the set becomes an Abelian group and under multiplication, the set with the zero element removed is also an Abelian group, multiplication should distribute over addition. Examples of fields:Rationals, reals, complex numbers, finite fields of the form $\{0, 1, ..., p-1\}$ with p a prime.

2. Definition of a vector space: A set whose elements are called vectors along with a field. The set has an addition operation under which with a specified element called the zero vector as the identity, the set becomes an Abelian group. The elements of the field can be multiplied with vectors in the usual way and we call this operation as scalar multiplication. Scalar multiplication should distribute over vector addition and should satisfy :$c_1(c_2\mathbf{x}) = (c_1c_2)\mathbf{x}$. if 0 is the additive identity of the scalar field and $\mathbf{0}$, the zero vector, then the identity $0.\mathbf{x} = \mathbf{0}$ should be valid.

3. Linear superposition: Given vectors $v_1, ..., v_k$ in a vector space V and scalars $c_1, ..., c_n$ in the underlying field, we define the linear superposition

$$v = \sum_{i=1}^{n} c_i v_i$$

If the $v_i's$ are linearly independent, then this combination can be the zero vector when and only when all the $c_i's$ are zero. If further, all vectors $v \in V$ are covered by such a superposition by varying the $c_i's$, then the $v_i's$ are said to form a basis for the vector space V. n is then the dimension of the vector space. It can be shown that every basis has the same number of elements and hence the dimension of the vector space is uniquely determined.

4. Linear transformations on a vector space:Let V, W be vector spaces over the same field $\mathbb{F}$. A mapping $T : V \to W$ is called a linear transformation if for all $x, y \in V$ and all scalars $c_x, c_y \in \mathbb{F}$, one has

$$T(c_x x + c_y y) = c_x T(x) + c_y T(y)$$

41.

1.Let X be a normed linear space and $T : X \to X$ a compact linear operator. Then for all $\lambda \neq 0$, we have that $\mathcal{R}(T - \lambda.I)$ is closed.

2.Let X be a normed linear space and $T : X \to X$ a compact linear operator, ie, if B is the open unit ball in X, then $ClT(B)$ is compact as a metric space. Show that compactness of T implies boundedness of T.

Ans:Suppose B is unbounded. Then there exists a sequence $x_n, n = 1, 2, ...$ such that $x_n \in B$ but $T(x_n), n = 1, 2, ...$ has the property:$\| Tx_n \| > n, n = 1, 2, ...$ It follows that $\{Tx_n\}$ cannot have any convergent subsequence, for if $Tx_n \to y, n \to \infty$, then for all sufficiently large n, we must have

$$\| Tx_n - y \| \leq 1$$

and hence

$$\| Tx_n \| \leq 1 + \| y \|$$

which is a contradiction. This means that T cannot be compact and we are done.

3.Let X, Y be the normed spaces and $T : X \to Y$ be compact, ie, if B is the unit open ball in X, then $ClT(B)$ is compact in Y. Then, show that if M is any bounded subset of X, $ClT(M)$ is compact in Y. For let $M \subset X$ be such that for all $x \in M$ we have $\| x \| < K < \infty$ where K is fixed. Then, $K^{-1}M \subset B$ and hence,

$$K^{-1}ClT(M) = ClT(K^{-1}M) \subset ClT(B)$$

By definition of compactness of T, $ClT(B)$ is compact and hence $K^{-1}ClT(M)$ is compact (A closed subset of a compact metric space is compact). It follows that $ClT(M)$ is compact. We are done and this furnishes us with an alternative definition of compactness: If X, Y are normed linear spaces and

$T : X \to Y$ is a linear map, then T is said to be compact if $ClT(M)$ is compact for every bounded set $M \subset X$.

4. Let X_1, X_2, X_3, X_4 be normed linear spaces and $T_1 : X_1 \to X_2, T_2 : X_2 \to X_3, T_3 : X_3 \to X_4$ be bounded linear operators. If T_2 is compact, then show that $T_3T_2T_1$ is compact. For let $x_n, n = 1, 2, ...$ be any sequence in X_1 such that $\| x_n \| < 1$ for all n. Then, $\{T_1(x_n)\}$ is bounded since T_1 is bounded. It follows from the definition of compactness that $\{T_2T_1(x_n)\}$ has a convergent subsequence, say $T_2T_1(x_{n_k})$. Since T_3 is bounded, it is also continuous and hence $T_3T_2T_1(x_{n_k})$ is a convergent sequence in X_3. It follows that $T_3T_2T_1$ is compact and we are done.

5. Let $\mathcal{H}$ be a Hilbert space and $T : \mathcal{H} \to \mathcal{H}$ a bounded linear operator and T^* the adjoint of T. Then T is compact iff T^*T is compact. For let T be compact and $\{x_n\}$ a bounded sequence. Then, Tx_n has a convergent subsequence say Tx_{n_k}. T^* is bounded since for all $x, y \in \mathcal{H}$, we have

$$| < T^*x, y > | = | < x, Ty > | \leq \| T \| \| x \| \| y \|$$

and hence

$$\| T^*x \|^2 \leq \| T \| \| T^*x \| \| x \|$$

or equivalently,

$$\| T^* \| \leq \| T \|$$

Hence, $T^*Tx_{n_k}$ is convergent proving compactness of T^*T. Conversely suppose T^*T is compact. Let x_n be a bounded sequence and let $T^*Tx_{n_k} \to x$. For simplicity of notation, set $y_k = x_{n_k}$. Then

$$\| Ty_n - Ty_m \|^2 = < Ty_n, Ty_n > + < Ty_m, Ty_m > - < Ty_n, Ty_m > - < Ty_m, Ty_n >$$

Now,

$$| < Ty_n, Ty_n > - < x, y_n > | = | < T^*Ty_n - x, y_n > | \leq \| T^*Ty_n - x \|$$

Likewise,

$$| < Ty_m, Ty_m > - < x, y_m > | = | < T^*Ty_m - x, y_m > | \leq \| T^*Ty_m - x \|$$

$$| < Ty_n, Ty_m > - < x, y_m > | = | < T^*Ty_n - x, y_m > | \leq \| T^*Ty_n - x \|$$

$$| < Ty_m, Ty_n > - < x, y_n > | = | < T^*Ty_m - x, y_n > | \leq \| T^*Ty_m - x \|$$

so combining all these we get

$$\| Ty_n - Ty_m \| \leq 2 \| T^*Ty_n - x \| + 2 \| T^*Ty_m - x \|$$

proving that $\{Ty_n\}$ is Cauchy in $\mathcal{H}$ and hence convergent.

6. Notion of spectrum of an operator: Let X be a normed linear space and $T : X \to X$ a bounded linear operator. For any $\lambda \in \mathbb{C}$, we define $T_\lambda = T - \lambda.I$. Consider three properties: (1) T_λ is not one-one, (2) $R(T_\lambda)$ is not dense in X, (3) T_λ^{-1} is not bounded. Suppose (1) holds, then we say that λ belongs to the point spectrum of T. Suppose (1) does not hold but (2) holds. Then we say that λ belongs to the residual spectrum of T. Finally, suppose (1) and (2) do not hold but (3) holds. Then, we say that λ belongs to the continuous spectrum of T. We denote the point spectrum by $\sigma_p(T)$, the continuous spectrum by $\sigma_c(T)$ and the residual spectrum by $\sigma_r(T)$. The spectrum of T is the set

$$\sigma(T) = \sigma_p(T) \cup \sigma_c(T) \cup \sigma_r(T)$$

We note that $\sigma_p(T), \sigma_c(T), \sigma_r(T)$ are pairwise disjoint and their union is the spectrum of T. We write $\rho(T) = \sigma(T)^c$. $\rho(T)$ is called the resolvent set of T. Thus, $\lambda \in \rho(T)$, iff T_λ is injective, has a dense range and its inverse defined on its range is a bounded linear operator. In a finite dimensional normed linear space, the continuous and residual spectra of all linear operators are empty. To see this, suppose X is finite dimensional and $T : X \to X$ is linear. Then let $\lambda \in \mathbb{C}$. Suppose T_λ is one-one. Then T_λ is also onto (because X finite dimensional). Hence, the residual spectrum of T is empty. Moreover any linear operator on a finite dimensional normed linear space is bounded and hence λ cannot belong to the continuous spectrum:Suppose $B = (e_1, ..., e_n)$ is a basis for X and let L be a linear operator on X. Then,

$$Le_i = \sum_{j=1}^{n} L_{ji} e_j$$

For $x = \sum x_i e_i \in X$, define

$$\| x \| = \sum |x_i|$$

Then, writing L for $[L]_B$, we have

$$(Lx)_i = \sum_j L_{ij} x_j$$

and hence

$$\| Lx \| \leq max_j \sum_i |L_{ij}|). \sum_j |x_j|$$

showing that

$$\| L \| \leq max_j \sum_i |L_{ij}|$$

which is finite.

Elements in the point spectrum of T are called eigenvalues of T. In a finite dimensional normed linear space,

Example: Let $\mathcal{H} = l^2(\mathbb{Z}_+)$ and define $T : \mathcal{H} \to \mathcal{H}$ by the equation

$$T(x_1, x_2, ...) = (0, x_1, x_2, ...)$$

Then, T is one-one. Let Y be the subspace of $\mathcal{H}$ consisting of all sequences with first element vanishing. Then Y is not dense in $\mathcal{H}$ and hence 0 belongs to the continuous spectrum of T.

Closed linear operator:Let X, Y be normed linear spaces and $T : \mathcal{D}(T) \to Y$ a linear operator with $\mathcal{D}(T) \subset X$. T is called a closed linear operator if

$$Gr(T) = \{(x, Tx) : x \in \mathcal{D}(T)\}$$

is closed in $X \times Y$. Here, the norm on $X \times Y$ is defined by

$$\| (x, y) \| = \| x \| + \| y \|$$

Closed graph theorem: Suppose X, Y are Banach spaces and $T : \mathcal{D}(T) \rightarrow Y$ is a closed linear operator. If $\mathcal{D}(T)$ is closed in X, then T is bounded.

Proof: $X \times Y$ under the above norm and usual addition and scalar multiplication, is a Banach space. For suppose (x_n, y_n) is Cauchy in $X \times Y$. Then

$$\| x_n - x_m \| + \| y_n - y_m \| \rightarrow 0, n, m \rightarrow \infty$$

Hence, x_n is Cauchy in X and y_n Cauchy in Y. Hence, $x_n \rightarrow x, y_n \rightarrow y$ for some $(x, y) \in X \times Y$. It follows that

$$\| (x_n, y_n) - (x, y) \| = \| x_n - x \| + \| y_n - y \| \rightarrow 0$$

proving completeness of $X \times Y$. since $Gr(T)$ is closed in $X \times Y$ and $\mathcal{D}(T)$ is closed in X, it follows that $Gr(T)$ and $\mathcal{D}(T)$ are Banach spaces. Consider the map

$$P : Gr(T) \rightarrow \mathcal{D}(T)$$

defined by

$$(x, Tx) \rightarrow x$$

This map is linear. Further, P is bounded because

$$\| Px \| = \| x \| \leq \| x \| + \| Tx \| = \| (x, Tx) \|$$

Further, P is bijective with the inverse map given by

$$P^{-1}(x) = (x, Tx)$$

The bounded inverse theorem (corollary of the open mapping theorem) then implies that P^{-1} is bounded. Thus,

$$\| Tx \| \leq \| x \| + \| Tx \| = \| P^{-1}x \| \leq \| P^{-1} \| . \| x \|$$

proving boundedness of T.

Theorem: Let X, Y be normed spaces and $T : \mathcal{D}(T) \rightarrow Y$ a linear operator. Then, T is closed iff $x_n \rightarrow x$ and $Tx_n \rightarrow y$ imply $x \in \mathcal{D}(T)$ and $y = Tx$.

Proof:First assume that T is closed. Let $x_n \rightarrow x$ and $Tx_n \rightarrow y$. Then, $(x_n, Tx_n) \rightarrow (x, y)$. Since $Gr(T)$ is closed and $(x_n, Tx_n) \in Gr(T)$, it follows that $(x, y) \in Gr(T)$, ie, $x \in \mathcal{D}(T)$ and $y = Tx$ and we are through with the only if part. Conversely, suppose $x_n \rightarrow x, Tx_n \rightarrow y$ imply $x \in \mathcal{D}(T)$ and $y = Tx$. We must show that $Gr(T)$ is closed. So suppose $z_n \in Gr(T)$ and $z_n \rightarrow z$. Then, $z_n = (x_n, Tx_n)$ for some $x_n \in \mathcal{D}(T)$. Write $z = (x, y)$. Then we have by hypothesis,

$$0 = lim \| z_n - z \| = \| x_n - x \| + \| Tx_n - y \|$$

and hence

$$x_n \rightarrow x, Tx_n \rightarrow y$$

It follows by hypothesis that $x \in \mathcal{D}(T)$ and $y = Tx$, ie, $(x, y) \in Gr(T)$ completing the proof of the if part.

42.Rajeev's Ph.d problem: Consider the wave equation with random forcing. The motion corresponds to a string stretched between the points $x = 0$ and $x = L$. The displacement at time t at the point x, $u(t,x)$ satisfies the one dimensional wave equation with random forcing $F(t,x)$:

$$u_{tt}(t,x) - u_{xx}(t,x) = F(t,x)$$

We expand u, F as a half wave Fourier series:

$$u(t,x) = \sum_{n=1}^{\infty} c_n(t).sin(n\pi x/L), F(t,x) = \sum_{n=1}^{\infty} f_n(t).sin(n\pi x/L)$$

Then the wave equation gives

$$c_n''(t) + (n\pi/L)^2 c_n(t) = f_n(t), n = 1, 2, ...$$

The energy of the string at time t can be expressed as

$$E(t) = \frac{\lambda}{2}\int_0^L u_x(t,x)^2 dx + \frac{1}{2}\int_0^L u_t(t,x)^2 dx$$

where λ is a constant related to the Young's modulus of the string. We express this energy in terms of the Fourier coefficients:

$$u_x = \sum c_n(n\pi/L)cos(n\pi x/L)$$

$$\int_0^L u_x^2 dx = \frac{L}{2}\sum(n\pi/L)^2 c_n^2 = \sum_n c_n^2(n^2\pi^2/2L)$$

$$u_t = \sum_n c_n'.sin(n\pi x/L)$$

$$\int_0^L u_t^2 dx = \frac{L}{2}\sum c_n'^2$$

so that

$$E(t) = \sum_n c_n^2(n^2\pi^2/4L) + c_n'^2 L/4$$

We shall use the simplified notation

$$E(t) = A\sum_n n^2 c_n(t)^2 + B\sum_n c_n(t)'^2$$

where

$$A = \pi^2/4L, B = L/4$$

Assume that the Fourier series has been truncated to N coefficients. We assume that $f_n(t) = \sigma_n dB_n(t)/dt$, where $B_n's$ are standard Brownian motion processes independent of each other. Then the equations of motion are

$$dc_n(t) = v_n(t)dt, dv_n(t) = \sigma_n dB_n(t) - (n\pi/L)^2 c_n(t)dt$$

The solution to c_n, v_n can actually be derived directly from the differential equations:

$$c_n''(t) + (n\pi/L)^2 c_n(t) = f_n(t)$$

Taking Laplace transforms gives

$$s^2 C_n(s) + (n\pi/L)^2 C_n(s) = F_n(s)$$

so that

$$C_n(s) = \frac{F_n(s)}{s^2 + (n\pi/L)^2}$$

or

$$c_n(t) = \int_0^t h_n(\tau) f_n(t-\tau) d\tau$$

where

$$h_n(t) = \frac{L}{n\pi}.sin(n\pi t/L)$$

We can write the solution as

$$c_n(t) = \sigma_n \int_0^t h_n(t-\tau) dB_n(\tau)$$

The displacement is then given by

$$u_n(t,x) = \sum_{n=1}^{N} \sigma_n.sin(n\pi x/L) \int_0^t h_n(t-\tau) dB_n(\tau) = \sum_{n=1}^{N} \int_0^t K_n(t-\tau, x) dB_n(\tau)$$

where

$$K_n(t,x) = (L\sigma_n/n\pi).sin(n\pi x/L) sin(n\pi t/L)$$

An alternate way to solve this is to write a state variable like first order differential equation for the two vectors $(c_n(t), v_n(t))$:

$$d(c_n(t), v_n(t))^T = \begin{pmatrix} 0 & 1 \\ -\omega_n^2 & 0 \end{pmatrix} (c_n(t), v_n(t))^T + (0, \sigma_n)^T dB_n(t)$$

where

$$\omega_n = n\pi/L$$

Write

$$A_n = \begin{pmatrix} 0 & 1 \\ -\omega_n^2 & 0 \end{pmatrix}$$

$$(sI - A_n)^{-1} = (s^2 + \omega_n^2)^{-1} \begin{pmatrix} s & 1 \\ -\omega_n^2 & s \end{pmatrix}$$

so that

$$exp(tA_n) = \begin{pmatrix} cos(\omega_n t) & \omega_n^{-1} sin(\omega_n t) \\ -\omega_n.sin(\omega_n t) & cos(\omega_n t) \end{pmatrix}$$

Thus, the solution to the above state variable equations is given by

$$(c_n(t), v_n(t))^T = \int_0^t exp((t-\tau)A_n)(0,\sigma_n)^T dB_n(\tau)$$

$$= \sigma_n \int_0^t (\omega_n^{-1} sin(\omega_n(t-\tau)), cos(\omega_n(t-\tau)))dB_n(\tau)$$

or equivalently,

$$c_n(t) = \sigma_n \omega_n^{-1} \int_0^t sin(\omega_n(t-\tau))dB_n(\tau),$$

$$v_n(t) = \sigma_n \int_0^t cos(\omega_n(t-\tau))dB_n(\tau)$$

The expression for v_n may also have been obtained directly from the formula

$$v_n(t) = c_n'(t) = \sigma_n \int_0^t h_n'(t-\tau)dB_n(\tau)$$

We can write

$$u_t(t,x) = \sum_n c_n'(t) sin(n\pi x/L) = \sum_n \sigma_n.sin(n\pi x/L) \int_0^t cos(\omega_n(t-\tau))dB_n(\tau)$$

$$= \sum_{n=1}^N \int_0^t F_n(t-\tau, x)dB_n(\tau)$$

where

$$F_n(t,x) = \sigma_n.cos(\omega_n t) sin(\omega_n x)$$

Note that

$$K_n(t,x) = \sigma_n.\omega_n^{-1}.sin(\omega_n t) sin(\omega_n x)$$

We are now in a position to calculate (1) space-time correlations in the displacement and velocity field of the vibrating string, mean energy and energy fluctuations of the string.

43. Application of Laurent Schwarz and Sobolev's theory of distributions to signal theory. Ω is an open subset of $\mathbb{R}^n$ and $\mathcal{D}(\Omega)$ is the space of infinitely differentiable functions on Ω having compact support. More precisely, for any compact subset K of Ω $\mathcal{D}_K$ is the space of infinitely differentiable functions on K and $\mathcal{D}(\Omega)$ is the union of all the $\mathcal{D}_K's$. The support of any function f on a topological space is the closure of the set $\{x : f(x) \neq 0\}$. Choose compact sets $K_i, i = 1, 2, ...$ such that $K_i \subset int(K_{i+1})$ for all i. For $f \in C^\infty(\Omega)$ and $N = 1, 2, ...,$ define

$$p_N(f) = max(|D^\alpha f(x)| : x \in K_N, |\alpha| \leq N\}$$

These seminorms define a metrizable locally convex topology on $C^\infty(\Omega)$. We can show that $\{p_N : N = 1, 2, ...\}$ is a separating family of seminorms on $C^\infty(\Omega)$. Indeed, suppose $f \neq 0$. Then there exists $x \in \Omega$

such that $f(x) \neq 0$. Choose N such that $x \in K_N$ and then $p_N(f) > |f(x)| > 0$ proving the claim. If $f, g \in \Omega$ and $f \neq g$, then the argument is applied to $f - g$ resulting in an N such that $p_N(f - g) > 0$.

Result: Suppose X is a vector space and $\mathcal{P}$ is a separating family of seminorms. For each $p \in \mathcal{P}$, define the set

$$V(p, n) = \{x : p(x) < 1/n\}$$

Let $\mathcal{B}$ be the collection of all finite intersections of sets of the form $V(p, n)$. Say that a set $A \subset X$ is open iff A is a union of translates of members of $\mathcal{B}$. Then we get a topology for X. $\mathcal{B}$ is a local base for this topology. Each element of $\mathcal{B}$ is convex and balanced. To prove convexity, suppose $x, y \in V(p_1, n_1) \cap ... \cap V(p_k, n_k)$. Then, $p_j(x) < 1/n_j, p_j(y) < 1/n_j$, so that if $a \in [0, 1]$, then

$$p_j(ax + (1 - a)y) \leq ap_j(x) + (1 - a)p_j(y) < 1/n_j$$

proving that

$$ax + (1 - a)y \in V(p_1, n_1) \cap ... \cap V(p_j, n_j)$$

This proves convexity of the elements of $\mathcal{B}$. Suppose

$$x \in \bigcap_{j=1}^{k} V(p_j, n_j)$$

and $|a| \leq 1$, then

$$p_j(ax) = |a|p_j(x) < |a|/n_j \leq 1/n_j$$

proving that

$$ax \in \bigcap_{j=1}^{k} V(p_j, n_j)$$

Using this result applied to $C^\infty(\Omega)$, this becomes a locally convex topological vector space. Further this space is metrizable since in the above result, we may define

$$d(x, y) = \sum_{i=1}^{\infty} 2^{-i} p_i(x - y)/(1 + p_i(x - y))$$

Then d is a metric. The fact that $d(x, y) > 0$ if $x \neq y$ follows from the fact that $\{p_i\}$ is a separating family of seminorms. If K is a compact subset of Ω, then $\mathcal{D}_K$, the set of all $f \in C^\infty(\Omega)$ with support K is closed in $C^\infty(\Omega)$. This follows from the fact that for any $x \in \Omega$, the map $f \rightarrow f(x)$ from $C^\infty(\Omega)$ into $\mathbb{C}$ is continuous. Indeed, suppose $f_n \rightarrow f$ in $C^\infty(\Omega)$. Let $x \in \Omega$ be arbitrary and choose N sufficiently large so that $x \in K_N$. Then since $f_n \rightarrow f$, it follows that $p_N(f_n - f) \rightarrow 0$ and since

$$p_N(f_n - f) = max(|D^\alpha(f_n - f)(u)| : u \in K_N, |\alpha| \leq N) \geq |f_n(x) - f(x)|$$

it follows that

$$f_n(x) \rightarrow f(x), n \rightarrow \infty$$

To prove that D_K is closed, it suffices therefore to show that D_K is the intersection of the nullspaces of the maps $f \to f(x)$ as x ranges over K^c. Indeed, suppose f is in this intersection. Then, $f(x) = 0$ for all $x \in K^c$. It follows that $supp(f)$ is contained in the closure of K which is K itself. On the other hand, suppose $f \in D_K$. Then, $f(x) = 0$ for all $x \notin K^c$ and hence f is in this intersection. We next show that a local base for $C^\infty(\Omega)$ is given by the sets

$$V_N = \{f : p_N(f) < 1/N\}, N = 1, 2, ...$$

Indeed, since $p'_N s$ form a separating family of seminorms, it follows from the result just proved that the sets of the form $V(p_{i_1}, N_1) \cap ... \cap V(p_{i_k}, N_k)$ form a local base for the topology. Let $M = max(i_1, ..., i_k) = i_r$ say and let $N = max(M, N_r)$. Then, $p_N(f) < 1/N$ implies $p_N(f) < 1/N_j$ for all j which implies $p_{i_j}(f) < 1/N_j$ for all j. Hence,

$$V_N \subset V(p_{i_1}, N_1) \cap ... \cap V(p_{i_k}, N_k)$$

proving the claim.

44. Prove that convolution in $\mathbb{R}^n$ is (a) associative, (b) commutative, (c) distributes over addition. Hint:

$$f * g(\mathbf{x}) = \int f(\mathbf{x}-\mathbf{y})g(\mathbf{y})d^n y = \int f(\mathbf{y})g(\mathbf{x}-\mathbf{y})d^n y = g * f(\mathbf{x})$$

$$(f * g) * h(\mathbf{x}) = \int f * g(\mathbf{x}-\mathbf{y})h(\mathbf{y})d^n y = \int h(\mathbf{y})d^n y. \int f(\mathbf{x}-\mathbf{z}-\mathbf{y})g(\mathbf{z})d^n z$$

$$= \int f(\mathbf{x}-\mathbf{z}-\mathbf{y})h(\mathbf{y})g(\mathbf{z})d^n y d^n z$$

45. Show that if $D_i = \frac{\partial}{\partial x^i}, i = 1, 2, ..., n$, then the solution to the pde

$$\Pi_{i=1}^n D_i^{m_i} f(\mathbf{x}) = g(\mathbf{x})$$

can be expressed as

$$f(\mathbf{x}) = (2\pi)^{-n} \int F(\mathbf{u}) exp(j(\mathbf{u}, \mathbf{x}))d^n u$$

where

$$F(\mathbf{u}) = G(\mathbf{u})/\Pi_{i=1}^n (ju_i)^{m_i}$$

$$G(\mathbf{u}) = \int g(\mathbf{x}) exp(j(\mathbf{u}, \mathbf{x}))d^n x$$

Show that

$$g(\mathbf{x}) = \int K(\mathbf{x}-\mathbf{y})f(\mathbf{y})d^n y$$

and determine the kernel K.

46. Consider two kernels K, H on $\mathbb{R}^n$. Show that if f is defined on $\mathbb{R}^n$, then

$$K * (H * f) = H * (K * f)$$

what is the system theoretic interpretation of this identity.

47. Show that the system of Maxwell equations with charge density and current density as input and the electric and magnetic fields as output define a linear shift invariant multiinput-multioutput system for four dimensional signals. Determine the impulse response and transfer function of this system. Hint:Start with the retarded potential formula for the electric potential $V(t, \mathbf{x})$ and magnetic vector potential $\mathbf{A}(t, \mathbf{x})$ in terms of the charge and current density:

$$V(t, \mathbf{x}) = \frac{1}{4\pi\epsilon_0} \int \rho(t - |\mathbf{x} - \mathbf{y}|/c, \mathbf{y}) d^3y / |\mathbf{x} - \mathbf{y}|$$

$$\mathbf{A}(t, \mathbf{x}) = \frac{\mu_0}{4\pi} \int \mathbf{J}(t - |\mathbf{x} - \mathbf{y}|/c, \mathbf{y}) d^4y / |\mathbf{x} - \mathbf{y}|$$

Show using these formulae that

$$V(t, \mathbf{x}) = H(t, \mathbf{x}) * \rho(t, \mathbf{x}), \mathbf{A}(t, \mathbf{x}) = \mu_0\epsilon_0 H(t, \mathbf{x}) * \mathbf{J}(t, \mathbf{x})$$

where

$$H(t, \mathbf{x}) = \frac{1}{4\pi\epsilon_0} \delta(t - |\mathbf{x}|/c) / |\mathbf{x}|$$

Show that if $\hat{V}, \hat{\mathbf{A}}$ are respectively the four dimensional Fourier transforms of $V, \mathbf{A}$, then

$$\hat{V}(k) = \hat{H}(k)\hat{\rho}(k)$$

with

$$\hat{H}(k) = \int H(t, \mathbf{x}) exp(-i(k_0 t - \mathbf{k}.\mathbf{x})) dt d^3x, k = (k_0, \mathbf{k})$$

Note that

$$\hat{V}(k) = \int V(t, \mathbf{x}) exp(-i(k_0 t - \mathbf{k}.\mathbf{x})) dt d^3x, \hat{\rho}(k) = \int \rho(t, \mathbf{x}) exp(-i(k_0 t - \mathbf{k}.\mathbf{x})) dt d^3x$$

Determine the explicit formula for $\hat{H}(k)$. This is easier to do if we start directly from the wave equation satisfied by $V, \mathbf{A}$:

$$\nabla^2 V - \frac{1}{c^2}\frac{\partial^2 V}{\partial t^2} = -\rho(t, \mathbf{x})/\epsilon_0$$

Taking Fourier transforms, we get taking $c = 1$,

$$\hat{V}(k)(-\mathbf{k}^2 + k_0^2) = -\hat{\rho}(k)/\epsilon_0$$

so that

$$H(k) = -\frac{1}{\epsilon_0(k_0^2 - \mathbf{k}^2)}$$

We use the shorthand notation

$$k.k = k^2 = k_0^2 - \mathbf{k}^2$$

so that

$$H(k) = -\frac{1}{\epsilon_0 k^2}$$

Explain how you would determine $h(t, \mathbf{x})$ using $H(k)$ by choosing the contour for k_0 appropriately.

48. DSP implementation of the creation and annihilation operators for bosons.

49. Minkowski functional on a topological vector space (construction of seminorms)

Let X be a vector space and $A \subset X$ an absorbing convex subset. By absorbing, we mean that given any $x \in X$, we have $x \in tA$ for some $t = t(x) > 0$. If X is a topological vector space, then we know that every nhood of zero is absorbing. In fact, we know that if V is a nhood of zero and r_n is a sequence of positive reals increasing to infinity, then $X = \bigcup_n r_n V$. It is easy to see that A must contain the zero vector. For by definition of absorbing, $0 \in tA$ for some $t > 0$. it follows that $0 = ty$ for some $y \in A$. Since $t > 0$, it follows that $y = 0$ and hence $0 \in A$.

Then define

$$\mu_A(x) = inf(t > 0 : t^{-1}x \in A\} = \inf(t > 0 : x \in tA)$$

Then, $\mu_A : X \to [0, \infty)$ is a well defined map. First, we want to show that

$$\mu_A(x + y) \leq \mu_A(x) + \mu_A(y), x, y \in X$$

For $x \in X$, define

$$H_A(x) = \{t > 0 : t^{-1}x \in A\} = \{t > 0 : x \in tA\}$$

Obviously

$$\mu_A(x) = inf(t : t \in H_A(x))$$

Let $t \in H_A(x)$ and $s > t$. Since $0 \in A$ and A is convex, it follows that $s \in H_A(x)$. Here is the proof. $t \in H_A(x)$ implies $t^{-1}x \in A$. Since $0 \leq t/s < 1$ and A is convex, we get

$$s^{-1}x = (t/s)(t^{-1}x) + (1 - t/s)0 \in A$$

and hence $s \in H_A(x)$. It follows that $H_A(x) = [\mu_A(x), \infty)$. Let $\mu_A(x) < s, \mu_A(y) < t$ and $u = s + t$. Then, $s^{-1}x \in A, t^{-1}y \in A$ and since A is convex,

$$u^{-1}(x + y) = (s/u)s^{-1}x + (t/u)t^{-1}y \in A$$

so that

$$\mu_A(x + y) \leq u$$

Setting $s = \mu_A(x) + \delta, t = \mu_A(y) + \delta$ with $\delta > 0$, it follows then that

$$\mu_A(x + y) \leq \mu_A(x) + \mu_A(y) + 2\delta$$

and letting $\delta \downarrow 0$ in this inequality gives

$$\mu_A(x+y) \leq \mu_A(x) + \mu_A(y)$$

We next want to show that $\mu_A(tx) = t\mu_A(x)$ if $t \geq 0$. We have for $t > 0$,

$$H_A(tx) = \{s > 0 : s^{-1}tx \in A\} = \{s > 0 : (st^{-1})^{-1}x \in A\} = t\{st^{-1} : s > 0 : (st^{-1})^{-1}x \in A\}$$

$$= tH_A(x)$$

which on taking infimum, gives

$$\mu_A(tx) = t\mu_A(x)$$

Suppose $t = 0$. Then,

$$H_A(tx) = H_A(0) = \{s > 0 : s^{-1}0 \in A\} = \{s > 0 : 0 \in A\} = (0, \infty)$$

since $0 \in A$. Taking infimum, we get

$$\mu_A(tx) = \mu_A(0) = 0 = 0.\mu_A(x)$$

Finally, we wish to show that apart from being convex and absorbing, if A is balanced, then $\mu_A(.)$ is a seminorm. This will be done if we show that for $t < 0$, $\mu_A(tx) = |t|\mu_A(x)$. But since A is balanced, we have $-A = A$ (By balanced, we mean that $tA \subset A$ whenever $|t| \leq 1$). It follows that

$$H_A(tx) = \{s > 0 : s^{-1}tx \in A\} = \{s > 0 : s^{-1}|t|x \in -A\} = \{s > 0 : s^{-1}|t| \in A\} = |t|H_A(x)$$

and taking infimum, we get

$$\mu_A(tx) = |t|\mu_A(x)$$

completing the proof. We next want to show that if

$$B = \{x : \mu_A(x) < 1\}, C = \{x : \mu_A(x) \leq 1\}$$

then

$$B \subset A \subset C, \mu_A = \mu_B = \mu_C$$

For if $x \in B$, then $\mu_A(x) < 1$ and hence, there exists $t < 1$ such that $t^{-1}x \in A$, so that $x \in tA$. Thus, $x = tu$ for some $u \in A$ and $0 \leq t < 1$. Since A is convex and contains 0, we get that

$$x = tu = tu + (1-t)0 \in A$$

proving that $B \subset A$. Suppose $x \in A$. Then, by definition of μ_A, we have $\mu_A(x) \leq 1$ and hence $x \in C$. This proves that $A \subset C$.

50. Definition of a Banach algebra: A complex algebra A is a vector space over $\mathbb{C}$ with an additional operation of multiplication $(a_1, a_2) \to a_1a_2$ from $A \times A \to A$ that is associative and distributes over addition, ie,

$$a_1(a_2a_3) = (a_1a_2)a_3, a_1(a_2 + a_3) = a_1a_2 + a_1a_3, a_i \in A, i = 1, 2, 3$$

Further, scalar multiplication $(c,a) \rightarrow c.a$ from $\mathbb{C} \times A \rightarrow A$ satisfies $c(a_1a_2) = (ca_1)a_2 = a_1(ca_2)$ where $c \in \mathbb{C}, a_1, a_2 \in A$. An example of a non-commutative algebra is the space of all 2×2 complex matrices with addition and multiplication being specified by matrix addition and matrix multiplication. A Banach algebra is an algebra with a norm such that under this norm A becomes a complex normed linear space and in addition, the norm function satisfies

$$\| a_1a_2 \| \leq \| a_1 \| . \| a_2 \|, a_1, a_2 \in A$$

An example of a commutative Banach algebra is the space of all complex valued functions on a set X with norm defined by

$$\| f \| = sup(|f(x)| : x \in X), f \in A$$

where

$$A = \{f : X \rightarrow \mathbb{C}\}$$

We verify that

$$\| f + g \| =$$

$$sup(|f(x) + g(x)| : x \in X) \leq sup(|f(x)| + |g(x)| : x \in X) \leq sup(|f(x)| : x \in X) + sup(|g(x)| : x \in X)$$

$$= \| f \| + \| g \|$$

We have

$$|f(x)| \leq \| f \|, |g(x)| \leq \| g \|, x \in X$$

so that

$$|f(x)g(x)| = |f(x)||g(x)| \leq \| f \| \| g \|$$

and taking sup over all x gives

$$\| fg \| < \| f \| \| g \|$$

51. Wavelet based system identification with advantage as lesser data storage.

Consider the problem of identification of parameters of a second order Volterra system using wavelets. The input-output equation for the system is given by

$$y(t) \approx \int_{-\infty}^{\infty} h(u,\theta)x(t-u)du + \int_{-\infty}^{\infty} g(u,v,\theta)x(t-u)x(t-v)dudv$$

Here, $\theta = (\theta_1, ..., \theta_p)$ is the parameter vector to be identified. For each value of θ in a finite set say $\Omega = \{\theta_i, i = 1, 2, ..., p\}$, we calculate the wavelet coefficients $c(n,k|\theta)$ and $d(n,k,m,r|\theta)$ of respectively of the first order kernel $h(u,\theta)$ and the second order kernel $g(u,v,\theta)$,ie, if

$$\psi_{n,k}(t) = 2^{n/2}\psi(2^n t - k), n, k \in \mathbb{Z}$$

are the wavelet functions, then

$$h(u,\theta) = \sum_{n,k} c(n,k|\theta)\psi_{n,k}(u), g(u,v,\theta) = \sum d(n,k,m,r|\theta)\psi_{n,k}(u)\psi_{m,r}(v)$$

so that the Volterra model may also be represented as

$$y(t) = \sum_{n,k} c(n,k|\theta) \int x(t-u)\psi_{n,k}(u)du + \sum_{n,k,m,r} d(n,k,m,r|\theta) \int x(t-u)x(t-v)\psi_{n,k}(u)\psi_{m,r}(v)dudv$$

Let $f(n,k)$ denote the wavelet coefficients of the output. Then from the above model,

$$f(p,q) = \int y(t)\psi_{p,q}(t)dt = \sum_{n,k} c(n,k|\theta) \int x(t-u)\psi_{p,q}(t)\psi_{n,k}(u)dtdu$$

$$+ \sum_{n,k,m,r} d(n,k,m,r|\theta) \int x(t-u)x(t-v)\psi_{n,k}(u)\psi_{m,r}(v)\psi_{p,q}(t)dtdudv$$

Set

$$A(p,q,n,k) = \int x(t-u)\psi_{p,q}(t)\psi_{n,k}(u)dtdu, B(p,q,n,k,m,r)$$

$$= \int x(t-u)x(t-v)\psi_{n,k}(u)\psi_{m,r}(v)\psi_{p,q}(t)dtdudv$$

Then, the above equation can be expressed as

$$f(p,q) = \sum_{n,k} A(p,q,n,k)c(n,k|\theta) + \sum_{n,k,m,r} B(p,q,n,k,m,r)d(n,k,m,r|\theta)$$

We note that the ranges of n,k,m,r are to be selected so that the truncated wavelet expansion is a good approximation to the given first and second order kernels. To that end, it suffices to determine the duration and maximum frequency of variation of the kernels h, g for all choices of the parameter vector in the range and then determine θ so that the sum of error squares

$$E(\theta) = \sum_{(p,q)\in E} (f(p,q) - \sum_{(n,k)\in D} A(p,q,n,k)c(n,k|\theta) - \sum_{(n,k,m,r)\in D\times D} B(p,q,n,k,m,r)d(n,k,m,r|\theta))^2$$

is minimized. Such a minimization may be achieved using the gradient search algorithm. However, to apply the gradient search algorithm, we need to know the parametric form of $c(n,k|\theta), d(n,k,m,r|\theta)$ as functions of θ. The search algorithm involves selecting that θ_i for which $E(\theta_i)$ is a minimum.

52. Types of topological vector spaces:

[a]. X is a topological vector space with topology τ. Thus, every point of X is a closed set and vector space operations are continuous with respect to this topology. The latter means that if $x \in X$ and $a \in \Phi$, where Φ is the field then if U is an open set in X containing the point ax, then there exists a $\delta > 0$ and an open set V in X containing x such that $(a-\delta, a+\delta)V \subset U$.

[b]. A local base for a topological vector space X is a collection $\mathcal{B}$ of nhoods of zero such that given any open set W containing 0, there exists a $B \in \mathcal{B}$ such that $B \subset W$. Suppose $x \in X$ and let W be a nhood of x. Then, $W - x$ is a nhood of zero (since the vector space operations are continuous). It follows that there exists a $B \in \mathcal{B}$ such that $B \subset W - x$, or equivalently, $x + B \in W$. Thus, $x + \mathcal{B}$ is a local base for the topology at x.

Example: Let $X = \mathbb{R}^n$ with its usual topology, Then, let

$$B(\delta) = (-\delta, \delta)^n$$

with $\delta > 0$. It is easy to see that $\{B(\delta) : \delta > 0\}$ is a local base for X. Another example of a local base for $\mathbb{R}^n$ is the collection of all sets of the form

$$S(\delta) = \{x \in \mathbb{R}^n : \sum_{i=1}^{n} x_i^2 < \delta\}$$

Suppose X is a normed vector space. Then, a local base for the topology is the set of all open spheres with zero as centre.

[c] If X is a topological vector space such that there exists a local base $\mathcal{B}$, such that every element of $\mathcal{B}$ is convex, then X is said to be a locally convex topological vector space. A set $E \subset X$ is convex if for every $x, y \in E$ and $0 < a < 1$, we have $ax + (1-a)y \in E$. For example, let X be a normed vector space and let $\mathcal{B}$ be the local base consisting of all open spheres with origin as the centre. Then, every element of $\mathcal{B}$ is convex since $x, y \in S(\delta)$ and $0 < p < 1$ imply

$$\| px + (1-p)y \| \leq p \| x \| + (1-p) \| y \| < p\delta + (1-p)\delta = \delta$$

Thus, every normed space is a locally convex topological vector space.

[d] X is called a locally compact space if 0 has a nhood whose closure is compact. For example, suppose X is a finite dimensional normed linear space. Then it is one of the important theorems of functional analysis that the closed unit ball is compact in X . Hence, every finite dimensional normed linear space is locally compact.

[e] Let X be a topological vector space and let K, C be disjoint subsets of X with K compact and C closed. Then there is a nhood V of 0 such that

$$(K + V) \cap (C + V) = \phi$$

Choose $x \in K$. Since C^c is open and $x \in C^c$, we can choose a symmetric nhood V_x of zero such that

$$V_x + V_x + V_x \subset C^c - x$$

or

$$x + V_x + V_x + V_x \subset C^c$$

It follows that $(x + V_x + V_x) \cap (C + V_x) = \phi$ (since $-V_x = V_x$). $\{x + V_x : x \in K\}$ is an open covering for K, hence there is a finite subcovering $\{x_i + V_{x_i} : i = 1, 2, ..., n\}$. Put $V = \bigcap_{i=1}^{n} V_{x_i}$. Then,

$$K + V \subset \bigcup_{i=1}^{n} (x_i + V_{x_i} + V) \subset \bigcup_{i} (x_i + V_{x_i} + V_{x_i})$$

Since

$$(x_i + V_{x_i} + V_{x_i}) \cap (C + V) = \phi$$

it follows that

$$(K+V)\cap(C+V)=\phi$$

and we are done.

Remark:If W is an open nhood of 0, then since vector addition is continuous, there exists a nhood U of 0 such that $U+U\subset W$. Then define $V=U\cap(-U)$. Then, V is a symmetric nhood of zero and $V+V\subset W$. By replacing W with V in this argument, there exists a symmetric nhood S of zero such that $S+S+S+S\subset W$. Let now $x\in X$ be arbitrary and let W be a nhood of x. Then $W-x$ is a nhood of zero and hence there exists a symmetric nhood U of zero such that $U+U+U+U\subset W-x$. This implies in particular that $x+U+U+U\subset W$, a fact that has been made use of in the above proof.

Topology: Let X be a topological space and $E\subset X$. $int(E)$ is the set of all $x\in E$ for which there exists an open nhood U such that $x\in U\subset X$. In other words,

$$int(E)=\bigcup\{U:U open, U\subset E\}$$

A subset E of X is said to be nowhere dense if $int(\bar{E})=\phi$. In that case, we may show that $\bar{E}^c$ is dense in X. For any subset E, $int(\bar{E})\subset\bar{E}$ implies $\bar{E}^c\subset int(\bar{E})^c$. it follows that

$$Cl(\bar{E}^c)\subset int(\bar{E})^c$$

since $int(\bar{E})^c$ is closed. Conversely, suppose $x\notin Cl(\bar{E}^c)$. Then, since $Cl(\bar{E}^c)$ is closed, its complement is open and hence there exists an open set U such that $x\in U\subset Cl(\bar{E}^c)^c$. Thus,

$$U\cap Cl(\bar{E}^c)=\phi$$

Hence,

$$U\cap\bar{E}^c=\phi$$

and hence

$$U\subset\bar{E}$$

Since U is open, it follows that

$$U\subset int(\bar{E})$$

and hence

$$x\notin int(\bar{E})^c$$

Thus we have proved that for any subset E of X, we have

$$Cl(\bar{E}^c)=int(\bar{E})^c$$

It follows that if $int(\bar{E})=\phi$, then $Cl(\bar{E}^c)=X$, or equivalently that $\bar{E}^c$ is dense in X.

Let A be a commutative Banach algebra and J a proper closed ideal in A. Define the quotient map $\pi:A\to A/J$ so that $\pi(x)=x+J, x\in A$. Define

$$\|\pi(x)\|=inf\{\|x+y\|:y\in J\}$$

Then A/J equipped with this norm is a Banach space.

Proof: First we must show that the norm is well defined. Let $\pi(x) = \pi(y)$. Then, $z = y - x \in J$. Since then $z + J = J$, it easily follows that

$$inf(\| x+u \| : u \in J) = inf(\| y+z+u \| : u \in J) = inf(\| y+u \| : u \in J)$$

showing that $\| \pi(x) \|$ is well defined for all $x \in A$. Now, let $x, y \in A$. Let $\delta > 0$ be arbitrary. Choose $u, v \in J$ so that $\| x+u \| \leq \| \pi(x) \| + \delta$ and $\| y+v \| \leq \pi(y) + \delta$. Then $u+v \in J$ and

$$\| \pi(x) + \pi(y) \| = \| \pi(x+y) \| \leq \| x+y+u+v \| \leq \| x+u \| + \| y+v \| \leq \| \pi(x) \| + \| \pi(y) \| + 2\delta$$

and letting $\delta \downarrow 0$ gives the triangle inequality:

$$\| \pi(x) + \pi(y) \| \leq \| \pi(x) \| + \| \pi(y) \|$$

Now for the Completeness property. Let $x_n, n = 1, 2, ...$ be a sequence in A such that $\pi(x_n), n = 1, 2, ...$ is Cauchy in A/J. This means that $\| \pi(x_n) - \pi(x_m) \| \to 0, n, m \to \infty$. It follows that there is a subsequence n_i of $\mathbb{N}$ such that $\| \pi(x_{n_{i+1}}) - \pi(x_{n_i}) \| < 1/2^i$. In particular, $\| \pi(x_{n_2}) - \pi(x_{n_1}) \| < 1/2$. So we can choose a $u_1 \in J$ so that

$$\| x_{n_2} - x_{n_1} + u_2 \| < 1/2$$

Since $\pi(x_{n_2} + u_2) = \pi(x_{n_2})$, we can choose $u_3 \in J$ so that

$$\| x_{n_3} - x_{n_2} - u_2 + u_3 \| < 1/2^2$$

and continuing this process, we get a sequence $u_i, i = 1, 2, ...$ in J such that

$$\| x_{n_{i+1}} + u_{i+1} - x_{n_i} - u_i \| < 1/2^i, i = 1, 2, ...$$

In other words, $x_{n_i} + u_i, i = 1, 2, ...$ is Cauchy in A. It follows that there exists a $y \in A$ such that $x_{n_i} + u_i \to y$ as $i \to \infty$. Since

$$\| \pi(x_{n_i}) - \pi(y) \| \leq \| x_{n_i} + u_i - y \|$$

it follows that $\pi(x_{n_i}) \to \pi(y)$ in the normed space A/J. Since $\pi(x_n)$ is Cauchy, it follows then that $\pi(x_n) \to \pi(y)$ proving that A/J is a Banach space.

53. Let X be a topological vector space and Y a locally compact subspace of X. Then, Y is closed in X.

1.Define a bounded linear functional on a normed linear space X. Show that $\mathcal{M}$ is a closed subspace of X and f is a linear functional on M that satisfies $|f(x)| \leq \| x \|$ for all $x \in M$, then there exists a linear functional g on X such that $g = f$ on M and $|g(x)| \leq \| x \|$ for all $x \in X$. Hint:Use the Hahn-Banach theorem. Deduce that if $x \in X$, then there exists an $f \in X^*$ (X^* is the class of all bounded linear functionals on X), such that $\| f \| = \| x \|$

2. Define a convex subset of a vector space and give examples. Specifically show that if $f_1, ..., f_k$ are linear functionals on a vector space X and $c_1, ..., c_k$ are arbitrary scalars, then the set

$$\{x \in X : f_i(x) \leq c_i, i = 1, 2, ..., k\}$$

is a convex set. What region does this correspond to in the case $X = \mathbb{R}^2$.

3. Let X be a topological vector space, ie a vector space over a field endowed with a topology such that scalar multiplication and vector addition are continuous operations and singletons are closed. Let V be an open nhood of the zero vector. Since scalar multiplication is continuous, there exists a nhood W of zero and a $\delta > 0$ such that $[-\delta, \delta]W \subset V$. Define

$$W_0 = \bigcup_{|r| \leq \delta} r.W$$

Then W_0 is an open nhood of zero, $W_0 \subset V$ and $rW_0 \subset W_0$ for all $|r| \leq 1$. In other words, every open nhood of zero contains a balanced nhood of zero.

54.Least mean square algorithm applied to trajectory tracking of point masses from measurements of the gravitational potential: Let $\rho(\mathbf{r})$ be the spatial density of matter and $\Phi(\mathbf{r})$ the gravitational potential. Then, Newton's inverse square law of gravitation can be expressed as

$$\Phi(\mathbf{r}) = -G(4\pi)^{-1} \int_{\mathbb{R}^3} \rho(\mathbf{r}') |\mathbf{r} - \mathbf{r}'|^{-1} d^3 r'$$

Using the property of the Green's function

$$\nabla^2 |\mathbf{r} - \mathbf{r}'|^{-1} = -4\pi\delta(\mathbf{r} - \mathbf{r}')$$

we get Poisson's equation for the gravitational potential:

$$\nabla^2 \Phi(\mathbf{r}) = -4\pi G \rho(\mathbf{r})$$

Suppose that we are given point masses $M_1, ..., M_n$ located at points $\mathbf{r}_1(t), ..., \mathbf{r}_n(t)$ which are functions of time t. Assume these postitions to be slowly time varying and we wish to estimate, or more precisely, track these trajectories based on measurements of the gravitational potential at a given point $\mathbf{R}$. We then make a match

$$Z(t) = \Phi(t, \mathbf{R}) \approx -G \sum_{i=1}^{n} \frac{M_i}{|\mathbf{R} - \mathbf{r}_i(t)|}$$

The matching is done so that

$$\delta \hat{\mathbf{r}}_i(t) = \hat{\mathbf{r}}_i(t + \delta t) - \hat{\mathbf{r}}_i(t) = -\mu \delta t. \nabla_{\mathbf{r}_i} (Z(t) + \sum_{i=1}^{n} \frac{GM_i}{|\mathbf{R} - \hat{\mathbf{r}}_i(t)|})^2$$

Note that

$$\nabla_{\mathbf{r}} |\mathbf{R} - \mathbf{r}|^{-1} = (\mathbf{R} - \mathbf{r}) / |\mathbf{R} - \mathbf{r}|^3$$

This gives the algorithm

$$\delta \hat{\mathbf{r}}_i(t) = 2\mu \delta t. e(t). (GM_i (\mathbf{R} - \mathbf{r}_i(t))) / |\mathbf{R} - \mathbf{r}_i(t)|^3$$

where

$$e(t) = Z(t) + \sum_{i=1}^{n} GM_i / |\mathbf{R} - \mathbf{r}_i(t)|$$

is the error signal.

55. Invariant functions: Let X be a set on which a group G acts and let $\mathcal{F}(X)$ denote the set of all functions $f : X \to \mathbb{C}$. G acts on $\mathcal{F}(X)$ in the natural way:

$$(g.f)(x) = f(g^{-1}x)$$

A function $f \in \mathcal{F}(X)$ is said to be G-invariant if $g.f = f$. Note that if $f_1, f_2 \in \mathcal{F}(X)$ and $c_1, c_2 \in \mathbb{C}$, then

$$(g.(c_1 f_1 + c_2 f_2))(x) = c_1 f_1(g^{-1}x) + c_2 f_2(g^{-1}x) = (c_1 g_1.f + c_2 g_2.f)(x)$$

so that each $g \in G$ acts linearly on the algebra $\mathcal{F}(X)$. Also each $g \in G$ preserves the multiplication operation in $\mathcal{F}(X)$ as may be seen from

$$(g.(f_1 f_2))(x) = (f_1 f_2)(g^{-1}x) = f_1(g^{-1}x) f_2(g^{-1}x) = ((g.f_1).(g.f_2))(x)$$

The set of G-invariant functions on X forms an algebra as can be seen from $g \in G$, $g.f_1 = f_1, g.f_2 = f_2$ implies $g.(c_1 f_1 + c_2 f_2) = c_1 g.f_1 + c_2 g.f_2 = c_1 f_1 + c_2 f_2$ and $g.(f_1 f_2) = (g.f_1)(g.f_2) = f_1 f_2$.

56. Bilinear forms:Let V be a vector space over $\mathbb{R}$ of dimension $n < \infty$ and let $f : V \times V \to \mathbb{R}$ be a symmetric non-degenerate bilinear form, ie, f is linear in both of its arguments, $f(u, v) = f(v, u)$ for all $u, v \in V$ and $f(u, v) = 0$ for all $v \in V$ iff $u = 0$. Choose a basis $B = \{e_1, ..., e_n\}$ for V and consider the matrix

$$A = ((f(e_i, e_j))$$

A is a real Hermitian matrix and hence, by the spectral theorem, it can be represented as

$$A = UDU^T$$

where U is a real orthogonal matrix and D is a real diagonal matrix. Let $x, y \in V$ and set $x = \sum x_i e_i, u = \sum y_i e_i$ so that $x_i, y_i \in \mathbb{R}$. Then,

$$f(x, y) = \sum_{i,j} x_i y_j f(e_i, e_j) = x^T Ay = [x]_B^T UDU^T [y]_B$$

There exists a basis B' of V, such that $[x]_{B'} = U^T [x]_B$ for all $x \in B$. To see this, suppose $B' = \{f_1, ..., f_n\}$ so that $x = \sum x'_i f_i$ with $[x]_{B'} = (x'_i)$, then

$$\sum x_j e_j = x = \sum x'_i f_i = \sum (U^T [x]_B)_i f_i = \sum u_{ji} x_j f_i$$

so that the new basis is given by

$$\sum_i u_{ji} f_i = e_j$$

or

$$f_i = \sum_j u_{ji} e_j$$

using orthogonality of U. Thus,

$$f(x, y) = [x]_{B'}^T D[y]_{B'}$$

which means that

$$f(f_i, f_j) = d_i \delta_{ij}$$

By applying a permutation to the basis elements (f_i), we may assume without loss of generality that $d_i > 0$ for $t = 1, 2, ..., r$, and < 0 for $i = r+1, ..., n$. Note that since we are assuming that f is non-degenerate, no d_i equals zero. Then letting $g_i = |d_i|^{-1/2} f_i$, it follows that $((f(g_i, g_j)))$ has the diagonal form

$$C = \begin{pmatrix} I_r & 0 \\ 0 & -I_{n-r} \end{pmatrix}$$

This is the canonical form of a symmetric, non-degenerate real bilinear form. Now we shall derive the canonical form for the antisymmetric bilinear form. Let V be a real vector space and $f : V \times V \to \mathbb{R}$ be an antisymmetric, non-degenerate bilinear form. By antisymmetric, we mean that $f(u, v) = -f(v, u)$ for all $u, v \in V$. Then choose a basis $B = (e_i)$ as before and define $A = ((f(e_i, e_j))$. A is then a real antisymmetric matrix and iA is a Hermitian matrix. We can diagonalize iA relative to an orthonormal basis for $\mathbb{C}^n$. Thus, we choose vectors $\xi_1, ..., \xi_n$ in $\mathbb{C}^n$ so that

$$iA\xi_k = c_k \xi_k, k = 1, 2, ..., n$$

$c_k's$ are all real and ξ_k are complex vectors. If The eigenvalues c_k appear in pairs as $(c, -c)$ for taking the complex conjugate gives us $-iA\bar{\xi}_k = \bar{c}_k \bar{\xi}_k$, None of the eigenvalues is zero since the form is non-degenerate. Writing $\xi_k = u_k + iv_k$ where u_k, v_k are real, we get

$$iA(u_k + iv_k) = c_k(u_k + iv_k)$$

and equating real and imaginary parts, we get

$$Au_k = c_k v_k, Av_k = -c_k u_k$$

Thus, $sp\{u_k, v_k\}$ is an invariant subspace for V and the restriction of A relative to this space is

$$\begin{pmatrix} 0 & -c_k \\ c_k & 0 \end{pmatrix}$$

By appropriately permuting the elements u_k, v_k, and applying a scaling, we may obtain a real basis for V relative to which the form has the matrix

$$\begin{pmatrix}) & I_r \\ -I_r & 0 \end{pmatrix}$$

where $r = 2n$. This means that a non-degenerate bilinear skew symmetric form on a real vector space exists iff the space has even dimension.

57. Let G be a group and let H, K be subgroups of G. Consider the G action on the set $(G/H) \times (G/K)$. This action is defined by the equation

$$g(g_1 H, g_2 K) = (gg_1 H, gg_2 K), g, g_1, g_2 \in G$$

Let $O(g_1H, g_2K)$ denote the orbit of (g_1H, g_2K) under G-action. Corresponding to this orbit define the element $Hg_1^{-1}g_2K$ of $\{HgK : g \in G\}$. This is a one-one correspondence. First to see that this mapping is well defined, we observe that if $(g_1'H, g_2'K)$ is any element in the orbit $O(g_1H, g_2K)$, then $g_1'H = gg_1H, g_2'K = gg_2K$ for some $g \in G$ and then

$$Hg_1'^{-1}g_2'K = Hg_1^{-1}g^{-1}gg_2K = Hg_1^{-1}g_2K$$

showing that the map is well defined. We next show that this map is one-one. Indeed, suppose $Hg_1^{-1}g_2K = Hg_1'^{-1}g_2'K$. Then we must show that $O(g_1H, g_2K) = O(g_1'H, g_2'K)$. But the equation implies the existence of $h \in H, k \in K$ such that $g_1'^{-1}g_2' = hg_1^{-1}g_2k$, so that

$$g_2'K = g_1'hg_1^{-1}g_2K$$

$$g_1'H == g_1'hH$$

so that

$$(g_1'H, g_2'K) = g_1'h(H, g_1^{-1}g_2K) = g_1'hg_1^{-1}(g_1H, g_2K)$$

proving that

$$O(g_1'H, g_2'K) = O(g_1H, g_2K)$$

ie the two orbits coincide. Finally, it is trivial to see that the map is onto. Since as g_1, g_2 range over G, $Hg_1^{-1}g_2K$ ranges over the entire set $\{HgK : g \in G\}$.

Let G be a group and consider its group algebra that formally consists of elements of the form $a = \sum_g a(g)g$. The product of two elements a, b in the group algebra is given by

$$a.b = \sum_{g,h \in G} a(g)b(h)gh = \sum_{g,h} a(gh^{-1})b(h)g$$

so that the function $a.b$ is nothing but the group convolution of the functions a and b. Let V be a vector space on which the group G acts as a representation T, ie, $g.v = T(g)v$ for $g \in G, v \in V$. Then given $v \in V$ an element a in the group algebra acts on v in accord with the rule

$$a.v = \sum_g a(g)g.v = \sum_g a(g)T(g)v$$

Denote the group algebra by $F(G)$. Then under this action, the representation space V becomes an $F(G)$ module.

58. Use the Monte-Carlo simulation estimate and plot the error probability for a binary communication system employing the matched filter.

Assume that under the hypothesis H_1 when 1 is the bit to be transmitted, we transmit the signal $s_1(t), 0 \le t \le T$ and under the hypothesis H_0 when 0 is the bit to be transmitted, we transmit the signal $s_0(t), 0 \le t \le T$. The received signal is $s_i(t) + w(t), t \in [0, T]$ under the hypothesis H_i and $w(.)$ is white Gaussian noise with autocorrelation

$$\mathbb{E}(w(t)w(s)) = N_0\delta(t-s)/2$$

If $x(t)$ is the received signal, the likelihood ratio for performing the decision is given by

$$\Lambda(x) = exp(-\int_0^T (x(t) - s_1(t))^2/N_0)/exp(-\int_0^T (x(t) - s_0(t))^2/N_0)$$

If Λ is greater than 1, we decide in favour of H_1 while if Λ is smaller than one, we decide in favour of H_0. Here, we are assuming the two hypotheses to be equally probable apriori. The likelihood ratio test can be simplified by defining

$$l(x) = \int_0^T x(t)s_1(t)dt - \int_0^T x(t)s_0(t)dt$$

If $l(x) > 0$, decide H_1 while if $l(x) < 0$, decide H_0. Define two impulse response functions $h_i(t) = s_i(T-t), i = 1, 0$. Then

$$\int_0^T x(t)s_i(t)dt = \int_0^T h_i(\tau)x(T-\tau)d\tau = \int_0^T h_i(T-\tau)x(\tau)d\tau, i = 1, 2$$

So the decision process can be viewed as the construction of two filters having impulse responses $h_i, i = 1, 0$. We pass the received signal over the duration $[0, T]$ through each of these filters and select the maximum. To implement this using MATLAB, we do the following. Generate two discrete time signals $s_i[n], n = 0, 1, 2, ..., N-1$. For example, we can take s_i to be one and minus one, or one and zero. We also generate white Gaussian noise $w[n], n = 0, 1, ..., N-1$ using the randn command followed by an amplitude scaling. Then select $i = 0, 1$ at random and generate $x[n] = s_i[n] + w[n], n = 0, 1, ..., N-1$. Using this x generate the numbers $\sum_{n=0}^{N-1} x[n]s_i[n], i = 1, 0$ and choose the maximum. Repeat this process K times for the same $s_i's$ but different realizations of the noise $w[.]$ and count the proportion of times that an error has occurred. Plot this error propagability as a function of the signal to noise ratio $\sum s_i[n]^2/\mathbb{E}w[n]^2$. Here we are assuming that $s_i's$ have been amplitude scaled appropriately so that

$$\sum s_1[n]^2 = \sum s_0[n]^2$$

59. Implement an adaptive equalizer based on LMS algorithm and study the effect of step size on the MSE.

Choose $p << N$ real numbers (filter tap weights) $h(0), h(1), ..., h(p)$. Generate an input sequence $x(n), n = 0, 1, 2, ..., N-1$. Generate white Gaussian noise $w[n], n = 0, 1, ..., N-1$ and generate the output process

$$y[n] = \sum_{k=0}^{p} h(k)x(n-k) + w(n), n = p, p+1, ..., N$$

Choose an initial guess for the tap weights $g_0[k], k = 0, 1, 2, ..., p$ and generate the tap weights in accord with the LMS algorithm:

$$g_{n+1}[k] = g_n[k] - \mu\frac{\partial}{\partial g_n[k]}(y[n] - \sum_{m=0}^{p} g_n[k]x[n-k])^2$$

or

$$g_{n+1}[k] = g_n[k] + 2\mu e[n]x[n-k], k = 0, 1, ..., p$$

where

$$e[n] = y[n] - \sum_{m=0}^{p} g_n[k]x[n-k]$$

is the error at time n. μ is the adaptation weight. The aim is to show that this algorithm for sufficiently small μ guarantees convergence of the weights $g_n[k]$ to $h[k]$ and also guarantees good tracking in case the $h's$ vary slowly with time. This gives an implementation of an adaptive system identifier. For an adaptive equalizer, we pass the input through an fir filter $\{h(n)\}$ and then letting $y(n)$ denote the output, we apply the above algorithm taking as input $y(n)$ and output $x(n)$. The resulting adaptive filter $\{g_n(k)\}$ is then expected to converge to an inverse of the filter $\{h(k)\}$. We may take $h(n)$ as an IIR filter of the form

$$H(z) = \frac{1}{1 - az^{-1}}$$

The process $y(n)$ is thus generated by the algorithm

$$z(n) + az(n-1) = x(n), y(n) = z(n) + w(n)$$

The adaptive filter output is

$$\hat{x}(n) = \sum_{k=0}^{p} g_n(k)y(n-k)$$

so that

$$g_{n+1}(k) = g_n(k) + 2\mu e(n)y(n-k), e(n) = x(n) - \sum_{k-0}^{p} g_n(k)y(n-k)$$

If the noise is small, we then expect

$$G_n(z) \approx 1 - az^{-1}$$

for large n. Check this.

60. Monte carlo simulation of a QPSK communication system. Generate a sequence of ones and zeroes. Pair this sequence so that we get a string of elements of the form $(ij), i, j = 0, 1$. To transmit a (00) we use the signal $cos(\omega_c n)$, to transmit (01), we transmit $cos(\omega_c n + 2\pi/4)$, to transmit (10), we transmit $cos(\omega_c n + \pi/2)$ and finally to transmit (11), we transmit $cos(\omega_c n + 3\pi/4)$. We add white Gaussian noise $w(n)$ to the transmitted signal with a variance of 1/10. Denote the resulting received signal by $x(n), n = 0, 1, 2, ..., N-1$. Take $\omega_c = 2\pi/100$ and $N = 100$, or more generally, $\omega_c = 2\pi/N$ so that the transmitted signal consists of one complete cycle of the carrier signal. To decode at the receiver end, we use the correlation method. Specifically compute

$$((2/N)\sum_{n=0}^{N-1} x(n)cos(\omega_c n), (2/N)\sum_{n=0}^{N-1} x(n)sin(\omega_c n))$$

Compute the distance of this pair from the four points $(1, 0)$, $(cos(2\pi/4), -sin(2\pi/4)), (cos(\pi/2) - sin(\pi/2))$ and $(cos(3\pi/4), -sin(3\pi/4))$ and if the minimum distance is to the first pair, then decide (00), if it is to the second pair, decide (01), if it is to the third pair, then decide (10) and finally if the minimum distance is to the third pair, decide (11). In this way the QPSK detector is constructed.

61. Banach Algebras: [a] Let A be a Banach algebra and $x \in A$ with $\| x \| < 1$. Then $e - x$ is invertible. Indeed define

$$y_n = e + x + x^2 + ... + x^n$$

Then,

$$\| y_{n+p} - y_n \| \leq \sum_{k=n+1}^{n+p} \| x \|^k \to 0$$

as $n \to \infty$ since $\sum_{k=0}^{\infty} \| x \|^k < \infty$. It follows that $\| y^n$ is a Cauchy sequence in A and hence converges to some y. We have

$$(e - x)y_n = (e - x)(e + x + x^2 + ... + x^n) = e - x^{n+1}$$

It follows that

$$\| (e - x)y - e \| = \| (e - x)(y - y_n) + (e - x)y_n - e \| = \| (e - x)(y - y_n) - x^{n+1} \|$$

$$\leq \| e - x \| . \| y - y_n \| + \| x \|^{n+1} \to 0, n \to \infty$$

from which we conclude that

$$y = (e - x)^{-1}$$

We now show that

$$\| (e - x)^{-1} - e - x \| \leq \| x \|^2 / (1 - \| x \|)$$

Indeed, since $y_n \to (e - x)^{-1}$, it follows that

$$\| (e - x)^{-1} - e - x \| = lim \| y_n - e - x \|$$

$$= lim \| e + x + x^2 + ... + x^n - e - x \| = lim \| x^2 + ... + x^n \|$$

$$= lim \| x^2 (e + x + ... + x^{n-2}) \| \leq \| x \|^2 .lim \sum_{k=0}^{n-2} \| x \|^k$$

$$= \| x \|^2 / (1 - \| x \|)$$

Finally, we want to show that if ϕ is any complex homomorphism on A, then $|\phi(x)| < 1$. First we note that $z \in A$ is invertible, then $\phi(x) \neq 0$. For then

$$1 = \phi(zz^{-1}) = \phi(z)\phi(z^{-1})$$

Now let x be as above and consider any $\lambda \in \mathbb{C}$ with $|\lambda| \geq 1$. Then $e - \lambda^{-1}x$ invertible and hence $\phi(1 - \lambda^{-1}x) \neq 0$. It follows that $\phi(x) \neq \lambda$ and hence $|\phi(x)| < 1$. This settles the claim and shows that any homomorphism on a Banach algebra is a bounded linear functional (or a continuous linear functional) on A.

62. Commutative Banach algebras:Let A be a commutative Banach algebra and Δ the set of all complex homomorphisms of A. If M is a maximal ideal of A, then $M = kerh$ for some homomorphism h of A and conversely if h is a homomorphism of A, then $Kerh$ is a maximal ideal of A.

Lemma: If A is a Banach algebra in which every nonzero element is invertible, then A is isometrically isomorphic to the complex field.

For choose any two distinct complex numbers $\lambda_i, i = 1, 2$. Then at most one of the elements $\lambda_i e - x, i = 1, 2$ is zero and hence at least one of them is invertible. Since the spectrum $\sigma(x)$ is non-empty, it follows that $\sigma(x)$ has exactly one point say $\lambda(x)$ for each $x \in A$. $\lambda(x)e - x$ is not invertible and hence it is zero. It is easy to see that λ is a linear map from A into $\mathbb{C}$ and that

$$|\lambda(x)| = \| \lambda(x)e \| = \| x \|$$

proving isometry.

Note:We have used an important result that the spectrum of any element in a Banach algebra is nonempty. For suppose $x \in A$ where A is a Banach algebra. Define $g : \mathbb{C} \to A$ by $g(\lambda) = \lambda.e - x$. g is a continuous function. However the set $G(A)$ of invertible elements in A is an open set (Let y be invertible. Then $z = y - (y - z)$ gives

$$z^{-1} = (y - (y - z))^{-1} = (1 - y^{-1}(y - z))^{-1} y^{-1}$$

exists if $\| y^{-1}(y - z) \| < 1$ which is in turn true if $\| y - z \| < \| y^{-1} \|^{-1}$. Note that $\| y \| \| y^{-1} \| \geq 1$. Thus, we have shown that z^{-1} exists for all z inside an open ball of sufficiently small radius with y as centre.). It follows that $g^{-1}(G(A))$ is an open set. But $g^{-1}(G(A)) = \sigma(x)^c$, ie, the resolvent set of x. The resolvent set is open and hence the spectrum is closed. We now show that the spectrum is bounded. For let x be an element of the Banach algebra and let λ be a complex number with $|\lambda| > \| x \|$. Then, $e - \lambda^{-1}x$ is invertible and hence $\lambda \in \sigma(x)^c$. It follows that if $\lambda \in \sigma(x)$, then $|\lambda| \leq \| x \|$. This proves that the spectrum of x is bounded by $\| x \|$. Since then $\sigma(x)$ is a closed and bounded subset of $\mathbb{C}$, it is compact. Define $f : \sigma(x)^c \to A$ by

$$f(\lambda) = (\lambda.e - x)^{-1}$$

We now make use of the inequality if $x \in G(A)$ and $h \in A$ is such that $\| h \| < \frac{1}{2} \| x^{-1} \|^{-1}$, then $x + h \in G(A)$ (this we have already proved above) and

$$L = \| (x + h)^{-1} - x^{-1} + x^{-1}hx^{-1} \| \leq 2 \| x^{-1} \|^3 \| h \|^2$$

Indeed,

$$(x + h)^{-1} = (x(1 + x^{-1}h))^{-1} = (1 + x^{-1}h)^{-1}x^{-1} = (x^{-1} - x^{-1}hx^{-1} + \sum_{n=2}^{\infty}(-1)^n(x^{-1}h)^n x^{-1}$$

giving

$$L \leq \sum_{n=2}^{\infty} \| x^{-1} \|^{n+1} \| h \|^n$$

$$= \| h \|^2 \| x^{-1} \|^3 / (1 - \| x^{-1} \| . \| h \|)$$

$$\leq 2 \| x^{-1} \|^3 . \| h \|^2$$

It follows that if $|\mu - \lambda|$ is sufficiently small, then

$$\| f(\mu) - f(\lambda) + (\mu - \lambda) f(\lambda)^2 \| \leq 2 f(\lambda)^3 |\mu - \lambda|^2$$

From which we deduce that f is strongly differentiable at λ and is hence holomorphic on $\sigma(x)^c$. If Γ is a circle in $\mathbb{C}$ with origin as center and radius $r > \| x \|$, then we have the result that for all $\lambda \in \Gamma$, the series

$$f(\lambda) = \lambda^{-1} \sum_{n=0}^{\infty} \lambda^{-n} x^n$$

is uniformly convergent and hence term by term integration gives

$$x^n = (2\pi i)^{-1} \int_{\Gamma} \lambda^n f(\lambda) d\lambda, n = 0, 1, 2, ...$$

Suppose $\sigma(x)$ is empty. Then, f is holomorphic in the entire complex plane $\mathbb{C}$ and hence the above integrals are zero for all n which contradict the fact that $x^0 = e$. This proves that $\sigma(x)$ is non-empty.

63. Parseval's theorem: Let G be a locally compact Abelian group, for example $\mathbb{R}^n$ and for $g \in G$, suppose $\chi(g)$ is a complex number of unit magnitude such that $\chi(g + g') = \chi(g)\chi(g')$ for all, $g, g' \in G$, in other words, $\chi : G \to \mathbb{T}$ where $\mathbb{T}$ is the unit torus, is a homomorphism. Then χ is called a character of G. For example, if $G = \mathbb{R}^n$, then for any $(t_1, ..., t_n) \in \mathbb{R}^n$, we may define

$$\chi_{t_1,...,t_n}(x_1, ..., x_n) = exp(i \sum_{a=1}^{n} t_a x_a)$$

or equivalently,

$$\chi_{\mathbf{t}}(\mathbf{x}) = exp(i < \mathbf{t}, \mathbf{x} >)$$

We easily verify that $|\chi_{\mathbf{t}}(\mathbf{x})| = 1$ and

$$\chi_{\mathbf{t}}(\mathbf{x} + \mathbf{y}) = \chi_{\mathbf{t}}(\mathbf{x}) + \chi_{\mathbf{t}}(\mathbf{y}), \mathbf{x}, \mathbf{y} \in \mathbb{R}^n$$

The Fourier transform theorem on $\mathbb{R}^n$ tells us that any square integrable function on $\mathbb{R}^n$ can be expanded as a superposition of the characters, ie,

$$f(\mathbf{x}) = \int \hat{f}(\mathbf{t}) \chi_{\mathbf{t}}(\mathbf{x}) d\mathbf{t}$$

where

$$\hat{f}(\mathbf{t}) = (2\pi)^{-n} \int_{\mathbb{R}^n} f(\mathbf{x}) \bar{\chi}_{\mathbf{t}}(\mathbf{x}) d\mathbf{x}$$

The Parseval theorem then gives for two functions $f, g \in L^2(\mathbb{R}^n)$,

$$< f, g >= (2\pi)^n < \hat{f}, \hat{g} >$$

This equation was generalized by A.S.Pontryagin to arbitrary locally compact Abelian groups. On G, we first introduce an invariant measure μ so that $\mu(x+E)=\mu(E), \forall x \in G$ and $E \in \mathcal{B}(G)$. On $\hat{G}$, the space of characters, we can introduce a dual measure $\hat{\mu}$ that is also invariant. Note that $\hat{G}$ is also an Abelian group. We then have the following Fourier and inverse Fourier relations:

$$f(g) = \int_{\hat{G}} \hat{f}(\chi)\chi(g)d\hat{\mu}(\chi),$$

$$\hat{f}(\chi) = \int f(g)\bar{\chi}(g)d\mu(g)$$

For these two to be consistent, we need that

$$f(g) = \int f(g')\bar{\chi}(g')\chi(g)d\mu(g)d\hat{\mu}(\chi)$$

which symbolically can be written as

$$\delta(g,g') = \int \chi(g)\bar{\chi}(g')d\hat{\mu}(\chi)$$

and the corresponding dual relation

$$\hat{f}(\chi) = \int \hat{f}(\chi')\chi'(g)\bar{\chi}(g)d\hat{\mu}(\chi')d\mu(g)$$

or equivalently,

$$\delta(\chi,\chi') = \int \chi(g)\bar{\chi}'(g)d\mu(g)$$

Here, $\delta(g,g')$ is a function that satisfies

$$\int f(g)\delta(g,g')d\mu(g) = f(g')$$

and likewise,

$$\int f(\chi)\delta(\chi,\chi') = f(\chi')$$

64.

1. Action plan for research and development:

(1). Developing specialized software for determining the modes of propagation of electromagnetic waves inside a waveguide of arbitrary cross-section having non-uniform and possibly anisotropic permittivity. The inputs to this programme will be the shape of the waveguide boundary in pixel format and the permittivity tensor at each point inside the guide and the outputs will be the propagation modes. The computations will be based on the finite element/moment method.

(2). Developing specialized software for simulating stochastic systems. The software will consist of simulating: (a) random variables/sequences with arbitrary prescribed probability distributions, (b)

Computing the statistics of the output of systems described by differential and difference equations with Gaussian input , white or colored, (c) Simulating nonlinear filters for trajectory tracking problems.

(3). Developing software for simulating quantum mechanical systems. These include: (a) Solving Schrodinger's equation for various kinds of potential numerically, (b) Applying independent and time dependent perturbation theory for computing the energy levels of many electron atoms and transition probabilities for atoms excited by the radiation field, (c) Numerically evaluating the Feynman path integral for computing the propagation for various kinds of quantum states e.g. harmonic oscillator and hydrogen atom (Coulamb potential), (d) Numerical evaluation of the amplitude for various kinds of scattering processes based on Feynman diagrams.

Requirements: Ph.D. students and M. Tech. students and 5 computers with sitting space and facilities.

Lab Development: A lab adjoining the DSP lab is to be developed for carrying out M. Tech. projects.

Requirements: 10 computer tables and 10 chairs, 1 printer table, 1 staff table and chair, 5 computers and a printer cum scanner cum copier.

65.**Proposal for music laboratory**

Main features of the laboratory

1.Installation of musical instruments like flute, Tabla, Veena etc.

2. Development of Signal processing toolboxes like Fourier and Wavelet techniques for the analysis of the music.

For example: let $x(t)$ be the music coming from an instrument. Suppose that $x(t)$ consists of the discrete frequencies $\omega_k, k = 1, 2, ..., n$. Then,

$$x(t) = \sum_{k=1}^{n} A_k cos(\omega_k t + \phi_k)$$

and its Fourier transform is given by

$$X(\omega) = \int_{-\infty}^{\infty} x(t)exp(-i\omega t) = \pi \sum_{k=1}^{n} [exp(i\phi_k)\delta(\omega - \omega_k) + exp(-i\phi_k)\delta(\omega + \omega_k)]$$

The frequencies $\{\omega_k\}$ can be estimated by simply forming the periodogram of $x(t)$ which is a finite data set based approximation of the Fourier transform and then locating its peaks. The problem of estimating the phases ϕ_k is harder. Spectral analysis suppresses the phase information. Bispectral analysis can then be used to estimate the phases. The periodogram of the signal based on the data set $x(t), t \in [0, T]$ is defined by

$$X_T(\omega) = \int_{-T}^{T} x(t)exp(-i\omega t)dt$$

$$= \frac{1}{2} \sum_{k=1}^{n} exp(i\phi_k) \int_{-T}^{T} [exp(i(\omega_k - \omega)t)dt + exp(-i\phi_k) \int_{-T}^{T} exp(-i(\omega_k + \omega)t)]dt$$

$$= \sum_{k=1}^{n} exp(i\phi_k)\frac{sin((\omega-\omega_k)T)}{(\omega-\omega_k)} + exp(-i\phi_k)\frac{sin((\omega+\omega_k)T)}{(\omega+\omega_k)}$$

$X_T(\omega)$ shows smeared peaks at the points $\{\omega_k\}$ and can be used to obtain good estimates for the frequencies $\{\omega_k\}$. For doing local spectral analysis, we may use the STFT (short time Fourier transform). To this end, let $h(t)$ be a window function. The STFT of the signal $x(.)$ with respect to this window is defined by

$$S_h(t,\omega) = \int_{-\infty}^{\infty} h(\tau - t)x(\tau)exp(-i\omega\tau)d\tau$$

For a given value of t, the peaks of this function with respect to the frequency variable ω give the dominant frequencies present in the vicinity of the time t. Let

$$h(t) = \frac{1}{2\pi}\int_{-\infty}^{\infty} H(\omega)exp(i\omega t)d\omega$$

Then,

$$S_h(t,\omega) = (2\pi)^{-1}\int H(\theta)x(\tau)exp(i\theta(\tau-t))exp(-i\omega\tau)d\tau d\theta$$

$$= \int_{-\infty}^{\infty} H(\theta)X(\omega-\theta)exp(-i\theta t)d\theta$$

3.Bispectral analysis: Suppose $x(t), t \in \mathbb{R}$ is a signal. Assume that this signal is passed through a linear time invariant filter having impulse response $h(t), t \in \mathbb{R}$. The output is given by

$$y(t) = \int_{-\infty}^{\infty} h(\tau)x(t-\tau)d\tau$$

The third order moment function of the input signal with calculations based on time averages are given by

$$C_x(u,v) = \int_{-\infty}^{\infty} x(t)x(t+u)x(t+v)dt$$

Its bivariate Fourier transform equals

$$B_x(\omega_1,\omega_2) = \int_{-\infty}^{\infty}\int_{-\infty}^{\infty} C_x(u,v)exp(-j(\omega_1 u + \omega_2 v))dudv$$

$$= X(\omega_1)X(\omega_2)\bar{X}(\omega_1+\omega_2)$$

Suppose $\phi_x(\omega) = ArgX(\omega), \phi_y(\omega) = ArgY(\omega)$ and $\phi_h(\omega) = ArgH(\omega)$. Then we have

$$ArgB_x(\omega_1,\omega_2) = \phi_x(\omega_1) + \phi_x(\omega_2) - \phi_x(\omega_1+\omega_2)$$

This formula can be used to estimate the signal phase spectrum from bispectral measurements. In musical notes the relative phase difference between two harmonics plays a fundamental role. For example, suppose that there are just two frequencies.

4.Applications of eigenstructure methods to frequency determination. If the music signal $x(t)$ consists of discrete frequencies $\omega_i, i = 1, 2, ..., k$, then the problem of estimating these frequencies in a background white noise can be achieved through the high resolution eigenstructure techniques like MUSIC and ESPRIT algorithm. The MUSIC algorithm involves determining the autocorrelation function of a signal consisting of a sum of sinusoids and white noise, arranging the elements of this autocorrelation function in the form of a matrix resulting in the following structure

$$\mathbf{R}_{xx} = \mathbf{ADA}^* + \sigma^2\mathbf{I}$$

where $\mathbf{D}$ is a diagonal matrix whose diagonal entries are the signal powers and $\mathbf{A}$ is a rectangular matrix whose columns are the steering vectors corresponding to the different signal frequencies. σ^2 is the noise variance and can be extracted out as the minimum eigenvalue of $\mathbf{R}_{xx}$. Having done so the signal part of the autocorrelation matrix can be determined and using the signal eigenvectors, the frequencies of the signal are extracted. This algorithm is particularly appropriate to extraction of the frequencies of the music signal in the presence of noise. Typically, owing to the nonstationary character of music, the autocorrelation function must be calculated over different time slots using a window and the MUSIC algorithm applied to each slot. Specifically, suppose $\{x(t) : t \in [0, T]\}$ is the signal.

5. Applications of the theory of stochastic calculus to the analysis of music. Part of this research is presently being conducted by one of my Ph.D students, Rajeev. This involves modeling the motion of the vibrating string and vibrating membrane with random forces using stochastic differential equations. A piece of music is generated when the string is subject to a definite deterministic pattern of forcing. The same is true for a vibrating membrane, like the Tabla. Such partial differential equations can be cast in the form

$$\Delta u(t, \mathbf{x}) - c^{-2}u_{tt}(t, \mathbf{x}) = f(t, \mathbf{x})$$

where the spatial variable $\mathbf{x}$ is two dimensional if it is a vibrating membrane, and is one dimensional if it is a vibrating string. This equation can be solved using Fourier series and the harmonic structure of the musical notes determined. In typical applications, over a finite period of time, we can treat the forcing term f as a sum of impulses at different space-time points plus noise and analyze the statistical properties of the displacement $u(t, \mathbf{x})$. Stochastic calculus can be used to analyze the statistics of the signals produced by a vibrating string even when nonlinearities in the dynamics are taken into account. For example, if the displacement of the string is large, then we have to model the motion using nonlinear corrections to the wave equation. Such an equation on discretization gives rise to a sequence of nonlinear coupled stochastic differential equations expressible in the general Ito format:

$$dx_i(t) = \mu_i(x_1(t), ..., x_n(t), t)dt + \sum_{j=1}^{d} \sigma_{ij}(x_1(t), ..., x_n(t), t)dB_j(t), i = 1, 2, ..., n$$

Here if $u(t, \xi)$ is the displacement of the string at a distance ξ from one endpoint, then $x_i(t) = u(t, iL/N)$, $i = 0, 1, ..., N-1$ represent the positions obtained after discretizing the motion with respect to the spatial argument. When the music is processed by a signal processing algorithm, then such a processing amounts to applying a nonlinear operation on the discretized variables $\{x_1(t), ..., x_n(t)\}$. If such a processing corresponds to a nonlinear operation then we can represent the output of the algorithm by an equation of the form

$$y(t) = \phi(x_1(t), ..., x_n(t))$$

We can analyze the mean and fluctuations of the process $\{y(t)\}$ using the Ito differential rule:

$$\frac{d}{dt}\mathbb{E}y(t) = \mathbb{E}[\nabla\phi(\mathbf{x}(t))^T\mu(\mathbf{x}(t))] + \frac{1}{2}Tr[\mathbb{E}\sigma(\mathbf{x}(t))\sigma(\mathbf{x}(t))^T\nabla\nabla^T\phi(\mathbf{x}(t))]$$

The mean of the process $y(.)$ is a good estimate of the music itself and fluctuations must be minimized for they correspond to noisy disturbances. To reduce fluctuations, we must use nonlinear filtering theory. This involves taking measurements of some other functions of the displacement process $\mathbf{x}(.)$ and solving Kushners' filter for the conditional probability density of the displacement process given the measurements. Another important kind of disturbance that enters into musical tones generated by stringed instruments is spikes. Such noise can be modeled using the Poisson process or the compound Poisson process in which the time difference between two spikes has the exponential distribution. If $N(t), t \geq 0$ is a Poisson process with arrival times $\tau_1 < \tau_2 < ...$ and the spikes have random amplitudes represented by independent random variables $X_1, X_3, ...$, then the spiky noise can be represented by the process

$$Z(t) = \sum_{i=1}^{N(t)} X_i$$

Such a process satisfies the generalized Ito rule:

$$df(Z(t)) = [f(Z(t) + X_{N(t)+1}) - f(Z(t))].dN(t)$$

6. Study of music produced by a flute. The governing equations are obtained by considering pressure waves inside a tube. These are longitudinal waves. Assume that the length of the tube is L. Consider the infinitesimal element $[x, x+dx]$ of air inside the pipe at time $t = 0$. At time t, the point x gets displaced by a small amount $\xi(t, x)$, so that the column of air now occupies the length $[x + \xi, x + dx + d\xi] = [x + \xi, x + dx(1 + \frac{\partial\xi}{\partial x})]$. The density of air in this new region ρ is is obtained by applying the mass conservation equation

$$\rho(t, x) = \rho_0 dx/(dx + d\xi) \approx \rho_0(1 - \frac{\partial\xi}{\partial x})$$

The air inside the tube is assumed to be adiabatic, so that the pressure is given by

$$p = p_0(\rho/\rho_0)^\gamma \approx p_0(1 - \gamma\frac{\partial\xi}{\partial x})$$

The equation of motion of this element of air is given by

$$\rho_0 dx\frac{\partial^2\xi}{\partial t^2} = -dx.\frac{\partial p}{\partial x}$$

or

$$\xi_{tt} = c^2\xi_{xx}, c^2 = \gamma p_0/\rho_0$$

A flute is a long pipe of air with holes along its length. Suppose we take a typical example in which the flute extends between the points $x = 0$ and $x = L$ and there is a single open hole at $x = L/2$, ie, midway. The blow is applied at $x = 0$ and we assume that this blow produces a pressure signal $p(t, 0)$. The end $x = L$ is open and hence $p(t, L) = 0$. Also $p(t, L/2) = 0$ since the hole at $x = L/2$ is open. The problem is then to obtain a complete solution to the wave equation for the pressure with thse boundary conditions. This problem can be extended to include a flute with more than one hole.

66. Numerical integration: Consider the Banach space $X = C[0, 1]$ and for $x \in X$, define

$$f(x) = \int_0^1 x(t)dt$$

f is a bounded linear functional on X, ie, $f \in X^*$. Choose points $0 = t_0 < t_1 < ... < t_n = 1$ and $\alpha(k), k = 0, 1, ..., n$ and define

$$g(x) = \sum_{k=0}^{n} \alpha(k)x(t_k)$$

Then $g \in X^*$. We want to approximate f by g. This is achieved by choosing the coefficients $\{\alpha(k)\}$ such that $g(x) = f(x)$ for all polynomials x having degree $\leq n$. This means that

$$f(t^m) = g(t^m), m = 0, 1, ..., n$$

or

$$\frac{1}{m+1} = \sum_{k=0}^{n} \alpha(k)t_k^m, m = 0, 1, ..., n$$

This is a system of $n + 1$ linear equations for the $n + 1$ unknowns $\alpha(k), k = 0, 1, ..., n$ and can be solved readily by MATLAB.

(Taken from Kreyszig, Introductory Functional Analysis).

67. Let $\mathcal{H}$ be an inner product space and let T be a Hermitian operator in $\mathcal{H}$. Then

$$\| T \| = sup(\| Tx \| : \| x \| \leq 1) = sup(| < Tx, x > | : \| x \| = 1)$$

Note that $< Tx, x >$ is real so the modulus sign on the inner product on the right can be removed.

Proof: Let

$$K = sup(| < Tx, x > | : \| x \| = 1)$$

Then the Schwarz inequality implies

$$K \leq \| T \|$$

Choose z such that $\| z \| = 1$ but $Tz \neq 0$ and define $v = \| Tz \|^{1/2} z$ and $w = \| Tz \|^{-1/2} Tz$. Then

$$\| v \| = \| w \| = \| Tz \|^{1/2}$$

Define

$$y_1 = v + w, y_2 = v - w$$

Then,

$$< Ty_1, y_1 > - < Ty_2, y_2 > = < Tv, v > + < Tv, w > + < Tw, v > + < Tw, w >$$

$$-(< Tv, v > - < Tv, w > - < Tw, v > + < Tw, w >)$$
$$= 2(< Tv, w > + < Tw, v >) = 2(< Tz, Tz > + << T^2 z, z >) = 4 \parallel Tz \parallel^2$$

Also if $y \neq 0$, then

$$< T(y/ \parallel y \parallel), y/ \parallel y \parallel > \leq K$$

so that

$$< Ty, y > \leq K \parallel y \parallel^2$$

It follows that

$$| < Ty_1, y_1 > - < Ty_2, y_2 > | \leq | < Ty_1, y_1 > | + | < Ty_2, y_2 > | \leq K(\parallel y_1 \parallel^2 + \parallel y_2 \parallel^2)$$
$$= 2K(\parallel v \parallel^2 + \parallel w \parallel^2) = 4K \parallel Tz \parallel$$

Combining this with the previous identity gives $\parallel Tz \parallel \leq K$, thereby completing the proof.

68. Review of a paper on control theory, given by Dr.Tarun

The paper starts by writing down the differential equation of motion for a rigid spacecraft. The dynamical variables are (a) the Gibbs vector that defines the altitude of the craft and (b) the angular velocity vector. The driving input process that appears in these differential equations is the control torque. Apart from this input, there is an external disturbing torque. that produces erraticity in the motion of the craft. It would be nice if the authors could present in their paper a brief derivation of these equations of motion starting from the general principles of classical mechanics explaining how the structure of the matrices $T(\xi)$ and H appears. They may for this purpose make use of the basic equations for describing the motion of a body in a rotating frame in terms of the coriolis and centrifugal forces. The aim of the paper as stated at the beginning is to design a controlling input torque $u(t)$ subject to certain constraints such that after a sufficiently long time, the dynamical variables converge to certain desired functions of time, ie, asymptotic tracking is achieved. In order to derive the tracking law, the authors first introduce a function $s(t)$ which is a weighted sum of errors in the angular velocity and in the Gibbs vector. The control input is decomposed into the sum of two components. One component is designed to preserve the nominal motion and the second component is designed to overcome erraticity in the dynamics. The erraticity in the motion results in a term δ in the equations of motion of the error. The authors assume that this δ is bounded by certain non-negative scalar functions. The authors then display two feedback control laws for the second input component. This feedback guarantees that the error does not increase greatly. Perhaps it would do good if the authors mention the word feedback. If the error s is large, the feedback is negative and vice versa. The authors observe that the two control laws displayed as given in the existing literature are not smooth functions and hence have drawbacks like excitation of high frequency dynamics. The authors suggest an improved control law and analyze its performance. The performance analysis carried out by the authors is a neat piece of applied mathematics. Certain bounds on the perturbed moment of inertia matrix are assumed in the course of the performance analysis. These need some explanation To derive the performance, the authors introduce a region of phase space in which the error between the true and the desired trajectories decays exponentially with time and then they prove that under mild conditions, the state will enter this region in finite time. The error, as mentioned, has two components, one for the angular velocity and two, for the Gibbs vector. The authors then prove another theorem which states that the error in the Gibbs vector alone also converges to zero. The paper is well written and may be published after incorporating the suggestions made.

69. Wavelet based estimation of Volterra kernels

Sudipta's thesis work: Assume that the input output equation in continuous time is given by

$$y(t) = \int h_1(u)x(t-u)du + \int h_2(u,v)x(t-u)x(t-v)dudv$$

We wish to estimate the first and second order Volterra kernels $h_1(.)$ and $h_2(.,.)$ from input-output data. Assume that the kernels h_1 and h_2 are respectively of duration $[0,T]$ and $[0,T] \times [0,T]$ and that the maximum frequency of h_1 is f_1 and the maximum frequency of h_2 along both the two directions is f_2. We expand the kernels in terms of a wavelet basis. Specifically, let ψ be the mother wavelet function that is concentrated over the band $[a,b]$. Then set $\psi_{n,k}(t) = 2^{n/2}\psi(2^n t - k)$ with $n,k \in \mathbb{Z}$. The wavelet function $\psi_{n,k}$ is concentrated over all t for which $2^n t - k \in [a,b]$ or equivalently, over all t such that $t \in [(k+a)/2^n, (k+b)/2^n]$. The duration of this wavelet function is given by $2^{-n}(b-a)$ and this must be set to $1/f_1$. Thus, the maximum value of the resolution n_1 for the first order component of the impulse response is given by $2^{-n_1}(b-a) = 1/f_1$, or equivalently,

$$n_1 = log_2(f_1(b-a))$$

The minimum value of the resolution n is n_2 and may be chosen so that ψ_{nk} is focussed over an interval of length T. Thus, $(b-a)/2^{n_2} = T$, or equivalently,

$$n_2 = log_2((b-a)/T)$$

For a given value of the resolution n, the translation variable k must vary in such a way so that $\psi_{n,k}$ covers the entire range $[0,T]$. This means that if $k_1(n) \le k \le k_2(n)$, then we must have

$$(k_1+a)/2^n = 0, (k_2+b)/2^n = T$$

or

$$k_1 = -a, k_2 = 2^n T - b$$

Thus, we must seek an expansion of $h_1(.)$ in the form

$$h_1(t) \approx \sum_{n_1 \le n \le n_2, k_1(n) \le k \le k_2(n)} c(n,k)\psi_{n,k}(t)$$

We denote this range of the ordered doublet (n,k) by D. In other words,

$$h_1(t) \approx \sum_{(n,k)\in D} c(n,k)\psi_{n,k}(t)$$

Likewise, we choose a domain E for the resolution and translation indices for expanding the second order Volterra kernel:

$$h_2(u,v) \approx \sum_{(n,k,p,q)\in E} d(n,k,p,q)\psi_{n,k}(u)\psi_{p,q}(v)$$

Specifically, n, p each varies over $[log_2((b-a)/T), log_2(f_2(b-a))]$ and for a given n, p, k, q respectively vary over $[-a, k_2(n)]$ and $[-a, k_2(p)]$. This domain, we denote by E. The output data can then be represented as

$$y(t) = \sum_{(n,k)\in D} c(n,k) \int x(t-u)\psi_{n,k}(u)du + \sum_{(n,k,p,q)\in E} d(n,k,p,q)\psi_{n,k}(u)\psi_{p,q}(v)x(t-u)x(t-v)dudv$$

Assume that $y(t)$ has a maximum frequency f_3 and a duration T', so that in accord with the above method, we can choose a maximum and minimum resolution index n_3, n_4 and a corresponding translation variable range. Specifically,

$$n_4 = log_2((b-a)/T), n_3 = log_2(f_3(b-a))$$

and for a given resolution index n, the translation index k varies in the range $[-a, 2^n T' - b]$. We denote this range of indices by F. Thus, the wavelet coefficients of $\{y(t)\}$ are given by

$$f(r,s) = \int y(t)\psi_{r,s}(t)dt = \sum_{(n,k)\in D} A(r,s|n,k)c(n,k) + \sum_{(n,k,p,q)\in E} B(r,s|n,k,p,q)d(n,k,p,q), (r,s) \in F$$

where

$$A(r,s|n,k) = \int x(t-u)\psi_{r,s}(t)\psi_{n,k}(u)dtdu, B(r,s|n,k,p,q) = \int x(t-u)x(t-v)\psi_{n,k}(u)\psi_{p,q}(v)\psi_{r,s}(t)dudvdt$$

This system of equations can be solved for the wavelet coefficients c, d of the two Volterra kernels provided the number of points in the set F is larger than the sum of the number of points in the sets D and E. For any finite set A, let $\mu(A)$ denote the number of elements in A. Then the condition for solvability is that

$$\mu(F) \geq \mu(D) + \mu(E)$$

We order the elements of D, E, F and construct matrices $\mathbf{A}, \mathbf{B}$ respectively of sizes $\mu(D) \times \mu(D)$ and $\mu(E) \times \mu(E)$. Then the above set of linear equations can be expressed in vectorial form as

$$\mathbf{f} = \mathbf{Ac} + \mathbf{Bd} = [\mathbf{A}|\mathbf{B}] \begin{pmatrix} \mathbf{c} \\ \mathbf{d} \end{pmatrix}$$

Suppose that the kernels h_1 and h_2 are known except for some parameters θ. We write $h_1(t,\theta)$ and $h_2(u,v,\theta)$. The problem is to estimate these parameters. Suppose $\mu(F) >> \mu(D) + \mu(E)$. Then, we can use a least squares matching procedure combined with the gradient algorithm to estimate θ. This is achieved by defining the energy function

$$E(\theta) = \| \mathbf{f} - \mathbf{Ac}(\theta) - \mathbf{Bd}(\theta) \|^2$$

where $\mathbf{c}(\theta)$ and $\mathbf{d}(\theta)$ are respectively obtained by arranging $c(n,k,\theta)$ and $d(n,k,p,q,\theta)$ as vectors with

$$c(n,k,\theta) = \int h_1(u,\theta)\psi_{n,k}(u)du, d(n,k,p,q,\theta) = \int h_2(u,v,\theta)\psi_{n,k}(u)\psi_{p,q}(v)dudv$$

Equivalently,

$$E(\theta) = \sum_{(r,s)\in F} (f(r,s) - \sum_{(n,k)\in D} A(r,s|n,k)c(n,k,\theta) - \sum_{(n,k,p,q)\in E} B(r,s|n,k,p,q)d(n,k,p,q,\theta))^2$$

To minimize this function, we may use the gradient algorithm:

$$\theta[k+1] = \theta[k] - \mu\nabla_\theta E(\theta[k]), k = 0, 1, 2, ...$$

Note that the optimal equations are

$$\nabla E(\theta) = 0$$

which gives

$$\sum_{(r,s)\in F} (f(r,s) - \sum_{(n,k)\in D} A(r,s|n,k)c(n,k,\theta) - \sum B(r,s|n,k,p,q)d(n,k,p,q,\theta))$$

$$\times(\sum_{(n',k')\in D} A(r,s|n',k')\nabla_\theta c(n',k',\theta) - \sum_{(n',k',p',q')\in E} B(r,s|n',k',p',q')\nabla_\theta d(n',k',p',q',\theta)) = 0$$

This expression is complicated and not amenable to closed form solutions.

Comparison with a direct least squares based identification: If we were to use directly the least squares method for determining the filter kernels h_1, h_2, we would have to replace the integrals by sums over samples and likewise the integral of the squared error by the sum of the squared errors of the samples and this approximation could lead to serious errors. The wavelet based method inherently accounts for the resolution and localization of the kernels and hence the sum of error squares over the wavelet coefficients reflects better the exact error energy of the signals. In other words, in the direct least squares approach, we are replacing $\int_0^T e^2(t)dt$ by $\sum_{n=0}^{N-1} e^2(n\Delta)$ while in the wavelet method we are replacing it by $\sum_{k=1}^p |< e, \phi_k>|^2$ where $\phi_k, k = 1, 2, ..., p$ is a truncated orthonormal basis for the L^2 space and

$$< e, \phi_k >= \int_0^T e(t)\phi_k(t)dt$$

When the basis functions ϕ_k are chosen appropriately (wavelet basis in accord with resolution and duration), then the latter can give even for small values of p a very good approximation to $\int e^2 dt$ that what the former gives with even large values of N. Thus, the wavelet method is superior in performance.

70.**Compact operators**: Let $\mathcal{H}$ be a Hilbert space and let $T_n, n = 1, 2, ...$ be a sequence of compact operators on $\mathcal{H}$ and let T be a bounded linear operator in $\mathcal{H}$ such that

$$\| T_n - T \|\to 0, n \to \infty$$

Then, T is compact. For let $(v_m : m = 1, 2, ...)$ be a sequence in $\mathcal{H}$ with $\| v_m \|= 1$ for all m. We want to show that $Tv_m, m = 1, 2, ...$ has a convergent subsequence. Choose a subsequence $(s_1(m))$ of (m) such that $T_1 v_{s_1(m)}$ converges. Then choose a subsequence $(s_2(m))$ of $(s_1(m))$ such that $T_2 v_{s_2(m)}$ converges. In general, $(s_i(m))$ is a subsequence of $s_{i-1}(m)$ with the property $T_i v_{s_i(m)}$ converges. Define $w(m) = s_m(m)$. We shall show that $Tv_{w(m)}$ is Cauchy. Let $n \geq m$. Then, $w(n) = s_n(n) = s_m(k)$ for some $k \geq n$. We have

$$\| T(v_{w(n)}) - T(v_{w(m)}) \|\leq$$

$$\leq\| T(v_{w(n)}) - T_m(v_{w(n)}) \| + \| T_m(v_{w(n)}) - T_m(v_{w(m)}) \| + \| T_m(v_{w(m)}) - T(v_{w(m)}) \|$$

$$\leq 2 \| T - T_m \| + \| T_m(v_{w(n)}) - T_m(v_{w(m)}) \|$$

$$= 2 \parallel T - T_m \parallel + \parallel T_m(v_{s_m(k)}) - T_m(v_{s_m(m)}) \parallel$$

From this, the result readily follows.

71. Quantum probability

1.Let $v_1, ..., v_n$ be vectors in $\mathcal{C}^N$. Write a program to compute

$$v_1 \otimes ... \otimes v_n \in \mathbb{C}^{nN}$$

Use the definition of the Kronecker tensor product. Write a program to compute the vector in the symmetric tensor product $\mathbb{C}^{N \otimes_s n}$ obtained by projecting $v_1 \otimes ... \otimes v_n$ onto this space, ie,

$$S(v_1 \otimes ... \otimes v_n) = \frac{1}{n!} \sum_{\sigma \in S_n} v_{\sigma 1} \otimes ... \otimes v_{\sigma n}$$

Take a basis $\{e_1, ..., e_N\}$ for $\mathbb{C}^N$ and form the vectors

$$e_1^{\otimes h_1} \otimes e_2^{\otimes h_2} \otimes ... \otimes e_N^{\otimes h_N}$$

where for any $v \in \mathbb{C}^N$, we have

$$v^{\otimes k} = v \otimes ... \otimes v (kfoldtensorproduct)$$

For $h_i \geq 0$ and $\sum_{i=1}^N h_i = n$, define

$$e_1^{h_1} ... e_N^{h_N} = S(e_1^{\otimes h_1} \otimes ... \otimes e_N^{h_N})$$

Then these vectors as the ordered N-tuple $(h_1, ..., h_N)$ ranges over all non-negative numbers with the constraint $\sum h_i = n$ form an orthogonal basis for the n-fold symmetric tensor product of $\mathbb{C}^N$ with itself. We shall compute the length of such a vector. It is easy to see that

$$< \Pi_{i=1}^N e_i^{h_i}, \Pi_{i=1}^N e_i^{h_i} >$$

$$= \frac{1}{(n!)^2} (\frac{n!}{h_1! ... h_N!})(h_1! ... h_N!)^2 = \frac{h_1! ... h_N!}{n!}$$

from which we deduce that the vectors

$$e(h_1, ..., h_n) = (\frac{n!}{h_1! ... h_N!})^{1/2} \Pi_{i=1}^N e_i^{h_i}$$

as $(h_1, ..., h_N)$ ranges over all the ordered partitions of n forms an orthonormal basis for the symmetric space $S((\mathbb{C}^N)^{\otimes n})$. Write a program to compute the matrix of the symmetrizing operator S relative to the standard basis for $\mathbb{C}^{Nn}$ and use this formula to compute and tabulate the basis vectors $e(h_1, ..., h_N)$ for $C^{N \otimes_s n}$. Check orthonormality of this basis. Verify from your formula for S that $S^2 = S$. The number of ways (ordered partition) in which n is expressible as the sum of N non-negative integers $h_1, ..., h_N$ is given by the coefficient of n in the expression

$$(1 + x + x^2 + ...)^N = (1 - x)^{-N}$$

This coefficient is clearly

$$\binom{-N}{n}(-1)^n = \binom{N+n-1}{n}$$

Verify that this dimension is the same as $Tr(S)$.

2.Quantum Markov process (MATLAB simulation)(Simulation of a star unital homomorphism): Let $\mathcal{H}_0$ be a Hilbert space and $\mathcal{H}$ another Hilbert space. Let $\theta : \mathcal{B}(H_0) \to \mathcal{B}(H_0) \otimes \mathcal{B}(H)$ be an algebra homomorphism that preserves the star operation and $\theta(1) = 1$. Choose an orthonormal basis $(e_0, ..., e_{d-1})$ for $\mathcal{H}$ and define $\theta_{ij} : \mathcal{B}(H_0) \to \mathcal{B}(H_0)$ by the equation

$$\theta_{ij}(X) = \mathbb{E}_{|e_j><e_i|}(\theta(X))$$

This notation means that for any $u, v \in \mathcal{B}(H_0)$, we have

$$< u, \theta_{ij}(X)v > = < u \otimes e_i, \theta(X)(v \otimes e_j) >, i, j = 0, 1, ..., d-1$$

Then we have

$$< u, \theta_{ij}(1)v > = < u, \mathbb{E}_{|e_j><e_i|}(\theta(1))v > = < u \otimes e_i, \theta(1)(v \otimes e_j) >$$

$$= < u \otimes e_i, v \otimes e_j > = < u, v > \delta_{ij}$$

so that

$$\theta_{ij} = \delta_{ij}.I$$

Also

$$< u, \theta_{ij}(X^*)v > = < u \otimes e_i, \theta(X^*)(e_j \otimes v) >$$

$$= < u \otimes e_i, \theta(X)^*(e_j \otimes v) > = < \theta(X)(u \otimes e_i), e_j \otimes v >$$

$$= < e_j \otimes v, \theta(X)(u \otimes e_i) >^* = < v, \theta_{ji}(X)u >^* = < \theta_{ji}(X)u, v > = < u, \theta_{ji}(X)^* v >$$

so that

$$\theta_{ij}(X^*) = \theta_{ji}(X)^*$$

Finally, let $\{u_r\}$ be an orthonormal basis for H_0. Then,

$$< u, \theta_{ij}(XY)v > = < u \otimes e_i, \theta(XY)(v \otimes e_j) >$$

$$= < u \otimes e_i, \theta(X)\theta(Y)(v \otimes e_j) > = \sum_{r,s} < u \otimes e_i, \theta(X)(u_r \otimes e_s) >< (u_r \otimes e_s), \theta(Y)(v \otimes e_j) >$$

$$= \sum_{r,s} < u, \theta_{is}(X)u_r >< u_r, \theta_{sj}(Y)v > = \sum_s < u, \theta_{is}(X)\theta_{sj}(Y)v >$$

$$= < u, \sum_s \theta_{is}(X)\theta_{sj}(Y)v >$$

or equivalently,

$$\theta_{ij}(XY) = \sum_s \theta_{is}(X)\theta_{sj}(Y)$$

Example:Let (S, μ) be a measure space and let $\phi_j : S \to S, j = 0, 1, 2, ..., d-1$ be measurable maps and let $p_j : S \to [0, 1], j = 0, 1, ..., d-1$ be measurable functions such that $p_j(x) \geq 0$ and $\sum_{j=0}^{d-1} p_j(x) = 1$. Let $H_0 = L^2(S, \mu)$ and $H = \mathbb{C}^d$. Define the map $T : \mathcal{B}(H_0) \to \mathcal{B}(H_0)$ so that

$$Tf(x) = \sum_{j=0}^{d-1} p_j(x) f(\phi_j(x))$$

The physical interpretation of T is as follows. Let $X_n, n = 0, 1, 2, ...$ be a Markov chain with state space S such that conditioned on the event $X_n = x$, we have $X_{n+1} = \phi_j(x)$ with probability $p_j(x)$. Thus,

$$Pr(X_{n+1} = \phi_j(x) | X_n = x) = p_j(x), j = 0, 1, ..., d-1, x \in S$$

Then for any $f \in L^2(\mu)$, we have

$$\mathbb{E}[f(X_{n+1}) | X_n = x] = Tf(x)$$

Define $\mathcal{B}(H_0)$ consists of complex valued functions on H_0, $H_0 \otimes H$ consists of $\mathbb{C}^d$ valued functions on S. $\mathcal{B}(H_0) \otimes \mathcal{B}(H)$ consists of $d \times d$ complex matrix valued functions on S Define $\theta : \mathcal{B}(H_0) \to \mathcal{B}(H_0) \otimes \mathcal{B}(H)$ by

$$\theta(f) = U.diag[fo\phi_0, ..., fo\phi_{d-1}]U^*$$

where U is a unitary $d \times d$ matrix valued function on S. Then, θ is a star unital homomorphism. Indeed, we have for two functions f, g on S,

$$\theta(f)\theta(g) = \theta(fg)$$

We have

$$\theta_{ij}(f) = \sum_r u_{ir} \bar{u}_{jr} fo\phi_r$$

We can arrange the matrix U so that

$$|u_{0r}(x)|^2 = p_r(x)$$

In that case, we get the equation

$$\theta_{00}(f) = T(f)$$

the transition generator of the Markov chain. This idea provides the means for realizing a classical Markov process via quantum Markov processes, ie, via, star unital homomorphisms.

Note: Let $\mathcal{H}_n, n = 0, 1, 2, ...$ be a sequence of Hilbert spaces. For each $n = 0, 1, 2, ...$, define

$$\mathcal{H}_{n]} = \mathcal{H}_0 \otimes ... \otimes \mathcal{H}_n, n = 0, 1, 2, ...$$

and also define

$$\mathcal{H}_{[n} = \mathcal{H}_n \otimes \mathcal{H}_{n+1} \otimes ...$$

with respect to a stabilizing sequence $\phi_n \in H_n, n = 0, 1, 2, ...$ This means that every vector in $\mathcal{H}_{[n}$ has the form $u \otimes \phi_k \otimes \phi_{k+1} \otimes ...$ for some $k > n$ and

$$u \in \mathcal{H}_n \otimes ... \otimes H_k$$

Let $\mathcal{B}_{n]}$ denote the Von-Neumann algebra of all bounded operators in $\mathcal{H}_{n]}$ and $\mathcal{B}_{[n}$ the Von-Neumann algebra of all bounded operators in $\mathcal{H}_{[n}$. Note that we can identify the elements of $\mathcal{B}_{n]}$ with elements of the form $B \otimes I_{[n+1}$ where B is a bounded operator in $\mathcal{H}_{n]}$. Let $\mathcal{H} = \mathcal{H}_0 \otimes \mathcal{H}_1 ...$ with respect to the given stabilizing sequence. This means that

$$\mathcal{H} = \mathcal{H}_{n]} \otimes \mathcal{H}_{[n+1}, n = 0, 1, 2, ...$$

Let $\mathcal{B} = \mathcal{B}(H)$, ie, the algebra of all bounded operators in H. For $B \in \mathcal{B}$, we then define $\mathcal{E}_{n]}B \in \mathcal{B}_{n]}$ by the formula that if $u, v \in \mathcal{H}_{n]}$, then

$$< u, \mathbb{E}_{n]}(B)v >=< u \otimes \phi_{[n+1}, B(v \otimes \phi_{[n+1}) >$$

where

$$\phi_{[n+1} = \phi_{n+1} \otimes \phi_{n+2} \otimes ...$$

Remark: $\mathcal{H}_{n]}, n = 0, 1, 2, ...$ in the quantum theory plays the role of a filtration in discrete time while the Hermitian operators in $\mathcal{B}_{n]}$ play the role of random variables measurable with respect to $\mathcal{F}_n$, the filtration at time n.

72. Problem:(Ph.D thesis problem of Dr.Rajbir Yaduvanshi working under guidance of Dr.Dhananjay and H.Parthasarathy) Implementation of Largrangian and Hamiltonian systems on DSP processors:Assume that the Lagrangian has a kinetic energy term quadratic in the velocities, ie,

$$T(q_1, ..., q_n, q_1', ..., q_n') = \frac{1}{2}\sum_{i,j=1}^{n} a_{ij}(q_1, ..., q_n)q_i'q_j'$$

The Lagrangian for the system of particles moving under the potential energy $U(q_1, ..., q_n)$ is then given by

$$\mathcal{L}(q_1, ..., q_n, q_1', ..., q_n') = T(q, q') - U(q)$$

and the Euler-Lagrange equations are

$$\frac{d}{dt}\frac{\partial T(q, q')}{\partial q_i'} = \frac{\partial T(q, q')}{\partial q_i} - \frac{\partial U(q)}{\partial q_i}$$

We compute

$$\frac{\partial T}{\partial q_i'} = \sum_j a_{ij}(q)q_j'$$

and

$$\frac{d}{dt}\frac{\partial T}{\partial q_i'} = \sum_{j,k} a_{ij,k}(q)q_k'q_j' + \sum_j a_{ij}(q)q_j''$$

$$\frac{\partial T}{\partial q_i} = \sum_{k,m} a_{km,i}(q)q_k'q_m'$$

and hence the equations of motion can be written down from here:

$$\sum_{j,k} a_{ij,k}(q)q'_k q'_j + \sum_j a_{ij}(q)q''_j = \sum_{k,m} a_{km,i}(q)q'_k q'_m + U_{,i}$$

These equations must be implemented after discretization. The Hamiltonian can be computed from here: The canonical momenta are

$$p_i = \frac{\partial \mathcal{L}}{\partial q'_i} = \sum_j a_{ij}(q)q'_j$$

or equivalently in vectorial notation,

$$\mathbf{p} = \mathbf{A}(\mathbf{q})\mathbf{q}'$$

The Lagrangian is

$$\mathcal{L} = \frac{1}{2}\mathbf{q}'^T\mathbf{A}(\mathbf{q})\mathbf{q}' - U(\mathbf{q})$$

so the Hamiltonian is

$$H(\mathbf{q},\mathbf{p}) = (\mathbf{p},\mathbf{q}') - \mathcal{L} = \frac{1}{2}\mathbf{q}'T\mathbf{A}(\mathbf{q})\mathbf{q}' + U(\mathbf{q})$$

$$= \frac{1}{2}\mathbf{p}^T\mathbf{A}(\mathbf{q})^{-1}\mathbf{p} + U(\mathbf{q})$$

Define the matrix

$$\mathbf{B}(\mathbf{q}) = \mathbf{A}(\mathbf{q})^{-1}$$

Then the Hamiltonian can be expressed as

$$H(\mathbf{q},\mathbf{p}) = \frac{1}{2}\mathbf{p}^T\mathbf{B}(\mathbf{q})\mathbf{p} + U(\mathbf{q})$$

We have

$$\mathbf{q}' = \nabla_{\mathbf{p}}H = \mathbf{B}(\mathbf{q})\mathbf{p},$$

$$\mathbf{p}' = -\nabla_{\mathbf{q}}H = -\frac{1}{2}\mathbf{p}^T\nabla_{\mathbf{q}}\mathbf{B}(\mathbf{q})\mathbf{p} - \nabla_{\mathbf{q}}U(\mathbf{q})$$

where the ith component of the vector $\mathbf{p}^T\nabla_{\mathbf{q}}\mathbf{B}(\mathbf{q})\mathbf{p}$ is defined by

$$\sum_{k,m} b_{km,i}(\mathbf{q})p_k p_m$$

The DSP implmentation of this system involves discretization once again.

73. Least mean phase algorithm in continuous time for continuous time processes.

$$d(t) = W^T Z(t) + n(t), \hat{d}(t) = W(t)^T Z(t)$$

where $W, W(t), Z(t)$ are complex vectors. The LMP algorithm reads

$$dW(t) = -\mu.dt.\nabla_{W(t)}(Arg(d(t)) - Arg(\hat{d}(t)))^2$$

$$= 2\mu.dt.(\nabla_{W(t)} Arg(\hat{d}(t))).(Arg(d(t)) - Arg(\hat{d}(t)))$$

$$Arg(d(t)) = Imlog(d(t)) = Imlog(W^T Z(t) + n(t)) \approx Im[log(W^T Z(t)) + n(t)/(W^T Z(t))]$$
$$= Arg(W^T Z(t)) + Im(n(t)/(W^T Z(t)))$$

Write

$$Z(t) = Z_1(t) + iZ_2(t), W = W_1 + iW_2$$

so that

$$W^T Z(t) = W_1^T Z_1(t) - W_2^T Z_2(t) + i(W_1^T Z_2(t) + W_2^T Z_1(t))$$

It follows that

$$Arg(W^T Z(t)) = tan^{-1} \frac{(W_1^T Z_2(t) - W_2^T Z_1(t))}{(W_1^T Z_1(t) - W_2^T Z_2(t))}$$

Also writing

$$n(t) = n_1(t) + in_2(t)$$

we have

$$\frac{n(t)}{W^T Z(t)} = \frac{n_1 + in_2}{A + iB}$$

where

$$A = Re(W^T Z(t)) = W_1^T Z_1 - W_2^T Z_2, B = W_1^T Z_2 + W_2^T Z_1 = Im(W^T Z(t))$$

giving

$$Im(\frac{n(t)}{W^T Z(t)}) = \frac{An_1 - Bn_2}{A^2 + B^2}$$

$$= \frac{(W_1^T Z_1 - W_2^T Z_2)n_1 - (W_1^T Z_2 + W_2^T Z_1)n_2}{|W^T Z|^2}$$

Likewise, we have

$$Arg(W^T(t)Z(t)) = Imlog(W^T(t)Z(t)) = Imlog(W_1(t)^T Z_1(t) - W_2(t)^T Z_2(t)$$
$$+ i(W_1(t)^T Z_2(t) + W_2(t)^T Z_1(t)))$$

so that

$$\nabla_{W(t)} Arg(W(t)^T Z(t)) = \nabla_{W_1(t)} Arg(W(t)^T Z(t)) + i\nabla_{W_2(t)} Arg(W(t)^T Z(t))$$

where

$$\nabla_{W_1(t)} Arg(W(t)^T Z(t)) = Im((Z_1(t) + iZ_2(t))/W(t)^T Z(t)) = Im(Z(t)/W(t)^T Z(t))$$

$$\nabla_{W_2(t)} Arg(W(t)^T Z(t)) = Im\frac{(-Z_2(t) + iZ_1(t))}{W(t)^T Z(t)} = Re(Z(t)/W(t)^T Z(t))$$

This gives us

$$\nabla_{W(t)} Arg(W(t)^T Z(t)) = Im(Z(t)/W(t)^T Z(t)) + i.Re(Z(t)/W(t)^T Z(t)) = i.(Z(t)/W(t)^T Z(t))^*$$

Thus,

$$dW(t) = 2\mu.dt.i(Z(t)/W(t)^T Z(t))^*(Arg(W^T Z(t)) - Arg(W(t)^T Z(t)) + (An_1(t) - Bn_2(t))/|W^T Z|^2)$$

or in full expanded form,

$$dW(t) = 2\mu.dt.i(Z(t)/W(t)^T Z(t))^*(Arg(W^T Z(t)) - Arg(W(t)^T Z(t))$$

$$+|W^Z(t)|^{-2}(Re(W^T Z(t))n_1(t) - Im(W^T Z(t))n_2(t)))$$

74.Concluding remarks for a paper by Tarun and Myself: The paper starts by developing a differential equation model for the motion of a plane under the influence of noise. If there were no noise, the plane would execute motion with uniform velocity. The presence of noise causes the plane to turn by small random amounts. Equations for the evolution of mean and covariance of the position and velocity variables are derived using approximation techniques. These equations are valid in between observations which are taken at discrete time instants. The evolution of the conditional mean and covariance just after an observation has been made are obtained. In filtering theory terminology, this method of estimating the trajectory of plane is called the continuous-discrete filter. The observations are made on the range and bearing of the plane. The methods used in this paper can be applied to any mechanical problem, for example estimating the trajectory of the Foucault pendulum used for determining the angular velocity of the earth's rotation. Any mechanical system is generally described by Newton's equations of motion in the presence of a potential field and frictional forces. In addition, to incorporate uncertainties in the model like random impacts on the moving object, we include white noise terms resulting thereby in a system of stochastic differential equations describing the motion. Measurements may be taken on any set variables to which we have access, for example, if we take a movie photograph of the object from one angle, we are then measuring a projection of the trajectory (position and velocity) on a plane, for example, one component of the particle displacement. Or we may be able to measure the energy of the object which is a nonlinear function of its position and displacement. The point is that measurements are usually discrete and then estimating a continuous trajectory from discrete measurements becomes important. Another example where the same techniques can be used is in non-linear circuit theory. For example, suppose we are given an LCR circuit with R a nonlinear function of the current with some unknown parameters. We write $R(i, \theta)$. We can excite this circuit with white Gaussian noise and write down the state variable equations for the current and its derivative. We may then extend the state by writing a sde for the unknown parameter θ with noise incorporated and from noisy measurements on the current through the circuit taken at discrete time instants, use the continuous discrete filter to estimate the parameter. The general philosophy is to write down the dynamical equations for the states of a system having unknown parameters, and then write down an equation for the parameter variation. In case the parameters are constant, the dynamical equation for the parameter variation will involve only noise and the parameters can be estimated by taking discrete measurements on some functions of the system state using the continuous discrete filter.

75. Product topology, subspace topology: Let $(X_i, \tau_i), i = 1, 2$ be topological spaces and let $\mathcal{C}$ be the collection of all subsets of $X_1 \times X_2$ having the form $E_1 \times E_2$ where $E_i \in \tau_i, i = 1, 2$. Then, let $\tau_1 \times \tau_2$ be the topology on $X_1 \times X_2$ generated by $\mathcal{C}$. $(X_1 \times X_2, \tau_1 \times \tau_2)$ is then a topological space called the product of the two topological spaces $(X_i, \tau_i), i = 1, 2$. Let $(X_i, \tau_i), i = 1, 2$ be topological spaces and let $f : X_1 \to X_2$ be a map such that $f^{-1}(\mathcal{E})) \subset \tau_1$ where $\mathcal{E}$ is any collection of subsets of X_2 that generates

τ_2. Then f is continuous. To see this, let

$$\mathcal{D} = \{E \subset X_2 : f^{-1}(E) \in X_1\}$$

Then $\mathcal{D}$ is easily verified to be a topology on X_2 and $\mathcal{E} \subset \mathcal{D}$ by hypothesis. Hence, $\mathcal{D}$ contains τ_1 proving that f is continuous.

76. Sudipta's Ph.D. problem: This problem involves estimating using wavelets the parameters of a simple pendulum. The equation of motion of the pendulum is given by

$$\theta''(t) = -a.sin(\theta(t))$$

where $\theta(t)$ is the deflection angle of the pendulum at time t and $a = g/L$ with g as the gravitational constant and L the length of the pendulum. The problem is to identify the parameter a by giving a noisy input. The approximate equation of motion of the pendulum (for small deflections) is given by

$$\theta''(t) = -a\theta(t) + a\theta^3(t)/6 + w(t)$$

The problem is to calculate the wavelet coefficients of $\theta(t)$ using a second order Volterra approximation to the motion. To this end we write

$$\theta(t) = \theta_0(t) + \theta_1(t)$$

where θ_0 is the deflection obtained if only the linear component of the pendulum is taken into consideration and $\theta_1(t)$ is the correction produced by the nonlinear component θ^3 in the motion. We have

$$\theta_0''(t) = -a\theta_0(t) + w(t),$$

$$\theta_1''(t) = -a\theta_1(t) + a\theta_0^3(t)$$

The first gives on Laplace transforming,

$$(s^2 + a)\hat{\theta}_0(s) = W(s)$$

so that

$$\theta_0(t) = a^{-1/2} \int_0^t sin((t-u)\sqrt{a})w(u)du$$

The second gives on Laplace transforming,

$$(s^2 + a)\hat{\theta}_1(s) = aL(\theta_0^3)(s)$$

so that

$$\theta_1(t) = a^{1/2} \int_0^t sin((t-u)\sqrt{a})\theta_0^3(u)du$$

$$= a^{-1} \int_{0<u_1,u_2,u_3<u<t} sin((t-u)\sqrt{a})sin((u-u_1)\sqrt{a})sin((u-u_2)\sqrt{a})sin((u-u_3)\sqrt{a})w(u_1)w(u_2)w(u_3)$$

$$du_1du_2du_3du$$

For a given input $w(.)$ we simulate the system and evaluate the wavelet coefficients of the output and then compare these with the theoretical expressions for the wavelet coefficients obtained using the above integral formula. Let $\psi(t)$ denote the mother wavelet function and $\psi_{n,k}(t) = 2^{n/2}\psi(2^n t - k), n, k \in \mathbb{Z}$ the wavelet basis functions. Then the wavelet coefficients of the unperturbed motion θ_0 are

$$c(n,k) = \int \theta_0(t)\psi_{n,k}(t)dt$$

The wavelet coefficients of the perturbation to the angular displacement are

$$d(n,k) = \int \theta_1(t)\psi_{n,k}(t) =$$

$$a^{-1}\int sin((t-u)\sqrt{a})sin((u-u_1)\sqrt{a})sin((u-u_2)\sqrt{a})sin((u-u_3)\sqrt{a})$$

$$\times w(u_1)w(u_2)w(u_3)\psi_{n,k}(t)du_1du_2du_3dudt$$

The wavelet coefficients upto first order correction terms of the total angular displacement of the simple pendulum are

$$e(n,k) = c(n,k) + d(n,k), n, k \in \mathbb{Z}$$

Some algebraic simplifications can be achieved here.

$$sin((u-u_1))\sqrt{a}).sin((u-u_2)\sqrt{a}) = \frac{1}{2}(cos((u_1-u_2)\sqrt{a}) - cos((2u-u_1-u_2)\sqrt{a}))$$

$$sin((u-u_1)\sqrt{a}).sin((u-u_2)\sqrt{a})sin((u-u_3)\sqrt{a})$$

$$= \frac{1}{4}(sin((u+u_1-u_2-u_3)\sqrt{a}) + sin((u-u_1+u_2-u_3)\sqrt{a})$$

$$-sin((3u-u_1-u_2-u_3)\sqrt{a}) - sin((-u-u_1-u_2+u_3)\sqrt{a}))$$

And finally,

$$sin((t-u)\sqrt{a})sin((u-u_1)\sqrt{a})sin((u-u_2)\sqrt{a})sin((u-u_3)\sqrt{a})$$

$$= \frac{1}{8}(sin((t+u_1-u_2-u_3)\sqrt{a}) + sin(t-2u-u_1+u_2+u_3)\sqrt{a})$$

$$+sin((t-u_1+u_2-u_3)\sqrt{a}) + sin((t-2u+u_1-u_2+u_3)\sqrt{a})$$

$$-sin((t+2u-u_1-u_2-u_3)\sqrt{a}) - sin((t-4u+u_1+u_2+u_3)\sqrt{a})$$

$$-sin((t-2u-u_1-u_2+u_3)\sqrt{a}) - sin((t+2u+u_1+u_2-u_3)\sqrt{a}))$$

Steps for identification of the parameter a:

1.Simulate white noise $\{w(t)\}$.

2.Generate the motion $\theta(t)$ of the simple pendulum by discretizing the differential equation.

3.Calculate and store the wavelet coefficients of the output corresponding to appropriate resolution and translation variables.

4.Calculate and store the wavelet coefficients of the theoretically expected result $\theta(t)$ using the above third order Volterra formula. Do this calculation for various choices of the parameter a.

5.Match the wavelet coefficients of the simulated output with the theoretical expressions for different $a's$ and determine the optimum choice of a.

77. Justification for the procurement of items for the DSP laboratory.

1.High speed scanner: Currently, a lot of literature exists in the form of unpublished reprints and books that are out of print in the field of "Group Representations in Signal and Image Processing". The authors who have contributed much to the field of group representations are mainly eminent mathematicians like I.Schur, Burnside, Frobenius, H.Weyl, Harish-Chandra, E.Wigner and more recently, Barry Simon, Zelobenko, Sternberg, Procesi and V.S.Varadarajan. The manuscripts written by these authors contain chiefly mathematical ideas like the construction of irreducible representations of semi-simple Lie algebras that are not accessible to the engineering student owing to the absence of such books in the institute library. For this purpose, the scanner will be a useful tool. We can borrow the books and reprints from well known mathematical institutions like the Indian Statistical Institute, The Tata Institute of Fundamental Research, scan these and store the manuscript on the PC. The engineering student will then be exposed to the mathematical ideas (like Harish-Chandra's use of maximal ideals in the construction of both finite and infinite dimensional irreducible representations of semisimple Lie groups) and they can directly apply it to image processing problems. The current literature available in image processing deals only with the rotation group $SU(2), SO(3)$ and the Lorentz group which can be regarded as $SL(2, \mathbb{C})$. For other semisimple Lie groups, one must take resort to the theory developed by Harish-Chandra and the antique papers of this famous mathematician will be available directly to the student if a scanner is purchased.

2.Lap-top computer: Mechanics is the most ancient of all the physical sciences. Ever since Sir Isaac Newton laid down the fundamental laws of mechanics in the form of the three laws, mechanics has taken rapid advances. After Newton formulated his laws of mechanics and the inverse square law of gravitation and used these two laws and calculus (which he called fluxions) to arrive at Kepler's laws of planetary motion, there have been several developments. After Newton, Joseph Louis Lagrange derived the laws of mechanics from a variational principle according to which the trajectories of particles between two time instants can be obtained by minimizing a cost functional of the path. Lagrange was able to formulate equations of motion for more complex systems like the simple pendulum double pendulum and other systems without having to introduce auxiliary forces like normal reaction, tension in a rod etc. His method involved merely the introduction of generalized coordinates to describe the motion of the system and then write down an expression for the difference between the kinetic and potential energies of the system in terms of these generalized coordinates and by minimizing the integral of this difference over a path, he arrived at the equations of motion. After Lagrange, came Hamilton who taught us that by applying the Legendre transformation to the Lagrangian system, one can describe the equations of motion in terms of first order coupled non-linear differential equations for the positions and momenta of the system. Recently, there has been a surge of interest in the formulation of the equations of mechanics for generalized systems using the tool of differentiable manifolds. The manifold can be any topological space with a differentiable structure. This has led to unexpected discoveries by mathematicians such as Stephen Smale, Rene Thom, V.I.Arnold etc. In short, mechanics has become a big industry for obtaining qualitative characteristics of the motion, like stablity of many particle systems, chaos in dynamical systems and so on. The calculus of differential forms plays a fundamental role in these investigations and we engineers must borrow the relevant research write ups from the mathematicians library. The calculus of forms first of all introduces the Lie derivative of a function defined on the manifold and the directional derivative of function along a given vector field, then it introduces the Lie derivative acting on vector fields, one forms and more generally tensor fields. These tools are eventually used in obtaining integrals of the three body problem. Currently, in the DSP lab at the NSIT, we are interested in studying such qualitative characteristics of the motion and in order to effect a free flow of ideas from all branches of science into mechanics, a lap top computer is required. For example, the research workers in the DSP lab need to interact with mechanical engineers and control theorists who specialize in mechanical problems. A lap top computer is therefore an essential tool. The lap top computer is also needed for the lab staff to keep track of the software purchased, the software condemned and so on.

3.DSP starter kit: In the DSP lab, currently many lab courses are being conducted. These courses are both at the B.E. and M.E.levels. The experiments involve (1) Simulating FIR and IIR filters via assembly language programs. The programs for FIR filters involve convolving two sequences and this can be achieved using a simple "For loop". The programs for IIR filters involve using the For loop to solve recursive difference equations. (2) Writing programs to implement system identification algorithms. The DSP starter kit will also involve writing programs to solve nonlinear differential equations with applications to mechanics. For example, in mechanics, we come across the notion of a Hamiltonian vector field on the cotangent manifold. Thc point on thc original manifold is denoted by $\mathbf{q}$ while the momentum of the system when the position is $\mathbf{q}$ is defined by a linear functional $\mathbf{p}$ on $TM_{\mathbf{q}}$. The equations of motion are 2n first order Hamiltonian differential equations and the solution for n large can be achieved by appropriate programming of the starter kit. This can be used for galactic simulation problems.

4.Pentium duo core pc:This will be used for simulating complex systems of statistical mechanics. Examples include computation of the canonical and grand canonical partition functions of a gas and hence derive the equation of state of the gas taking into account molecular interactions. We also will simulate the Boltzmann kinetic transport equation that is used for studying approach to equilibrium of a gas which has been perturbed slightly. High speed computation is required for simulating systems of statistical mechanics and this pc is appropriate for this purpose.

78.Exploration of the ground of Mars, signal processing techniques: The primary tool available for exploring the ground of mars is electromagnetic waves. Suppose that the ground of mars has like that of the earth, several layers numbered $1, 2, 3, ...N$. We transmit an electromagnetic wave $s(t)$ from the surface of mars and after reflection from the ith layer, the signal received back on the ground has suffered an attenuation $h(i)$ and a time delay $\tau(i)$. The signal received at the surface can then be expressed as $x(t) = \sum_i h(i)s(t - \tau(i))$. Some information about the structure of the layers can be obtained if we can recover accurate estimates of the parameters $\{h(i), \tau(i)\}$. Taking the Fourier transform of the equation for $x(t)$ gives us

$$X(\omega) = \sum_i h(i)exp(-j\omega\tau(i))S(\omega)$$

and hence the transfer function of the ensemble of layers is given by

$$H(\omega) = \frac{X(\omega)}{S(\omega)} = \sum_i h(i)exp(-j\omega\tau(i))$$

If we assume that the signals have been discretized, then the corresponding model for the received signal can be expressed as

$$x(n) = \sum_{kh=1}^{p} (k)s(n \quad k)$$

Assuming that $s(n)$ is white non-Gaussian noise, we can recover the magnitude transfer function

$$|\sum_n h(n)exp(-j\omega n)|^2$$

from a spectral analysis of $\{x(n)\}$, ie by estimating the autocorrelation function of $x(.)$ followed by an FFT. The phase of the transfer function can be recovered by performing a bispectral analysis of $x(n)$, ie, estimate the third order moment sequence followed by a bivariate FFT. The spectrum of the received signal has the form $\sigma^2|H(\omega)|^2$ while the bispectrum has the form $\gamma.H(\omega_1)H(\omega_2)\bar{H}(\omega_1+\omega_2)$. The phase can be recovered from the bispectrum by a simple recursive algorithm. On can generalize this technique by sending an electromagnetic signal distributed over a spatial domain. Assume that the transmitted signal on the surface is $E(x, y, t)$. Let $\tau(i, x, y)$ be the delay suffered by the signal at a layer below the point (x, y) and $h(i, x, y)$ the corresponding attenuation. Then, the received space-time signal has the form

$$F(t, x, y) = \sum_i h(i, x, y)E(x, y, t - \tau(i, x, y))$$

Signal processing can be performed on such signals to receive information about the layer structure distributed under a region of space. The above equation decouples the received signal at different points (x, y). More generally, we can have coupling.

79. Vireshwar's M.Tech disseration synopsis: Suppose $x(n, m)$ is an image obtained after discretization using pixels. Assume that it is blurred by noise. Then the blurred image can be represented as $y(n, m) = x(n, m)+w(n, m)$ where x is the unblurred image, w is the blurring noise and y is the blurred image. The problem of removing the blurring noise can be achieved by passing this two dimensional signal through an FIR filter having impulse response $h(n, m)$. The output is

$$\hat{x}(n, m) = \sum_{i,j} h(i, j)y(n - i, m - j)$$

with $h(i, j)$ chosen so that

$$\mathbb{E}(x(n, m) - \sum_{i,j} h(i, j)y(n - i, m - j))^2$$

is a minimum. The optimization can be carried out by setting to zero the derivative of the above energy with respect to the filter coefficients $\{h(i, j)\}$. The resulting optimal normal equations are

$$\mathbb{E}(x(n, m)y(n - i, m - j)) = \sum_{p,q} h(p, q)\mathbb{E}(y(n - p, m - q)y(n - i, m - j))$$

These equations are solved to obtain the optimal filter. In practice, we will be given N copies of original images $x_i(n,m)$ and their blurred versions $y_i(n,m)$. Then the impulse response of the deblurring filter will be determined by solving the system of linear equations

$$\sum_i x_i(n,m)y_i(n-r,m-s) = \sum_{p,q} h(p,q) \sum_i y_i(n-p,m-q)y_i(n-r,m-s)$$

This method of removing blurs can be extended to situations when there is a distorting linear transformation. In this case, the original image $x(n,m)$ and its distorted cum blurred version $y(n,m)$ are related via an equation of the form

$$y(n,m) = g(n,m) * x(n,m) + w(n,m)$$

where $*$ stands for convolution. In the ideal case of known statistics, the optimal Wiener filter $h(n,m)$ that takes as input $y(n,m)$ and outputs an optimal estimate $\hat{x}(n,m)$ of the unblurred image can be designed easily. In our work, we shall explore the applicability of second order Volterra filters for image deblurring. Such a deblurring filter will have the form

$$\hat{x}(n,m) = \sum h_1(p,q)x(n-p,m-q) + \sum h_2(p_1,q_1,p_2,q_2)x(n-p_1,m-q_1)x(n-p_2,m-q_2)$$

with the kernels h_1, h_2 designed so that

$$\mathbb{E}(x(n,m) - \hat{x}(n,m))^2$$

is a minimum. In practice, the expectation will be replaced by a sum over prototype unblurred and blurred images. The effectiveness of the second order Volterra filters in edge smoothing will also be investigated. We shall also explore the applicability of recursive Volterra filters in image modeling. Specifically, given an image $y(n,m)$, we shall try to determine parameters $h_1(k)$ and $h_2(k,m)$ so that if

$$\epsilon(n,m) = y(n,m) + \sum_{p,q} h_1(p,q)y(n-p,m-q) + \sum_{p_1,q_1,p_2,q_2} h_2(p_1,q_1,p_2,q_2)y(n-p_1,m-q_1)y(n-p_2)y(m-q_2)$$

then $\sum_{n,m} \epsilon^2(n,m)$ is a minimum. We shall then explore the possibility of using such a model to recursively generate the given image. Such a model will provide and efficient and accurate parametrization of the given image.

80. Replies to referees comments for Tarun's paper:Under what conditions can be model the inductance as $L + w(t)$ where L is a constant and $w(t)$ is a correlated random process. First note that by application of Ampere's law, the magnetic flux through the inductance is given by $N\Phi = N^2AI/d$ where d is the length of the inductance, I the current flowing through the coil and N the number of turns. A is the cross sectional area of the inductance. The inductance is therefore $L = N\Phi/I = N^2A/d$. While actually computing the inductance, the parameters N, A, d are subject to random measurement errors $\delta N, \delta A, \delta d$. It follows that the first order error in the inductance computation is given by

$$\delta L = (2NA/d)\delta N + (N^2/d)\delta A - (N^2A/d^2)\delta d$$

This expression gives us a method for computing the statistics of the random fluctuation in the inductance. However, this fluctuation is a static measurement error, it is not a random process. The random process error $w(t)$ arises from the BH curve according to which the magnetic field B follows the magnetic field intensity $H = NI/L$ in accord with a memory scheme, ie, B depends upon the entire history of H. This curve will also have random fluctuations. Writing $B(t) = B(H(s), s \leq t) = B(NI(s)/L, s \leq t)$, it follows that the ideal model for B will be a correlated random process, the correlation arising from the memory of the $B - H$ curve. For example, if the curve has the form

$$B(t) = \int_0^t F(NI(s)/L)ds$$

then it can be modeled by the differential equation

$$dB(t) = F(NI(t)/L)dt$$

and random fluctuations in $I(t)$ will contribute to random fluctuations in $B(t)$. Thus, writing $I(t) = I_0(t) + \delta I(t)$, we find that

$$dB(t)/dt = F(NI_0(t)/L) + F'(NI_0(t)/L)\delta I(t)$$

and this provides the required differential equation model for $B(t)$ and hence for the instantaneous inductance $NB(t)A/d.I_0(t)$ The fact that the $B - H$ curve has memory can also be viewed from the stand-point of magnetic permeability. The magnetic permeability $\mu = B/H$ depends on the entire history of the magnetic field intensity by which the material came to its current state. The same arguments can be given for a capacitor. The dielectric constant $\epsilon = D/E$ of a capacitor depends on the entire history of the electric field. This can be seen from the fact that if we model the electron bound to the atom by a harmonic potential, then the equation of motion for its displacement from the equilibrium is given by

$$mx''(t) + \gamma x'(t) + m\omega_0^2 x(t) = -eE(t)$$

Solving this differential equation for $x(t)$ gives us the displacement $x(t)$ as a function of the entire past of the elecric field, ie, $x(t)$ is a functional of $E(s), s \leq t$. Hence the dipole moment of the atom which is $-ex(t)$ is also a functional of the electric field history and hence so is the polarization of the material (dipole moment per unit volume). It follows that the electric displacement D is also a functional of the history of the electric field and hence so is the permittivity. This suggests that the capacitance $\epsilon A/d$ should have a fluctuation that is modeled as a correlated random process.

Chapter 3

Quantum Mechanics

1.Path integral for the harmonic oscillator; A discrete Fourier transform based approach. The action for the particle in traveling from $(x(0),0)$ to $(x(T),T)$ is given by

$$\mathcal{A}(x) = \int_0^T (Mx'(t)^2/2 - M\omega^2 x(t)^2/2)dt$$

we write

$$x(t) = x_{cl}(t) + \delta x(t)$$

where $x_{cl}(t)$ is the action for the classical trajectory. We have

$$x''_{cl}(t) = -\omega^2 x_{cl}(t)$$

whose solution is given by

$$x_{cl}(t) = A.cos(\omega t) + B.sin(\omega t)$$

The boundary conditions are

$$x_{cl}(0) = x(0) = A, x_{cl}(T) = A.cos(\omega T) + B.sin(\omega T)$$

so that

$$A = x(0), B = (x(T) - x(0).cos(\omega T))/sin(\omega T)$$

The classical action for motion between the two end points can now be evaluated:

$$Mx'^2/2 - M\omega^2 x^2/2 = (M/2)(-A\omega.sin(\omega t) + B\omega.cos(\omega t))^2 - (M\omega^2/2)(A.cos(\omega t) + B.sin(\omega t))^2$$

$$= -(MA^2\omega^2/2)cos(2\omega t) + (MB^2\omega^2/2)cos(2\omega t) - MAB\omega^2 sin(2\omega t)$$

giving

$$\mathcal{A}_{cl}(x(0), x(T), T) = (M(A^2 - B^2)\omega/4)(1 - sin(2\omega T)) + (MAB\omega/2)(cos(2\omega T) - 1)$$

By virtue of the equation of motion for x_{cl}, it follows that

$$\mathcal{A}(x) = \mathcal{A}_{cl}(x(0), x(T), T) + \int_0^T M(\delta x'(t))^2/2 - M\omega^2(\delta x(t))^2/2)dt$$

Now observe that $\delta x(t) = 0$ when $t = 0$ and $t = T$. We can therefore expand δx over $[0, T]$ as a half wave Fourier sine series:

$$\delta x(t) = \sum_{m=1}^{\infty} \hat{x}[m] sin(m\pi t/T), t \in [0, T]$$

We pick up samples at intervals of $\epsilon = T/N$ setting $t_n = n\epsilon, n = 0, 1, ..., N$ giving

$$\delta x(t_n) = \sum_{m=1}^{N} \hat{x}[m] sin(\pi mn/N)$$

It is preferable to add a normalization factor:

$$\delta x(t_n) = (2/N)^{1/2} \sum_{m=1}^{N} \hat{x}[m] sin(\pi mn/N)$$

The transformation from the variables $\{\delta x(t_n)\}$ to the variables $\{\hat{x}[m]\}$ is unitary and the path integral can be evaluated easily using this transform. Note that the integrals

$$\int_0^T (\delta x'(t))^2 dt, \int_0^T (\delta x(t))^2 dt$$

can be expressed in terms of $\hat{x}[m]$. The details of the evaluation of the path integral can be found in the treatise by Kleinert.

2.Dirac field:

$$\psi(x) = u(p) exp(-ip.x), p^2 = m^2, p = (p^0, \mathbf{p}), p^2 = p^{02} - (\mathbf{p}, \mathbf{p})$$

The Dirac equation

$$(i\gamma^\mu \partial_\mu - m)\psi(x) = 0$$

implies

$$(\gamma^\mu p_\mu - m)u(p) = 0$$

or for short

$$(\gamma.p - m)u(p) = 0$$

Note that

$$\gamma^0 = \begin{pmatrix} 0 & 1 \\ 1 & 0 \end{pmatrix}, \gamma^i = \begin{pmatrix} 0 & \sigma^i \\ \sigma^i & 0 \end{pmatrix}$$

Pauli matrix vector is

$$\sigma^\mu = (1, \sigma), \bar{\sigma}^\mu = (1, -\sigma), \sigma = (\sigma^1, \sigma^2, \sigma^3)$$

With $p^0 = E$, we claim that

$$\sqrt{p.\bar{\sigma}} = \sqrt{E+p^3}(1-\sigma^3)/2 + \sqrt{E-p^3}(1+\sigma^3)/2$$

$$\sqrt{p.\bar{\sigma}} = \sqrt{E+p^3}(1+\sigma^3)/2 + \sqrt{E-p^3}(1-\sigma^3)/2$$

Indeed, the square of the right side of the first equation is

$$((E+p^3)/4)(1+1-2\sigma^3) + ((E-p^3)/4)(1+1+2\sigma^3) + sqrtE^2 - p^{32}(1-\sigma^{32})/2$$

the last term is zero since $\sigma^{32} = 1$ and the above expression simplifies to

$$((E+p^3)/2)(1-\sigma^3) + ((E-p^3)/2)(1+\sigma^3) = E - \sigma^3 p^3$$

so that if $p = (E, 0, 0, p^3)$, then the formula for the square root of $p.\sigma$ is correct. Likewise, we find that

$$\sqrt{p.\bar{\sigma}} = \sqrt{p_\mu \bar{\sigma}^\mu} = \sqrt{E+p^3}(1+\sigma^3)/2 + \sqrt{E-p^3}(1-\sigma^3)/2$$

Let ξ be any two component spinor. Then, we wish to show that if

$$u(p) = \begin{pmatrix} \sqrt{p.\sigma}\xi \\ \sqrt{p.\bar{\sigma}}\xi \end{pmatrix}$$

then

$$(\gamma.p - m)u(p) = 0$$

Assume $p^1 = p^2 = 0$ so that $p = p^\mu = (E, 0, 0, p^3)$. Then

$$\gamma.p = \gamma^0 E - \gamma^3 p^3 = E\begin{pmatrix} 0 & 1 \\ 1 & 0 \end{pmatrix} - \begin{pmatrix} 0 & \sigma^3 \\ -\sigma^3 & 0 \end{pmatrix} p^3$$

$$= \begin{pmatrix} 0 & E - p^3\sigma^3 \\ E + p^3\sigma^3 & 0 \end{pmatrix}$$

This gives

$$\gamma.p - m = \begin{pmatrix} -m & E - p^3\sigma^3 \\ E + p^3\sigma^3 & -m \end{pmatrix}$$

The two 2×1 block components of $(\gamma.p - m)u(p)$ are then evaluated as

$$a = (-m)\sqrt{p.\sigma}\xi + (E - p^3\sigma^3)\sqrt{p.\bar{\sigma}}\xi$$

$$= (-m)[\sqrt{E+p^3}(1-\sigma^3)/2 + \sqrt{E-p^3}(1+\sigma^3)/2]\xi + (E-p^3\sigma^3)[\sqrt{E+p^3}(1+\sigma^3)/2 + \sqrt{E-p^3}(1-\sigma^3)/2]\xi$$

The second term above can be expanded as

$$(E-p^3\sigma^3)[\sqrt{E+p^3}(1+\sigma^3)/2 + \sqrt{E-p^3}(1-\sigma^3)/2]\xi$$

$$[0.5\sqrt{E+p^3}(E-p^3\sigma^3)(1+\sigma^3) + 0.5\sqrt{E-p^3}(E-p^3\sigma^3)(1-\sigma^3)]\xi$$

$$= 0.5[\sqrt{E+p^3}(E-p^3)(1+\sigma^3) + \sqrt{E-p^3}(E+p^3)(1-\sigma^3)$$

$$= 0.5[m\sqrt{E-p^3}(1+\sigma^3) + m\sqrt{E+p^3}(1-\sigma^3)]$$

since

$$E^2 - p^{32} = m^2$$

It follows that $a = 0$. Likewise, $b = 0$.

Suppose γ^μ are the Dirac matrices. They obey the anticommutation rule

$$\{\gamma^\mu, \gamma^\nu\} = 2g^{\mu\nu}$$

Define

$$S^{\mu\nu} = \frac{i}{4}[\gamma^\mu, \gamma^\nu]$$

We wish to verify that these matrices represent the Lorentz algebra. Observe that in view of the above anticommutation relation,

$$S^{\mu\nu} = \frac{i}{2}(\gamma^\mu\gamma^\nu - g^{\mu\nu}) = iT^{\mu\nu}/2$$

say, where

$$T^{\mu\nu} = \gamma^\mu\gamma^\nu - g^{\mu\nu}$$

We compute

$$[S^{\mu\nu}, S^{\rho\sigma}] = -\frac{1}{4}[\gamma^\mu\gamma^\nu, \gamma^\rho\gamma^\sigma]$$

We have

$$[\gamma^\mu\gamma^\nu, \gamma^\rho\gamma^\sigma]$$
$$= \gamma^\mu\gamma^\nu\gamma^\rho\gamma^\sigma - \gamma^\rho\gamma^\sigma\gamma^\mu\gamma^\nu$$
$$= \gamma^\mu(2g^{\nu\rho} - \gamma^\rho\gamma^\nu)\gamma^\sigma - \gamma^\rho\gamma^\sigma\gamma^\mu\gamma^\nu$$
$$= 2g^{\nu\rho}\gamma^\mu\gamma^\sigma - (2g^{\mu\rho} - \gamma^\rho\gamma^\mu)\gamma^\nu\gamma^\sigma$$
$$-\gamma^\rho\gamma^\sigma\gamma^\mu\gamma^\nu$$
$$= 2g^{\nu\rho}\gamma^\mu\gamma^\sigma - 2g^{\mu\rho}\gamma^\nu\gamma^\sigma$$
$$+\gamma^\rho\gamma^\mu(2g^{\nu\sigma} - \gamma^\sigma\gamma^\nu) - \gamma^\rho\gamma^\sigma\gamma^\mu\gamma^\nu$$
$$= 2g^{\nu\rho}\gamma^\mu\gamma^\sigma - 2g^{\mu\rho}\gamma^\nu\gamma^\sigma$$
$$+2g^{\nu\sigma}\gamma^\rho\gamma^\mu - \gamma^\rho\gamma^\mu\gamma^\sigma\gamma^\nu - \gamma^\rho\gamma^\sigma\gamma^\mu\gamma^\nu$$
$$= 2g^{\nu\rho}\gamma^\mu\gamma^\sigma - 2g^{\mu\rho}\gamma^\nu\gamma^\sigma + 2g^{\nu\sigma}\gamma^\rho\gamma^\mu$$
$$-2g^{\mu\sigma}\gamma^\rho\gamma^\nu$$

This formula can be expressed as a linear expression in the $T^{\mu\nu}$ and the correct Lorentz algebra structure constants result. It is the $S^{\mu\nu}$ which must be used as the generators of the Lorentz transformation applied to bispinor wave functions, for example, the bispinor wave function appearing in the Dirac equation. Note that the above expression equals

$$2g^{\nu\rho}(g^{\mu\sigma} + T^{\mu\sigma}) - 2g^{\mu\rho}(g^{\nu\sigma} + T^{\nu\sigma}) + 2g^{\nu\sigma}(g^{\mu\rho} + T^{\mu\rho})$$

$$-2g^{\mu\sigma}(g^{\nu\rho}+T^{\nu\rho})$$

$$=2(g^{\nu\rho}T^{\mu\sigma}-g^{\mu\rho}T^{\nu\sigma}+g^{\nu\sigma}T^{\mu\rho}-g^{\mu\sigma}T^{\nu\rho})$$

which is the correct Lie algebra commutation relation for the Lorentz group.

3.Consider the Klein-Gordon field. The field can be expanded in terms of the creation and annihilation operator fields $a(p)$ and $a(p)^*$. Thus,

$$\phi(x)=(2\pi)^{-3}\int d^3p(2E(p))^{-1/2}(a(p)exp(ip.x)+a(p)^*exp(-ip.x))$$

where $a(p), a(p)^*$ satisfy the standard Boson commutation relations, ie,

$$[a(p),a(p)^*]=C.\delta(p-p')$$

Note that here p is a three momentum since the fields are evaluated at time $t=0$. Let $|0>$ be the vacuum state, ie, the state in which for every three momentum value p, there is no boson in the field. We have $a(p)^*(|0>=|p>$, ie, the operator $a(p)^*$ creates a boson having momenum p. Also $a(p)|0>=0$, since if the field contains no boson, then nothing can be annihilated. It follows that

$$\phi(x)|0>=(2\pi)^{-3}\int d^3p(2E(p))^{-1}exp(-ip.x)|p>$$

We also find that

$$<0|\phi(x)|p>=(2\pi)^{-3}\int(2E(p'))^{-1}<p'|p>exp(-ip'.x)d^3p'=exp(-ip.x)$$

where the normalization of the states $|p>$ has been chosen so that

$$<p'|p>=\delta(p-p')$$

It follows that the operator $\phi(x)$ can be regarded as the creation of a single particle at position x. Note that our convention for normalization is that

$$a(p)^*|0>=(2E(p))^{-1/2}|p>$$

The reason for this normalization can be understood as follows: Assume that p,q are three momenta and that E is the energy corresponding to p. This means that (E,p) is a four momentum vector. Consider a Lorentz transformation from $p\to p'$ corresponding to a boost of velocity β along the x^3 direction. Then, p_1,p_2 remain unchanged while p_3 changes to $p_3'=\gamma(p_3+\beta.E(p))$ where $E(p)=\sqrt{p^2+m^2}$. We thus find that

$$\delta_3(p-q)=\delta_3(p'-q)dp_3'/dp_3=\delta_3(p'-q)\gamma(1+\beta.dE/dp_3)$$

Now,

$$dE/dp_3=p_3/E$$

so that

$$\delta_3(p-q)=\delta_3(p'-q)\gamma(1+\beta p_3/E)$$

Now, the energy transformation law under the Lorentz boost along x_3 axis is given by

$$E' = \gamma(E + \beta.p_3)$$

so that the above formula can be expressed as

$$\delta_3(p-q) = \delta_3(p'-q)E'/E$$

which proves that $E(p)\delta_3(p-q)$ is a Lorentz invariant. We therefore define

$$|p>= \sqrt{2E(p)}a(p)^*|0>$$

This gives

$$<p'|p>= 2E(p) <0|a(p')a(p)^*|0>= 2E(p) <0|a(p)^*a(p') + C.\delta(p-p')|0>= 2E(p)\delta(p-p')$$

which is a Lorentz invariant. In other words, the inner product between two momentum states of the field is a Lorentz invariant as it should be. Suppose $|p>$ is the state of a Klein-Gordon field in which there is exactly one particle with three momentum p. Then if Λ is a Lorentz transformation and $U(\Lambda)$ is the corresponding unitary transformation acting on the quantum field, we must have $U(\Lambda)|p>= |\Lambda p>$ It follows that

$$|\Lambda p>= (2E(\Lambda p))^{1/2}a(\Lambda.p)^*|0>= (2E(\Lambda p))^{1/2}a(\Lambda.p)U(\Lambda)|0>$$

on the one hand, and on the other,

$$|\Lambda p>= U(\Lambda)|p>= U(\Lambda)(2E(p))^{1/2}a(p)^*|0>= (2E(p))^{1/2}U(\Lambda)a(p)^*|0>$$

so that

$$U(\Lambda)a(p)^*U(\Lambda)^{-1} = (E(\Lambda p)/E(p))^{1/2}a(\Lambda p)$$

Klein-Gordon propagator: We have

$$\phi(x) = \int (2E(p))^{-1/2}(a(p)exp(-ip.x) + a(p)^*exp(ip.x))d^3p$$

where now

$$p.x = p^\mu x_\mu = Et - (\mathbf{p}, \mathbf{x})$$

We evaluate

$$[\phi(x), \phi(y)] = \int (2E(p))^{-1/2}(2E(q))^{-1/2}[a(p)exp(-ip.x) + a(p)^*exp(ip.x)$$

$$, a(q)exp(-iq.y) + a(q)^*exp(iq.y)]d^3pd^3q$$

$$= \int (2E(p))^{-1/2}(2E(q))^{-1/2}[a(p), a(q)^*](exp(i(q.y - p.x)) + [a(p)^*, a(q)]exp(i(p.x - q, y))d^3pd^3q$$

$$= \int (2E(p))^{-1}(exp(ip.(y-x)) + exp(p.(x-y)))d^3p$$

except for a constant of proportionality. We define

$$D(x) = \int d^3p(2E(p))^{-1}exp(-ip.(x-y))$$

We now observe that the vacuume field correlation function is given by

$$< 0|\phi(x)\phi(y)|0 >= \int (2E(p))^{-1/2}(2E(q))^{-1/2} < 0|(a(p)exp(-ip.x)+$$

$$a(p)^*exp(ip.x))(a(q)exp(-iq.y) + a(q)^*exp(iq.y))|0 > d^3pd^3q$$

Now

$$< 0|a(p)a(q)|0 >= 0, < 0|a(p)a(q)^*|0 >=< 0|a(q)^*a(p) + \delta(p-q)|0 >= \delta(p-q),$$
$$< 0|a(p)^*a(q)|0 >= 0, < 0|a(p)^*a(q)^*|0 >= 0$$

so that

$$< 0|\phi(x)\phi(y)|0 >= \int (2E(p))^{-1}exp(ip.(y-x))d^3p$$

We denote this by $D(x-y)$, ie.,

$$D(x) = \int (2E(p))^{-1}exp(-ip.x)d^3p$$

We have

$$(\partial^\mu\partial_\mu + m^2)D(x) = \int (2E(p))^{-1}(-p^2 + m^2)exp(ip.x)d^3p = 0$$

since $p^2 - m^2 = 0$. We now consider the four dimensional integral

$$I = \int \frac{exp(-ip.x)}{p^2 - m^2}d^4p$$

We can write this expression as

$$I = \int \frac{exp(-ip.x)}{p^{02} - E(p)^2}d^4p$$

The integral over p^0 can be evaluated by contour integration. We shift the poles $\pm E(p)$ slightly above the p^0 axis and complete the contour using the upper half semicircle. Then standard residue calculus gives

$$I = \int (2E(p))^{-1}[exp(-i(E(p)t - (\mathbf{p}, \mathbf{x}))) - exp(i(E(p)t + (\mathbf{p}, \mathbf{x})))]d^3p$$

This can be expressed as

$$I = \int (2E(p))^{-1}[exp(-ip.x) - exp(ip.x)]d^3p = D(x) - D(-x) =< 0|[\phi(x), \phi(y)]|0 >$$

Suppose we define

$$I' = \int \frac{exp(-ip.x)}{p^2 - m^2} d^4p$$

Noting that $p^2 - m^2 = p^{02} - E(p)^2$, the integration over p^0 can be closed using the semicircle in the lower half complex plane with p^0 as the real axis. The two poles $\pm E(p)$ can be pushed to $\pm E(p) - i\epsilon$ for $x^0 > 0$. This condition guarantees that the integral over the semicircle is convergent. Thus, by the standard residue theorem, we have

$$I' = \int (2E(p))^{-1}(exp(-ip.x) - exp(ip.x))d^3p, x^0 > 0$$

We thus define

$$D_R(x) = \theta(x^0)(2\pi i)^{-1} \int \frac{exp(-ip.x)}{p^2 - m^2} d^4p$$

with the understanding that both the poles $\pm E(p)$ involved in the integration over p^0 are pushed downwards by a small amount. It follows from the above remarks that for $x^0 > y^0$, we have

$$D_R(x) = D(x) - D(-x)$$

Define

$$D_0(x) = (2\pi i)^{-1} \int exp(-ip.x)d^4p/(p^2 - m^2)$$

for all x with the above understanding of pole pushes and contour. Then

$$D_R(x) = \theta(x^0)D_0(x)$$

Further, let K denote the Klein-Gordon operator

$$K = \partial^\mu \partial_\mu + m^2 = \partial^2 + m^2$$

We have

$$K.D_R(x) = (\partial^2 + m^2)D_R(x) = (\partial^2\theta(x^0))D_0(x) + \theta(x^0)\partial^2 D_0(x) + 2(\partial_\mu\theta(x^0))\partial^\mu D_0(x) + m^2\theta(x^0)D_0(x)$$

$$= \delta'(x^0)D_0(x) + \theta(x^0)(2\pi i)^{-1} \int (-p^2 + m^2)exp(-ip.x)d^4p/(p^2 - m^2) + 2\delta(x^0)D_{0,t}(x)$$

$$= \delta'(x^0)D_0(x) + i\theta(x^0)\delta^4(x) + 2\delta(x^0)D_{0,t}(x)$$

Now observe that

$$\int f(u)(\delta'(u)g(u) + 2\delta(u)g'(u))du = -(fg)'(0) + 2f(0)g'(0) = f(0)g'(0) - g(0)f'(0)$$

Thus,

$$\delta'(u)g(u) + 2\delta(u)g'(u) = g'(0)\delta(u) + g(0)\delta'(u)$$

It follows that

$$KD_R(x) = D_{0,t}(x^0 = 0)\delta(x^0) + D_0(x^0 = 0)\delta'(x^0) + i\theta(x^0)\delta^4(x)$$

Now,

$$D_0(x^0 = 0) = 0$$

since for $x^0 \geq 0$, we have $D_0(x) = D(x) - D(-x)$. Also,

$$D_{0,t}(x^0 = 0) = (-i/2)\int (exp(-ip.x) + exp(ip.x))d^3p = -i\delta^3(x)$$

Thus, we get

$$K.D_R(x) = -i\delta^4(x)$$

Note that

$$\theta(x^0)\delta^4(x) = 0$$

It follows that $D_R(.)$ is the retarded Green's function for the Klein-Gordon operator. The Feynman propagator is defined by

$$D_F(x-y) =< 0|T(\phi(x).\phi(y))|0 >$$

where T is the time ordering operator. We have for $x^0 > y^0$,

$$D_F(x-y) =< 0|\phi(x)\phi(y)|0 >= D(x-y)$$

and for $x^0 < y^0$, we have

$$D_F(x-y) =< 0|\phi(y)\phi(x)|0 >= D(y-x)$$

Now define

$$D_1(x) = \int d^4p.exp(-ip.x)/(p^2 - m^2 + i\epsilon)$$

the contour being taken over the lower half semicircle such that the pole $E(p)$ is included but the pole $-E(p)$ is excluded when $x^0 > 0$ but $-E(p)$ is included and $E(p)$ is excluded when $x^0 < 0$. In the latter case, the contour is closed by the upper half semicircle. We then get for $x^0 > 0$,

$$D_1(x) = \int d^3pexp(-ip.x)/2E(p), x^0 > 0,$$

$$D_1(x) = \int d^3pexp(ip.x)/2E(p), x^0 < 0$$

In other words,

$$D_1(x) = D_F(x)$$

for all $x \in \mathbb{R}^4$. The Feynman propagator plays a fundamental role in calculating transition probabilities for scattering processes, the reason being that the evolution operator is calculated by adopting the time ordering scheme for the time varying Hamiltonian operator.

Dirac propagator: First we recall the quantization of the Dirac field:

$$\psi(x) = \int d^3p(2E(p))^{-1}(a_s(p)u_s(p)exp(-ip.x) + b_s(p)^*v_s(p)exp(ip.x))d^3p$$

and

$$\bar{\psi}(x) = \psi(x)^*\gamma^0 = \sum d^3p(2E(p))^{-1}(a_s(p)^*\bar{u}_s(p)exp(ip.x) + b_s(p)\bar{v}_s(p)exp(-ip.x))$$

where s denotes the polarization states. s takes two values $1, 2$. We have

$$u_s(p) = (\sqrt{p.\sigma}\xi_s, \sqrt{p.\bar{\sigma}}\xi_s)$$

u_s is an eigenvector of the algebraic Dirac equation, ie,

$$(\gamma.p - m)u_s(p) = 0, s = 1, 2$$

$\xi_1 = (1, 0)$ and $\xi_2 = (0, 1)$. We have

$$u^*u = \xi^*(p.\sigma)\xi + \xi^*(p, \bar{\sigma})\xi$$

For $\xi = \xi_1$ and ξ_2, it is easy to see that

$$\xi^*(p, \sigma)\xi = E(p)$$

We thus get

$$u^*u = 2E(p)\xi^*\xi$$

Note that ξ is a spinor and hence $\xi^*\xi$ is invariant under $SU(2)$ action. It follows that u^*u is not invariant since $E(p)$ is not invariant. To construct an invariant inner product for bispinors, we set

$$\bar{u} = u^*\gamma^0$$

Then,

$$\bar{u}u = (\xi^*(p, \sigma), \xi^*(p, \bar{\sigma}))\gamma^0((p, \sigma)\xi, (p, \bar{\sigma})\xi)$$
$$= \xi^*\sqrt{(p, \sigma)(p, \bar{\sigma})}\xi + \xi^*\sqrt{(p, \bar{\sigma})(p, \sigma)}\xi = 2m\xi^*\xi$$

since

$$(p, \sigma)(p, \bar{\sigma}) = p^2 = m^2$$

Thus, $\bar{u}u$ is a Lorentz invariant. We can also verify the relations

$$\sum_s u_s(p)\bar{u}_s(p) = \gamma.p + m$$

Recall that $u_s(p)$ is a solution to the algebraic Dirac equation, ie,

$$(\gamma.p - m)u_s(p) = 0$$

$u_s(p)exp(-ip.x)$ is a solution to the Dirac equation

$$(i\gamma.\partial - m)u_s(p)exp(-ip.x) = (\gamma.p - m)u_s(p)exp(-ip.x) = 0$$

Corresponding negative frequency solutions (positrons) to the Dirac equation are

$$v_s(p)exp(ip.x)$$

They satisfy

$$(i\gamma.\partial - m)v_s(p)exp(ip.x) = 0$$

or equivalently,

$$(\gamma.p + m)v_s(p) = 0$$

It may be shown by direct substitution that

$$v_s(p) = (\sqrt{p.\sigma}\eta_s, -\sqrt{p.\bar{\sigma}}\eta_s)$$

We have correspondingly

$$\sum_s v_s(p)\bar{v}_s(p) = \gamma.p - m$$

It follows that

$$\sum_s u_s(p)u_s(p)^* = (\gamma.p - m)\gamma^0 =$$

It is easily verified that

$$u_r(p)^*u_s(p) = 2E(p)\delta_{rs} = v_r(p)^*v_s(p)$$

Now recall the structure of the quantized Dirac field:

$$\psi_a(x) = \int d^3p(2E(p))^{-1/2}(a_s(p)u_{s,a}(p)exp(-ip.x) + b_s(p)^*v_{s,a}(p)exp(ip.x))$$

$$\psi_b(x)^* = \int d^3p(2E(p))^{-1/2}(a_s(p)^*u_{s,b}(p)^*exp(ip.x) + b_s(p)v_{s,b}(p)^*exp(-ip.x))d^3p$$

giving

$$\{\psi_a(x), \psi_b(y)^*\} = \int d^3pd^3q(2E(p))^{-1/2}(2E(q))^{-1/2}(\{a_s(p), a_s(q)^*\}u_{s,a}(p)u_{s,b}(p)^*exp(i(q.y - p.x))$$

$$+\{b_s(p)^*, b_s(q)\}v_{s,a}(p)v_{s,b}(p)^*)exp(i(p.x - q.y))$$

$$= \int d^3p(2E(p))^{-1}(u_{s,a}(p)u_{s,b}(p)^*exp(ip.(y - x)) + v_{s,a}(p)v_{s,b}(p)^*exp(ip.(x - y)))$$

Summation over the repeated index s is implied. Further,

$$\sum_s u_{s,a}(p)u_{s,b}(p)^* = \sum_(su_s(p)u_s(p)^*)_{ab}$$

Now recall that

$$\sum_a u_{r,a}(p)^*u_{s,a}(p) = 2E(p)\delta_{rs}$$

Define the matrix A by

$$A(a, s) = u_{s,a}(p)$$

Then the above equation can be written as

$$\sum_a A(a, r)^*A(a, s) = 2E(p)\delta_{rs}$$

or in matrix notation,

$$A^*A = 2E(p) > I$$

It follows that,

$$AA^* = 2E(p)I$$

that is,

$$\sum A(a,r)A(b,r)^* = 2E(p)\delta_{a,b}$$

or

$$\sum u_{r,a}(p)u^*_{r,b} = 2E(p)\delta_{a,b}$$

A similar equation holds for $v_s(p)$. From these substitutions, it follows that

$$\{\psi_a(x), \psi_b(y)^*\} = \delta_{a,b}\frac{1}{2}\int (exp(ip.(y-x)) + exp(ip.(x-y))d^3p$$

It follows that if $x^0 = y^0 = 0$, then we must have

$$\{\psi_a(x), \psi_b(y)^*\} = \delta_{a,b}\delta(x-y)$$

which is the desired anticommutation relation.

4. Non-Abelian gauge theories. The Yang-Mills field equations. Let $\sigma_a, a = 1,2,3$ be the three Pauli spin matrices. For each $\mu = 0,1,2,3$ and $a = 1,2,3$, assume that we have a field function $A_{\mu,a}(x)$. Define the covariant derivative

$$D_\mu = \partial_\mu + igA_{\mu,a}(x)\sigma_a$$

summation over the repeated index a being implied. The bracket operation of D_μ is obtained as

$$F_{\mu\nu} - [D_\mu, D_\nu] = ig\partial_\mu A_{\nu,a}\sigma_a - ig\partial_\nu A_{\mu,a}\sigma_a - 2ig^2\epsilon_{abc}\sigma_c A_{\mu,a}(x)A_{\nu,b}(x)$$

From this expression, we derive the form of the field tensor:

$$F_{\mu\nu} = ig(\partial_\mu A_{\nu,a} - \partial_\nu A_{\mu,a})\sigma_a - 2ig^2 A_{\mu,a}A_{\nu,b}\epsilon_{abc}\sigma_c$$

The Lagrangian density for the field is given by

$$Tr(F_{\mu\nu}.F^{\mu\nu})$$

Compute this Lagrangian and hence by varying $A_{\mu,a}$, determine the field equations. Note that

$$F^{\mu\nu} = ig(\partial^\mu A^\nu_a - \partial^\nu A^\mu_a)\sigma_a - 2ig^2 A^\mu_a A^\nu_b \epsilon_{abc}\sigma_c$$

Define the scalar fields

$$F_{\mu\nu a} = \partial_\mu A_{\nu,a} - \partial_\mu A_{\nu,a}$$

and also

$$G_{\mu\nu ab} = A_{\mu,a}A_{\nu,b}$$

Then,

$$F_{\mu\nu} = ig.F_{\mu\nu a}\sigma_a - 2ig^2 G_{\mu\nu ab}\epsilon_{abc}\sigma_c$$
$$= (igF_{\mu\nu a} - 2ig^2 G_{\mu\nu bc}\epsilon_{bca})\sigma_a$$

Note that summation over the repeated indices bc and a is implied. Using $\sigma_a\sigma_b = i\epsilon_{abc}\sigma_c$, we get

$$F_{\mu\nu}F^{\mu\nu} = (ig.F_{\mu\nu a} - 2ig^2 G_{\mu\nu bc}\epsilon_{bca})^2$$

$$+i\epsilon_{adf}(igF_{\mu\nu a} - 2ig^2 G_{\mu\nu bc}\epsilon_{bca})(iF_{\mu\nu d} - 2ig^2 G_{\mu\nu bc}\epsilon_{bcd})\sigma_f$$

where $(C_{\mu\nu})^2$ is understood to be $C_{\mu\nu}C^{\mu\nu}$. Thus, noting that $Tr(\sigma_a) = 0$, it follows that the Lagrangian density of the Yang-Mills field is given by

$$L = Tr(F_{\mu\nu}.F^{\mu\nu}) = 2(ig.F_{\mu\nu a} - 2ig^2 G_{\mu\nu bc}\epsilon_{bca})^2$$

5. Computation of the S-matrix in quantum field theory. $|\mathbf{k}>$ denotes a one particle state of the field in which there is a Boson having momentum $\mathbf{k}$ The field can be expressed as

$$\psi(x) = \int (2E(k))^{-1/2}(a(k)exp(ik.x) + a(k)^* exp(-ik.x))d^3k$$

Let $\phi(x)$ be a function on $\mathbb{R}^3$ and $\phi(\mathbf{k})$ its three dimensional spatial Fourier transform. The corresponding wave packet at time $t = 0$ is given by

$$|\phi >= \int (2E(k))^{-1/2}\phi(\mathbf{k})|\mathbf{k}>$$

We have

$$|\mathbf{k} >= (2E(\mathbf{k}))^{1/2}a(\mathbf{k})^*|0>$$

and also

$$a(\mathbf{k})|0 >= 0$$

It follows that

$$|\phi >= [\int \phi(\mathbf{k})a(\mathbf{k})^* + \phi(\mathbf{k})^* a(\mathbf{k})]|0>$$

The normalization condition can be worked out

$$1 =< \phi|\phi >= \int \phi(k)\phi(k')^* < 0|a(\mathbf{k}')a(\mathbf{k})^*|0 > d^3k d^3k'$$

Now use the commutation relation

$$a(\mathbf{k})a(\mathbf{k}')^* = a(\mathbf{k}')^* a(\mathbf{k}) + \delta(\mathbf{k} - \mathbf{k}')$$

and $a(\mathbf{k})|0 >= 0$ to get

$$1 =< \phi|\phi >= \int |\phi(k)|^2 d^3k$$

Note that the wave packet $|\phi>$ corresponds to a linear superposition of states in which in each term, there is one particle in a definite momentum eigenstate. Let $|\phi_A,\phi_B>$ denote a state in which the system consists of two wavepackets in the far past. Let $|\phi_1,\phi_2...>$ denote a state of several wave packets one for each final state particle in the far future. The probability of scattering of the two initial wave packets to the final wave packets is given by

$$P = |<\phi_1,\phi_2...|\phi_A,\phi_B>|^2$$

We assume that the wavepacket B is incident on the wave packet A with an impact parameter defined by the vector $\mathbf{b}$. Now spatial translation by $\mathbf{b}$ corresponds to multiplication of the momentum space wave function by $exp(-i\mathbf{b}.\mathbf{k}_B)$ abd hence the initial state should be represented by the equation

$$|\phi_A\phi_B>= \int d^3k_A.d^3k_B(2E_A)^{-1/2}(2E_B)^{-1/2}\phi_A(k_A)\phi_B(k_B)exp(-ib.k_B)|k_Ak_B>_{in}$$

Similarly, we can write the out state as

$$<\phi_1,\phi_2,...| = \Pi_f\int d^3p_f(\phi_f(p_f)/\sqrt{2E_f})_{out} <p_1,p_2,...|$$

It follows that

$$(out)<\phi_1,\phi_2,...|\phi_A,\phi_B>(in) = \int \Pi_f(d^3p_f)d^3k_A.d^3k_B\Pi_f(\phi_f(p_f)/$$

$$\sqrt{2E_f})(\phi_A(k_A)\phi_B(k_B)/\sqrt{2E_A2E_B})(out)<p_1,p_2,...|k_A,k_B>(in)$$

The S-matrix appears in the integrand. It is given by

$$(out)<p_1,p_2,...|k_A,k_B>(in) = lim_{T\to\infty}<p_1,p_2,...|exp(-iHT)|k_A,k_B>$$

We write this relation as

$$(out)<p_1,p_2,...|k_A,k_B>(in) =<p_1,p_2,...|S|k_A,k_B>$$

where formally,

$$S = lim_{T\to\infty}exp(-iTH)$$

This limit makes sense only when its matrix elements are taken into account, on the one side, the in states and on the other the out states.

6.Feynman rules in quantum electrodynamics. The Dirac field is given by

$$\psi(x) = \int (2E(p))^{-1/2}d^3p(a_s(p)u_s(p)exp(-ip.x)+b_s(p)^*v_s(p)exp(ip.x))$$

and its modified conjugate is

$$\bar{\psi}(x) = \psi(x)^*\gamma^0 = \int (2E(p))^{-1/2}d^3p(a_s(p)\bar{u}_s(p)exp(ip.x)+b_s(p)\bar{v}_s(p)exp(-ip.x))$$

The equal time anticommutation rules read

$$\{a_s(p), a_r(q)^*\} = \delta_{rs}\delta_3(p-q) = \{b_s(p), b_r(q)^*\}$$

The electronic current operator is given by

$$J^\mu(x) = \bar{\psi}(x)\gamma^\mu\psi(x)$$

Assume that $A_\mu(x)$ is the external electromagnetic four potential. Assuming that this field is external, we can assume A_μ to be an ordinary function and not an operator. Then the interaction Hamiltonian between the electromagnetic field and the electron field is given by

$$H = \int J^\mu(x)A_\mu(x)d^3x = \int \bar{\psi}(x)\gamma^\mu\psi(x)A_\mu(x)d^3x$$

Here, $x = (x^0, x^1, x^2, x^3) = (t, \mathbf{x})$ and $d^3x = d\mathbf{x} = dx_1dx_2dx_3$. We can write $A_\mu(x)$ as a four dimensional Fourier transform, that is,

$$A_\mu(x) = \int A_\mu(p)exp(ip.x)d^4p$$

where

$$p.x = p^\mu x_\mu = p^0x^0 - \sum_{r=1}^{3} p^r x^r = p^0t - (\mathbf{p}, \mathbf{x})$$

Let us compute to first order, the amplitude for an electron with momentum p to be scattered into an electron with momentum p' via the electromagnetic field. This amplitude is proportional to

$$< p'|\int_{-\infty}^{\infty} Hdt|p> = < p'|(\int J^\mu(x)A_\mu(x)d^4x)|p>$$

Now,

$$\int Hdt = \int J^\mu(x)A_\mu(x)d^4x = \int \bar{\psi}(x)\gamma^\mu\psi(x)A_\mu(x)d^4x$$

$$= \int (2E(p).2E(p'))^{-1/2}(a_s(p)^*\bar{u}_s(p)exp(-ip.x)$$

$$+b_s(p)\bar{v}_s(p)exp(ip.x))\gamma^\mu(a_r(p')u_r(p')exp(ip'.x)+b_r(p')^*v_r(p')exp(-ip'.x))A_\mu(p'')exp(ip''.x)d^4xd^3pd^3p'd^4p''$$

$$= \int (2E(p).2E(p'))^{-1/2}[a_s(p)^*a_r(p')\bar{u}_s(p)\gamma^\mu u_r(p')exp(i(p''-p+p',x))$$

$$+a_s(p)^*b_r(p')^*\bar{u}_s(p)\gamma^\mu v_r(p')exp(i(p''-p-p',x))$$

$$+b_s(p)a_r(p')\bar{v}_s(p)\gamma^\mu u_r(p')exp(i(p''+p+p',x))$$

$$+b_s(p)b_r(p')^*\bar{v}_s(p)\gamma^\mu v_r(p')exp(i(p''+p-p',x))]A_\mu(p'')d^4xd^3pd^3p'd^4p''$$

$$= \int (2E(p).2E(p'))^{-1/2}[a_s(p)^*a_r(p')\bar{u}_s(p)\gamma^\mu u_r(p')\delta(p''-p+p') + a_s(p)^*b_r(p')^*\bar{u}_s(p)\gamma^\mu v_r(p')\delta(p''-p-p')$$

$$+b_s(p)a_r(p')\bar{v}_s(p)\gamma^\mu u_r(p')\delta(p''+p+p') + b_s(p)b_r(p')^*\bar{v}_s(p)\gamma^\mu v_r(p')\delta(p''+p-p')]A_\mu(p'')d^4p''d^3pd^3p'$$

$$= \int (2E(p).2E(p'))^{-1/2}[a_s(p)^* a_r(p')\bar{u}_s(p)\gamma^\mu u_r(p')A_\mu(p-p') + a_s(p)^* b_r(p')^* \bar{u}_s(p)\gamma^\mu v_r(p')A_\mu(p+p')$$

$$+b_s(p)a_r(p')\bar{v}_s(p)\gamma^\mu u_r(p')A_\mu(-p-p') + b_s(p)b_r(p')^*\bar{v}_s(p)\gamma^\mu v_r(p')A_\mu(p'-p)]d^3pd^3p'$$

We also use the fact that if $|0>$ is the vacuum, then $a_s(p)^*|0>= (2E(p))^{-1/2}|p, s, e>$ where $|p, s, e>$ is the state of the electron field with one electron having momentum p and polarization s. Thus, We drop the index e with the understanding that $|p, s>$ is the state of the electron field with one electron having momentum p and polarization s. Then

$$< p_2 s_2| \int Hdt|p_1 s_1 >$$

is the required amplitude. We observe that

$$< p_2 s_2|a_s(p)^* a_r(p')|p_1 s_1 >= (2E(p_2).2E(p_1))^{1/2} < 0|a_{s_2}(p_2)a_s(p)^* a_r(p')a_{s_1}(p_1)^*|0 >$$

Now,

$$< 0|a_{s_2}(p_2)a_s(p)^* a_r(p')a_{s_1}(p_1)^*|0 >$$

$$= \delta_{s_2 s}\delta_{s_1 r}\delta(p-p_2)\delta(p'-p_1)$$

so that

$$< p_2 s_2|a_2(p)^* a_r(p')|p_1 s_1 >= (2E(p).2E(p'))^{1/2}\delta_{s_2 s}\delta_{s_1 r}\delta(p-p_2)\delta(p'-p_1)$$

$$< p_2 s_2|a_s(p)^* b_r(p')^*|p_1 s_1 >= 0$$

because except for a normalization factor, $b_r(p')^*|p_1 s_1 >$ is a state in which there is one positron of momentum p' and one electron of momentum p_1 while $a_s(p)|p_2 s_2 >$ is a state in which there is no positron. These two states are orthogonal. Another way to see this is to note that but for a normalization factor,

$$< p_2 s_2|a_s(p)^* b_r(p')^*|p_1 s_1 >=< 0|a_{s_2}(p_2)a_s(p)^* b_r(p')^* a_{s_1}(p_1)^*|0 >$$

$$=< 0|b_r(p')^* a_{s_2}(p_2)a_s(p)^* a_{s_1}(p_1)^*|0 >= 0$$

since b_r anticommutes with a_s and $b_r(p')|0 >= 0$. Also by similar reasoning,

$$< p_2 s_2|b_s(p)a_r(p')|p_1 s_1 >= 0$$

$$< p_2 s_2|b_s(p)a_r(p')|p_1 s_1 >= 0,$$

$$< p_2 s_2|b_s(p)b_r(p')|p_1 s_1 >= 0$$

the last following from the fact that $|p_1, s_1 >$ is a state in which there is no positron and hence if the positron annihilation operator $b_r(p')$ acts on such a state, the result is zero. Combining all these facts, we finally get

$$< p_2 s_2| \int Hdt|p_1 s_1 >= \bar{u}_{s_2}(p_2)\gamma^\mu u_{s_1}(p_1)A_\mu(p_2-p_1)$$

This result can be depicted by a Feynman diagram.

7.Interaction between electrons and bosons. The bosonic field with the boson having mass m satisfies the Klein-Gordon equation. The quantized bosonic wave field is given by

$$\phi(x) = \int (2E(p))^{-1/2}(a(p)exp(ip.x) + a(p)^* exp(-ip.x))d^3p$$

Here,

$$E(p) = \sqrt{m^2 + p^2}$$

where m is the mass of the boson. $a(p)$ annihilates a boson having momentum p while $a(p)^*$ creates a boson with momentum p. The commutation relations are

$$[a(p), a(p')^*] = \delta(p - p')$$

and all other commutators between $\{a(p), a^*(p) : \mathbf{p} \in \mathbb{R}^3\}$ are zero. Then

$$\phi(x)\phi(y) = \int (2E(p).2E(p'))^{-1/2}(a(p)exp(ip.x)$$

$$+a(p)^* exp(-ip.x))(a(p')exp(ip.y) + a(p')^* exp(-ip'.y))d^3pd^3p'$$

Further,

$$< 0|a(p)a(p')|0 >= 0, < 0|a(p)a(p')^*|0 >= \delta(p - p'), < 0|a(p)^* a(p')^*|0 >= 0$$

so

$$< 0|\phi(x)\phi(y)|0 >= \int (2E(p))^{-1} exp(ip.(x - y))d^3p = D(x - y)$$

say. Define the Feynman propagator

$$D_F(x) = \int \frac{exp(ip.x)}{p^2 - m^2 + i\epsilon} d^4p = \int \frac{exp(ip.x)}{p^{02} - E(p)^2 + i\epsilon}$$

Consider the integral with respect to p_0. The integrand has poles at $p^0 = \pm(E(p) - i\epsilon)$. Assume that $x^0 > 0$. Then if we close the contour of integration with respect to p^0 by an upper semicircle, the integral will be convergent since $ip.x$ contains the summand ip^0x^0 and in the upper semicircle, $Imp^0 > 0$ so that ip^0x^0 has a negative real part causing $exp(ip.x)$ to have exponential decay to zero as p^0 reaches the upper semicircle. This contour includes only one pole namely $-E(p) + i\epsilon$ and hence by residue calculus

$$D_F(x) = 2\pi i \int \frac{exp(ip.x)}{-2E(p)} d^3p = D(x)$$

for $x^0 > 0$. Likewise for $x^0 < 0$, we can close the contour below and get

$$D_F(x) = D(-x)$$

In other words, the Feynman propagator is given by

$$D_F(x) = D(x)\theta(x^0) + D(-x)\theta(-x^0)$$

8.

1.The fields are $\phi_i, i = 1, 2, ..., N$. An infinitesimal gauge transformation of these fields has the form

$$\phi_i \to (\delta_{ij} + ig\alpha_a t^a_{ij})\phi_j$$

summation over the repeated indices a, j being implied. $T^a = (t^a_{ij})$ are Hermitian matrices and α_a are small real numbers. We can write the above gauge transformation as

$$\phi \to (I + ig\alpha_a T^a)\phi$$

This is an infinitesimal unitary transformation since

$$I + ig\alpha_a T^a = exp(ig\alpha_a T^a) + O(|\alpha|^2)$$

Let A^a_μ be fields and assume that the covariant derivative is defined by

$$D_\mu = \partial_\mu + \lambda.A^a_\mu T^a$$

Assume that under the gauge transformation, the fields A^a_μ transform as

$$A^a_\mu \to A^a_\mu + X^a_\mu$$

where X^a_μ are linear functions of α^a and its partial derivatives. In other words, X^a_μ is like the $\{\alpha^a\}$ of the first order of smallness. Then $D_\mu\phi$ under the Gauge transformation transforms as

$$D_\mu\phi = (\partial_\mu + \lambda.A^a_\mu T^a)\phi$$

$$\to (\partial_\mu + \lambda.(A^a_\mu + X^a_\mu)T^a)(I + ig\alpha_b T^b)\phi$$

$$= (I + ig\alpha_b T^b)(\partial_\mu + \lambda.(A^a_\mu + X^a_\mu T^a))\phi + [\partial_\mu + \lambda.(A^a_\mu + X^a_\mu)T^a, I + ig\alpha_b T^b]\phi$$

With neglect of quadratic terms in α (noting that $X's$ are linear in α), this is the same as

$$(I + ig\alpha_a T^a)(\partial_\mu + \lambda.A^a_\mu)\phi + \lambda X^a_\mu T^a\phi + (ig(\partial_\mu\alpha_b)T^b + ig\alpha_b\lambda.A^a_\mu[T^a, T^b])\phi$$

We want only the first term in this sum to remain. That would guarantee that $(D_\mu\phi)^*(D^\mu\phi)$ is an invariant. This happens provided that

$$\lambda X^a_\mu + ig\partial_\mu\alpha_a + ig\alpha_c\lambda.A^b_\mu c(bca) = 0$$

where $c(abc)$ are the structure constants of the Lie algebra generated by $\{T^a\}$.

2. Spontaneous symmetry breaking: There is just one field ϕ and its field equations are determined from the Lagrangian density

$$\mathcal{L} = \frac{1}{2}(\partial_\mu\phi)^2 - m^2\phi^2/2 - \lambda\phi^4/24$$

where

$$(\partial_\mu\phi)^2 = (\partial_\mu\phi)(\partial^\mu\phi) = (\partial_t\phi)^2 - (\nabla\phi)^2$$

We find that after ignoring a perfect divergence,

$$\delta\mathcal{L} = (\partial^\mu\phi)(\delta\partial_\mu phi) - m^2\phi\delta\phi - \lambda\phi^3\delta\phi/6$$

$$= (\delta\phi).(-\partial^2\phi - m^2\phi - \lambda\phi^3/6)$$

giving the field equations

$$\partial^2\phi + m^2\phi + \lambda\phi^3/6 = 0$$

The potential energy of the field is given by

$$V(\phi) = m^2\phi^2/2 + \lambda\phi^4/24$$

and it attains its minimum at ϕ_0 where $V'(\phi_0) = 0$, ie,

$$\mu^2\phi_0 = \lambda\phi_0^3/6$$

the nonzero solution to which is

$$\phi_0 = \pm\mu(6/\lambda)^{1/2}, \mu^2 = -m^2$$

We define

$$v = \mu(6/\lambda)^{1/2}$$

and set

$$\phi(x) = v + \sigma(x)$$

Then,

$$\mathcal{L} = \frac{1}{2}(\partial_\mu\sigma)^2 + \mu^2(v+\sigma)^2/2 - \lambda(v+\sigma)^4/24$$

$$= \frac{1}{2}(\partial_\mu\sigma)^2 + \mu^2(v^2 + \sigma^2 + 2v\sigma)/2 - (\lambda/24)(v^4 + 4v^3\sigma + 6v^2\sigma^2 + 4v\sigma^3 + \sigma^4)$$

$$= \frac{1}{2}(\partial_\mu\sigma)^2 + \mu^2v^2 + (\mu^2/2 - v^2\lambda/4)\sigma^2 - (\lambda v/6)\sigma^3 - \lambda\sigma^4/24$$

Now μ^2v^2 is a constant and can be ignored.

$$\mu^2/2 - v^2\lambda/4 = \mu^2/2 - (3/2)\mu^2 = -\mu^2$$

Also,

$$\lambda.v/6 = (\lambda/6)\mu(6/\lambda)^{1/2} = (\lambda/6)^{1/2}$$

so

$$\mathcal{L} = \frac{1}{2}(\partial_\mu\sigma)^2 - \frac{1}{2}(2\mu^2)\sigma^2/2 - (\lambda/6)^{1/2}\sigma^3 - \lambda\sigma^4/4$$

We see that the Lagrangian consists of three parts. The first is $\frac{1}{2}(\partial_\mu\sigma)^2 - \frac{1}{2}(2\mu^2)\sigma^2$ which corresponds to a massive boson field of mass $\mu\sqrt{2}$ and the second and third parts involve σ^3 and σ^4 terms. The symmetry $\phi \to -\phi$ under which the orignal Lagrangian was invariant is no longer true for $\sigma \to -\sigma$. In other words, this is a spontaneously broken symmetry.

9.**Quantum Computation**

1.Definition of a state vector for m qubits. The state vector has size $N = 2^m$, For $m = 1$, $N = 2$ and we get a one qubit state.

2. Different kinds of gates on a quantum system. C-NOT gate, Hadamard gate, Toffoli gate, Fredkin gate, phase gate, $\pi/8$ gate. A gate is simply a unitary matrix that takes a state as input and outputs another state. When the states are pure, the gate action is given by $|\psi> \to U|\psi>$ while when the states are mixed, the action is given by $\rho \to U\rho U^*$.

3.Notion of generalized measurement: Let ρ be a state of the quantum system. It may be pure or mixed. If pure, then there exists a unit vector $|\psi>$ such that $\rho = |\psi><\psi|$. If mixed, then there exists an onb $|\psi_i>, i = 1, 2, ...$ for the Hilbert space and non-negative numbers $p_1, p_2, ...$ such that $\sum p_i = 1$ and

$$\rho = \sum_i |\psi_i > p_i < \psi_i|$$

A generalized measurement is a countable family of operators $\{M_m, m = 1, 2, ...\}$ such that $\sum_m M_m^* M_m = I$ and after the measurement if the outcome is unknown, the state of the system collapses to

$$M(\rho) = \sum_m M_m.\rho.M_m^*$$

It is easily verified that $Tr(M(\rho)) = 1$ and that $M(\rho)$ as defined here is non-negative definite so that it is a valid density operator.

10.Feynman propagator for the Dirac field:

$$\psi_a(x) = \int (2E(p))^{-1/2}(u_{s,a}(p)a_s(p)exp(-ip.x) + b_s(p)^* v_{s,a}(p)exp(ip.x))d^3p$$

$$\bar{\psi}_a(x) = \int (2E(p))^{-1/2}(\bar{u}_{s,a}(p)a_s(p)^* exp(ip.x) + b_s(p)\bar{v}_{s,a}(p)exp(-ip.x))d^3p$$

Then,

$$\psi_a(x)\bar{\psi}_b(y) = \int (2E(p))^{-1/2}(2E(p'))^{-1/2}[u_{s,a}(p)\bar{u}_{r,b}(p')a_s(p)a_r(p')^* exp(-ip.x + ip'.y)$$

$$+u_{s,a}(p)\bar{v}_{r,b}(p')a_s(p)b_r(p')exp(-ip.x - ip'.y) + v_{s,a}(p)\bar{u}_{r,b}(p')b_s(p)^* a_s(p')^* exp(ip.x + ip'.y)$$

$$+v_{s,a}(p)\bar{v}_{r,b}(p')b_s(p)^* b_r(p')exp(ip.x - ip'.y))]d^3pd^3p'$$

Also

$$< 0|a_s(p)a_r(p')^*|0 >= \delta(p - p')$$

since

$$\{a_s(p), a_r(p')^*\} = \delta_{sr}\delta(p - p')$$

and

$$a_s(p)|0 >= 0$$

Thus,

$$< 0|\psi_a(x)\bar{\psi}_b(y)|0 >= \int (2E(p))^{-1} \sum_s (u_{s,a}(p)\bar{u}_{s,b}(p)exp(-ip.(x - y))d^3p$$

Now

$$\sum_s u_{s,a}(p)\bar{u}_{s,b}(p) = (\gamma.p + m)_{a,b}$$

Thus,

$$< 0|\psi_a(x)\bar{\psi}_b(y)|0 >= (i\gamma.\partial + m)_{ab} \int (2E(p))^{-1} exp(-ip.(x-y))d^3p$$

Likewise, we may show that

$$< 0|\bar{\psi}_b(y)\psi_a(x) >= -(i\gamma.\partial + m)_{ab} \int (2E(p))^{-1} exp(-ip.(x-y))d^3p$$

Define the Feynman propagator for the Fermion field:

$$S_F(x) = \int d^4p[(i\gamma.p + m)/(p^2 - m^2 + i\epsilon)]exp(-ip.x)$$

For $x_0 > 0$, the integral over p^0 is closed by a lower half semicircle. The corresponding pole that falls inside the contour is given by $p^0 = E(p) - i\epsilon$ and we get

$$S_F(x) = \int d^3p((i\gamma.p + m)/2E(p))exp(-ip.x)$$

In other words, for $x^0 > y^0$

$$(S_F(x-y))_{ab} =< 0|\psi_a(x)\bar{\psi}_b(y)|0 >$$

and likewise, we may show that for $x^0 < y^0$,

$$(S_F(x-y))_{ab} = -< 0|\bar{\psi}_b(y)\psi(x)|0 >$$

In other words, the Feynman propagator is given by

$$S_F(x-y) =< 0|T(\psi(x)\bar{\psi}(y)|0 >$$

where T is the Fermionic Chronological ordering operator.

11. Scattering matrix computation: Let $|k_A, k_B >$ denote the state in which there are two incident particles, one having momentum k_A and the other having momentum k_B. This is the initial state. The final state is denoted by $|p_1, p_2, ... >$ consisting of particles having momenta $p_1, p_2, ...$ H is the interaction Hamiltonian operator The S-matrix is simply the limit of $exp(-iHt)$ for large t. Thus,

$$< p_1, p_2, ...|S|k_A, k_B >= lim_{T\to\infty} < p_1, p_2, ...|exp(-iHT)|k_A, k_B >$$

Let $|\Omega >$ denote the ground state of H, the total Hamiltonian. The unperturbed Hamiltonian operator is H_0 whose ground state is $|0 >$. Let $|n >$ denote the eigenstate of H corresponding to the eigenvalue E_n. We have

$$|0 >= \sum_n |n >< n|0 >= |\Omega >< \Omega|0 > + \sum_{n\neq\Omega} |n >< n|0 >$$

$$exp(-iHT)|0>= exp(-iE_0T)|\Omega><\Omega|0> + \sum_{n\neq\Omega>} exp(-iE_nT)|n><n|0>$$

E_0 is the ground state energy level of H. We have $E_n > E_0$ and

$$<\Omega|0>= exp(-iHT)|0>(exp(-iE_0T)<\Omega|0>)^{-1} - (<\Omega|0>)^{-1}\sum_{n\neq\Omega} exp(i(E_0-E_n)T)<n|0>$$

It follows that since $E_0 < E_n$,

$$|\Omega>= lim_{T\to\infty} exp(-iH(T-i\epsilon))|0>(exp(-iE_0T)<\Omega|0>)^{-1}$$

When the presence of ϵ is understood, we shall abbreviate the above formula as

$$|\Omega>= lim_{T\to\infty}(exp(-iE_0T)<\Omega|0>)^{-1}exp(-iHT)|0>$$

The eventual aim is to construct the transition probability amplitudes using Feynman diagrams. Consider a quantum field $\phi(x)$. We assume that this is a bosonic field so that it has the expansion

$$\phi(x) = \int (2E(p))^{-1/2}(a(p)exp(-ip.x) + a(p)exp(ip.x))d^3p$$

where if m is the boson mass, then

$$E(p) = \sqrt{m^2+p^2}$$

We write

$$\phi^+ = \int (2E(p))^{-1/2}a(p)exp(-ip.x)d^3p, \phi^- = \int (2E(p))^{-1/2}a(p)^*exp(ip.x)d^3p$$

so that

$$[\phi^+(x), \phi^-(y)] = \int (2E(p))^{-1}exp(ip.(x-y))d^3p$$

We define the contraction $C(\phi(x), \phi(y))$ of the field variables $\phi(x)$ and $\phi(y)$ to be $[\phi^+(x), \phi^-(y)]$ for $x^0 > y^0$ and $[\phi^+(y), \phi^-(x)]$ if $y^0 > x^0$. Then from the above, we get

$$C(\phi(x), \phi(y)) = D_F(x-y)$$

where D_F is the Feynman propagator. If $\phi_1, ..., \phi_m$ are field variables, ie expressible as superpositions of $a(p)$ and $a(p)^*$, we define the normal ordering of the product $\phi_1...\phi_m$ to be obtained from the ordinary product by pushing all terms involving $a^*(p)$ to the left and all terms involving $a(p)$ to the right. To put this in more clear language, we write

$$\phi_i = \phi_i^+ + \phi_i^-$$

where ϕ_i^+ is a superposition of only the annihilation operators and ϕ_i^- is a superposition of only the creation operators. Suppose $x^0 > y^0$. Then,

$$T(\phi(x)\phi(y)) = \phi(x)\phi(y) =$$

$$(\phi(x)^+ + \phi(x)^-)(\phi(y)^+ + \phi(y)^-) =$$

$$= \phi(x)^+\phi(y)^+\phi(x)^+\phi(y)^- + \phi(x)^-\phi(y)^+ + \phi(x)^-\phi(y)^-$$

so that

$$T(\phi(x)\phi(y)) = \phi(x)^+\phi(y)^+ + \phi(y)^-\phi(x)^+ + \phi(x)^-\phi(y)^+ + \phi(x)^-\phi(y)^-$$
$$+[\phi(x)^+, \phi(y)^-]$$
$$= N(\phi(x)\phi(y)) + C(\phi(x), \phi(y))$$

More generally, Wick's theorem states that

$$T(\phi(x_1)\phi(x_2)...\phi(x_m)) = N(\phi(x_1)...\phi(x_m) + allpossiblecontractions)$$

If $x^{10} > x^{20} > x^{30}$, then prove this equation by explicit expansion.

$$T(\phi_1\phi_2\phi_3) =$$

$$(\phi_1^+ + \phi_1^-)(\phi_2^+ + \phi_2^-)(\phi_3^+ + \phi_3^-)$$
$$= \phi_1^+\phi_2^+\phi_3^+ + ...$$

We leave the details of manipulation to the interested reader.

12. Write down the total Lagrangian density of the Maxwell field interacting with the Dirac field and derive the Maxwell-Dirac field equations.

Hint:

$$\mathcal{L} = K_1(F_{\mu\nu})^2 + K_2\bar{\psi}(\gamma^\mu((i\partial_\mu - eA_\mu) - m)\psi$$

where γ^μ are the Dirac matrices, and $F_{\mu\nu} = \partial_\mu A_\nu - \partial_\nu A_\mu$ is the antisymmetric electromagnetic field tensor. The total Lagrangian for the Maxwell-Dirac field can be expressed as

$$\mathcal{L} = -\frac{1}{4}(F_{\mu\nu})^2 + \bar{\psi}(i\gamma.\partial - m_0)\psi - e_0\bar{\psi}\gamma^\mu\psi A_\mu$$

13. Let $\phi(x)$ be a scalar field defined for $x \in \mathbb{R}^4$. Assume that the Lagrangian density of the field is given by

$$\mathcal{L} = \frac{1}{2}\partial_\mu\phi\partial^\mu\phi - m^2\phi^2/2$$

Give an algorithm to compute the path integral

$$\int exp(\frac{i}{h}S[\phi])D\phi$$

where the path integral is over the space $\{\phi(x) : \mathbf{x} \in \mathbb{R}^3, t_a \le t \le t_b\}$ with $\phi_a = \phi(., t_a)$ and $\phi_b = \phi(., t_b)$ given. Note that the path integral can be expressed in three dimensional Fourier space by setting

$$\hat{\phi}(\mathbf{k}, t) = \int \phi(x)exp(i\mathbf{k}.\mathbf{x})d^3x$$

We leave the problem of expressing the action $S[\phi]$ in Fourier space.

Hint:

$$\int (\partial_\mu \phi . \partial^\mu \phi) d^3 x = - \int k_{\mu} k^\mu \phi(\mathbf{k}, t)^2 d^3 k$$
$$= - \int k^2 \phi(\mathbf{k}, t) d^3 k$$

14. Quantization of the electromagnetic field using path integrals. The photon propagator has the form $-ig_{\mu\nu}/(k^2 + i\epsilon)$. The Lagrangian density for the free electromagnetic field is given by $-\frac{1}{4}(F_{\mu\nu})^2$. The action function of the field is

$$S[A] = -\frac{1}{4} \int (F_{\mu\nu})^2 d^4 x$$

Recall that $F_{\mu\nu} = \partial_\mu A_\nu - \partial_\nu A_\mu$ and the path integral to be evaluated is given by

$$\int \mathcal{D}A . exp(iS[A])$$

Note that

$$(F_{\mu\nu})^2 = (A_{\nu,\mu} - A_{\mu,\nu})(A_{\nu,\mu} - A_{\mu,\nu}) = 2A_{\nu,\mu} A_{\nu,\mu} - 2A_{\nu,\mu} A_{\mu,\nu}$$
$$= -2A_{\nu,\mu\mu} A_\nu + 2A_{\nu,\mu\nu} A_\mu$$

on ignoring a perfect divergence. We can thus write

$$S[A] = \frac{1}{2} \int (A_\nu A_{\nu,\mu\mu} - A_\mu A_{\nu,\mu\nu}) d^4 x$$

$$= \frac{1}{2 \int ((-k^2)(A_\nu[k])^2 + k_\mu k_\nu A_\mu[k] A_\nu[k]}) d^4 k$$

This formula can be used in the evaluation of the Feynman path integral for the em field.

Gauge fixed Lagrangian:

$$L = -\frac{1}{4}(F_{\mu\nu})^2 + |D_\mu \phi|^2 - V(\phi)$$

where $F_{\mu\nu}$ is the antisymmetric field tensor of the electromagnetic field, D_μ is the covariant derivative defined by

$$D_\mu = \partial_\mu + ieA_\mu$$

and ϕ is a complex scalar field defined by

$$\phi = (\phi^1 + i\phi^2)/\sqrt{2}$$

The problem is to formulate the field equations using classical Lagrangian field theory and then quantize it using the path integral. Setting $\phi^1 = v + h$ where v is a constant deduce that by adding Gauge term to the Lagrangian, we arrive at a mass term for the Goldstone Boson field ϕ^2.

For the quantized electromagnetic field, deterime the propagator $< 0|T(A_\mu(x)A_\nu(y))|0 >$ from first principles. Hint: Expand the vector potential in terms of the annihilation and creation opeartors and use the commutation relations between these operators. Show that the propagator is proportional to the four dimensional Fourier transform of $g_{\mu\nu}/(k^2 + i\epsilon)$.

15. $U(x)$ is a spatially dependent unitary operator. We define a two component complex field

$$\phi(x) = 2^{-1/2}U(x)(0, v + h(x))^T$$

Consider the Lagrangian

$$L = |D_\mu \phi|^2 + \mu^2 \phi^* \phi - \lambda(\phi^* \phi)^2$$

where

$$D_\mu \phi = (\partial_\mu - ieA_{\mu,a}(x)\sigma_a)\phi$$

$A_{\mu,a}(x)$ are some appropriately defined field variables. We calcluate

$$\partial_\mu \phi = 2^{-1/2}(\partial_\mu U)(0, v + h)^T + 2^{-1/2}U(0, \partial_\mu h)^T$$

so that

$$D_\mu \phi = 2^{-1/2}(\partial_\mu U)(0, v + h)^T + 2^{-1/2}U(0, \partial_\mu h)^T - 2^{-1/2}ieA_\mu U(0, v + h)^T$$

Consider a term of the form $(D_\mu \phi)^*(D^\mu \phi)$ in the Lagrangian. We can eliminate the matrix $U(x)$ by a gauge transformation. Suppose for example that $U(x)$ is infinitesimal. We write

$$U(x) = I + i\alpha_a(x)\sigma_a$$

where α_a are small, ie, of the first order of smallness. The gauge transformation involves changing $A_{\mu,a}$ and ψ simultaneously by small amounts defined by the functions α_a to remove the effect of α_a on ψ. Specifically changing ψ to $(I + i\beta_a\sigma_a)\psi$ and $A_{\mu,a}$ to $A_{\mu,a} + X_{\mu,a}$ where β_a, α_a and $X_{\mu,a}$ are all of the first order of smallness, changes $D_\mu(U\psi)$ to

$$(\partial_\mu - ie(A_{\mu,a} + X_{\mu,a})\sigma_a)(I + i(\alpha_b + \beta_b)\sigma_b)\psi$$

$$= (I + i(\alpha_b + \beta_b)\sigma_b)(\partial_\mu - ie(A_{\mu,a} + X_{\mu,a})\sigma_a)\psi$$

$$+[\partial_\mu - ie(A_{\mu,a} + X_{\mu,a})\sigma_a, I + i(\alpha_b + \beta_b)\sigma_b]\psi$$

$$= (I + i(\alpha_b + \beta_b)\sigma_b)(\partial_\mu - ieA_{\mu,a}\sigma_a)\psi$$

$$+[-ieX_{\mu,a}\sigma_a + i.\partial_\mu(\alpha_b + \beta_b)\sigma_b + ieA_{\mu,a}(\alpha_b + \beta_b)\epsilon_{abc}\sigma_c]\psi$$

The coefficient of σ_a in the last bracket is given by

$$-ieX_{\mu,a} + i\partial_\mu(\alpha_a + \beta_a) + ieA_{\mu,b}(\alpha_c + \beta_c)\epsilon_{bca}$$

This can be made to vanish by appropriate choice of the functions $X_{\mu,a}$ in terms of β_a and when this happens, it is clear that $(D_\mu \phi).(D^\mu \phi^*)$ goes transforms to $(D_\mu \psi)^*(D^\mu \psi$ with the covariant derivatives defined by A_μ and not by $A_\mu + X_\mu$.

16. Consider the reaction in which an electron having four momentum p interacts with a positron having four momentum p' to produce first a photon having four momentum $q = p + p'$ and this photon then disintegrates into a muon and an antimuon having respectively momenta k and k'. The transition probability amplitude for this process is obtained as the matrix element

$$< k, k'|T(\int d^4xd^4y\bar{\psi}(x)\gamma^\mu\psi(x)\bar{\psi}(y)\gamma^\nu\psi(y)A_\mu(x)A_\nu(y))|p, p' >$$

The terms $A_\mu(x)A_\nu(y)$ contract to give the photon propagator $\Delta_{\mu\nu}(x - y)$ where

$$\Delta_{\mu\nu}(x) = \int exp(-iq.x)d^4q/(q^2 + i\epsilon)$$

The term $\bar{\psi}(y)$ contains an antifermion annihilation operator with a coefficient of $\bar{v}$ and hence can be contracted with $|p>$ to give $\bar{v}(p)exp(-ip.y)$. The term $\bar{\psi}(y)$ also contains a Fermion creation operator with a coefficient of $\bar{u}$ and hence can be contracted with $< k|$ to give $exp(ik.y)\bar{u}(k)$. We select this term for the computation of the amplitude. Likewise, the term $\psi(y)$ contains an Antifermion creation operator with a coefficient of v and hence can be contracted with $< k'|$ to give $exp(ik'.y)v(k')$. This gives one term in the product as $exp(i(k + k', y))\bar{u}(k)\gamma^\nu v(k')$. Likewise the other term in the product evaluates to $exp(-i(p + p', x))\bar{v}(p')\gamma^\mu u(p)$. The total amplitude is therefore given by

$$\int exp(i((k + k', y) - (p + p', x)))\bar{u}(k)\gamma^\nu v(k')\bar{v}(p')\gamma^\mu u(p)\Delta_{\mu\nu}(x - y)d^4xd^4y$$

$$= \int exp(i(k + k' - p - p', y))exp(-i(p + p', \xi))\bar{u}(k)\gamma^\nu v(k')\bar{v}(p')\gamma^\mu u(p)\Delta_{\mu\nu}(\xi)d^4\xi d^4y$$

$$= \delta(k + k' - p - p')\bar{u}(k)\gamma(\nu)v(k')\bar{v}(p')\gamma^\mu u(p)\hat{\Delta}_{\mu\nu}(p + p')$$

The coefficient of $\delta(k + k' - p - p')$ is the desired scattering amplitude. It equals

$$\bar{u}(k)\gamma^\nu v(k')\bar{v}(p')\gamma^\mu u(p)g_{\mu\nu}/(p + p')^2$$

We write q for $p + p'$, the total momentum.

Reference:[1] Peskin and Schroeder, An introduction to quantum field theory, [2] Steven Weinberg, Quantum Field Theory.

17.Dirac equation for the free electron

$$(i\gamma^\mu\partial_\mu - m)\psi = 0$$

Setting $\psi(x) = u(p)exp(-ip.x)$ gives

$$(\gamma.p - m)u(p) = 0$$

which may be termed as the algebraic Dirac equation. We take

$$\gamma^0 = \begin{pmatrix} 0 & 1 \\ 1 & 0 \end{pmatrix}, \gamma^r = \begin{pmatrix} 0 & -\sigma_r \\ \sigma_r & 0 \end{pmatrix}, r = 1, 2, 3$$

Note that $\gamma^{02} = i, \gamma^{r2} = 1$,

$$\gamma^r\gamma^s + \gamma^s\gamma^r = 0, r \neq s$$

and

$$\gamma^0\gamma^r = -\gamma^r\gamma^0, r = 1, 2, 3$$

These equations can be summarized as

$$\gamma^\mu\gamma^\nu + \gamma^\nu\gamma^\mu = 2g^{\mu\nu}$$

We have taking the conjugate transpose of the algebraic Dirac equation

$$u(p)^*(\gamma^{\mu *}p_\mu - m) = 0$$

where star denotes conjugate transpose. Now observe that

$$\gamma^{0*} = \gamma^0, \gamma^{r*} = -\gamma^r, r = 1, 2, 3$$

The above equation can be cast as

$$\bar{u}(p)(\gamma^0\gamma^{\mu *}p_\mu - \gamma^0 m) = 0$$

or

$$\bar{u}(p)(\gamma^0\gamma^{\mu *}\gamma^0 p_\mu - m) = 0$$

Now observe that

$$\gamma^0\gamma^{0*}\gamma^0 = \gamma^0,$$

$$\gamma^0\gamma^{r*}\gamma^0 = -\gamma^0\gamma^r\gamma^0 = \gamma^r, r = 1, 2, 3$$

Thus the above equation gives on postmultiplying by γ^0,

$$\bar{u}(p)(\gamma^\mu p_\mu - m) = 0$$

or equivalently,

$$\bar{u}(p)(\gamma.p - m) = 0$$

which can be called the conjugate of the algebraic Dirac equation.

18.Interaction picture representation of Schrodinger dynamics. Suppose that the Hamiltonian of the system is given by

$$H(t) = H_0 + V(t)$$

where H_0 is independent of time and $V(t)$ is a time dependent perturbation. The wave equation to be solved is

$$id\psi/dt = H(t)\psi_t$$

We define

$$\phi_t = exp(itH_0)\psi_t$$

and then we have

$$id\phi_t/dt = -H_0 exp(itH_0)\psi_t + exp(itH_0)H(t)\psi_t = exp(itH_0)V(t)\psi_t = V_{int}(t)\phi_t$$

where

$$V_{int}(t) = exp(itH_0)V(t)exp(-itH_0)$$

is the perturbing energy in the interaction representation. Suppose that the initial state of the system is $|\psi_i>$ (at time $t = t_i$). We wish to determine the probability that after time $t_f - t_t$, the state of the system will by $|\psi_f>$. We assume that at time $t < t_i$ the perturbing energy was switched off and so also at time $t > t_f$. Then $|\psi_i> = |\phi_i>$ and $|\psi_f> = |\phi_f>$. Upto linear orders in the interaction energy, we have

$$\phi_{t_f} = \phi_i - i\int_{t_i}^{t_f} V_{inft}(t)\phi_i dt$$

so the transition amplitude from the state $\psi_i = \phi_i$ to the state $\psi_f = \phi_f$ is given by

$$-i\int_{t_i}^{t_f} <\phi_f|V_{int}(t)|\phi_i> dt$$

Note that

$$<\phi_f|V(t)|\phi_i> = <\phi_f|exp(itH_0)V(t)exp(-itH_0)|\phi_i>$$

As an example, suppose that the unperturbed system is an an electron bound to the nucleus of an atom and that $V(t)$ is a weak time varying electric field acting on the electron, ie,

$$V(t) = -e(\mathbf{r}, \mathbf{E}(t))$$

Then

$$exp(itH_0)V(t)exp(-itH_0) = -e(\mathbf{r}(t), \mathbf{E}(t))$$

where $\mathbf{r}(t)$ is the position of the electron after time t under the influence of the atomic Hamiltonian only. Thus we can write

$$<\phi_f|V_{int}(t)|\phi_i> = -e <\phi_f|(\mathbf{r}(t), \mathbf{E}(t))|\phi_i>$$

and the transition probability amplitude from the initial state ϕ_i at time t_i to the final state ϕ_f at time t_f is proportional to the integral of the above quantity over the duration $[t_i, t_f]$ (Landau and Lifshitz, Quantum mechanics).

19. relationship between two and four component spinors.

1.By considering the algebraic Dirac equation, we assume that the bispinor wave function ψ can be expressed in block vector form (ϕ, χ) where ϕ and χ are two spinors. Then derive the relationship between ϕ and χ in the standard representation. Hint:

$$\gamma^0 = \begin{pmatrix} 0 & I \\ I & 0 \end{pmatrix}, \gamma^r = \begin{pmatrix} 0 & \sigma_r \\ -\sigma_r & 0 \end{pmatrix}, r = 1, 2, 3$$

Then observe that

$$\gamma.p = \gamma^\mu p_\mu = \begin{pmatrix} 0 & -(\sigma, p) \\ (\sigma, p) & 0 \end{pmatrix}$$

The Dirac equation

$$(\gamma.p - m)\psi = 0$$

then reads

$$E\chi - (\sigma, p)\chi - m\phi = 0, E\phi + (\sigma, p)\phi - m\chi = 0$$

We can change over from the spinor representation (ϕ, χ) to the standard representation by setting

$$u = (\phi + \chi)/\sqrt{2}, v = (\phi - \chi)/\sqrt{2}$$

so that

$$\phi = (u + v)/\sqrt{2}, \chi = (u - v)/\sqrt{2}$$

plugging this into the spinor equations (the Weyl equation) then gives us

$$E(u - v) - (\sigma, p)(u - v) - m(u + v) = 0, E(u + v) + (\sigma, p)(u + v) - m(u - v) = 0$$

so that

$$Eu + (\sigma, p)v - mu = 0, Ev + (\sigma, p)u + mv = 0$$

or

$$(E - m)u + (\sigma, p)v = 0, (E + m)v + (\sigma, p)u = 0$$

which implies

$$v = -(\sigma, p)u/(E + m), u = -(\sigma, p)v/(E - m)$$

The two equations are consistent since from the first, we get

$$(\sigma, p)v = -p^2 u/(E + m) = -(E^2 - m^2)u/(E + m) = -(E - m)u$$

which is precisely the second equation. The normal convention is to set

$$u = \sqrt{E + mw}$$

so that

$$v = -\sqrt{E + m}(\sigma, p)w$$

and the bispinor for the free Dirac particle can then be expressed as

$$\psi = (\sqrt{E + mw}, -\sqrt{E + m}(\sigma, p)w)$$

20. Write down the radial wave equation for the hydrogen atom and determine the unbounded states. Assume that the state consists of an incident wave of particles that interact with the nucleus and then get scattered to ∞. Write down the wave equation for the scattered system.

Hint: Let $U(r)$ be any radial potential. Assuming Azimuthal symmetry, we can write the wave function as

$$\psi(r, \theta) = \sum_{l-0}^{\infty} R_l(r).P_l(cos\theta)$$

Inserting this into the stationary state Schrodinger equation gives us

$$-\frac{h^2}{2mr}(r.R_l(r))'' + l(l+1)h^2R_l(r)/2mr^2 = ER_l(r)$$

Let

$$E = h^2k^2/2m$$

Then for different choices of $U(r)$, determine the asymptotical wave functions $R_l(r)$ by numerical methods. These wavefunctions will be parametrized by l, k. In the continuous spectrum (unbounded states), k can assume a continuum of values in a given range. In the discrete spectrum (bounded case), k assumes discrete values. DSP implementation of this proceeds by discretizing the differential equation.

21. Atom in interaction with radiation. Let the vector potential be $\mathbf{A}(\mathbf{r})$. The momentum operator of the atom is $\mathbf{p} = -ih\nabla$. The interaction energy between the atom and the radiation is obtained as the cross term in $(\mathbf{p} - e\mathbf{A})^2/2m$. This cross term equals $H_{int} = -e(\mathbf{A}, \nabla)/m$. Note that this is the non-relativistic picture of the interaction. Assuming that $\mathbf{A}$ is a plane wave, ie,

$$\mathbf{A}(\mathbf{r}) = \mathbf{A}_0 exp(i\mathbf{k}.\mathbf{r}) + \mathbf{A}_0^* exp(-i\mathbf{k}.\mathbf{r})$$

we get for the interaction energy,

$$H_{int} = \frac{ihe}{m} exp(i\mathbf{k}.\mathbf{r})(\mathbf{A}_0, \nabla) - \frac{ihe}{m} exp(-i\mathbf{k}.\mathbf{r})(\mathbf{A}_0^*, \nabla)$$

More precisely, we should include time dependence by assuming the magnetic vector potential to have the form

$$\mathbf{A}(\mathbf{r}, t) = 2\mathbf{A}_0 cos(\omega t - \mathbf{k}.\mathbf{r} + \delta) = \mathbf{A}_0 exp(i(\omega t - \mathbf{k}.\mathbf{r})) + \mathbf{A}_0^* exp(-i(\omega t - \mathbf{k}.\mathbf{r}))$$

where on the rightmost, $\mathbf{A}_0$ is a phasor having the same amplitude as the real vector $\mathbf{A}_0$ but with a phase angle δ. The interaction energy operator is given by

$$V(t) = \frac{ihe}{m}(\mathbf{A}(\mathbf{r}, t), \nabla)$$

The matrix element of this interaction energy determines upto first order, the probability of transition of the atom between two stationary states $|\psi_i>$ and $|\psi_f>$ per unit time. This element is given by

$$V_{fi}(t) =<\psi_f|V(t)|\psi_i>$$

$$= exp(i\omega t)\int \psi_f(\mathbf{r})^*(\mathbf{A}_0, \nabla)\psi_i(\mathbf{r}) exp(-i\mathbf{k}.\mathbf{r})d^3r$$

$$+exp(-i\omega t)\int \psi_f(\mathbf{r})^*(\mathbf{A}_0^*, \nabla)\psi_i(\mathbf{r}) exp(i\mathbf{k}.\mathbf{r})d^3r$$

Suppose ω_{fi} is the difference between the energy of the final and initial state, ie,

$$\omega_{fi} = (E_f - E_i)/h$$

Then the transition probability amplitude in duration $[0, t]$ is given by

$$\int_0^t V_{fi}(t)exp(-i\omega_{fi}t)dt$$

22.Atom in interaction with radiation; relativistic description. The Dirac Hamiltonian is given by

$$H = (\alpha, \mathbf{p} + e\mathbf{A}(\mathbf{r}, t)) + \beta mc^2 + e\Phi$$

The interaction part of the Hamiltonian, that is the part that describes the interaction between the atom and the electromagnetic radiation field is given by

$$V(t) = e(\alpha, \mathbf{A}(\mathbf{r}, t))$$

Note that $\Phi(\mathbf{r})$ is the electrostatic potential of the atom and $\mathbf{A}(\mathbf{r}, t)$ is the magnetic vector potential of the radiation field that impinges on the atom. Assume that ψ_i and ψ_f are two stationary states of the atom. These satisfy the stationary Dirac equation:

$$((\alpha, \mathbf{p}) + \beta mc^2 + e\Phi)\psi_a = E_a\psi_a, a = i, f$$

Then under the influence of radiation, the probability amplitude per unit time for transition from state ψ_i to state ψ_f is given by the matrix element of the interaction energy:

$$V_{fi}(t) =< \psi_f|V(t)|\psi_i >= e\int \psi_f(\mathbf{r})^*(\alpha, \mathbf{A}(\mathbf{r}, t))\psi_i(\mathbf{r})d^3r$$

where $\alpha = (\alpha_i, i = 1, 2, 3)$ are the Dirac α matrices.

23. Electron wave function in the Dirac formalism:

$$\psi(x) = C.u(p)exp(ip.x)$$

where $u(p)$ is a bispinor satisfying the Dirac equation

$$(\gamma.p - m)u(p) = 0$$

We assume the normalization

$$\bar{u}(p)u(p) = 2m$$

where

$$\bar{u}(p) = u(p)^*\gamma^0$$

Then, the algebraic Dirac equation implies

$$\bar{u}(p)\gamma u(p).p = m\bar{u}(p)u(p) = 2m^2 = 2p^2$$

so that

$$\bar{u}(p)\gamma u(p) = 2p$$

The Dirac four current density is given by

$$j(x) = \bar{\psi}(x)\gamma\psi(x) = |C|^2\bar{u}(p)\gamma u(p) = 2p|C|^2$$

If our normalization is such that there is one charged particle in a box of volume V, then the current density must be given by $(1, \mathbf{v})$ where $\mathbf{v} = \mathbf{p}/\epsilon$ with ϵ the energy. Note that $\epsilon = \sqrt{m^2 + \mathbf{p}^2}$. Thus we must have $j(x) = p/\epsilon$ and this implies that $2|C|^2 = 1/\epsilon$, or equivalently, $C = (2\epsilon)^{-1/2}$..

Ref:Landau, Berseteski, Lifshitz, Quantum Electrodynamics.

24. Let $\phi(t, \mathbf{r}) = \phi(x)$ denote the massive Boson scalar field. Assuming that the Boson has mass m, it follows that $\phi(x)$ satisfies the Klein-Gordon equation

$$\partial_\mu\partial^\mu\phi + m^2\phi = 0$$

or expressed explicitly in terms of space and time coordinates,

$$(\partial_t^2 - \nabla^2 + m^2)\phi = 0$$

We try a plane wave solution

$$\phi(x) = A.exp(ip.x)$$

where

$$p.x = p^0x^0 - \sum_{r=1}^{3} p^r x^r = p^\mu x_\mu = p_\mu x^\mu$$

Plugging this into the Klein-Gordon equation gives us

$$p^\mu p_\mu = m^2$$

or in the general short hand notation of field theory,

$$p^2 = m^2$$

The solution of this equation is

$$p^0 = \pm\sqrt{m^2 + \mathbf{P}^2}$$

The positive sign corresponds to positive energy particles while the negative sign corresponds to negative energy particles. The general solution to Klein-Gordon equation is given by a sum of positive energy and negative energy parts:

$$\phi(x) = \int (a(\mathbf{p})exp(ip.x) + a(\mathbf{p})^*exp(-ip.x))d^3p$$

After second quantization, $a(\mathbf{p})$ becomes the annihilation operator of a boson having momentum $\mathbf{p}$ and $a(\mathbf{p})^*$ becomes the creation operator of a boson having momentum $\mathbf{p}$. The Lagrangian density of the field is given by

$$\mathcal{L}(\phi) = \frac{1}{2}(\partial_\mu\phi)(\partial^\mu\phi) - \frac{1}{2}m^2\phi^2$$

We have

$$\delta\mathcal{L} = \partial_\mu\phi\delta\partial^\mu\phi - m^2\phi\delta\phi$$
$$= \partial^\mu(\partial_\mu\phi\delta\phi) - \delta\phi.\partial^\mu\partial_\mu\phi - \delta\phi.m\phi$$

On integrating this over the whole space-time manifold $\mathbb{R}^4$, the first term, namely the perfect divergence cancels out and since the variation $\delta\phi$ is arbitrary, we end up with the pde

$$\partial^\mu\partial_\mu\phi + m^2\phi = 0$$

which is the Klein-Gordon equation. The Hamiltonian density can be calculated from this expression and then by substituting for the field expression in terms of the creation and annihilation operators, we can obtain an expression for the Hamiltonian in terms of the creation and annihilation operators. The procedure is as follows:

$$\pi = \frac{\partial L}{\partial\phi_t} = \partial_t\phi$$

Then the Hamiltonian density equals

$$\mathcal{H} = \pi\partial_t\phi - L = \frac{1}{2}((\partial_t\phi)^2 + (\nabla\phi)^2 + m^2\phi^2)$$

and the Hamiltonian of the field is

$$H = \int \mathcal{H}d^3x$$

We leave it as an exercise to evaluate the Hamiltonian in terms of the creation and annihilation operator fields.

25. Compute the Feynman propagator for the massive boson field. Specialize to the case of the massless boson field.

Hint: Let $\phi(x)$ be the second quantized Boson field. Express ϕ in terms of the creation and annihilation operators. Choose two space-time points x, x' such that $x^0 > x^{0'}$. The propagator is given by

$$D(x - x') = < 0|T(\phi(x)\phi(x')|0 > = < 0|\phi(x)\phi(x')|0 >$$

Write

$$\phi(x) = \phi_+(x) + \phi_-(x), \phi(x') = \phi_+(x') + \phi_-(x')$$

where ϕ_+ consists only of boson annihilation operators and ϕ_- consists of only boson creation operators. Now observe that

$$\phi(x)\phi(x') = \phi_+(x)\phi_+(x') + \phi_+(x)\phi_-(x') + \phi_-(x')\phi_+(x') + \phi_-(x)\phi_-(x')$$

Also

$$\phi_+(z)|0 >= 0, < 0|\phi_-(z) = 0$$

and hence

$$D(x - x') = < 0|\phi_+(x)\phi_-(x')|0 > = [\phi_+(x), \phi_-(x')]$$

since the latter is a c-number. This commutator can be easily evaluated.

Reference:Bjorken and Drell, Relativistic Quantum Fields.

26. Derive the form of the photon propagator from first principles.

Hint: Let $A_\mu(x)$ be the electromagnetic four potential. Expand this in terms of the creation and annihilation operators of the field, ie,

$$A_\mu(x) = \sum_{s=1}^{2} \int (2E(p))^{-1/2}(a_s(p)\epsilon(s,\mu)exp(-ip.x) + a_s(p)^*\epsilon(s,\mu)exp(ip.x))d^3p$$

where $s = 1,2$ are the polarization states of the photon and $\epsilon(s,\mu)$ are the polarization four vectors. They satisfy the orthogonality relation $k^\mu\epsilon(s,\mu) = 0$. Introduce the commutation relation

$$[a_s(p), a_{s'}(p')^*] = \delta_{s,s'}\delta(\mathbf{p}-\mathbf{p}')$$

and use the facts that

$$a_s(p)|0>= 0, < 0|a_s(p)^* = 0$$

to arrive at the form of the photon propagator

$$D_{\mu\nu}(x-x') =< 0|T(A_\mu(x).A_\nu(x'))|0>$$

where T is the time ordering operator.

27. Suppose $u_s(p), s = 1,2$ satisfy the Dirac equation

$$(\gamma.p - m)u_s(p) = 0$$

Define

$$\bar{u}_s(p) = u_s(p)^*\gamma^0$$

where γ^μ are the Dirac matrices. Show that

$$\bar{u}_s(p)(\gamma.p - m) = 0, s = 1,2$$

Show that the amplitudes u_s can be normalized so that

$$\bar{u}_s(p)u_{s'}(p) = 2m\delta_{s,s'}$$

Deduce that

$$\sum_s u_s(p)\bar{u}_s(p) = \gamma.p + m$$

Hint: We first solve

$$(\gamma.p - m)u_s(p) = 0$$

Write

$$\gamma^0 = \begin{pmatrix} 0 & I \\ I & 0 \end{pmatrix}, \gamma^r = \begin{pmatrix} 0 & \sigma_r \\ -\sigma_r & 0 \end{pmatrix}, r = 1,2,3$$

so that

$$\begin{pmatrix} -m & p^0 - (\sigma, \mathbf{p}) \\ p^0 + (\sigma, \mathbf{p}) & -m \end{pmatrix} u_s(p) = 0$$

Write

$$u_s(p) = \begin{pmatrix} \xi \\ \eta \end{pmatrix}$$

giving

$$-m\xi + (p^0 - (\sigma, \mathbf{p}))\eta = 0, (p^0 + (\sigma, \mathbf{p}))\xi - m\eta = 0$$

The two equations are consistent for on multiplying the former by $p^0 + (\sigma, \mathbf{p})$ and using $p^{02} - \mathbf{p}^2 = m^2$, we get

$$m^2\eta - m(p^0 + (\sigma, \mathbf{p}))\xi = 0$$

which is precisely the latter. Thus, the solution can be written as

$$u(p) = (m\xi, (E + (\sigma, \mathbf{p}))\xi)$$

where $E = p^0 = \sqrt{\mathbf{p}^2 + m^2}$. Now, we can choose $\xi = \xi_1$ so that it is an eigenvector of $(\sigma, \mathbf{p})$ with eigenvalue $|\mathbf{p}|$ and we can choose $\xi = \xi_2$ so that it is an eigenvector of the same with eigenvalue $-|\mathbf{p}|$. it is clear that $\xi_i, i = 1, 2$ are orthogonal, ie, $\xi_1^*\xi_2 = 0$ since they are both eigenvectors of a Hermitian operator $(\sigma, \mathbf{p})$ corresponding to distinct eigenvalues $\pm|\mathbf{p}|$. Then let $u_i(p)$ be $u(p)$ with $\xi = \xi_i, i = 1, 2$. Then observe that

$$\gamma^0 u(p) = ((E + (\sigma, \mathbf{p}))\xi, m\xi)$$

$$\bar{u}_1(p)u_1(p) = u_1(p)^*\gamma^0 u_1(p) = 2m\xi_1^*(E + \sigma, \mathbf{p}))\xi_1 = 2m(E + |\mathbf{p}|)\xi_1^*\xi_1$$

$$\bar{u}_1(p)u_2(p) = 2m\xi_1^*(E + (\sigma, \mathbf{p}))\xi_2 = 0$$

since ξ_1 and ξ_2 are orthogonal and ξ_2 is eigen to the operator $E + (\sigma, \mathbf{p})$. The remaining properties can be proved along similar lines.

28. Derivation of the Feynman rules from path integrals: The Hamiltonian of the scalar field is given by

$$H = \int d^3x[\pi^2/2 + (\nabla\phi)^2/2 + V(\phi)]$$

The classical field equations are obtained by setting $\pi = \partial_t\phi$ and these equations are

$$\partial_t\phi = \frac{\delta H}{\delta\pi} = \pi,$$

$$\partial_t\pi = -\frac{\delta H}{\delta\phi} = \nabla^2\phi - V'(\phi)$$

Combining the two, we get

$$\partial_t^2\phi - \nabla^2\phi + V'(\phi) = 0$$

Let $\phi_a(\mathbf{x})$ denote the state of the field at time t_a and $\phi_b(\mathbf{x})$ the state at time $t_b = t_a + T$. Under the quantum dynamics of the field, the transition probability amplitude from the former to the latter, is, according to the Feynman path integral formulation of quantum mechanics, given by

$$< \phi_b(\mathbf{x})|exp(-iHT)|\phi_a(\mathbf{x}) >$$

$$= \int \mathcal{D}\phi\mathcal{D}\pi.exp(i\int_{[0,T]\times\mathbb{R}^3} d^4x(\pi\partial_t\phi - \pi^2/2 - (\nabla\phi)^2/2 - V(\phi)))$$

where the path integral over the momentum field is the entire region and that over the position field is between the times t_a and t_b with the constraints on the endpoints $\phi_a(\mathbf{x}), \phi_b(\mathbf{x})$. The momentum integral can be transformed into a Gaussian integral using

$$\pi\partial_t\phi - \pi^2/2 = -\frac{1}{2}(\pi - \partial_t\phi)^2 + (\partial_t\phi)^2/2$$

Thus, we obtain for the transition amplitude,

$$< \phi_b(\mathbf{x})|exp(-iHT)|\phi_a(\mathbf{x}) >= \int \mathcal{D}\phi.exp(i\int_{[0,T]\times\mathbb{R}^3} ((\partial_t\phi)^2 - V(\phi))d^4x)$$

Correlation functions: Consider the computation of the correlation function

$$\int \mathcal{D}\phi(x).\phi(x_1)\phi(x_2)exp(i\int_{[-T,T]} \mathcal{L}(\phi)d^4x)$$

with the boundary condition

$$\phi_a(\mathbf{x}) = \phi(-T, \mathbf{x}), \phi_b(\mathbf{x}) = \phi(T, \mathbf{x})$$

Here, x_1, x_2 are fixed points in space time and $x_1^0 < x_2^0$. By writing

$$\int_{[-T,T]} \mathcal{L}(\phi)d^4x = (\int_{[-T,x_1^0]} + \int_{[x_1^0,x_2^0]} + \int_{[x_2^0,T]})\mathcal{L}d^4x$$

it easily follows that the above correlation function is the same as

$$\int \mathcal{D}\phi_1(\mathbf{x})\mathcal{D}\phi_2(\mathbf{x})\phi_1(\mathbf{x}_1)\phi_2(\mathbf{x}_2) < \phi_b|exp(-iH(T - x_2^0))|\phi_2 >$$

$$< \phi_2|exp(-iH(x_2^0 - x_1^0))|\phi_1 >< \phi_1|exp(-iH(x_1^0 + T))|\phi_a >$$

Let $\phi_S(\mathbf{x}_1)$ be the Schrodinger operator defined by

$$\phi_S(\mathbf{x}_1)|\phi_1 >= \phi_1(\mathbf{x}_1)|\phi_1 >$$

and likewise, $\phi_S(\mathbf{x}_2)$. Then, the above correlation function can be expressed as

$$< \phi_b|exp(-iH(T - x_2^0))\phi_S(\mathbf{x}_2)exp(-iH(x_2^0 - x_1^0))\phi_S(\mathbf{x}_1)exp(-iH(x_1^0 + T))|\phi_a >$$

This follows from the spectral resolution

$$\phi_S(\mathbf{x}_1) = \int |\phi_1 > \phi_1(\mathbf{x}_1) < \phi_1|\mathcal{D}\phi_1(\mathbf{x})$$

and likewise for $\phi_S(\mathbf{x}_2)$. Thus, we get for the above correlation function the above formula in terms of Canonical Schrodinger theory. $\phi_S(\mathbf{x}_1)$ is a Schrodinger variable. It evolves after time t under the Hamiltonian H to the Heisenberg variable $\phi_H(x_1)$ according to the rule

$$\phi_H(x_1) = exp(iHx_1^0)\phi_S(\mathbf{x}_1)exp(-iHx_1^0)$$

Likewise,

$$\phi_H(x_2) = exp(iHx_2^0)\phi_S(\mathbf{x}_2)exp(-iHx_2^0)$$

Thus, the above correlation can be expressed in terms of Heisenberg operators as

$$< \phi_b|exp(-iHT)\phi_H(x_2)\phi_H(x_1)exp(iHT)|\phi_a >$$

Let $|\Omega >$ denote the vacuum state with energy E_0 and let $|n >$ denote the state of the field with energy $E_n, n \geq 1$. Note that $E_0 < E_n, n = 1, 2, ...$ The state $|phi_a >$ can be expressed as

$$|phi_a >= |\Omega >< \Omega|\phi_a > + \sum_{n\geq 1} |n >< n|\phi_a >$$

so that

$$exp(iH(T(1+i\epsilon))|\phi_a >= |\Omega >< \Omega|\phi_a > exp(iE_0T(1+i\epsilon)) + \sum_{n\geq 1} exp(iE_nT(1+i\epsilon))|n >< n|\phi_a >$$

Here we have replaced T by $T(1 + i\epsilon)$ where ϵ is a small positive real number. As $T \to \infty$, we have $exp(-\epsilon E_nT) << exp(-\epsilon E_0T)$ and hence the above can be approximated for large T as

$$|\Omega >< \Omega|\phi_a > exp(iE_0T(1+i\epsilon))$$

Likewise, $< \phi_b|exp(-iHT(1-i\epsilon))$ in this limit, can be replaced by $< \phi_b|\Omega >< \Omega|exp(-iE_0T(1-i\epsilon))$. With these modifications, the above correlation function becomes in the limit as $T \to \infty$ so that $exp(iHT)|\phi_a >\to |\Omega >$ and $exp(iHT)|\phi_b >\to |\Omega >$,

$$< \phi_b|\Omega >< \Omega|\phi_a >< \Omega|\phi_H(x_2)\phi_H(x_1)|\Omega >$$

If we remove the terms $\phi(x_1)$ and $\phi(x_2)$ from the correlations, then the last factor will be replaced by unity. This means that we have the following elegant expression for the two point correlation function or the propagator:

$$< \Omega|\phi(x_1)\phi(x_2)|\Omega >$$

$$= \frac{\int \mathcal{D}\phi(x)\phi(x_1)\phi(x_2)exp(i\int_{[-T,T]}\mathcal{L}(\phi)d^4x)}{\int \mathcal{D}\phi(x)exp(i\int_{[-T,T]}\mathcal{L}(\phi)d^4x)}$$

This is the fundamental equation that is used to transform quantities related to the propagator to functional integrals.

Ref:Peskin and Schroeder, An introduction to Quantum Field Theory.

29.Feynman diagrams for ϕ^4-theory. The aim is to develop computational techniques for evaluating the n-point correlation function Suppose $H_I(t)$ is the interaction Hamiltonian. Under the interaction Hamiltonian $H_I(t)$, the evolution operator is given by the formula

$$U(t_2,t_1) = T\{exp(-i\int_{t_1}^{t_2} H_I(t)dt)\}$$

$$= I - i\int_{t_1}^{t_2} H_I(t)dt + (-i)^2\int_{0<t_1<t_2<t} H_I(t_2)H_I(t_1)dt_1dt_2 + ...$$

$$= I + \sum_{n=1}^{\infty}(-i)^n\int_{0<t_1<...<t_n<t} H_I(t_n)H_I(t_{n-1})...H_I(t_1)dt_1...dt_n$$

The evolution operator satisfies the Schrodinger equation corresponding to the interaction Hamiltonian:

$$i\frac{\partial U(t,t')}{\partial t} = H_I(t)U(t,t')$$

Note that

$$H = H_0 + H_{int}, H_I(t) = exp(iH_0(t-t_0))H_{int}exp(-iH_0(t-t_0))$$

We claim that

$$U(t,t') = exp(iH_0(t-t_0))exp(-iH(t-t'))exp(-iH_0(t'-t_0))$$

Indeed differentiating this expression with respect to time gives

$$i\frac{\partial U(t,t')}{\partial t} = exp(iH_0(t-t_0))(H-H_0)exp(-iH(t-t'))exp(-iH_0(t'-t_0))$$

$$= exp(iH_0(t-t_0))H_{int}exp(-iH(t-t'))exp(-iH_0(t'-t_0))$$

$$= exp(iH_0(t-t_0))H_{int}exp(-iH_0(t-t_0))exp(iH_0(t-t_0))exp(-iH(t-t'))exp(-iH_0(t'-t_0))$$

$$= H_I(t)U(t,t')$$

proving the claim.

30.

1.Two state quantum systems and qubits: The state here at time t can be represented as $\psi(t) = (c_1(t), c_2(t))^T$ where $c_1(t)$ and $c_2(t)$ are complex numbers satisfying $|c_1(t)|^2 + |c_2(t)|^2 = 1$ for all t. At time t, the probability that the system will be found in the upspin state $|1>$ is then $|c_1(t)|^2$ and the probability that it will be found in the downspin state is $|c_2(t)|^2$. The evolution dynamics of the state is governed by Schrodinger's equation.

2. n qubit states as vectors of size 2^n. Representation of such states is given by $|x_1x_2...x_n>$ where each x_i is either a one or a zero. A general state of an n-qubit system has the form

$$\psi = \sum_{x_1,...,x_n=0,1} c(x_1x_2...x_n)|x_1x_2...x_n>$$

where $c(x_1...x_n)$ are complex numbers satisfying

$$\sum_{x_1...x_n} |c(x_1...x_n)|^2 = 1$$

3. Quantum gates: A quantum gate on an n qubit system is an $2^n \times 2^n$ unitary matrix U. Examples of single qubit gates are (a) NOT (b) Hadamard. Examples of two qubit gates are swap and CNOT.

4.Quantum system simulation: Involves simulation of Schrodinger and Heisenberg dynamics. To simulate the Schrodinger dynamics in a prescribed potential $U(x, y, z)$, the partial differential equation needs to be discretized while to simulate the Heisenberg dynamics, the infinite matrices are to be truncated to get finite matrices and the first order differential equation for the truncated observables must be discretized in time thereby yielding a discrete time first order difference equation. Specifically, if $X(t)$ is the observable at time t.

31. Treatment of a collection of Bosons in radiation theory: Let $|\alpha_1 >, |\alpha_2 >, ..., |\alpha_m >$ be orthogonal states. There are N particles in the system numbered $1, 2, ..., N$. The state $|\alpha(i, a_i) >$ corresponds to the ith particle being in the state $|\alpha_{a_i} >$. Thus, the state of the entire system is given by

$$|\alpha(1, a_1) > |\alpha(2, a_2) > ...|\alpha(N, a_N) >$$

This states corresponds to the situation when for each $i = 1, 2, ..., N$, the ith particle is in the state $|\alpha_{a_i} >$. This situation is okay as long as the particles are distinguishable. If they are indistinguishable, then we must apply the symmetrizing operator to the above ket. We then get the state

$$|\alpha(a_1)\alpha(a_2)...\alpha(a_N) >= S|\alpha(1, a_1) > ...|\alpha(N, a_N) >$$

The symmetrizing operator S can be expressed as

$$S = (n!)^{-1/2} \sum P$$

where P runs over all the permutations of the particles. More precisely, we have

$$|\alpha(a_1)\alpha(a_2)...\alpha(a_N) >= (n!)^{-1/2} \sum_{\sigma} |\alpha(\sigma 1, a_1) > ...|\alpha(\sigma N, a_N) >$$

Suppose that in the sequence $a_1, ..., a_N$, k appears n_k times, so that $n_1 + ... + n_k = N$. Then it is easy to see that the norm of the above symmetrized ket is given by

$$n_1!n_2!...n_k!$$

It follows that the normalized symmetrized ket is given by

$$(n_1!...n_k!)^{-1/2}|\alpha(a_1)...\alpha(a_N) >$$

We can represent these states (Dirac, Principles of quantum mechanics) using the standard oscillator format. Corresponding to each ket $|\alpha_i >$, introduce an annihilation operator η_i^* and a creation operator η_i. In addition, introduce the vacuum state $|0 >$. Then $\eta_{a_1}\eta_{a_2}...\eta_{a_N}|0 >$ is the state $|\alpha(a_1)...\alpha(a_N) >$.

This state is the same as $\eta_1^{n_1}\eta_2^{n_2}...\eta_k^{n_k}|0>$. Note that the $\eta_i's$ all commute amongst themselves. The normalized state $|n_1, n_2, ..., n_k>$ in which n_i bosons are in the state $|\alpha_i>$ thus has the form

$$|n_1 n_2 ... m_k >= (n_1!...n_k!)^{-1/2}\eta_1^{n_1}...\eta_k^{n_k}|0>$$

We introduce the observable n_i by the equation

$$n_i + 1 = \eta_i^*\eta_i, \eta_i\eta_i^* = n_i$$

This is a postulate, ie

$$[\eta_i^*, \eta_i] = 1$$

Then

$$\eta_i|n_1...n_k >= (n_1!...n_k!)^{-1/2}\eta_1^{n_1}...\eta_i^{n_i+1}...\eta_k^{n_k}|0>$$

$$= (n_1!...(n_i+1)!...n_k!)^{1/2}/(n_1!...n_k!)^{1/2}\eta_1^{n_1}...\eta_i^{n_i+1}...\eta_k^{n_k}|0>$$

$$= (n_i+1)^{1/2}|n_1...(n_i+1)...n_k>$$

Likewise,

$$\eta_i^*|n_1...n_k >= (n_1!...n_k!)^{-1/2}\eta_1^{n_1}...\eta_i^*\eta_i.\eta_i^{n_i-1}...\eta_k^{n_k}|0>$$

$$= (n_1!...n_k!)^{-1/2}(\eta_i^*\eta_i).\eta_1^{n_1}...\eta_i^{n_i-1}...\eta_k^{n_k}|0>$$

$$= (n_1!...n_k!)^{-1/2}(n_1!...(n_i-1)!...n_k!)^{1/2}(\eta_i^*\eta_i)|n_1...(n_i-1)...n_k>$$

$$= \sqrt{n_1}|n_1...(n_i-1)...n_k>$$

Now suppose that the Hamiltonian of the field of bosons interacting with some other objec has the form

$$H = \sum_i(\mu(i)\eta_i + \bar{\mu}(i)\eta_i^*)$$

$\mu(i)$ are complex scalars or more accurately, these are functions of observables related to the object with which the radiation field interacts. Then

$$H|n_1...n_k >= \sum_i \mu(i)(n_i+1)^{1/2}|n_1...(n_i+1)...n_k> +\bar{\mu}(i)(n_i)^{1/2}|n_1...(n_i-1)...n_k>$$

so that

$$< n_1...(n_i+1)...n_k|H|n_1...n_k >= \mu(i)(n_i+1)^{1/2}$$

This means that the probability of the emission of a boson into the ith state under the interaction is proportional to n_i+1. Likewise, the probability of absorption of a boson from the ith state is proportional to n_i. The study of the entire radiation field with a variable number of particles is accomplished by the introduction of the Fock space. To define this space, we first define the space $\mathcal{H}_N$ as being spanned by the kets $|\alpha(a_1)...\alpha(a_N)>$. The Hilbert space $\mathcal{H}_N$ is then the N-particle space. The direct sum of $\mathcal{H}_N$ as N assumes values $0, 1, 2, ...$ is the Fock space $\Gamma(\mathcal{H})$ where $\mathcal{H} = \mathcal{N}_1$. The Fock space has been used extensively by Hudson and Parthasarathy in the development of quantum Brownian motion and quantum stochastic calculus.

32.Simulation of quantum radiation field: First the massive scalar field is simulated. This involves Fourier resolution. $\phi(\mathbf{x}, t)$, the scalar field satisfies the Klein-Gordon equation

$$(\nabla^2 - \frac{\partial^2}{\partial t^2} - m^2)\psi(\mathbf{x}, t) = 0$$

or in the language of four vectors,

$$(\partial_\mu \partial^\mu + m^2)\phi = 0$$

In the quantum theory of radiation, $\phi(x)$ becomes an operator, ie, it has the form

$$\phi(x) = \int (2(m^2 + \mathbf{p}^2))^{-1/2}[a(\mathbf{p})exp(-ip.x) + a(\mathbf{p})^* exp(ip.x)]d^3p$$

where $a(\mathbf{p})$ is the annihilation operator for a particle of momentum $\mathbf{p}$ and $a(\mathbf{p})^*$ is the creation operator for a particle of momentum $\mathbf{p}$. They satisfy the commutation rules

$$[a(\mathbf{p}), a(\mathbf{p}')^*] = \delta^3(\mathbf{p} - \mathbf{p}')$$

The quantization of the electromagnetic field is harder. We have to take into account the fact that the polarization of the radiation is perpendicular to the direction of the wave propagation. Suppose $A_\mu(x)$ is the covariant four potential. Here, $x = (\mathbf{x}, t)$ with $\mathbf{x} = (x_1, x_2, x_3)$ the position vector and t the time variable. it satisfies the wave equation

$$(\nabla^2 - \frac{\partial^2}{\partial t^2})A_\mu(x) = 0$$

This equation is supported by the gauge condition

$$\frac{\partial A_\mu}{\partial x_\mu} = 0$$

Hence, we can expand the four potential as a superposition of plane waves:

$$A_\mu(x) = \int [\hat{A}_\mu(k)exp(ik.x) + \hat{A}_\mu(k)^* exp(-ik.x)]d^3k$$

where

$$k = (k^\mu) = (\omega, \mathbf{k}), \omega = |\mathbf{k}|$$

and

$$k.x = \omega t - (\mathbf{k}, \mathbf{x}) = k^\mu x_\mu = k_\mu x^\mu$$

The gauge condition implies that

$$k^\mu A_\mu = 0$$

We let the roman indices $r, s, ...$ assume values $1, 2, 3$ in contrast to the greek indices $\mu, \nu, ...$ which assume values $0, 1, 2, 3$. Write $A^0 = V$. Then the gauge condition gives

$$k_\mu A^\mu = 0$$

or

$$k^0 V = k^r A^r$$

Equivalently,

$$V = k^r A^r / k^0 = (\mathbf{k}, \mathbf{A})/\omega$$

Here, $\mathbf{A} = (A^r)_{r=1}^3$. We decompose the vector potential $\mathbf{A}$ into a transverse part and a longitudinal part. The transverse part, $\mathcal{A}^r$ is given by

$$\hat{\mathcal{A}}^r = \hat{A}^r - k^r k^s \hat{A}^s/\omega^2$$

The longitudinal part is evidently given by

$$\hat{L}^r = k^r k^s \hat{A}^s/\omega^2$$

Thus the longitudinal part is the projection of the vector potential along the direction of propagation and the transverse part is obtained after removing the longitudinal part from the complete potential (Dirac, Principles of Quantum Mechanics, last chapter on "Quantum Electrodynamics"). We note that in view of the gauge condition, the scalar potential V is expressible completely in terms of the longitudinal part of the vector potential. We shall obtain using the Fourier resolution of the vector potential and the gauge condition, the Fourier resolution of the electric and magnetic field. The electric field components are given by

$$E^r = -A^r_{,0} - A^0_{,r}$$

Now,

$$\nabla.\mathbf{A} = A^r_{,r} = \int [-ik^r \hat{A}^r(\mathbf{k})exp(ik.x) + ik^r \hat{A}^r(\mathbf{k})exp(-ik.x)]d^3k$$

so that

$$V = -\int [\nabla.\mathbf{A}]dt = \int [(k^r/\omega)(\hat{A}^r(\mathbf{k})exp(ik.x) - \hat{A}^r(\mathbf{k})exp(-ik.x))]d^3k$$

Then,

$$V,r = -i\int (k^s k^r/\omega)(\hat{A}^r(\mathbf{k})exp(ik.x) - \hat{A}^r(\mathbf{k})exp(-ik.x))d^3k$$

$$A^r_{,0} = \int i\omega(\hat{A}^r(\mathbf{k})exp(ik.x) - \hat{A}^r(\mathbf{k})exp(-ik.x))d^3k$$

so that

$$E^r = -V_{,r} - A^r_{,0} = -i\int \omega(\hat{A}^r - k^r k^s \hat{A}^s/\omega^2)exp(ik.x) - (\hat{A}^{r*} - k^r k^s \hat{A}^{s*})exp(-ik,x))d^3k$$

This equation shows that the electric field is transverse, ie it depends only on the transverse components of the vector potential. This fact could also have been derived directly using $\nabla.\mathbf{E} = 0$.

Quantization of the electromagnetic field is harder. It involves expressing the energy of the radiation field as a quadratic functional of the magnetic vector potential in the spatial frequency domain thereby making it possible to assign a quantum harmonic oscillator to each spatial frequency mode by assigning canonical commutation relations between the field Fourier components.

33.Fermion coupled to a spontaneously broken gauge theory through a chiral interaction.

We let A_μ be the gauge boson field and ϕ the complex field of the Goldstone boson. The Lagrangian function is given by

$$\mathcal{L} = \bar{\psi}_L(i.D)\psi_L + \bar{\psi}_R(i\partial)\psi_R - \lambda(\bar{\psi}_L\phi\psi_R + \bar{\psi}_R\phi^*\psi_L)$$

where $iD = \gamma^\mu(i\partial_\mu - e.A_\mu)$ and $\partial = \gamma^\mu\partial_\mu$. Here, if $\psi(x)$ denotes the electron field, we have

$$\bar{\psi}(x) = \psi(x)^*\gamma^0, \gamma^0 = \begin{pmatrix} 0 & 1 \\ 1 & 0 \end{pmatrix}$$

$$\psi(x) = \begin{pmatrix} \psi_L(x) \\ \psi_R(x) \end{pmatrix}$$

$$\bar{\psi}_L\psi_R = (\psi_L^*, 0)\gamma^0(0, \psi_R)^T = \psi_L^*\psi_R$$

Define the projection operator

$$P = \begin{pmatrix} I & 0 \\ 0 & 0 \end{pmatrix}$$

P projects the wave function onto the left handed component. $I - P$ projects the wave function onto the right handed component. We can write

$$\bar{\psi}_L\psi_R = \psi^*P\gamma^0(I-P)\psi = \bar{\psi}\gamma^0P\gamma^0(I-P)\psi$$

and likewise,

$$\bar{\psi}_R\psi_L = \psi^*(I-P)\gamma^0P\psi = \bar{\psi}\gamma^0(I-P)\gamma^0P\psi$$

Define the matrices

$$A = \gamma^0P\gamma^0(I-P), B = (I-P)\gamma^0(I-P)\gamma^0P$$

Then the interaction term between the electron field and the Goldstone boson field can be expressed as

$$\phi.(\bar{\psi}.A.\psi) + \phi^*\bar{\psi}.B.\psi$$

The probability of the state $|p, k>$ consisting of two electrons with four momenta p, k getting scattered to the final state $|p', k'>$ consisting of two electrons with four momenta p', k' with an exchange of a Goldstone boson is given upto second order by the matrix element

$$< p', k'|(\int (\phi(x)\bar{\psi}(x)A.\psi(x) + \phi^*(x)\bar{\psi}(x).B.\psi(x))d^4x)2|p, k>$$

This matrix element can be evaluated by expanding ψ in terms of the electron annihilation operators and $\bar{\psi}$ in terms of the electron creation operators, the positron part can be omitted since it does not contribute to scattering. We then make use of the propagator of the Goldstone boson field and arrive at concrete expressions for the scattering amplitude. For example, the cross term gives

$$\int \phi(x)\phi^*(y)[\bar{\psi}(x)A\psi(x)].[\bar{\psi}(y)B\psi(y)]d^4xd^4y$$

$\phi(x)$ can be paired with $\phi^*(y)$ to get $D_\phi(x-y)$, the propagator of the Goldstone boson field. $\bar{\psi}(x)$ has a term in its integrand $a(p_1)^*\bar{u}(p_1)exp(ip_1.x)$. This pairs with $< p'|$ to give $\bar{u}(p')exp(ip'.x)$. $\psi(x)$ has a term $a(p_1)u(p_1)exp(-ip_1.x)$. This pairs with $|p>$ to give $u(p)exp(-ip.x)$. Thus the term involving the matrix A gives $\bar{u}(p')Au(p)exp(i(p'-p,x))$. Likewise, the term $\bar{\psi}(y)$ has a term $a(p_1)^*\bar{u}(p_1)exp(ip_1.y)$ which pairs with the state $< k'|$ to give $\bar{u}(k')exp(ik'.y)$. The term $\psi(y)$ has a term $a(p_1)u(p_1)exp(-ip_1.y)$ which pairs with the state $|k>$ to give $u(k)exp(-ik.y)$. Thus, the term involving the matrix B gives $\bar{u}(k')Bu(k)exp(i(k'-k,y))$ and the total amplitude for one of the Feynman diagrams is given by

$$\int D(x-y)[\bar{u}(p')Au(p)].[\bar{u}(k')Bu(k)]exp(i((p'-p,x)+(k'-k,y)))d^4xd^4y$$

This integral can be evaluated easily :

$$\int D(x-y)exp(i(a.x+b.y)d^4xd^4y = \int D(\xi)exp(i(a.(\xi+y)+b.y))d^4\xi d^4y = \hat{D}(a)\delta(a+b)$$

thus the required amplitude is given by

$$[\bar{u}(p')Au(p)][\bar{u}(k')Bu(k)]\hat{D}(p'-p)\delta(p'-p+k'-k)$$

34.Moller scattering. The initial state is $(u_1,p_1),(u_2,p_2)$ for the two electrons. $p_i, i=1,2$ are the momentum values of the two electrons and u_1,u_2 are wave functions (solution to the Dirac equation) that define the polarizations of the two electrons. Thus, the electron (p_1,u_1) has wave function $u_1.exp(ip_1.x)$ where $p_1.x$ is a Lorentz scalar product. We have $p^{10} = \sqrt{m^2+\mathbf{p_1}^2}$ and likewise for the second electron. Let $a(p,u)$ be the electron annihilation operator for momentum p and polarization u. Then, the second quantized electron wave function is given by

$$\psi(x) = \sum_{p,u} a(p,u).u.exp(ip.x)$$

The matrix element of this second quantized wave function between the vacuum state and the state corresponding to the first electron is given by

$$< 0|\psi(x)|u_1,p_1 >=< 0|\psi(x)a^*(p_1,u_1)|0 >=< 0|\sum_{p,u} u.exp(ip.x)a(p,u)a^*(p_1,u_1)|0 >= u_1.exp(ip_1.x)$$

where we have used the commutation relation

$$[a(p,u),a^*(p',u')] = \delta_{p,p'}\delta_{u,u'}$$

The interaction energy is given by

$$H = \int J^\mu(x)A_\mu(x)d^3x$$

where

$$J^\mu(x) = \bar{\psi}(x)\gamma^\mu\psi(x)$$

Thus, the second order contribution to the evolution operator $P(exp(i\int_{-\infty}^{\infty} Hdt)$ is proportional to $\int P(H(t_1)H(t_2))dt_1dt_2$. P is the chronological ordering operator. Letting $\{.,.\}$ denote the anticommutator, in order to calculate the scattering of two electrons from the initial state $[(p_1u_1),(p_2u_2)]$ to the final state $[(p_2u_2),(p_2'u_2')]$, we find that the second order contribution to the scattering operator can be expressed as

$$U_2 = \int d^4x_1d^4x_2P\{\bar{\psi}(x_1)(\gamma^\mu A_\mu(x_1))\psi(x_1),\bar{\psi}(x_2)(\gamma^\mu A_\mu(x_2)\psi(x_2)\}$$

Note that a factor of one half must be inserted before U_2. For simplicity of notation, we write $\gamma.A$ in place of $\gamma^\mu A_\mu$. We have

$$\psi(x) = \sum_{p,u} a(p,u).u.exp(ip.x) + \sum_{p,v} b^*(p,v).v.exp(-ip.x)$$

where a is the electron annihilation operator, while b^* is the positron creation operator. The sum is over the mass shell, ie. p such that $p^2 = m^2$, or equivalently, $p^0 = \sqrt{m^2+\mathbf{p}^2}$. We note that

$$\bar{\psi}(x) = \sum_{p,u} a^*(p,u).\bar{u}.exp(-ip.x) + \sum_{p,v} b(p,v).\bar{v}.exp(ip.x)$$

We use the notation $\gamma.A$ for $\gamma^\mu A_\mu$. Then,

$$U_2 =$$

$$\int_{x10>x20} A_\mu(x_1)A_\nu(x_2)\gamma^\mu_{ab}\gamma^\nu_{cd}(\bar{\psi}_a(x_1)\psi_b(x_1)\bar{\psi}_c(x_2)\psi_d(x_2))dx_1dx_2$$

$$= \int_{x10>x20} A_\mu(x_1)A_\nu(x_2)\gamma^\mu_{ab}\gamma^\nu_{cd}(\bar{\psi}_a(x_1)\psi_b(x_1)\bar{\psi}_c(x_2)\psi_d(x_2))dx_1dx_2$$

$$+ \int_{x10<x20} A_\nu(x_2)A_\mu(x_1)\gamma^\mu_{ab}\gamma^\nu_{cd}(\bar{\psi}_c(x_2)\psi_d(x_2)\bar{\psi}_a(x_1)\psi_b(x_1))dx_1dx_2$$

We now have

$$< p_1'u_1', p_2'u_2'|\bar{\psi}_a(x_1)\psi_b(x_1)\bar{\psi}_c(x_2)\psi_d(x_2)|p_1u_1,p_2u_2>$$

$$=< p_1'u_1', p_2'u_2'|(\sum_{p,u} a^*(p,u).\bar{u}_a.exp(-ip.x_1) + \sum_{p,v} b(p,v).\bar{v}_a.exp(ip_1.x_1))\times$$

$$(\sum_{p,u} a(p,u)u_b exp(ip.x_1) + \sum_{p,v} b^*(p,v)v_b.exp(-ip.x_1))\times$$

$$(\sum_{p,u} a^*(p,u)\bar{u}_c exp(-ip.x_2) + \sum_{p,v} b(p,v)\bar{v}_c.exp(ip.x_2))\times$$

$$(\sum_{p,u} a(p,u)u_d.exp(ip_2.x_2) + \sum_{p,v} b^*(p,v)v_d.exp(-ip.x_2))|p_1u_1p_2u_2>$$

This evaluates to

$$\bar{u}'_{1a}\bar{u}'_{2c}u_{1b}u_{2d}exp(-i(p_1'.x_1+p_2'.x_2)).exp(i(p_1.x_1+p_2.x_2))$$

+similar terms. One of the similar terms is

$$\bar{u}'_{2a}\bar{u}'_{1c}u_{2b}u_{1d}exp(-ip_2'.x_1 - ip_1'.x_2 + ip_2.x_1 + ip_1.x_2)$$

We thus get for the corresponding amplitude

$$M_1 = \int_{x^{10} > x^{20}} < (A_\mu(x_1)A_\nu(x_2)) > exp(i((p_1 - p_1').x_1 + (p_2 - p_2').x_2))dx_1dx_2\bar{u}_1'\gamma^\mu u_1\bar{u}_2'\gamma^\nu \bar{u}_2$$

plus other terms. One of the other terms corresponding to the above is given by

$$M_2 = \int_{x^{10} > x^{20}} < A_\mu(x_1)A_\nu(x_2) > exp(i(p_2 - p_2').x_1 + (p_1 - p_1').x_2)dx_1dx_2\bar{u}_2'\gamma^\mu u_2\bar{u}_1'\gamma^\nu u_1$$

$$= \int_{x^{20} > x^{10}} < A_\mu(x_2)A_\nu(x_1) > exp(i(p_1 - p_1').x_1 + (p_2 - p_2').x_2)dx_1dx_2\bar{u}_1'\gamma^\nu u_1\bar{u}_2'\gamma^\mu u_2$$

Writing

$$D_{\mu\nu}(x_1 - x_2) =< P(A_\mu(x_1)A_\nu(x_2)) >$$

and noting that $D_{\mu\nu} = D_{\nu mu}$, we get for the sum of these two

$$M = M_1 + M_2 == \int_{\mathbb{R}^4\times\mathbb{R}^4} D_{\mu\nu}(x_1 - x_2)exp(i(p_1 - p_1').x_1 + (p_2 - p_2').x_2)dx_1dx_2\bar{u}_1'\gamma^\mu u_1\bar{u}_2'\gamma^\nu u_2$$

Reference:Advanced Quantum Mechanics, by Freeman Dyson.

Recent advances in signal processing:Finite groups and their properties, group homomorphism, Cayley's theorem, direct product of finite groups, permutation groups, normal subgroup, compact groups, representation, unitary representation, Schur's orthogonality lemmas, Peter-Weyl theorem, application of group representation theory to image processing (Wiener filter, matched filter, feature extraction, estimation of the group element connecting two images, introduction to Lie groups and Lie algebras, Lie groups acting on differentiable manifolds.

35. Lorentz transformation of one particle quantum states: p denotes the four momentum, σ denotes the collection of all the other dynamical variables that commute with the four momentum. The state of the quantum system in which the four momentum has a definite value p and these other dynamical variables have the values σ is denoted by $\psi_{p,\sigma}$. Let P^μ denote the operator of four momentum of the system. Then obviously,

$$P^\mu\psi_{p,\sigma} = p^\mu\psi_{p,\sigma}$$

We are assuming that the quantum system comprises a single particle. If there are N particles, then the state will have the form $\psi_{p_1,...,p_N,\sigma_1,...\sigma_N}$. If Λ denotes a Lorentz transformation of space time coordinates and a denotes a four vector corresponding to translations, then (Λ, a) is an element of the Poincare group. Its action on a four vector x is given by

$$(\Lambda, a)x = \Lambda x + a$$

The composition rule for the Poincare group is obtained as

$$(\Lambda_1, a_1).(\Lambda_2, a_2)x = (\lambda_1, a_1)(\Lambda_2 x + a_2) = \Lambda_1(\Lambda_2 x + a_2) + a_1 = \Lambda_1\Lambda_2 x + \Lambda_1 a_2 + a_1$$

$$= (\Lambda_1\Lambda_2, \Lambda_1 a_2 + a_1)x$$

so that

$$(\Lambda_1, a_1)(\Lambda_2, a_2) = (\Lambda_1\Lambda_2, \Lambda_1 a_2 + a_1)$$

This defines the composition law for the Poincare group. The Poincare group comprises Lorentz transformations (spatial rotations and boosts and their combination) along with spatial inversion, time inversion, spatial translations and time translations. Let $(\Lambda, a) \rightarrow U(\Lambda, a)$ denote the representation of the Poincare group acting on the space of quantum states. Then $U(1, a)$ is a translation of space-time. Since the four momentum operator P^μ is the generator of space-time translations, it follows that

$$U(1, a)\psi_{p,\sigma} = exp(-iP^\mu a_\mu)\psi_{p,\sigma} = exp(-ip.a)\psi_{p,\sigma}$$

Let Λ^μ_ρ denote the matrix elements of the Lorentz transformation Λ. Its action on a four vector x^μ is given by

$$\bar{x} = \Lambda.x$$

or in terms of components,

$$\bar{x}^\mu = \Lambda^\mu_\rho x^\rho$$

Let $\eta_{\mu\nu}$ be the metric tensor. Then

$$\eta_{\mu\nu}\bar{x}^\mu\bar{x}^\nu = \eta_{\mu\nu}x^\mu x^\nu$$

or

$$\eta_{\mu\nu}\Lambda^\mu_\rho\Lambda^\nu_\sigma = \eta_{\rho\sigma}$$

This can be expressed as

$$\eta = \Lambda^T\eta\Lambda$$

so that

$$\Lambda^{-1} = \eta^{-1}\Lambda^T\eta$$

or in terms of components,

$$(\Lambda^{-1})^\mu_\nu = \eta^{\mu\rho}\Lambda^\sigma_\rho\eta_{\nu\sigma}$$

or equivalently,

$$(\Lambda^{-1})^\mu_\nu = \eta_{\nu\sigma}\eta^{\mu\rho}\Lambda^\sigma_\rho$$

Now observe that

$$P^\mu\psi_{p,\sigma} = p^\mu\psi_{p,\sigma}$$

and hence,

$$U(\Lambda)P^\mu\psi_{p,\sigma} = p^\mu U(\Lambda)\psi_{p,\sigma} = p^\mu\psi_{\Lambda p,\sigma}$$

This can be expressed as

$$U(\Lambda)P^\mu U(\Lambda)^{-1}U(\Lambda)\psi_{p,\sigma} = p^\mu\psi_{\Lambda p,\sigma}$$

or

$$U(\Lambda)P^\mu U(\Lambda)^{-1}\psi_{\Lambda.p,\sigma} = p^\mu\psi_{\Lambda p,\sigma}$$

On the other hand,

$$P^\mu\psi_{\Lambda p,\sigma} = (\Lambda p)^\mu\psi_{p,\sigma}$$

so that

$$(\Lambda^{-1}P)^{\mu}\psi_{\Lambda p,\sigma} = p^{\mu}\psi_{\Lambda p,\sigma}$$

where

$$(\Lambda^{-1}P)^{\mu} = (\Lambda^{-1})^{\mu}_{\rho}P^{\rho}$$

so that

$$U(\Lambda)P^{\mu}U(\Lambda)^{-1} = (\Lambda^{-1}P)^{\mu} = (\Lambda^{-1})^{\mu}_{\rho}P^{\rho}$$

$$= \Lambda^{\mu}_{\rho}P^{\rho}$$

Replacing Λ by Λ^{-1} gives us

$$U(\Lambda)^{-1}P^{\mu}U(\Lambda) = (\Lambda^{-1})^{\mu}_{\rho}P^{\rho} = \Lambda^{\mu}_{\rho}P^{\rho}$$

In this derivation, we are assuming that

$$U(\Lambda)\psi_{\Lambda p,\sigma}$$

More generally, we can derive the equation

$$U(\Lambda)^{-1}P^{\mu}U(\Lambda) = \Lambda^{\mu}_{\rho}P^{\rho}$$

from the equation

$$U(\Lambda)\psi_{p,\sigma} = \sum_{\sigma'} C_{\sigma',\sigma}(\Lambda,p)\psi_{\Lambda p,\sigma'}$$

For then,

$$P^{\mu}U(\Lambda)\psi_{p,\sigma} = P^{\mu}\sum_{\sigma'} C_{\sigma',\sigma}(\Lambda,p)\psi_{\Lambda p,\sigma'}$$

$$= \sum_{\sigma'} C_{\sigma',\sigma}(\Lambda,p)P^{\mu}\psi_{\Lambda p,\sigma'}$$

$$= \sum_{\sigma'} C_{\sigma',\sigma}(\Lambda,p)(\Lambda.p)^{\mu}\psi_{\Lambda p,\sigma'}$$

on the one hand and on the other,

$$P^{\mu}U(\Lambda)\psi_{p,\sigma} = U(\Lambda)U(\Lambda)^{-1}P^{\mu}U(\Lambda)\psi_{p,\sigma}$$

so that

$$U(\Lambda)^{-1}P^{\mu}U(\Lambda)\psi_{p,\sigma}$$

$$= \sum_{\sigma'} C_{\sigma',\sigma}(\Lambda,p)(\Lambda.p)^{\mu}U(\Lambda)^{-1}\psi_{\Lambda p,\sigma'}$$

$$= \sum_{\sigma',\sigma''}(\Lambda.p)^{\mu}C_{\sigma',\sigma}(\Lambda,p)C_{\sigma'',\sigma'}(\Lambda^{-1},\Lambda p)\psi_{p,\sigma''}$$

The derivation of the Lorentz transformation law of P^{μ} will be complete once we note that

$$\sum_{\sigma'} C_{\sigma',\sigma}(\Lambda,p)\psi_{\Lambda p,\sigma'} = U(\Lambda)\psi_{p,\sigma},$$

$$\psi_{p,\sigma} = U(\Lambda^{-1})U(\Lambda)\psi_{p,\sigma}$$

$$= \sum_{\sigma'} C_{\sigma',\sigma}(\Lambda,p)U(\Lambda^{-1})\psi_{\Lambda p,\sigma'}$$

$$= \sum_{\sigma'} C_{\sigma',\sigma}(\Lambda,p)C_{\sigma'',\sigma'}(\Lambda^{-1},\Lambda p)\psi_{p,\sigma''}$$

We restrict our study to proper orthochronous Lorentz transformations, ie, Lorentz transformations having unit determinant and that preserve the sign of time, ie, Λ is a proper orthochronous Lorentz transformation iff $det\Lambda = 1$ and if $x = (x^\mu)$ is such that $x^0 > 0$, then $\bar{x} = \Lambda x$ is such that $\bar{x}^0 > 0$. Choose a standard four momentum k^μ and let $L^\mu_\nu(p)$ be a Lorentz transformation such that

$$p^\mu = L^\mu_\nu(p)k^\nu$$

ie

$$p = L(p)k$$

where $L(p)$ is a proper orthochronous Lorentz transformation. Let $\Lambda \to U(\Lambda)$ be a unitary representation of the Lorentz group that act on the quantum states $\psi_{p,\sigma}$. With $N(p)$ a numerical factor we define

$$\psi_{p,\sigma} = N(p)U(L(p))\psi_{k,\sigma}$$

Then, for an proper orthochronous Lorentz transformation, we have

$$U(\Lambda)\psi_{p,\sigma} = N(p)U(\Lambda)U(L(p))\psi_{k,\sigma}$$

$$= N(p)U(\Lambda.L(p))\psi_{k,\sigma} = N(p)U(L(\Lambda p))U(L(\lambda p)^{-1}\Lambda L(p))\psi_{k,\sigma}$$

Consider the Lorentz transformation

$$W(\Lambda,p) = L(\Lambda p)^{-1}\Lambda L(p)$$

We have

$$W(\Lambda,p)k = L(\Lambda p)^{-1}\Lambda p = k$$

so that $W(\Lambda,p)$ belongs to the little group consisting of all Lorentz transformations that fix $k = (k^\mu)$. For any W belonging to this little group, we have

$$U(W)\psi_{k,\sigma} = \sum_{\sigma'} D_{\sigma',\sigma}(W)\psi_{k,\sigma'}$$

where $((D_{\sigma',\sigma}(W)))$ is a unitary matrix. It follows that

$$U(\Lambda)\psi_{p,\sigma} = N(p)U(L(\Lambda p))U(W(\Lambda,p))\psi_{k,\sigma}$$

$$= N(p)U(L(\Lambda p))\sum_{\sigma'} D_{\sigma',\sigma}(\Lambda,p)\psi_{k,\sigma'}$$

$$= N(p)\sum_{\sigma'} D_{\sigma',\sigma}(\Lambda,p)U(L(\Lambda p))\psi_{k,\sigma'}$$

$$= (N(p)/N(\Lambda p))\sum D_{\sigma',\sigma}(\Lambda,p)\psi_{\Lambda p,\sigma'}$$

Note that $W \to D(W) = ((D_{\sigma',\sigma}(W)))$ is a unitary representation of the little group. The problem of determining the quantum Lorentz transformations thus, apart from a normalization factor, reduces to determining the unitary representations of the little group.

36.Commutation relations for the Poincare algebra: Consider an infinitesimal Poincare transformation having matrix elements (Λ, a) where

$$\Lambda^\mu_\nu = \delta^\mu_\nu + \omega^\mu_\nu, a^\mu = \epsilon^\mu$$

ω^μ_ν and ϵ^μ are both infinitesimal quantities (ie, of the first order of smallness). We have

$$\eta_{\rho\sigma} = \eta_{\mu\nu}(\delta^\mu_\rho + \omega^\mu_\rho)(\delta^\nu_\sigma + \omega^\nu_\sigma)$$

$$= \eta_{\rho\sigma} + \eta_{\nu\rho}\omega^\nu_\sigma + \eta_{\mu\sigma}\omega^\mu_\rho$$

$$= \eta_{\rho\sigma} + \omega_{\rho\sigma} + \omega_{\sigma\rho}$$

Thus $(\omega_{\mu\nu})$ is an antisymmetric real matrix. This is a necessary and sufficient condition for the matrix $(\delta^\mu_\nu + \omega^\mu_\nu)$ to be an infinitesimal Lorentz transformation. The rule for lowering and contracting indices is through the metric tensor $\eta_{\mu\nu}$. Let $(\Lambda, a) \to U(\Lambda, a)$ be a unitary representation of the Poincare group. Its infinitesimal form is given by

$$U(1+\omega,\epsilon) = I + i\omega_{\mu\nu}J^{\mu\nu} - i\epsilon_\mu P^\mu$$

where since ω is antisymmetric, we can assume that $J^{\mu\nu} = -J^{\nu\mu}$. The operators $J^{\mu\nu}$ and P^μ are Hermitian matrices since U is unitary. Note that

$$2iJ^{\mu\nu} = \frac{\partial U(1+\omega,0)}{\partial\omega_{\mu\nu}}|_{\omega=0}$$

and

$$-iP^\mu = \frac{\partial U(1,\epsilon)}{\partial\epsilon_\mu}|_{\epsilon=0}$$

We shall evaluate

$$Q = U(\Lambda,a)U(1+\omega,\epsilon)U(\Lambda,a)^{-1}$$

First note that

$$(\Lambda,a)^{-1}x = y$$

implies

$$\Lambda y + a = x$$

so that

$$y = \Lambda^{-1}(x-a) = (\Lambda^{-1}, -\Lambda^{-1}a)$$

and this gives

$$Q = U((\Lambda, a).(1+\omega, \epsilon).(\Lambda^{-1}, -\Lambda^{-1}a))$$

Now,

$$(\Lambda, a).(1+\omega, \epsilon).(\Lambda^{-1}, -\Lambda^{-1}a)x$$
$$= (\Lambda, a)(1+\omega, \epsilon)(\Lambda^{-1}x - \Lambda^{-1}a)$$
$$= (\Lambda, a)((1+\omega)(\Lambda^{-1}x - \Lambda^{-1}a) + \epsilon)$$
$$= \Lambda((1+\omega)(\Lambda^{-1}x - \Lambda^{-1}a) + \epsilon) + a$$
$$= \Lambda(1+\omega)\Lambda^{-1}x - \Lambda(1+\omega)\Lambda^{-1}a + \Lambda\epsilon + a$$
$$= \Lambda(1+\omega)\Lambda^{-1}x - \Lambda\omega\Lambda^{-1}a + \Lambda\epsilon$$

It follows that

$$Q = U(\Lambda(1+\omega)\Lambda^{-1}, -\Lambda\omega\Lambda^{-1}a + \Lambda\epsilon)$$

Thus,

$$U(\Lambda, a)(I + i\omega_{\mu\nu}J^{\mu\nu} - i\epsilon_\mu P^\mu)U(\Lambda, a)^{-1}$$
$$= I + i(\Lambda\omega\Lambda^{-1})_{\mu\nu}J^{\mu\nu} + i(\Lambda\omega\Lambda^{-1}a - \Lambda\epsilon)_\mu P^\mu$$

so that

$$\omega_{\rho\sigma}U(\Lambda, a)J^{\rho\sigma}U(\Lambda, a)^{-1} = \Lambda^\rho_\mu\omega_{\rho\sigma}(\Lambda^{-1})^\sigma_\nu J^{\mu\nu}$$
$$+\Lambda^\rho_\mu\omega_{\rho\sigma}(\Lambda^{-1})^\sigma_\nu a^\nu P^\mu$$

or

$$U(\Lambda, a)J^{\rho\sigma}U(\Lambda, a)^{-1} = \Lambda^\rho_\mu\Lambda^\sigma_\nu J^{\mu\nu}$$
$$+\frac{1}{2}\Lambda^\rho_\mu\Lambda^\sigma_\nu(a^\nu P^\mu - a^\mu P^\nu)$$

Likewise,

$$-i\epsilon_\mu U(\Lambda, a)P^\mu U(\Lambda, a)^{-1} = -i(\Lambda\epsilon)_\mu P^\mu$$
$$= -i\Lambda_\mu^{rho}\epsilon_\rho P^\mu = -i\epsilon_\mu\Lambda^\mu_\rho P^\rho$$

so that

$$U(\Lambda, a)P^\mu U(\Lambda, a)^{-1} = \Lambda^\mu_\rho P^\rho$$

37. DSP implementation of Lorentz transformation of quantum fields.

Consider an operator field like $\mathcal{H}(x) = (\mathbf{x}, t)$. for each $x \in \mathbb{R}^4$, $\mathcal{H}(x)$ is an operator on the Hilbert space of quantum states. We assume that $\mathcal{H}(x)$ is the Hamiltonian of the field. For example, if $\phi(x)$ is the Klein-Gordon field, then

$$\mathcal{H}(x) = \frac{1}{2}(\partial_t\psi(x))^2 + \frac{1}{2}(\nabla\phi(x))^2 + m\phi(x)^2/2$$

The corresponding Lagrangian density is given by

$$\mathcal{L} = \frac{1}{2}(\partial_t \phi)^2 - \frac{1}{2}(\nabla\phi)^2 - m\phi^2/2$$

We may easily verify that $\mathcal{H}(x)$ is obtained by applying a Legendre transformation to the Lagrangian density. The canonical momentum density is given by

$$\pi = \frac{\partial \mathcal{L}}{\partial(\partial_t \phi)} = \partial_t \phi$$

so that

$$\mathcal{H} = \pi\partial_t\phi - \mathcal{L} = \frac{1}{2}(\partial_t\phi(x))^2 + \frac{1}{2}(\nabla\phi(x))^2 + m\phi(x)^2/2$$

as written above. In the quantized Klein-Gordon theory, for each $x \in \mathbb{R}^4$, $\phi(x)$ becomes an operator acting on a Hilbert space, more precisely on the symmetric Fock space. For any Poincare transformation (Λ, a), with Λ a Lorentz transformation and a a space-time translation, in order that the Hamiltonian $\mathcal{H}$ transform like a scalar, we require that

$$U(\Lambda, a)\mathcal{H}(x)U(\Lambda, a)^{-1} = \mathcal{H}(\Lambda x + a)$$

Suppose that the field consists of components $\psi_l(x), l = 1, 2, ...$ We can decompose ψ_l into the sum of an annihilation part and a creation part. The annihilation part can be expressed as a superposition of the annihilation operators $a(\mathbf{p}, \sigma)$ while the creation part can be expressed as a superposition of the creation operators $a(\mathbf{p}, \sigma)^*$. Thus,

$$\psi_l(x) = \psi_l^+(x) + \psi_l^-(x)$$

where

$$\psi_l^+(x) = \sum_\sigma \int g_l(\mathbf{p}, \sigma)a(\mathbf{p}, \sigma)exp(-ip.x)d^3\mathbf{p}$$

$$\psi_l^-(x) = \int_\sigma \int g_l^*(\mathbf{p}, \sigma)a^*(\mathbf{p}, \sigma)exp(ip.x)d^3\mathbf{p}$$

Note that if the particle has mass m, then $p^0 = \sqrt{m^2 + \mathbf{p}^2}$ and in view of this relationship, the integration need be taken only over three momentum space. The Lorentz transformation of these annihilation and creation fields can be expressed as

$$U(\Lambda, a)\psi_l^+(x)U(\Lambda, a)^{-1} = \sum_{l'} D_{l,l'}(\Lambda)\psi_{l'}^+(\Lambda.x + a)$$

and

$$U(\Lambda, a)\psi_l^-(x)U(\Lambda, a) = \sum_{l'} \bar{D}_{l,l'}(\Lambda)^*\psi_{l'}^-(\Lambda.x + a)$$

$$= \sum_{l'} D_{l',l}(\Lambda)\psi_{l'}^-(\Lambda.x + a)$$

Here, for each Lorentz transformation Λ, $(D_{l,l'}(\Lambda))$ is a unitary matrix. We are assuming that the fields ψ_l before being quantized form a the substratum of a vector space on which the matrices D act as representations of the Lorentz group. Suppose that the Hamiltonian density has the form

$$H(x) = \sum_{l_1,...,l_N,m_1,...,m_K} h(l_1,...,l_N,m_1,...,m_K)\psi^+_{l_1}(x)...\psi^+_{l_N}(x)\psi^-_{m_1}(x)...\psi^-_{m_N}(x)$$

Then we get

$$U(\Lambda,a)H(x)U(\Lambda,a)^{-1} = \sum h(l_1,...,l_N,m_1,...,m_K)D_{l_1,l_1'}(\Lambda)...D_{l_N,l_N'}(\Lambda)D_{m_1',m}(\Lambda)...D_{m_K',m_K}(\Lambda)$$
$$\times\psi^+_{l_1'}(\Lambda x+a)...\psi^+_{l_N'}(\Lambda x+a)\psi^-_{m_1'}(\Lambda x+a)...\psi^-_{m_K}(\Lambda x+a)$$

For H to transform like a scalar, this must equal

$$H(\Lambda x+a) = \sum h(l_1,...,l_N,m_1,...,m_K)\psi^+_{l_1}(\Lambda x+a)...\psi^-_{m_K}(\Lambda x+a)$$

and hence

$$h(l_1',...,m_K') = \sum_{l_1,...,m_K} D_{l_1,l_1'}(\Lambda)...D_{l_N,l_N'}(\Lambda)D_{m_1',m_1}(\Lambda)...D_{m_K',m_K}(\Lambda)h(l_1,...,m_K)$$

ie, h must be an invariant of the tensor representation of the Lorentz group (Steven Weinberg, Quantum Field Theory, Foundations).

37.Let $\Phi_{q_1,...,q_n}$ be a symmetric n particle quantum state. Thus, if σ is any permutation of $\{1,2,...,n\}$, then

$$\Phi_{q_{\sigma 1},q_{\sigma 2},...,q_{\sigma n}} = \Phi_{q_1,q_2,...,q_n}$$

The action of a creation operator $a(q)^*$ on such a state is given by

$$a(q)^*\Phi_{q_1,...,q_n} = \Phi_{q,q_1,...,q_n}$$

We focus on $n=1$ for simplicity.

$$a(q)^*\Phi_{q_1} = \Phi_{q,q_1}$$
$$<\Phi_{q_1',q_2'},\Phi_{q_1,q_2}> = \delta(q_1-q_1')\delta(q_2-q_2')+\delta(q_1-q_2')\delta(q_2-q_1')$$

Then,

$$<\Phi_{q_1',q_2'},a(q)^*\Phi_{q_1}> = <\Phi_{q_1',q_2'},\Phi_{q,q_1}>$$
$$= \delta(q-q_1')\delta(q_1-q_2')+\delta(q-q_2')\delta(q_1-q_1')$$

on the one hand, while on the other, this must be the same as

$$<a(q)\Phi_{q_1',q_2'},\Phi_{q_1}>$$

Suppose that

$$a(q)\Phi_{q_1,q_2} = \delta(q-q_1)\Phi_{q_2}+\delta(q-q_2)\Phi_{q_1}$$

Then

$$< a(q)\Phi_{q_1',q_2'}, \Phi_{q_1} >=< \delta(q-q_1')\Phi_{q_2'} + \delta(q-q_2')\Phi_{q_1'}, \Phi_{q_1} >$$
$$= \delta(q-q_1')\delta(q_1-q_2') + \delta(q-q_2')\delta(q_1-q_1')$$

which agrees with the above result. More generally, we have the result that if $\Phi_{q1,...,q_n}$ is an n-particle bosonic state, then

$$a(q)\Phi_{q1,...,q_n} = \sum_{i=1}^{n} \delta(q-q_i)\Phi_{q1,...,q_{i-1},q_{i+1},...,q_n}$$

This result can also be adapted to Fermionic states as follows. First we define an n-particle Fermionic state $\Phi_{q1,...,q_n}$ in such a way that it is antisymmetric, ie, if σ is any permutation of $\{1, 2, ..., n\}$, then

$$\Phi_{q\sigma 1,...,q\sigma n} = sgn(\sigma)\Phi_{q1,...,q_n}$$

Then we define the creation operator $a(q)^*$ so that

$$a(q)^*\Phi_{q1,...,q_n} = \Phi_{q,q1,...,q_n}$$

Then, it is easy to show that

$$a(q)\Phi_{q1,...,q_n} = \sum_{r=1}^{n} (-1)^r \delta(q-q_r)\Phi_{q1,...,q_{r-1},q_{r+1},...,q_n}$$

Indeed, assuming that this relation is correct, we get the correct expression for

$$< a(q)\Phi_{q1,...,q_n}, \Phi_{q_1',...,q_{n-1}'} >$$

as can be verified readily by writing all the terms.

Free fields:

39.$U(\Lambda, a)$ is a Poincare transformation corresponding to a Lorentz transformation Λ and a space-time translation a. U is a unitary representation of the Poincare group. $H(x)$ is the Hamiltonian density of the field. We can write it as $H(\mathbf{x}, t)$ where $x = (\mathbf{x}, t)$ are the space-time coordinates of an event. We require that H be a scalar in the sense that

$$U(\Lambda, a)H(x)U(\Lambda, a)^{-1} = H(\Lambda x + a)$$

We have noted earlier that if $\psi_{p,\sigma}$ is a one particle state corresponding to four momentum p and helicity σ, then its Lorentz transformation rule is given by

$$U(\Lambda)\psi_{p,\sigma} = ((\Lambda p^0)/p_0)^{1/2} \sum_{\sigma'} D_{\sigma',\sigma}(W(\Lambda, p))\psi_{\Lambda p,\sigma'}$$

where $U(\Lambda) = U(\Lambda, 0)$. If in addition, we apply a space-time translation a, then we have using

$$U(1, a)U(\Lambda, 0) = U((1, a).(\Lambda, 0)) = U(\Lambda, a)$$

and the fact that

$$U(1,a) = exp(-ia_\mu P^\mu)$$

with $P^\mu = (\mathbf{P}, H)$ the four momentum operator (that generates space time translations, note that energy H is the generator of time translations and the three momentum $\mathbf{P}$ is the generator of spatial translations),

$$U(\Lambda,a)\psi_{p,\sigma} = ((\Lambda p)^0/p^0)^{1/2}\sum D_{\sigma',\sigma}(W(\Lambda,p))U(1,a)\psi_{\Lambda p,\sigma'}$$

$$= ((\Lambda p)^0/p^0)^{1/2} exp(-ia_\mu(\Lambda p)^\mu)\sum D_{\sigma',\sigma}(W(\Lambda,p))\psi_{\Lambda p,\sigma'}$$

More generally, the action of the Poincare element $U(\Lambda,a)$ on an N-particle state $\psi_{p1,\sigma_1,...,p_N,\sigma_N}$ is given by

$$U(\Lambda,a)\psi_{p1,\sigma_1,...,p_N,\sigma_N} = \Pi_{i=1}^N[(\Lambda p_i)^0/p_i^0)^{1/2}]exp(-ia_\mu(\Lambda.\sum p_i)^\mu)$$

$$\sum_{\sigma'_1,...,\sigma'_N} \Pi_{j=1}^n D_{\sigma'_j,\sigma_j}(W(\Lambda,p_j))\psi_{\Lambda p1,\sigma'_1,...\Lambda p_N,\sigma'_N}$$

More precisely, if the ith particle has spin j_i, we should use the representation D^{j_i} of the little group. Thus, the transformation law should be expressed as

$$U(\Lambda,a)\psi_{p1,\sigma_1,...,p_N,\sigma_N} = (\Pi(\Lambda p_i)^0/p_i^0)^{1/2}\sum \Pi_{k=1}^N D^{(j_k)}_{\sigma'_k,\sigma_k}(W^{(j_k)}(\Lambda,p_k))\psi_{\Lambda p1,\sigma'_1,...,\Lambda p_N,\sigma'_N}$$

The Poincare transformation of the annihilation operator $a(p,\sigma)$ for a single particle can be deduced easily from this. We start with the fact that if Φ_0 denotes the vacuum state (in which there is no particle), then

$$\Phi_{p1,\sigma_1,...,p_N,\sigma_N} = a(p_1,\sigma_1)^*...a(p_N,\sigma_N)^*\Phi_0$$

In particular a one particle state is given by

$$\Phi_{p,\sigma} = a(p,\sigma)^*\Phi_0$$

Now,

$$U(\Lambda,a)a(p,\sigma)^*\Phi_0 = U(\Lambda,a)\Phi_{p,\sigma} = ((\Lambda p)^0/p^0)^{1/2} exp(-ia.Lambdap)\sum_{\sigma'} D_{\sigma',\sigma}(W(\Lambda,p))\Phi_{\Lambda p,\sigma'}$$

$$= ((\Lambda p)^0/p^0))^{1/2} exp(-ia.\Lambda p)\sum D_{\sigma',\sigma}(W(\Lambda,p))a(\Lambda p,\sigma')^*\Phi_0$$

We also note that from physical considerations, a Poincare transformation of the vacuum is the vacuum itself, ie,

$$U(\Lambda,a)\Phi_0 = \Phi_0$$

Thus,

$$U(\Lambda,a)a(p,\sigma)^*U(\Lambda,a)^{-1} = ((\Lambda p)^0/p^0)^{1/2} exp(-ia.\Lambda p)\sum D_{\sigma',\sigma}(W(\Lambda,p))a(\Lambda p,\sigma')^*$$

We can verify this formula for two particle states. We have, assuming the above transformation rule for the creation operator a^*,

$$U(\Lambda,a)\Phi_{p1,\sigma_1,p2,\sigma_2} = U(\Lambda,a)a(p_1,\sigma_1)^*a(p_2,\sigma_2)^*\Phi_0$$

$$= U(\Lambda, a)a(p_1, \sigma_1)U(\Lambda, a)^{-1}U(\Lambda, a)a(p_2, \sigma_2)U(\Lambda, a)^{-1}\Phi_0$$
$$= [((\Lambda p_1)^0/p_1^0)(\Lambda p_2)^0/p_2^0)]^{1/2} exp(-ia.(\Lambda(p_1 + p_2))$$
$$\sum D_{\sigma_1', \sigma_1}(W(\Lambda, p_1))D_{\sigma_2', \sigma_2}(W(\Lambda, p_2))a(\Lambda p_1, \sigma')^* a(\Lambda p_2, \sigma_2')^* \Phi_0$$
$$= [((\Lambda p_1)^0/p_1^0)(\Lambda p_2)^0/p_2^0)]^{1/2} exp(-ia.(\Lambda(p_1 + p_2))$$
$$\sum_{\sigma_1', \sigma_2'} D_{\sigma_1', \sigma_1}(W(\Lambda, p_1))D_{\sigma_2', \sigma_2}(W(\Lambda, p_2))\Phi_{\Lambda p_1, \sigma_1', \Lambda p_2, \sigma_2'}$$

40. Schwinger's action principle: $\mathcal{L}$ is the Lagrangian density. It is a function of the field variables $\phi_\alpha(x)$ and its partial derivatives $\phi_{\alpha,\mu}(x)$. We write

$$\mathcal{L}(\phi_\alpha, \phi_{\alpha,\mu})$$

The field is distributed over the four volume Ω. The action functional is given by

$$I(\Omega) = \int_\Omega \mathcal{L} d^4x$$

We wish to compute the change in the action caused by (a) the change in the fields $\Delta\phi_\alpha$ at each point, and (b) the change in the coordinates δx^μ of the boundary. The total change in the fields caused by these two is given by

$$\delta\phi_\alpha = \Delta\phi_\alpha + \phi_{\alpha,\mu}\delta x^\mu$$

The corresponding change in the action is given by

$$\delta I(\Omega) = \int [\frac{\partial \mathcal{L}}{\partial \phi_\alpha}\delta\phi_\alpha + \frac{\partial \mathcal{L}}{\partial \phi_{\alpha,\mu}}\delta\phi_{\alpha,\mu}]d^4x$$

We use the field equations

$$\frac{\partial}{\partial x^\mu}\frac{\partial \mathcal{L}}{\partial \phi_{\alpha,\mu}} = \frac{\partial \mathcal{L}}{\partial \phi_\alpha}$$

Let Ω_1 be the original volume and Ω_2 the new volume following the infinitesimal transformation of coordinates. Likewise, let ϕ_1 be the original field and ϕ_2 the transformed field. Thus,

$$\phi_{2\alpha}(x) - \phi_{1\alpha}(x) = \Delta\phi_\alpha(x)$$

The change in the action is then given by

$$\int_{\Omega_2} \mathcal{L}(\phi_{2\alpha}(x), \phi_{2\alpha,\mu}(x))d^4x - \int_{\Omega_1} \mathcal{L}(\phi_{1\alpha}(x), \phi_{1\alpha,\mu}(x))d^4x$$

$$= \int_\Omega n_\mu \delta x^\mu \mathcal{L}(\phi_\alpha, \phi_{\alpha,\mu})d^4x + \int_\Sigma \Delta\phi_\alpha \frac{\partial \mathcal{L}}{\partial \phi_{\alpha,\mu}} n_\mu d\sigma$$

For simplicity of notation, we set

$$\pi_\alpha = n_\mu \frac{\partial \mathcal{L}}{\partial \phi_{\alpha,\mu}}$$

Then since

$$\Delta\phi_\alpha = \delta\phi_\alpha - \phi_{\alpha,\mu}\delta x^\mu$$

it follows that

$$\delta I(\Omega) = \int_\Omega n_\mu \delta x^\mu \mathcal{L} d^4x + \int_\Sigma \pi^\alpha \Delta\phi_\alpha d\sigma$$

$$= \int_\Sigma \pi^\alpha \delta\phi_\alpha d\sigma + \int (n_\mu \mathcal{L} - \pi^\alpha \phi_{\alpha,\mu})\delta x^\mu d\sigma$$

Now let ϕ_α be the second quantized field and let σ be a surface. Let O be an operator function of the fields. $|\phi''_\alpha \sigma>$ denotes a state of the field in which the operators ϕ_α on the surface σ have eigenvalues ϕ''_α. We are interested in the matrix element

$$<\phi'_\alpha, \sigma|O|\phi''_\alpha, \sigma>$$

We want to compute the change in this matrix element caused by a small change in the fields, the surface σ and the Lagrangian $\mathcal{L}$. We have

$$\delta I(\Omega) = \int \delta\mathcal{L} d^4x + \int (\pi_\alpha \delta\phi_\alpha + (n_\mu \mathcal{L} - \phi_{\alpha,\mu}\pi_\alpha)\delta x_\mu) d\sigma$$

Let σ_0 be the initial surface (in the remote past). $<\phi'_\alpha, \sigma|O|\phi''_\alpha, \sigma>$ will be denoted by $<\sigma'|O|\sigma''>$. We have

$$<\sigma'|O|\sigma''> = \sum_{\phi'_o, \phi''_o} <\sigma'|\sigma'_o><\sigma'_o|O|\sigma''_o><\sigma''_o|\sigma''>$$

so that

$$\delta <\sigma'|O|\sigma''> = \sum (\delta <\sigma'|\sigma'_o>) <\sigma'_o|O|\sigma''_o><\sigma''_o|\sigma''>$$

$$+ \sum <\sigma'|\sigma'_o><\sigma'_o|O|\sigma''_o> \delta <\sigma''_o|\sigma''>$$

Note how the transition amplitude is defined (according to Feynman): If σ_1, σ_2 are two surfaces and Ω is the four volume enclosed by these surfaces, let $I(\Omega)$ be the action in this region. Then,

$$<\phi'_{1\alpha}, \sigma_1|\phi'_{2\alpha}, \sigma_2>$$

$$= \sum exp(i.I(\Omega))$$

the sum being over all fields in the region with prescribed boundary conditions, ie, the fields in each summand assume values $\phi'_{1\alpha}$ on σ_1 and $\phi'_{2\alpha}$ on σ_2. If $O(x)$ is any operator field, then

$$<\phi'_{1\alpha}\sigma_1|O(x)|\phi'_{2\alpha}\sigma_2>$$

$$= \sum O(x).exp(i.I(\Omega))$$

Hence,

$$\delta <\phi'_{1\alpha}\sigma_1|\phi'_{2\alpha}\sigma_2>$$

$$= i\sum \delta I(\Omega).exp(iI(\Omega)) = i <\phi'_{1\alpha}\sigma_1|\delta I(\Omega)|\phi'_{2\alpha}\sigma_2>$$

The change in the action $I(\Omega)$ caused by shift in the bounding surfaces and the field values on these surfaces is denoted by $\delta I_{\sigma_1-\sigma_2}$. We have the obvious relation $\delta I_{\sigma_2-\sigma_1} = -\delta I_{\sigma_1-\sigma_2}$. We thus get

$$\delta < \sigma'|O|\sigma'' >=$$

$$\sum < \sigma'|i\delta I_{\sigma-\sigma_o}|\sigma_o' >< \sigma_o'|O|\sigma_o'' >< \sigma_o''|\sigma'' >$$

$$+\sum < \sigma'|\sigma_o' >< \sigma_o'|O|\sigma_o'' >< \sigma_o''|i\delta I_{\sigma 0-\sigma}|\sigma'' >$$

$$= i < \sigma'|[\delta I_{\sigma-\sigma_o}, O]|\sigma'' >$$

or in terms of shorthand notation as

$$\delta < \sigma'|O|\sigma'' >= i < \sigma'|[\delta I, O]|\sigma'' >$$

41.Quantum electrodynamics on the DSP processor

1. Let $a(p,\sigma)$ denote the annihilation operator of a particle having momentum p and azimuthal spin quantum number σ, $b(p,\sigma)$ is the corresponding annihilation operator for the antiparticle. The postulated commutation relations are

$$[a(p,\sigma), a^*(p',\sigma')] == \delta(\mathbf{p}-\mathbf{p}')\delta_{\sigma\sigma'}$$

The quantum field within a box is given by

$$\Psi(x) = \int [a(p,\sigma)u(p,\sigma)exp(ip.x) + b(p,\sigma)^* v(p,\sigma)exp(-ip.x)]d^3p$$

For the hermitian conjugate, we then have

$$\Psi^*(x) = \int [a(p,\sigma)^* u^*(p,\sigma)exp(-ip.x) + b(p,\sigma)v^*(p,\sigma)exp(ip.x)]d^3p$$

where the four vector p is on the mass shell, ie, $p^2 = p^{02} - |\mathbf{p}|^2 = m^2$. The commutator is given by

$$[\Psi(x), \Psi^*(y)] = \sum [u(p,\sigma)u^*(p,\sigma)exp(ip.(x-y)) - v(p,\sigma)v^*(p,\sigma)exp(-ip.(x-y))]d^3p$$

To simulate the quantum field on the DSP processor, we need to operate in finite position and momentum domains. Our spatial variables k, m, r take values $0, 1, ..., N-1$ and the time variable n is also assumed to take the same values. Let Δ denote the discretization step size of the spatial variables. Then let $a[p,q,s,\sigma]$ denote the annihilation operator for the particles and $b[p,q,s,\sigma]$ the annihilation operator for antiparticles having momentum $(p\Delta, q\Delta, s\Delta)$. We set $x = 2\pi k/N\Delta, y = 2\pi m/N\Delta, z = 2\pi r/N\Delta$ and hence we have the expansion

$$\Psi[t,k,m,r] = \sum_{p,q,s=0}^{N-1} \{a[p,q,s,\sigma]u[p,q,s,\sigma]exp(2\pi i(t\Delta\sqrt{p^2+q^2+s^2}$$

$$-(kp+mq+rs)/N)) + b[p,q,s,\sigma]^* v[p,q,s,\sigma] exp(-2\pi i(t\Delta\sqrt{p^2+q^2+s^2} - (kp+mq+rs)/N))\}$$

so that

$$\Psi^*[t,k,m,r] = \sum_{p,q,s=0}^{N-1} \{a^*[p,q,s,\sigma]u^*[p,q,s,\sigma]exp(-2\pi i(t\Delta\sqrt{p^2+q^2+s^2} - (kp+mq+rs)/N))$$

$$+b^*[p,q,s,\sigma]v^*[p,q,s,\sigma]exp(2\pi i(t\Delta\sqrt{p^2+q^2+s^2} - (kp+mq+rs)/N))$$

In this formalism, the field is discretized in space but not in time. We postulate the commutation relations

$$[a[p,q,s,\sigma], a^*[p',q',s',\sigma'] = \delta[\sigma - sigma']\delta[p-p']\delta[q-q']\delta[s-s']$$

and likewise for b. All other commutators equal zero. Then the equal time commutation relations for the field Ψ are

$$[\Psi[t,k,m,r], \Psi^*[t,k',m',r']]$$

$$\sum u[p,q,s,\sigma]u^*[p,q,s,\sigma]exp(-2\pi i(k(p-p') + m(q-q') + r(s-s'))/N)$$

$$-v[p,q,s,\sigma]v^*[p,q,s,\sigma]exp(2\pi i(k(p-p') + m(q-q') + r(s-s'))/N)$$

Note that $\Psi[t,k,m,r]$ is required to satisfy a discrete version of the Klein-Gordon equation, discrete with respect to the spatial indices and continuous with respect to the time index. This means that the exponential function that we have used is not appropriate. Specifically, consider the following discrete Klein-Gordon equation:

$$\frac{d^2}{dt^2}\Phi[t,k,m,r] - \frac{1}{\Delta^2}(\Phi[t,k+1,m,r] + \Phi[t,k-1,m,r]$$

$$+\Phi[t,k,m+1,r]+\Phi[t,k,m-1,r]+\Phi[t,k,m,r+1]+\Phi[t,k,m,r-1]-6\Phi[t,k,m,r])+M^2\Phi[t,k,m,r] = 0$$

We write

$$\Phi[t,k,m,r] = exp(2\pi i(kp+mq+rs)/N)\eta(t)$$

Plugging this into the differential equation gives us

$$\eta''(t) - \Delta^{-2}(2.cos(2\pi k/N) + 2.cos(2\pi m/N) + 2.cos(2\pi r/N) - 6)\eta(t) + M^2\eta(t) = 0$$

It follows that

$$\eta(t) = A.exp(ip^0 t) + B.exp(-ip^0 t)$$

where

$$p^0 = p^0(k,m,r) = (M^2 + 2\Delta^{-2}(3 - cos(2k\pi/N) - cos(2\pi m/N) - cos(2\pi r/N)))^{1/2}$$

In order to compare this with the standard Klein-Gordon formula, suppose N is large and we make the approximation $cos(\xi) \approx 1 - \xi^2/2$, then p^0 becomes approximately

$$p^0 \approx (M^2 + (4\pi^2/N^2\Delta^2)(k^2+m^2+r^2))^{1/2}$$

if we set

$$p^1 = 2\pi k/N\Delta, p^2 = 2\pi m/N\Delta, p^3 = 2\pi r/N\Delta$$

then we recover the standard Klein-Gordon formula

$$p^0 = \sqrt{M^2 + \mathbf{p}^2}, \mathbf{p} = (p^1, p^2, p^3)$$

2. Let $\psi_l, l = 0, 1, 2, 3$ denote the Dirac field and let A_μ denote the four potential of the electromagnetic field. The current density is given by

$$J^\mu = \bar{\psi}\gamma^\mu\psi$$

where

$$\bar{\psi} = \psi^*\gamma^0$$

γ^μ are the Dirac matrices. The Lagrangian density for the free Dirac field is given by

$$\mathcal{L}_D = i\bar{\psi}\gamma^\mu\partial_\mu\psi - m\bar{\psi}\psi = \bar{\psi}(i\gamma^\mu\partial_\mu - m)\psi$$

To check this, we form the variation of $\mathcal{L}_D$ treating ψ and $\bar{\psi}$ as independent variables. Thus,

$$\delta\mathcal{L}_D = i\delta\bar{\psi}\gamma^\mu\partial_\mu\psi + i\bar{\psi}\gamma^\mu\partial_\mu\delta\psi$$
$$-m\delta\bar{\psi}\psi - m\bar{\psi}\delta\psi$$

so that setting the column vector coefficient of $\delta\bar{\psi}$ to zero gives us

$$i\gamma^\mu\partial_\mu\psi - m\psi = 0$$

or

$$(i\gamma^\mu\partial_\mu - m)\psi = 0$$

which is Dirac's equation. The term involving $\partial_\mu\delta\psi$ can be integrated by parts, to give the row vector coefficient of $\delta\psi$ as

$$0 = -i(\partial_\mu\bar{\psi})\gamma^\mu - m\bar{\psi} = 0$$

This is the same as

$$i\partial_\mu\psi^*\gamma^0\gamma^\mu + m\psi^*\gamma^0 = 0$$

or

$$i\partial_0\psi^* + i\partial_j\psi^*\gamma^0\gamma^j + m\psi^*\gamma^0 = 0$$

Using the identities

$$\gamma^{0*} = \gamma^0, \gamma^{i*} = -\gamma^i, i = 1, 2, 3$$

we get on taking conjugate transpose of the above equation,

$$-i\partial_0\psi + i\gamma^j\gamma^0\partial_j\psi + m\gamma^0\psi = 0$$

Premultiply by γ^0 to get

$$-i\gamma^0\partial_0\psi + i\gamma^0\gamma^j\gamma^0\partial_j\psi + m\psi = 0$$

and since

$$\gamma^0\gamma^j\gamma^0 = -\gamma^j, j = 1, 2, 3$$

we get

$$-i\gamma^0\partial_0\psi - i\gamma^j\partial_j\psi + m\psi = 0$$

or

$$(i\gamma^\mu\partial_\mu - m)\psi = 0$$

which is once again Dirac's equation. The Dirac equation in the presence of the electromagnetic field is given by

$$\gamma^\mu(i\partial_\mu - A_\mu)\psi - m\psi = 0$$

and this can be derived from the Lagrangian density

$$\mathcal{L}_1 = \bar{\psi}.\gamma^\mu(i\partial_\mu - A_\mu)\psi = \mathcal{L}_D + \mathcal{L}_I$$

where

$$\mathcal{L}_I = -\bar{\psi}\gamma^\mu\psi.A_\mu = -J^\mu A_\mu, J^\mu = \bar{\psi}\gamma^\mu\psi$$

$\mathcal{L}_I$ represents the interaction between the Dirac field and the electromagnetic field. Note that in classical electromagnetism, the Lagrangian that represents the interaction between charged matter and the electromagnetic field is given by $-J^\mu A_\mu = -\rho(V - (\mathbf{v}, \mathbf{A}))$, where ρ is the charge density, V is the electrostatic potential and $\mathbf{A}$ is the magnetic vector potential.

42.The method of self consistent fields: Let the Hamiltonian of a two particle quantum system be given by

$$H = H_1 + H_2 + V_{12}$$

where H_1 is a function of only the coordinates and momenta of particle one, H_2 is a function of only the coordinates and momenta of the second particle and V_{12} represents the interaction energy between the two particles. It is a function of the coordinates and momenta of both the particles. We try a wave function for the total system in separable form, ie

$$\psi(r_1, r_2) = \psi_1(r_1)\psi_2(r_2)$$

where ψ_1, ψ_2 have unit length and are chosen so that $< \psi|H|\psi >$ is a minimum. Now,

$$< \psi|H|\psi > = < \psi_1|H_1|\psi_1 > + < \psi_2|H_2|\psi_2 > + < \psi_1 \otimes \psi_2|V_{12}|\psi_1 \otimes \psi_2 >$$

$$= \int \bar{\psi}_1(r_1)H_1\psi_1(r_1)d^3r_1 + \int \bar{\psi}_2(r_2)H_2\psi_2(r_2)d^3r_2 + \int \bar{\psi}_1(r_1)\bar{\psi}_2(r_2)V_{12}\psi_1(r_1)\psi_2(r_2)d^3r_1d^3r_2$$

Setting the variation of this quantity with respect to $\bar{\psi}_1$ to $E_1\psi_1$ gives us the equation

$$H_1\psi_1(r_1) + (\int \bar{\psi}_2(r_2)V_{12}\psi_2(r_2)d^3r_2)\psi_1(r_1) = E_1\psi_1(r_1)$$

Likewise, setting the variation of the same with respect to ψ_2 to zero gives us

$$H_2\psi_2(r_2) + (\int \bar{\psi}_1(r_1)V_{12}\psi_1(r_1))\psi_2(r_2) = E_2\psi_2(r_2)$$

These equations are called the Hartree-Fock equations. They decouple the two dimensional problem into two one dimensional problems. The solution is only approximate since we are assuming the wave function to have the separable form.

Let us try antisymmetric wave functions in place of products. We then automatically take care of the Pauli exclusion principle. Specifically, we try

$$\psi(r_1,r_2)=\frac{1}{\sqrt{2}}(\psi_1(r_1)\psi_2(r_2)-\psi_2(r_1)\psi_1(r_2))$$

Then we find that

$$\begin{aligned}<\psi|H|\psi>=&\frac{1}{2}<(\psi_1\otimes\psi_2-\psi_2\otimes\psi_1)|(H_1\otimes I_2+I_1\otimes H_2\\ &+V_{12})|(\psi_1\otimes\psi_2-\psi_2\otimes\psi_1)>\\ =&\frac{1}{2}(<\psi_1|H_1|\psi_1>-<\psi_1|H_1|\psi_2><\psi_2|\psi_1>\\ &-<\psi_2|H_1|\psi_1><\psi_1|\psi_2>+|<\psi_2|H_1|\psi_2>+<\psi_2|H_2|\psi_2>\\ &-<\psi_1|\psi_2><\psi_2|H_2|\psi_1>-<\psi_2|\psi_1><\psi_1|H_2|\psi_2>+<\psi_1|H_2|\psi_1>)\\ &+<\psi_1\otimes\psi_2|V_{12}|\psi_1\otimes\psi_2>-<\psi_1\otimes\psi_2|V_{12}|\psi_2\otimes\psi_1>-<\psi_2\otimes\psi_1|V_{12}|\psi_1\otimes\psi_2>\\ &+<\psi_2\otimes\psi_1|V_{12}|\psi_2\otimes\psi_1>)\end{aligned}$$

We are assuming that ψ_1 and ψ_2 are orthonormal so that

$$\begin{aligned}<\psi|H|\psi>=&\frac{1}{2}(<\psi_1|H_1|\psi_1>+<\psi_2|H_1|\psi_2>+<\psi_1|H_2|\psi_1>+<\psi_2|H_2|\psi_2>)\\ &<\psi_1\otimes\psi_2|V_{12}|\psi_1\otimes\psi_2>-<\psi_1\otimes\psi_2|V_{12}|\psi_2\otimes\psi_1>\\ &-<\psi_2\otimes\psi_1|V_{12}|\psi_1\otimes\psi_2>\\ &+<\psi_2\otimes\psi_1|V_{12}|\psi_2\otimes\psi_1>)\end{aligned}$$

Taking into account the orthonormality constraints, the functional to be extremized is given by

$$F(\psi_1,\bar{\psi}_1,\psi_2,\bar{\psi}_2,\eta_1,\eta_2,E_1,E_2)$$

43.Charge conjugation operation acting on quantum fields: Let $\psi(p_i,\sigma_i,n_i,i=1,2,...)$ denote a multiparticle quantum state in which the ith particle is of species n_i and has a four momentum p_i and azimuthal spin angular momentum quantum number σ_i. Let n_i^c denote the species of the antiparticle of the species n_i. Then, the charge conjugation operator is a unitary operator C acting on the multiparticle states with the effect

$$C\psi^{\pm}(\{p_i,\sigma_i,n_i\})=\Pi_i(\xi_{n_i})\psi^{\pm}(\{p_i,\sigma_i,n_i^c\})$$

where plus sign denotes the in state and minus sign the out state. Here, ξ_{n_i} is a complex number of unit magnitude. The scattering matrix is given by

$$S[\{p_i', \sigma_i', n_i'\}|\{p_i, \sigma_i, n_i\}] =< \psi^-(\{p_i', \sigma_i', n_i'\}), \psi^+(\{p_i, \sigma_i, n_i\}) >$$

$$= \Pi_i[\xi_{n_i'}^* \xi_{n_i}] S[\{p_i', \sigma_i', n_i^{c'}\}|\{p_i, \sigma_i, n_i^c\}]$$

since C is a unitary operator.

44.Scattering matrix: Let Φ_α be a free particle state corresponding to energy E_α. We have

$$H_0 \Phi_\alpha = E_\alpha \Phi_\alpha$$

Assume that in the presence of the interaction energy V this state becomes an instate ψ_α^+ and an outstate ψ_α^-. This means that in the remote past the state ψ_α^+ evolved as Φ_α and in the remote future, the state ψ_α^- evolves as Φ_α. By conservation of energy, both the state $\Psi_\alpha^\pm$ correspond to the energy E_α for the perturbed Hamiltonian $H = H_0 + V$. Thus,

$$H\psi_\alpha^\pm = E_\alpha \psi_\alpha^\pm$$

It follows that the limits of $\int exp(-iE_\alpha t) g(\alpha) \psi_\alpha^+ d\alpha$ and $\int exp(-iE_\alpha t) g(\alpha) \Phi_\alpha d\alpha$ as $t \to -\infty$ are equal and likewise, the limits of $\int exp(-iE_\alpha t) g(\alpha) \psi_\alpha^- d\alpha$ and $\int exp(-iE_\alpha t) g(\alpha) \psi_\alpha^+ d\alpha$ as $t \to \infty$ are equal. Define

$$\psi^+ = \int g(\alpha) \psi_\alpha^-, d\alpha, \psi^- = \int g(\alpha) \psi_\alpha, \Phi = \int g(\alpha) \Phi_\alpha d\alpha$$

These are time independent states and their evolutions in time are respectively given by

$$exp(-iHt)\psi^+, exp(-iHt)\psi^-, exp(-iH_0 t)\Phi$$

We thus have,

$$\psi^+ = lim_{t \to -\infty} exp(iHt) exp(-iH_0 t)\Phi, \psi^- = lim_{t \to \infty} exp(iHt) exp(-iH_0 t) Phi$$

Define

$$\Omega(t) = exp(iHt) exp(-iH_0 t)$$

Then,

$$\psi^+ = \Omega(-\infty)\Phi, \psi^- = \Omega(\infty)\Phi$$

and we get for the scattering matrix elements

$$S_{\beta\alpha} =< \psi_\beta^-, \psi_\alpha^+ >=< \Phi_\beta, \Omega(\infty)^* \Omega(-\infty)\Phi_\alpha >$$

which means that the scattering matrix relative to the basis of unperturbed energy eigenstates is given by

$$S = \Omega(\infty)^* \Omega(-\infty)$$

We claim that (The Lippman-Schwinger equation)

$$\psi_\alpha^+ = \Phi_\alpha + (E_\alpha - H_0 + i\epsilon)^{-1} V \psi_\alpha^+$$

Indeed, this equation gives

$$(E_\alpha - H + i\epsilon)\psi_\alpha^+ = (E_\alpha - H_0 + i\epsilon)\Phi_\alpha = i\epsilon.\Phi_\alpha$$

The ϵ on the right side may be ignored and we thus get

$$H\psi_\alpha^+ = (E_\alpha + i\epsilon)\psi_\alpha^+$$

so that

$$exp(-iHt)\psi_\alpha^+ = exp(\epsilon t)exp(-iE_\alpha t)\psi_\alpha^+$$

or

$$exp(-\epsilon t)exp(-iHt)\psi_\alpha^+ = exp(-iE_\alpha t)\psi_\alpha^+$$

The limit on the left as $t \to \infty$ exists owing to the exponential decaying factor and that confirms the Lippman-Schwinger equation. A better proof would proceed along the following lines. Let g be a smooth function of the variable α that parametrizes the energy levels. Define

$$\Phi_g(t) = \int g(\alpha)\Phi_\alpha(t)d\alpha$$

and

$$\psi_g^-(t) = \int g(\alpha)\psi_\alpha^-(t)d\alpha$$

where $\psi_\alpha^-(t)$ is given by the Lippman-Schwinger equation. Specifically, at time $t = 0$, we have

$$\psi_\alpha^- = \Phi_\alpha + (E_\alpha - H_0 - i\epsilon)^{-1}V\psi_\alpha^-$$

We have using this equation,

$$\psi_g^-(t) = \int g(\alpha)exp(-iE_\alpha t)\psi_\alpha^- d\alpha$$

$$= \int g(\alpha)exp(-iE_\alpha t)\Phi_\alpha d\alpha + \int (E_\alpha - H_0 - i\epsilon)^{-1}V\psi_\alpha^- exp(-iE_\alpha t)d\alpha$$

$$= \Phi_g(t) + \int (E_\alpha - E_\beta - i\epsilon)^{-1}T_{\beta\alpha}exp(-iE_\alpha t)\Phi_\beta d\alpha d\beta$$

where

$$T_{\beta\alpha} = < \Phi_\beta|V|\psi_\alpha^- >$$

We want to show that

$$lim_{t\to\infty}[\psi_g^-(t) - \Phi_g(t)] = 0$$

It suffices therefore to show that

$$lim_{t\to\infty}\int (E_\alpha - E_\beta - i\epsilon)^{-1}T_{\beta\alpha}exp(-iE_\alpha t)d\alpha = 0$$

Assume the integral with respect to α is taken over the real axis and then over the large semicircle in the lower half of the complex plane. Such an extension is valid since if we consider $-i(E_\alpha - i\delta)t$, for $\delta > 0$, has a large negative real part $-\delta t$ as $t \to \infty$ and hence, in this limit, the integral over the semicircle will become zero. The resulting integral now becomes a contour integral over the lower half semicircle and if we assume that $T_{\beta\alpha}$ has no pole in α that falls in the interior of this semicircle, then the resulting contour integral is zero since the only pole of the integrand is at $E_\alpha = E_\beta + i\epsilon$ which is in the upper half part of the complex plane. Thus,

$$lim_{t\to\infty}[\psi_g^-(t) - \Phi_g(t)] = 0$$

and we are through.

45. Quantum probability

1. Let $\mathcal{H}$ be a Hilbert space and for $u \in \mathcal{H}$, $e(u)$ is the exponential vector in $\mathcal{H}$. Thus,

$$e(u) = \bigoplus_{n=0}^{\infty} u^{\otimes n}/\sqrt{n}!$$

Then,

$$< e(u), e(v) >= exp(< u, v >)$$

$W(u, U)$ is the Weyl operator in $\mathcal{H}$ corresponding to the unitary operator U in $\mathcal{H}$ and $u \in \mathcal{H}$. Its action on the exponential vector is given by

$$W(u, U)e(v) = exp(- \parallel u \parallel^2 /2- < u, Uv >)e(Uv + u)$$

It is a simple matter to check that $W(u, U)$ is a unitary operator. In fact write

$$W(u, U)e(v) = \lambda(u, U, v)e(Uv + u)$$

Then,

$$< W(u, U)e(v_1), W(u, U)e(v_2) >= \bar{\lambda}(u, U, v_1).\lambda(u, U, v_2) < e(Uv_1 + u), e(Uv_2 + u) >$$

$$= \bar{\lambda}(u, U, v_1)\lambda(u, U, v_2)exp(< Uv_1 + u, Uv_2 + u >)$$

$$= \bar{\lambda}(u, U, v_1)\lambda(u, U, v_2)exp(\parallel u \parallel^2 + < v_1, v_2 > + < Uv_1, u > + < u, Uv_2 >)$$

It is clear that if we take

$$\lambda(u, U, v) = exp(- \parallel u \parallel^2 /2- < u, Uv >)$$

then the above will coincide with $exp(< v_1, v_2 >)$, ie, with $< e(v_1), e(v_2) >$. We shall evaluate the product $W(u_1, U_1)W(u_2, U_2)$ where $u_1, u_2 \in \mathcal{H}$ and $U_1, U_2 \in \mathcal{U}(\mathcal{H})$, the unitary group in $\mathcal{H}$. We have

$$W(u_2, U_2)e(v) = exp(- \parallel u_2 \parallel^2 /2- < u_2, U_2v >)e(U_2v + u_2)$$

so that

$$W(u_1, U_1)W(u_2, U_2)e(v) = exp(- \parallel u_2 \parallel^2 /2- < u_2, U_2v >)W(u_1, U_1)e(U_2v + u_2)$$

$$= exp(- \parallel u_2 \parallel^2 /2- < u_2, U_2v >).exp(- \parallel u_1 \parallel^2 /2- < u_1, U_1(U_2v + u_2) >)e(U_1(U_2v + u_2) + u_1)$$

$$= exp(- \parallel u_2 \parallel^2 /2- \parallel u_1 \parallel^2 /2- < u_2, U_2v > - < u_1, U_1U_2v > - < u_1, U_1u_2 >)e(U_1U_2v + U_1u_2 + u_1)$$

Note that the group G consisting of all elements of the from (u, U) with $u \in \mathcal{H}$ and $U \in \mathcal{U}(\mathcal{H})$ acts on $\mathcal{H}$ in the following way.

$$(u, U).v = Uv + u$$

Thus,

$$(u_1, U_1).(u_2, U_2).v = (u_1, U_1)(U_2 v + u_2) = U_1(U_2 v + u_2) + u_1 = U_1 U_2 v + U_1 u_2 + u_1 = (U_1 u_2 + u_1, U_1 U_2).v$$

In other words, the composition law in G is given by

$$(u_1, U_1).(u_2, U_2) = (U_1 u_2 + u_1, U_1 U_2)$$

We have

$$W((u_1, U_1).(u_2, U_2))e(v) = W(U_1 u_2 + u_1, U_1 U_2)e(v)$$

$$= exp(- \| U_1 u_2 + u_1 \|^2 /2 - < U_1 u_2 + u_1, U_1 U_2 v >)e(U_1 U_2 v + U_1 u_2 + u_1)$$

$$= exp(- \| u_1 \|^2 /2 - \| u_2 \|^2 /2 - Re < u_1, U_1 u_2 > - < u_2, U_2 v > - < u_1, U_1 U_2 v >)e(U_1 U_2 v + U_1 u_2 + u_1)$$

It follows that

$$W(u_1, U_1)W(u_2, U_2)e(v) = exp(-i.Im < u_1, U_1 u_2 >)W((u_1, U_1).(u_2, U_2))$$

or equivalently,

$$W((u_1, U_1).(u_2, U_2)) = exp(i.Im < u_1, U_1 u_2 >)W(u_1, U_1)W(u_2, U_2)$$

Thus, $(u, U) \rightarrow W(u, U)$ from $G \rightarrow \mathcal{U}(\mathcal{H})$ is a projective unitary representation. Consider $u \in \mathcal{H}$ and define $W(u) = W(u, I)$. Then,

$$W(u)e(v) = exp(- \| u \|^2 /2 - < u, v >)e(u + v)$$

so that

$$W(tu)W(su)e(v) = exp(-s^2 \| u \|^2 - s < u, v > /2)W(tu)e(su + v)$$

$$= exp(-s^2 \| u \|^2 /2 - s < u, v > -t^2 \| u \|^2 /2 - t < u, su + v >)e((t + s)u + v)$$

$$= exp(-s^2 \| u \|^2 /2 - t^2 \| u \|^2 /2 - ts \| u \|^2 -(t + s) < u, v >)e((t + s)u + v)$$

$$= W((t + s)u)e(v)$$

It follows that $\{W(tu) : t \in \mathbb{R}\}$ is a one parameter unitary subgroup of $\mathcal{U}(\mathcal{H})$. We denote its generator by $-ip(u)$. Thus,

$$exp(-itp(u)) = W(tu), t \in \mathbb{R}, u \in \mathcal{H}$$

We have

$$W(tu)e(v) = exp(- \| u \|^2 t^2/2 - t < u, v >)e(v + tu)$$

so that

$$< e(w), W(tu)e(v) >= exp(- \| u \|^2 t^2/2 - t < u, v >)exp(< w, v + tu >)$$

$$= exp(- \| u \|^2 t^2/2 - t < u, v >)exp(t < w, u >)exp(< w, v >)$$

it follows that

$$-i < e(w), p(u)e(v) >= \frac{d}{dt} < e(w), W(tu)e(v) > |_{t=0}$$

$$= (- < u, v > + < w, u >)exp(< w, u >) = (< w, u > - < u, v >) < e(w), e(v) >$$

or equivalently,

$$< e(w), p(u)e(v) >= i(< w, u > - < u, v >) < e(w), e(v) >$$

It follows that

$$< e(w), p(iu)e(v) >= (- < w, u > + < u, v >) < e(w), e(v) >$$

Thus,

$$< e(w), (p(u) + ip(iu))e(v) >= -2i < u, v >< e(w), e(v) >$$

implying that

$$(-(p(u) + ip(iu))/2i)e(v) =< u, v > e(v)$$

Define

$$a(u) = \frac{i}{2}(ip(iu) + p(u))$$

Then we get

$$a(u)e(v) =< u, v > e(v)$$

It follows that

$$a(u)e(tv) = t < u, v > e(tv)$$

Pick the coefficient of t^n on both sides to get

$$a(u)v^{\otimes n}/\sqrt{n}! =< u, v > v^{\otimes(n-1)}/\sqrt{(n-1)!}$$

or

$$a(u)v^{\otimes n} = \sqrt{n} < u, v > v^{\otimes(n-1)}$$

This equation shows that $a(u)$ annihilates one particle in the n-particle state to give an $n-1$ particle state.

46. Quantization of fields. The wave function $\psi(r)$ becomes an operator field for bosons, we have the commutation relations

$$[\psi(r), \psi^*(r')] = \delta(r - r'), [\psi(r), \psi(r')] = 0, [\psi^*(r), \psi^*(r')] = 0$$

The number operator is defined by

$$N = \int \psi^*(r)\psi(r)d^3r$$

We have

$$[\psi(r), N] = \int [\psi(r), \psi^*(r')]\psi(r')d^3r' + \int \psi^*(r')[\psi(r), \psi(r')]d^3r'$$

$$= \int \delta(r - r')\psi(r')d^3r' + 0 = \psi(r)$$

or equivalently,

$$\psi(r)(N-1) = N\psi(r)$$

which is the same as

$$\psi(r)N = (N+1)\psi(r)$$

Suppose $|v>$ is an eigenstate of N with eigenvalue n, ie, $N|v>= n|v>$. Then we have

$$n\psi(r)|v>= \psi(r)N|v>= (N+1)\psi(r)|v>$$

or

$$N\psi(r)|v>= (n-1)\psi(r)|v>$$

that is $\psi(r)|v>$ is an eigenstate of N with eigenvalue $n-1$. That is why $\psi(r)$ corresponds to annihilation of a boson at the location r. Introduce the Hamiltonian operator

$$H = K + \Omega$$

where K is the kinetic energy operator defined as

$$K = -\frac{h^2}{2M}\int \psi^*(r)\nabla^2\psi(r)d^3r$$

and if $U(r)$ is an ordinary function of r, then

$$\Omega = \int \bar{\psi}(r)U(r)\psi(r)$$

We get

$$[K, \psi(r)] = -\frac{h^2}{2m}\int [\psi^*(r'), \psi(r)]\nabla'^2\psi(r')d^3r = \frac{h^2}{2m}\nabla^2\psi(r)$$

$$[K, \psi^*(r)] = -\frac{h^2}{2m}\int \psi^*(r')\nabla'^2[\psi(r'), \psi^*(r)]d^3r'$$

$$= -\frac{h^2}{2m}\int \psi^*(r')\nabla'^2\delta(r'-r)d^3r' = -\frac{h^2}{2m}\nabla^2\psi^*(r)$$

47. Quantum probability, sample computations:

1. $a(u)e(v) =< u, v > e(v)$,

$$a^*(u)e(v) = \frac{d}{dt}e(v+tu)|_{t=0}$$

Note that

$$< e(w), a^*(u)e(v) >= \frac{d}{dt} < e(w), e(v+tu) > |_{t=0} = \frac{d}{dt}exp(< w, v+tu >)|_{t=0} =< w, u >< e(w), e(v) >$$

on the one hand, while on the other,

$$< a(u)e(w), e(v) >=< u, w >^*< e(w), e(v) >=< w, u >< e(w), e(u) >$$

and both expressions tally with each other.

48. Formation of the covalent bond between molecules:Assume that we are studying the covalent bond of the H_2 molecule. Each H atom consists of a single positive charge $+e$ and a negative charge $-e$ bound to this positive charge. The energy levels of the atom are obtained from the Schrodinger equation, ie by solving computing the eigenvalues of the Hamiltonian

$$-\frac{h^2}{2m}\nabla^2 - e^2/r$$

where ∇^2 is the Laplacian corresponding to a single atom. Specifically, we should number the two electrons as one and two and likewise the two protons as one and two. The position vector of the first proton is $\mathbf{0}$ and the position vector of the second proton is $\mathbf{a}$. $\mathbf{a}$ is a fixed vector. The position vector of the first electron is $\mathbf{r}_1$ and that of the second electron is $\mathbf{r}_2$. The Total Hamiltonian of the system is given by

$$H = -\frac{h^2}{2m}\nabla_1^2 - \frac{h^2}{2m}\nabla_2^2 - \frac{e^2}{|\mathbf{r}_1|} - \frac{e^2}{|\mathbf{r}_1 - \mathbf{a}|} - \frac{e^2}{|\mathbf{r}_2|} - \frac{e^2}{|\mathbf{r}_2 - \mathbf{a}|} + \frac{e^2}{|\mathbf{r}_1 - \mathbf{r}_2|}$$

$$= H_1 + H_2 + H_{12}$$

where

$$H_1 = -\frac{h^2}{2m}\nabla_1^2 - e^2/r_1 - e^2/|\mathbf{r}_1 - \mathbf{a}|$$

$$H_2 = -\frac{h^2}{2m}\nabla_2^2 - e^2/r_2 - e^2/|\mathbf{r}_2 - \mathbf{a}|$$

$$H_{12} = e^2/|\mathbf{r}_1 - \mathbf{r}_2|$$

Note that H_1 and H_2 are identical Hamiltonians provided that we interchange the indices one and two. The first step is to determine numerically the eigenvalues and eigenfunctions of H_1. This problem is best attacked in the cylindrical coordinate system. Suppose we assume that $\mathbf{a}$ is much larger in magnitude as compared with the Bohr radius. Then the last term in the Hamiltonian

$$W = -\frac{h^2}{2m}\nabla^2 - e^2/r - e^2/|\mathbf{r} - \mathbf{a}|$$

can be regarded as a small perturbation to the energy of the Hydrogen atom and the first order correction to the wave function can be determined from that. For applying perturbation theory, we use the expansion

$$1/|\mathbf{r} - \mathbf{a}| = (r^2 + a^2 - 2arcos\theta)^{-1/2} = a^{-1}(1 + r^2/a^2 - 2rcos(\theta)/a)^{-1/2}$$

$$= \sum_{l=0}^{\infty}(r^l/a^{l+1})P_l(cos(\theta))$$

and retain just the first few terms in the expansion. Suppose that $u_{nlm}(r, \theta, \phi)$ is the wave function of the unperturbed Hydrogen atom, the terms will for example be if we retain $l = 0, 1$,

$$a^{-1} + (r/a^2)cos(\theta)$$

Then the first order shift in the energy level E_n in this state is given by

$$\Delta E_{nlm} = \int |u_{nlm}(r,\theta,\phi)|^2 (a^{-1} + (r/a^2)cos(\theta))sin(\theta)r^2 dr d\theta d\phi$$

49. Quantum mechanical system with slowly time varying parameters (Ref:Arno Bohm, Quantum Mechanics) Let $H(\theta)$ be the Hamiltonian of a system dependent upon parameters θ. For each choice of θ, we determine the energy eigenvalues $E_n(\theta)$ and eigenfunctions $|n,\theta>, n = 1,2,...,$ so that

$$H(\theta)|n,\theta> = E_n(\theta)|n,\theta>, n = 1,2,..., <n,\theta|m,\theta> = \delta_{nm},$$

$$\mathbf{I} = \sum_{n=1}^{\infty} |n,\theta><n,\theta|$$

We now let θ vary with time. Then the Hamiltonian H becomes time dependent, ie, $H(\theta(t))$. We assume that the corresponding time dependent wave functions are given by $|\psi(n,t)> = |n,\theta(t)>\phi(t)$ where ϕ is a function of time. Plugging this into the time dependent Schrodinger equation,

$$ih\frac{d}{dt}|\psi(n,t)> = H(\theta(t))|\psi(n,t)>$$

gives us

$$ih\phi'(t)|n,\theta(t)> + ih\phi(t)\frac{d}{dt}|n,\theta(t)> = H(\theta(t))|n,\theta(t)>\phi(t) = \phi(t)E_n(\theta(t))|n,\theta(t)>$$

Take the inner product on both sides with $<m,\theta(t)|$ to get

$$ih\phi'(t)\delta_{mn} + ih\phi(t)<m,\theta(t)|\frac{d}{dt}|n,\theta(t)> = \phi(t)E_n(\theta(t))\delta_{mn}$$

It will in general be impossible to choose ϕ so that this equation is satisfied for all m. However taking $m = n$ gives us

$$ih\phi'(t) - E_n(\theta(t))\phi(t) + ih\phi(t)<n,\theta(t)|\frac{d}{dt}|n,\theta(t)> = 0$$

of e

$$ih\phi'(t)/\phi(t) = E_n(\theta(t)) = -<n,\theta(t)|\frac{d}{dt}|n,\theta(t)>$$

The solution to this equation is readily obtained by integration:

$$ih.log(\phi(t)) = \int_0^t E_n(\theta(u))du - \int_0^t <n,\theta(u)|\frac{d}{du}|n,\theta(u)>du$$

or

$$log(\phi(t)) = -\frac{i}{h}\int_0^t E_n(\theta(u))du + \frac{i}{h}\int_0^t <n,\theta(u)|\frac{d}{du}|n,\theta(u)>du$$

This gives an approximate solution to the time dependent problem. Let $\theta = (\theta_1,...,\theta_p)$ so that

$$\nabla_\theta = (\frac{\partial}{\partial\theta_i}, i = 1,2,...,m)$$

Then

$$\frac{d}{dt}|n,\theta(t)>=\sum_{i=1}^{m}\int_0^t \frac{\partial}{\partial\theta_i}|n,\theta(t)>d\theta_i(t)$$

$$=\int_0^t \nabla_\theta|n,\theta(t)>.d\theta(t)$$

It follows that

$$log(\phi(t))=-\frac{i}{h}\int_0^t E_n(\theta(u))du+\frac{i}{h}\int_0^t d\theta(u).\nabla_\theta|n,\theta(u)>$$

It follows from this formula that if the system parameters vary over a duration $[0,T]$ such that $\theta(T)=\theta(0)$, ie, so that at the end of the duration, the parameters return back to their starting values, then $\phi(t).exp((i/h)\int_0^t E_n(\theta(u))du)$ need not return to zero. At the end of the period, the wave function may acquire a phase shift. This phase shift is called the Berry phase named after its discoverer Michael Berry.

Let e_3 denote the unit vector along the z axis, in conventional language $e_3=\hat{z}$. Suppose that J_1, J_2, J_3 are the angular momentum generators of a quantum system. $|k,e_3>$ denotes the eigenvector of J_3 corresponding to the eigenvalue kh, or setting $h=1$, the eigenvalue is simply k. Thus, $J_3|k,e_3>=k|k,e_3>$. We may without loss of generality assume that all the states are eigenvectors of $J^2=\sum_{i=1}^3 J_i^2$ with a definite eigenvalue $j(j+1)$. We define the operator

$$U(\theta,\phi)=exp(-i\phi J_3)exp(-i\theta J_2)exp(i\phi J_3)$$

Suppose that $R\rightarrow\pi(R)$ is the underlying representation of the rotation group $SO(3)$. It follows that

$$\pi(R_3(\phi))=exp(iJ_3\phi),\pi(R_2(\theta))=exp(iJ_2\theta),\pi(R_1(\phi))=exp(iJ_2\phi)$$

In other words, the Lie algebra of $\pi(SO(3))$ is generated by $iJ_k, k=1,2,3$. Define

$$\mathbf{R}(\theta,\phi)=\mathbf{e}_1cos(\phi)sin(\theta)+\mathbf{e}_2sin(\phi)sin(\theta)+\mathbf{e}_3cos(\theta)$$

Then

$$\mathbf{R}(\theta,\phi).\mathbf{J}=J_1cos(\phi)sin(\theta)+J_2sin(\phi)sin(\theta)+J_3cos(\theta)$$

Define

$$|k,\theta,\phi>=U(\theta,\phi)|k,e_3>$$

We then want to show that

$$\mathbf{R}(\theta,\phi).\mathbf{J}|k,\theta,\phi>=k|k,\theta,\phi>$$

This amounts to proving that

$$U(\theta,\phi)^{-1}(\mathbf{R}(\theta,\phi).\mathbf{J})U(\theta,\phi)|k,e_3>=k|k,e_3>$$

Now observe that

$$U(\theta,\phi)^{-1}(\mathbf{R}(\theta,\phi).\mathbf{J})U(\theta,\phi)=$$

$$exp(-i\phi J_3).exp(i\theta J_2)exp(i\phi J_3)(\mathbf{R}(\theta,\phi).\mathbf{J})exp(-i\phi J_3)exp(-i\theta_2 J_2)exp(i\phi J_3)$$

To evaluate this, we need the following:

$$exp(i\phi J_3)J_i exp(-i\phi J_3), i = 1, 2, 3$$

$$exp(i\theta J_2)J_i exp(-i\theta J_2), i = 1, 2, 3$$

Observe that

$$exp(i\phi J_3)J_i exp(-i\phi J_3) = exp(i\phi ad(J_3))J_i = J_i + \sum_{n=1}^{\infty}((i\phi)^n/n!)ad(J_3)^n J_i$$

where

$$ad(X)(Y) = [X, Y]$$

Now,

$$ad(J_3)(J_1) = [J_3, J_1] = iJ_2, ad(J_3)^2(J_1) = i[J_3, J_2] = J_1$$

In general,

$$ad(J_3)^{2n}(J_1) = J_1, ad(J_3)^{2n+1} iJ_2$$

so that

$$exp(i\phi J_3)J_1 exp(-i\phi J_3) = \sum_{n=0}^{\infty}((i\phi)^{2n}/(2n)!)J_1 + iJ_2 \sum_{n=0}^{\infty}(i\phi)^{2n+1}/(2n+1)!$$

$$= J_1 cos(\phi) - J_2 sin(\phi)$$

$$exp(i\phi J_3)J_2 exp(-i\phi J_3) = \sum_{n=0}^{\infty}((i\phi)^n/n!)ad(J_3)^n(J_2)$$

Now,

$$ad(J_3)(J_2) = [J_3, J_2] = -iJ_1, ad(J_3)^2(J_2) = -i[J_3, J_1] = J_2$$

and in general,

$$ad(J_3)^{2n}(J_2) = J_2, ad(J_3)^{2n+1}(J_2) = -iJ_1$$

so that

$$exp(i\phi J_3)J_2 exp(-i\phi J_3) = J_2 cos(\phi) + J_1 sin(\phi)$$

Finally, we trivially have

$$exp(i\phi J_3)J_3 exp(-i\phi J_3) = J_3$$

Next,

$$exp(i\theta J_2)J_1 exp(-i\theta J_2) = \sum_{n=0}^{\infty}((i\theta)^n/n!)(adJ_2)^n(J_1)$$

Now,

$$ad(J_2)(J_1) = [J_2, J_1] = -iJ_3, ad(J_2)^2(J_1) = -iad(J_2)(J_3) = -i[J_2, J_3] = J_1$$

and in general,

$$ad(J_2)^{2n}(J_1) = J_1, ad(J_2)^{2n+1}(J_1) = -iJ_3$$

so that

$$exp(i\theta J_2)J_1exp(-i\theta J_2) = J_1cos(\theta) + J_3sin(\theta),$$

$$exp(i\theta J_2)J_3exp(-i\theta J_2) = \sum_{n=0}^{\infty}((i\theta)^n/n!)ad(J_2)^n(J_3)$$

and using

$$ad(J_2)(J_3) = [J_2, J_3] = iJ_1, ad(J_2)^2(J_3) = iad(J_2)(J_1) = i[J_2, J_1] = J_3$$

so that

$$exp(i\theta J_2)J_3exp(-i\theta J_2) = J_3cos(\theta) - J_1sin(\theta)$$

Now we are in a position to combine all the formulas:

$$U(\theta,\phi)^{-1}J_1U(\theta,\phi) =$$

$$exp(-i\phi J_3).exp(i\theta J_2)exp(i\phi J_3)J_1exp(-i\phi J_3)exp(-i\theta J_2)exp(i\phi J_3) =$$
$$exp(-i\phi J_3)exp(i\theta J_2)(J_1cos(\phi) - J_2sin(\phi))exp(-i\theta J_2)exp(i\phi J_3)$$
$$= exp(-i\phi J_3)(cos(\phi)(J_1cos(\theta) + J_3sin(\theta)) - sin(\phi)J_2)exp(i\phi J_3)$$
$$= cos(\phi)cos(\theta)exp(-i\phi J_3)J_1exp(i\phi J_3) + J_3cos(\phi)sin(\theta) - sin(\phi)exp(-i\phi J_3)J_2exp(i\phi J_3)$$
$$= cos(\phi)cos(\theta)(J_1cos(\phi) + J_2sin(\phi)) + J_3cos(\phi)sin(\theta) - sin(\phi)(J_2cos(\phi) - J_1sin(\phi))$$
$$= J_1(cos^2(\phi)cos(\theta) + sin^2(\phi)) + J_2(sin(\phi)cos(\phi)cos(\theta) - sin(\phi)cos(\phi)) + J_3cos(\phi)sin(\theta)$$

Likewise,

$$U(\theta,\phi)^{-1}J_2U(\theta,\phi) =$$
$$= exp(-i\phi J_3).exp(i\theta J_2)exp(i\phi J_3)J_2exp(-i\phi J_3)exp(-i\theta J_2)exp(i\phi J_3)$$
$$= exp(-i\phi J_3)exp(i\theta J_2)(J_2cos(\phi) + J_1sin(\phi))exp(-i\theta J_2)exp(i\phi J_3)$$
$$= exp(-i\phi J_3)(cos(\phi)J_2 + sin(\phi)(J_1cos(\theta) + J_3sin(\theta))exp(i\phi J_3)$$
$$= cos(\phi)exp(-i\phi J_3)J_2exp(i\phi J_3) + cos(\theta)sin(\phi)exp(-i\phi J_3)J_1exp(i\phi J_3) + sin(\phi)sin(\theta)J_3$$
$$= cos(\phi)(J_2cos(\phi) - J_1sin(\phi)) + cos(\theta)sin(\phi)(J_1cos(\phi) + J_2sin(\phi)) + sin(\theta)sin(\phi)J_3$$
$$= J_1(sin(\phi)cos(\phi)cos(\theta) - sin(\phi)cos(\phi)) + J_2(cos^2(\phi) + sin^2(\phi)cos(\theta)) + sin(\theta)sin(\phi)J_3$$

Finally,

$$U(\theta,\phi)^{-1}J_3U(\theta,\phi) =$$
$$exp(-i\phi J_3)exp(i\theta J_2)exp(i\phi J_3)J_3exp(-i\phi J_3)exp(-i\theta J_2)exp(i\phi J_3)$$
$$= exp(-i\phi J_3)exp(i\theta J_2)J_3exp(-i\theta J_2)exp(i\phi J_3)$$
$$= exp(-i\phi J_3)(J_3cos(\theta) - J_1sin(\theta))exp(i\phi J_3)$$
$$= cos(\theta)J_3 - sin(\theta)(J_1cos(\phi) + J_2sin(\phi))$$
$$= -cos(\phi)sin(\theta)J_1 - sin(\phi)sin(\theta)J_2 + cos(\theta)J_3$$

Thus,

$$U(\theta,\phi)^{-1}(\mathbf{R}(\theta,\phi).\mathbf{J})U(\theta,\phi)$$
$$= cos(\phi)sin(\theta)U(\theta,\phi)^{-1}J_1U(\theta,\phi) + sin(\phi)sin(\theta)U(\theta,\phi)^{-1}J_2U(\theta,\phi)$$
$$+cos(\theta)U(\theta,\phi)^{p-1}J_3U(\theta,\phi)$$
$$= cos(\phi)sin(\theta)(J_1(cos^2(\phi)cos(\theta) + sin^2(\phi)) + J_2(sin(\phi)cos(\phi)cos(\theta) - sin(\phi)cos(\phi)) + J_3cos(\phi)sin(\theta))$$
$$+sin(\phi)sin(\theta)(J_1(sin(\phi)cos(\phi)cos(\theta) - sin(\phi)cos(\phi)) + J_2(cos^2(\phi) + sin^2(\phi)cos(\theta)) + sin(\theta)sin(\phi)J_3)$$
$$+cos(\theta)(-cos(\phi)sin(\theta)J_1 - sin(\phi)sin(\theta)J_2 + cos(\theta)J_3)$$
$$= a_1J_1 + a_2J_2 + a_3J_3$$

where

$$a_1 = cos(\phi)sin(\theta)(cos^2(\phi)cos(\theta) + sin^2(\phi)) + sin^2(\phi)cos(\phi)sin(\theta)(cos(\theta) - 1)$$
$$-sin(\theta)cos(\theta)cos(\phi) = 0$$
$$a_2 = sin(\phi)cos^2(\phi)sin(\theta)(cos(\theta) - 1) + sin(\phi)sin(\theta)(cos^2(\phi) + sin^2(\phi)cos(\theta)) - sin(\phi)sin(\theta)cos(\theta)$$
$$= 0$$
$$a_3 = cos^2(\phi)sin^2(\theta) + sin^2(\phi)sin^2(\theta) + cos^2(\theta) = 1$$

Thus,

$$U(\theta,\phi)^{-1}(\mathbf{R}(\theta,\phi).\mathbf{J})U(\theta,\phi) = J_3$$

and the proof is complete.

50. Problem:Let $Q_i, P_i, i = 1,2,3$ be the position and momentum operators in a quantum mechanical system. The angular momentum operators are $L_i = \epsilon_{ijk}Q_jP_k$ with summation over the repeated indices j,k being understood. Consider the operator $F = L\times P + P\times L$. Show that $F.L = 0$. This identity plays a fundamental role in deriving the spectrum of the Hydrogen atom from purely group theoretic considerations (see the book Quantum mechanics, foundations and applications by Arno Bohm).

Proof of the identity:

$$F.L = F_1L_1 + F_2L_2 + F_3L_3 = L.L\times P + L.P\times L$$

$$L.L\times P = L_1(L_2P_3 - L_3P_2) + L_2(L_3P_1 - L_1P_3) + L_3(L_1P_2 - L_2P_1)$$
$$= [L_1,L_2]P_3 + [L_2,L_3]P_1 + [L_3,L_1]P_2 = i(L_1P_1 + L_2P_2 + L_3P_3) = iL.P = 0$$

since

$$L_1P_1 + L_2P_2 + L_3P_3 = (Q_2P_3 - Q_3P_2)P_1 + (Q_3P_1 - Q_1P_3)P_2 + (Q_1P_2 - Q_2P_1)P_3$$
$$= Q_1(P_2P_3 - P_3P_2) + Q_2(P_3P_1 - P_1P_3) + Q_3(P_1P_2 - P_2P_1) = 0$$

as the $P's$ commute amongst themselves. We will be through therefore if we can show that $L.P\times L = 0$ Now observe that

$$P_2L_3 - P_3L_2 = L_3P_2 - L_2P_3 + [P_2,L_3] - [P_3,L_2]$$

But,

$$[P_2,L_3] = [P_2,Q_1P_2 - Q_2P_1] = iP_1,$$

$$[P_3, L_2] = [P_3, Q_3P_1 - Q_1P_3] = -iP_1$$

etc. Thus

$$P_2L_3 - P_3L_2 = L_3P_2 - L_2P_3 + 2iP_1$$

and likewise, for the other components. Now use the fact that

$$L.P = 0$$

and the previous identity to arrive at the desired conclusion.

51. Classical Kepler problem for the Hydrogen atom:

$$E = p^2/2m - e^2/r$$

is the total energy. It is a constant of the motion. Let **L** denote the total angular momentum of the particle. It is a constant of the motion. The Runge Lenz vector is defined by

$$\mathbf{a} = -\mathbf{p} \times \mathbf{L}/m + e^2\hat{r}$$

It is also a constant of the motion. Proof:

$$d\mathbf{a}/dt = \frac{\mathbf{L}}{m} \times d\mathbf{p}/dt + e^2\frac{d}{dt}(\hat{r}/r)$$

Now,

$$d\mathbf{p}/dt = -e^2\hat{r}/r^2$$

so the first term on the rhs in the above equation is

$$-e^2(\mathbf{L}/m) \times \hat{r}/r^2) = e^2(\hat{r} \times \mathbf{L})/mr^2$$

The second term is evaluated as follows:

$$(d/dt)\hat{r} = (d/dt)(\mathbf{r}/r) = \mathbf{r}'/r - \mathbf{r}(\mathbf{r}', \mathbf{r})/r^3$$

$$= r^{-3}(r^2\mathbf{r}' - (\mathbf{r}', \mathbf{r})\mathbf{r}) = r^{-3}\mathbf{r} \times (\mathbf{r}' \times \mathbf{r}) = -r^{-3}\mathbf{r} \times (\mathbf{L})/m$$

These equations imply the desired result, namely constancy of the vector **a** in time. The equation of the elliptic orbit can be determined from here.

$$\mathbf{a.r} = -L^2/m + e^2r$$

Consider in addition the equations

$$\mathbf{L.r} = 0, \mathbf{a.L} = 0$$

These are obvious from the defining relations for **a**, **L**. The equation $\mathbf{L.r} = 0$ implies that motion takes place in a plane perpendicular to **L**. The equation $\mathbf{a.L} = 0$ implies that the vector **a** falls in the plane of motion. Assume that **r** makes at time t an angle ϕ with **a**. Then we get from the above

$$arcos(\phi) = -L^2/m + e^2r$$

or

$$r = (L^2/m)/(e^2 - a.cos(\phi))$$

The maximum value of r is given by setting $\phi = 0$ and we get

$$r_{max} = (L^2/m)/(e^2 - a)$$

The minimum value is obtained by setting $\phi = \pi$ and is

$$r_{min} = (L^2/m)/(e^2 + a)$$

The semimajor axis is

$$A = (r_{max} + r_{min})/2 = (e^2L^2/m)/(e^4 - a^2)$$

We have

$$a^2 = p^2L^2/m^2 + e^4 - 2e^2L^2/mr = e^4 + (2L^2/m)(p^2/2m - e^2/r) = e^4 + 2L^2E/m$$

From this equation, we calculate the energy as

$$E = m(a^2 - e^4)/2L^2 = -e^2/2A$$

The eccentricity of the orbit can now be calculated: From the canonical equation of an ellipse in polar coordinates $r = l/(1 - \epsilon.cos\phi)$, one deduces that the eccentricity is given by

$$\epsilon = a/e^2$$

so the minor axis B is given by

$$B^2 = A^2(1 - \epsilon^2) = \frac{e^4}{4E^2}(1 - a^2/e^4) = \frac{1}{4E^2}(-2L^2E/m) = -L^2/2mE$$

52. Gauge theories: Let $A_i(x,t), i = 1,2,3$ be the magnetic vector potential and $A_0(x,t)$ the electrostatic potential. Consider the operators

$$\mathbf{D} = \nabla + ie\mathbf{A}, D_0 = \partial_t - ieA_0$$

Let $\psi(x,t)$ be the wave function of the electron. It satisfies the equation

$$-\frac{h^2}{2m}D^2\psi = iD_0\psi$$

Assume that ψ suffers a Gauge transformation

$$\psi \to \psi' = exp(i\alpha)\psi$$

where $\alpha(x,t)$ is a scalar function. We wish to determine the transformation law of $\mathbf{A}, A_0$ that would guarantee that $D\psi$ transforms into $exp(i\alpha)D\psi$ and $D_0\psi$ transforms into $exp(i\alpha)D_0\psi$. Assume that the transformed $\mathbf{A}$ and A_0 are respectively $\mathbf{A}'$ and A_0'. Then,

$$(D\psi)' = (\nabla + ie\mathbf{A}')\psi' = (\nabla + ie\mathbf{A}')exp(i\alpha)\psi = exp(i\alpha)(\nabla\psi + ie\mathbf{A}' + i\nabla\alpha)\psi$$

For this to equal

$$exp(i\alpha)D\psi = exp(i\alpha)(\nabla + ie\mathbf{A})\psi$$

we must necessarily have

$$e\mathbf{A}' + \nabla\alpha = e\mathbf{A}$$

or equivalently,

$$\mathbf{A}' = \mathbf{A} - e^{-1}\nabla\alpha$$

Also,

$$(D_0\psi)' = (\partial_t - ieA_0')exp(i\alpha)\psi = exp(i\alpha)(\partial_t - ieA_0' + i\partial_t\alpha)\psi$$

For this to equal

$$exp(i\alpha)D_0\psi = exp(i\alpha)(\partial_t - ieA_0)\psi$$

we must necessarily have

$$-ieA_0' + i\partial_t\alpha = -ieA_0$$

or

$$A_0' = A_0 + e^{-1}\partial_t\alpha$$

We call this transformation of the wave function and the electromagnetic four potential, a gauge transformation. This is the $U(1)$ gauge. The Schrodinger equation is invariant under this transformation. Specifically, we consider the Schrodinger equation for a nonrelativistic particle in an electromagnetic field:

$$-\frac{h^2}{2m}D^2\psi = ihD_0\psi$$

Under the Gauge transformation of the electromagnetic fields and the wave function, we have

$$(D\psi)' = D'\psi' = exp(i\alpha)D\psi, (D_0\psi)' = D_0'\psi' = exp(i\alpha)D_0\psi$$

so the gauge transformed wave function ψ' satisfies the Schrodinger equation corresponding to the gauge transformed four potentials. Note that $D'\psi' = (D\psi)' = exp(i\alpha)D\psi$ so that

$$D'^2\psi' = D'D'\psi' = D'(D\psi)' = D'exp(i\alpha)D\psi = exp(i\alpha)D^2\psi$$

implying that

$$-\frac{h^2}{2m}D'^2\psi' = iD_0'\psi'$$

in view of the Schrodinger equation for ψ.

53. Born-Oppenheimer method for a diatomic molecule: $\mathbf{P}_1$ is the momentum of the first nucleus and $\mathbf{P}_2$ the momentum of the second nucleus. The total kinetic energy of the nuclei is given by

$$T_{nuc} = \mathbf{P}_1^2/2m_1 + \mathbf{P}_2^2/2m_2$$

where m_1, m_2 are respectively the masses of the two nuclei. Let there be N_e electrons in the field of the two nuclei and label their momenta by $\mathbf{p}_i, i = 1, 2, ..., N_e$. The total kinetic energy of the electrons is then given by

$$T_{elec} = \frac{1}{2m_e}\sum_{i=1}^{N_e}\mathbf{p}_i^2$$

The potential energy of interaction of the two nuclei is given by

$$U_{nuc} = \frac{Z_1 Z_2 e^2}{|\mathbf{R}_1 - \mathbf{R}_2|}$$

where $\mathbf{R}_i, i = 1, 2$ are the position vectors of the two nuclei and $Z_1 e, Z_2 e$ are the nuclear charges. The potential energy of interaction between the electrons and nuclei is

$$U_{nuc-elec} = -\sum_{i=1}^{N_e} \sum_{j=1}^{2} \frac{Z_i e^2}{|\mathbf{r}_i - \mathbf{R}_j|}$$

Finally, the potential energy of interaction between the electrons is given by

$$U_{elec} = \sum_{i \neq j, i,j=1}^{N_e} \frac{e^2}{|\mathbf{r}_i - \mathbf{r}_j|}$$

The dynamical variables that describe the total Hamiltonian H of the molecule are $\mathbf{r}_i, \mathbf{R}_j, \mathbf{p}_i, \mathbf{P}_j, i = 1, 2, ..., N_e, j = 1, 2$ and we have

$$H = T_{nucl} + T_{elec} + U_{nuc} + U_{nuc-elec} + U_{elec}$$

The stationary state Schrodinger equation to be solved is given by

$$H|\psi_E >= E|\psi_E >$$

The position space representation of the stationary state wave function $|\psi_E >$ corresponding to the energy eigenvalue E is given by

$$\psi_E(\mathbf{r}_1, ..., \mathbf{r}_{N_e}, \mathbf{R}_1, \mathbf{R}_2) =< \mathbf{r}_1, ..., \mathbf{r}_{N_e}, \mathbf{R}_1, \mathbf{R}_2 | \psi_E >$$

The nuclei are heavy and if they are assumed to be static, we can to a good approximation regard the electrons as constituting a system moving in the static electrostatic field generated by the nuclei and the energy of the electronic system is then given by

$$E_{elec} = T_{elec} + U_{nuc-elec} + U_{elec}$$

An interesting problem is to consider $N_e - 1$ electrons in the ground state and one electron in an excited state moving in electrostatic field produced by the nuclei and the other $N_e - 1$ electrons. Let $\mathbf{r}$ denote the coordinate of the single excited electron. It effective potential energy of interaction with the smeared out charge of the other electrons and the two almost static nuclei is represented by $U(\mathbf{r}) = U(\mathbf{r}, \mathbf{R})$. The total energy of this excited electron is given by

$$E_e = \mathbf{p}_e^2/2m_e + U(\mathbf{r}, \mathbf{R})$$

We now proceed to describe the motion from the standpoint of the nuclei. $E_{n,e}(\mathbf{R})$ denotes the average of $U_{nuc-elec}$ over the electronic variables. The average potential experienced by the nuclei is thus

$$U_n(\mathbf{R}) = E_{n,e}(\mathbf{R}) + U_{nuc}(\mathbf{R})$$

The total energy of relative motion of the nuclei is then given by

$$E_{nuc} = T_{nuc} + U_n(\mathbf{R}) = T_{nuc} + E_{n,e}(\mathbf{R}) + U_{nuc}(\mathbf{R})$$

where T_{nuc} is the kinetic energy of relative motion of the nuclei. Note that the total nuclear kinetic energy has been decomposed into the kinetic energy of translation of the center of mass and the kinetic energy of relative motion and the kinetic energy of the center of mass has been removed from the picture. Letting $\mathbf{P}$ denote the relative nuclear momentum, ie, difference between the momenta of the two nuclei, we have

$$T_{nuc} = \mathbf{P}^2/2\mu, \mu = m_1 m_2/(m_1 + m_2)$$

Note that the total nuclear kinetic energy is given by

$$T_{nuc,tot} = \mathbf{P}_1^2/2m_1 + \mathbf{P}_2^2/2m_2$$

Writing

$$\mathbf{P}_{cm} = \mathbf{P}_1 + \mathbf{P}_2, \mathbf{P}_{rel} = \mu(\mathbf{P}_2/m_2 - \mathbf{P}_1/m_1) = (m_1\mathbf{P}_2 - m_2\mathbf{P}_1)/(m_1 + m_2)$$

we find that

$$T_{nuc,tot} = \mathbf{P}_{cm}^2/2(m_1 + m_2) + \mathbf{P}_{rel}^2/2\mu$$

The first term does not give any contribution to the bound states. The total energy of the molecule corresponding to the model in which just one electron is occupying an excited state is given by

$$H = E_{nuc} + E_e = T_{nuc} + U_n(\mathbf{R}) + E_e = \mathbf{P}^2/2\mu + U_n(\mathbf{R}) + \mathbf{p}^2/2m_e + U(\mathbf{r}, \mathbf{R})$$

We set

$$V(\mathbf{r}, \mathbf{R}) = U_n(\mathbf{R}) + U(\mathbf{r}, \mathbf{R})$$

and

$$h(\mathbf{p}, \mathbf{r}, \mathbf{R}) = \mathbf{p}^2/2m_e + V(\mathbf{r}, \mathbf{R})$$

and then we get

$$H = \mathbf{P}^2/2\mu + h(\mathbf{p}, \mathbf{r}, \mathbf{R})$$

The variables $\mathbf{r}, \mathbf{p}$ describe the fast degrees of freedom, namely those of the electron while the variables $\mathbf{P}, \mathbf{R}$ describe the slow degrees of freedom, namely those of the nuclei and are called the slow variables. We want to solve the Schrodinger equation

$$H\psi_E = E\psi_E$$

The slow position operators R_i and the fast position operators r_i have joint eigenfunctions $|x_i, r_i>$:

$$\hat{R}_i|x_i, r_i >= x_i|x_i, r_i >, \hat{r}_i|x_i, r_i >= r_i|x_i, r_i >$$

We now regard the slow position variables $\mathbf{R} = (R_i)$ as parameters in the fast Hamiltonian $h(\mathbf{p}, \mathbf{r}, \mathbf{R})$ and denote the eigenstates of this Hamiltonian by $|x_i, n>$. Note that x_i is a parameter in this eigenstate and n is a quantum number. The eigenvalue equation reads with the notation

$$h = h(\mathbf{p}, \mathbf{r}, \mathbf{R}) = h(\mathbf{R}),$$

$$h(\mathbf{R})|x_i, n>= \epsilon_n(x)|x_i, n>$$

$\epsilon_n(x)$ are the energy levels of the fast electron with the positions x of the nuclei fixed. These positions therefore appear as parameters in the energy levels of the fast electron. The state $|x, n>$ describes a state of the molecule in which the nuclei are at fixed locations x while the electrons are in the energy state $\epsilon_n(x)$. We have the obvious orthogonality relation

$$< x_i', n'|x_i, n>= \delta(x - x')\delta_{n',n}$$

The overall wave function of the molecule is an element of $\mathcal{H}_{slow} \otimes \mathcal{H}_{fast}$ and $|x, n>$ which is a wave function in this space can be expressed as

$$|x, n>= |x> \otimes|n(x)>$$

where $|x>$ is in $\mathcal{H}_{slow}$ and represents a state of the nuclei in which they are located at position x. $|n(x)>$ is a state of the electrons in which the electrons energy $\epsilon_n(x)$. The operator $\mathbf{P}$ acts only in $\mathcal{H}_{slow}$. Let us compute its action on $|\psi_E>$ using the representation $|x, n>= |x> \otimes|n(x)>$. We have

$$< x, n|\mathbf{P}|\psi_E>= \sum_{n'} \int d^3x' < x, n|\mathbf{P}|x', n'>< x', n'|\psi_E>$$

$$= \sum_{n'} \int d^3x' < n(x)| < x|\mathbf{P}|x'> |n'(x')>< x', n'|\psi_E>$$

The operator $< x|\mathbf{P}|x'>$ equals $-i\nabla_{x'}\delta(x - x')$ and it acts on all the terms to its right. Thus, we get

$$< x, n|\mathbf{P}|\psi_E>= -i\sum_{n'} \int d^3x' < n(x)|\nabla_{x'}\delta(x - x')|n'(x')>< x', n'|\psi_E>$$

$$= i\sum_{n'} \int d^3x'\delta(x - x') < n(x)|\nabla_{x'}|n'(x')>< x', n'|\psi_E>$$

$$= i\sum_{n'} < n(x)|\nabla_x(|n'(x)>< x, n|\psi_E>)$$

$$= i\sum_{n'}(< n(x)|\nabla_x|n'(x)>) < x, n|\psi_E> +i < n(x)|n'(x)> \nabla_x < x, n|\psi_E>$$

$$= A_{n,n'}(x) < x, n|\psi_E> +i\delta_{n,n'}\nabla_x < x, n|\psi_E>$$

where

$$A_{n,n'}(x) = i < n(x)|\nabla_x|n'(x)>$$

Thus the Schrodinger equation for the slow part can be expressed as with $\psi_{E,n}(x) =< x, n|\psi_E>$,

$$(2\mu)^{-1}\sum_l(\sum_m(-i\delta_{n,m}\nabla_x - A_{n,m}(x))(-i\delta_{m,l}\nabla_x - A_{m,l}(x)) + \epsilon_n(x)\delta_{nl})\psi_{E,l}(x) = E\psi_{E,n}(x)$$

In the Born-Oppenheimer approximation, we neglect the action of ∇_x on $|n'(x)>$. This approximation is justified in view of the slow condition, ie, the nuclear momentum is small in magnitude. With this approximation, we have $A_{n,n'}(x) = 0$ and the above equation simplifies to

$$-(2\mu)^{-1}\nabla^2\psi_{E,n}(x) + \epsilon_n(x)\psi_{E,n}(x) = E\psi_{E,n}(x)$$

54.Gauge theory (Reference:A.Bohm, Quantum mechanics, foundations and applications): Let $A_i(x), i = 1,2,3$ be an $N \times N$ matrix valued function of spatial position $x = (x_1, x_2, x_3)$. This will be called a gauge field provided that it satisfies certain transformation properties. We define the covariant derivative

$$D = \nabla - igA$$

where ∇ is the ordinary gradient $(\partial_i, i = 1,2,3)$ and g is a constant. Let $\psi(x)$ be an $N \times 1$ vector valued function of spatial location. $U(x)$ is a unitary $N \times N$ matrix valued function of position. The gauge transformation of the wave function is

$$\psi(x) \to U(x)\psi(x)$$

or in terms of components,

$$\psi_i(x) \to \sum_{j=1}^{N} U_{ij}(x)\psi_j(x), i = 1,2,...,N$$

We wish to define the gauge transformation of the gauge fields A so that $D\psi$ transforms into $UD\psi$. This means that if $A \to A'$, then

$$D\psi = (\nabla - igA)\psi \to (\nabla - igA')\psi' = (\nabla - igA')U\psi = UD\psi = U(\nabla - igA)\psi$$

or

$$U(\nabla - igA)\psi = (\nabla U)\psi + U\nabla\psi - igA'\psi$$

so that

$$-igUA = \nabla U - igA'$$

or

$$A' = A - (i/g)U^{-1}\nabla U$$

The stationary Schrodinger equation

$$(-\frac{1}{2m}D^2 + V)\psi = E\psi$$

under the gauge transformation gives

$$(-\frac{1}{2m}UD^2 + UV)\psi = EU\psi$$

or

$$-\frac{1}{2m}D'2\psi' + UVU^{-1}\psi' = E\psi'$$

Thus the gauge transformed scalar potential V is given by

$$V' = UVU^{-1}$$

Let us define the field tensor

$$F_{ij} = [D_i, D_j] = [\partial_i - igA_i, \partial_j - igA_j]$$

Note that $D_i\psi$ transforms to $UD_i\psi$. Writing this as $D_i'\psi' = D_i'U\psi$, we find that $D_i' = UD_iU^{-1}$. It follows that

$$F_{ij}' = [D_i', D_j'] = [UD_iU^{-1}, UD_jU^{-1}] = U[D_i, D_j]U^{-1} = UF_{ij}U^{-1}$$

This is the transformation law of the field tensor.

55.Quantum field theory: Suppose $I[\psi]$ is the action functional of a set of fields $\psi = (\psi_l)$. Suppose that this functional is invariant under infinitesimal transformations of the form

$$\psi_l(x) \to \psi_l(x) + \epsilon.F_l(x)$$

where ϵ is a small real parameter. Then we can write

$$0 = \delta I = \epsilon \int (\delta I[\psi]/\delta\psi_l)(x)\delta\psi_l(x)d^4x$$

This is called a global symmetry. Consider now transformations of the form

$$\psi_l(x) \to \psi_l(x) + \epsilon(x)F_l(x)$$

where $\epsilon(.)$ is an infinitesimal function of x. The variation in the action caused by such a transformation has the form

$$\delta I = \int J^\mu(x)\partial_\mu\epsilon(x)d^4x$$

in order that I satisfy the global symmetry, ie, in order that $\delta I = 0$ when ϵ is a constant. Suppose now that ψ_l satisfy the field equations $\delta I[\psi] = 0$. Then it follows from the arbitrariness of $\epsilon(x)$ that

$$\partial_\mu J^\mu = 0$$

and we thus obtain a conservation law. Thus we have proved the fundamental theorem of Noether:If the action is invariant under infinitesimal transformations of the above kind, then we have a conservation law. The conservation law implies that

$$0 = dF/dt, F = \int J^0 d^3x$$

Suppose that the Lagrangian density of the field is $\mathcal{L}(\psi, \partial_\mu\psi)$. The Lagrangian is then given by

$$L(t) = \int \mathcal{L}d^3x$$

The action is given by

$$I = \int L(t)dt = \int \mathcal{L}d^3x$$

Suppose that I has the above mentioned global symmetry, ie, invariance under

$$\psi_l(x) \to \psi_l(x) + \epsilon.F_l(x)$$

Assume that $L(t)$ is also invariant under the above transformation. Then we shall calculate δI under transformations of the form

$$\psi_l(x) \to \psi_l(x) + \epsilon(t)F_l(x)$$

$$\delta I = \int dt \int \delta\mathcal{L}d^3x$$

$$= \int dt \int [\frac{\delta\mathcal{L}}{\delta\psi_l}\epsilon(t)F_l(x) + (\delta\mathcal{L}/\delta\psi_l')(\partial/\partial t(\epsilon(t)F_l(x))]d^3x$$

$$= \int dt \int d^3x[(\delta\mathcal{L}/\delta\psi_l)F_l + (\delta\mathcal{L}/\delta\psi_l')(\partial F_l/\partial t)]\epsilon(t)d^3x$$

$$+ \int dt \int d^3x\epsilon'(t)F_l(x)(\delta\mathcal{L}/\delta\psi_l')$$

The first integral is zero since by virtue of invariance of $\int \mathcal{L}d^3x$ under the transformation

$$\psi_l(x) \to \psi_l(x) + \epsilon.F_l(x)$$

we have

$$(\delta\mathcal{L}/\delta\psi_l)F_l + (\delta\mathcal{L}/\delta\psi_l')F_{l,t} = 0$$

Thus,

$$\delta I = \int d^4x\epsilon'(t)F_l(x)(\delta\mathcal{L}/\delta\psi_l')$$

If the fields ψ_l satisfy the field equations, then we can set $\delta I = 0$ and obtain on integrating by parts, the conservation equation

$$\frac{dF(t)}{dt} = 0$$

where

$$F(t) = \int F_l(x)(\delta\mathcal{L}/\delta\psi_l')d^3x$$

Thus in the case that the action as well as the Lagrangian (integral of the Lagrangian density over space) are invariant under the transformations with ϵ constant, we obtain an explicit formula for the conserved charge $\int J^0 d^3x$. Suppose now that this transformation leaves also the Lagrangian density invariant. Then we can obtain an explicit formula for the conserved current density J^μ. The procedure is to note that under the transformation (local transformation)

$$\psi_l(x) \to \psi_l(x) + \epsilon(x)F_l(x)$$

we have

$$\delta I = \int (\frac{\partial \mathcal{L}}{\partial \psi_l})\epsilon(x)F_l(x) + (\frac{\partial \mathcal{L}}{\partial \partial_\mu \psi_l})\partial_\mu(\epsilon(x)F_l(x)))d^4x$$

and since $\mathcal{L}$ is invariant under the global transformation, (ϵ constant), we must have

$$\frac{\partial \mathcal{L}}{\partial \psi_l}F_l + \frac{\partial \mathcal{L}}{\partial \partial_\mu \psi_l}\partial_\mu F_l = 0$$

and hence we deduce that

$$\delta I = \int (\frac{\partial \mathcal{L}}{\partial \partial_\mu \psi_l}F_l \partial_\mu \epsilon)d^4x$$

It follows that if ψ_l satisfy the field equations, then $\delta I = 0$ giving the conservation law $\partial_\mu J^\mu = 0$ with

$$J^\mu = \frac{\partial \mathcal{L}}{\partial \partial_\mu \psi_l}F_l$$

56.Lagrangian and Hamiltonian formulation of the Dirac field:

Consider an infinitesimal Lorentz transformation

$$\Lambda^\mu_\nu = \delta^\mu_\nu + \omega^\mu_\nu$$

The metric tensor is

$$(\eta_{\mu\nu}) = diag[1, 1, 1, -1]$$

ie

$$\eta_{ii} = 1, i = 1, 2, 3, \eta_{00} = -1, \eta_{\mu\nu} = 0, \mu \neq \nu$$

We set

$$\omega_{\mu\nu} = \eta_{\alpha\mu}\omega^\alpha_\nu$$

The equation

$$\eta_{\alpha\beta}(\delta^\alpha_\mu + \omega^\alpha_\mu)(\delta^\beta_\nu + \omega^\beta_\nu)$$
$$= \eta_{\mu\nu}$$

gives on retaining terms upto first order in ω:

$$\eta_{\mu\beta}\omega^\beta_\nu + \eta_{\nu\alpha}\omega^\alpha_\mu = 0$$

or

$$\omega_{\mu\nu} + \omega_{\nu\mu} = 0$$

ie, $(\omega_{\mu\nu})$ is an antisymmetric matrix. If the action for the fields $I[\psi]$ is invariant under these transformations, then there is a set of conserved currents which we denote by $M^{\rho\mu\nu}$. The conservation law is

$$\partial_\rho M^{\rho\mu\nu} = 0$$

It follows that

$$dJ^{\mu\nu}/dt = 0, J^{\mu\nu} = \int M^{0\mu\nu} d^3x$$

Assume that the fields (ψ_l) fall in a representation space of the Lorentz group so that under infinitesimal Lorentz transformations, they transform as

$$\delta\psi^l = \frac{i}{2}\omega^{\mu\nu}(I_{\mu\nu})^l_m\psi^m$$

$I^{\mu\nu}$ are the generators of the Lie algebra of the representation of the Lorentz group. They satisfy the standard commutation relations. The field derivatives transform as

$$\delta\partial_k\psi_l = \frac{i}{2}\omega^{\mu\nu}(I_{\mu\nu})^m_l\partial_k\psi^m - \omega^\mu_k\partial_\mu\psi_l$$

Note that we have used the fact that if $x \to x'$ under the infinitesimal Lorentz transformation $x'\mu - x^\mu = \delta x^\mu = x^\mu + \omega^\mu_\nu x^\nu$, then the derivatives transform as

$$\partial_{\mu'} = \frac{\partial x^\alpha}{\partial x^{\mu'}}\partial_\alpha = -\omega^\alpha_\mu\partial_\alpha$$

Assume that the Lagrangian density L is invariant under the Lorentz transformation of the fields. Then,

$$0 = \frac{\partial L}{\partial\psi^l}\delta\psi^l + \frac{\partial\mathcal{L}}{\partial\partial_k\psi^l}\delta(\partial_k\psi^l)$$

We can substitute for $\delta\psi^l$ and $\delta(\partial_k\psi^l)$ and derive a conservation law for a tensor called the Belinfante tensor, closely related to the energy momentum tensor $T^{\mu\nu}$. The details can be found in Weinberg's book, "The quantum theory of fields".

57. Path integrals for solving quantum mechanical problems:Assume that the Hamiltonian is given by $H(q,p) = p^2/2m + V(q)$. The Lagrangian is then given by $L = pq' - H = pq' - p^2/2m - V(q)$. Consider the Path integral

$$\int exp(i\int_0^T(pq' - p^2/2m - V(q))dt)DpDq = \int exp(i\int_0^T(pq' - H(q,p))dt)DpDq$$

The integral with respect to p is a Gaussian integral and is easily evaluated by discretization:

$$\int_0^T(pq' - p^2/2m)dt \approx \sum_i(p_i(\Delta q_i/\Delta t) - p_i^2/2m)\Delta t$$

and

$$\int exp(i\sum_i p_i(\Delta q_i/\Delta t) - p_i^2/2m)\Delta t\Pi_i dp_i = (2\pi m/i)^{N/2}exp(i\sum_i m(\Delta q_i)^2/2\Delta t)$$

This means that but for a multipicative constant(complex), the path integral is given by

$$\int exp(i\int_0^T(m(dq(t)/dt)^2/2 - V(q))dt)Dq$$

This is the classic path integral originally introduced into non-relativistic quantum mechanics by Feynman.

58. Evolution of Heisenberg observables:

$$Q_a(t) = exp(itH)Q_a exp(-itH), P_a(t) = exp(itH)P_a exp(-itH)$$

Define

$$|q,t> = exp(itH)|q>, |p,t> = exp(itH)|p>$$

Then we have

$$Q_a(t)|q,t> = q_a|q,t>, P_a(t)|p,t> = p_a|p,t>$$

We also have

$$<q',t|q,t> = <q'|exp(-itH)exp(itH)|q> = <q'|q> = \delta(q'-q)$$

and likewise,

$$<p',t|p,t> = \delta(p'-p)$$

$$\int |q,t> dq < q,t| = exp(itH)\int |q> dq < q|exp(-itH) = exp(itH).I.exp(-itH) = I$$

and likewise,

$$\int |p,t> dp < p,t| = I$$

Further,

$$<q,t|p,t> = <q|p> = exp(i\sum q_a p_a)/\sqrt{2\pi}$$

Next,

$$<q',\tau + d\tau|q,\tau> = <q'|exp(-iHd\tau)|q> = <q',\tau|exp(-iHd\tau)|q,\tau>$$

Note that since the Hamiltonian commutes with itself, it follows that

$$H(Q,P) = exp(itH)H(Q,P)exp(-itH) = H(Q(t),P(t))$$

Assume that the quantum Hamiltonian is such that all the $Q's$ appear to the left of all the $P's$. We then have

$$<q',\tau + d\tau|q,\tau> =$$

$$\int \Pi dp_a < q',\tau|exp(-iH(Q(\tau),P(\tau))d\tau)|p,\tau >< p,\tau|q,\tau>$$

$$= \int (\Pi dp_a/\sqrt{2\pi})exp(-iH(q',p)d\tau) < q',\tau|p,\tau >< p,\tau|q,\tau>$$

$$= \int (\Pi dp_a/\sqrt{2\pi})exp(-iH(q',p)d\tau).exp(i\sum(q'_a - q_a)p_a)$$

We then use with $\tau_0 < \tau_1 < ... < \tau_N$,

$$<q_N,\tau_N|q_0,\tau_0> = \int <q_N,\tau_N|q_{N-1},\tau_{N-1}> dq_{N-1} < q_{N-1},\tau_{N-1}|q_{N-2},\tau_{N-2}>$$

$$dq_{N-2}...dq_1 < q_1, \tau_1 | q_0, \tau_0 >$$

and taking the limit $N \to \infty, max(\tau_{i+1} - \tau_i) \to 0, \tau_0, \tau_N$ fixed gives us the path integral formula

$$< q', t_2 | q, t_1 > = < q_N | exp(-iH(\tau_N - \tau_0)) | q_0 >$$

$$= \int exp(i \int_{t_1}^{t_2} (p(t)q'(t) - H(q(t), p(t)))dt) DqDp$$

59. Computation of the vertex function: First we must develop expressions for the Dirac Fermion propagator.

$$\psi_l(x) = \sum_\sigma \int [a(p,\sigma)u_l(p,\sigma)exp(-ip.x) + b(p,\sigma)^* v_l(p,\sigma)exp(ip.x)]d^3p$$

where $a(p,\sigma)$ is the operator that annihilates an electron of four momentum p and spin component σ while $b(p,\sigma)$ annihilates a positron of four momentum p and spin component σ. Here $p.x = p^0 t - \mathbf{p.x}$ and p falls on the mass shell, ie, $p^0 = \sqrt{m^2 + \mathbf{p}^2}$. We can write this in vector notation as

$$\psi(x) = \sum_\sigma \int [a(p,\sigma)\mathbf{u}(p,\sigma)exp(-ip.x) + b(p,\sigma)\mathbf{v}(p,\sigma)exp(ip.x)]d^3p$$

Weinberg in his book "The quantum theory of fields" uses the notation $a(p,\sigma,n)$ and $a(p,\sigma,n^c)$ respectively for $a(p,\sigma)$ and $b(p,\sigma)$. Since $\psi(x)$ satisfies the free particle Dirac equation, we must have

$$(i\gamma^\mu \partial_\mu + m)\psi(x) = 0$$

This gives us

$$(\gamma^\mu p_\mu + m)\mathbf{u}(p,\sigma) = \mathbf{0},$$

$$(\gamma^\mu p_\mu - m)\mathbf{v}(p,\sigma) = \mathbf{0}$$

or using the shorthand notation,

$$(\gamma.p + m)\mathbf{u}(p,\sigma) = \mathbf{0}, (\gamma.p - m)\mathbf{v}(p) = \mathbf{0}$$

The anticommutation relations read

$$[a(p,\sigma), a^*(p',\sigma')] = \delta^3(p - p')\delta_{\sigma,\sigma'},$$

$$[b(p,\sigma), b^*(p',\sigma')] = \delta^3(p - p')\delta_{\sigma,\sigma'}$$

Thus,

$$[\psi(x), \psi^*(y)] = \int d^3p\{\sum_\sigma \mathbf{u}(p,\sigma)\mathbf{u}^*(p,\sigma)exp(-ip.(x-y)) + \sum_\sigma \mathbf{v}(p,\sigma)\mathbf{v}^*(p,\sigma)exp(ip.(x-y))\}$$

60. Problem:Discuss one loop radiative corrections to the electron propagator. Hint Let Δ be the photon propagator without corrections and Π the one loop correction due to electron terms. We have for the exact photon propagator:

$$\Delta' = \Delta + \Delta\Pi\Delta + \Delta\Pi\Delta\Pi\Delta + ... = \Delta(1 - \Pi\Delta)^{-1}$$

61. A heuristic derivation of the form of the exact photon propagator: Let $A_\mu(x)$ be the quantized vector potential of the electromagnetic field and J_μ the electronic current, ie,

$$J^{\mu(x)} = \psi^*(x)\gamma^\mu\psi(x)$$

The Lagrangian density for the electromagnetic field interacting with matter has the form

$$\mathcal{L} = a.F_{\mu\nu}(x)F^{\mu\nu}(x) + b.J_\mu(x)A^\mu(x)$$

where

$$F_{\mu\nu} = A_{\nu,\mu} - A_{\mu,\nu}$$

Written in full

$$\mathcal{L} = a.(\partial_\nu A_\mu - \partial_\mu A_\nu).(\partial^\nu A^\mu - \partial^\mu A^\nu)$$

$$= 2a.(\partial_\nu A_\mu)(\partial^\nu A^\mu) - 2a.(\partial_\nu A_\mu)(\partial^\mu A^\nu)$$

The free photon propagator can be expressed as a Feynman path integral:

$$\Delta_{\mu\nu}(x-y) = \int A_\mu(x)A_\nu(y).exp((ia/h)\int F_{\mu\nu}(w)F^{\mu\nu}(w)d^4w)DA_\mu$$

We can express the action integral in the momentum domain as follows:

$$A_\mu(x) = \int A_\mu(p)exp(ip.x)d^4p$$

$$\partial_\nu A_\mu(x) = \int ip_\nu A_\mu(p).exp(ip.x)d^4p$$

$$\int(\partial_\nu A_\mu(x))(\partial^\nu A^\mu(x))d^4x$$

$$= -\int p^2 A_\mu(p)A_\mu(p)d^4p,$$

$$\int(\partial_\nu A_\mu(x))(\partial^\mu A^\nu(x))d^4x$$

$$= -\int p_\nu p^\mu A_\mu(p)A^\nu(p)d^4p$$

after dropping constant proportionality factors. Thus,

$$\int F_{\mu\nu}(x)F^{\mu\nu}(x)d^4x = \int(p^2\eta_{\mu\nu} - p_\mu p_\nu)A^\mu(p)A^\nu(p)d^4p$$

We find that

$$\Delta_{\mu\nu}(x-y) = \int A_\mu(x)A_\nu(y).exp((ia/h)\int F_{\mu\nu}(x)F^{\mu\nu}(x)d^4x)DA_\mu$$

$$= \int exp(i(p.x+p'.y))d^4pd^4p' \int A_\mu(p)A_\nu(p')exp((2ia/h)\int R^{\mu\nu}(p)A_\mu(p)A_\nu(p)d^4p)DA_\mu$$

The path integral in the momentum space representation is like evaluating the covariance of a Gaussian random vector. The answer is the inverse of the matrix $R^{\mu\nu}(p)$, ie,

$$\Delta_{\mu\nu}(x-y) = \int exp(ip.(x-y))[R^{\mu\nu}(p)]^{-1}d^4p$$

where

$$R^{\mu\nu}(p) = p^2\eta^{\mu\nu} - p^\mu p^\nu$$

Here, $\eta^{\mu\nu}$ is the space time metric. Let us solve the equation

$$R^{\mu\nu}x_\nu = y^\mu$$

for x. We can express this equation as

$$(p^2\eta^{\mu\nu} - p^\mu p^\nu)x_\nu = y^\mu$$

so that

$$p^2x^\mu - p^\mu(p.x) = y^\mu$$

or

$$p^2x - p(p.x) = y$$

so that

$$x = (p(p.x)+y)/p^2$$

and hence,

$$p.x = (1/p^2)(p^2(p.x)+p.y)$$

There is no solution, ie, the matrix $(R^{\mu\nu}(p))$ is singular. To get over this problem in the evaluation of the bare photon propagator, consult the book by Weinberg.

62. Dirac formalism: The Lorentz metric is given by

$$(\eta_{\mu\nu}) = diag[1,1,1,-1]$$

An infinitesimal Lorentz transformation has the form

$$\Lambda^\mu_\nu = \delta^\mu_\nu + \omega^\mu_\nu$$

where if

$$\omega_{\mu\nu} = \eta_{\mu\alpha}\omega^\alpha_\nu$$

then

$$\omega_{\mu\nu} = -\omega_{\nu\mu}$$

Let $\Lambda \to D(\Lambda)$ be a unitary representation of the Lorentz group. Then for Λ as above, we have

$$D(\Lambda) = I + \frac{i}{2}\omega_{\mu\nu}I^{\mu\nu}$$

where $I^{\mu\nu}$ are the generators of the representation D. Note the commutation relations satisfied by the $I^{\mu\nu}$: Let $\omega_{\mu\nu}, \theta_{\mu\nu}$ be two infinitesimal elements of the Lorentz group. Let $L^\mu_\nu = \delta^\mu_\nu + \omega^\mu_\nu$. Then we have

$$D(\Lambda)D(L)D(\Lambda^{-1}) = D(\Lambda.L.\Lambda^{-1})$$

Now

$$D(\Lambda)D(L)D(\Lambda^{-1}) = (I + (i/2)\omega_{\mu\nu}I^{\mu\nu})(I + (i/2)\theta_{ab}I^{ab})(I - (i/2)\omega_{pq}I^{pq})$$
$$= [I + (i/2)(\omega_{\mu\nu} + \theta_{\mu\nu})I^{\mu\nu} - (1/4)\omega_{\mu\nu}\theta_{ab}I^{\mu\nu}I^{ab})][I - (i/2)\omega_{pq}I^{pq}]$$
$$= I + (i/2)\theta_{\mu\nu}I^{\mu\nu} - (1/4)\omega_{\mu\nu}\theta_{ab}I^{\mu\nu}I^{ab} + (1/4)\omega_{pq}(\omega_{\mu\nu} + \theta_{\mu\nu})I^{\mu\nu}I^{pq}$$
$$= I + (i/2)\theta_{\mu\nu}I^{\mu\nu} + (1/4)\omega_{\mu\nu}\theta_{ab}(I^{ab}I^{\mu\nu} - I^{\mu\nu}I^{ab}) + O(\omega^2)$$

On the other hand,

$$\Lambda.L.\Lambda^{-1} = (I + \omega)(I + \theta)(I - \omega) = (I + \omega + \theta + \omega\theta)(I - \omega)$$
$$= I + \theta + \omega\theta - \theta\omega + O(\omega^2)$$

It follows that

$$D(\Lambda.L.\Lambda^{-1}) = I + (i/2)\theta_{\mu\nu}I^{\mu\nu} + (i/2)[\omega, \theta]_{\mu\nu}I^{\mu\nu}$$

Equating terms of the same degree gives us

$$(1/4)\omega_{\mu\nu}\theta_{ab}[I^{ab}, I^{\mu\nu}]$$
$$= (i/2)[\omega, \theta]_{\mu\nu}I^{\mu\nu}$$

Now,

$$[\omega, \theta]_{\mu\nu} = \omega_{\mu\alpha}\theta^\alpha_\nu - \theta_{\mu\alpha}\omega^\alpha_\nu$$

Note that from this, it follows that

$$[\omega, \theta]_{\nu\mu} = \omega_{\nu\alpha}\theta^\alpha_\mu - \theta_{\nu\alpha}\omega^\alpha_\mu$$
$$= -[\omega, \theta]_{\mu\nu}$$

as it should be. We have

$$[\omega, \theta]_{\mu\nu}I^{\mu\nu}$$
$$= (\omega_{\mu\alpha}\theta^\alpha_\nu - \theta_{\mu\alpha}\omega^\alpha_\nu)I^{\mu\nu}$$
$$= \omega_{\rho\sigma}\theta_{ab}(\eta^{a\sigma}I^{\rho b} - \eta^{b\rho}I^{a\sigma})$$

Using antisymmetry of ω and θ, this equals

$$\frac{1}{2}\omega_{\rho\sigma}\theta_{ab}(\eta^{a\sigma}I^{\rho b} - \eta^{b\rho}I^{a\sigma} - \eta^{a\rho}I^{\sigma b} + \eta^{b\sigma}I^{a\rho})$$

Note that the generators $I^{\mu\nu}$ of the representation D have been defined so that $I^{\mu\nu} = -I^{\nu\mu}$. The term inside the bracket above is thus manifestly antisymmetric with respect to interchange of (ρ, σ) and of (a, b). This yields the desired commutation relations for the Lie algebra of the representation:

$$i[I^{ab}, I^{\rho\sigma}] =$$

$$\eta^{b\rho} I^{a\sigma} - \eta^{a\rho} I^{b\sigma} - \eta^{\sigma a} I^{\rho b} + \eta^{\sigma b} I^{\rho a}$$

Problem:Write down the structure constants corresponding to these commutation relations.

63. Computation of the propagator for bosons:

$$\Delta_{lm}(x,y) = < 0|T(\psi_l(x)\psi_m^*(y))|0 >$$

$$\psi(x) = \psi^+(x) + \psi^-(x)$$

where

$$\psi^+(x) = \sum_\sigma \int u(p,\sigma)a(p,\sigma)exp(ip.x)d^3p,$$

$$\psi^-(x) = \sum_\sigma \int v(p,\sigma)a^*(p,\sigma)exp(-ip.x)d^3p$$

If $x^0 > y^0$, then

$$\Delta_{lm}(x,y) = < 0|\psi_l(x)\psi_m^*(y)|0 > = < 0|\psi_l^{+*}(x)\psi_m^-(y)|0 >$$

while if $y^0 > x^0$m, then

$$\Delta_{lm}(x,y) = < 0|\psi_m^*(y)\psi_l(x)|0 > = < 0|\psi_m^{-*}(y)\psi_l^-(x)|0 >$$

We find that

$$\Delta_{lm}(x,y) = \theta(x^0 - y^0) \int u_l(p,\sigma)u_m^*(p,\sigma)exp(ip.(x-y))d^3p$$

$$+\theta(y^0 - x^0) \int v_l(p\, sigma)v_m^*(p\, sigma)exp(ip.(x-y))d^3p$$

This expression holds for bosonic fields. Note that in case antiparticles exist, we have

$$\psi^+(x) = \int u(p,\sigma)a(p,\sigma,n)exp(ip.x)d^3p,$$

$$\psi^-(x) = \int v(p,\sigma)a^*(p,\sigma,n^c)exp(-ip.x)d^3p$$

$$\psi(x) = \psi^+(x) + \psi^-(x) = \int u(p,\sigma)a(p,\sigma,n)exp(ip.x)d^3p + \int v(p,\sigma)a^*(p,\sigma,n^c)exp(-ip.x)d^3p$$

with the commutation relations

$$[a(p,\sigma,n), a^*(p',\sigma',n)] = \delta^3(p-p')\delta_{\sigma,\sigma'}$$

$$[a(p,\sigma,n^c),a^*(p',\sigma',n^c)]=\delta^3(p-p')\delta_{\sigma,\sigma'}$$

$$[a(p,\sigma,n),a(p',\sigma',n)]=0,$$

$$[a(p,\sigma,n^c),a(p',\sigma',n^c)]=\delta^3(p-p')\delta_{\sigma,\sigma'}$$

$$[\psi(x),\psi^*(y)]=[\int[a(p,\sigma,n)u(p,\sigma)exp(ip,x)$$

$$+a^*(p,\sigma,n^c)v(p,\sigma)exp(-ip.x)]d^3p,\int[a^*(p,\sigma,n)u^*(p,\sigma)exp(-ip.y)+a(p,\sigma,n^c)v^*(p,\sigma)exp(ip.y)]d^3p]$$

$$=\sum[u(p,\sigma,n)u^*(p,\sigma)exp(ip.(x-y))-v(p,\sigma)v^*(p,\sigma)exp(-ip.(x-y))]d^3p$$

It is a theorem (see Wienberg, "Quantum theory of fields") that

$$\sum_\sigma u_l(p,\sigma)u^*_m(p,\sigma)=P_{lm}(\mathbf{p},\sqrt{\mathbf{p}^2+m^2})/\sqrt{\mathbf{p}^2+m^2}$$

where P_{lm} is a polynomial. This result is true for particles of arbitrary spin. Note that P_{lm} has been defined on the mass shell, ie for all four vectors p satisfying $p^0=\pm\sqrt{\mathbf{p}^2+m^2}$. We now define a polynomial $P^L_{lm}(p)$ for all four vectors p, not necessarily on the mass shell so that (1) P^L_{lm} is linear in p^0 and (2) P^L_{lm} agrees with P_{lm} on the mass shell. We can thus write

$$P^L_{lm}(p)=P^{(0)}_{lm}(\mathbf{p})+p^0P^{(1)}_{lm}(\mathbf{p})$$

Define

$$\Delta_+(x)=\int(p^0)^{-1}exp(ip.x)d^3p$$

where p falls on the mass shell. Then,

$$P_{lm}(-i\nabla)\Delta_+(x)=\int p^{0-1}P_{lm}(p)exp(ip.x)d^3p=\int\sum u(p,\sigma)u^*(p,\sigma)exp(ip.x)d^3p$$

$$P_{lm}(-i\nabla)\Delta_+(-x)=\int p^{0-1}P_{lm}(-p)exp(-ip.x)d^3p=\int\sum v(p,\sigma)v^*(p,\sigma)exp(-ip.x)d^3p$$

so that the propagator is given by

$$\Delta_{lm}(x)=\theta(x^0)P_{lm}(-i\nabla)\Delta_+(x)+\theta(-x^0)P_{lm}(-i\nabla)\Delta_+(-x)$$

We have

$$P^L_{lm}(i\partial_\mu)(\theta(x^0)\Delta_+(x))=(P^{(0)}_{lm}(-i\nabla)+i\partial_0P^{(1)}_{lm}(-i\nabla))(\theta(x^0)\Delta_+(x))$$

$$=\theta(x^0)P^{(0)}_{lm}(-i\nabla)\Delta_+(x)+i\delta(x^0)P^{(1)}_{lm}(-i\nabla)\Delta_+(x)+\theta(x^0)i\partial_0P^{(1)}_{lm}(-i\nabla)\Delta_+(x)$$

$$=i\delta(x^0)P^{(1)}_{lm}(-i\nabla)\Delta_+(x)+\theta(x^0)P^L_{lm}(i\partial_\mu)\Delta_+(x)$$

We now observe that

$$P^L_{lm}(i\partial_\mu)\Delta_+(x)=\int(P^L_{lm}(p)/p^0)exp(ip.x)d^3p=\int(P_{lm}(p)/p^0)exp(ip.x)d^3p=P_{lm}(-i\nabla)\Delta_+(x)$$

since in the integral expression for $\Delta_+(x)$, the four vector p falls on the mass shell. Thus,

$$P^L_{lm}(i\partial_\mu)(\theta(x^0)\Delta_+(x)+\theta(-x^0)\Delta_+(-x))$$

$$i\delta(x^0)P^{(1)}_{lm}(-i\nabla)(\Delta_+(x)-\Delta_-(-x))+\theta(x^0)P_{lm}(-i\nabla)\Delta_+(x)+\theta(-x^0)P_{lm}(-i\nabla)\Delta_+(-x)$$

Also when $x^0=0$, we have $\Delta_+(x)=-\Delta_+(-x)$. It follows that

$$P^L_{lm}(i\partial_\mu)(\theta(x^0)\Delta_+(x)+\theta(-x^0)\Delta_+(-x))=\Delta_{lm}(x)$$

We usually set

$$\Delta_F(x)=\theta(x^0)\Delta_+(x)+\theta(-x^0)\Delta_+(-x)$$

and call Δ_F the Feynman propagator. We have thus obtained a concrete expression for the propagator for a particle of arbitrary spin. For fermions, there is a minus sign involved in the expressions for the propagator.

Causality requirement: The commutator

$$[\psi(x),\psi^*(y)]$$

for bosons and the anticommutator $\{\psi(x),\psi^*(y)\}$ for fermions should vanish when $x-y$ is space like. Note that when $x-y$ is space-like, we can apply a Lorentz transformation that brings $x-y$ to the form $(\mathbf{u},0)$, ie, in the transformed reference frame, $x-y$ has zero time component. Consider the situation for bosons first.

$$\psi(x)=\int[a(p,\sigma)u(p,\sigma)exp(ip.x)+a^*(p,\sigma)v(p,\sigma)exp(-ip.x)]d^3p$$

$$\psi^*(x)=\int[a^*(p,\sigma)u^*(p,\sigma)exp(-ip.x)+a(p,\sigma)v^*(p,\sigma)exp(ip.x)]d^3p$$

We have

$$[\psi(x),\psi^*(y)]=\int[u(p,\sigma)u^*(p,\sigma)exp(ip.(x-y))-v(p,\sigma)v^*(p,\sigma)exp(ip.(y-x))]d^3p$$

We must prove that this commutator is zero when $x-y$ is space-like. We use the relation

$$\sum u(p,\sigma)u^*(p,\sigma)=P(\mathbf{p},\sqrt{\mathbf{p}^2+m^2})/\sqrt{\mathbf{p}^2+m^2}$$

$$\sum v(p,\sigma)v^*(p,\sigma)=P(-\mathbf{p},-\sqrt{\mathbf{p}^2+m^2})/\sqrt{\mathbf{p}^2+m^2}$$

This can be shown to be zero along the following lines (Weinberg). Consider for simplicity the case of scalar fields. We have two components $\phi^+(x)$ and $\phi^-(x)$, where

$$\phi^+(x)=\int(2p^0)^{-1/2}a(p)exp(ip.x)d^3p,$$

$$\phi^-(x)=\int(2p^0)^{-1/2}a^*(p)exp(-ip.x)$$

Define

$$\phi(x) = K_1\phi^+(x) + K_2\phi^-(x)$$

Then,

$$[\phi(x), \phi^*(y)] = |K_1|^2 \int [(2p^0)^{-1} exp(ip.(x-y))d^3p - |K_2|^2 \int (2p^0)^{-1} exp(ip.(y-x))]d^3p$$

Consider the function

$$\Delta_+(x) = \int (2p^0)^{-1} exp(ip.x)d^3p$$

giving

$$[\phi(x), \phi^*(y)] = |K_1|^2\Delta_+(x-y) - |K_2|^2\Delta_+(y-x)$$

Suppose x is space-like. Then relative to an appropriate frame we can set $x^0 = 0$ and in this frame, we have

$$\Delta_+(x) = \int (2p^0)^{-1} exp(-i\mathbf{p}.\mathbf{x})d^3p$$

In this frame, we get

$$\Delta_+(-x) = \int (2p^0)^{-1} exp(i\mathbf{p}.\mathbf{x})d^3p = \Delta_+(x)$$

ie relative to this special frame, Δ_+ is an even function of x ($x^0 = 0$). It follows that in this frame, the commutator evaluates to

$$[phi(x), \phi^*(y)] = (|K_1|^2 - |K_2|^2)\Delta_+(x-y)$$

This is zero iff $|K_1| = |K_2|$. This is the condition for causality:Commutators must vanish at space-like separations. Let us now look at the case of causal vector fields having spin one. Here,

$$v^\mu(x) = K_1\phi^{\mu+}(x) + K_2\phi^{\mu-}(x)$$

where

$$\phi^{\mu+}(x) = \int (2p^0)^{-1/2} e^\mu(p,\sigma)a(p,\sigma)exp(ip.x)d^3p$$

$$\phi^{\mu-}(x) = \int (2p^0)^{-1/2} e^{\mu*}(p,\sigma)a^*(p,\sigma)exp(-ip.x)d^3p$$

We can regard this particle as a photon propagating along the direction $\mathbf{p}$. σ takes values $1, 0, -1$ and $e^\mu(p,1), e^\mu(p,-1)$ correspond respectively to left and right circular polarizations (perpendicular to $\mathbf{p}$). $e^\mu(p,0)$ is directed along the $\mathbf{p}$ direction. Assuming for simplicity $\mathbf{p} = (0,0,1,0)$, we may take

$$e^\mu(p,1) = \frac{1}{\sqrt{2}}(1,i,0,0), e^\mu(p,-1) = \frac{1}{\sqrt{2}}(1,-i,0,0),$$

$$e^\mu(p,0) = (0,0,1,0)$$

All vectors are normalized to have unit length as complex vectors. Then, form the matrix

$$\Pi^{\mu\nu}(p) = \sum_\sigma e^\mu(p,\sigma)\bar{e}^\nu(p,\sigma)$$

The matrix $\Pi^{\mu\nu}(p)$ is Hermitian, ie,

$$\bar{\Pi}^{\nu\mu}(p) = \Pi_{\mu\nu}(p)$$

It is clear that $\Pi(p)$ is an orthogonal projection matrix onto the range of the vectors $(e^{\mu}(p,\sigma)), \sigma = 1, 0, -1$. However we need to be a little careful here. The matrix Π acts on a four vector a^{μ} to give the four vector $\Pi^{\mu\nu}(p)a_{\nu} = \eta_{\nu\sigma}\Pi^{\mu\sigma}(p)a^{\nu}$. We see that Π has rank three and if a^{μ} is along the time direction, then $\Pi^{\mu\nu}(0)a_{\nu} = 0$. Thus, $(\Pi^{\mu\nu}(0))$ is the projection matrix on the subspace orthogonal to the time direction. Let $L^{\mu}_{\nu}(p)$ be a Lorentz transformation matrix that maps the the vector $(0,0,0,m)$ onto the four vector p^{μ}. Note that $p_{\mu}p^{\mu} = -m^2$. Then, it follows that

$$e^{\mu}(p,\sigma) = L^{\mu}_{\nu}(p)e_{\nu}(0,\sigma)$$

It follows that

$$\Pi^{\mu\nu}(p) = L^{\mu}_{\rho}(p)L^{\nu}_{\sigma}(p)\Pi^{\rho\sigma}(0)$$

We have,

$$\Pi^{\mu\nu}(p)p_{\nu} = L^{\mu}_{\rho}(p)L^{\nu}_{\sigma}(p)p_{\nu}\Pi^{\rho\sigma}(0)$$

We have

$$p = L(p)u, u = (0,0,0,m)$$

and

$$p^T\eta p = -m^2$$

Thus,

$$u^T L(p)^T \eta L(p)u = -m^2$$

Note that

$$L(p)^T \eta L(p) = \eta$$

so that

$$L(p)^T = \eta L(p)^{-1}\eta = \eta L(-p)\eta$$

This equation means that

$$L^{\nu}_{\sigma}(p)p_{\nu} = L^{\nu}_{\rho}(-p)p_{\rho} = (0,0,0,-m) = u_{\sigma}$$

It follows that

$$\Pi^{\mu\nu}(p)p_{\nu} = L^{\mu}_{\rho}(p)u_{\sigma}\Pi^{\rho\sigma}(0) = 0$$

since

$$\Pi^{\rho\sigma}(0)u_{\sigma} = 0$$

It follows that $\Pi(p)$ is the projection onto the subspace orthogonal to the p^{μ}-direction. Consider now the matrix

$$P^{\mu\nu} = \eta^{\mu\nu} + p^{\mu}p^{\nu}/m^2$$

This is a rank three matrix and its action on the four vector p^{μ} is given by

$$P^{\mu\nu}p_{\nu} = p^{\mu} - p^{\mu} = 0$$

since $p^{\mu}p_{\mu} = -m^2$. It follows that

$$\Pi^{\mu\nu}(p) = \eta^{\mu\nu} + p^{\mu}p^{\nu}/m^2$$

upto a constant of proportionality. Then,

$$[v^\mu(x), v^{\nu*}(y)] = [K_1\phi^{\mu+}(x) + K_2\phi^{\mu-}(x), K_1\phi^{\nu*}(y) + K_2\phi^{\nu*}(y)]$$

$$= |K_1|^2 \int \Pi^{\mu\nu}(p)(2p^0)^{-1} exp(ip.(x-y)d^3p - |K_2|^2 \int \Pi^{\mu\nu}(p)(2p^0)^{-1} exp(ip.(y-x))d^3p$$

Now observe that

$$\int \Pi^{\mu\nu}(p)(2p^0)^{-1} exp(ip.x)d^3p = \int (\eta^{\mu\nu} + p^\mu p^\nu/m^2)(2p^0)^{-1} exp(ip.x)d^3p$$

$$= (\eta^{\mu\nu} - \partial^\mu\partial^\nu/m^2)\Delta_+(x)$$

Thus,

$$[v^\mu(x), v^{\nu*}(y)] = (\eta^{\mu\nu} - \partial^\mu\partial^\nu/m^2)(|K_1|^2\Delta_+(x-y) - |K_2|^2\Delta_+(y-x))$$

If $x - y$ is space like, then $\Delta_+(x-y) = \Delta_+(y-x)$ and the above will evaluate to zero iff $|K_1| = |K_2|$.

64. Bosonic and Fermionic states: The bosonic state corresponding to particles at $q_1, ..., q_n$ may be represented as

$$\Phi[q_1, ..., q_n] = \sum_\sigma |q_{\sigma 1}, ..., q_{\sigma n} >$$

where σ runs over all the permutations of $\{1, 2, ..., n\}$. The Fermionic state on the other hand has the form

$$\Phi[q_1, ..., q_n] = \sum_\sigma sgn(\sigma)|q_{\sigma 1}, ..., q_{\sigma n} >$$

For bosonic states, we define the creation operator at q $a^*(q)$ by its action:

$$a^*(q)\Phi[q_1, ..., q_n] = \Phi[q, q_1, ..., q_n]$$

and the annihilation operator $a(q)$ by

$$a(q)\Phi[q_1, ..., q_n] = \sum_{r=1}^{n} \delta(q - q_r)\Phi[q_1, ..., q_{r-1}, q_{r+1}, ..., q_n]$$

It is easy to check that $a^*(q)$ is the adjoint of $a(q)$. We do the calculation for $n = 2$:

$$a(q)\Phi[q_1, q_2] = \delta(q - q_1)\Phi[q_2] + \delta(q - q_2)\Phi[q_1]$$

so that

$$< \Phi[q_3], a(q)\Phi[q_1, q_2] >=< \Phi[q_3], \Phi[q_2] > \delta(q - q_1)+ < \Phi[q_3], \Phi[q_1] > \delta(q - q_2)$$

$$= \delta(q_2 - q_3)\delta(q - q_1) + \delta(q_1 - q_3)\delta(q - q_2)$$

on the one hand while

$$a^*(q)\Phi[q_3] = \Phi[q, q_3]$$

so that

$$< a^*(q)\Phi[q_3], \Phi[q_1, q_2] >=< \Phi[q, q_3], \Phi[q_1, q_2] >$$

$$= (< qq_3| + < q_3q|)(|q_1q_2 > + |q_2q_1 >) = < qq_3|q_1q_2 > + < qq_3|q_2q_1 > + < q_3q|q_1q_2 > + < q_3q|q_2q_1 >$$

$$= \delta(q-q_1)\delta(q_3-q_2) + \delta(q-q_2)\delta(q_3-q_1) + \delta(q_3-q_1)\delta(q-q_2) + \delta(q_3-q_2)\delta(q-q_1)$$

$$= 2(\delta(q-q_1)\delta(q_3-q_2) + \delta(q-q_2)\delta(q_3-q_1)$$

and after normalization by the factor 1/2, we get back the previous result. Note that for Fermions, the creation operator has the same action while the annihilation operator has the form

$$a(q)\Phi[q_1, ..., q_n] = \sum_{r=1}^{n}(-1)^{r-1}\delta(q-q_r)\Phi[q_1, ..., q_{r-1}, q_{r+1}, ..., q_n]$$

65.Computation of the exact electron propagator. The Dirac Lagrangian for the free field has the form

$$\mathcal{L}_D(\psi, \psi^*) = \psi^*(x)(i\gamma^\mu\partial_\mu + m)\psi(x)$$

Note that the covariant four momentum operator is

$$p_\mu = i\partial_\mu$$

so that

$$p^r = -p_r = -i\partial_r, r = 1, 2, 3, p^0 = i\partial_0 = p_0$$

The Dirac Lagrangian in the presence of an external radiation field described by the electromagnetic four potential $A_\mu(x)$ is given by

$$\mathcal{L}_D + \mathcal{L}_{int} = \psi^*(x)\gamma^\mu(i\partial_\mu + eA_\mu(x) + m)\psi(x)$$

Note that the interaction Lagrangian is therefore given by

$$\mathcal{L}_{int} = e\psi^*(x)\gamma^\mu\psi(x)A_\mu(x)$$

The electromagnetic four current vector is given by

$$J^\mu(x) = e\psi^*(x)\gamma^\mu\psi(x)$$

and in view of the Dirac equation it is conserved:

$$\partial_\mu J^\mu = 0$$

Indeed, the Dirac equation reads

$$\gamma^\mu(i\partial_\mu + eA_\mu + m)\psi = 0$$

We can take

$$\gamma^0 = \begin{pmatrix} I & 0 \\ 0 & -I \end{pmatrix}$$

$$\gamma^r = \begin{pmatrix} 0 & \sigma_r \\ -\sigma_r & 0 \end{pmatrix}, r = 1, 2, 3$$

We easily verify the anticommutation relations

$$\gamma^\mu\gamma^\nu + \gamma^\nu\gamma^\mu = 2g^{\mu\nu}$$

where

$$(g^{\mu\nu}) = diag[1, -1, -1, -1]$$

We see that γ^0 is Hermitian while γ^r is skew Hermitian. We write the Dirac equation as

$$[\gamma^0(i\partial_0 + eA_0) + \gamma^r(i\partial_r + eA_r) + m]\psi = 0$$

and taking complex conjugate find

$$-i(\partial_0\psi^*)\gamma^0 + eA_0\psi^*\gamma^0 + i(\partial_r\psi^*)\gamma^r - eA_r\psi^*\gamma^r + m\psi^* = 0$$

Now note that

$$\gamma^0\gamma^r\gamma^0 = -\gamma^r$$

and hence from the above we get

$$-i\partial_0\psi^*\gamma^0\gamma^0 + eA_0\psi^*\gamma^0\gamma^0 - i\partial_r\psi^*\gamma^0\gamma^r + eA_r\psi^*\gamma^0\gamma^r + m\psi^*\gamma^0 = 0$$

Define

$$\bar{\psi} = \psi^*\gamma^0$$

Then the above can be arranged as

$$(i\partial_0 - eA_0)\bar{\psi}\gamma^0 + (i\partial_r - eA_r)\bar{\psi}\gamma^r - m\bar{\psi} = 0$$

Equivalently,

$$(i\partial_\mu - eA_\mu)\bar{\psi}\gamma^\mu - m\bar{\psi} = 0$$

We thus arrive at the equation

$$\bar{\psi}\gamma^\mu(i\partial_\mu + eA_\mu + m)\psi + (i\partial_\mu - eA_\mu)\bar{\psi}\gamma^\mu\psi = 0$$

or

$$\partial_\mu(\bar{\psi}\gamma^\mu\psi) = 0$$

which is the desired charge conservation equation.

66. Model for Fermion-Boson scattering (Wienberg, Quantum theory of fields): $\phi_k(x)$ are the bosonic fields and $\psi_l(x)$ are the fermionic fields. The interaction Hamiltonian is given by

$$H = \int \mathcal{H}(x)d^3x$$

where the $\mathcal{H}(x)$ is the interaction Hamiltonian density defined by

$$\mathcal{H}(x) = \sum_{lmk} g_{lmk}\psi_l^*(x)\psi_m(x)\phi_k(x)$$

$(p_i, \sigma_i, n_i), i = 1, 2$ are respectively the specifications of the initial fermion and boson while (p'_i, σ'_i, n'_i) $i = 1, 2$ are respectively the specifications of the final fermion and boson. The scattering amplitude for this process is

$$S[p'_1, \sigma'_1, n'_1, p'_2, \sigma'_2, n'_2 | p_1, \sigma_1, n_1, p_2, \sigma_2, n_2]$$

$$= < p'_i, \sigma'_i, n'_i, i = 1, 2 | T(exp(-i \int \mathcal{H}(x) d^4x)) | p_i, \sigma_i, n_i, i = 1, 2 >$$

where T is the usual time ordering operator. In the desired amplitude, we retain only upto terms second order in the interaction Hamiltonian. Thus, the desired amplitude to be evaluated is given by(apart from a proportionality factor)

$$< p'_i, \sigma'_i, n'_i, i = 1, 2 | \int T\{\mathcal{H}(x)\mathcal{H}(y)\} d^4x d^4y | p_i, \sigma_i, n_i, i = 1, 2 >$$

We denote the states $|p_i, \sigma_i, n_i, i = 1, 2 >$ and $|p'_i, \sigma'_i, n'_i, i = 1, 2 >$ respectively by $|p', \sigma', n' >$ and $|p, \sigma, n >$. Then,

$$< p', \sigma', n' | T(\mathcal{H}(x)\mathcal{H}(y)) | p, \sigma, n >$$

$$= \sum_{lmkl'm'k'} g_{lmk} g_{l'm'k'} < p', \sigma', n' | T(\psi^*_{l'}(x)\psi_{m'}(x)\phi_{k'}(x)\psi^*_l(y)\psi_m(y)\phi_k(y)) | p, \sigma, n >$$

We consider a specific pairing in the computation of the above. First note that $\psi_m(y)$ pairs with $\psi^*_{l'}(x)$ yielding the Fermionic propagator

$$< T(\psi_m(y)\psi^*_{l'}(x)) >= \Delta_{ml'}(y - x)$$

The other terms are paired as follows: $\psi^*_l(y)$ contains a term $a^*(p, \sigma, n) u^*_l(p, \sigma, n)$ exp(-ip.y) in its Fourier integral and this can pair with the final state Fermion $p'_1 \sigma'_1 n'_1$ to give

$$u^*_l(p'_1, \sigma'_1, n'_1) exp(-ip'_1.y)$$

We are assuming $n'_1 = n$. Likewise, $\psi_{m'}(x)$ contains a term $a(p, \sigma, n) u_m(p, \sigma, n) exp(ip.x)$ and this can be paired with the initial state Fermion p_1, σ_1, n_1 to give

$$u^*_m(p_1, \sigma_1, n_1) exp(ip_1.x)$$

$\phi_k(y)$ contains factors $a(p, \sigma, n) u_k(p, \sigma, n) exp(ip.y)$ and $a^*(p, \sigma, n) u^*_k(p, \sigma, n) exp(-ip.y)$. The first and likewise for $\phi_{k'}(x)$. Thus, $\phi_k(y)$ can be paired with the initial state boson and $\phi_{k'}(x)$ with the final state boson to give the factor $u_k(p_2, \sigma_2, n_2) exp(ip_2.y) u^*_{k'}(p'_2, \sigma'_2, n'_2) exp(-ip'_2.x)$ or $\phi_k(y)$ can be paired with the final state boson and $\phi_{k'}(x)$ with the initial state boson to give the factor $u^*_k(p'_2, \sigma'_2, n'_2) exp(-ip'_2.y) u_{k'}(p_2, \sigma_2, n_2)$ $exp(ip_2.x)$. The total amplitude for the Fermion-Boson scattering process can now be written down.

67.An introduction to quantum field theory, by Peskin and Schroeder.

1. Derive the Hamiltonian of the Klein-Gordon field from first principles.

ans: The Klein-Gordon equation for the field $\phi(t, \mathbf{x}) = \phi(x)$ reads

$$(\partial_t^2 - \nabla^2 + m^2)\phi = 0$$

This can be expressed in relativistic covariant notation as

$$(\partial^\mu \partial_\mu + m^2)\phi = 0$$

Let

$$L(\phi, \partial_\mu \phi) = \frac{1}{2}(\partial^\mu \phi \partial_\mu \phi - m^2 \phi^2)$$

Then the action is

$$S[\phi] = \int L d^4 x$$

and the action principle $\delta S[\phi] = 0$ yields the Klein-Gordon equation. We can write

$$L = \frac{1}{2}(\partial_t \phi)^2 - \frac{1}{2}|\nabla \phi|^2 - m^2 \phi^2/2$$

Then the canonical momentum density is

$$\pi = \frac{\partial L}{\partial \partial_t \phi} = \partial_t \phi$$

and the Hamiltonian density is

$$\mathcal{H} = \pi \partial_t \phi - L = \frac{1}{2}(\pi^2 + |\nabla \phi|^2) + m^2 \phi^2/2$$

The Hamilton equations of motion for the classical field are

$$\partial_t \phi = \delta \mathcal{H}/\delta \pi = \pi,$$

$$\begin{aligned} -\partial_t \pi &= \delta \mathcal{H}/\delta \phi = \partial \mathcal{H}/\partial \phi - (\nabla, \partial \mathcal{H}/\partial \nabla \phi) \\ &= m^2 \phi - \nabla^2 \phi \end{aligned}$$

and hence, combining both of the above equations gives us

$$\partial_t^2 \phi - \nabla^2 \phi + m^2 \phi = 0$$

which is the Klein-Gordon equation.

Noether's theorem: Let L be a Lagrangian function of the field ϕ and its partial derivatives $\partial_\mu \phi$. Then consider an infinitesimal change in the fields

$$\phi \to \phi + \alpha \Delta \phi$$

where α is a constant and $\Delta \phi$ is a function of the space-time point. The change in L under this transformation is given by

$$\Delta L = (\partial L/\partial \phi)\alpha \Delta \phi + (\partial L/\partial \partial_\mu \phi)\alpha \partial_\mu \Delta \phi$$

since

$$\Delta \partial_\mu \phi = \partial_\mu \Delta \phi$$

This can be expressed as

$$\Delta L = (\alpha\Delta\phi)(\partial L/\partial\phi - \partial_\mu(\partial L/\partial\partial_\mu\phi))$$
$$+\partial_\mu(\alpha(\partial L/\partial\partial_\mu\phi)\Delta\phi)$$

Suppose that the action is invariant under such a transformation. Then it follows that there is a vector field τ^μ such that

$$\Delta L = \partial_\mu\tau^\mu$$

Suppose then that ϕ satisfies the field equations. It then would follow from the above that

$$\partial_\mu(\alpha(\partial L/\partial\partial_\mu\phi)\Delta\phi - \tau^\mu) = 0$$

which is Noether's conservation equation.

Example: Suppose $L = L(\phi(x), \partial_\mu\phi(x))$, ie L does not depend explicitly on x. Then under infinitesimal translations of the field, we have

$$\phi(x) \to \phi(x+a) = \phi(x) + a^\mu\partial_\mu\phi(x)$$

and obviously under this transformation, the Lagrangian transforms as

$$L \to L + a^\mu\partial_\mu L = L + \partial_\mu(a^\mu L)$$

ie, L changes by a perfect divergence. It follows from the above that

$$\partial_\mu J^\mu = 0$$

where

$$J^\mu = (\partial L/\partial_\mu\phi)a^\nu\partial_\nu\phi - La^\mu$$

Letting

$$a^\mu = \delta^\mu_\rho$$

gives us

$$J^\mu = T^\mu_\rho = (\partial L/\partial_\mu\phi)\partial_\rho\phi - L\delta^\mu_\rho$$

T^μ_ρ is precisely the energy-momentum tensor and Noether's theorem leads to the conservation law for this tensor. This conservation law could equivalently have been derived by just taking the partial derivative of L.

3. Quantization of the Klein-Gordon field:Let $E(\mathbf{p}) = \sqrt{m^2 + \mathbf{p}^2}$ and let $a(\mathbf{p}), \mathbf{p} \in \mathbb{R}^3$ be an operator field satisfying the commutation relations

$$[a(\mathbf{p}), a^*(\mathbf{p}')] = (2\pi)^3\delta(\mathbf{p} - \mathbf{p}')$$

Define the operator field

$$\phi(x) = (2\pi)^{-3}\int(2E(\mathbf{p}))^{-1/2}[a(\mathbf{p})exp(-ip.x) + a^*(\mathbf{p})exp(ip.x)]d^3p$$

where

$$x = (t, \mathbf{x}), p = (p^0, \mathbf{p}) = (E(\mathbf{p}), \mathbf{p})$$

so that

$$p.x = p^0 t - \mathbf{p}.\mathbf{x} = E(\mathbf{p})t - \mathbf{p}.\mathbf{x}$$

Let

$$\phi(\mathbf{x}) = \phi(0, \mathbf{x})$$

Set

$$\pi(\mathbf{x}) = \pi(0, \mathbf{x}) = \partial_t \phi(t, \mathbf{x})|_{t=0}$$

Then

$$\pi(\mathbf{x}) = (2\pi)^{-3}(-i)\int 2^{-1/2}E(\mathbf{p})^{1/2}(a(\mathbf{p})exp(i\mathbf{p}.\mathbf{x}) - a^*(\mathbf{p})exp(-i\mathbf{p}.\mathbf{x}))d^3p$$

Let us calculate the commutator $[\phi(\mathbf{x}), \pi(\mathbf{y})]$:

$$[\phi(\mathbf{x}), \pi(\mathbf{y})] = (2\pi)^{-6}(-i)\int (4E(\mathbf{p})E(\mathbf{p}')^{-1})^{-1/2}(-[a(\mathbf{p}), a^*(\mathbf{p}')]exp(i(\mathbf{p}.\mathbf{x} - \mathbf{p}'.\mathbf{y}))$$

$$+[a^*(\mathbf{p}), a(\mathbf{p}')]exp(i(\mathbf{p}'.\mathbf{y} - \mathbf{p}.\mathbf{x})))d^3pd^3p'$$

$$= (2\pi)^{-3}(-i)\int 2^{-1}(-\delta(\mathbf{p} - \mathbf{p}')exp(i(\mathbf{p}.\mathbf{x} - \mathbf{p}'.\mathbf{y})) - \delta(\mathbf{p} - \mathbf{p}')exp(i(\mathbf{p}'.\mathbf{y} - \mathbf{p}.\mathbf{x})))d^3pd^3p'$$

$$= (2\pi)^{-3}(-i)\int 2^{-1}(-2)exp(i\mathbf{p}.(\mathbf{x} - \mathbf{y}))d^3p$$

$$= i\delta(\mathbf{x} - \mathbf{y})$$

This is the correct commutation relation for canonical position and momentum fields as expected in the quantum theory of fields. Let us now calculate the energy

$$H = \frac{1}{2}\int ((\partial_t\phi)^2 + (\nabla\phi)^2 + m^2\phi^2)d^3x$$

68. Landau and Lifshitz, Quantum Electrodynamics:

Atom interacting with radiation, second order perturbation theory is needed. Two kinds of transitions are possible under the interaction between matter and radiation. Let V be the interaction energy. With $\mathbf{d}$ denoting the dipole moment operator of the atom and $\mathbf{E}$ the electric field operator of the incident radiation, we find that the Fourier component at the frequency ω of the electric field is given by

$$\mathbf{E}_\omega = -j\omega\mathbf{A}_\omega$$

where the magneric vector potential Fourier component $\mathbf{A}_\omega$ at the frequency ω has the form

$$\mathbf{A}_\omega = \mathbf{e}exp(-i\omega t)a(\omega)/\sqrt{2\omega} + \mathbf{e}^*exp(i\omega t)a^*(\omega)/\sqrt{2\omega}$$

where **e** is the polarization vector and $a(\omega), a^*(\omega)$ are respectively the operators of annihilation and creation of a photon of frequency ω. The quantum system comprises atom in interaction with radiation. The interaction potential is **d.E**. The initial state $|1>$ consists of the atom in the energy state $|E_1>$ along with a photon of frequency ω. We denote this state by $|E_1, \omega>$. After the perturbation, is turned on, the atom jumps to and intermediate state $|E_n>$ and the photon is absorbed. This intermediate state consisting of the atom having energy E_n in conjunction with the absorbed photon is denoted by $|E_n, 0>$. The corresponding matrix element is $<E_n, 0|V|E_1, \omega>$. The photon part contributes $<0|a(\omega)|\omega>= 1$ since $a(\omega)|\omega>= |0>$. (The annihilation operator $a(\omega)$ annihilates a photon of frequency ω. Thus,

$$<E_n, 0|V|E_1, \omega>= -j\sqrt{\omega}\mathbf{e}.\mathbf{d}_{n1}$$

where

$$\mathbf{d}_{nm} =< E_n|\mathbf{d}|E_m>$$

is the matrix element of the dipole moment of the atom between two energy states of the atom. The first order correction to the wave function of the atom cum photon corresponding to absorption of the photon is from standard perturbation theory given by

$$|\psi>= \sum_n |E_n, 0>< E_n, 0|V|E_1, \omega> /(E_1 + \omega - E_n)$$

After the interaction, the probability amplitude of a photon of frequency ω' being emitted and the atom moving to the state $|E_2>$ is given by

$$<E_2, \omega'|V|\psi>= \sum <E_2, \omega'|V|E_n, 0>< E_n, 0|V|E_1, \omega> /(E_1 + \omega - E_n)$$

Now observe that

$$<E_2, \omega'|V|E_n, 0>= -j\sqrt{\omega'}\mathbf{e}^*.\mathbf{d}_{2n}$$

where we have taken into account

$$<\omega'|a^*(\omega')|0>= 1$$

This gives the amplitude of the atom and photon interacting, the photon getting absorbed and then emitted with a different frequency as

$$-\sqrt{\omega\omega'}sum(\mathbf{e}^*.\mathbf{d}_{2n})(\mathbf{e}.\mathbf{d}_{n1})/(E_1 + \omega - E_n)$$

There is another possible mechanism of transition in which a photon of different frequency ω' is emitted. In this transition, the atom first makes a transition from $|E_1>$ to $|E_n>$ and a photon of frequency ω' is emitted. In this intermediate state, there are two photons of frequencies ω and ω' respectively. After this, the atom makes a transition to $|E_2>$ and the photon of frequency ω is absorbed. We leave it as an exercise to write down the amplitude for this process. The amplitude can be computed along the following lines. The initial state of the atom and photons is given by $|E_1, \omega, 0>$ where E_1 stands for the atom being in the energy state E_1, ω stands for the first photon having frequency ω and 0 for the second photon nonexistent. The intermediate state is given by $|E_n, \omega, \omega'>$ in which the second photon has frequency ω' and the first photon stays as it is. The corresponding transition probability amplitude is given by

$$<E_n, \omega, \omega'|V|E_1, \omega, 0>=< E_n, \omega'|V|E_1, 0>$$

Now the component of V that contributes to this matrix element is given by $-j\sqrt{\omega'}\mathbf{d}.\mathbf{e}^*a^*(\omega')$ and since $<\omega'|a^*(\omega')|0>=1$, this amplitude evaluates to

$$-i\sqrt{\omega'}<E_n|\mathbf{d}.\mathbf{e}^*|E_1>=\mathbf{d}_{n1}.\mathbf{e}^*$$

The wave function after this intermediate interaction is given by

$$\sum_n |E_n,\omega,\omega'>\mathbf{d}_{n1}.\mathbf{e}^*/(E_1+\omega-E_n-\omega-\omega')$$

$$=\sum_n |E_n,\omega,\omega'>\mathbf{d}_{n1}.\mathbf{e}^*/(E_1-E_n-\omega')$$

Again, there is an interaction causing atom to reach the state $|E_2>$ with the photon of frequency ω absorbed. The final state is thus, $|E_2,0,\omega'>$ and the total probability amplitude for the resulting process is therefore given by

$$\sum_n <E_2,0,\omega'|V|E_n,\omega,\omega'>\mathbf{d}_{n1}.\mathbf{e}^*/(E_1-E_n-\omega')$$

Now observe that

$$<E_2,0,\omega'|V|E_n,\omega,\omega'>=<E_2,0|V|E_n,\omega>=-i\sqrt{\omega}\mathbf{d}_{2n}.\mathbf{e}$$

The total amplitude is then given by the formula

$$-\sqrt{\omega\omega'}\sum(\mathbf{d}_{n1}.\mathbf{e}^*)(\mathbf{d}_{2n}.\mathbf{e})/(E_1-E_n-\omega')$$

69. Quantum mechanical analysis of antenna radiation pattern: The current through the wire of an antenna for a given applied electrostatic potential is obtained by applying the rules of quantum mechanics. If we assume nonrelativistic quantum mechanics and spin is neglected, then the wave functions satisy the time dependent Schrodinger wave equation

$$ih\partial_t\psi=-\frac{h^2}{2m}\nabla^\psi+U\psi$$

where U is the potential field produced by the applied voltage. The charge density is given by $\rho=e|\psi|^2=e\psi\psi^*$ and the conjugate wave function satisfies the pde

$$-ih\partial_t\psi^*=-\frac{h^2}{2m}\nabla^2\psi^*+U\psi^*$$

Here we are assuming U to be real. It follows by multiplying the Schrodinger equation by ψ^* and the conjugate Schrodinger equation by ψ and subtracting that

$$ih\partial_t(\psi^*\psi)=(-h^2/2m)div(\psi^*\nabla\psi-\psi\nabla\psi^*)$$

The charge density for the single electron is given by

$$\rho=-e\psi^*\psi$$

and hence we get

$$\partial_t \rho = (-ieh/2m)div(\psi^* \nabla\psi - \psi\nabla\psi^*)$$

In view of the charge conservation equation

$$\partial_t \rho = -div\mathbf{J}$$

it follows that the current density is given by the formula

$$\mathbf{J} == (ieh/2m)(\psi^* \nabla\psi - \psi\nabla\psi^*)$$

The magnetic vector potential produced by this current density assuming classical rules is given by

$$\mathbf{A}(t, \mathbf{r}) = \frac{ieh\mu}{8\pi m}\int(\psi^*(t - |\mathbf{r} - \mathbf{r}'|/c, \mathbf{r}')\nabla\psi(t - |\mathbf{r} - \mathbf{r}'|/c, \mathbf{r}') - cc)exp(-jk|\mathbf{r} - \mathbf{r}'|)d^3r'/|\mathbf{r} - \mathbf{r}'|$$

Dirac equation, expression for the current density. Define the γ matrices

$$\gamma^0 = \begin{pmatrix} 0 & I \\ I & 0 \end{pmatrix},$$

$$\gamma^r = \begin{pmatrix} 0 & \sigma_r \\ -\sigma_r & 0 \end{pmatrix}, r = 1, 2, 3$$

where $\sigma_r, r = 1, 2, 3$ are the Pauli spin matrices. They are given by

$$\sigma_1 = \begin{pmatrix} 0 & 1 \\ 1 & 0 \end{pmatrix},$$

$$\sigma_2 = \begin{pmatrix} 0 & -i \\ i & 0 \end{pmatrix},$$

$$\sigma_3 = \begin{pmatrix} 1 & 0 \\ 0 & -1 \end{pmatrix}$$

The anticommutation relations are standard:

$$\sigma_i\sigma_j + \sigma_j\sigma_i = 2\delta_{ij}, 1 \leq i, j \leq 3$$

This leads to the anticommutation relations

$$\gamma^\mu\gamma^\nu + \gamma^\nu\gamma^\mu = 2g^{\mu\nu}$$

where

$$g^{00} = 1, g^{rr} = -1, r = 1, 2, 3, g^{\mu\nu} = 0, \mu \neq \nu$$

Note that if star denotes conjugate transpose, then

$$\gamma^{r*} = -\gamma^r, r = 1, 2, 3, \gamma^{0*} = \gamma^0$$

Also,

$$\gamma^0\gamma^r\gamma^0 = -\gamma^r, r = 1, 2, 3$$

The Dirac equation reads

$$(i\gamma^\mu \partial_\mu - m)\psi = 0$$

From this, we infer that

$$(i\gamma^\nu \partial_\nu + m)(i\gamma^\mu \partial_\mu - m)\psi = 0$$

Now observe that in view of the anticommutation relations between the γ matrices,

$$\gamma^\mu \gamma^\nu \partial_\mu \partial_\nu = g^{\mu\nu} \partial_\mu \partial_\nu$$

which is the wave operator and hence the above becomes

$$(\nabla^2 - \partial_t^2 - m^2)\psi = 0$$

which is the Klein-Gordon equation. Thus, any wave function stratifying the Dirac equation also satisfies the Klein-Gordon equation.

70.Selection rules in quantum mechanics:

Let L_x, L_y, L_z denote the three components of the angular momentum of a quantum mechanical particle. We wish to derive the rules for dipole transition from one state to another. Let $E_0\hat{z}$ be the applied electric field. Then, if $-e$ denotes the charge of the electron, the interaction energy between the electron and the field is given by $-eE_0 z$. The probability amplitude of the atom making a transition from an initial state $|\psi_i>$ to a final state $|\psi_f>$ is given by

$$<\psi_f| - eE_0 z|\psi_i> = -eE_0 <\psi_f|z|\psi_i>$$

The electric dipole selection rules determine constraints on the initial and final states that guarantee this transition probability amplitude to be non-zero. Suppose $|\psi_i>$ is the state $|l, m>$ in which the squared angular momentum observable L^2 has eigenvalue $l(l+1)$ and L_z has eigenvalue m, $|\psi_f>$ is the state $|l, m'>$ in which the squared angular momentum has the same eigenvalue l while L_z has eigenvalue m'. Then, we seek conditions for $<m'|z|m>$ to be non-zero. To derive these conditions, we use the following commutation relations:

$$[L_z, z] = [xp_y - yp_x, z] = 0$$

and hence

$$0 =< l, m'|[L_z, z]|l, m >=< l, m'|(L_z z - zL_z|l, m >= (m' - m) < l, m'|z|l, m >$$

and hence the required selection rule is $m = m'$. Again suppose that the electric field is $E_0\hat{x}$, ie, along the x direction. Then the conditions must be derived for $< l, m'|x|l, m > \neq 0$. We use the commutation relations

$$[L_z, x] = [xp_y - yp_x, x] = iy$$

so that

$$[L_z, [L_z, x]] = i[L_z, y] = i[xp_y - yp_x, y] = x$$

or when expanded in full notation,

$$L_z(L_z x - xL_z) - (L_z x - xL_z)L_z = x$$

or

$$L_z^2 x - 2L_z x L_z + x L_z^2 = x$$

It follows from this that

$$< l, m'|x|l, m >= (m'^2 - 2mm' + m^2) < l, m'|x|l, m >$$

and this implies that $< l, m'|x|l, m >$ is nonzero when and only when

$$m'^2 - 2mm' + m^2 - 1 = 0$$

or

$$(m' - m)^2 - 1 = 0$$

or

$$(m' - m - 1)(m' - m + 1) = 0$$

or

$$m' - m = \pm 1$$

Likewise the condition for $< l, m'|y|l, m >$ to be nonzero is that $m' - m = \pm 1$. We thus conclude that for any constant electric field, electric dipole interaction with the electron leads to nonzero transition probabilities iff the change in the azimuthal angular momentum quantum number is $0, \pm 1$.

71. Wave functions for particles having spin one. There are two parts of the wave function, one ψ_μ which by taking the analogy of the electromagnetic four potential for photons, is a four vector and then a tensor $\psi_{\mu\nu}$ which is derived from ψ_μ so that

$$i\psi_{\mu\nu} = p_\mu \psi_\nu - p_\nu \psi_\mu$$

The wave equation from tensor symmetry must have the form

$$p^\nu \psi_{\mu\nu} = im\psi_\mu$$

Here, $p_\mu = i\partial_\mu$. For photons, $m = 0$ and the second equation reduces to Maxwell's equations in free space, ie, the divergence of the antisymmetric field tensor vanishes. The problem is to derive these wave equations from an action principle. Consider

$$\mathcal{L}_1 = \psi^{\mu\nu*}(\partial_\nu \psi_\mu - \partial_\mu \psi_\nu) + \psi^{\mu\nu}(\partial_\nu \psi_\mu^* - \partial_\mu \psi_\nu^*)$$

The coefficient of $\delta\psi^{\mu\nu*}$ in $\delta\mathcal{L}_1$ is

$$\partial_\nu \psi_\mu - \partial_\mu \psi_\nu$$

and this must be $\psi_{\mu\nu}$. So we add a term

$$\mathcal{L}_2 = \psi^{\mu\nu*}\psi_{\mu\nu} = \psi_{\mu\nu}^* \psi^{\mu\nu}$$

to the Lagrangian. Then the coefficient of $\delta\psi^{\mu\nu}$ in $\delta(\mathcal{L}_1 + \mathcal{L}_2)$ becomes

$$\partial_\nu \psi_\mu^* - \partial_\mu \psi_\nu^* - \psi_{\mu\nu}^*$$

which is once again zero by the field equations. Finally, the coefficient of $\delta\psi_\mu^*$ in $\delta(\mathcal{L}_1+\mathcal{L}_2)$ after ignoring total partial differentials is given by

$$-\partial_\nu\psi^{\mu\nu}+\partial_\nu\psi^{\nu\mu}=-2\partial_\nu\psi^{\mu\nu}$$

by virtue of the antisymmetry of $\psi^{\mu\nu}$. This must equal $-2m\psi^\mu$ by virtue of the wave equations. So we must add a term

$$\mathcal{L}_3=-2m\psi^\mu\psi_\mu^*$$

The Lagrangian density is therefore given by

$$\mathcal{L}=\mathcal{L}_1+\mathcal{L}_2+\mathcal{L}_3=$$

$$\psi^{\mu\nu*}(\partial_\nu\psi_\mu-\partial_\mu\psi_\nu)+\psi^{\mu\nu}(\partial_\nu\psi_\mu^*-\partial_\mu\psi_\nu^*)$$

$$\psi^{\mu\nu*}\psi_{\mu\nu}-2m\psi^\mu\psi_\mu^*$$

We leave it as an exercise to carry out the variation of the action functional defined by this Lagrangian density with respect to $\psi_{\mu\nu},\psi_{\mu\nu}^*,\psi_\mu,\psi_\mu^*$ and write down the field equations and check that these tally with the given field equations we started with.

72. Quantum mechanics

Scattering theory problems:

1.Explain what you understand by the unitary group of a free particle, ie, the free particle propagator. Explain with reference to the solution to the heat equation. Hint: Schrodinger's equation for the free particle reads

$$ih\psi_{,t}=-\frac{h^2}{2m}\psi_{,xx}$$

The solution can be expressed as

$$\psi_t(x)=exp((ih^2/2m)d^2/dx^2)\psi_0(x)$$

The explicit form of the integral kernel corresponding to the linear operator $exp((ih^2/2m)d^2/dx^2)$ may be obtained using the Fourier transformation: Let

$$\hat{\psi}(t,k)=\int\psi(t,x)exp(-ikx)dx$$

so that

$$ih\hat{\psi}_{,t}=(h^2k^2/2m)\hat{\psi}$$

giving the solution

$$\hat{\psi}(t,k)=exp(-ihk^2/2m)\hat{\psi}(0,k)$$

Fourier inversion of this equation gives the required free particle kernel $K_t(x-y)$, ie

$$\psi(t,x)=\int K_t(x-y)\psi(0,y)dy$$

2.Define the resolvent of a Hermitian operator in a Hilbert space and derive its integral expression in terms of the one parameter unitary group generated by the operator.

Hint: Let A be a Hermitian operator and consider $U_t = exp(itA)$, the one parameter unitary group generated by A. We have

$$\int_0^\infty exp(-t(x+iA))dt = (x+iA)^{-1} = R_x(A), Rex > 0$$

This is the resolvent of the operator A. An integral expression for the resolvent of an operator can be obtained in terms of the spectral measure associated with the operator. Specifically, suppose that the spectral representation of A is given by

$$A = \int_{-\infty}^{\infty} x.dE(x)$$

Then,

$$exp(itA) = \int_{-\infty}^{\infty} exp(itx).dE(x)$$

and then for $Rex > 0$, we have

$$R_x(A) = \int_{-\infty}^{\infty} (x+iu)^{-1} dE(u)$$

Suppose that $U_t = exp(it(H_0+V))$ and $V_t = exp(itH_0)$. The Moller operator Ω_+ is defined as the limit as $t \to -\infty$ of $U_t^* V_t$ restricted to $M_\infty(H_0)$. Now observe that for $\eta > 0$,

$$\eta \int_0^\infty exp(-\eta t)\Phi(t)dt$$

$$= -\Phi(t)exp(-\eta t)|_0^\infty + \int_0^\infty \Phi'(t)exp(-\eta t)dt$$

$$= \Phi(0) + \int_0^\infty \Phi'(t)exp(-\eta t)dt$$

which converges to

$$\Phi(0) + \Phi(\infty) - \Phi(0) = \Phi(\infty)$$

as $\eta \to 0$. It follows that the Moller operator (after appropriate restriction) is the limit of $\Omega_{\eta+}$ as $\eta \to 0$ where

$$\Omega_{\eta+} = \eta \int_0^\infty exp(-\eta t) U_t^* V_t dt$$

Let A, B, C be three Hermitian operators and define

$$X(t) = exp(-itA), Y(t) = exp(-itB), Z(t) = exp(itC)$$

Let $E(.)$ be the spectral measure of C. Then,

$$\int_{-\infty}^{0} exp(\eta t)X^*(t)dt \int_{-\infty}^{0} exp(\delta s)Y^*(s)TZ(s+t)ds$$

$$= \int exp((\eta - iA + i\lambda)t)exp((\delta + iB + i\lambda)s)TdE(\lambda)dtds$$

over the appropriate ranges. This evaluates for $\eta, \delta > 0$ to

$$\int (\eta + iA + i\lambda)^{-1}(\delta + iB + i\lambda)^{-1}TdE(\lambda)$$

$$= \int R_{\eta+i\lambda}(A)R_{\delta+i\lambda}(B)TdE(\lambda)$$

This is a fundamental formula in scattering theory. Note that

$$\eta \int_{0}^{\infty} X^*(t)Y(t)exp(-\eta t)dt$$

as $\eta \to 0$ is the Moller operator associated with the unitary groups $X(t), Y(t)$.

73. Problems in quantum computation:

(a) Quantum teleportation. Alice can communicate one qubit of information to Bob by transmitting just two classical bits of information. This is done as follows. Alice and Bob share an epr pair, ie, The state shared by Alice and Bob is $(|00> + |11>)/\sqrt{2}$. Alice has access to the first qubit while Bob has access to the second. Let the state that Alice wishes to communicate to Bob be $|\psi> = \alpha|0> + \beta|1>$. Then the state initial state of the system owned by Alice and Bob is given by

$$|\psi> (|00> + |11>)/\sqrt{2} = \frac{1}{\sqrt{2}}(\alpha|0> + \beta|1>)(|00> + |11>)$$

$$= \frac{1}{\sqrt{2}}(\alpha|000> + \alpha|011> + \beta|100> + \beta|111>)$$

Alice has access here to the first two qubits while Bob has access to the third qubit. Alice sends her qubits through a CNOT gate and then after doing so, the state of the overall system gets changed to

$$\frac{1}{\sqrt{2}}(\alpha|000> + \alpha|011> + \beta|110> + \beta|101>)$$

Alice then sends her first qubit through a Hadamard gate resulting in the state of the overall system as

$$\frac{1}{2}(\alpha(|0> + |1>)|00> + \alpha(|0> + |1>)|11> + \beta(|0> - |1>)|10> + \beta(|0> - |1>)|01>)$$

$$= \frac{1}{2}(|00> (\alpha|0> + \beta|1>) + |01> (\alpha|1> + \beta|0>) + |10> (\alpha|0> - \beta|1>) + |11> (\alpha|1> - \beta|0>)$$

Alice then makes a measurement on her two qubits in the standard basis $|00>, |01>, |10>, |11>$. If she measures $|00>$, she communicates two classical bits 00 to Bob. The state of the overall system after Alice's measurement is $|00> (\alpha|0> +\beta|1>)$ and Bob reads the correct state $|\psi>$ from this which was originally owned by Alice. If Alice measures $|01>$, she communicates this to Bob again via two classical bits. In this case, the state of the overall system collapses to $|01> (\alpha|1> +\beta|0>)$ and Bob's state is then $\alpha|1> +\beta|0>$. Bob applies the NOT gate to his state giving $\alpha|0> +\beta|1>$. The fact that Bob has selected this gate is because Alice has communicated the classical bits 01 to Bob. If Alice measures $|10>$, then Bob's state becomes $\alpha|0> -\beta|1>$ and Bob recovers $|\psi>$ by applying the phase change gate to his state. Note that the phase change gate is $|0>\rightarrow |0>, |1>\rightarrow -|1>$. Finally, if Alice measures $|11>$, Bob's state becomes $\alpha|1> -\beta|0>$. Bob on receiving the classical bits 11 from Alice, first applies the NOT gate giving his resulting state as $\alpha|0> -\beta|1>$ and then he applies the phase change gate giving his state as $\alpha|0> +\beta|1>$ which is precisely $|\psi>$. The gate that Bob chooses to apply to his state is based on the two classical bits communicated by Alice to Bob.

(b) Superdense coding. Here, Alice is able to send two classical bits of information to Bob by transmitting just a single qubit. This is achieved by Alice and Bob initially sharing an EPR pair. The initial state of Alice and Bob is $(|00> +|11>)/\sqrt{2}$. Alice owns the first qubit and Bob owns the second qubit. The If Alice wishes to transmit the classical bit sequence 00 to Bob, she does nothing other than simply communicate her qubit to Bob. Then, Bob's state becomes $(|00> +|11>)/\sqrt{2}$. If Alice wishes to communicate 01, she applies a NOT gate to her qubit and transmits this qubit to Bob. Bob's state then becomes $(|10> +|01>)/\sqrt{2}$. If Alice wishes to communicate 10 to Bob, she applies a phase change gate and communicate her qubit to Bob and the resulting state of Bob becomes $(|00> -|11>)/\sqrt{2}$. If Alice wishes to communicate 11 to Bob, she applies a NOT gate followed by a phase change gate and transmits her qubit to Bob whose state then becomes $(|01> -|10>)/\sqrt{2}$. Bob then makes a measurement in the basis

$$\mathcal{B} = \{(|00> +|11>)/\sqrt{2}, (|01> +|10>)/\sqrt{2}, (|00> -|11>)/\sqrt{2}, (|01> -|10>)/\sqrt{2}\}$$

and decides base on the measurement which classical bit pair was communicated by Alice to him.

(c) Quantum Fourier Transform: The orthonormal basis $|k>, k = 0, 1, ..., n-1$ is transformed into the orthonormal basis

$$U|k>= N^{-1/2}\sum_{m=0}^{N-1} exp(2\pi ikm/N)|m>$$

We write

$$m = m_0 + 2m_1 + 2^2m_2 + ... + 2^{N-1}m_{N-1}, m_j = 0, 1$$

and assume that

$$|m>= |m_0m_1...m_{r-1}>= |m_0>|m_1>...|m_{r-1}>, N = 2^r$$

Note that this defines an orthonormal basis in $\mathbb{C}^{2r}$. Then write

$$k = k_0 + 2k_1 + ... + 2^{r-1}k_{r-1}$$

and hence

$$U|k_0k_1...k_{r-1}>=2^{-r/2}\sum_{m_0,...,m_{r-1}=0,1} exp(2\pi ik(m_0+2m_1+...+2^{r-1}m_{r-1})/2^r)|m_0m_1...m_{r-1}>$$

$$=2^{-r/2}\sum_{m_0=0,1} exp(2\pi ikm_0/2^r)|m_0>.\sum_{m_1=0,1} exp(2\pi i2km_1/2^r)|m_1>...$$

$$\sum_{m_{r-1}=0,1} exp(2\pi i2^{r-1}km_{r-1}/2^r)|m_{r-1}>$$

Now,

$$k/2^r=(k_0+2k_1+...+2^{r-1}k_{r-1})/2^r=.k_{r-1}k_{r-2}...k_0$$

as a binary fraction. Likewise,

$$2k/2^r=k_{r-1}+.k_{r-2}k_{r-3}...k_0$$

and in general,

$$2^jk/2^r=integer+.k_{r-j-1}k_{r-j-2}...k_0$$

Thus,

$$U|k_0k_1...k_{r-1}>=2^{-r/2}(|0>+exp(2\pi i.k_{r-1}k_{r-2}...k_0)|1>)(|0>+exp(2\pi i.k_{r-2}k_{r-3}...k_0)|1>)...$$
$$(|0>+exp(2\pi i.k_0)|1>)$$

This operator can be depicted diagrammatically.

Ref:Quantum Computation and quantum information by Nielsen and Chuang and Notes by Vishal Sahni.

Optical cavity quantum computer: Let a, a^* denote respectively the operations of annihilation and and creation of a photon in the radiation field. The electric field corresponding to this can be expressed as

$$E=iE_0(a.exp(ik.r)-a^*exp(-ik.r))$$

The integral of E^2 over the cavity gives the Hamiltonian of the field apart from a numerical factor. The terms $exp(\pm 2ik.r)$ integrate out to zero giving the classical Hamiltonian as proportional to a^*a. With ω denoting the frequency of the photon, the Hamiltonian of the radiation field enclosed in the cavity is therefore given by $\frac{1}{2}h\omega.a^*a$. The commutation relations introduced are $[a,a^*]=1$. The dipole moment of the atom is to a good degree of approximation represented by the Pauli matrix Y. The interaction Hamiltonian between the radiation field and the atom equals the scalar product of the atomic dipole moment and the electric field at its site. This is proportional to $Y(a-a^*)$. X, Y, Z are the Pauli spin matrices. They are in the standard representation, given by

$$X=\begin{pmatrix} 0 & 1 \\ 1 & 0 \end{pmatrix}, Y=\begin{pmatrix} 0 & -i \\ i & 0 \end{pmatrix},$$

$$Z=\begin{pmatrix} 1 & 0 \\ 0 & -1 \end{pmatrix}$$

We introduce the Ladder operators for the atom

$$\sigma_+ = (X + iY)/2, \sigma_- = (X - iY)/2$$

Let $|0>$ denote the eigen state of Z corresponding to the eigenvalue -1 and $|1>$ the eigenstate of Z corresponding to the eigenvalue $+1$. Thus $Z|0>= -|0>, Z|1>= |1>$. Then $X|0>= |1>, X|1>= |0>$ and $Y|0>= -i|1>, Y|1>= i|0>$. It follows that

$$\sigma_+|0>= (X + iY)|0> /2 = |1>, \sigma_+|1>= (X + iY)|1> /2 = 0,$$

$$\sigma_-|0>= (X - iY)|0> /2 = 0, \sigma_-|1>= (X - iY)|1> /2 = |0>$$

In other words, σ_+ raises the eigenvalue of an eigenstate of Z by unity and σ_- lowers the eigenvalue of an eigenstate of Z by unity. We have $Y = -i(\sigma_+ - \sigma_-)$ so that the interaction Hamiltonian between the atom and the photon field is given by

$$(\sigma_+ - \sigma_-)(a - a^*) = \sigma_+ a + \sigma_- a^* - \sigma_+ a^* - \sigma_- a$$

The atom also interacts with the external magnetic field and this interaction can be represented as $\delta.Z$ where δ is a scalar. This formula stems from the fact that the magnetic field is applied along the z direction and the component of the magnetic moment in the z direction is proportional to Z. Consider the atom alone interacting with the magnetic field. We find that

$$d\sigma_+/dt = i[\delta Z, \sigma_+] = (i\delta/2)[Z, X + iY] = (i\delta/2)(2iY + 2X) - i\delta(X + iY) = 2i\delta\sigma_+$$

so that σ_+ varies with time as $exp(2i\delta t)$. Likewise,

$$d\sigma_-/dt = i[\delta Z, \sigma_-] = (i\delta/2)[Z, X - iY] = (i\delta/2)(2iY - 2X) = -2i\delta\sigma_-$$

so that σ_- varies with time as $exp(-2i\delta t)$. Now consider the radiation field alone, ie not interacting with the atom. We find that

$$da/dt = i[h\omega a^* a/2, a] = (ih\omega/2)[a^*, a]a = -ih\omega a/2$$

which means that a varies with time as $exp(-i\omega t/2)$. Likewise, a^* varies with time as $exp(i\omega t/2)$. It follows that in the absence of interaction, $\sigma_+ a^*$ and $\sigma_- a$ vary respectively with time as $exp(\pm i(2\delta + \omega/2)t)$. In the rotating wave approximation, we set these to zero since the averaging a high frequency physical quantity over even a small duration gives zero.

74. Scattering matrix in quantum theory: The initial state of the system of particles is denoted by i and the final state by f. The scattering matrix can be written as the sum of a unit matrix corresponding to no interactions between the particles an an interaction part. Thus,

$$S_{fi} = \delta_{fi} + i\delta(P_f - P_i)T_{fi}$$

where P_i is the four momentum of the system of particles before the interaction and P_f is the same after the interaction. This formula can be justified along the following lines. Assuming that the interaction potential is V, we then apply standard time dependent perturbation theory. Let $|n>$ be the unperturbed eigenstates and $H = H_0 + V$ the total Hamiltonian. The wave function after perturbation is $\psi(t)$. We set

$$\psi(t) = \sum_n a_n(t)|n>$$

Then

$$ih\psi'(t) = ih\sum_n a_n'(t)|n> = (H_0 + V)\psi(t) = \sum_n a_n(t)E_n|n> + \sum_n a_n(t)V|n>$$

giving

$$iha_n'(t) = E_n a_n(t) + \sum_m a_m(t) < n|V|m>$$

so that if $|i>$ is the initial state, and f the final state the transition probability amplitude in duration $[0,t]$ approximately satisfies the d.e

$$iha_f'(t) = V_{fi}exp(-iE_it) + E_f a_f(t)$$

so that

$$a_f(t) = exp(-iE_ft)\int_0^t exp(i(E_f - E_i)s)V_{fi}ds$$

where $f \neq i$. For large times t, the integral becomes $\delta(E_f - E_i)$ signifying energy conservation. In the general case of particles, there are other delta functions corresponding to momentum conservation. Let $u_i, i = 1, 2, ...$ be the wave functions of the initial particles and $u_i', i = 1, 2, ...$ the wave functions of the final particles. They are solutions of the momentum space Dirac equation. Then the form of the scattering amplitude T_{fi} is given by

$$T_{fi} = (\otimes_i u_i'^*)Q(\otimes_i u_i)$$

where Q is some matrix. Let p_a be the initial particle three momenta and p_a' the final three momenta. Then the differential transition probability of scattering of the initial particles into the three momentum element $\Pi d^3p_a'$ has the form

$$dw = \delta(P_f - P_i)|T_{fi}|^2\Pi d^3p_a'$$

provided that we assume a unit volume. In the general case, we must multiply this expression by the volume V in which the particles are enclosed. Suppose we make the computation of the transition probabilities using the normalized Dirac wave functions u where the normalization is given by $u^*\gamma^0u = 2m$ or equivalently using the notation $\bar{u} = u^*\gamma^0$, $\bar{u}u = 2m$. In this case, the resulting amplitude is $M_{fi} = T_{fi}(\Pi\epsilon\epsilon')^{1/2}$. This can be proved on elementary grounds by considering the wave function of a photon corresponding to one photon in a unit volume. See Lifshitz, Bersetski and Pitaevski, "Quantum Electrodynamics". It follows that the differential probability amplitude for a particle of energy ϵ to decay into particles having energies ϵ_a' and three momenta in the range d^3p_a' is given by

$$dw = \delta(P_f - P_i)|M_{fi}|^2\epsilon^{-1}\Pi\epsilon_a'^{-1}d^3p_a'$$

75. Suppose u is the initial state of an electron and u' the final state. These are assumed to be solutions to the Dirac equation corresponding to four momenta p and p'. Then the scattering matrix element between these two states has the form

$$M_{fi} = \bar{u}'Au$$

where A is some matrix and $\bar{u}' = u'^{*}\gamma^0$. The corresponding transition probability is given by

$$|M_{fi}|^2 = |\bar{u}' A u|^2$$

We note that

$$M_{fi} = u'^{*}\gamma^0 A u$$

so that

$$M_{fi}^* = u^* A^* \gamma^{0*} u' = \bar{u}\gamma^0 A^* \gamma^{0*} u'$$

Define

$$\bar{A} = \gamma^0 A^* \gamma^{0*}$$

Then,

$$M_{fi}^* = \bar{u}\bar{A}u'$$

whence,

$$|M_{fi}|^2 = (\bar{u}' A u)(\bar{u}\bar{A}u')$$

and when expressed in terms of components,

$$|M_{fi}|^2 = \bar{u}'_i A_{ij} u_j \bar{u}_k \bar{A}_{km} u'_m = u'_m \bar{u}'_i A_{ij} u_j \bar{u}_k \bar{A}_{km}$$

Note the average of $u_i\bar{u}_j$ over all polarizations is ρ_{ij}, where ρ is the initial density matrix and likewise, the average of $u'_i\bar{u}'_j$ is ρ'_{ij}, where ρ' is the final density matrix. Thus, after averaging out, the probability of the electron getting scattered from an initial state ρ to a final state ρ' is given by

$$|M_{fi}|^2 = Tr(\rho' A \rho \bar{A})$$

76. Steven Weinberg, The quantum theory of fields, Volume I.

Problem in groups: On $\mathbb{R}^4$, define the bilinear form

$$\phi(\xi,\eta) = tt' - xx' - yy' - zz', \xi = (t,x,y,z), \xi' = (t',x',y',z')$$

Then a linear transformation L on $\mathbb{R}^4$ that preserves ϕ is called a Lorentz transformation, ie,

$$\phi(L\xi, L\eta) = \phi(\xi,\eta), \xi,\eta \in \mathbb{R}^4$$

Let Q denote the quadratic form associated to the bilinear form ϕ, ie,

$$Q(\xi) = \phi(\xi,\xi)$$

Let $\mathcal{H}$ denote the vector space of all 2×2 Hermitian matrices. Consider the linear mapping $\Phi : \mathbb{R}^4 \to \mathcal{H}$ defined by

$$\Phi(\xi) = \begin{pmatrix} t+z & x-iy \\ x+iy & t-z \end{pmatrix}$$

Φ is an onto map. We then have

$$det\Phi(\xi) = Q(\xi)$$

Let $A \in SL(2,\mathbb{C})$. Then for any $\xi \in \mathbb{R}^4$, $\Phi(\xi)$ is a Hermitian matrix and hence $A\Phi(\xi)A^*$ is also a Hermitian matrix. Hence, $\eta = \Phi^{-1}(A\Phi(\xi)A^*) \in \mathbb{R}^4$ is well defined. It follows that

$$Q(\eta) = det\Phi(\eta) = det(A\Phi(\xi)A^*) = detA.det\Phi(\xi).detA^* = det\Phi(\xi) = Q(\xi)$$

Writing $\eta = L\xi$, it thus follows that L is a Lorentz transformation on $\mathbb{R}^4$.

Problem: For a given $A \in SL(2,\mathbb{C})$, determine explicitly, the matrix $L = L(A)$.

Chapter 4

Electromagnetics

[1] Write down the basic transmission line partial differential equations for the voltage and current propagation along the line. Assume that the capacitance, inductance, resistance and conductance per unit length are constants. Derive in the process formulae for the characteristic impedance and propagation constant of the signals along the line. For a source excitation $v_s(t)$ when the internal resistance of the source is R_s and a load impedance $R(j\omega) + iX(j\omega)$, give explicit integral formulae for the voltage and current along the line.

[2] Explain the use of the Smith chart in the computation of the input impedance of a transmission line for any given complex load impedance.

[3] If two transmission lines having characteristic impedances $Z_{0i}, i = 1, 2$ are connected at the load end with a common load of Z_L, the transmission lines have source phasors $V_i, i = 1, 2$ at the same frequency and propagation constants $\beta_i, i = 1, 2$, then determine the current and voltage along the two lines.

[4] Display the analogy between transmission lines and waveguides.

Hint: Assume first that the waveguide has infinite breadth along one direction and write down the Maxwell equations satisfied by the electric and magnetic fields. For example, suppose that the waveguide has a length along the z axis and breadth a along the x direction and infinite breadth along the y direction. Assume the propagation of a transverse electromagnetic wave. Let the electric field be given by $E_x(z, t)$ and the magnetic field by $H_y(z, t)$. The Maxwell equation $\nabla \times E = -\mu \frac{\partial H}{\partial t}$ gives

$$\frac{\partial E_x}{\partial z} = -\mu \frac{\partial H_y}{\partial t}$$

and the Maxwell equation

$$\nabla \times H = \sigma E + \epsilon \frac{\partial E}{\partial t}$$

gives

$$-\frac{\partial H_y}{\partial z} = \sigma E_x + \epsilon \frac{\partial E_x}{\partial t}$$

Compare these equations to the transmission line equations for voltage and current propagation.

[5] Assuming that in a rectangular waveguide having axis along the z direction and breadths a, b respectively along the x and y directions, that the dependence on z of the electric and magnetic fields has the form $exp(-\gamma z)$, write down the Maxwell equations $\nabla \times E = -i\omega\mu H$ and $\nabla \times H = (\sigma + j\omega\epsilon)E$ in component form. Using these equations, express the transverse components of the electric and magnetic fields E_x, E_y, H_x, H_y in terms of the partial derivatives of E_z, H_z with respect to x and y and hence derive the Helmholtz equations for E_z and H_z.

[6] Write down the expression for the curl operator in the cylindrical coordinate system and use this to formulate the Maxwell equations for a cylindrical waveguide. Express the transverse components of the field $E_\rho, H_\rho, E_\phi, H_\phi$ in terms of partial derivatives of E_z, H_z with respect to ρ, ϕ and hence derive the Helmholtz equation for E_z, H_z in cylindrical coordinates.

[7] Consider the $TE_{n,m}$ and $TM_{n,m}$ modes of electromagnetic wave propagation inside a rectangular waveguide. Obtain analytic expressions for the field lines at a given value of z say $z = 0$ and plot them.

Hint: The transverse components of the electric and magnetic field phasors have the form $(E_x(x,y), E_y(x,y), H_x(x,y), H_y(x,y)).exp(-\gamma z)$. It should be noted that the functions E_x, E_y, H_x, H_y are real valued functions. The differential equations for the electric field lines in the plane $z = 0$ are then given by

$$\frac{dx}{E_x(x,y)} = \frac{dy}{E_y(x,y)}$$

and likewise for the magnetic field

$$\frac{dx}{H_x(x,y)} = \frac{dy}{H_y(x,y)}$$

These equations can readily be integrated since the functions E_x, E_y, H_x, H_y have the separable form, ie, separable into the product of a function of x and a function of y.

[8] A transmission line has inductance, capacitance, conductance and resistance per unit length varying slowly along the line. By considering these parameters to be the sum of large constants and a small perturbation that is line length dependent, determine the transmission line equations and solve them using first order perturbation theory. Assume sinusoidal excitation Hint: Consider the linear differential equation

$$\frac{d}{dz}\mathbf{X}(z) = \mathbf{A}\mathbf{X}(z) + \epsilon.\mathbf{B}(z)\mathbf{X}(z)$$

where ϵ is a small parameter, $\mathbf{A}$ is a constant matrix and $\mathbf{B}(z)$ is a matrix that varies along the line. Here, $\mathbf{X}(z)$ for each z is a $d \times 1$ vector and $\mathbf{A}, \mathbf{B}$ are matrices of size $d \times d$, apply first order perturbation theory to this equation and determine $\mathbf{X}(z)$ upto $O(\epsilon)$:

$$\mathbf{X}(z) = \mathbf{X}_0(z) + \epsilon.\mathbf{X}_1(z)$$

giving

$$\mathbf{X}_0'(z) = \mathbf{A}\mathbf{X}_0(z), \mathbf{X}_1'(z) = \mathbf{A}\mathbf{X}_1(z) + \mathbf{B}(z)\mathbf{X}_0(z)$$

so that

$$\mathbf{X}_0(z) = exp(z\mathbf{A})\mathbf{s}$$

and

$$\mathbf{X}_1(z) = \int_0^z exp((z-u)\mathbf{A})\mathbf{B}(u)\mathbf{X}_0(u)du$$

Show that the transmission line equations can be cast in this framework.

[9] Consider a waveguide of arbitrary cross section S. Let $\Gamma = \partial S$ denote the boundary of S. Express the electric and magnetic fields as the sum of a longitudinal and a transverse component, plug these expressions into the Maxwell equations and hence derive formulae for the transverse components of the field in terms of the axial components. Formulate the wave equation for the longitudinal components with appropriate boundary conditions and explain how you would numerically determine the modes. You may use either the method of moments or the finite element method for determining the propagation modes.

10.State whether the electromagnetic four potential is a four vector or not. Assuming that it is a four vector, write down the Lorentz transformation equations that give the four potential field in a reference frame S' moving relative to the reference frame S with a uniform velocity v along the direction specified by the unit vector $\mathbf{n}$.

Ans:Let the four potential be $(V, \mathbf{A})$ in S and $(V', \mathbf{A}')$ in S'. Then the transformation equations are

$$V' = \gamma(V - (\mathbf{v}, \mathbf{A}), A'_{\parallel} = \gamma(A_{\parallel} - vV), \mathbf{A}'_{\perp} = \mathbf{A}_{\perp}$$

where $A_{\parallel}$ and $A_{\perp}$ denote respectively the components of the magnetic vector potential along the direction of $\mathbf{v}$ and perpendicular to the $\mathbf{v}$.

11.If $g_{\mu\nu}(x)$ is the metric tensor of space-time, and A^{μ}, B^{μ} are four vectors, then $g_{\mu\nu}A^{\mu}B^{\nu} = A_{\mu}B^{\mu}$ is an invariant, ie, a scalar. Use this fact to derive the formula for parallel displacement of a covariant vector in terms of the formula for the parallel displacement of a contravariant vector.

12. Write down the electromagnetic field equations for a medium having anisotropic homogeneous permittivity and derive the modified wave equations for the fields.

Hint: Choose the principal axes of the permittivity as the x, y, z axes and use the constitutive relations

$$D_x = \epsilon_x.E_x, D_y = \epsilon_y.E_y, D_z = \epsilon_z E_z$$

The relevant Maxwell equations are

$$div(\mathbf{D}) = 0$$

or in component form,

$$\epsilon_x \frac{\partial D_x}{\partial x} + \epsilon_y \frac{\partial E_y}{\partial y} + \epsilon_z \frac{\partial E_z}{\partial z} = 0$$

$$\nabla \times \mathbf{H} = \frac{\partial \mathbf{D}}{\partial t}$$

or in component form,

$$\frac{\partial H_z}{\partial y} - \frac{\partial H_y}{\partial z} = \epsilon_x.\frac{\partial E_x}{\partial t}$$

with similar equations for the y, z components, and finally,

$$\nabla \times \mathbf{E} = -\mu_0 \frac{\partial \mathbf{H}}{\partial t}$$

which can also be expressed in component form.

13. Consider a wave consisting of one frequency but two wavelengths. What does the amplitude of this wave look like when the wavelengths are very closely separated. Assume both components have the same amplitude.

Hint:Consider the wave motion along the x axis. Then

$$F(x,t) = cos(\omega t k_1 x) + cos(\omega t - k_2 x) = 2cos(\omega t - (k+1+k_2)x/2)cos((k_1 - k_2)x/2)$$

The amplitude of this wave is $cos((k_1 - k_2)x/2)$ which varies slowly in space and is independent of time.

14. Determine a formula for the amplitude and phase of a superposition of n plane waves, all having the same frequency but different wave numbers. Assume the waves propagate all along the x direction.

Hint:

$$F(t,x) = Re[\sum_{i=1}^{n} A_i exp(j\phi_i) exp(j(\omega t - k_i x))]$$

Now show that

$$|\sum_{i=1}^{n} A_i exp(j\theta_i)| = (\sum_i A_i^2 + 2\sum_{i<j} A_i A_j cos(\theta_i - \theta_j))^{1/2}$$

and

$$Arg(\sum_{i=1}^{n} A_i exp(j\theta_i)) = tan^{-1}\frac{(\sum_i A_i sin(\theta_i))}{\sum_i A_i cos(\theta_i)}$$

where A_i, ϕ_i, θ_i are all real.

15. Energy momentum tensor of the electromagnetic field

Assume first that the field functions are $\phi_\alpha, \alpha = 1, 2, ..., N$ so that $\phi_\alpha \mathbb{R}^4 \to \mathbb{R}$ are well defined differentiable functions. The Lagrangian density is given by $\mathcal{L}(\phi_\alpha, \phi_{\alpha,\mu})$ and the action for the entire field is

$$S[\{\phi_\alpha\}] = \int \mathcal{L}(\phi_\alpha(x), \phi_{\alpha,\mu}(x)) d^4x$$

The field equations obtained from the action principle

$$\delta S[\phi_\alpha] = 0$$

are

$$\frac{\partial}{\partial x^\mu}\frac{\partial \mathcal{L}}{\partial \phi_{\alpha,\mu}} = \frac{\partial \mathcal{L}}{\partial \phi_\alpha}$$

We consider

$$\frac{\partial \mathcal{L}}{\partial x^\mu} = \frac{\partial \mathcal{L}}{\partial \phi_\alpha}\phi_{\alpha,\mu} + \frac{\partial \mathcal{L}}{\partial \phi_{\alpha,\nu}}\phi_{\alpha,\nu\mu}$$

$$= (\frac{\partial}{\partial x^\nu}\frac{\partial \mathcal{L}}{\partial \phi_{\alpha,\nu}})\phi_{\alpha,\mu} + \frac{\partial \mathcal{L}}{\partial \phi_{\alpha,\nu}}\phi_{\alpha,\nu\mu}$$

$$= \frac{\partial}{\partial x^\nu}(\frac{\partial \mathcal{L}}{\partial \phi_{\alpha,\nu}}\phi_{\alpha,\mu})$$

It follows that the conservation law

$$\frac{\partial T^\nu_\mu}{\partial x^\nu} = 0$$

holds where

$$T^\nu_\mu = \frac{\partial \mathcal{L}}{\partial \phi_{\alpha,\nu}}\phi_{\alpha,\mu} - \mathcal{L}\delta^\nu_\mu$$

We may interpret T^ν_μ as the energy momentum tensor of the field. The problem with this definition is that T may not be a symmetric tensor. However in most problems, it can be symmetrized as follows. Suppose we consider a tensor $\psi^{\mu\nu\rho}$ that is antisymmetric with respect to the last two indices, ie, $\psi^{\mu\nu\rho} = -\psi^{\mu\rho\nu}$. Then

$$\frac{\partial^2 \psi^{\mu\nu\rho}}{\partial x^\nu \partial x^\rho} = 0$$

and hence if

$$T^{\mu\nu}_{,\nu} = 0$$

and we define

$$S^{\mu\nu} = T^{\mu\nu} + \psi^{\mu\nu\rho}_{,\rho}$$

then

$$S^{\mu\nu}_{,\nu} = 0$$

In other words, if T satisfies the conservation law then S will also satisfy the conservation law. Hence, S may equally well be used as the energy-momentum tensor of the field. In a majority of cases, the tensor ψ can be selected so that S becomes a symmetric tensor. For example, consider the electromagnetic field for which the field functions are A^μ the four potential. The covariant potential components are $A_\mu = g_{\mu\nu}A^\nu$. The antisymmetric field tensor is given by

$$F_{\mu\nu} = A_{\nu,\mu} - A_{\mu,\nu}$$

and the Lagrangian density is

$$\mathcal{L} = F_{\mu\nu}F^{\mu\nu}$$

To compute the tensor T for this field, we need the partial derivatives

$$\frac{\partial \mathcal{L}}{\partial A_{\mu,\nu}}$$

To obtain this, we consider the change in $\mathcal{L}$ induced by a small change δA_μ in the potentials. We get

$$\delta\mathcal{L} = 2F^{\mu\nu}\delta F_{\mu\nu} = 2F^{\mu\nu}(\delta A_{\nu,\mu} - \delta A_{\mu,\nu})$$

$$= 4F^{\mu\nu}\delta A_{\nu,\mu}$$

since

$$F^{\mu\nu} = -F^{\nu\mu}$$

This gives

$$\frac{\partial \mathcal{L}}{\partial A_{\nu,\mu}} = 4F^{\mu\nu}$$

Hence,

$$T^{\mu}_{\rho} = \frac{\partial \mathcal{L}}{\partial A_{\nu,\mu}} A_{\nu,\rho} - \mathcal{L}\delta^{\mu}_{\rho}$$

$$= 4F^{\mu\nu}A_{\nu,\rho} - F_{\alpha\beta}F^{\alpha\beta}\delta^{\mu}_{\rho}$$

or equivalently,

$$T_{\mu\rho} = 4F_{\mu\nu}A^{\nu}_{,\rho} - F_{\alpha\beta}F^{\alpha\beta}g_{\mu\rho}$$

$$= 4F^{\nu}_{\mu}A_{\nu,\rho} - F_{\alpha\beta}F^{\alpha\beta}g_{\mu\rho}$$

This is not symmetric. However,

$$T_{\mu\rho} - 4F^{\nu}_{\mu}A_{\rho,\nu} = 4F^{\nu}_{\mu}F_{\rho\nu} - F_{\alpha\beta}F^{\alpha\beta}g_{\mu\rho}$$

is symmetric and further,

$$(F^{\mu\nu}A_{\nu,\rho})_{,\mu} = F^{\mu\nu}_{,\mu}A_{\nu,\rho} + F^{\mu\nu}A_{\nu,\rho\mu} = 0$$

by virtue of the Maxwell equations in free space and the antisymmetry of $F^{\mu\nu}$.

16.Hertz potential:Consider an em field in free space. The electric field is **E** and the magnetic field is **H**. They satisfy the equations

$$\nabla.E = 0, \nabla \times E = -i\omega\mu H, \nabla.H = 0, \nabla \times H = i\epsilon\omega E$$

The first equation implies the existence of a vector field F such that

$$E = \nabla \times F$$

So from the fourth, we get

$$\nabla \times H = i\epsilon\omega\nabla \times F$$

or equivalently,

$$\nabla \times (H - i\epsilon\omega F) = 0$$

Thus there is a scalar function ϕ such that

$$H = i\epsilon\omega F + \nabla\phi$$

and plugging this into the second Maxwell equation gives us

$$\nabla \times (\nabla \times F) = -i\omega\mu(i\epsilon\omega F + \nabla\phi)$$

or

$$\nabla(\nabla.F) - \nabla^2 F = \epsilon\mu\omega^2 F - i\omega\mu\nabla\phi$$

or

$$\nabla(\nabla.F + i\omega\mu\phi) = \nabla^2 F + \omega^2\epsilon\mu F$$

We can choose a gauge so that

$$\nabla.F + i\omega\mu\phi = 0$$

and then we obtain the Helmholtz equation for F:

$$(\nabla^2 + \omega^2\epsilon\mu)F = 0$$

17. Complex Poynting theorem: Assume that the fields have sinusoidal dependence. We write for the electric field

$$\mathbf{E}(t, \mathbf{r}) = \frac{1}{2}(E(\mathbf{r})exp(i\omega t) + E(\mathbf{r})^* exp(-i\omega t))$$

and likewise for the magnetic field

$$\mathbf{H}(t, \mathbf{r}) = \frac{1}{2}(H(\mathbf{r})exp(i\omega t) + H(\mathbf{r})^* exp(-i\omega t))$$

The Poynting vector is given by

$$\mathbf{S}(t, \mathbf{r}) = \frac{1}{4}(E \times H.exp(2i\omega t) + E^* \times H^* exp(-2i\omega t) + E \times H^* + E^* \times H)$$

and the energy density of the electric field is given by

$$U_E(t, \mathbf{r}) = \frac{\epsilon_0}{8}(E^2 exp(2i\omega t) + E^{*2} exp(-2i\omega t) + 2EE^*)$$

The energy density of the magnetic field is

$$U_H(t, \mathbf{r}) = \frac{1}{8\mu_0}(H^2 exp(2i\omega t) + H^{*2} exp(-2i\omega t) + 2HH^*)$$

The current density can also be expressed as

$$\mathbf{J}(t, \mathbf{r}) = \frac{1}{2}(J(\mathbf{r})exp(i\omega t) + J^*(\mathbf{r})exp(-i\omega t))$$

Plugging these expressions into the energy conservation equation

$$\frac{d}{dt}\int (U_E + U_H)dV = -\int \mathbf{S}.\mathbf{n}dA - \int \mathbf{J}.\mathbf{E}dV$$

gives a relationship between the complex phasor quantities. Determine this relationship.

18. Arti's Ph.D thesis work:

Maxwell's equations in a rectangular waveguide having anisotropic but homogeneous permittivity; variational formulation of the problem. The permittivity tensor is assumed to have the form

$$\epsilon = \begin{pmatrix} \epsilon_{xx} & \epsilon_{xy} & 0 \\ \epsilon_{yx} & \epsilon_{yy} & 0 \\ 0 & 0 & \epsilon_z \end{pmatrix}$$

The Maxwell equations for the guide corresponding to frequency ω are

$$\nabla \times E = -i\omega\mu H, \nabla \times H = i\omega\epsilon.E$$

where μ is a scalar while ϵ is a 3×3 tensor of the type just described. We have in component form assuming $exp(-\gamma z)$ dependence of the fields on z,

$$E_{z,y} + \gamma E_y = -i\omega\mu H_x, -\gamma E_x - E_{z,x} = -i\omega\mu H_y,$$

$$E_{y,x} - E_{x,y} = -i\omega\mu H_z,$$
$$H_{z,y} + \gamma H_y = i\omega(\epsilon_{xx}E_x + \epsilon_{xy}E_y),$$
$$-\gamma H_x - H_{z,x} = i\omega(\epsilon_{yx}E_x + \epsilon_{yy}E_y),$$
$$H_{y,x} - H_{x,y} = i\omega\epsilon_z E_z$$

These equations can be arranged in matrix form as

$$\mathbf{Af} = \mathbf{g}$$

where

$$\mathbf{f} = [E_x, E_y, H_x, H_y]^T, \mathbf{g} = [-E_{z,x}, -E_{z,y}, H_{z,y}, -H_{z,x}]^T$$

and

$$\mathbf{A} = \begin{pmatrix} \gamma & 0 & 0 & -i\omega\mu \\ 0 & \gamma & i\omega\mu & 0 \\ i\omega\epsilon_{xx} & i\omega\epsilon_{xy} & 0 & -\gamma \\ i\omega\epsilon_{yx} & i\omega\epsilon_{yy} & \gamma & 0 \end{pmatrix}$$

Let (A_{ij}) be the cofactor matrix of A, ie, A_{ij} is the determinant of A after deleting the ith row and j^{th} column. ELementary algebra gives the following:

$$A_{11} = det \begin{pmatrix} \gamma & i\omega\mu & 0 \\ i\omega\epsilon_{xy} & 0 & -\gamma \\ i\omega\epsilon_{yy} & \gamma & 0 \end{pmatrix}$$

$$= \gamma(\gamma^2 + \omega^2\mu\epsilon_{yy})$$

$$A_{12} = det \begin{pmatrix} 0 & i\omega\mu & 0 \\ i\omega\epsilon_{xx} & 0 & -\gamma \\ i\omega\epsilon_{yx} & \gamma & 0 \end{pmatrix}$$

$$= \omega^2\mu\gamma\epsilon_{yx}$$

$$A_{13} = det \begin{pmatrix} 0 & \gamma & 0 \\ i\omega\epsilon_{xx} & i\omega\epsilon_{xy} & -\gamma \\ i\omega\epsilon_{yx} & i\omega\epsilon_{yy} & 0 \end{pmatrix}$$

$$= -i\omega\gamma^2\epsilon_{yx}$$

$$A_{14} = det \begin{pmatrix} 0 & \gamma & i\mu\omega \\ j\omega\epsilon_{xx} & i\omega\epsilon_{xy} & 0 \\ j\omega\epsilon_{yx} & j\omega\epsilon_{yy} & \gamma \end{pmatrix}$$

$$= -j\omega\gamma^2\epsilon_{xx} - j\omega^3\mu d = -j\omega(\omega^2\mu d + \gamma^2\epsilon_{xx})$$

where

$$d = \epsilon_{xx}\epsilon_{yy} - \epsilon_{xy}\epsilon_{yx}$$

$$A_{21} = det \begin{pmatrix} 0 & 0 & j\omega\mu \\ j\omega\epsilon_{xy} & 0 & -\gamma \\ j\omega\epsilon_{yy} & \gamma & 0 \end{pmatrix}$$

$$= -\omega^2\mu\gamma\epsilon_{xy}$$

$$A_{22} = det \begin{pmatrix} \gamma & 0 & -j\omega\mu \\ j\omega\epsilon_{xx} & 0 & -\gamma \\ j\omega\epsilon_{yx} & \gamma & 0 \end{pmatrix}$$

$$= \gamma(\gamma^2 + \omega^2\mu\epsilon_{xx})$$

$$A_{23} = det \begin{pmatrix} \gamma & 0 & -j\omega\mu \\ j\omega\epsilon_{xx} & j\omega\epsilon_{xy} & -\gamma \\ j\omega\epsilon_{yx} & j\omega\epsilon_{yy} & \omega \end{pmatrix}$$

$$= j\omega(\gamma^2\epsilon_{yy} + \omega^2\mu d)$$

where

$$d = \epsilon_{xx}\epsilon_{yy} - \epsilon_{xy}\epsilon_{yx}$$

$$A_{24} = det \begin{pmatrix} \gamma & 0 & 0 \\ j\omega\epsilon_{xx} & j\omega\epsilon_{xy} & 0 \\ j\omega\epsilon_{yx} & j\omega\epsilon_{yy} & \gamma \end{pmatrix}$$

$$= j\omega\gamma^2\epsilon_{xy}$$

$$A_{31} = det \begin{pmatrix} 0 & 0 & -j\omega\mu \\ \gamma & j\omega\mu & 0 \\ j\omega\epsilon_{yy} & \gamma & 0 \end{pmatrix}$$

$$= -j\omega\mu(\gamma^2 + \omega^2\mu\epsilon_{yy})$$

$$A_{32} = det \begin{pmatrix} \gamma & 0 & -j\omega\mu \\ 0 & j\omega\mu & 0 \\ j\omega\epsilon_{yx} & \gamma & 0 \end{pmatrix}$$

$$= -j\omega^3\mu^2\epsilon_{yx}$$

$$A_{33} = det\begin{pmatrix} \gamma & 0 & -j\omega\mu \\ 0 & \gamma & 0 \\ j\omega\epsilon_{yx} & j\omega\epsilon_{yy} & 0 \end{pmatrix}$$
$$= -\omega^2\mu\gamma\epsilon_{yx}$$

$$A_{41} = det\begin{pmatrix} 0 & 0 & -j\gamma\omega \\ \gamma & j\omega\mu & 0 \\ j\omega\epsilon_{xy} & 0 & -\gamma \end{pmatrix}$$
$$= -j\omega^3\gamma\mu\epsilon_{xy}$$

$$A_{42} = det\begin{pmatrix} \gamma & 0 & -j\omega\mu \\ 0 & j\omega\mu & 0 \\ j\omega\epsilon_{xx} & 0 & -\gamma \end{pmatrix}$$
$$= -j\omega\mu(\gamma^2 + \omega^2\mu\epsilon_{xx})$$

$$A_{43} = det\begin{pmatrix} \gamma & 0 & -j\omega\mu \\ 0 & \gamma & 0 \\ j\omega\epsilon_{yx} & j\omega\epsilon_{yy} & 0 \end{pmatrix}$$
$$= -\omega^2\mu\gamma\epsilon_{yx}$$

and finally,

$$A_{44} = det\begin{pmatrix} \gamma & 0 & 0 \\ 0 & \gamma & j\omega\mu \\ j\omega\epsilon_{xx} & j\omega\epsilon_{xy} & 0 \end{pmatrix}$$
$$= \omega^2\gamma\mu\epsilon_{xy}$$

To calculate A^{-1}, we also need the determinant of A. It evaluates to

$$detA = \gamma.A_{11} + j\omega\mu.A_{14}$$
$$= \gamma(\gamma(\gamma^2 + \omega^2\mu\epsilon_{yy})) + j\omega\mu(-j\omega\gamma^2\epsilon_{xx} - j\omega^3\mu d)$$
$$= \gamma^2(\gamma^2 + \omega^2\mu\epsilon_{yy}) + \omega^2\mu\gamma^2\epsilon_{xx} + \omega^4\mu^2 d$$
$$= \gamma^2(\gamma^2 + \omega^2\mu\epsilon_{yy} + \omega^2\mu\epsilon_{xx}) + \omega^4\mu^2 d$$
$$= \gamma^2(\gamma^2 + \omega^2\mu(\epsilon_{xx} + \epsilon_{yy})) + \omega^4\mu^2 d$$
$$\mathbf{A}^{-1} = (detA)^{-1}(((-1)^{i+j}A_{ji}))_{1\le i,j\le 4}$$
$$= \begin{pmatrix} A_{11} & -A_{21} & A_{31} & -A_{41} \\ -A_{12} & A_{22} & -A_{32} & A_{42} \\ A_{13} & -A_{23} & A_{33} & -A_{43} \\ -A_{14} & A_{24} & -A_{34} & A_{44} \end{pmatrix}$$

Thus, from the equation

$$\mathbf{f} = \mathbf{A}^{-1}\mathbf{g}$$

we get with the notation $\Delta = detA$,

$$E_x = \Delta^{-1}(-A_{11}E_{z,x} + A_{21}E_{z,y} + A_{31}H_{z,y} - A_{41}H_{z,x})$$

$$E_y = \Delta^{-1}(A_{12}E_{z,x} - A_{22}E_{z,y} + A_{32}H_{z,y} - A_{42}H_{z,x})$$

We shall henceforth assume that the modes that propagate inside the waveguide are the TM modes so that $H_z = 0$. This gives

$$E_x = \Delta^{-1}(-A_{11}E_{z,x} + A_{21}E_{z,y}), E_y = \Delta^{-1}(A_{12}E_{z,x} - A_{22}E_{z,y})$$

Note that

$$\frac{A_{11}}{\Delta} = \frac{\gamma(\gamma^2 + \omega^2\mu\epsilon_{yy})}{\gamma^2(\gamma^2 + \omega^2\mu(\epsilon_{xx} + \epsilon_{yy})) + \omega^4\mu^2 d}$$
$$= -F_1(\omega, \gamma)$$

say.

$$\frac{A_{21}}{\Delta} = \frac{-\omega^2\mu\gamma\epsilon_{xy}}{\gamma^2(\gamma^2 + \omega^2\mu(\epsilon_{xx} + \epsilon_{yy})) + \omega^4\mu^2 d} = F_2(\omega, \gamma)$$

say.

$$\frac{A_{12}}{\Delta} = \frac{\omega^2\mu\gamma\epsilon_{yx}}{\gamma^2(\gamma^2 + \omega^2\mu(\epsilon_{xx} + \epsilon_{yy})) + \omega^4\mu^2 d} = F_3(\omega, \gamma)$$

say.

$$\frac{A_{22}}{\Delta} = \frac{\gamma(\gamma^2 + \omega^2\mu\epsilon_{xx})}{\gamma^2(\gamma^2 + \omega^2\mu(\epsilon_{xx} + \epsilon_{yy})) + \omega^4\mu^2 d}$$
$$= -F_4(\omega, \gamma)$$

say. With these substitutions, we get

$$E_x = F_1E_{z,x} + F_2E_{z,y}, E_y = F_3E_{z,x} + F_4E_{z,y}$$

Let us continue the calculations for general fields, ie, without making the TM or TE assumption. The final aim is to obtain the field equations from a variational principle as this would enable the finite element method to be applicable.

$$\mathbf{f} = \mathbf{A}^{-1}\mathbf{g}$$

so that

$$f_1 = \Delta^{-1}(A_{11}g_1 - A_{21}g_2 + A_{31}g_3 - A_{41}g_4)$$

$$f_2 = \Delta^{-1}(-A_{12}g_1 + A_{22}g_2 - A_{32}g_3 + A_{42}g_4)$$

$$f_3 = \Delta^{-1}(A_{13}g_1 - A_{23}g_2 + A_{33}g_3 - A_{43}g_4)$$

$$f_4 = \Delta^{-1}(-A_{14}g_1 + A_{24}g_2 - A_{34}g_3 + A_{44}g_4)$$

19. Millman and Halkias, Integrated Electronics:The initial chapters deal with the properties of a simple pn junction diode. Later on in the book, transistor circuits are analyzed, then operational

amplifiers and finally in the appendix, the book discusses the physics of semiconductors. If E is the electric field inside a semiconductor and μ the mobility of the electrons/holes, then the steady state velocity of the carriers is given by $v = \mu E$. This is also called the drift velocity. If τ is the average time in between collisions of two carriers, then by Newton's law, the acceleration of the particles is eE/m where m is the effective mass of the carrier. It follows that the drift velocity, which is taken to be the average velocity of the carrier between two successive collisions is given by $eE\tau/2m$ and therefore the mobility of the carrier is $\mu = e\tau/2m$. Apart from drift velocity, there is also a diffusion velocity for minority carriers. Suppose holes (p) are the minority carriers. Then, if $p(t,x)$ denotes their concentration and D_p the diffusion constant for the holes, the flux of the holes is given by $-D_p\nabla p$ the corresponding diffusion hole current density is $J_{pd} = -eD_p\nabla p$. The total hole current is the sum of the diffusion and drift currents ie,

$$J_p = -eD_p\nabla p + \mu epE$$

Similarly for electrons diffusing as minority carriers in the sea of holes, the current density is given by

$$J_n = eD_n\nabla n + \mu enE$$

Under conditions of equilibrium, we have $J_p = 0$ and hence we get

$$D_p\nabla(log(p)) = -\mu\nabla\Phi$$

where Φ is the electrostatic potential. Integrating this equation and using the fact that both p and Φ vanish at ∞ gives us

$$D_p.log(p) = -\mu\Phi$$

so that

$$p = C.exp(-\mu\Phi/D_p)$$

However under equilibrium, we have from Boltzmann's law,

$$p = C.exp(-e\Phi/kT)$$

Thus we obtain Einstein's formula for D_p:

$$e/kT = \mu/D_p$$

or

$$D_p = \mu kT/e$$

Suppose that p_0 is the equilibrium hole conservation then from charge conservation and using $p_{,t} = (p_0 - p)/\tau$, we have

$$D_p p'' = (p_0 - p)/\tau$$

which has an oscillatory solution with characteristic length

$$L_p = (\tau/D_p)^{1/2}$$

Consider now the more general problem of determining the hole and electron concentrations in a semiconductor. The relevant equations to be solved are

$$J_p = -eD_p\nabla p + p\mu eE,$$

$$J_n = eD_n\nabla n + n\mu eE,$$

$$\nabla.J_p + ep_{,t} = 0,$$

$$\nabla.J_n - en_{,t} = 0,$$

$$\nabla.E = e(p-n)/\epsilon,$$

$$E = -\nabla\Phi$$

These are twelve equations for the twelve unknowns

$$E(3components), p, n, J_p(threecomponents), J_n(threecomponents), \Phi$$

The boundary condition is that the difference between Φ at the two ends of the semiconductor should be the applied voltage.

20. Balanis, Antenna Theory:

Salient features: Polarization of a plane wave. The direction of propagation of the wave is $\hat{z}$. The x and y components of the electric field at a given time have the form

$$E_x = A.cos(kz + \phi_x), E_y = B.cos(kz + \phi_y)$$

Linear polarization occurs iff $\phi_x - \phi_y = n\pi$ where n is an integer. In this case, we have $E_y/E_x = \pm B/A$. If $\phi_x - \phi_y = 2n\pi \pm \pi/2$, and $A = B$, we have circular polarization. In that case, E_x, E_y are related via the equation

$$E_x^2 + E_y^2 = A^2$$

In all the other cases, we have elliptic polarization. Expressions for the semimajor and semiminor axes of the ellipse in terms of the parameters A, B, ϕ_x, ϕ_y have been derived. One way to calculate the semimajor and semiminor axes is to first derive the equation of the ellipse. It is obtained by eliminating the variable z from the two equations. This can be obtained as follows:

$$E_y = B.cos(kz + \phi_x + \phi_y - \phi_x) = B.(cos(kz + \phi_x).cos(\phi_y - \phi_x) - sin(kz + \phi_x).sin(\phi_y - \phi_x))$$

$$= B.cos(\phi_y - \phi_x)E_x/A - B.sin(\phi_y - \phi_x)\sqrt{1 - E_x^2/A^2}$$

so that

$$(E_y - B.cos(\phi_y - \phi_x)E_x/A)^2 = B^2 sin^2(\phi_y - \phi_x)^2(1 - E_x^2/A^2)$$

This equation can be put in the form

$$\alpha.E_x^2 + \beta.E_y^2 + 2\gamma E_x E_y = 1$$

and then a rotation can be applied to the vector (E_x, E_y) to bring it into standard form

$$\xi^2/a^2 + \eta^2/b^2 = 1$$

where

$$E_x = \xi.cos(\theta) - \eta.sin(\theta), E_y = \xi.sin(\theta) + \eta.cos(\theta)$$

Different kinds of effective areas of an antenna: V_T is the voltage developed at the terminals of the antenna. R_r is the radiation resistance, R_L is the load resistance and R_T is the resistance at the terminals. The total resistance presented by the antenna is given by $R_r+R_T+R_L$ and the total reactance is X_A+X_T where X_A is the aperture reactance. The current at the antenna terminals is given by

$$I_T = \frac{V_T}{R_r + R_T + R_L + j(X_A + X_T)}$$

and if W_i denotes the incident power density, then the effective aperture area of the antenna is given by the ratio of the power developed at its terminals to the incident power density:

$$A_e = |I_T|^2 R_T / 2W_i = \frac{|V_T|^2}{2W_i(((R_r + R_T + R_L)^2 + (X_A + X_T)^2)}$$

$R_r + R_L$ is the aperture resistance. The effective aperture area is a maximum when matching takes place, ie, when $X_A = -X_T$ and $R_r + R_L = R_T$. The maximum aperture area is therefore given by

$$A_{em} = \frac{|V_T|^2}{8(R_r + R_L)}$$

The loss area of the antenna is the power dissipated in the load resistance R_L divided by the incident power. It equals

$$A_l = \frac{|V_T|^2 R_L}{2((R_r + R_T + R_L)^2 + (X_A + X_T)^2)}$$

Its maximum value occurs when matching takes place and it is given by

$$A_{lm} = \frac{|V_T|^2 R_L}{8(R_r + R_L)^2}$$

Finally, the scattering area is given by

$$A_s = \frac{|V_T|^2 R_r}{2((R_r + R_T + R_L)^2 + (X_A + X_T)^2)}$$

and its maximum value is

$$A_{sm} = \frac{|V_T|^2 R_r}{8(R_r + R_L)^2}$$

The capture area is the sum of the three areas. Suppose a transmitter antenna emits power P_t. If it is an isotropic source, then the power density at a distance R from the antenna is given by $\frac{P_t}{4\pi R^2}$. If the source is not isotropic, then it will have a directivity $D_t(\theta, \phi)$ in the direction (θ, ϕ). Note that $D_t(\theta, \phi)$ equals the actual power density in the given direction divided by the averaged power density over all the directions. Suppose $U(\theta, \phi)$ is the power per unit solid angle. Note that U is obtained by multiplying the Poynting vector by R^2. Then,

$$D_t(\theta, \phi) = \frac{4\pi U(\theta, \phi)}{\int_0^{\pi} \int_0^{2\pi} U(\theta, \phi) sin(\theta) d\theta d\phi}$$

The power density at the receiver located at a distance R from the transmitter is given by

$$\frac{P_t D_t}{4\pi R^2}$$

and if the receiver has an effective area A_r, the power collected by it is given by

$$P_r = \frac{P_t D_t A_r}{4\pi R^2}$$

Thus,

$$\frac{P_r}{P_t} = \frac{D_t A_r}{4\pi R^2}$$

If the places of the transmitter and receiver are interchanged, then we get

$$\frac{P_r}{P_t} = \frac{D_r A_t}{4\pi R^2}$$

The ratio is the same in both the cases. Note that in the latter case, the antenna that was originally operating as the receiver, emits a power P_t and the antenna that was originally operating as the transmitter collects power P_r. Thus, we get the fundamental reciprocity equation

$$D_r A_t = D_t A_r$$

or

$$\frac{D_r}{A_r} = \frac{D_t}{A_t}$$

In case the transmitter antenna is isotropic, we have $D_t = 1$ and we get for the effective area of an isotropic source

$$A_{is} = \frac{A_r}{D_r}$$

This formula gives us a method for computing A_{is}. We can compute A_r and D_r for an infinitesimal dipole by the usual method. Note that for an infinitesimal dipole, the power density is proportional to $sin^2(\theta)$. Its average value is given by

$$\frac{1}{4\pi}.2\pi \int_0^\pi sin^3(\theta)d\theta$$

$$= \frac{1}{2}\int_0^\pi (3.sin(\theta) - sin(3\theta))d\theta/4 = \frac{1}{8}(6 - 2/3) = 16/24 = 2/3$$

and its maximum value is 1 so that its maximum directivity is $3/2 = 1.5$. The radiation resistance R_r for the infinitesimal dipole can also be computed easily. Let E be the electric field incident on the dipole. Then, the incident power density is $E^2/2\eta$ where η is the impedance of space. With L denoting the length of the dipole, the voltage developed at its terminals is EL and hence the power collected by the antenna in the receiver mode is $E^2L^2/2R_r$. This gives $A_{rm} = (E^2L^2/2R_r)/(E^2/2\eta) = L^2\eta/R_r$

$$A_{rm}/D_{rm} = L^2\eta/1.5R_r$$

and this is the effective area of an isotropic source.

The concept of electric vector potential **F** and the magnetic current density **M** has been given. These are respectively the duals of the magnetic vector potential **A** and the electric current density **J**. In the presence of magnetic current density **M** and electric current density **J** but no electric or magnetic charges, the Maxwell equations can be expressed as

$$\nabla.E = 0, \nabla.H = 0, \nabla \times E = -\mu\frac{\partial H}{\partial t} - M, \nabla \times H = \epsilon\frac{\partial E}{\partial t} + J$$

The first two equations imply the existence of two vector fields F, A respectively called the electric and magnetic vector potential such that

$$E = -\frac{1}{\epsilon}\nabla \times F, H = \frac{1}{\mu}\nabla \times A$$

We now solve for the fields under two cases (a) $M = 0, J \neq 0$ and (b) $J = 0, M \neq 0$. In the first case, we get from the third Maxwell equation

$$\nabla \times E = -\nabla \times \frac{\partial A}{\partial t}$$

or

$$\nabla \times (E + \frac{\partial A}{\partial t}) = 0$$

and this implies the existence of a scalar field Φ_e called the electric scalar potential such that

$$E = -\nabla\Phi_e - \frac{\partial A}{\partial t}$$

In the second case, we have

$$\nabla \times H = \frac{\partial}{\partial t}\nabla \times F$$

so that

$$\nabla \times (H + \frac{\partial F}{\partial t}) = 0$$

implying the existence of a scalar field Φ_m, called the magnetic scalar potential such that

$$H = -\nabla\Phi_m - \frac{\partial F}{\partial t}$$

In the presence of both electric and magnetic current densities, the total electric and magnetic fields can thus be expressed using the superposition principle as

$$E = -\nabla\Phi_e - \frac{\partial A}{\partial t} - \frac{1}{\epsilon}\nabla \times F$$

$$H = \frac{1}{\mu}\nabla \times A - \nabla\Phi_m - \frac{\partial F}{\partial t}$$

Antenna arrays:This topic has been discussed in considerable depth by Balanis. The Schelnukoff synthesis method has been stressed upon. Linear arrays have been considered here. Specifically, if sensor elements are located at points $(0,0,md), m = 0,1,...,N-1$ along the z axis and excited by currents $I_n exp(jn\psi), n = 0,1,...,N-1$, then with θ denoting the usual elevation angle, ie, the angle between the z axis and the point where the field is being evaluated, the expression for the field (each sensor is assumed to be an isotropic radiator) is given by

$$E = \sum_{m=0}^{N-1} I_m exp(jm(kd.cos(\theta) + \psi))$$

Consider the polynomial

$$P(z) = \sum_{m=0}^{N-1} I_m z^m$$

Then, the field amplitude pattern can be expressed as

$$E(\theta, \phi) = P(exp(j(kd.cos(\theta) + \psi))$$

Suppose $P(z)$ has zeros on the unit circle at the points $exp(j\phi_k), k = 1,2,...,r$. Then, the nulls of the antenna pattern will be given at the values of θ satisfying

$$kd.cos(\theta) + \psi = \phi_k$$

or

$$\theta = cos^{-1}((\lambda/2\pi d)(\phi_k - \psi)), k = 1,2,...,r$$

Suppose that the polynomial $P(z)$ has a zero $r.exp(j\alpha)$ where r is close to unity but not necessarily unity. Then a null of the antenna pattern will be close to θ given by $kd.cos(\theta) + \psi = \alpha$.

Computation of the radiation field pattern of a circular loop carrying sinusoidal current. Let I denote the current phasor. Assume that the radius of the loop is a and that the loop is on the xy plane with center coinciding with the origin. The far field vector potential phasor is given by

$$\mathbf{A}(\mathbf{r}) = \frac{\mu I.a.exp(-jkr)}{4\pi r} \int_0^{2\pi} (-sin(\phi')\hat{x} + cos(\phi')\hat{y})exp(-jka.sin(\theta)cos(\phi' - \phi))d\phi'$$

We have used the general formula for a thin wire"

$$\mathbf{A}(\mathbf{r}) = \frac{\mu.exp(-jkr)}{4\pi r} \int_\Gamma I(\mathbf{r}')exp(jk\hat{r}.\mathbf{r}')d\mathbf{r}'$$

where $\mathbf{r}$ is the point where the field is being evaluated and $\mathbf{r}'$ is a point in the current source. Note that for the circular loop,

$$\hat{r} = \hat{x}.cos(\phi)sin(\theta) + \hat{y}.sin(\phi)sin(\theta) + \hat{z}.cos(\theta)$$

$$\mathbf{r}' = a.cos(\phi')\hat{x} + a.sin(\phi')\hat{y}$$

so that

$$\hat{r}.\mathbf{r}' = a.sin(\theta).cos(\phi' - \phi)$$

Let (ρ,ϕ,z) denote cylindrical coordinates. Then we have from the above,

$$A_\rho = A_x.cos(\phi) + A_y.sin(\phi) =$$

$$\frac{\mu Ia.exp(-jkr)}{4\pi r}\int_0^{2\pi} sin(\phi-\phi')exp(-jka.sin(\theta).cos(\phi'-\phi)d\phi' = 0$$

as follows from by replacing the integration variable ϕ' by $\xi = cos(\phi'-\phi)$. It is clear from the formula that $A_z = 0$. Thus, the only non-vanishing component of the vector potential in cylindrical coordinates is A_ϕ and this component is given by

$$A_\phi = -A_x.sin(\phi) + A_y.cos(\phi) =$$

$$\frac{\mu Ia.exp(-jkr)}{4\pi r}\int_0^{2\pi} cos(\phi-\phi')exp(-jka.sin(\theta)cos(\phi'-\phi))d\phi'$$

$$= \frac{\mu Ia.exp(-jkr)}{4\pi r} int_0^{2\pi} cos(u).exp(-jka.sin(\theta).cos(u))du$$

The integral equals a constant times $J_1(ka.sin(\theta))$. It follows that

$$A_\phi(r,\theta) = \frac{\mu Ia}{4\pi r}exp(-jkr)J_1(ka.sin(\theta))$$

Letting $F(\theta) = \frac{\mu Ia}{4\pi}J_1(ka.sin(\theta))$, it follows that

$$A_\phi(r,\theta) = F(\theta)exp(-jkr)/r$$

The far field electric and magnetic field patterns can immediately be calculated from this. We start with the expression for curl in spherical polar coordinates.

$$curl\mathbf{F} = \frac{1}{r^2.sin(\theta)}((r.sin(\theta)F_\phi)_{,\theta} - (rF_\theta)_{,\phi})\hat{r} + \frac{1}{r.sin(\theta)}(F_{r,\phi} - (r.sin(\theta)F_\phi)_{,r})\hat{\theta}$$

$$+\frac{1}{r}((r.F_\theta)_{,r} - F_{r,\theta}))$$

Suppose that the only nonvanishing component of $\mathbf{F}$ is F_ϕ. Then,

$$curl\mathbf{F} = \frac{1}{r^2.sin(\theta)}((r.sin(\theta)F_\phi)_{,\theta}\hat{r} - \frac{1}{r.sin(\theta)}(r.sin(\theta)F_\phi)_{,r}\hat{\theta}$$

In our case, we replace $\mathbf{F}$ by $\mathbf{A}$ and note from the above formula that the $\hat{r}$ component does not have any term proportional to $1/r$ and hence the radiation component ($1/r$ dependence) of the magnetic field is given by

$$\mathbf{H} = curl\mathbf{A}/\mu = -\frac{1}{\mu r}(rA_\phi)_{,r}\hat{\theta}$$

$$= -\frac{1}{\mu}F(\theta)(exp(-jkr)/r)\hat{\theta}$$

In other words, the only non-vanishing component of the magnetic field is given by

$$H_\theta = -\frac{1}{\mu}F(\theta)exp(-jkr)/r$$

We get the corresponding only nonvanishing component of the electric field as

$$E_\phi = -\eta H_\theta = \frac{\eta.exp(-jkr)}{\mu.r}F(\theta)$$

Evaluation of $F(\theta)$: The relevant integral is

$$\int_0^{2\pi} cos(u)exp(jv.cos(u))du = \int_0^{\pi} cos(u).exp(jv.cos(u))du + \int_{\pi}^{2\pi} cos(u).exp(jv.cos(u))du$$

The second integral on the right becomes after replacing the integration variable by $t = 2\pi - u$,

$$-\int_{\pi}^{0} cos(t)exp(jv.cos(t))dt$$

and hence

$$\int_0^{2\pi} cos(u).exp(jv.cos(u))du = 2\int_0^{\pi} cos(u).exp(jv.cos(u))du$$

We wish to show that the function

$$f(v) = \int_0^{\pi} cos(nu)exp(jv.cos(u))du$$

satisfies the Bessel equation of order n where n is any non-negative integer.

$$f'(v) = j\int_0^{\pi} cos(u)cos(nu)exp(jv.cos(u))du,$$

$$f''(v) = -\int_0^{\pi} cos^2(u)cos(nu)exp(jv.cos(u))du$$

The method of images has been discussed for antenna radiation problems in good detail. The simplest example illustrating this method is the problem of computing the radiation fields produced by an infinitesimal vertical dipole located at a height h above an infinite conducting plane that coincides with the xy plane. Assume that the dipole has length dl. We may assume that the z coordinate of the dipole extends from h to $h+dl$. Since current flows along the positive z direction, we can assume that the point on the dipole with z coordinate h has negative charge $-Q(t)$ and that the point on the dipole with z coordinate $h+dl$ has charge $+Q(t)$. The current through the dipole is $i(t) = Q'(t)$. The corresponding image dipole must be constructed below the $z = 0$ plane in such a way as to guarantee zero tangential electric field on the $z = 0$ plane. This happens provided that the image dipole has z coordinates extending from $z = -h$ upto $z = -h - dl$. From image theory, the point with $z = -h$ will have charge $+Q(t)$ and the point with $z = -h - dl$ will have charge $-Q(t)$. This means that current in the image dipole of amount $i(t) = Q'(t)$ will also flow in the positive z direction. Let r_1 and r_2 denote respectively the distances from the point where the field is being calculated to the two dipoles, the original one and the image one. Then we have approximately for $r >> h$,

$$r_1 \approx r - h.cos(\theta), r_2 = r + h.cos(\theta)$$

so that the vector potential is $A = A_z\hat{z}$ where

$$A_z(r,\theta) \approx \frac{\mu I.exp(-jkr)}{4\pi r}(exp(jkh.cos(\theta)) + exp(-jkh.cos(\theta)))$$

$$= \frac{\mu I.exp(-jkr)}{2\pi r}cos(kh.cos(\theta))$$

This is actually a consequence of the pattern multiplication formula applied to the two element array consisting of two dipoles. The total pattern of the array is a product of the individual pattern and the array factor. The electric field is then given by

$$E_\theta = (j\omega\mu I.exp(-jkr)/2\pi r)cos(kh.cos(\theta))sin(\theta)$$

Suppose $h = \lambda/2$. Then the electric field pattern is proportional to $cos(\pi.cos(\theta)).sin(\theta)$ and its nulls appear at $\theta = 0, \pi, \pi/3$.

One of the problems deals with the computation of the field pattern produced by a circular loop placed above a perfectly conducting plane. Two configurations are possible here, one horizontal and two vertical. In the horizontal configuration, the image loop has current flowing in the opposite direction. Using the pattern multiplication formula, the electric field has the form

$$E_\phi(r,\theta) = C.J_1(ka.sin(\theta))cos(kh.cos(\theta))exp(-jkr)/r$$

and this formula can be used to evaluate the radiation resistance of the antenna. Let us evaluate the field pattern when the loop is vertical. First we compute the vector potential for a verically located circular loop. Assume that the loop equation is given by $y = 0, x^2 + (z-h)^2 = R^2$. Any point on the loop can be represented as

$$\mathbf{r}' = (x', y', z') = (R.cos(\alpha), 0, h + R.sin(\alpha))$$

The length element is then given by

$$d\mathbf{r}' = d(x', y', z') = R(-sin(\alpha), 0, cos(\alpha))d\alpha$$

and the unit radius vector of the point where the field is being computed is given by

$$\hat{r} = \hat{x}.cos(\phi)sin(\theta) + \hat{y}.sin(\phi)sin(\theta) + \hat{z}.cos(\theta)$$

so that

$$\hat{r}.\mathbf{r}' = R(cos(\alpha)cos(\phi)sin(\theta) + sin(\alpha)cos(\theta)) + h.cos(\theta)$$

so that the far field vector potential has the form

$$\mathbf{A}(\mathbf{r}') = \frac{\mu I.R.exp(-jkr)}{4\pi r}exp(jkh.cos(\theta))\int_0^{2\pi} exp(jkR(cos(\alpha)cos(\phi)sin(\theta) + sin(\alpha)cos(\theta)))$$

$$(-\hat{x}.sin(\alpha) + \hat{z}.cos(\alpha))d\alpha$$

In fact, this pattern can also be obtained by simply applying a rotation and a translation to the pattern produced by a horizontal circular loop located in the xy plane.

One of the problems deals with connecting a transmission line to a linear antenna and obtaining the VSWR of the transmission line at the input terminals under matching. Suppose R_0 is the characteristic impedance of the line. Then resonance occurs when the characteristic impedance of the line. The total impedance presented by the antenna is the sum of the radiation resistance, the loss resistance and the load resistance. Under matching condition, there is no reflected wave. Consider a transmission line of length L terminated by a load impedance Z_L. The voltage and current along the line are respectively given by the formulas

$$V(z) = V_+exp(-\gamma z) + V_-exp(\gamma z), I(z) = Z_0^{-1}(V_+exp(-\gamma z) + V_-exp(\gamma z))$$

so that at $z = L$, we have with $r = V_+/V_-$,

$$(r.exp(-2\gamma L) + 1)/(r.exp(-2\gamma L) - 1) = Z_L/Z_0$$

giving

$$r.exp(-2\gamma L)(1 - Z_L/Z_0) = -(Z_L/Z_0 + 1)$$

For no reflected wave to exist we must have $V_- = 0$ and that gives $Z_L = Z_0$, ie, the terminating impedance must be the same as the characteristic impedance. In this case, we have

$$VSWR = 1$$

Equivalent sources: J_1, M_1 are the sources of electric current and magnetic current in region one where the conductivity is σ_1, the permittivity is ϵ_1 and the permeability μ_1. Let E_1, H_1 denote the electric and magnetic fields in this region. Then we have

$$\nabla \times E_1 = -j\omega\mu_1 H_1 - M_1, \nabla \times H_1 = (\sigma_1 + j\omega\epsilon_1)E_1 + J_1$$

We define

$$Z_1 = j\omega\mu_1, Y_1 = \sigma_1 + j\omega\epsilon_1$$

Then in region one, we have

$$\nabla \times E_1 = -Z_1 H_1 - M_1, \nabla \times H_1 = Y_1 E_1 + J_1$$

Region two is the volume inside a body characterized by $\epsilon_2, \mu_2, \sigma_2$. Letting Z_2, Y_2 denote the corresponding impedance and admittance, we have in this region since $J_2 = 0, M_2 = 0$,

$$\nabla \times E_2 = -Z_2 H_2, \nabla \times H_2 = Y_2 E_2$$

We can write these equations as

$$\nabla \times E_2 = -Z_1 H_2 - M_2, \nabla \times H_2 = Y_1 E_2 + J_2$$

where

$$M_2 = (Z_2 - Z_1)H_2, J_2 = (Y_2 - Y_1)E_2$$

We can view the body as being replaced by these equivalent sources M_2, J_2 and calculate the fields from them. (See Jordan and Balmain, Electromagnetic waves and radiating systems). The above equations tell us how to compute E_2, H_2 given J_2, M_2. However these formulae are not practicably applicable since J_2, M_2 themselves depend on E_2, H_2. Therefore one uses the Born approximation in which J_2, M_2 are computed using E_1, H_1 in place of E_2, H_2. This approximation is valid provided $Z_2 - Z_1, Y_2 - Y_1$ are small.

Far field computation for aperture antennas are based on introducing equivalent sources J_S, M_S on the aperture surface and then

$$A(r) = \frac{\mu.exp(-jkr)}{4\pi r}\int_S J_S(r')exp(jk\hat{r}.r')dS(r'),$$

$$F(r) = \frac{\epsilon}{4\pi.exp(-jkr)}4\pi r\int_S M_S(r')exp(jk\hat{r}.r')dS(r')$$

the integrals being evaluated over the surface of the body. Then, the distant electric and magnetic fields due to the electric current source has components

$$E_\theta = -j\omega A_\theta, H_\phi = E_\theta/\eta,$$

$$E_\phi = -j\omega A_\phi, H_\theta = -E_\phi/\eta$$

and those due to the magnetic current source are

$$H_\theta = -j\omega F_\theta, E_\phi = -\eta H_\theta$$

$$H_\phi = -j\omega F_\phi, E_\theta = -j\omega\eta F_\phi$$

The total electric field due to both the electric and magnetic current sources is then

$$E_\theta = -j\omega A_\theta + j\omega\eta F_\theta$$

$$E_\phi = -j\omega A_\phi + j\omega\eta F_\phi$$

$$H_\phi = E_\theta/\eta, H_\theta = -E_\phi/\eta$$

with $\eta = \sqrt{\mu/\epsilon}$.

Linear current source for synthesis of patterns. Assume that the line source is of length L and carries a current $I(z)exp(jk_z z), |z| \leq L/2$. The space factor of the pattern is then given by

$$SF(\theta) = \int_{-L/2}^{L/2} I(z')exp(jk_z z')exp(jkz'cos(\theta))dz'$$

In terms of the Fourier transform

$$\hat{I}(u) = \int_{-L/2}^{L/2} I(z')exp(juz')dz'$$

of $I(.)$, we can write

$$SF(\theta) = \hat{I}(k_z + k.cos(\theta))$$

The technique of design is that given the space factor of a pattern function $F(\theta)$, we define a function

$$f(u) = F(cos^{-1}((u - k_z)/k)), |u - k_z| \leq k$$

and compute

$$I(z') = \frac{1}{2\pi} \int_{k_z-k}^{k_z+k} f(u)exp(-juz')du$$

and this completes the design. Consider the special case when the current amplitude $I(z)$ is a constant I_0. Then,

$$SF(\theta) = I_0 \int_{-L/2}^{L/2} exp(jz'(k_z + k.cos(\theta)))dz'$$

$$= 2I_0.\{sin[(k_z + k.cos(\theta))L/2]/(k_z + k.cos(\theta))\}$$

For the special case when $k_z = 0$, the pattern satisfies the symmetry condition $SF(\pi - \theta) = SF(\theta)$. We consider now a generalization of this for more general problems of pattern synthesis. Suppose that the current distribution has the form

$$I(z) = \sum_{m=1}^{N} b(m)exp(j\alpha(m)z)$$

where $\{b(m), \alpha(m)\}$ are real numbers to be determined from the given pattern. The space factor for this current distribution is given by

$$SF(\theta) = \sum_{m=1}^{N} b(m) \int_{-L/2}^{L/2} exp(j(\alpha(m) + k.cos(\theta))z')dz'$$

This evaluates to

$$SF(\theta) = \sum_{m=1}^{N} 2b(m)\frac{sin((L/2)(\alpha(m) + k.cos(\theta)))}{(\alpha(m) + k.cos(\theta))}$$

When $cos(\theta) = -\alpha(m)/k$, it follows that $SF(\theta) = b(m)L$ provided that $sin((L/2)(\alpha(n) - \alpha(m))) = 0$ for all $n \neq m$. This happens provided $\alpha(n) = 2\pi n/L$ and hence we get a method for matching a given amplitude pattern by choosing an appropriate current distribution.

Aperture antenna analysis using Fourier transform: The electric field in space produced by a two dimensional aperture can be expressed as a superposition of plane waves:

$$\mathbf{E}(x, y, z) = \int_{-\infty}^{\infty} \int_{-\infty}^{\infty} \mathbf{f}(k_x, k_y)exp(j(k_x x + k_y y + k_z z))dk_x dk_y$$

The electric field must satisfy the Helmholtz equation at frequency $\omega = kc = k/\sqrt{\epsilon\mu}$ and hence

$$k_z = \sqrt{k^2 - k_x^2 - k_y^2}$$

The other condition is that

$$div\mathbf{E} = 0$$

and that gives

$$k_x f_x + k_y f_y + k_z f_z = 0$$

so that

$$f_z = -(k_x f_x + k_y f_y)/k_z = -(k_x f_x(k_x, k_y) + k_y f_y(k_x, k_y))/\sqrt{k^2 - k_x^2 - k_y^2}$$

We have

$$\mathbf{f}(k_x, k_y) = (2\pi)^{-1} \int \mathbf{E}(x, y, 0) exp(-j(k_x x + k_y y)) dx dy$$

and hence the electric field in space can be completely determined from $E_x(x, y, 0), E_y(x, y, 0)$, ie, from the tangential components of the electric field in the plane of the aperture ($z = 0$ plane):

Dolph-Chebysheff array: Recursion relation

$$T_m(z) = 2zT_{m-1}(z) - T_{m-2}(z)$$

and the solution is

$$T_m(z) = cos[m.cos^{-1}(z)]$$

We shall show that T_m satisfies the recurrence relation

$$T_m(z) = 2zT_{m-1}(z) - T_{m-2}(z)$$

Indeed, let $z = cos(u)$. Then,

$$T_m(z) = cos[m.cos(u)] = cos[(m-1)cos(u)]cos(u) - sin[(m-1)cos(u)]sin(u)$$

$$= zT_{m-1} - \sqrt{1 - z^2} P_{m-1}$$

where

$$P_m(z) = sin[(m-1)z]$$

Also

$$P_{m-1}(z) = sin[(m-2)z]cos(u) + cos[(m-2)z]sin(u) = zP_{m-2}(z) + T_{m-2}(z)\sqrt{1 - z^2}$$

This only gives a joint recursion for the pair $[T_m, P_m]$. From the first equation, we get

$$P_{m-1} = (zT_{m-1} - T_m)/\sqrt{1 - z^2}$$

and plugging this into the second gives us

$$(zT_{m-1} - T_m)/\sqrt{1 - z^2} = z(zT_{m-2} - T_{m-1})/\sqrt{1 - z^2} + T_{m-2}\sqrt{1 - z^2}$$

which on simplification gives

$$zT_{m-1} - T_m = z(zT_{m-2} - T_{m-1}) + T_{m-2}(1 - z^2) = T_{m-2} - zT_{m-1}$$

or

$$T_m = 2zT_{m-1} - T_{m-2}$$

This formula gives us a recursive method for calculating the Chebyshev polynomials. Note that for $-1 \le z \le 1$, we have defined $T_m(z)$. This definition can be extended for $z > 1$ and $z < -1$. by replacing cos with $cosh$. Indeed, $cos(iu) = cosh(u)$ so that $-i.cos^{-1}(z) = cosh^{-1}(z)$ and

$$cosh(m.cosh^{-1}(z)) = cosh(-im.cos^{-1}(z)) = cos(m.cos^{-1}(z))$$

which justifies the extension

$$T_m(z) = cosh(m.cosh^{-1}(z)), z > 1, z < -1$$

To design the Cheybshev array, we first choose an array factor of the form $\sum a(m).cos(mu)$ where $u = \pi.d.cos(\theta)/\lambda$ and expand $cos(mu)$ in powers of $cos(u)$. Choose an order m (positive integer) and determine z_0 so that $T_m(z_0) = R_0$. This z_0 corresponds to the location of the main lobe of the antenna pattern. Specifically, if z_1 denotes the null of $T_m(z)$ closest to $z = 1$, then the mainlobe is focussed over the region $[z_1, z_0]$. We then substitute $cos(u) = z/z_0$ in the array factor of the original polynomial and equate this to $T_m(z)$, ie,

$$\sum_k a(k)T_k(z/z_0) = T_m(z)$$

and determine $a(k)'s$ from this equation. The given pattern is then

$$T_m(z_0.cos(u)) = T_m(z_0.cos(\pi d.cos(\theta))/\lambda)$$

. Note that this is also the same as $\sum_k a(k)T_k(cos(\pi d.cos(\theta)/\lambda))$. This pattern has a mainlobe centred at $\theta = \pi/2$ and extends upto θ_1 where $\pi.d.cos(\theta_1)/\lambda = z_1/z_0$. The number m must be one less than the total number of elements of the designed array.

Rectangular microstrip antenna:The slot is rectangular of height h along the x direction, length a along the y direction and length b along the z direction. The walls of the cavity are assumed to be perfect magnetic conductors so that the tangential components of the magnetic fields on these surfaces vanish. Note that the walls are the surfaces $y = 0, a$ and $z = 0, b$. The top and bottom surfaces of the cavity are perfect electric conductors so that the tangential components of the electric fields on these surfaces vanish. Note that these surfaces are $z = 0, h$. The only non-vanishing component of the vector potential is A_x. This leads to $H_x = 0$ which means that only transverse magnetic field TM modes exist inside the conductor. The boundary conditions read

$$H_y|_{z=0,b} = H_z|_{y=0,a} = 0,$$

$$E_y|_{x=0,h} = E_z|_{x=0,h} = 0$$

A_x satisfies the Helmholtz equation which reads

$$\nabla^2 A_x + \omega^2 \mu\epsilon.A_x = 0$$

The electric scalar potential is obtained from the Lorentz gauge condition:

$$\frac{\partial A_x}{\partial x} = -j\omega\epsilon\mu V$$

so that

$$V = (j/\omega\epsilon\mu)\frac{\partial A_x}{\partial x}$$

and it follows that

$$E_x = -\frac{\partial V}{\partial x} - j\omega A_x$$

$$= -(j/\omega\epsilon\mu)(\frac{\partial^2 A_x}{\partial x^2} + k^2 A_x)$$

$$E_y = -\frac{\partial V}{\partial y} = -(j/\omega\epsilon\mu)\frac{\partial^2 A_x}{\partial x \partial y}$$

$$E_z = -\frac{\partial V}{\partial z} = -(j/\omega\epsilon\mu)\frac{\partial^2 A_x}{\partial x \partial z}$$

The magnetic field components are

$$H_x = 0, H_y = \frac{1}{\mu}\frac{\partial A_x}{\partial z}, H_z = -\frac{1}{\mu}\frac{\partial A_x}{\partial y}$$

We can verify that the solution to the vector potential

$$A_x = C.cos(k_x x)cos(k_y y)cos(k_z z), k_x = m\pi/h, k_y = n\pi/a, k_z = p\pi/b$$

where m, n, p are integers satisfies the Helmholtz equation with

$$k^2 = \omega^2\mu\epsilon = (m\pi/h)^2 + (n\pi/a)^2 + (p\pi/b)^2$$

the resonant frequencies are therefore discrete. To check the boundary conditions, we observe that

$$E_x = -(j/\omega\epsilon\mu)(k^2 - k_x^2)C.cos(k_x x)cos(k_y y)cos(k_z z)$$

$$E_y = -(j\omega\epsilon\mu)(-k_x)(k_y)C.sin(k_x x)sin(k_y y)cos(k_z z)$$

and this is zero at $x = 0, h$. Likewise,

$$E_z = -(j/\omega\epsilon\mu).(-k_x)(k_z)C.sin(k_x x)cos(k_y y)sin(k_z z)$$

and this is zero at $x = 0, h$. We now verify the boundary conditions for the magnetic field.

$$H_y = (1/\mu)(-k_z)C.cos(k_x x)cos(k_y y)sin(k_z z)$$

and this is zero at $z = 0, b$. Finally,

$$H_z = (1/\mu)(-k_y)C.cos(k_x x)sin(k_y y)cos(k_z z)$$

and this is zero at $y = 0, a$. We now calculate the surface electric and magnetic current densities. On the surface $x = 0$, the surface electric current density is given by

$$\mathbf{J}_s(x = 0, y, z) = \hat{x} \times \mathbf{H}|_{x=0} = -(Ck_z/\mu)cos(k_y y)sin(k_z z)\hat{z} + (Ck_y/\mu)sin(k_y y)cos(k_z z)\hat{y}$$

On the surface $x = b$. Likewise on the surface $x = h$, we have

$$\mathbf{J}_s(x = h, y, z) = -\hat{x} \times \mathbf{H}|_{x=h} = -(Ck_z/\mu)(-1)^p cos(k_y y)sin(k_z z)\hat{z} + (Ck_y/\mu)(-1)^p sin(k_y y)cos(k_z z)\hat{y}$$

The magnetic current densities on the side walls are given by

$$\mathbf{M}_s(x, y = 0, z) = -\hat{y} \times \mathbf{E}|_{y=0} = \hat{z}E_x(x, y = 0, z) - \hat{x}E_z(x, y = 0, z)$$

$$= \hat{z}(-j/\omega\epsilon\mu)(k^2 - k_x^2)C.cos(k_x x)cos(k_z z) - \hat{x}(jk_x k_z C/\omega\mu\epsilon).sin(k_x x)sin(k_z z)$$

$$\mathbf{M}_s(x, y = a, z)$$

These expressions can be used to derive far field expressions for the electric vector potential $\mathbf{F}$. Specifically

$$\mathbf{F} = \frac{\mu.exp(-jkr)}{4\pi.r} \int \mathbf{M}_s(\mathbf{r}')exp(jk\hat{r}.\mathbf{r}')dS(\mathbf{r}')$$

Define the integral

$$\mathbf{L} = \frac{\mu}{4\pi} \int \mathbf{M}_s(\mathbf{r}')exp(jk\hat{r}.\mathbf{r}')dS(\mathbf{r}')$$

so that

$$\mathbf{F}(r, \theta, \phi) = \mathbf{L}(\theta, \phi)exp(-jkr)/r$$

Then, the magnetic scalar potential Φ_m is obtained from

$$div\mathbf{F} = -j\omega\epsilon\mu\Phi_m$$

so that Φ_m is $O(1/r)$. Let us calculate H_r, H_θ, H_ϕ.

$$H_r = -j\omega F_r - \frac{\partial\Phi_m}{\partial r}$$

We show that H_r is $O(1/r^2)$. Indeed, with neglect of $O(1/r^2)$, we have

$$div\mathbf{F} = -jkL_r exp(-jkr)/r$$

so that with neglect of $O(1/r^2)$,

$$\Phi_m = (j/\omega\epsilon\mu)(-jkL_r)exp(-jkr)/r = (k/\omega\epsilon\mu)L_r exp(-jkr)/r$$

so that with neglect of $O(1/r^2)$,

$$H_r = -j\omega L_r exp(-jkr)/r + jk^2/\omega\epsilon\mu)L_r exp(-jkr)/r = 0$$

proving the claim. We then get with neglect of $O(1/r^2)$,

$$H_\theta = -j\omega L_\theta exp(-jkr)/r, H_\phi = -j\omega L_\phi exp(-jkr)/r$$

- Magnetic frill generator. This model is often used in computing the incident field E_z^i which is used in Pocklington's integral equation to determine the distribution of current on the wire surface. The dipole is fed with a voltage source at $z = 0$. The dipole length stretches from $z = -L/2$ to $z = L/2$ and the dipole surface is $\rho = a$. The incident field is produced by the voltage source which can be replaced by a magnetic frill generator. The assumption is that the magnetic frill generator is an annulus of inner radius a and outer radius b located in the xy plane with centre at the origin. The wire is a conducting cylinder of radius a. The electric field on the annulus is given by

$$\mathbf{E}_f = \frac{V_i}{2.\rho.log(b/a)}\hat{\rho}, a \leq \rho \leq b$$

The potential difference between the points $\rho = a$ and $\rho = b$ on the xy plane is given by

$$\int_a^b \mathbf{E}_f.\hat{\rho}.d\rho = \frac{V_i}{2}$$

as desired. We wish to calculate the electric field on the surface $\rho = a$. First note that the magnetic current density on the frill is given by

$$\mathbf{M}_f = -2\hat{z} \times \hat{\rho}.E_\rho = 2E_\rho\hat{\phi} = \frac{V_i}{log(b/a)}\hat{\phi}$$

The electric vector potential at the point $\rho = a, \phi = 0, z = 0$ is calculated by a simple integration:

$$\mathbf{F} = \frac{\epsilon}{4\pi}\int \mathbf{M}_f(\rho)\frac{exp(-jk((\rho.cos(\phi) - a)^2 + (\rho.sin(\phi))^2 + z^2)^{1/2})}{((\rho.cos(\phi) - a)^2 + (\rho.sin(\phi))^2 + z^2)^{1/2}}\rho.d\rho.d\phi$$

where the region of integration is $a \leq \rho \leq b, 0 \leq \phi \leq 2\pi$. This can be simplified as

$$\mathbf{F} = \frac{\epsilon}{4\pi}\int M_s(\rho)(-\hat{x}.sin(\phi) + \hat{y}.cos(\phi))\rho.d\rho.d\phi.exp(-jk(\rho^2 + z^2 - 2a\rho.cos(\phi))^{1/2})$$

$$/(\rho^2 + z^2 + a^2 - 2a\rho.cos(\phi))^{1/2}$$

Clearly $F_z = 0$. Also $F_x = 0$ and the only non-vanishing component of $\mathbf{F}$ is F_y. For a general point on the cylinder $\rho = a$, it follows that the only non-vanishing component of $\mathbf{F}$ is F_ϕ. We have

$$F_\phi(z, \rho = a) = \frac{\epsilon}{4\pi}\int_0^a\int_0^{2\pi} M_s(\rho,\phi).cos(\phi).\rho.d\rho.d\phi.exp(-jk(\rho^2 + z^2 + a^2 - 2a\rho.cos(\phi))^{1/2})$$

$$/(\rho^2 + z^2 + a^2 - 2a\rho.cos(\phi))^{1/2}$$

For a general ρ, we have

$$F_\phi(\rho, z) = \frac{\epsilon}{4\pi}\int_a^b\int_0^{2\pi} M_s(\rho').cos(\phi').\rho.d\rho'.d\phi'.exp(-jk\sqrt{\rho^2 + \rho'^2 + z^2 - 2\rho\rho'.cos(\phi')})$$

$$/\sqrt{\rho^2+\rho'^2+z^2-2\rho\rho'.cos(\phi')}$$

From this, we can compute E_z. We have

$$E_z(\rho,z)=\frac{1}{\rho}(\rho.F_\phi)_{,\rho}$$

This field is used as the incident field in Pocklington's integral equation to determine the current distribution on the surface of the cylindrical antenna.

Generalization of Pocklington's integral equation to an arbitrary surface. Let the surface equation be given by $\mathbf{r}(u,v)$ where u,v are free parameters. Define the following tangent vectors on the surface:

$$e_u(u,v)=\frac{\partial\mathbf{r}}{\partial u}, e_v(u,v)=\times\frac{\partial\mathbf{r}}{\partial v}$$

The surface element is given by

$$dS(u,v)=e_u\times e_v dudv$$

The surface current density can be expressed as

$$\mathbf{J}_s(u,v)=J_u(u,v)e_u(u,v)+J_v(u,v)e_v(u,v)$$

It follows that the vector potential produced by this surface current density is given by

$$\mathbf{A}(\mathbf{r})=\frac{\mu}{4\pi}\mathbf{J}_s(u,v).exp(-jkR(u,v))dS(u,v)/R(u,v)$$

Here

$$R(u,v)=|\mathbf{r}-\mathbf{r}(u,v)|$$

We calculate the scalar potential from this making use of the Lorentz gauge condition:

$$\Phi=(j/\omega\epsilon\mu)div\mathbf{A}$$

Now,

$$div\mathbf{A}=\frac{\mu}{4\pi}\int(\mathbf{J}_s(u,v),\nabla R(u,v))G(\mathbf{r},u,v)dS(u,v)$$

where

$$\nabla R(u,v)=\nabla|\mathbf{r}-\mathbf{r}(u,v)|=\mathbf{n}(u,v)$$

where

$$\mathbf{n}(u,v)=(\mathbf{r}-\mathbf{r}(u,v))/|\mathbf{r}-\mathbf{r}(u,v)|=\mathbf{R}(u,v)/R(u,v)$$

and

$$G(\mathbf{r},u,v)=\frac{d}{dR}(exp(-jkR)/R)=-(jk/R+1/R^2)exp(-jkR)$$

It follows that

$$\Phi=\frac{j}{4\pi\omega\epsilon}\int(\mathbf{J}_s,\mathbf{n})(u,v)G(\mathbf{r},u,v)dS(u,v)$$

so that

$$\mathbf{E} = -\nabla\Phi - j\omega\mathbf{A} = -\frac{j}{4\pi\omega\epsilon}\int(\mathbf{J}_s,\mathbf{n})\nabla G(\mathbf{r},u,v)dS(u,v)$$

$$-(j\omega\mu/4\pi)\int \mathbf{J}_s(u,v)H(\mathbf{r},u,v)dS(u,v)$$

where

$$H(\mathbf{r},u,v) = exp(-jkR(u,v))/R(u,v)$$

This electric field, we denote by $\mathbf{E}_s(\mathbf{r})$, the subscript s standing for "scattered". We can write in component form as

$$E_{sk} = \sum_{m=1}^{3}\int \psi_{km}(\mathbf{r},u,v)J_m(u,v)dudv, k = 1,2,3$$

where

$$\psi_{km}(\mathbf{r},u,v) = (-j/4\pi\omega\epsilon)G_{,k}(\mathbf{r},u,v)n_m(u,v) - (j\omega\mu/4\pi)H(\mathbf{r},u,v)\delta_{km}$$

The tangential components of the scattered electric field on the given surface (respectively along the e_u and e_v directions) are then

$$(e_u, E_s) = \sum_{k=1}^{3} x_{k,u}E_{sk}, (e_v, E_s) = \sum_{k=1}^{3} x_{k,v}E_{sk}$$

These can be used to derive generalization of the Pocklington integral equation.

Yagi-Uda array:Finite straight wire dipoles are located at points $(x_n, y_n), n = 1, 2, ..., N$ in the xy plane. The nth dipole stretches from $z = -L_n/2$ upto $z = +L_n/2$. The $n = 1$ dipole is fed at the centre with a current source I_0. This generates a current distribution in the feeder element (the $n = 0$ dipole) which produces a scattered electromagnetic field that induces currents in the other dipoles. Let $I_n(\xi), |\xi| \leq L_n/2$ be the current in the nth dipole. Then we need to derive integral equations for these functions. These integral equations are approximated by matrix linear equations by considering expansions of the form

$$I_n(z) = \sum_{m=0}^{M-1} I_{nm}cos((2m+1)z/L_n), |z| \leq L_n/2$$

Define

$$R_n(x,y,z,\xi) = \sqrt{(x-x_n)^2 + (y-y_n)^2 + (z-\xi)^2}$$

Then, the magnetic vector potential field produced by the current in the nth dipole is given by

$$A_n(x,y,z) = \hat{z}\frac{\mu}{4\pi}\int_{-L_n/2}^{L_n/2} I_n(\xi)exp(-jkR_n(x,y,z,\xi))d\xi/R_n(x,y,z,\xi)$$

and the corresponding z component of the electric field is given by

$$E_{nz} = \frac{-j}{\omega\epsilon\mu}(d^2A_n/dz^2 + k^2A_n)$$

$$= (-j/4\pi\omega\epsilon)\int_{-L_n/2}^{L_n/2} I_n(\xi)(\frac{d^2}{dz^2}G_n(x,y,z,\xi) + k^2 G_n(x,y,z,\xi))d\xi$$

Integration by parts using $G_{n,zz} = G_{n,\xi\xi}$ gives

$$\int_{-L_n/2}^{L_n/2} I_n(\xi)\frac{d^2}{dz^2}G_n d\xi = \int_{-L_n/2}^{L_n/2} I_n(\xi)G_{n,\xi\xi}d\xi$$

$$= I_n(\xi)G_{n,\xi}|_{-L_n/2}^{L_n/2} - \int_{-L_n/2}^{L_n/2} I'_n(\xi)G_{n,\xi}d\xi$$

$$= -\int_{-L_n/2}^{L_n/2} I'_n(\xi)G_{n,\xi}d\xi$$

since $I_n(\pm L_n/2) = 0$. Another integration by parts shows that

$$\int_{-L_n/2}^{L_n/2} I_n(\xi)G_{n,zz}d\xi$$

$$= -I'_n(\xi)G_n|_{-L_n/2}^{L_n/2} + \int_{-L_n/2}^{L_n/2} I''_n(\xi)G_n d\xi$$

$$= I'_n(-L_n/2)G_{n-} - I'_n(L_n/2)G_{n+} + \int_{-L_n/2}^{L_n/2} I''_n(\xi)G_n d\xi$$

where

$$G_{n-} = G_n|_{\xi=-L_n/2} = exp(-jkR_{n-})/R_{n-}, G_{n+} = G_n|_{\xi=L_n/2} = exp(-jkR_{n+})/R_{n+}$$

where

$$R_{n-} = \sqrt{(x-x_n)^2 + (y-y_n)^2 + (z+L_n/2)^2}, R_{n+} = \sqrt{(x-x_n)^2 + (y-y_n)^2 + (z-L_n/2)^2}$$

We thus obtain the following formula for the total z-component of the electric field due to all the dipoles in space:

$$E_z = \sum_{n=1}^{N}(-j/\omega\epsilon\mu)\int_{-L_n/2}^{L_n/2} I_n(\xi)(G_{n,zz} + k^2 G_n)d\xi$$

$$= (-j/\omega\epsilon\mu)(\sum_{n=1}^{N} I'_n(-L_n/2)G_{n-} - I'_n(L_n/2)G_{n+} + \int_{-L_n/2}^{L_n/2}(I''_n(\xi) + k^2 I_n(\xi))G_n d\xi)$$

Let

$$G_{n+}|_{x=x_m,y=y_m} = G_{nm+}, G_{n-}|_{x=x_m,y=y_m} = G_{nm-}$$

In other words, defining

$$R_{nm-}(z) = \sqrt{(x_n-x_m)^2 + (y_n-y_m)^2 + (z+L_n/2)^2}, R_{nm+}(z)$$
$$= \sqrt{(x_n-x_m)^2 + (y_n-y_m)^2 + (z-L_n/2)^2}$$

we have

$$G_{nm-}(z) = exp(-jkR_{nm-}(z))/R_{nm-}(z), G_{nm+}(z) = exp(-jkR_{nm+}(z))/R_{nm+}(z)$$

We also define

$$G_{nm}(z,\xi) = G_n|_{x=x_m,y=y_m} = exp(-jkR_{nm}(z,\xi))/R_{nm}(z,\xi)$$

where

$$R_{nm}(z,\xi) = \sqrt{(x_n - x_m)^2 + (y_n - y_m)^2 + (z-\xi)^2}$$

Then

$$E_z(x_m, y_m, z) = (-j/\omega\epsilon\mu)\sum_{n=1}^{N}[I'_n(-L_n/2)G_{nm-}(z) - I'_n(L_n/2)G_{nm+}(z)$$

$$+\int_{-L_n/2}^{L_n/2}(I''_n(\xi) + k^2 I_n(\xi))G_{nm}(z,\xi)d\xi]$$

This is zero for all $|z| \leq L_m/2$ provided $m \neq 0$. For $m = 0$, this is zero at all z except at $z = 0$ where a source is present creating a singularity. The current distribution in all the elements can be solved for by taking into account this singularity using the above expansion for $I_n(z)$. We divide the length of each dipole into M segments with centres at $z_0, z_1, z_2, ..., z_{M-1}$ where $z_0 = 0$. Then the above equation is evaluated at $z_k, k = 0, 1, 2, ..., M-1$ to zero at all $m = 2, 3, ..., N$ and at $m = 0$, we evaluate it to zero at $z_1, ..., z_{M-1}$ giving in all a set of $(N-1)M + M - 1 = NM - 1$ equations for the NM unknowns $I_{nm}, n = 1, 2, ..., N, m = 0, 1, ..., M-1$. The last equation is obtained as

$$I_0 = \sum_{m=0}^{M-1} I_{1,m}$$

and a computer program can be developed for solving this system of NM linear equations.

Circular microstrip antenna: The cavity is cylindrical in shape of radius R and height h. The electric vector potential is zero while the magnetic vector potential is $A_z(\rho, \phi, z)\hat{z}$. The field satisfies Helmholtz equation

$$\nabla^2 A_z + k^2 A_z = 0$$

which in cylindrical coordinates translates to

$$\frac{1}{\rho}\frac{\partial}{\partial\rho}\rho\frac{\partial A_z}{\partial\rho} + \frac{1}{\rho^2}\frac{\partial^2 A_z}{\partial\phi^2} + \frac{\partial^2 A_z}{\partial z^2} + k^2 A_z = 0\,.$$

This equation can be expressed as

$$A_{z,\rho\rho} + \frac{1}{\rho}A_{z,\rho} + \frac{1}{\rho^2}A_{z,\phi\phi} + A_{z,zz} + k^2 A_z = 0$$

The side walls of the cylinder are perfect magnetic conductors while the top and bottom surfaces are perfect electric conductors. Thus the boundary conditions are that the tangential components of the magnetic field vanish on the side walls while the tangential components of the electric field vanish on the top and bottom surfaces. We note that since $\mathbf{A} = A_z\hat{z}$, it follows that

$$\mathbf{H} = \nabla A_z \times \hat{z} = A_{z,\rho}\hat{\rho} \times \hat{z} + \rho^{-1}A_{z,\phi}\hat{\phi} \times \hat{z}$$

$$= -A_{z,\rho}\hat{\phi} + \rho^{-1}A_{z,\phi}\hat{\rho}$$

so that

$$H_\rho = \rho^{-1}A_{z,\phi}, H_\phi = -A_{z,\rho}, H_z = 0$$

Further, the electric field can be calculated using

$$\nabla \times H = j\omega\epsilon E$$

Now,

$$\nabla \times H = \frac{1}{\mu}\nabla \times \nabla \times A = \mu^{-1}(\nabla(\nabla.A) - \nabla^2 A)$$

so that

$$E = (-j/\omega\epsilon\mu)(\nabla(\nabla.A) + k^2 A)$$

giving

$$E_\rho = (-j/\omega\epsilon\mu)A_{z,z\rho}, E_\phi = (-j/\omega\epsilon\mu\rho)A_{z,z\phi},$$

We do not need E_z for the computation of the modes since E_z is the normal component of the electric field on the top and bottom surfaces of the cylinder. Apply a separation of variables to solve the Helmholtz equation for A_z:

$$A_z(\rho, \phi, z) = R(\rho).\Phi(\phi).Z(z)$$

Then,

$$R''/R + R'/\rho R + \Phi''/\rho^2\Phi + k^2 = -Z''/Z$$

The left side is a function of ρ, ϕ only while the right side a function of z only. Hence, both sides must equal a constant, say α^2. We get

$$Z(z) = A_1.cos(\alpha z) + A_2 sin(\alpha z)$$

We get further,

$$\rho^2 R''/R + \rho R'/R + (k^2 - \alpha^2)\rho^2 = -\Phi''/\Phi$$

The left side is a function of ρ only while the right side of ϕ only. Hence, both sides must equal a constant, say m^2. It follows that

$$\Phi(\phi) = B_1.cos(m\phi) + B_2.sin(m\phi)$$

and for single-valuedness, m must be an integer, which may be assumed to be non-negative. Then,

$$\rho^2 R'' + \rho R' + ((k^2 - \alpha^2)\rho^2 - m^2)R = 0$$

This has solution $J_m(\beta r), \beta = \sqrt{k^2 - \alpha^2}$ where J_m satisfies the differential equation

$$\rho^2 R'' + \rho R' + (\rho^2 - m^2)R = 0$$

Note that $J_m(\alpha r)$ is the Bessel function of order m. The complete solution can now be written as

$$A_z(\rho,\phi,z) = J_m(\beta\rho).(B_1.cos(m\phi) + B_2.sin(m\phi)).(A_1.cos(\alpha z) + A_2 sin(\alpha z))$$

We find that since H_ϕ is proportional to $A_{z,\rho}$, the condition for H_ϕ to be zero when $\rho = a$ is that

$$J_m(\beta.a) = 0$$

and if $\xi_{mr}, r = 1, 2, ...$ are the zeros of J_m, then β can assume values $\xi_{mr}/a, r = 1, 2, ...$ Thus,

$$R(\rho) = J_m(\xi_{mr}\rho/a)$$

and

$$\xi_{mr}^2/a^2 + \alpha^2 = k^2$$

The condition that the tangential components of the electric field vanish when $z = 0, h$ give $\alpha = q\pi/h$ where q is an integer. Combining these facts we get the following expression for the possible resonant frequencies inside the cavity resonator:

$$k^2 = \omega^2\mu\epsilon = \xi_{mr}^2/a^2 + q^2\pi^2/h^2$$

where m, r, q assume non-negative integral values.

Helical antenna analysis: The equation of the helix is

$$z = c\phi, \rho = a$$

or equivalently in terms of cartesian coordinates,

$$\mathbf{r} = \hat{x}.a.cos(\phi) + \hat{y}.a.sin(\phi) + c.\phi$$

The line element is

$$d\mathbf{r} = d\mathbf{l} = d\phi(-a.sin(\phi)\hat{x} + a.cos(\phi)\hat{y} + c\hat{z})$$

and assuming a constant current I through the helix, the magnetic vector potential field is given by

$$\mathbf{A}(x,y,z) = \frac{\mu I}{4\pi}\int_0^{2n\pi}(-a.sin(\phi')\hat{x} + a.cos(\phi')\hat{y} + c\hat{z})d\phi'.$$

$$exp(-jk\sqrt{(x - a.cos(\phi'))^2 + (y - a.sin(\phi'))^2 + (z - c\phi')^2})$$

where n is the total number of turns of the helix. In terms of cylindrical coordinates,

$$\mathbf{A}(\rho,\phi,z) = \frac{\mu I}{4\pi}\int_0^{2n\pi}(-a.sin(\phi')\hat{x} + a.cos(\phi')\hat{y} + c\hat{z})d\phi'.$$

$$exp(-jk\sqrt{\rho^2 + a^2 - 2\rho a.cos(\phi - \phi') + (z - c\phi')^2})$$

giving

$$A_\rho = A_x.cos(\phi) + A_y.sin(\phi) =$$

$$\frac{\mu Ia}{4\pi}\int_0^{2n\pi} sin(\phi-\phi')d\phi'.exp(-jk\sqrt{(\rho^2+a^2-2\rho.a.cos(\phi-\phi')+(z-c.\phi')^2)}$$

$$A_\phi = -A_x.sin(\phi)+A_y.cos(\phi)=\frac{\mu Ia}{4\pi}\int_0^{2n\pi} cos(\phi-\phi').d\phi'.$$

$$exp(-jk\sqrt{(\rho^2+a^2-2\rho a.cos(\phi-\phi')+(z-c.sin(\phi')^2)}$$

$$A_z=\frac{\mu Ic}{4\pi}\int_0^{2n\pi} d\phi'.exp(-jk\sqrt{(\rho^2+a^2-2\rho.a.cos(\phi'-\phi)+(z-c\phi')^2)}$$

The electric and magnetic fields can be determined from these expressions by elementary differentiations using the formulae for gradient and curl in cylindrical coordinates.

Reciprocity theorem: Let E_i, H_i be the fields produced by sources (J_i, M_i) respectively where $i = 1, 2$. Then,

$$curlE_i = -j\omega\mu H_i - M_i, curlH_i = j\omega\epsilon E_i + J_i, i = 1, 2$$

Thus,

$$div(E_1 \times H_2) = (curlE_1, H_2) - (E_1, curlH_2) = (-j\omega\mu H_1 - M_1, H_2) - (E_1, j\omega\epsilon E_2 + J_2)$$

$$div(E_2 \times H_1) = (-j\omega\mu H_2 - M_2, H_1) - (E_2, j\omega\epsilon E_1 + J_1)$$

Taking the difference gives us

$$div(E_1 \times H_2) - div(E_2 \times H_1) = (M_2, H_1) - (M_1, H_2) + (J_1, E_2) - (J_2, E_1)$$

Let V be a volume bounded by the surface S. Then, it follows from this equation and Gauss' theorem that

$$\int_S (E_1 \times H_2 - E_2 \times H_1, n)dS = \int_V [(M_2, H_1) - (M_1, H_2) + (J_1, E_2) - (J_2, E_1)]dV$$

Consider now two dipoles numbered 1, 2 oriented along the z axis. Assume that dipole 1 is excited by a voltage source V_1. This voltage source can be replaced by an equivalent magnetic surface current M_1. The field produced by this induces a current J_2 in dipole two. Thus, $J_1 = 0, M_2 = 0$ and the above equation gives taking V to be the entire space and S to be the surface of a sphere of infinite radius,

$$\int_{V_1} (M_1, H_2)dV = -\int_{V_2} (J_2, E_1)dV$$

We can write E_1 as E_{21} in the second integral, where E_{21} is the electric field produced in the second dipole due to the voltage in the first. The right side is then $\int_{-L_2/2}^{L_2/2} E_{21}(z)I_2(z)dz$. Note that V_i is the volume bounded by the ith dipole, $i = 1, 2$. From this equation, we can show that the mutual impedance between the two antennae is given by

$$Z_{21} = \frac{1}{I_1(0)I_2(0)}\int_{-L_2/2}^{L_2/2} E_{21}(z)I_2(z)dz$$

Not clear how this is derived. M_1 is a surface magnetic current of value $(V_1/2)\hat{\rho}/log(b/a)$ and $H_2 = I_2(0)/2\pi d$ where d is the distance between the two dipoles. Let us determine an algorithm for computing the mutual impedance between two dipoles. The first dipole is along the z axis at the origin, ie, its location is $\rho = 0, |z| \leq L_1/2$ while the second dipole is at $\rho = d, \phi = \pi/2$, ie, its cartesian coordinates are $x = 0, y = d, |z| \leq L_2/2$. A voltage V_1 is applied at the origin to dipole one. This source can be replaced by an equivalent magnetic frill generator producing a surface magnetic current $\mathbf{M}_f = -2\hat{z} \times \hat{\rho}\frac{V_1}{2\rho.log(b/a)}, a \leq \rho \leq b$. This evaluates to $-V_1\hat{\phi}/(\rho.log(b/a))$. Denote this by $M_f(\rho)\hat{\phi}$. The electric vector potential corresponding to this is

$$\mathbf{F} = \frac{\epsilon}{4\pi}\int_a^b\int_0^{2\pi} M_f(\rho')(-\hat{x}.sin(\phi') + \hat{y}.cos(\phi'))\frac{exp(-jk\sqrt{(\rho^2+\rho'^2-2\rho\rho'cos(\phi-\phi')+z^2)}}{\sqrt{\rho^2+\rho'^2-2\rho\rho'cos(\phi-\phi')+z^2}}\rho'd\rho'd\phi'$$

It follows that

$$F_z = 0,$$

$$F_\rho = F_x cos(\phi) + F_y sin(\phi) =$$

$$C.\int M_f(\rho')sin(\phi-\phi')exp(-jk\sqrt{\rho^2+\rho'^2-2\rho\rho'cos(\phi-\phi')+z^2})\rho'd\rho'd\phi'$$

$$/\sqrt{\rho^2+\rho'^2-2\rho\rho'cos(\phi-\phi')+z^2}$$

$$= 0$$

$$F_\phi = \frac{\epsilon}{4\pi}\int M_f(\rho')cos(\phi-\phi')\frac{exp(-jk\sqrt{\rho^2+\rho'^2-2\rho\rho'cos(\phi-\phi')+z^2})}{\sqrt{\rho^2+\rho'^2-2\rho\rho'cos(\phi-\phi')+z^2}}\rho'd\rho'd\phi'$$

The electric field produced by the frill can now be determined:

$$E = \nabla \times (F_\phi\hat{\phi})$$

giving

$$E_z = (\rho F_\phi)_{,\rho}$$

We denote this field by $E_z^f(x,y,z)$, the superscript f standing for "frill". The z component of the electric field produced by the current distribution in the two dipoles is given by

$$E_z^d(x,y,z) = (j\mu/4\pi\omega\epsilon)\int_{-L_1/2}^{L_1/2} I_1(\xi)(\frac{\partial^2 G(x,y,z-\xi)}{\partial z^2} + k^2 G(x,y,z-\xi))d\xi +$$

$$(j\mu/4\pi\omega\epsilon)\int_{-L_2/2}^{L_2/2} I_2(\xi)(\frac{\partial^2 G(x,y-d,z-\xi)}{\partial z^2} + k^2 G(x,y-d,z-\xi))d\xi$$

where

$$G(x,y,z) = exp(-jk\sqrt{x^2+y^2+z^2})/\sqrt{x^2+y^2+z^2}$$

Biconical antenna: $E_\theta(r,\theta), H_\phi(r,\theta)$ are the nonzero components of the electric and magnetic fields. These are substituted into the equation $curlE = -j\omega\mu H, curlH = j\omega\epsilon E$. We get

$$curlE = (\hat{\phi}/r)(rE_\theta)_{,r}$$

$$curlH = (\hat{r}/r^2 sin(\theta))(r.sin\theta H_\phi)_{,\theta} - (\hat{\theta}/r.sin\theta)(rsin(\theta)H_\phi)_{,r}$$

The equations

$$curlE = -j\omega\mu H, curlH = j\omega\epsilon E$$

give

$$H_\phi = C.exp(-jkr)/r.sin(\theta), E_\theta = \eta H_\phi$$

Take this as an exercise and prove it. Note that $curlH = j\omega\epsilon E$ gives

$$H_\phi = f(r)/sin\theta$$

where f is some function of the radial coordinate. By substituting into the other two equations, it easily follows that f satisfies the radial Helmholtz equation

$$\frac{1}{r}(rf)'' + k^2 f = 0$$

which has the solution

$$f(r) = exp(-ikr)/r$$

Planar array: Assume that the array elements are located on the xy plane at the points (md_x, nd_y), $m = 0,1,...,M-1, n = 0,1,...,N-1$. Let the current sources in the array elements have the separable form $I(m,n) = I_1(m)I_2(n)$. Then the array amplitude pattern is given by

$$A(\theta,\phi) = \sum_{m=0}^{M-1}\sum_{n=0}^{N-1} I(m,n)exp(jk(md_x cos(\phi)sin(\theta) + nd_y sin(\phi)sin(\theta)))$$

$$= A_1(\theta,\phi).A_2(\theta,\phi)$$

where

$$A_1(\theta,\phi) = \sum_{m=0}^{M-1} I_1(m)exp(jkmd_x cos(\phi)sin(\theta))$$

$$A_2(\theta,\phi) = \sum_{n=0}^{N-1} I_2(n)exp(jknd_y sin(\phi)sin(\theta))$$

We assume that

$$I_1(m) = |I_1(m)|exp(jm\psi_x), I_2(n) = |I_2(n)|exp(jn\psi_y)$$

and then conclude that the maximum occurs when

$$kd_x cos(\phi)sin(\theta) + \psi_x = 0, kd_y sin(\phi)sin(\theta) + \psi_y = 0$$

These two equations can be solved for θ and ϕ. Assume now that $|I_1(n)|$ and $|I_2(m)|$ are constants, say unity. Then,

$$A_1(\theta, \phi) = \sum_{m=0}^{M-1} exp(jkmd_x cos(\phi) sin(\theta) = sin(kMd_x cos(\phi) sin(\theta)/2)$$

Array synthesis:

First consider an array with an odd number of elements. The amplitude pattern is given by

$$AF(\theta) = \sum_{m=-M}^{M} a(m) exp(jm\psi), \psi = kd.cos(\theta) + \beta$$

From this equation, we get

$$a(m) = \frac{1}{2\pi} \int_{-\pi}^{\pi} AF(\psi) exp(-jm\psi) d\psi$$

Suppose that the amplitude pattern is specified over the range $\theta \in [\theta_1, \theta_2]$. We assume that $kd(cos(\theta_1) - cos(\theta_2)) \leq 2\pi$. Then setting $\psi_i = kd.cos(\theta_i) + \beta, i = 1, 2$, we get for the current weights

$$a(m) = \frac{1}{2\pi} \int_{\psi_2}^{\psi_1} AF(\psi) exp(-jm\psi) d\psi$$

where $AF(\psi)$ is obtained by replacing θ in $AF(\theta)$ with $cos^{-1}((\psi - \beta)/kd)$.

Woodward-Lawson method:

$$I(z') = \sum_{m=1}^{M} b(m) exp(-jkz'.cos(\theta_m)$$

is the assumed current distribution. The amplitude pattern is then given by

$$A(\theta) = \int_{-l/2}^{l/2} I(z') exp(jkz'.cos(\theta)) dz'$$

$$= \sum_{m=1}^{M} b(m) sin(kl(cos(\theta) - cos(\theta_m))/2)/k(cos(\theta) - cos(\theta_m))$$

Assume that

$$kl(cos(\theta_{m+1}) - cos(\theta_m))/2 = \pi$$

so that

$$cos(\theta_m) = 2m\pi/kl = m\lambda/l, m = 1, 2, ..., N$$

Then,

$$A(\theta_m) = lb(m), m = 1, 2, ..., M$$

and this suggests the method of calculating the coefficients $b(m)$ of the current distribution by sampling the given amplitude pattern. This method is important for antenna pattern synthesis.

Method of stationary phase: The electric field is assumed to have the form

$$\mathbf{E}(x,y,z) = \int \mathbf{f}(k_x,k_y)exp(j(k_x x + k_y y + k_z z))dk_x dk_y$$

where

$$k_z = \sqrt{k^2 - k_x^2 - k_y^2}$$

The contribution to the integral is maximum at that (k_x, k_y) for which the phase $\mathbf{k.r}$ is stationary, ie,

$$\frac{\partial}{\partial k_x}(\mathbf{k.r}) = 0, \frac{\partial}{\partial k_y}(\mathbf{k.r}) = 0$$

This gives

$$x - k_x z/k_z = 0, y - k_y z/k_z = 0$$

or setting

$$x = r.cos(\phi)sin(\theta), y = r.sin(\phi)sin(\theta), z = r.cos(\theta)$$

we get

$$cos(\phi)sin(\theta) - k_x cos(\theta)/k_z = 0, sin(\phi)sin(\theta) - k_y cos(\theta)/k_z = 0$$

giving

$$k_x = k_z.cos(\phi)sin(\theta)/cos(\theta), k_y = k_z.sin(\phi)sin(\theta)/cos(\theta),$$

so that

$$k^2 = k_x^2 + k_y^2 + k_z^2 = k_z^2/cos^2(\theta)$$

giving

$$k_x = k.cos(\phi)sin(\theta), k_y = k.sin(\phi)sin(\theta), k_z = k.cos(\theta)$$

We expand the phase $\mathbf{k.r}$ about this stationary point (k_{x0}, k_{y0}) retaining upto terms quadratic in $(k_x - k_{x0}, k_y - k_{y0})$. Specifically,

$$\mathbf{k.r} = \mathbf{k}_0.\mathbf{r} + A\xi^2 + C\eta^2 + 2B\xi\eta$$

where

$$A = \frac{1}{2}\frac{\partial^2}{\partial k_x^2}(\mathbf{k.r})|_{\mathbf{k}=\mathbf{k}_0}, C = \frac{1}{2}\frac{\partial^2}{\partial k_y^2}(\mathbf{k.r})|_{\mathbf{k}=\mathbf{k}_0},$$

$$B = \frac{1}{2}\frac{\partial^2}{\partial k_x \partial k_y}(\mathbf{k.r})|_{\mathbf{k}=\mathbf{k}_0}$$

The method of stationary phase can be applied to the approximate evaluation of integrals of the form $\int f(u,v)exp(j\phi(u,v))dudv$. Let (u_0, v_0) satisfy the equation

$$\frac{\partial\phi(u,v)}{\partial u} = \frac{\partial\phi(u,v)}{\partial v} = 0$$

Then setting

$$A = \frac{\partial^2\phi}{\partial u^2}|_{(u_0,v_0)}, C = \frac{\partial^2\phi}{\partial v^2}|_{(u_0,v_0)},$$

$$B = \frac{\partial^2\phi}{\partial u\partial v}|_{(u_0,v_0)}$$

Then

$$\phi(u,v) \approx \phi_0 + \frac{1}{2}(A(u-u_0)^2 + 2B(u-u_0)(v-v_0) + C(v-v_0)^2)$$

where $\phi_0 = \phi(u_0, v_0)$, so that

$$I = \int f(u,v)exp(j\phi(u,v))dudv \approx f(u_0,v_0)exp(j\phi_0)\int exp(j(A\xi^2 + 2B\xi\eta + C\eta^2)/2)d\xi d\eta$$

The Gaussian integral can be evaluated easily. Write

$$\mathbf{x} = (\xi,\eta)^T, \mathbf{S} = \begin{pmatrix} A & B \\ B & C \end{pmatrix}$$

so that

$$\int exp(j(A\xi^2 + 2B\xi\eta + C\eta^2)/2)d\xi d\eta = \int exp(j\mathbf{x}^T\mathbf{S}\mathbf{x}/2)d\mathbf{x}$$

$$= (2\pi)det(j\mathbf{S})^{-1/2} = -2\pi.j.det(\mathbf{S})^{-1/2}$$

In the current problem

$$\phi(u,v) = ux + vy + \sqrt{k^2 - u^2 - v^2}z$$

and taking

$$x = r.cos(\phi)sin(\theta), y = r.sin(\phi)sin(\theta), z = r.cos(\theta),$$

we have

$$(u_0, v_0) = k(cos(\phi)sin(\theta), sin(\phi)sin(\theta))$$

We need to calculate A, B, C, ie, the partial derivatives of ϕ at the stationary point.

$$\phi_{,u} = x - uz/\sqrt{k^2 - u^2 - v^2}$$

$$\phi_{,uu} = -z/\sqrt{k^2 - u^2 - v^2} - u^2z/(k^2 - u^2 - v^2)^{3/2}$$

so that

$$A = \phi_{,uu}(u_0,v_0) = -z/k.cos(\theta) - z.cos^2(\phi)sin^2(\theta)/k.cos^3(\theta) == -\frac{r}{k}(1 + cos^2(\phi)sin^2(\theta)/cos^2(\theta))$$

$$\phi_{,v} = y - zv/\sqrt{k^2 - u^2 - v^2}$$

$$\phi_{,vv} = -z/\sqrt{k^2 - u^2 - v^2} - zv^2/(k^2 - u^2 - v^2)^{3/2}$$

so that

$$C = \phi_{,vv}(u_0,v_0) == -z/k.cos(\theta) - z.sin^2(\phi)sin^2(\theta)/k.cos^3(\theta) = -\frac{r}{k}(1 + sin^2(\phi)sin^2(\theta)/cos^2(\theta))$$

$$\phi_{,uv} = -uvz/(k^2 - u^2 - v^2)^{3/2}$$

so that

$$B = \phi_{,uv}(u_0, v_0) = -z.cos(\phi)sin(\phi)sin^2(\theta)/k.cos^3(\theta) = -\frac{r}{k}cos(\phi)sin(\phi)sin^2(\theta)/cos^2(\theta)$$

We now note that

$$det(\mathbf{S}) = AC - B^2 = \frac{r^2}{k^2}[(1 + cos^2(\phi)sin^2(\theta)/cos^2(\theta))(1 + sin^2(\phi)sin^2(\theta)/cos^2(\theta)) \\ -cos^2(\phi)sin^2(\phi)sin^4(\theta)/cos^4(\theta)]$$

Problem in antenna theory(Computation of the mutual impedance between two finite wire dipoles): Let the first antenna be placed with centre at the origin and length along the z axis from $z = -L/2$ to $z = L/2$. The centre of this dipole is at the origin $(0, 0, 0)$. The second dipole has its centre at $(a, 0, 0)$ and stretches from $z = -L/2$ to $z = L/2$. Assume that the wires can be modeled as cylinders of radius a. The first antenna is excited at the origin by a voltage V and we assume that the gap over which the excitation is done has a length Δ. Then the electric field of this excitation is $E = V\hat{z}/\Delta$ and the equivalent magnetic current density is given by $M_s = -2\hat{\rho} \times E = 2V\hat{\phi}/\Delta$. This surface current density is spread out over a cylinder of length Δ centred at the origin and having a radius R. The electric vector potential F generated by M_s can be expressed as

$$F = \frac{\epsilon.R}{4\pi}\int_{-L/2}^{L/2}\int_0^{2\pi} dz'd\phi'(2V/\Delta)(-\hat{x}.sin(\phi') + \hat{y}.cos(\phi'))$$

$$\frac{exp(-jk\sqrt{(x - R.cos(\phi'))^2 + (y - R.sin(\phi'))^2 + (z - z')^2})}{\sqrt{(x - R.cos(\phi')^2 + (y - R.sin(\phi'))^2 + (z - z')^2}}$$

In terms of cylindrical coordinates,

$$x = \rho.cos(\phi), y = \rho.sin(\phi)$$

we have

$$(x - R.cos(\phi'))^2 + (y - R.sin(\phi'))^2 + (z - z')^2 = \rho^2 + R^2 - 2\rho R.cos(\phi - \phi') + (z - z')^2$$

so that

$$F_\rho = 0 = F_z,$$

$$F_\phi = -F_x.sin(\phi) + F_y.cos(\phi) =$$

$$\frac{\epsilon.R}{4\pi}\int_{-L/2}^{L/2}\int_0^{2\pi} dz'd\phi'(2V/\Delta)cos(\phi - \phi')exp(-jk\sqrt{\rho^2 + R^2 - 2\rho R.cos(\phi - \phi') + (z - z')^2})$$

$$/\sqrt{\rho^2 + R^2 - 2\rho.R.cos(\phi - \phi') + (z - z')^2}$$

Note that $r^2 = \rho^2 + z^2$ so that F_ϕ is a function of only ρ, z, ie, F_ϕ is independent of ϕ. The electric vector potential may be expressed as

$$F = F_\phi(\rho, z)\hat{\phi}$$

The electric field is given by

$$E = \frac{1}{\epsilon}\nabla \times F$$

$$= -\hat{\rho}.\rho^{-1}F_{\phi,z} + \hat{z}.\rho^{-1}F_{\phi,\rho}$$

In particular, for formulating the Pocklington integral equation, we need only E_z. We shall denote this by E_z^i, the superscript i corresponding to "incident field", ie the field generated by the source in antenna one. We have

$$E_z^i(\rho, z) = \rho^{-1}F_{\phi,\rho}$$

Let $I_1(z)$ be the current in the first antenna. Then, the magnetic vector potential due to this current is given by

$$A_1 = \hat{z}\frac{\mu}{4\pi}\int_{-L/2}^{L/2} I_1(z')\frac{exp(-jk\sqrt{x^2+y^2+(z-z')^2})}{\sqrt{(x^2+y^2+(z-z')^2}}dz'$$

The scalar potential is computed using the Lorentz gauge condition:

$$divA_1 = -j\omega\epsilon\mu\Phi_1$$

This gives

$$\Phi_1 = \frac{j}{\omega\epsilon}divA_1 = \frac{j}{\omega\epsilon\mu}A_{1z,z}$$

and the z component of the corresponding electric field is given by

$$E_{1z} = -\Phi_{1,z} - j\omega A_{1z}$$

$$= -\frac{j}{\omega\epsilon\mu}A_{1z,zz} - j\omega A_{1z}$$

$$= -\frac{j}{\omega\epsilon}(A_{1z,zz} + k^2 A_{1z})$$

Writing

$$A_{1z} = \int_{-L/2}^{L/2} I_1(z')G_1(x, y, z-z')dz'$$

we find that

$$E_{1z} = -\frac{j}{\omega\epsilon}\int_{-L/2}^{L/2} I_1(z')(G_{1,zz}(x, y, z-z') + k^2 G_1(x, y, z-z'))dz'$$

We express this in cylindrical coordinates giving

$$E_{1z}(\rho, \phi, z) = -\frac{j}{\omega\epsilon}\int_{-L/2}^{L/2} I_1(z')(G_{1,zz}(\rho, z-z') + k^2 G_1(\rho, z-z'))dz'$$

Note that

$$G_1(\rho, z) = \mu exp(-jk\sqrt{\rho^2+z^2})/4\pi\sqrt{\rho^2+z^2}$$

Similarly, let $I_2(z)$ be the current in the second antenna. The magnetic vector potential due to this current is given by

$$A_2 = \hat{z}\frac{\mu}{4\pi}\int_{-L/2}^{L/2} I_2(z')\frac{exp(-jk\sqrt{(x-a)^2+y^2+(z-z')^2})}{\sqrt{(x-a)^2+y^2+(z-z')^2}}dz'$$

The corresponding z component of the electric field is given by

$$E_{2z} = -\frac{j}{\omega\epsilon}\int_{-L/2}^{L/2} I_2(z')(G_{2,zz}(x,y,z-z')+k^2G_2(x,y,z-z'))dz'$$

where

$$G_2(x,y,z) = \mu.exp(-jk\sqrt{(x-a)^2+y^2+z^2})/4\pi\sqrt{(x-a)^2+y^2+z^2}$$

We can express G_2 using cylindrical coordinates:

$$G_2(\rho,\phi,z) = \frac{\mu}{4\pi}exp(-jk\sqrt{\rho^2+a^2-2\rho a.cos(\phi)+z^2})/\sqrt{\rho^2+a^2-2\rho a.cos(\phi)+z^2}$$

The z component of the electric field at any point in space is given by

$$E_z^t = E_{1z} + E_{2z} + E_z^i$$

and this is equated to zero at $\rho = 0$ and at $\rho = a, \phi = 0$. The result is a set of two integral equations for the currents $I_1(z)$ and $I_2(z)$.

Problem in antenna theory: Determine the width of the mainlobe of a uniform linear array having N elements.

Ans:

$$A(\theta) = \sum_{n=0}^{N-1} exp(jknd.cos(\theta)) = (exp(jkNd.cos(\theta))-1)/(exp(jkd.cos(\theta))-1)$$

$$|A(\theta)| = |sin(kNd.cos(\theta)/2)|/|sin(kd.cos(\theta)/2)|$$

The maximum occurs when $\theta = \pi/2$ and the first minimum when

$$kNd.cos(\theta)/2 = \pi$$

or equivalently, when

$$cos(\theta) = \lambda/Nd$$

or

$$\theta = cos^{-1}(\lambda/Nd) = \pi/2 - sin^{-1}(\lambda/Nd)$$

The width of the mainlobe, ie, angle between the first two nulls on either side of the maximum is given by

$$2.sin^{-1}(\lambda/Nd)$$

Another problem in antenna theory: Assume that an array consists of four isotropic elements with uniform excitation with the elements located at the origin and on the x, y, z axes at unit distance from the origin. Determine the array factor.

Ans:

$$A(\theta, \phi) = 1 + exp(jkd.cos(\phi)sin(\theta)) + exp(jkd.sin(\phi)sin(\theta)) + exp(jkd.cos(\theta))$$

Problem: Determine the equation satisfied by the direction in which the maximum occurs.

Hint: You've got to solve the equations

$$\frac{\partial A(\theta, \phi)}{\partial \theta} = 0 = \frac{\partial A(\theta, \phi)}{\partial \phi}$$

These are transcendental equations and can only be solved only by using numerical techniques.

Another problem:We want to design a three element linear array with perhaps non-uniform excitation so that the nulls of the array occur at $\theta = \theta_i, i = 1, 2$. Assume that the interelement spacing is d. The array factor is then given by

$$A(\theta) = I_0 + I_1.exp(jkd.cos(\theta)) + I_2.exp(j2kd.cos(\theta))$$

Let

$$z_i = exp(jkd.cos(\theta_i)), i = 1, 2$$

Set $u = I_1/I_0$ and $v = I_2/I_0$. Then the following equations for the unknowns (u, v) need to be solved:

$$1 + uz_i + vz_i^2 = 0, i = 1, 2$$

Problem in antenna theory:Consider a linear array with sensors located on the z axis at distances $d_n, n = 1, 2, ..., N$ from the origin. Let d_n be random with mean d_{n0} and variance σ_n^2 small. Determine the average amplitude and power pattern as well as the correlations between the amplitude patterns along two different directions.

Solution:

$$A(\theta) = \sum_{n=1}^{N} I_n exp(jkd_n cos(\theta))$$

Let

$$\Phi_n(\xi) = \mathbb{E}exp(j\xi d_n)$$

the characteristic function of d_n. Note that

$$\Phi_n(\xi) \approx 1 + j\xi d_{n0} - \xi^2(\sigma_n^2 + d_{n0}^2)/2$$

The average amplitude pattern is given by

$$< A(\theta) >= \sum_{n=1}^{N} I_n\Phi_n(k.cos(\theta))$$

and the amplitude correlations are

$$< A(\theta_1)A^*(\theta_2) >= \sum_n I_n^2\Phi_n(k(cos(\theta_1) - cos(\theta_2))) + \sum_{n\neq m} I_nI_m\Phi_n(k.cos(\theta_1))\Phi_m^*(k.cos(\theta_2))$$

The multiple correlations may also be evaluated. In particular, the average power pattern is given by

$$< |A(\theta)|^2 >= \sum I_n^2 + \sum_{n\neq m} I_nI_m\Phi_n(k.cos(\theta))\Phi_m^*(k.cos(\theta))$$

Suppose the sensors are moving along Brownian trajectories. Then calculate the time and ensemble averaged amplitude and power patterns.

$$A(\theta, t) = \sum_{n=1}^{N} I_n.exp(jkB_n(t)cos(\theta)$$

$$< A(\theta, t) >= \sum_{n=1}^{N} I_nexp(-ktcos^2(\theta)/2)$$

and the time and ensemble averaged amplitude pattern is given by

$$\frac{1}{T}\int_0^T < A(\theta, t) > dt = \sum_{n=1}^{N}(k.cos^2(\theta)/2))^{-1}I_n(1 - exp(-kT.cos^2(\theta)/2))$$

First mid-semester question paper on antenna theory:

[1] Derive from first principles, the far field electric and magnetic field radiation patterns for an infinitesimal dipole.

[2] Explain the notion of electric vector potential and magnetic current density.

[3] A circular loop of wire of radius R carries a current $I_0exp(j\omega t)$. Determine the first two terms in the expansion of the far field magnetic vector potential in powers of R.

[4] Explain the terms radiation resistance R_r, scattering area A_s, loss area A_l, loss resistance R_l, load resistance R_L and terminal resistance R_T.

[5] Suppose that an antenna has a gain function $G(\theta, \phi)$ in the direction θ, ϕ. Write down the expression for the antenna directivity. Calculate the directivity of a finite wire antenna of length L carrying current $I(z) = I_0(1 - 2|z|/L), |z| \leq L$

More problems in antenna theory:

1. Consider a spherical surface of radius R that carries a current density

$$J_\theta(\theta,\phi)\hat{\theta}+J_\phi(\theta,\phi)\hat{\phi}$$

Determine expressions for the far field magnetic vector potential and the far field electric and magnetic fields.

2. A parabola is defined as the locus of a point the ratio of whose distances from a point P and from a line is equal to unity. The point P is called the focus. Show that the equation of a parabola in the xy plane can be expressed as $y^2 = 4ax$. Show that the focus of such a parabola is given by $P = (a, 0)$. A paraboloid is a surface obtained by rotating a parabola about the normal to its apex. Show that the equation of a paraboloid can be expressed as

$$x^2+y^2=4az$$

or equivalently,

$$z=\rho^2/4a$$

The equation of the paraboloid can be expressed in parametric form as

$$\mathbf{r}(\rho,\phi)=\rho.cos(\phi)\hat{x}+\rho.sin(\phi)\hat{y}+(\rho^2/4a)\hat{z}$$

We calculate the tangent vectors on this surface relative to the parametric coordinates (ρ, ϕ). These are given by

$$\mathbf{e}_\rho=\frac{\partial\mathbf{r}}{\partial\rho}=\hat{x}.cos(\phi)+\hat{y}.sin(\phi)+(\rho/2a)\hat{z},$$

$$\mathbf{e}_\phi=\rho^{-1}\frac{\partial\mathbf{r}}{\partial\phi}=-sin(\phi)\hat{x}+cos(\phi)\hat{y}$$

The surface current density can be expressed as

$$\mathbf{J}(\rho,\phi)=J_\rho(\rho,\phi)\mathbf{e}_\rho+J_\phi(\rho,\phi)\mathbf{e}_\phi$$

$$=J_\rho(\hat{x}.cos(\phi)+\hat{y}.sin(\phi)+(\rho/2a)\hat{z})+J_\phi(-\hat{x}.sin(\phi)+\hat{y}.cos(\phi))$$

$$=[J_\rho cos(\phi)-J_\phi sin(\phi)]\hat{x}+[J_\rho sin(\phi)+J_\phi cos(\phi)]\hat{y}+(\rho/2a)J_\rho\hat{z}$$

The differential area element on the surface is given by

$$dS(\rho,\phi)=|\frac{\partial\mathbf{r}}{\partial\rho}\times\frac{\partial\mathbf{r}}{\partial\phi}|d\rho.d\phi$$

We have

$$\mathbf{e}_\rho\times\mathbf{e}_\phi=\hat{z}-(\rho/2a)sin(\phi)\hat{y}-(\rho/2a)cos(\phi)\hat{x}$$

so that

$$|\mathbf{e}_\rho\times\mathbf{e}_\phi|^2$$

$$=1+(\rho/2a)^2$$

and hence

$$dS(\rho, \phi) = \rho\sqrt{1 + \rho^2/4a^2}d\rho.d\phi$$

The magnetic vector potential can now be evaluated as

$$\mathbf{A}(r, \theta, \phi) = \frac{\mu.exp(-jkr)}{4\pi r} int\mathbf{J}(\rho', \phi').exp(jk(\rho'.sin(\theta)cos(\phi - \phi') + \rho'^2 cos(\theta)/4a))dS(\rho', \phi')$$

Remark on transmission lines: Consider a lossy transmission line with propagation constant $\gamma = \alpha + j\beta$. Let Z_0 be the characteristic impedance of the line. Assume the line to be of length L and terminated by a load impedance Z_L. Then from the basic differential equations for voltage and current along the line, we have

$$V(z) = V_+exp(-\gamma z) + V_-exp(\gamma z), I(z) = (V_+exp(-\gamma z) - V_-exp(\gamma z))/Z_0$$

Since

$$V(L) = I(L)Z_L,$$

we have with

$$r = V_-/V_+,$$

$$1 + r.exp(2\gamma L) = [1 - r.exp(2\gamma L)]Z_L/Z_0$$

so that

$$r.exp(2\gamma L)(Z_L/Z_0 + 1) = Z_L/Z_0 - 1$$

so that

$$r = \frac{Z_L - Z_0}{Z_L + Z_0} exp(-2\gamma L)$$

If the line is infinitely long, ie, $L \to \infty$, then we get

$$r = 0$$

in which case, $V_- = 0$ and the input impedance of the line is then given by

$$Z_{in} = V(0)/I(0) = Z_0(V_+ + V_-)/(V_+ - V_-) = Z_0$$

Thus an infinitely long line presents and impedance equal to the characteristic impedance of the line.

Problem: Consider a parallel wire transmission line in which each line has length l and is of radius R. Let d denote the distance between the centres of the two wires. Determine the characteristic impedance of such a line.

Hint: Let C, L denote respectively the capacitance and inductance of the line per unit length. Then the characteristic impedance is given by $Z_0 = \sqrt{L/C}$. Let I denote the current through the line. Then at a distance x from the axis of the first wire, the magnetic field in the plane of the two wires is given by

$$H = \frac{I}{2\pi}(\frac{1}{x} + \frac{1}{d - x}), R \leq x \leq d - R$$

and hence the flux of the magnetic field through the plane spanned by the wires between them is given by

$$\Phi = \int_R^{d-R} \mu H.l.dx = \frac{\mu I}{\pi} log((d-R)/R) \approx \frac{\mu I}{\pi} log(d/R)$$

This gives for the inductance of the line the formula

$$L = \Phi/I = \frac{\mu}{\pi} log(d/R)$$

Similarly, we can determine the capacitance of the line. Consider the complex potential

$$F(z) = A.log((z-a)/z-b))$$

where a, b are real numbers. Its real part is

$$u(x,y) = ReF(z) = A.log|(z-a)/(z-b)| = A.log(|z-a|/|z-b|)$$

Consider the level surface (equipotential)

$$u(x,y) = C$$

This is the same as

$$|z-a|/|z-b| = \alpha = exp(C/A)$$

when expanded, this gives

$$|z|^2 + a^2 - 2ax = |z-a|^2 = \alpha^2|z-b|^2 = \alpha^2(|z|^2 + b^2 - 2bx)$$

or

$$|z|^2(\alpha^2 - 1) + 2(a - b\alpha^2)x = a^2 - \alpha^2 b^2$$

Since

$$|z|^2 = x^2 + y^2$$

this is a circle with centre on the x axis. Thus, the equipotential surfaces are circles in the plane with centres on the x axis.

Charge in random motion, statistics of the radiation field produced by: Assume that a charge Q moves along the z axis in according to the law

$$x''(t) + \gamma x'(t) + \omega^2 x(t) = f(t) = \sigma.dB(t)/dt$$

where $B(.)$ is a Brownian motion process. This equation can be expressed in stochastic differential form as

$$dx(t) = v(t)dt, dv(t) = -\omega^2 x(t) - \gamma v(t)dt + \sigma dB(t)$$

or in vector-matrix form as

$$d\begin{pmatrix} x(t) \\ v(t) \end{pmatrix} = \begin{pmatrix} 0 & 1 \\ -\omega^2 & -\gamma \end{pmatrix} \begin{pmatrix} x(t) \\ v(t) \end{pmatrix} dt$$

$$+\sigma\begin{pmatrix}0\\1\end{pmatrix}dB(t)$$

Assuming zero initial conditions, the solution to this sde is given by

$$\begin{pmatrix}x(t)\\v(t)\end{pmatrix}=\sigma\int_0^t exp\{(t-s)\mathbf{A}\}\begin{pmatrix}0\\1\end{pmatrix}dB(s)$$

where

$$\mathbf{A}=\begin{pmatrix}0 & 1\\-\omega^2 & -\gamma\end{pmatrix}$$

The Laplace transform of $exp(t\mathbf{A})$ is given by (assuming appropriate region of convergence)

$$\int_0^\infty exp(-(sI-A)t)dt=(sI-A)^{-1}$$

Now,

$$(sI-A)=\begin{pmatrix}s & -1\\\omega^2 & s+\gamma\end{pmatrix}$$

so that

$$(sI-A)^{-1}=(s^2+\gamma s+\omega^2)^{-1}\begin{pmatrix}s+\gamma & 1\\-\omega^2 & s\end{pmatrix}$$

It follows on taking Laplace inverse that

$$exp(tA)=\begin{pmatrix}b_{11}(t) & b_{12}(t)\\b_{21}(t) & b_{22}(t)\end{pmatrix}$$

where

$$b_{11}(t)=\mathcal{L}^{-1}(\frac{s+\gamma}{s^2+\gamma s+\omega^2})$$

$$=\mathcal{L}^{-1}\frac{s+\gamma/2+\gamma/2}{(s+\gamma/2)^2+\omega^2-\gamma^2/4}$$

$$=exp(-\gamma t/2)(cos(t\sqrt{\omega^2-\gamma^2/4})+\frac{\gamma}{2\sqrt{\omega^2-\gamma^2/4}}sin(t\sqrt{\omega^2-\gamma^2/4}))$$

$$b_{12}(t)=\mathcal{L}^{-1}\frac{1}{s^2+\gamma s+\omega^2}=\frac{sin(\sqrt{\omega^2-\gamma^2/4t})}{\sqrt{\omega^2-\gamma^2/4}}exp(-\gamma t/2)$$

$$b_{21}(t)=-\frac{\omega^2}{\sqrt{\omega^2-\gamma^2/4}}sin(\sqrt{\omega^2-\gamma^2/4t})exp(-\gamma t/2)$$

$$b_{22}(t)=\mathcal{L}^{-1}\frac{s}{s^2+\gamma s+\omega^2}=\mathcal{L}^{-1}\frac{s+\gamma/2-\gamma/2}{s^2+\gamma s+\omega^2}$$

$$=[cos(t\sqrt{\omega^2-\gamma^2/4})-\frac{\gamma}{2\sqrt{\omega^2-\gamma^2/4}}sin(t\sqrt{\omega^2-\gamma^2/4})]exp(-\gamma t/2)$$

It follows that the solution to the sde describing harmonic motion in damping can be expressed as

$$x(t) = \sigma \int_0^t b_{12}(t-s)dB(s), v(t) = \sigma \int_0^t b_{22}(t-s)dB(s)$$

The position and velocity correlations can now be easily computed.

$$< x(t_1)x(t_2) >= \sigma^2 \int_0^{min(t_1,t_2)} b_{12}(t_1 - s)b_{12}(t_2 - s)ds,$$

$$< x(t_1)v(t_2) >= \int_0^{min(t_1,t_2)} b_{12}(t_1 - s)b_{22}(t_2 - s)ds,$$

$$< v(t_1)v(t_2) >= \int_0^{min(t_1,t_2)} b_{22}(t_1 - s)b_{22}(t_2 - s)ds$$

The vector potential produced by the charge in the far field approximation is approximately given by

$$A_z = \frac{\mu v(t)}{4\pi(x^2 + y^2 + (z - x(t))^2)^{1/2}}$$

The exact formula for the vector potential is obtained by replacing t by the retarded time and incoroporating a relativistic factor. We do not go into that here. We expand the vector potential in powers of $x(t)$ and retain only upto first order terms. We have

$$(x^2 + y^2 + (z - x(t))^2)^{-1/2} = (r^2 - 2zx(t))^{-1/2} = r^{-1} + r^{-3}zx(t)$$

$$= r^{-1} - r^{-2}cos(\theta)x(t)$$

and hence

$$A_z \approx \frac{\mu}{4\pi}v(t)(r^{-1} - r^{-2}x(t).cos(\theta))$$

From this formula, the electromagnetic field can be computed and therefrom, the correlations in the fields at different space-time points may be evaluated. We find for the vector potential correlations

$$< A_z(r_1, \theta_1, t_1)A_z(r_2, \theta_2, t_2) >$$

$$= \frac{\mu^2}{16\pi^2}[(r_1r_2)^{-1} < v(t_1)v(t_2) > +(r_1r_2)^{-2}cos(\theta_1)cos(\theta_2) < v(t_1)v(t_2)x(t_1)x(t_2) >]$$

the triple moments of the form $< v(t_1)v(t_2)x(t_i) >$ vanish since the processes $x(.)$ and $v(.)$ are zero mean Gaussian processes.

12.The Feynman lectures on physics, vols.II and III.

Vol. II is mainly about the story of electromagnetism. Feynman introduces the subject by discussing electrostatics problems appearing in nature such as transmission lines and slowly develops the subject of electrodynamics. Displacement current is introduced by him by considering the charging of capacitors. The Coulomb solution to electrostatics problems and the superposition principle for electric fields is discussed at length by him. The integral form of the Maxwell equations are discussed by Feynman and how they lead to the differential form by making use of Gauss' divergence theorem and Stokes theorem. The wave equation for the propagation of electromagnetic waves in space is discussed at length by Feynman and the complete solution to the wave equation in the form of retarded potentials is also explained without giving a detailed mathematical proof. Feynman explains by means of physical arguments how the acceleration term $\frac{1}{c^2}\frac{\partial^2\psi}{\partial t^2}$ leads to the retarded potential solution. The general solution to the wave equation is a superposition of the retarded and advanced potentials and for causality purposes, we retain only the retarded potential solution. This fact has not been explained in Feynman's lectures. The detailed derivation of the solution to the wave equation involves taking the spatio-temporal Fourier transform of the equation and then inverting the temporal part of the Fourier transform by means of contour integration. The spatial part of the Fourier transform is inverted then in the usual way. These detailed arguments can be found in the classic book by Jackson "Classical Electrodynamics". Feynman explains Ampere's law by means of the illustrative example of two parallel wires carrying current. The magnetic field produced by the first wire sets up a magnetic field which interacts with the current in the second wire thereby causing a force to be exerted on the second wire. Feynman observes that chronologically, Ampere's law was deduced from this force equation. Feynman also observes that the force exerted by the magnetic field of a current carrying wire on another does not follow the law of action and reaction. This fact is peculiar to the magnetic field. Feynman discusses the law of electromagnetic induction as originally discovered experimentally and formulated mathematically by Faraday and its various applications especially to the generation of electricity. Without this law, the modern era of electricity would not exist. Feynman, in his discussion on Maxwell's equations introduces several tools of vector calculus and gives intuitive ways for proving these results. For example, he proves the formula $div(A \times B) = (B, curlA) - (A, curlB)$, or rather gives a way for remembering this formula by using elementary properties of the algebraic vector triple product.

21.Landau and Lifshitz, Classical theory of fields: The book focuses around two of the great theories that have dominated physics, one, Maxwell's theory of electromagnetism and two, Einstein's general relativistic theory of gravitation. The special relativistic covariant formulation of Maxwell's field equations have been given. The Lorentz transformations of four vectors and tensors have been introduced. Using this string of ideas, the Lorentz transformation of the electromagnetic antisymmetric field tensor between two inertial frames has been obtained. Motion of charged particles in an electromagnetic field has been described via the Lagrangian formalism as well as the Hamiltonian formalism. The Lagrangian formalism involves first of all writing down the Lagrangian of a system of point charges in an external electromagnetic field and then penning down the Euler-Lagrange equations of motion. The result is the Newton's second law of motion with a driving Lorentz force. The Hamiltonian of the system of charged particles in an external electromagnetic field has been obtained using the Legendre transformation. It has the form

$$\mathcal{H}(\mathbf{r}, \mathbf{p}) = \sum_{i=1}^{N} (\mathbf{p}_i - \mathbf{A}(t, \mathbf{r}_i))^2/2m_i + \sum_{i=1}^{N} Q_i V(t, \mathbf{r}_i)$$

The equations of motion are then written down. The Hamilton-Jacobi method for charged particle motion has also been discussed in considerable detail. The method as outlined in one of the preceding volumes on mechanics, involves solving the partial differential equation

$$H(\mathbf{q}, \nabla_q S(\mathbf{q}), t) + \frac{\partial S(\mathbf{q}, t)}{\partial t} = 0$$

for the generating function S and using that to arrive at the equations of motion. The Lagrangian of a system of charged particles moving in the electromagnetic field generated by the charged particles themselves has also been obtained. Approximations to this Lagrangian have been obtained. Finally, application of the Legendre transformation to arrive at the Hamiltonian of the system of particles has been carried out. Before however doing so, the retarded potential solution to the Maxwell equations has been discussed. This solution has been obtained by using the Fourier transformation of the wave equation with source. Approximate versions of the retarded potential solution have been derived by expanding $\psi(t-r/c)$ in powers of $1/c$. Geometric optics and the Eikonal equation has been discussed in considerable detail. This involves plugging the form $\psi(t,q) = A(t,q)exp(iS(t,q))$ into the wave equation and making approximations regarding the behaviour of the functions A, S. Fraunhoffer and Fresnel diffraction theories have been discussed after that. The starting point for diffraction theory is the frequency domain version of the retarded potential solution. Consider an aperture on which is incident and electromagnetic field. The electric and magnetic field generate equivalent electric and magnetic surface current densities on the aperture surfaces. These are respectively given by $\mathbf{J}_S = \mathbf{n} \times \mathbf{H}$ and $\mathbf{M}_S = \mathbf{n} \times \mathbf{E}$. These equivalent sources are then used to compute the magnetic and electic vector potentials using aperture integration for the retarded potentials and thereby the near and far field patterns are obtained.

22. J.D.Jackson, Classical Electrodynamics:

1. The first two chapters focus on electrostatics. The highlights of these chapters is mainly techniques for computing the Green's function for various kinds of boundary value problems in electrostatics. As an example, the Green's function $G(x,y|x',y')$ for the two dimensional electrostatics problem with Dirichlet boundary conditions when the boundary is a rectangle $[0,a] \times [0,b]$ satisfies the differential equation

$$\frac{\partial^2 G(x,y|x',y')}{\partial x^2} + \frac{\partial^2 G(x,y|x',y')}{\partial y^2} = \delta(x-x')\delta(y-y')$$

for $0 < x, x' < a$ and $0 < y, y' < b$ with the boundary condition

$$G(x,y|x',y') = 0$$

when either $x = 0$ or $x = a$ and $0 < y < b$ or when $y = 0$ or $y = b$ and $0 < x < a$. The solution to this problem is given in the form of a Fourier series. The corresponding three dimensional problem reads

$$\Delta G(x,y,z|z',y',z') = \delta(x-x')\delta(y-y')\delta(z-z')$$

for $(x,y,z), (x',y',z') \in [0,a] \times [0,b] \times [0,c]$ with $G(x,y,z|x',y',z') = 0$ when either $x = 0$ or $x = a$ and $0 < y < b, 0 < z < c$ or $y = 0$ or $y = b$ and $0 < x < a, 0 < z < c$, or when $z = 0$ or $z = c$ and $0 < x < a$ and $0 < y < c$.

2. A problem:. Derive the dispersion relation for electromagnetic waves in a homogeneous medium having anisotropic refractive index.

ans:

$$\nabla \times E = -j\omega\mu H,$$

$$\nabla \times H = j\omega\epsilon.E$$

where μ is a scalar but ϵ is a 3×3 constant matrix. We also have Gauss' law $div(\epsilon.E) = 0$. Taking the curl of the first equation gives

$$\nabla(divE) - \nabla^2 E = -j\omega\mu curlH = -j\omega\mu(j\omega\epsilon.E)$$

$$= \omega^2\mu\epsilon.E$$

Note that this equation gives on taking divergence

$$div(\epsilon.E) = 0$$

and hence Gauss' law is superfluous. Assume a plane wave of the form

$$\mathbf{E} = \mathbf{E}_0 exp(i(\omega t - \mathbf{k.r}))$$

Plugging this into the above equation gives us

$$-\mathbf{k}(\mathbf{k.E}_0) + k^2\mathbf{E}_0 = \omega^2\mu\epsilon.\mathbf{E}_0$$

In matrix vector notation, $\mathbf{E}_0$ is a 3×1 column vector and so is $\mathbf{k}$. We can then arrange the above equation as

$$-\mathbf{kk}^T\mathbf{E}_0 + k^2\mathbf{E}_0 = \omega^2\mu\epsilon.\mathbf{E}_0 = 0$$

or

$$(\omega^2\mu\epsilon + \mathbf{kk}^T - k^2\mathbf{I})\mathbf{E}_0 = 0$$

which implies the dispersion relation

$$det(\omega^2\mu\epsilon + \mathbf{kk}^T - k^2\mathbf{I}) = 0$$

Note that ϵ is a 3×3 matrix.

3. Dipole radiation: The direction of radiation (beam direction) is the x axis. We $\mathbf{d}''$ is the rate of change of rate of change of the dipole moment. The radiated power is proportional to $|n \times d''|^2$. We can write this as

$$|n \times d''|^2 = d''^2 - (n, d'')^2$$

Let us average this expression over all directions perpendicular to the x direction. We have

$$d''^2 = d_x''^2 + d_y''^2 + d_z''^2$$

and its average is also denoted by d''^2.

Let us calculate the average of $(n, d'')^2$. Let θ denote the angle between $\mathbf{n}$ and the x axis. Then,

$$(n, d'')^2 = n_x^2 d_x''^2 + n_y^2 d_y''^2 + n_z^2 d_z''^2 + 2n_x n_y d_x'' d_y'' + 2n_y n_z d_y'' d_z'' + 2n_z n_x d_z'' d_x''$$

The average value of $d''_y d''_x$ is zero since d''_x is fixed. Likewise, the average of $d''_z d''_x$ is also zero. The average of $d''_y d''_z$ is also zero. We note that $n_x^2 = cos^2(\theta)$. We also note that the average of $d_y''^2$ equals $(d''^2 - d_x''^2)/2$ and this is also the average of $d_z''^2$. It follows that the average of $(n, d'')^2$ equals

$$cos^2(\theta)d_x''^2 + (d''^2 - d_x''^2)(n_y^2 + n_z^2)/2 = cos^2(\theta)d_x''^2 + (d''^2 - d_x''^2)sin^2(\theta)/2$$

$$= d_x''^2 cos^2(\theta) + (d''^2 - d_x''^2)(1 - cos^2(\theta))/2$$

and hence, the average of $|n \times d''|^2$ equals

$$d''^2 - d_x''^2 cos^2(\theta) - (d''^2 - d_x''^2)(1 - cos^2(\theta))/2$$

$$= (d''^2 + d_x''^2)/2 + cos^2(\theta)((d''^2 - d_x''^2)/2 - d_x''^2)$$

$$= (d''^2 + d_x''^2)/2 + (d''^2 - 3d_x''^2)cos^2(\theta)/2$$

(Ref:Landau and Lifshitz, Classical theory of fields).

4.Electromagnetics(Fourier analysis of radiation fields): Let $\rho(t, \mathbf{r})$ be the charge density in space. Consider its Fourier resolution:

$$\hat{\rho}(\omega, \mathbf{r}) = \frac{1}{2\pi}\int_{-\infty}^{\infty} \rho(t, \mathbf{r})exp(-i\omega t)dt$$

It follows that

$$\rho(t, \mathbf{r}) = \int_{-\infty}^{\infty} \hat{\rho}(\omega, \mathbf{r})exp(i\omega t)d\omega$$

Thus, the scalar potential is given by(Feynman, lectures on physics Vol.II and Landau and Lifshitz, Classical theory of fields)

$$\phi(t, \mathbf{r}) = \int \rho(t - |\mathbf{r} - \mathbf{r}'|/c, \mathbf{r}')d^3r'/|\mathbf{r} - \mathbf{r}'|$$

$$= \int \hat{\rho}(\omega, \mathbf{r}')exp(i\omega(t - |\mathbf{r} - \mathbf{r}'|/c))d^3r'd\omega/|\mathbf{r} - \mathbf{r}'|$$

$$= \int \hat{\phi}(\omega, \mathbf{r})exp(i\omega t)d\omega$$

where

$$\hat{\phi}(\omega, \mathbf{r}) = \int \hat{\rho}(\omega, \mathbf{r}')exp(-i\omega|\mathbf{r} - \mathbf{r}'|/c)d^3r'/|\mathbf{r} - \mathbf{r}'|$$

Likewise, if the current density has the Fourier resolution

$$\mathbf{j}(t, \mathbf{r}) = \int \hat{\jmath}(\omega, \mathbf{r})exp(i\omega t)$$

then the vector potential will have the resolution

$$\mathbf{A}(t, \mathbf{r}) = \int \hat{A}(\omega, \mathbf{r})exp(i\omega t)d\omega$$

where

$$\hat{A}(\omega,\mathbf{r})=\int \hat{j}(\omega,\mathbf{r}')exp(-i\omega|\mathbf{r}-\mathbf{r}'|/c)d^3r'/|\mathbf{r}-\mathbf{r}'|$$

Consider a moving charge whose position at time t is given by $\mathbf{R}(t)$. Then,

$$\rho(t,\mathbf{r})=Q.\delta(\mathbf{r}-\mathbf{R}(t))$$

and

$$\hat{\phi}(\omega,\mathbf{r})=\frac{Q}{2\pi}\int \delta(\mathbf{r}'-\mathbf{R}(t))exp(-i\omega t)exp(-i\omega|\mathbf{r}-\mathbf{r}'|/c)d^3r'dt/|\mathbf{r}-\mathbf{r}'|$$

$$=\frac{Q}{2\pi}\int exp(-i\omega t)exp(-i\omega|\mathbf{r}-\mathbf{R}(t)|)dt/|\mathbf{r}-\mathbf{R}(t)|$$

Likewise,

$$\mathbf{j}(t,\mathbf{r})=Q\mathbf{V}(t)\delta(\mathbf{r}-\mathbf{R}(t)),\mathbf{V}(t)=\mathbf{R}'(t)$$

and hence

$$\hat{A}(\omega,\mathbf{r})=\frac{Q}{4\pi}\int \mathbf{V}(t)exp(-i\omega(t+|\mathbf{r}-\mathbf{R}(t)|))dt/|\mathbf{r}-\mathbf{R}(t)|$$

5.Fields produced by a charge in uniform motion:Let $\mathbf{v}$ denote the velocity of the charge. Then the charge density is given by $\rho(t,\mathbf{r})=Q\delta(\mathbf{r}-\mathbf{v}t)$. The scalar potential $\phi(t,\mathbf{r})$ satisfies the wave equation with source

$$\Delta\phi-\frac{1}{c^2}\frac{\partial\phi}{\partial t^2}=-\rho/\epsilon_0$$

We define the spatial Fourier transforms:

$$\hat{\phi}(t,\mathbf{k})=\int \phi(t,\mathbf{r})exp(-i\mathbf{k}.\mathbf{r})d^3r,$$

$$\hat{\rho}(t,\mathbf{k})=\int \rho(t,\mathbf{r})exp(-i\mathbf{k}.\mathbf{r})d^3r$$

Then from the above wave equation, it follows that

$$\frac{\partial^2\hat{\phi}}{\partial t^2}=-k^2c^2\hat{\phi}+\hat{\rho}$$

Observe that

$$\hat{\rho}(t,\mathbf{k})=Q\int \delta(\mathbf{r}-\mathbf{v}t)exp(-i\mathbf{k}.\mathbf{r})d^3r=Q.exp(-i\mathbf{k}.\mathbf{v}t)$$

so that

$$\hat{\phi}_{tt}$$

6.Relationship between the electric displacement vector and the electric field vector relative to principal axes. In the xy plane, the electric displacement vector $\mathbf{D} = [D_x, D_y]^T$ is related to the electric field $\mathbf{E} = [E_x, E_y]^T$ via the matrix equation

$$D_x = \epsilon_{xx}E_x + \epsilon_{xy}E_y, D_y = \epsilon_{yx}E_x + \epsilon_{yy}E_y$$

If we assume that the permittivity matrix

$$\epsilon = \begin{pmatrix} \epsilon_{xx} & \epsilon_{xy} \\ \epsilon_{yx} & \epsilon_{yy} \end{pmatrix}$$

is symmetric, ie, $\epsilon_{xy} = \epsilon_{yx}$, then ϵ can be diagonalized using an orthogonal matrix T, ie we can write

$$\epsilon = T\Lambda T^t$$

where

$$\Lambda = diag[\epsilon_1, \epsilon_2]$$

with $\epsilon_i, i = 1, 2$ being the eigenvalues of ϵ and $\mathbf{T} = [\mathbf{v}_1, \mathbf{v}_2]$ are such that $\mathbf{v}_i, i = 1, 2$ are the eigenvectors of ϵ corresponding to the two eigenvalues $\epsilon_i, i = 1, 2$. $\mathbf{v}_1, \mathbf{v}_2$ are the corresponding eigenvectors. The eigenvalues $\epsilon_{1,2}$ are determined as roots of $det(\lambda.I - \epsilon) = 0$. This equation is the same as

$$(\lambda - \epsilon_{xx})(\lambda - \epsilon_{yy}) - \epsilon_{xy}^2 = 0$$

or

$$\lambda^2 - (\epsilon_{xx} + \epsilon_{yy})\lambda + (\epsilon_{xx}\epsilon_{yy} - \epsilon_{xy}^2) = 0$$

The roots are

$$\epsilon_{1,2} = \frac{\epsilon_{xx} + \epsilon_{yy}}{2} \pm \frac{1}{2}\sqrt{\epsilon_{xx} - \epsilon_{yy})^2 + 4\epsilon_{xy}^2}$$

The eigenvectors can be determined from this easily. Corresponding to the eigenvalue λ, let $\mathbf{v}_\lambda = [v_1, v_2]^T$ be the eigenvector. Then,

$$\epsilon_{xx}v_1 + \epsilon_{xy}v_2 = \lambda v_1$$

The second equation

$$\epsilon_{xy}v_1 + \epsilon_{yy}v_2 = \lambda v_2$$

will automatically be satisfied if the first is by virtue of the eigenvalue equation $det(\lambda.I - \epsilon) = 0$. Thus,

$$v_2 = (\lambda - \epsilon_{xx})v_1/\epsilon_{xy}$$

and the normalized eigenvector $\mathbf{v}_\lambda$ is given by

$$\mathbf{v}_\lambda = (\epsilon_{xy} + (\lambda - \epsilon_{xx})^2)^{-1/2}[\epsilon_{xy}, (\lambda - \epsilon_{xx})]^T$$

Subsituting into this expression $\lambda = \epsilon_1, \epsilon_2$ successively gives us both the eigenvectors $\mathbf{v}_i = \mathbf{v}_{\epsilon_i}, i = 1, 2$. it can be easily verified that the $\{\mathbf{v}_1, \mathbf{v}_2\}$ is an orthonormal set, ie, $\mathbf{v}_i^T\mathbf{v}_j = \delta_{ij}, i, j = 1, 2$. We can say that along the direction $\mathbf{v}_1$, the permittivity is ϵ_1 and along the direction $\mathbf{v}_2$, the permittivity is ϵ_2. To see this more explicitly, we define $\mathbf{T} = [\mathbf{v}_1, \mathbf{v}_2]$. Then, the equation

$$\mathbf{D} = \epsilon\mathbf{E}$$

gives

$$\mathbf{D} = \mathbf{T}\Lambda\mathbf{T}^{-1}\mathbf{E}$$

Since $T^{-1} = T^t$ ($\mathbf{TT}^t = I$ by virtue of orthonormality of the eigenvectors), it follows that

$$\mathbf{T}^t\mathbf{D} = \Lambda\mathbf{T}^t\mathbf{E}$$

Define

$$\tilde{D} = T^t D, \tilde{E} = T^t E$$

Then

$$\tilde{D} = [D_1, D_2]^T, \tilde{E} = [E_1, E_2]^T$$

where

$$D_1 = \mathbf{v}_1^T\mathbf{D}, D_2 = \mathbf{v}_2^T\mathbf{D}, E_1 = \mathbf{v}_1^T\mathbf{E}, E_2 = \mathbf{v}_2^T\mathbf{E}$$

and

$$D_i = \epsilon_i E_i, i = 1, 2$$

Chapter 5

Relativity

1.Finite element formulation of the electromagnetic field equations in a waveguide in the presence of a gravitational field:

The four potential is given by

$$(A^\mu) = (V, A_x, A_y, A_z)$$

The metric of space-time is assumed to be Newtonian with the gravitational potential U independent of z

$$d\tau^2 = (1 + 2U(x,y)/c^2)dt^2 - \frac{1}{c^2}(dx^2 + dy^2 + dz^2)$$

so that

$$g_{00} = 1 + 2U/c^2, g_{11} = g_{22} = g_{33} = -1/c^2$$

The electromagnetic field tensor is given by

$$F_{\mu\nu} = A_{\nu,\mu} - A_{\mu,\nu}$$

The Lagrangian density is given by

$$L = F_{\mu\nu}F^{\mu\nu} = g^{\mu\alpha}g^{\nu\beta}F_{\mu\nu}F_{\alpha\beta}$$

The derivatives with respect to z are replaced by multiplication with $-\gamma$ and the derivatives with respect to time are replaced by multiplication by $i\omega$ and a negative sign is added before these. To see the reason for the negative sign, we consider a simple case in which one of the components of the action functional is $\int V_{,z}^2 dxdydzdt$. The variation of this gives on integration by parts, $\int V_{,z}\delta V_{,z}dxdydzdt = -\int V_{,zz}\delta V dxdydzdt$. The coefficient of δV is $-V_{,zz} = -\gamma^2 V$. If in the original action, we replace $V_{,z}$ by $-\gamma.V$, then we get the term $\gamma^2 \int V^2 dxdydzdt$ which on variation gives $\gamma^2 \int V\delta V dxdydzdt$ and the coefficient of δV is $\gamma^2 V$. This shows that in the replacement of $V_{,z}$ by γV in the original action, a minus sign must be introduced to obtain the correct field equations. The same goes for derivative with respect to time. This can also be seen to be true for terms of the form $\int \psi.V_{,z}dxdydzdt$. Taking the variation with respect to V, integrating by parts and selecting the coefficient of δV gives us $-\psi_{,z}$ which is $\gamma\psi$. On the other hand, if we directly make the replacement $V_{,z} \rightarrow -\gamma V$ in the action, we get $-\gamma \int .\psi.V dxdydzdt$.

Taking the variation with respect to V and selecting the coefficient of δV gives us $-\gamma\psi$ and to get agreement with the former, a negative sign must be introduced. Now,

$$L = 2g^{00}g^{11}F_{01}^2 + 2g^{00}g^{22}F_{02}^2 + 2g^{00}g^{33}F_{03}^2 + 2g^{11}g^{22}F_{12}^2 + 2g^{22}g^{33}F_{23}^2 + 2g^{11}g^{33}F_{31}^2$$

The first term contains

$$F_{01}^2 = (A_{1,0} - A_{0,1})^2 = A_{1,0}^2 + A_{0,1}^2 - 2A_{1,0}A_{0,1}$$

As per the above rule, this term is to be replaced by

$$\omega^2 A_1^2 + A_{0,1}^2 + 2j\omega.A_1 A_{0,1} = \omega^2(g_{11}A_x)^2 + (g_{00}V)_{,x}^2 + 2j\omega.g_{11}A_x(g_{00}V)_{,x}$$

Likewise the second term contains F_{02}^2 and this must be replaced with

$$\omega^2(g_{22}A_y)^2 + (g_{00}V)_{,y}^2 + 2j\omega g_{11}A_y(g_{00}V)_{,y}$$

The third term contains

$$F_{03}^2 = (A_{3,0} - A_{0,3})^2 = A_{3,0}^2 + A_{0,3}^2 - 2A_{3,0}A_{0,3}$$

and as per the above prescription, this must be replaced with

$$\omega^2(g_{33}A_z)^2 - \gamma^2 g_{00}V^2 - 2j\omega\gamma.g_{33}g_{00}A_z V$$

The sign of the last term can be understood as follows. Consider the variation of $A_{3,0}A_{0,3}$ with respect to A_3. This gives on integrating by parts, $-A_{0,30}\delta A_3$ and the coefficient of δA_3 here is $-A_{0,30} = j\omega\gamma A_0$. On the other hand, replacing $A_{3,0}$ and $A_{0,3}$ respectively by $j\omega A_3$ and $-\gamma A_0$ reduces the term $A_{3,0}A_{0,3}$ to $-j\omega\gamma A_3 A_0$ and the variation with respect to A_3 gives $-j\omega\gamma A_0$. To get agreement, we must thus introduce an extra negative sign here. The fourth term contains F_{12}^2. This evaluates to

$$F_{12}^2 = (A_{2,1} - A_{1,2})^2 = ((g_{22}A_y)_{,x}^2 + (g_{11}A_x)_{,y}^2 - 2A_{y,x}A_{x,y}$$

No modification here. The fifth term contains F_{23}^2. This evaluates to

$$F_{23}^2 = (A_{3,2} - A_{2,3})^2 = A_{3,2}^2 + A_{2,3}^2 - 2A_{3,2}A_{2,3}$$

and this must in accord to the above rules be replaced with

$$(g_{33}A_z)_{,y}^2 - \gamma^2 A_y^2 - 2\gamma(g_{33}A_z)_{,y}g_{22}A_y$$

Finally, the sixth term contains F_{31}^2. This evaluates to

$$F_{31}^2 = (A_{1,3} - A_{3,1})^2 = A_{1,3}^2 + A_{3,1}^2 - 2A_{1,3}A_{3,1}$$

This must be replaced by

$$-\gamma^2(g_{11}A_x)^2 + (g_{33}A_z)_{,x}^2 - 2\gamma g_{11}A_x(g_{33}A_z)_{,x}$$

Taking $c = 1$ and substituting for metric coefficients, we get the modified Lagrangian density as

$$L = -2(1-2U)(\omega^2(A_x)^2 + ((1+2U)V)_{,x}^2 - 2j\omega.A_x((1+2U)V)_{,x})$$

$$-2(1-2U)(\omega^2(A_y)^2+((1+2U)V)^2_{,y}-2j\omega.A_y((1+2U)V)_{,y})$$
$$-2(1-2U)(\omega^2(A_z)^2-\gamma^2(1+2U)V^2+2j\omega\gamma.(1+2U)A_zV)$$
$$+2(A^2_{y,x}+A^2_{x,y}-2A_{y,x}A_{x,y})$$
$$+2(A^2_{z,y}-\gamma^2A^2_y-2\gamma A_{z,y}A_y)$$
$$+2(-\gamma^2A^2_x+A^2_{z,x}-2\gamma A_xA_{z,x})$$

Note that we have assumed the gravitational potential U to be weak, ie, $|U|<<1$ so that g^{00} can be taken as $1-2U$. We further simplify the above expression so that $O(U^2)$ terms do not appear. We then get

$$L=-2\omega^2(1-2U)A^2_x-2(1-2U)(V_{,x})^2-8(UV)_{,x}V_{,x}+4j\omega(1-2U)A_xV_{,x}+8j\omega A_x(UV)_{,x}$$
$$-2\omega^2(1-2U)A^2_y-2(1-2U)V^2_{,y}-8(UV)_{,y}V_{,y}+4j\omega(1-2U)A_yV_{,y}+8j\omega A_y(UV)_{,y}$$
$$-2\omega^2(1-2U)A^2_z-2\gamma^2(1-2U)V^2+4\gamma^2UV^2-4j\omega\gamma(1-2U)A_zV-4j\omega\gamma.U.A_z.V$$
$$+2(A^2_{y,x}+A^2_{x,y}-2A_{y,x}A_{x,y})+2(A^2_{z,y}-\gamma^2A^2_y-2\gamma A_{z,y}A_y)$$
$$+2(-\gamma^2A^2_x+A^2_{z,x}-2\gamma A_xA_{z,x})$$

We can write this equation as

$$L=F+UG$$

where F,G depend only on the electromagnetic four potentials. The Action principle involves

$$\delta\int L\sqrt{-g}dxdydt=0$$

and we have

$$L\sqrt{-g}=L(1+U)=(F+U.G)(1+U)=F+U(F+G)$$

with neglect of $O(U^2)$ terms. We divide the region into triangular elements and expand the four functions A_x,A_y,A_z,V as linear combination of their vertex values, calculate the integrals and carry out the minimization.

Remark: A term of the form $(UW)_{,x}F$ in the Lagrangian can be replaced by $-UWF_{,x}$ since the integral of the total derivative $(UWF)_{,x}$ is zero.

2.Determine the relativistic equations of hydrodynamics: Let $U^\mu=dx^\mu/d\tau$ be the four velocity vector of a fluid particle. Let $g_{\mu\nu}=diag[1,-1,-1,-1]$ be the covariant metric tensor. Then,

$$d\tau^2=g_{\mu\nu}dx^\mu dx^\nu=dt^2-dx^2-dy^2-dz^2=dt^2(1-v^2)$$

where

$$v^r=dx^r/dt,r=1,2,3,v^2=\sum_{r=1}^{3}(v^r)^2$$

Let ρ be the rest density (rest mass divided by rest volume) and p the pressure. Consider the tensor

$$T^{\mu\nu} = -pg^{\mu\nu} + (\rho + p)U^\mu U^\nu$$

In the rest frame, we have $U^0 = 1, U^r = 0, r = 1, 2, 3$ so that

$$T^{00} = -p + \rho + p = \rho, T^{rr} = p, r = 1, 2, 3$$

so that

$$T = diag[\rho, p, p, p]$$

The conservation equation

$$\frac{\partial T^{\mu\nu}}{\partial x^\nu} = 0$$

gives us

$$((\rho + p)U^\mu U^\nu)_{,\nu} = p_{,\nu}.g^{\mu\nu}$$

or equivalently,

$$((\rho + p)U^\nu)_{,\nu}U^\mu + ((\rho + p)U^\nu U^\mu_{,\nu} = p_{,\nu}g^{\mu\nu}$$

Multiply both sides by U_μ and use $U^\mu U_\mu = 1$ to get

$$((\rho + p)U^\nu)_{,\nu} = p_{,\nu}U^\nu$$

or equivalently,

$$(\rho U^\nu)_{,\nu} + pU^\nu_{,\nu} = 0$$

This is the relativistic modification of the equation of conservation of mass. For a nonrelativistic fluid, $p << \rho$ and the above approximates to

$$(\rho U^\mu)_{,\mu} = 0$$

which is the equation of mass conservation. Let n denote the number of fluid particles per unit volume. The number conservation equation reads

$$(nU^\alpha)_{,\alpha} = 0$$

This can be expressed as

$$n_{,\nu}U^\nu + nU^\nu_{,\nu} = 0$$

We shall combine this equation with the previous one which can be expressed as

$$(p + \rho)U^\nu_{,\nu} + \rho_{,\nu}U^\nu = 0$$

These two imply

$$-(p + \rho)(n_{,\nu}/n)U^\nu + \rho_{,\nu}U^\nu = 0$$

Now this equation is in fact the equation of entropy conservation. ρ is the mass per unit volume and n is the number of particles per unit volume. So ρ/n is the mass per particle. This is also the energy per particle. $1/n$ is the volume per particle and the energy equation in thermodynamics states that if s is the entropy per particle and T the temperature, then

$$T.ds = d(\rho/n) + p.d(1/n) = -pdn/n^2 - \rho.dn/n^2 + d\rho/n = -(p + \rho)dn/n^2 + d\rho/n$$

Also if X is any field in the fluid regime, its rate of change along the motion of a fluid particle is given by

$$dX = X_{,\mu}(dx^\mu/d\tau)d\tau = X_{,\mu}U^\mu d\tau$$

so that the above equation has the form

$$T.ds = 0$$

ie the entropy of each particle remains constant.(Weinberg, Gravitation and Cosmology). Let us look at the form assumed by the equation $T^{i\alpha}_{,\alpha} = 0, i = 1,2,3$. First observe that $v^i = dx^i/dt = (dx^i/d\tau)d\tau/dt$, or equivalently, $dx^i/d\tau = (dt/d\tau)v^i = U^0 v^i$. Note also that

$$U^0 = dt/d\tau = (1-v^2)^{-1/2} = \gamma(v)$$

Thus,

$$\begin{aligned}-p_{,i} &= ((p+\rho)U^iU^\nu)_{,\nu} = ((p+\rho)U^0v^iU^\nu)_{,\nu} = ((p+\rho)\gamma^2v^iv^j)_{,j} + ((p+\rho)\gamma^2v^i)_{,0} \\ &= (p+\rho)\gamma^2v^jv^i_{,j} + (p+\rho)\gamma^2v^j_{,j}v^i + (p+\rho)\gamma^2v^i_{,0} + ((p+\rho)\gamma^2)_{,0}v^i\end{aligned}$$

or

$$-[p_{,i}/\gamma^2(p+\rho)] = v^jv^i_{,j} + v^j_{,j}v^i + v^i_{,0} + ((p+\rho)\gamma^2)_{,0}v^i$$

This equation resembles the standard non-relativistic Navier-Stokes equation. Now recall the generalization of the mass conservation equation:

$$(p+\rho)U^\nu_{,\nu} + \rho_{,\nu}U^\nu = 0$$

Problem:Simplify the relativistic Navier-Stokes equation using this equation.

3.Define the Riemann Christoffel curvature tensor by considering the parallel transport of a vector around an infinitesimal closed curve. Show that if for any given infinitesimal closed curve around a point on the manifold, the parallel transport of any vector around the curve is zero, then the curvature tensor must vanish. Show that if the curvature tensor vanishes in a finite region of the manifold, then the metric on the manifold can be reduced to that of flat space.

4. Prove the Bianchi identity for the curvature tensor by passing to a geodesic reference frame.

5.Linearize the Riemann curvature tensor by considering a weak gravitational field. Hence, linearize the Einstein tensor and show that the Einstein field equations in this weak field limit yield Newton's theory of gravitation.

6. Linearize the Schwarzchild metric about the point (r,θ,ϕ) using a local cartesian coordinate system and formulate the Maxwell equations in a small neighbourhood of this point.

7.Metric does not depend on time and z coordinate. We can write

$$d\tau^2 = g_{00}dt^2 + g_{11}dx^2 + g_{22}dy^2 + g_{33}dz^2 + 2(g_{01}dtdx + g_{02}dtdy + g_{03}dtdz) + 2(g_{12}dxdy + g_{23}dydz + g_{13}dxdz)$$

where $g_{00}, g_{0i}, g_{ij}, i,j = 1,2,3$ are only functions of x,y, ie, they are independent of t,z. Let $g^{\mu\nu}$ denote the contravariant metric tensor, ie,

$$((g^{\mu\nu}(x,y))) = (((g_{\mu\nu}(x,y)))^{-1}$$

Let $A^\mu(x,y)$ be the contravariant four vector potential complex phasor. The complete potential is given by

$$A^\mu(t,x,y,z) = Re(A^\mu(x,y)exp(i\omega t - \gamma z)$$

The Lagrangian for the field inside the waveguide is given by

$$L = F_{\mu\nu}F^{\mu\nu} = g^{\mu\alpha}g^{\nu\beta}F_{\mu\nu}F_{\alpha\beta}$$

Set $A^0 = V, A^1 = A_x, A^2 = A_y, A^3 = A_z$. Then,

$$A_0 = g_{00}V + g_{01}A_x + g_{02}A_y + g_{03}A_z$$

$$A_{0,z} = -\gamma.A_0 = -\gamma(g_{00}V + g_{01}A_x + g_{02}A_y + g_{03}A_z)$$

In the action, this should be replaced with its negative as pointed above. We have then

$$L = g^{\mu\alpha}g^{\nu\beta}F_{\mu\nu}F_{\alpha\beta}$$

$$F_{01} = A_{1,0} - A_{0,1} = i\omega.(g_{10}V + g_{11}A_x + g_{12}A_y + g_{13}A_z) - (g_{00}V + g_{01}A_x + g_{02}A_y + g_{03}A_z)_{,x}$$

$$F_{02} = A_{2,0} - A_{0,2} = i\omega.(g_{20}V + g_{21}A_x + g_{22}A_y + g_{23}A_z) - (g_{00}V + g_{01}A_x + g_{02}A_y + g_{03}A_z)_{,y}$$

$$F_{03} = A_{3,0} - A_{0,3} = i\omega.(g_{30}V + g_{31}A_x + g_{32}A_y + g_{33}A_z) + \gamma.(g_{00}V + g_{01}A_x + g_{02}A_y + g_{03}A_z)$$

$$F_{12} = A_{2,1} - A_{1,2} = (g_{20}V + g_{21}A_x + g_{02}A_y + g_{03}A_z)_{,1} - (g_{10}V + g_{11}A_x + g_{12}A_y + g_{13}A_z)_{,2}$$

$$F_{23} = A_{3,2} - A_{2,3} = (g_{30}V + g_{31}A_x + g_{32}A_y + g_{33}A_z)_{,2} + \gamma.(g_{20}V + g_{21}A_x + g_{22}A_y + g_{23}A_z)$$

$$F_{31} = A_{1,3} - A_{3,1} = -\gamma.(g_{10}V + g_{11}A_x + g_{12}A_y + g_{13}A_z) - (g_{30}V + g_{31}A_x + g_{32}A_y + g_{33}A_z)_{,x}$$

Now,

$$F^{\mu\nu}F_{\mu\nu} = g^{\mu\alpha}g^{\nu\beta}F_{\mu\nu}F_{\alpha\beta}$$

$$= 2(g^{00}g^{11}F_{01}^2 + g^{00}g^{22}F_{02}^2 + g^{00}g^{33}F_{03}^2 + g^{11}g^{22}F_{12}^2 + g^{22}g^{33}F_{23}^2 + g^{11}g^{33}F_{13}^2)$$

From these expressions, formulate a finite element method for determine the fields.

8. Simplify the formula for the covariant divergence of a symmetric two tensor and for an antisymmetric two tensor.

Let $T^{\mu\nu}$ be a symmetric tensor. Then,

$$T^{\mu\nu}_{:\sigma} = T^{\mu\nu}_{,\sigma} + \Gamma^\mu_{\alpha\sigma}T^{\alpha\nu} + \Gamma^\nu_{\alpha\sigma}T^{\mu\alpha}$$

giving

$$T^{\mu\nu}_{:\nu} = T^{\mu\nu}_{,\nu} + \Gamma^\mu_{\alpha\nu}T^{\alpha\nu} + \Gamma^\nu_{\alpha\nu}T^{\mu\alpha}$$

Now use the identity

$$\Gamma^\nu_{\alpha\nu} = g^{-1/2}(g^{1/2})_{,\alpha}$$

to get

$$T^{\mu\nu}_{:\nu} = g^{-1/2}(T^{\mu\nu}_{,\nu}\sqrt{g})_{,\nu} + \Gamma^\mu_{\alpha\nu}T^{\alpha\nu}$$

This formula is valid for any two tensor. If T is an antisymmetric two tensor, the last term drops out and we get

$$T^{\mu\nu}_{:\nu} = g^{-1/2}(T^{\mu\nu}\sqrt{g})_{,\nu}$$

9. The scalar wave equation for a potential in a curved space time, ie, the Laplace-Beltrami equation has been studied by the authors. It can be obtained from an action principle which may be used to derive a finite element algorithm for determining the modes of propagation inside a wave-guide of arbitrary shape. Generalizing this idea, the Maxwell theory in a curved space time determines a set of linear partial differential equations for the electromagnetic four potential given the metric of space-time. These equations can also be derived from an action principle as was first demonstrated by Einstein when he formulated his field equations of gravitation. To study the propagation of electromagnetic waves inside a waveguide in the presence of an external gravitational field, we assume exponential dependence of the four potential on the z and time coordinate and modify the action principle slightly. This modified action will be quadratic in the electromagnetic four potential but will also involve the external gravitational field as a parameter. This action principle can be used to derive a finite element algorithm for obtaining the potentials. The idea is to partition the cross section of the guide into triangular elements, and assume that each component of the four potential inside a triangle can be expressed as a linear combination of the corresponding component at the vertices of the triangle, the interpolation functions being linear functions of the x and y coordinates. Using these interpolation rules, the action functional for the entire field is expressed as a quadratic function of $4N$ variables where N is the number of vertices and the factor of 4 arises due to the presence of four potential components. The modes of propagation can then be derived from the eigenvalues of the $4N \times 4N$ matrix associated with this quadratic form.

10. **Abstract of Arti's Ph.D thesis, general relativity and electromagnetism, numerical methods**

This thesis discussed numerical schemes for solving a variety of boundary value problems arising in electromagnetic and related fields. The methods discussed herein include the method of moments and the finite element method. These two methods have been in use for a very long time by applied mathematicians and engineers. To get a glimpse of the wide range and scope of the finite element method, one need only glance at the collection of IEEE papers edited by Pelosi. The method of moments in brief, involves the following. We are given a linear partial differential operator $\mathcal{L}$ acting on smooth functions f defined on an open subset U in $\mathbb{R}^n$. The pde to be solved is $\mathcal{L}(f) = g$ where g is the source function and f the response to be determined. Boundary conditions involve vanishing of f on the boundary ∂U of U. Note that ∂U is an $n-1$ dimensional manifold. The idea is to choose an appropriate set $\{\phi_1, ..., \phi_n\}$ of linearly independent smooth functions vanishing on ∂U and assume that f can be expanded as a linear combination of these test functions, ie, $f \approx \sum_{i=1}^{N} a_i\phi_i$ where $\{a_i\}$ are constants. Plugging this into the pde gives

$$\sum_{i=1}^{N} a_i\mathcal{L}(\phi_i) = g$$

The $a_i's$ are unknown constants constants and indeed this derived system may not have any solution at all. However, we form the inner product on both sides with the test functions $\{\phi_i\}$ and thereby derive a system of linear equations for the $\{a_i\}$:

$$\sum_{i=1}^{N} a_i < \mathcal{L}(\phi_i), \phi_j >=< g, \phi_j >, 1 \leq j \leq N$$

This system of linear equations for the $a_i's$ will have a unique solution provided that the matrix $((< \mathcal{L}(\phi_j), \phi_i >))_{1\leq i,j\leq N}$ is non-singular. In that case, we can write the solution as

$$(a_i) = ((< \mathcal{L}(\phi_j), \phi_i >))^{-1}(< g, \phi_i >)$$

It is clear that if N is large, we get a more accurate solution for the system. The scope of this method is quite large. Suppose, for example, the boundary condition on f is not that $f = 0$ on ∂U, but rather that $f = \psi$ on ∂U where ψ is a given function on ∂U. Then, this can be transformed into an equation involving vanishing boundary conditions on f by choosing a function ψ_0 on $\bar{U} = U \cup \partial U$ such that ψ_0 is smooth on U and $\psi_0|_{U=\partial U} = \psi$. Then define $h = f - \psi_0$. h will satisfy $\mathcal{L}(h) = \mathcal{L}(f) - \mathcal{L}(\psi_0)$ in U and $h|_{\partial U} = 0$. The method of moments can now be applied to h. Another extension of this method is to nonlinear partial differential equations. Suppose, for example, $\mathcal{L}_i, i = 1, 2, ..., k$ are linear partial differential operators and one desires to solve a quadratic nonlinear partial differential equation of the form

$$\sum \alpha_i \mathcal{L}_i(f) + \sum \beta_{ij} \mathcal{L}_i(f) \mathcal{L}_j(f) = g$$

where $\{\alpha_i, \beta_{ij}\}$ are constants. Then, approximate f as a linear combination of basis functions $f \approx \sum a_m \phi_m$ to get

$$\sum_{i,m} \alpha_i a_m \mathcal{L}_i(\phi_m) + \sum_{i,j,m,n} \beta_{ij} a_m a_n \mathcal{L}_i(\phi_m) \mathcal{L}_j(\phi_n) = g$$

and forming inner products with the $\phi_m, \phi_p \phi_q$ gives us a system of quadratic algebraic equations for the unknown coefficients a_m, ie,

$$\sum_m a_m \sum_i \alpha_i < \mathcal{L}_i(\phi_m), \phi_p > + \sum_{m,n} a_m a_n \sum_{i,j} \beta_{ij} < \mathcal{L}_i(\phi_m) \mathcal{L}_j(\phi_n), \phi_p >=< g, \phi_p >$$

$1 \leq p \leq N$,

$$\sum_m a_m \sum_i \alpha_i < \mathcal{L}_i(\phi_m), \phi_p \phi_q > + \sum_{m,n} a_m a_n \sum_{i,j} \beta_{ij} < \mathcal{L}_i(\phi_m) \mathcal{L}_j(\phi_n), \phi_p \phi_q >=< g, \phi_p \phi_q >$$

$1 \leq p, q \leq N$. It may easily be shown that the method of moments is nothing but a least squares approximation of the solution to the given partial differential equation as a linear combination of test functions, ie,

$$\hat{f} = \sum a_i \phi_i$$

with $a_i's$ chosen so that $\| \mathcal{L}(\hat{f}) - g \|$ is a minimum. The method of moments, as applied to electromagnetic problems involves determining approximate solutions to the Helmholtz equation. The Helmholtz equation describes the propagation of electromagnetic waves either in free space or in a waveguide or cavity resonator. When the waveguide has a nonuniform permittivity, we obtain the modified Helmholtz equation

$$\nabla^2 \psi(x, y) + (\gamma^2 + \omega^2 \mu_0 \epsilon_0 \epsilon_r(x, y)) \psi(x, y) = 0$$

where γ is the propagation constant along the z direction. This is an eigenvalue problem once we assume the vanishing boundary conditions for ψ on the waveguide boundary. The boundary condition forces γ to assume values in a countable set and associated with each γ value is an eigenfunction ψ. When ϵ_r is a constant and the waveguide has a rectangular cross section, the problem is easily solved in closed form. Any standard textbook on electromagnetism discusses these solutions. The are classified as the $TE_{m,n}, TH_{m,n}$ modes (transverse electric and transverse magnetic). When ϵ_r is not a constant, the method of moments is used to arrive at approximate solutions. The method can be generalized to the following: Let $\mathcal{L}$ be a partial differential operator acting on functions f defined on an open subset U of $\mathbb{R}^n$. We want solutions to the eigenvalue problem

$$\mathcal{L}(f(x)) - \lambda.g(x) = 0$$

in U with $f = 0$ on ∂U. To this end we choose test functions $\phi_i, i = 1, 2, ..., N$ that vanish on ∂U and assume that

$$f(x) \approx \sum_{i=1}^{N} a_i \phi_i(x)$$

Plugging this into the eigenvalue equation and forming inner products with ϕ_j, we get

$$\sum_{i=1}^{N} a_i < \mathcal{L}(\phi_i), \phi_j >= \lambda \sum_{j=1}^{N} a_j < g, \phi_j >, j = 1, 2, ..., N$$

Define the matrix

$$\mathbf{A} = ((< \mathcal{L}(\phi_j), \phi_i >))_{1 \le i,j \le N}$$

and the diagonal matrix

$$\mathbf{D} = diag[< g, \phi_j >, j = 1, 2, ..., N]$$

Then, the above linear equations can be cast as a matrix generalized eigenvalue problem

$$(\mathbf{A} - \lambda.\mathbf{D})\mathbf{a} = \mathbf{0}$$

The modes, ie possible values of λ are determined from the equation

$$det(\mathbf{A} - \lambda.\mathbf{D}) = 0$$

and this can be specialized to the Helmholtz problem of electromagnetic theory. In this thesis, we have applied this technique to the computation of the modes of propagation inside a rectangular waveguide having permittivity piecewise constant over little squares. The test functions chosen by us are the same sinusoidal functions that appear in the $TE_{m,n}$ and $TH_{m,n}$ expansions for the case when the permittivity is constant throughout the cross section of the waveguide.

The next problem that has been attempted in this thesis is the solution of the same, namely determination of the modes of propagation of em waves inside a waveguide having inhomogeneous permittivity using finite element methods. We start once again with the modified Helmholtz equation and divide the cross section of the guide into little triangles. the field inside each triangle is approximated by a linear combination of the fields at the vertices of the triangle using affine linear test functions, ie, functions of the form $ax + by + c$. The constants a, b, c depend on the coordinates of the triangle vertices. The next step is to derive the modified Helmholtz equation from an action principle, ie, by extremizing an appropriate functional of the field. Such a functional is well known in the physical literature under the heading "Action principle for classical field theories" and a good account of this can be found in Goldstein's "Classical Mechanics". The field in each triangular pixel is then substituted into the action integral and by summing the contributions of all the triangles, one obtains in place of the action functional, a quadratic functional of the vertex field values. Such a quadratic form on minimization yields a generalized eigenvalue equation that can be solved easily using MATLAB. While computing the action integral for the interpolated field, one needs to evaluate integrals of linear and quadratic functions of position over each triangle. Some simple computational methods have been developed in this thesis for this purpose. Having attempted these two problems, the next part of the thesis deals with the application of the finite element method to electromagnetic wave propagation problems in the presence of a weak gravitational field. According to Einstein's general theory of relativity, the gravitational field tends to curve the space-time manifold and the resulting field is described by Riemannian space-time metric. The metric being symmetric, has ten independent components, which in the general relativity literature, are referred to as the gravitational potentials. These potentials satisfy the Einstein field equations which are modifications of the Newtonian inverse square law of gravitation to covariant formalism. One first defines the notion of parallel displacement of vectors on a Riemannian manifold. This notion leads at once to the notion of covariant derivatives of vector and tensor fields on a Riemannian manifold. One then observes that the parallel displacement of a vector around a small closed loop in the manifold leads to a nonzero discrepancy in the resulting vector. This enables one to define the Riemann-Christoffel curvature tensor for a Riemannian manifold. If the space-time metric is flat, ie, Minkowskian, then the curvature tensor vanishes identically. The curvature tensor has four indices and contraction of two indices produces the Ricci tensor. The Einstein field equations of gravitation are formulated in terms of this Ricci tensor. More specifically, one derives the Einstein tensor from the Ricci tensor and proves that it has a vanishing divergence. In the presence of matter and electromagnetic radiation, the left side of the Einstein field equation contains the Einstein tensor which contains the metric tensor and its first and second partial derivatives while the right side contains the energy momentum tensor of matter plus radiation. The vanishing of the divergence of the Einstein tensor leads therefore to the Maxwell equations for the electromagnetic field. Suppose that the background gravitational field is specified and we wish to solve for the electromagnetic field in this background. Then we need to solve the usual Maxwell equations with ordinary partial derivatives replaced by covariant derivatives. The resulting Maxwell equations can also be derived from an action principle with the action a quadratic functional of the electromagnetic four potential and the metric of the gravitational field appearing herein as a parameter.

11.General relativity and gravitation

Problems:

1.Define the connection, Lie bracket, torsion and curvature.

X, Y are vector fields on a differentiable manifold. We denote the space of smooth vector fields on a manifold $\mathcal{M}$ by $\chi(\mathcal{M})$. The connection is a map $\nabla : \chi(\mathcal{M}) \times \chi(\mathcal{M}) \to \chi(\mathcal{M})$ such that (1) $\nabla_{fX}(Y) = f\nabla_X Y$ for all $f \in F(\mathcal{M})$, the space of smooth functions on $\mathcal{M}$, $\nabla_X(fY) = X(f)Y + f\nabla_X Y$ for all smooth functions f and all vector fields X, Y. $\mathcal{L}_X$ is the Lie derivative along the direction of the vector field X. It acts on vector fields, one forms and more generally all smooth tensor fields. The action of the Lie derivative on vector fields is given by

$$\mathcal{L}_X(Y) = [X, Y] = XY - YX$$

In terms of local coordinate bases $e_i = \frac{\partial}{\partial x^i}$, if we write $X = X^i e_i$ and $Y = y^i e_i$, then if f is any smooth function on the manifold,

$$XY(f) = X^j e_j Y^i e_i(f) = X^j e_j(Y^i) e_i(f) + X^j Y^i e_j e_i(f)$$

and

$$YX(f) = Y^j e_j(Y^i) e_i(f) + Y^j X^i e_j e_i(f)$$

and since $e_j e_i = e_i e_j$, it follows that

$$[X, Y] = (X^j e_j(Y^i) - Y^j e_j(X^i)) e_i$$

or

$$[X, Y]^i = X^j Y^i_{,j} - Y^j X^i_{,j}$$

Let ω be a one form and Y a vector field. Then $\omega \otimes Y$ is a tensor field. By definition of Lie derivative of a tensor field, we have that if X is any vector field, then

$$\mathcal{L}_X(\omega \otimes Y) = \mathcal{L}_X(\omega) \otimes Y + \omega \otimes \mathcal{L}_X Y$$

the contracted version of which is

$$\mathcal{L}_X(\omega(Y)) = (\mathcal{L}_X \omega)(Y) + \omega(\mathcal{L}_X Y)$$

and in terms of local coordinates,

$$X^j(\omega_i Y^i)_{,j} = (\mathcal{L}_X \omega)_i Y^i + \omega_i (\mathcal{L}_X Y)^i$$

so that

$$(\mathcal{L}_X \omega)_i Y^i = X^j \omega_{i,j} + \omega_j X^j_{,i}$$

Another useful identity:

$$\begin{aligned} &< \mathcal{L}_X \omega, Y > - Y < \omega, X > \\ &= (X^j \omega_{i,j} + \omega_j X^j_{,i}) Y_i - Y^i (\omega_j X^j)_{,i} \\ &= (X^j \omega_{i,j} + \omega_j X^j_{,i} - \omega_{j,i} X^j - \omega_j X^j_{,i}) Y^i \\ &= (\omega_{i,j} - \omega_{j,i}) X^j Y^i \end{aligned}$$

Now,

$$\omega = \omega_i dx^i$$

$$d\omega = \omega_{i,j} dx^j \wedge dx^i$$

so that

$$d\omega(X, Y) = \omega_{i,j}(X^j Y^i - Y^j X^i)/2 = \frac{1}{2}(\omega_{i,j} - \omega_{j,i}) X^j Y^i$$

Combining this with the preceding identity gives us

$$<\mathcal{L}_X\omega, Y> -Y <\omega, X>= 2.d\omega(X,Y)$$

Now,

$$<\mathcal{L}_X\omega, Y>= X(<\omega, Y>)- <\omega, \mathcal{L}_X Y>$$

giving

$$X(<\omega, Y>) - Y(<\omega, X>)- <\omega, [X,Y]>= 2d\omega(X,Y)$$

and this is a very important result.

Let (e_i) be a local basis for vector fields (not necessarily a coordinate basis) and let (e^i) be the dual basis for the cotangent manifold. The mapping $X \to e^i(\nabla_X e_j)$ is a one form and we denote this one form by ω^i_j. Thus,

$$\nabla_X e_j = \omega^i_j(X)e_i$$

It follows then that

$$\nabla_X Y = \nabla_X(Y^i e_i) = X(Y^i)e_i + Y^i\nabla_X e_i = X(Y^i)e_i + Y^i\omega^j_i(X)e_j = [X(Y^j) + Y^i\omega^j_i(X)]e_j$$

or equivalently,

$$(\nabla_X Y)^j = e^j(\nabla_X Y) = X(Y^j) + Y^i\omega^j_i(X)$$

The torsion is a tensor field defined by

$$T(X,Y) = \nabla_X Y - \nabla_Y X - [X,Y]$$

We have

$$T^j(X,Y) = e^j(T(X,Y)) = (\nabla_X Y)^j - (\nabla_Y X)^j - [X,Y]^j = [X(Y^j) + \omega^j_i(X)Y^i] - [Y(X^j) + \omega^j_i(Y)X^i] - [X,Y]^j$$

$$= X(<e^j, Y>) - Y(<e^j, X>) - e^j([X,Y]) + \omega^j_i(X)e^i(Y) - \omega^j_i(Y)e^i(X)$$

$$= 2de^j(X,Y) + (\omega^j_i \wedge e^i)(X,Y)$$

in view of the preceding identity. In other words, we have deduced "Cartan's first equation of structure"

$$\frac{1}{2}T^j = de^j + \omega^j_i \wedge e^i$$

In case the connection is derived from a metric, the torsion is zero and then the above reduces to

$$de^j = -\omega^j_i \wedge e^i$$

which is a ready formula for calculating the connection components ω^j_i relative to any basis for the space of vector fields.

The curvature tensor: For any three vector fields X, Y, Z, we define another vector field $R(X,Y,Z)$ via the equation

$$R(Z,X,Y) = \nabla_X\nabla_Y Z - \nabla_Y\nabla_X Z - \nabla_{[X,Y]}Z$$

We evaluate this in a basis (e_i) for the tangent space with (e^i) as the dual basis.

$$\nabla_Y Z = \nabla_Y(Z^i e_i) = Y(Z^i)e_i + Z^i \nabla_Y e_i = Y(< e^i, Z >)e_i + < e^i, Z > \omega_i^j(Y)e_j$$

$$= [Y(< e^i, Z >) + < e^j, Z > \omega_j^i(Y)]e_i$$

$$\nabla_X \nabla_Y Z = X[Y(< e^j, Z >) + < e^j, Z > \omega_j^i(Y)]e_i + [Y(< e^i, Z >) + < e^j, Z > \omega_j^i(Y)]\omega_i^k(X)e_k$$

$$= [XY(< e^j, Z >) + X(< e^j, Z >)\omega_j^i(Y) + < e^j, Z > X(\omega_j^i(Y)) + Y(< e^k, Z >)\omega_k^i(X)$$

$$+ < e^j, Z > \omega_j^k(Y)\omega_k^i(X)]e_i$$

Thus,

$$\nabla_X \nabla_Y Z - \nabla_Y \nabla_X Z$$

$$= [[X, Y](< e^j, Z >) + < e^j, Z > (X(\omega_j^i(Y)) - Y(\omega_j^i(X)) + \omega_j^k(Y)\omega_k^i(X) - \omega_j^k(X)\omega_k^i(Y)]e_i$$

Further,

$$\nabla_{[X,Y]} Z = \nabla_{[X,Y]}(Z^i e_i) = [X, Y](Z^i)e_i + Z^i \omega_i^j([X, Y])e_j$$

It follows that

$$R(Z, X, Y) =< e^j, Z > (X(\omega_j^i(Y)) - Y(\omega_j^i(X)) - \omega_j^i([X, Y]) + \omega_j^k(Y)\omega_k^i(X) - \omega_j^k(X)\omega_k^i(Y))e_i$$

$$=< e^j, Z > (2d\omega_j^i(X, Y)) + 2\omega_k^i \wedge \omega_j^k(X, Y)$$

or

$$\frac{1}{2}R(X, Y, Z) =< e^j, Z > (d\omega_j^i + \omega_k^i \wedge \omega_j^k)(X, Y)$$

Let ω be any one form. Then, it follows from the above that

$$\omega(R(Z, X, Y)) =< e^j, Z > (d\omega_j^i + \omega_k^i \wedge \omega_j^k)(X, Y)\omega(e_i)$$

Define a set of functions R^i_{jkm} so that $R^i_{jkm} = -R^i_{jmk}$ and

$$R^i_{jkm} e^k \wedge e^m = d\omega_j^i + \omega_k^i \wedge \omega_j^k$$

Then the above can be written as

$$\omega(R(Z, X, Y)) = R^i_{jkm} e^k \wedge e^m(X, Y) < e^j, Z > \omega(e_i)$$

Alternately, we define the functions R^i_{jkm} by this equation and term the former equation as Cartan's second equation of structure.

Covariant derivative of a one form: If f is a scalar, then we define its covariant derivative as $\nabla_X f = Xf$ for any vector field X. It follows that if ω is a one form and Y a vector field, then $\nabla_X(\omega(Y)) = X(\omega(Y))$. The covariant derivative acting on tensor fields is assumed to satisfy Leibniz rule. Thus, we get

$$X(\omega(Y)) = (\nabla_X \omega)(Y) + \omega(\nabla_X Y)$$

Thus,

$$(\nabla_X\omega)(Y) = X^j(\omega_{i,j}Y^j + \omega_i Y^i_{,j}) - \omega_i(\nabla_X Y)^i$$
$$= X^j(\omega_{i,j}Y^j + \omega_i Y^i_{,j}) - \omega_i(X(Y^i) + Y^j\omega^i_j(X))$$
$$= X^j\omega_{i,j}Y^i - \omega_i\omega^i_j(X)Y^j$$

so that

$$(\nabla_X\omega)_i = e_i(\nabla_X\omega) = X^j\omega_{i,j} - \omega_j\omega^j_i(X)$$
$$= X(\omega_i) - \omega_j\omega^j_i(X)$$

or

$$\nabla_X\omega = [X(\omega_i) - \omega_j\omega^j_i(X)]e^i$$

and in particular,

$$\nabla_X e^i = -\omega^i_j(X)e^j$$

Problem:Write down the expression for the components of the Riemann tensor in a local coordinate basis.

Problem:The Riemann curvature tensor in terms of components R_{ijkm} satisfies the symmetry relations $R_{jikm} = -R_{ijkm}, R_{ijmk} = -R_{ijkm}, R_{kmij} = R_{ijkm}$. The Ricci tensor is defined in terms of the Riemann tensor by the contraction $R_{ik} = g^{jm}R_{ijkm}$. The Weyl tensor C_{ijkm} is defined as the traceless part of the Riemann tensor. Write down the explicit expression for the Weyl tensor.

Metrical connection having zero torsion:Assume that relative to a basis (e^i) for the cotangent space of the manifold, the metric has the form $g = g_{ij}e^i \otimes e^j$. Symmetry of the metric, ie, $g(X,Y) = g(Y,X)$ for all vector fields X, Y implies that $g_{ij} = g_{ji}$. Suppose ω^i_j are the connection one forms, ie, $\nabla_X e^i = -\omega^i_j(X)e^j$. Assume that the connection satisfies the condition $\nabla_X g = 0$ for all vector fields X. Then we have

$$0 = \nabla_X g = X(g_{ij})e^i \otimes e^j + g_{ij}\nabla_X e^i \otimes e^j + g_{ij}e^i \otimes \nabla_X e^j$$
$$= X(g_{ij})e^i \otimes e^j - g_{ij}\omega^i_k(X)e^k \otimes e^j - g_{ij}\omega^j_k(X)e^i \otimes e^k$$

for all vector fields X and from this, it follows that

$$X(g_{ij}) - g_{kj}\omega^k_i(X) - g_{ik}\omega^k_j(X) = 0$$

Assume now that the dual basis (e_i) for the tangent space is a local coordinate basis, ie, $e_i = \frac{\partial}{\partial x^i}$ and $e^i = dx^i$. We write

$$\omega^i_k(e_j) = \Gamma^i_{kj}$$

Suppose that $T = 0$ for this connection. Then since for the local coordinate basis, $[e_i, e_j] = 0$, it follows that

$$\nabla_{e_i}e_j - \nabla_{e_j}e_i = 0$$

or

$$\omega^k_j(e_i) - \omega^k_i(e_j) = 0$$

or

$$\Gamma^k_{ji} = \Gamma^k_{ij}$$

From the above, we have further taking $X = e_k$,

$$g_{ij,k} - g_{rj}\Gamma^r_{ik} - g_{ir}\omega^r_{jk} = 0$$

or

$$g_{ij,k} - \Gamma_{jik} - \Gamma_{ijk} = 0$$

and using the symmetry of Γ_{jik} with respect to the last two indices, it follows that

$$\Gamma_{ijk} = \frac{1}{2}(g_{ij,k} + g_{ik,j} - g_{jk,i})$$

In other words, if we impose the two conditions on the connection (1) it is torsion free and (2) the covariant derivative of the metric is zero, then the connection is uniquely determined from the metric.

Weyl tensor: Assume that relative to a basis the Riemann curvature tensor in an n dimensional manifold has components R_{ijkl}. The symmetries of this tensor are $(1)R_{ijkl} = -R_{jikl}, (2)R_{ijkl} = -R_{ijlk}, (3)R_{ijkl} = R_{klij}$. The Ricci tensor is defined by $R_{ik} = g^{jl}R_{ijkl}$ and $R = g^{ij}R_{ij}$. Define

$$C_{ijkl} = R_{ijkl} - \frac{1}{n-2}(g_{ik}R_{jl} + g_{jl}R_{ik} - g_{jk}R_{il} - g_{il}R_{jk}) + \frac{1}{(n-1)(n-2)}(g_{ik}g_{jl} - g_{il}g_{jk})R$$

We have then

$$C_{ijkl} = -C_{jikl}, C_{ijkl} = -C_{ijlk}, C_{ijkl} = C_{klij}$$

Further, the trace of the Weyl tensor is zero:

$$g^{jl}C_{ijkl} = R_{ik} - \frac{1}{n-2}(g_{ik}R + nR_{ik} - R_{ik} - R_{ik}) + \frac{1}{(n-1)(n-2)}(ng_{ik} - g_{ik})R = 0$$

C_{ijkl} are the components of the Weyl tensor. This tensor has all the symmetries of the Riemann tensor but in addition it has a vanishing trace.

Tetrad formulation of the equations of general relativity: $g_{ij}(x)$ is the given metric. x has four components, the zeroth is time and first second and third are spatial components. Let $\mathbf{G}(x) = ((g_{ij}(x)))$. Let η be a constant symmetric matrix. Then we can write

$$\mathbf{G}(x) = \mathbf{E}(x)\eta\mathbf{E}^T(x)$$

or in terms of components,

$$g_{ij}(x) = \eta_{ab}e^a_i(x)e^b_j(x)$$

The $(ia)^{th}$ entry of $E(x)$ thus equals e^a_i. The numbers η_{ab} are all constant scalars. For each $a = 0, 1, 2, 3$, $(e^a_i)_{0\leq i\leq 3}$ is a covariant four vector. This can be seen from the transformation properties of the metric tensor and the assumption that η_{ab} are scalars. Assuming then that e_i^a is a covariant vector, its transformation law is given by

$$e_i^{a'}(x') = e_j^a(x')\frac{\partial x^j}{\partial x^{i'}}$$

Also since (η_{ab}) are scalars, their transformation laws are

$$\eta'_{ab} = \eta_{ab}$$

Thus, the transformation law of the metric tensor is given by

$$g'_{ij}(x') = (\eta_{ab}e_i^a e_j^b)' = \eta_{ab}e_i^{a'} e_j^{b'} = \eta_{ab}e_k^a\frac{\partial x^k}{\partial x^{i'}}e_m^b\frac{\partial x^m}{\partial x^{j'}}$$

$$= \eta_{ab}e_k^a e_m^b\frac{\partial x^k}{\partial x^{i'}}\frac{\partial x^m}{\partial x^{j'}}$$

$$= g_{km}(x)\frac{\partial x^k}{\partial x^{i'}}\frac{\partial x^m}{\partial x^{j'}}$$

which is the correct transformation law of the metric tensor. The four vector fields (e_i^a) form what is called a tetrad basis, or more precisely, a covariant tetrad basis. The corresponding contravariant tetrad basis is given by

$$e^{ai} = g^{ij}e_j^a$$

We compute

$$e^{ai}e_i^b = g^{ik}e_k^a e_i^b$$

This is the $(ab)^{th}$ entry of the matrix

$$E^T G^{-1}E = E^T E^{-T}\eta^{-1}E^{-1}E = \eta^{-1}$$

We define

$$(\eta^{ab}) = \eta^{-1}$$

Then, we have

$$\delta_b^a = \eta^{ac}\eta_{cb}$$

We also define the contravariant four vectors e_a^i so that

$$e_i^a e_b^i = \delta_b^a$$

In other words,

$$E^T F = I$$

where the $(ib)^{th}$ entry of F is e_b^i. It then follows from a result in elementary matrix theory that

$$FE^T = I$$

or

$$F_{ib}E_{jb} = \delta_{ij}$$

or

$$e_b^i e_j^b = \delta_j^i$$

Observe finally that

$$\eta_{ab} = g_{ij} e_a^i e_b^j$$

which in matrix notation reads

$$\eta = F^T G F = E^{-1} G E^{-T}$$

or

$$G = E \eta E^T$$

the original equation that we started off with.

Directional derivative, intrinsic derivative and spin coefficients.

Let A^i be a contravariant vector field so that $A_i = g_{ij} A^j$ is the corresponding covariant vector field. We choose a tetrad basis e_i^a and define the scalars

$$A_a = e_{ai} A^i = e_a^j A_j$$

Note that for each $a = 0, 1, 2, 3$, A_a is a scalar field. Its directional derivative along e_b is defined by the equation

$$A_{a,b} = e_b^i A_{a,i} = e_b^i \frac{\partial A_a}{\partial x^i}$$

Since A_a is a scalar field, its partial derivatives coincide with its covariant derivatives, ie,

$$A_{a,i} = A_{a:i} = (e_a^j A_j)_{:i} = A_{j:i} e_a^j + A_j e_{a:i}^j$$

This gives

$$A_{a,b} = A_{j:i} e_a^j e_b^i + A_j e_{a:i}^j e_b^i$$

The intrinsic derivatives are defined by

$$A_{a|b} = A_{j:i} e_a^j e_b^i$$

and more generally for tensors,

$$T^{a_1 \dots a_p}_{b_1 \dots b_q | c} = T^{i_1 \dots i_p}_{j_1 \dots j_q : m} e^{a_1}_{i_1} \dots e^{a_p}_{i_p} e^{j_1}_{b_1} \dots e^{j_q}_{b_q} e^m_c$$

The spin coefficients are defined by

$$\gamma_{abc} = e_a^i e_{bi:j} e_c^j$$

Now note that the spin coefficients are antisymmetric with respect to interchange of the first two indices. Indeed, $e_a^i e_{bi} = \eta_{ab}$ is a constant scalar and hence

$$0 = \eta_{ab:j} = (e_a^i e_{bi})_{:j} = e_a^i e_{bi:j} + e_{a:j}^i e_{bi}$$

This gives

$$\gamma_{abc} = -e_{bi} e_{a:j}^i e_c^j = -e_b^i e_{ai:j} e_c^j = -\gamma_{bac}$$

Thus, we have

$$A_j e^j_{a:i} e^i_b = A_c e^c_j e^j_{a:i} e^i_b = A_c \gamma_{uab} \eta^{uc} = \gamma_{uab} A^u$$

and hence,

$$A_{a,b} = A_{a|b} + \gamma_{uab} A^u$$

or

$$A_{a|b} = A_{a,b} - \gamma_{cab} A^c = A_{a,b} + \gamma_{acb} A^c$$

The directional derivative is the tetrad version of the ordinary directional derivative while the intrinsic derivative is the tetrad version of the covariant derivative. Let us now write down the Maxwell equations relative to a tetrad frame. The antisymmetric tensor is given in terms of ordinary components by

$$F_{ij} = A_{j,i} - A_{i,j} = A_{j:i} - A_{i:j}$$

The tetrad components of this tensor are

$$F_{ab} = F_{ij} e^i_a e^j_b$$

These are the Maxwell scalars. The homogeneous Maxwell equations in tetrad format are

$$F_{ab|c} + F_{bc|a} + F_{ca|b} = 0$$

These follow from the equations

$$F_{ij:k} + F_{jk:i} + F_{ki:j} = 0$$

on multiplying with $e^i_a e^j_b e^k_c$. The other Maxwell equations the absence of charges and currents are

$$F^{ij}_{:j} = 0$$

In tetrad format, these become

$$\eta^{bc} F_{ab|c} = 0$$

Indeed, we write down the original Maxwell equations as

$$g^{jk} F_{ij:k} = 0$$

which is the same as

$$\eta^{bc} e^j_b e^k_c e^i_a F_{ij:k} = 0$$

or

$$\eta^{bc} F_{ab|c} = 0$$

We now specialize the Maxwell equations to a space-time of sufficient generality that can be used to describe rotating blackholes. The metric is

$$ds^2 = exp(2\nu)(dt)^2 - exp(2\psi)(d\phi - \omega dt - q_r dr - q_\theta d\theta)^2 - exp(2\mu_2)dr^2 - exp(2\mu_3)d\theta^2$$

or equivalently writing

$$t = x^0, r = x^2, \theta = x^3 = \phi = x^1$$

$$ds^2 = exp(2\nu)(dx^0)^2 - exp(2\psi)(dx^1 - q_0 dx^0 - q_2 dx^2 - q_3 dx^3)^2 - exp(2\mu_2)(dx^2)^2 - exp(2\mu_3)(dx^3)^2$$

Note that setting

$$x^4 = ix^0, \mu_4 = \nu$$

we can write

$$ds^2 = -\sum_{A=2,3,4} exp(2\mu_A)(dx^A)^2 - exp(2\psi)(dx^1 - \sum_{A=2,3,4} q_A dx^A)^2$$

where μ_A, q_A are functions of x^2, x^3, x^4. Such a metric describes non-stationary axi-symmetric space-times. We introduce the following basis of one forms

$$\omega^A = exp(\mu_A)dx^A, A = 2,3,4, \omega^1 = exp(\psi)(dx^1 - \sum_{A=2,3,4} q_A dx^A)$$

Then the metric is given by

$$g = -\sum_{A=2}^{4} \omega^A \otimes \omega^A$$

Equivalently, we define

$$(\eta_{ab}) = diag[1, -1, -1, -1],$$

$$(e_{0i})_{i=0}^3 = (exp(\nu), 0, 0, 0), (e_{1i})_{i=0}^3 = (-q_0 exp(\psi), exp(\psi), -q_2 exp(\psi), -q_3 exp(\psi)),$$

$$(e_{2i}) = (0, 0, exp(\mu_2), 0), (e_{3i}) = (0, 0, 0, exp(\mu_3))$$

It follows that

$$e_{0i}dx^i = exp(\nu)dt, e_{1i}dx^i = exp(\psi)(d\phi - q_0 dt - q_2 dr - q_3 d\theta),$$

$$e_{2i}dx^i = exp(\mu_2)dr, e_{3i}dx^i = exp(\mu_3)d\theta$$

Thus,

$$ds^2 = g_{ij}dx^i dx^j = \eta^{ab} e_{ai} dx^i e_{bj} dx^j$$

so that

$$g_{ij} = \eta^{ab} e_{ai} e_{bj}$$

Note that if we define

$$\omega_a = e_{ai} dx^i$$

then

$$ds^2 = g = \eta^{ab} \omega_a \otimes \omega_b$$

Equivalently, let

$$\omega^a = \eta^{ab} \omega_a$$

so that

$$\omega^0 = \omega_0, \omega^i = -\omega_i, i = 1, 2, 3$$

then

$$g = \eta_{ab} \omega^a \otimes \omega^b$$

The Maxwell equations in tetrad notation read

$$\eta^{bc}F_{ab|c}=0$$

or

$$F_{a0|0}-F_{a1|1}-F_{a2|2}-F_{a3|3}=0$$

Now, if A^i, B^i are two four vector fields and A_a, B_b are the corresponding tetrad scalars, then

$$(A_aB_b)_{|c}=A_{a|c}B_b+A_aB_{b|c}=(A_{a,c}-\gamma_{dac}A^d)B_b+A_a(B_{b,c}-\gamma_{dbc}B^d)$$

$$=(A_aB_b)_{,c}-\gamma_{dac}A^dB_b-\gamma_{dbc}A_aB^d$$

from which we deduce

$$F_{ab|c}=F_{ab,c}-\gamma_{dac}\eta^{de}F_{eb}-\gamma_{dbc}\eta^{de}F_{ae}$$

Note that $F_{ab,c}$ should not be interpreted as the partial derivative of F_{ab} with respect to x^c but rather as the directional derivative of the scalar F_{ab} along the direction e_c. In Chandrasekhar's book "The mathematical theory of blackholes", to avoid this confusion, he uses the notation $F_{(a)(b),(c)}$ for this directional derivative with the notation $F_{(a)(b)}$ being used in place of F_{ab}.

Thus, the above Maxwell equation can be written as

$$e_0(F_{a0})-e_1(F_{a1})-e_2(F_{a2})-e_3(F_{a3})$$

$$-\gamma_{da0}\eta^{de}F_{e0}-\gamma_{d00}\eta^{de}F_{ae}$$

$$+\gamma_{da1}\eta^{de}F_{e1}+\gamma_{d11}\eta^{de}F_{ae}$$

$$+\gamma_{da2}\eta^{de}F_{e2}+\gamma_{d22}\eta^{de}F_{ae}$$

$$+\gamma_{da3}\eta^{de}F_{e3}+\gamma_{d33}\eta^{de}F_{ae}=0$$

Now observe the following:

$$\gamma_{da0}\eta^{de}F_{e0}=-\gamma_{1a0}F_{10}-\gamma_{2a0}F_{20}-\gamma_{3a0}F_{30},$$

$$\gamma_{d00}\eta^{de}F_{ae}=-\gamma_{100}F_{a0}-\gamma_{200}F_{a2}-\gamma_{300}F_{a3}$$

$$\gamma_{da1}\eta^{de}F_{e1}=\gamma_{0a1}F_{01}-\gamma_{2a1}F_{21}-\gamma_{3a1}F_{31}$$

$$\gamma_{d11}\eta^{de}F_{ae}=\gamma_{011}F_{a0}-\gamma_{211}F_{a2}-\gamma_{311}F_{a3}$$

$$\gamma_{da2}\eta^{de}F_{e2}=\gamma_{0a2}F_{02}-\gamma_{1a2}F_{12}-\gamma_{3a2}F_{32}$$

$$\gamma_{d22}\eta^{de}F_{ae}=\gamma_{022}F_{a0}-\gamma_{122}F_{a1}-\gamma_{322}F_{a3}=0$$

$$\gamma_{da3}\eta^{de}F_{e3}=\gamma_{0a3}F_{03}-\gamma_{1a3}F_{13}-\gamma_{2a3}F_{23}$$

$$\gamma_{d33}\eta^{de}F_{ae}=\gamma_{033}F_{a0}-\gamma_{133}F_{a1}-\gamma_{233}F_{a2}$$

To calculate e_0F for any function F, we need the contravariant vectors e_a^i rather than the covariant vectors e_{ai}. These can be found from the equations

$$e_{ai}e_b^i=\eta_{ab}$$

We write the various components of this equation and solve for e_b^i:

$$e_{ai}e_0^i = \eta_{a0}$$

gives

$$e_{0i}e_0^i = 1, e_{1i}e_0^i = 0, e_{2i}e_0^i = 0, e_{3i}e_0^i = 0$$

which translates to

$$exp(\nu)e_0^0 = 1, -q_0exp(\psi)e_0^0 + exp(\psi)e_0^1 - q_2exp(\psi)e_0^2 - q_3exp(\psi)e_0^3 = 0,$$
$$exp(\mu_2)e_0^2 = 0, exp(\mu_3)e_0^3 = 0$$

These result in

$$e_0^0 = exp(-\nu), e_0^2 = 0, e_0^3 = 0, e_0^1 = q_0e_0^0 = q_0exp(-\nu)$$

Thus,

$$(e_0^i) = (exp(-\nu), q_0exp(-\nu), 0, 0)$$

Likewise,

$$e_{0i}e_1^i = 0, e_{1i}e_1^i = -1, e_{2i}e_1^i = 0, e_{3i}e_1^i = 0$$

These give

$$exp(\nu)e_1^0 = 0, exp(\psi)(-q_0e_1^0 + e_1^1 - q_2e_1^2 - q_3e_1^3) = -1,$$
$$exp(\mu_2)e_1^2 = 0, exp(\mu_3)e_1^3 = 0$$

so that

$$e_1^0 = 0, e_1^2 = 0, e_1^3 = 0, e_1^1 = -exp(-\psi)$$

resulting into

$$(e_1^i) = (0, -exp(-\psi), 0, 0)$$

Next,

$$e_{0i}e_2^i = 0, e_{1i}e_2^i = 0, e_{2i}e_2^i = -1, e_{3i}e_2^i = 0$$

giving

$$exp(\nu)e_2^0 = 0, -q_0e_2^0 + e_2^1 - q_2e_2^2 - q_3e_2^3 = 0,$$
$$exp(\mu_2)e_2^2 = -1, exp(\mu_3)e_2^3 = 0$$

giving

$$e_2^0 = 0, e_2^2 = -exp(-\mu_2), e_2^3 = 0, e_2^1 = -q_2exp(-\mu_2)$$

so that

$$(e_2^i) = (0, -q_2exp(-\mu_2), -exp(-\mu_2), 0)$$

Finally,

$$e_{0i}e_3^i = 0, e_{1i}e_3^i = 0, e_{2i}e_3^i = 0, e_{3i}e_3^i = -1$$

giving

$$exp(\nu)e_3^0 = 0, -q_0e_3^0 + e_3^1 - q_2e_3^2 - q_3e_3^3 = 0,$$
$$exp(\mu_2)e_3^2 = 0, exp(\mu_3)e_3^3 = -1$$

so that

$$e_3^0 = 0, e_3^2 = 0, e_3^3 = -exp(-\mu_3), e_3^1 = -q_3 exp(-\mu_3)$$

or

$$(e_3^i) = (0, -q_3 exp(-\mu_3), 0, -exp(-\mu_3))$$

The directional derivatives are now easily computed:

$$e_a F = e_a^i F_{,i}$$

where now $F_{,i} = \frac{\partial F}{\partial x^i}$. Thus

$$e_0 F = exp(-\nu)(F_{,0} + q_0 F_{,1})$$

$$e_1 F = -exp(-\psi)F_{,1}, e_2 F = -exp(-\mu_2)(q_2 F_{,1} + F_{,2}),$$

$$e_3 F = -exp(-\mu_3)(q_3 F_{,1} + F_{,3})$$

We introduce the colon derivatives:

$$F_{:A} = F_{,A} + q_A F_{,1}$$

Note that the colon derivative is a derivation. Indeed, linearity follows at once and the product rule from

$$(FG)_{:A} = (FG)_{,A} + q_A (FG)_{,1} = F_{,A} G + F G_{,A} + q_A (F_{,1} G + F G_{,1})$$

$$= (F_{,A} + q_A F_{,1}) G + F(G_{,A} + q_A G_{,1}) = F_{:A} G + F G_{:A}$$

Then,

$$e_0(F) = exp(-\nu)F_{:0}, e_1(F) = -exp(-\psi)F_{,1}, e_2(F) = -exp(-\mu_2)F_{:2}, e_3(F) = -exp(-\mu_3)F_{:3}$$

We leave it as an exercise to expand the full Maxwell equations in the tetrad framc using these equations.

12. **Gravitation**

1. Sample computations of the Riemann curvature tensor using Cartan's equations of structure. Cartan's second structure equation reads

$$\frac{1}{2} R^j_{lkm} e^k \wedge e^m = d\omega^j_l + \omega^j_k \wedge \omega^k_m$$

where (e_i) is a basis for the tangent manifold (not necessarily a coordinate basis), (e^i) the dual basis for the cotangent manifold and ω^i_j are the connection one forms relative to this basis, ie, if ∇ is the connection,

$$\nabla_X e_i = \omega^j_i(X) e_j$$

or

$$\omega^j_i(X) = e^j(\nabla_X e_i)$$

We consider the metric

$$g = -\sum_{A=1,2,3,4} \omega^A \otimes \omega^A$$

where

$$\omega^A = exp(\mu_A)dx^A, A = 2,3,4, \omega^1 = exp(\psi)(dx^1 - \sum_{A=2,3,4} q_A dx^A)$$

where $q_A, \mu_A, A = 2,3,4, \psi$ are functions of $x^A, A = 1,2,3,4$, ie all the variables. $x^4 = it$, $x^1 = \phi, x^2 = r, x^3 = \theta$. We write $q_4 = -i\omega$, $\mu_4 = \nu$. Then,

$$dx^1 - \sum_{A=2,3,4} q_A dx^A = d\phi - q_2 dr - q_3 d\theta - \omega.dt$$

and the metric can be expressed as

$$g = exp(2\nu)dt^2 - exp(2\psi)(d\phi - \omega.dt - q_2 dr - q_3 d\theta)^2 - exp(2\mu_2)dr^2 - exp(2\mu_3)d\phi^2$$

$$= -\sum_{A=2,3,4} exp(2\mu_A)(dx^A)^2 - exp(2\psi)(dx^1 - \sum_{A=2,3,4} q_A dx^A)^2$$

The above metric can be used to describe not only stationary and nonstationary axisymmetric space-times (rotating blackholes) but also perturbations of such blackholes. Let us compute the Riemann curvature tensor relative to the dual basis $(\omega^A)_{A=1}^4$.

Suppose that relative to the dual basis $(\omega^A)_{A=1}^4$, we have computed the components of the Riemann tensor $R^(a)_{(b)(c)(d)}$. Let e_a^i denote the associated tetrad frame. Then we can write

$$\omega^a = e_i^a dx^i$$

We have from the definition of tensor components relative to a basis,

$$R^{(a)}_{(b)(c)(d)}(\omega^c \wedge \omega^d)(X,Y)\omega^b(Z)\theta(e_a) = \theta(R(Z,X,Y))$$

where (e_a) is the dual basis to ω^a, ie, $\omega^a(e_b) = \delta_b^a$, or writing $e_a = e_a^i \partial_i$,

$$\delta_b^a = \omega^a(e_b) = e_i^a dx^i(e_b^j \partial_j) = e_i^a e_b^i$$

The above equation for R can be expressed as

$$R^{(a)}_{(b)(c)(d)}(e_i^c dx^i \wedge e_j^d dx^j)(X,Y)e_k^b dx^k(Z)\theta_m dx^m(e_a) = \theta(R(Z,X,Y))$$

or

$$\frac{1}{2}R^{(a)}_{(b)(c)(d)}e_i^c e_j^d e_k^b e_a^m (X^i Y^j - X^j Y^i)Z^k \theta_m = \theta(R(Z,X,Y))$$

or setting

$$R^m_{kij} = R^{(a)}_{(b)(c)(d)}e_a^m e_k^b e_i^c e_j^d,$$

we get

$$\frac{1}{2}R^m_{kij}(X^i Y^j - X^j Y^i)Z^k i\theta_m = \theta(R(Z,X,Y))$$

or if we assume that R^m_{kij} is antisymmetric with respect to interchange of (i,j), we can write

$$\theta(R(Z,X,Y)) = R^m_{kij}\theta_m X^i Y^j Z^k$$

As another example, suppose T is a one-one tensor. Let (e_a) be a basis for the tangent space and (e^a) the corresponding dual basis. We set

$$T = T^{(a)}_{(b)} e_a \otimes e^b$$

We write

$$e_a = e^i_a \partial_i, e^b = e^b_i dx^i$$

so that

$$T = T^{(a)}_{(b)} e^i_a e^b_j \partial_i \otimes dx^j = T^i_j \partial_i \otimes dx^j$$

where

$$T^i_j = T^{(a)}_{(b)} e^i_a e^b_j$$

are the components of the tensor T relative to local coordinates (x^i), ie, relative to the local coordinate basis $(\partial_i = \frac{\partial}{\partial x^i})$. We now evaluate the Christoffel symbols using Cartan's first equation of structure which states that if the torsion vanishes which happens in this case since we are assuming the connection to be the metrical connection, then

$$d\omega^A + \omega^A_B \wedge \omega^B = 0$$

where

$$\omega^A_B(X)e_A = \nabla_X e_A$$

with (e_A) being the dual basis to ω^A, ie, $\omega^A(e_B) = \delta^A_B$. We start with the basic equations for our space-time

$$\omega^A = exp(\mu_A)dx^A, A = 2,3,4, \omega^1 = exp(\psi)(dx^1 - \sum_{A=2,3,4} q_A dx^A)$$

Thus,

$$d\omega^A = exp(\mu_A) \sum_{B=1,2,3,4} \mu_{A,B} dx^B \wedge dx^A,$$

$$= exp(\mu_A)\mu_{A,1} dx^1 \wedge dx^A + exp(\mu_A) \sum_{B=2,3,4} \mu_{A,B} dx^B \wedge dx^A, A = 2,3,4$$

From now on, the indices A, B will always be assumed to take values $2,3,4$ only unless otherwise specified. Now,

$$dx^A = exp(-\mu_A)\omega^A, dx^1 = exp(-\psi)\omega^1 + \sum_A q_A exp(-\mu_A)\omega^A$$

Thus,

$$d\omega^A = exp(\mu_A)\mu_{A,1}(exp(-\psi)\omega^1 + \sum_B q_B exp(-\mu_B)\omega^B) \wedge exp(-\mu_A)\omega^A$$

$$+exp(\mu_A)\sum_B \mu_{A,B} exp(-\mu_B)\omega^B \wedge exp(-\mu_A)\omega^A$$

$$= exp(-\psi)\mu_{A,1}\omega^1 \wedge \omega^A + \mu_{A,1}\sum_B q_B exp(-\mu_B)\omega^B \wedge \omega^A + \sum_B exp(-\mu_B)\mu_{A,B}\omega^B \wedge \omega^A$$

We also note that $\omega^A_B = -\omega^B_A$, as follows from the following argument. Suppose e_a form a tetrad basis so that the metric coefficients relative to this basis are constants, ie, $g = \eta_{ab}e^a \otimes e^b$ with η_{ab} constants. Then for any vector field X, assuming that the connection is the metrical one,

$$0 = \nabla_X g = X(\eta_{ab})e^a \otimes e^b + \eta_{ab}\nabla_X e^a \otimes e^b + \eta_{ab}e^a \otimes \nabla_X e^b$$

$$= 0 - \eta_{ab}\omega^a_c(X)e^c \otimes e^b - \eta_{ab}e^a \otimes \omega^b_c(X)e^c$$

so that

$$\eta_{bc}\omega^b_a(X)e^a \otimes e^c + \eta_{ab}\omega^b_c(X)e^a \otimes e^c = 0$$

or

$$\omega_{ca}(X)e^a \otimes e^c + \omega_{ac}(X)e^a \otimes e^c = 0$$

or

$$\omega_{ca} = -\omega_{ac}$$

In the present scenario, $\eta = diag[-1,-1,-1,-1]$, so we get $\omega_{ab} = -\omega^a_b$. Thus, it follows that

$$\omega^a_b = -\omega^b_a$$

We can write the equation for $d\omega^A$ as

$$d\omega^A = -exp(-\psi)\mu_{A,1}\omega^A \wedge \omega^1 - \mu_{A,1}q_B exp(-\mu_B)\omega^A \wedge \omega^B - exp(-\mu_B)\mu_{A,B}\omega^A \wedge \omega^B$$

$$= -exp(-\psi)\mu_{A,1}\omega^A \wedge \omega^1 - exp(-\mu_B)\sum \mu_{A:B}\omega^A \wedge \omega^B$$

where

$$\mu_{A:B} = \mu_{A,B} + a_B\mu_{A,1}$$

This must be the same as

$$-\omega^A_B \wedge \omega^B - \omega^A_1 \wedge \omega^1$$

We set

$$\omega^A_1 = a(A)\omega^A + b(A)\omega^1 + \sum_B c(A,B)\omega^B,$$

$$\omega^A_B = p(A,B)\omega^A + q(A,B)\omega^B + r(A,B)\omega^1$$

Then, from the Cartan structure equation,

$$-d\omega^A = (a(A)\omega^A + \sum c(A,B)\omega^B) \wedge \omega^1 + \sum_B (p(A,B)\omega^A + r(A,B)\omega^1) \wedge \omega^B$$

$$= a(A).\omega^A \wedge \omega^1 + \sum_B (c(A,B) - r(A,B))\omega^B \wedge \omega^1 + \sum_B p(A,B)\omega^A \wedge \omega^B$$

Comparing with the above equation, we get

$$a(A) = exp(-\psi)\mu_{A,1}, c(A,B) = r(A,B), p(A,B) = exp(-\mu_B)\mu_{A:B}$$

This means that

$$\omega_1^A = exp(-\psi)\mu_{A,1}\omega^A + b(A)\omega^1 + \sum c(A,B)\omega^B$$

$$\omega_B^A = exp(-\mu_B)\mu_{A:B}\omega^A + q(A,B)\omega^B + c(A,B)\omega^1$$

Since $\omega_B^A = -\omega_A^B$, this means that $c(A,B) = -c(B,A)$ and $q(B,A) = -exp(-\mu_B)\mu_{A:B}$. It remains to compute the functions $c(a,B), b(A)$. Now,

$$d\omega^1 = -\sum_B \omega_B^1 \wedge \omega^B = \sum_B \omega_1^B \omega^B = \sum_B b(B)\omega^1 \wedge \omega^B + \sum_{A,B} c(B,A)\omega^A \wedge \omega^B$$

On the other hand, using

$$\omega^1 = exp(\psi)(dx^1 - \sum_{A=2,3,4} q_A dx^A)$$

we find that

$$d\omega^1 = \sum_A exp(\psi)\psi_{,A} dx^A \wedge dx^1 - \sum_A (q_A exp(\psi))_{,1} dx^1 \wedge dx^A - \sum_{A,B} (exp(\psi)q_A)_{,B} dx^B \wedge dx^A$$

$$= \sum_A exp(\psi)(\psi_{,A} + \psi_{,1}q_A + q_{A,1})dx^A \wedge dx^1 - exp(\psi)\sum_{A,B}(q_{A,B} + \psi_{,B}q_A)dx^B \wedge dx^A$$

$$= exp(\psi)\sum_A (\psi_{:A} + q_{A,1})dx^A \wedge dx^1 - exp(\psi)\sum_{A,B}(q_{A,B} + \psi_{,B}q_A)dx^B \wedge dx^A$$

$$= \sum_A (\psi_{:A} + q_{A,1})exp(\psi - \mu_A)\omega^A \wedge (exp(-\psi)\omega^1 + \sum_A q_A exp(-\mu_A)\omega^A)$$

$$-exp(\psi)\sum_{A,B}(q_{A,B} + \psi_{,B}q_A)exp(-\mu_B - \mu_A)\omega^B \wedge \omega^A$$

$$= \sum_A (\psi_{:A} + q_{A,1})exp(-\mu_A)\omega^A \wedge \omega^1 + \sum_{A,B}(\psi_{:A} + q_{A,1})q_B exp(\psi - \mu_A - \mu_B)\omega^A \wedge \omega^B$$

$$-\sum_{A,B} exp(\psi - \mu_B - \mu_A)(q_{A,B} + q_A\psi_{,B})\omega^B \wedge \omega^A$$

$$= \sum_A (\psi_{:A} + q_{A,1})exp(-\mu_A)\omega^A \wedge \omega^1 + \sum_{A,B} exp(\psi - \mu_A - \mu_B)((\psi_{:A} + q_{A,1})q_B + q_{A,B} + q_A\psi_{,B})\omega^A \wedge \omega_B$$

Now

$$(\psi_{:A} + q_{A,1})q_B + q_{A,B} + q_A\psi_{,B} =$$

$$(\psi_{,A} + q_A\psi_{,1})q_B + q_{A,1}q_B + q_{A,B} + q_A\psi_{,B}$$

$$= q_{A:B} + \psi_{,A}q_B + \psi_{,B}q_A + \psi_{,1}q_Aq_B$$

and since the quantity $\psi_{,A}q_B + \psi_{,B}q_A + \psi_{,1}q_Aq_B$ is symmetric in the indices (A,B), it follows that after multiplying it by $\omega^A \wedge \omega^B$ and summing over A, B, we get zero. Thus,

$$d\omega^1 = \sum_A (\psi_{:A} + q_{A,1})exp(-\mu_A)\omega^A \wedge \omega^1 + \sum_{A,B} exp(\psi - \mu_A - \mu_B)q_{A:B}\omega^A \wedge \omega^B$$

Comparing this with the expression for $d\omega^1$ obtained using Cartan's equation of structure gives us

$$b(A) = -(\psi_{:A} + q_{A,1})exp(-\mu_A),$$

$$c(B,A) = \frac{1}{2}exp(\psi - \mu_A - \mu_B)(q_{A:B} - q_{B:A})$$

This completes our derivation of the connection forms:

$$\omega_1^A = exp(-\psi)\mu_{A,1}\omega^A - (\psi_{:A} + q_{A,1})exp(-\mu_A)\omega^1 - \frac{1}{2}\sum exp(\psi - \mu_A - \mu_B)(q_{A:B} - q_{B:A})\omega^B$$

$$\omega_B^A = exp(-\mu_B)\mu_{A:B}\omega^A - exp(-\mu_A)\mu_{B:A}\omega^B + \frac{1}{2}exp(\psi - \mu_A - \mu_B)(q_{B:A} - q_{A:B})\omega^1$$

We are now in a position to use Cartan's second equation of structure to derive expressions for the Riemann curvature tensor. First we introduce some notation:

$$Q_{AB} = q_{A:B} - q_{B:A}$$

Then,

$$\omega_1^A = exp(-\psi)\mu_{A,1}\omega^A - (\psi_{:A} + q_{A,1})exp(-\mu_A)\omega^1 - \frac{1}{2}\sum_B exp(\psi - \mu_A - \mu_B)Q_{AB}\omega^B$$

$$\omega_B^A = exp(-\mu_B)\mu_{A:B}\omega^A - exp(-\mu_A)\mu_{B:A}\omega^B - \frac{1}{2}exp(\psi - \mu_A - \mu_B)Q_{AB}\omega^1$$

Note that $\omega_A^1 = -\omega_1^A$. Now we derive an identity (Chandrasekhar): Let F be any function of $x^i, i = 1, 2, 3, 4$. The identity involves computing $d(F\omega^1)$ and $d(F\omega^A)$.

$$\omega^1 = exp(\psi)(dx^1 - \sum q_A dx^A), \omega^A = exp(\mu_A)dx^A$$

so that

$$d(F\omega^1) = d(Fexp(\psi)(dx^1 - \sum q_A dx^A))$$

$$= \sum(F.exp(\psi))_{,A}dx^A \wedge dx^1 - \sum_A(Fexp(\psi)q_A)_{,1}dx^1 \wedge dx^A - \sum_{A,B}(Fexp(\psi)q_A)_{,B}dx^B \wedge dx^A$$

Using

$$dx^A = exp(-\mu_A)\omega^A, dx^1 = exp(-\psi)\omega^1 + \sum q_A exp(-\mu_A)\omega^A$$

we get

$$d(F\omega^1) = \sum_A[(F.exp(\psi))_{,A} + (F.exp(\psi)q_A)_{,1}](exp(-\mu_A)\omega^A) \wedge (exp(-\psi)\omega^1 + \sum_B q_B exp(-\mu_B)\omega^B)$$

$$+ \sum_{A,B}(F.exp(\psi).q_A)_{,B}exp(-\mu_A - \mu_B)\omega^A \wedge \omega^B$$

$$= \sum[F_{,A} + F\psi_{,A} + (Fq_A)_{,1} + Fq_A\psi_{,1}]exp(-\mu_A)\omega^A \wedge \omega^1$$

$$+\sum_{A,B} exp(\psi-\mu_A-\mu_B)[(F_{,A}+F\psi_{,A}+(Fq_A)_{,1}+Fq_A\psi_{,1})q_B+(Fq_A)_{,B}+Fq_A\psi_{,B}]\omega^A\wedge\omega^B$$

Define the linear operator D_A so that

$$D_AF = F_{:A}+Fq_{A,1} = F_{,A}+F_{,1}q_A+Fq_{A,1} = F_{,A}+(Fq_A)_{,1}$$

Then,

$$D_A(F.exp(\psi)) = (Fexp(\psi))_{,A}+(F.exp(\psi)q_A)_{,1}$$

$$= exp(\psi)(F_{,A}+F\psi_{,A}+(Fq_A)_{,1}+Fq_A\psi_{,1})$$

Also,

$$(F_{,A}+F\psi_{,A}+(Fq_A)_{,1}+Fq_A\psi_{,1})q_B+(Fq_A)_{,B}+Fq_A\psi_{,B}$$

$$= [F_{,A}q_B+F_{,B}q_A+F(\psi_{,A}q_B+\psi_{,B}q_A)+(F_{,1}+F\psi_{,1})q_Aq_B]+F(q_{A,1}q_B+q_{A,B})$$

The totality of terms inside [.] are symmetric in the indices (A, B). It follows that the antisymmetric part of the right side of the above equation is precisely

$$\frac{1}{2}F(q_{A,1}q_B+q_{A,B}-q_{B,1}Q_A-q_{B,A}) = \frac{1}{2}F(q_{A:B}-q_{B:A}) = \frac{1}{2}FQ_{AB}$$

Thus,

$$d(F\omega^1) = \sum_A exp(-\psi)D_A(F.exp(\psi))exp(-\mu_A)\omega^A\wedge\omega^1+\frac{1}{2}\sum_{A,B} exp(\psi-\mu_A-\mu_B)FQ_{AB}\omega^A\wedge\omega^B$$

A similar computation can be performed for $d(F\omega^A)$:

$$d(F.\omega^A) = d(F.exp(\mu_A)dx^A) = (F.exp(\mu_A))_{,1}dx^1\wedge dx^A+\sum_B(F.exp(\mu_A)_{,B}dx^B\wedge dx^A$$

$$= (F.exp(\mu_A))_{,1}(exp(-\psi)\omega^1+\sum_B q_Bexp(-\mu_B)\omega^B)\wedge exp(-\mu_A)\omega^A$$

$$+\sum_B(F.exp(\mu_A))_{,B}exp(-\mu_A-\mu_B)\omega^B\wedge\omega^A$$

$$= (F.exp(\mu_A))_{,1}exp(-\psi-\mu_A)\omega^1\wedge\omega^A+\sum_B(F.exp(\mu_A))_{,1}q_B+(F.exp(\mu_A))_{,B})exp(-\mu_A-\mu_B)\omega^B\wedge\omega^A$$

$$= (F.exp(\mu_A))_{,1}exp(-\psi-\mu_A)\omega^1\wedge\omega^A+\sum_B(F.exp(\mu_A))_{:B}exp(-\mu_A-\mu_B)\omega^B\wedge\omega^A$$

Kerr metric

To derive the Kerr metric, we start with the general form

$$ds^2 = exp(2\nu)dt^2-exp(2\psi)(d\phi-\omega.dt)^2-exp(2\mu_2)dr^2-exp(2\mu_3)d\theta^2$$

where $\nu, \psi, \omega, \mu_2, \mu_3$ are functions of (r, θ) only. After writing down the Einstein field equations $R_{ij} = 0$ for this metric, and imposing a gauge condition on the functions μ_2, μ_3, we finally arrive at the Kerr metric (see Chandrasekhar's book "The mathematical theory of blackholes")

$$exp(2\nu) = \rho^2\Delta/\Sigma^2, exp(2\psi) = \Sigma^2.sin^2(\theta)/\rho^2, \omega = 2aMr/\Sigma^2, exp(2\mu_2) = \rho^2/\Delta, exp(2\mu_3) = \rho^2$$

where

$$\rho^2 = r^2 + a^2cos^2(\theta), \Sigma^2 = (r^2 + a^2)^2 - a^2\Delta\delta, \Delta = r^2 + a^2 - 2Mr, \delta = sin^2(\theta)$$

Note that setting $a = 0$ gives us

$$\rho = r, \Delta = r^2 - 2Mr, \Sigma^2 = r^4, exp(2\nu) = 1 - 2M/r, exp(2\mu_2) = r^2/(r^2 - 2Mr) = (1 - 2M/r)^{-1} exp(2\mu_3)$$
$$= r^2, exp(2\psi) = r^2.sin^2(\theta)$$
$$\omega = 0$$

and hence the metric becomes

$$ds^2 = (1 - 2M/r)dt^2 - (1 - 2M/r)^{-1}dr^2 - r^2(d\theta^2 + sin^2(\theta)d\phi^2)$$

ie, when $a = 0$, the Kerr metric becomes the Schwarzchild metric. We are now in a position to study geodesics and electromagnetic waves in the Kerr metric.

13.Gravitation:Developing multipole expansions for the gravitational potential produced by distribution of matter.

$$\Phi(r) = -G\int \rho(r')d^3r'/|r - r'|$$

$$|r - r'|^{-1} = (r^2 + r'^2 - 2r.r')^{-1/2} = r^{-1}\sum_{n=0}^{\infty} P_n(r.r'/rr')(r'/r)^n$$

Assume that the angle θ is measured from the direction of r. Then $r.r' = rr'cos(\theta)$ and

$$|r - r'|^{-1} = \sum_{n=0}^{\infty} P_n(cos(\theta))r'^n/r^{n+1}$$

giving the multipole expansion

$$\Phi(r, \theta) = \sum_{n=0}^{\infty} \phi_n(\theta)/r^{n+1}$$

where

$$\phi_n(\theta) = \int_0^{\infty}\int_0^{\pi} P_n(cos(\theta))r'^n\rho(r')r'^2 sin(\theta)dr'd\theta$$

Retaining upto the second order terms, we find that

$$\Phi(r, \phi) \approx -GM/r + \eta(\phi)/r^2$$

We are assuming the matter distribution to have cylindrical symmetry. The equations of motion are

$$r'' - r\phi'^2 = -\frac{\partial\Phi(r,\phi)}{\partial r}$$

$$r\phi'' + 2r'\phi' = -r^{-1}\frac{\partial\Phi(r,\phi)}{\partial\phi}$$

These two differential equations must be integrated numerically to determine the effect of small perturbations in the matter distribution of the central body on the motion. The term involving $\eta(\phi)$ is to be regarded as a first order perturbation to the dynamics:

$$r'' - r\phi'^2 = -GM/r^2 + 2\eta(\phi)/r^3,$$

$$r\phi'' + 2r'\phi' = 2\eta'(\phi)/r^4$$

An example where such perturbations are significant is the small perturbation in the shape of the body from spherical to ellipsoidal.

14. Effect of a weak gravitational field on the field pattern produced by an antenna. Assume a Newtonian metric:

$$d\tau^2 = (1 + 2U/c^2)dt^2 - \frac{1}{c^2}(dx^2 + dy^2 + dz^2)$$

where $|U| << c^2$. We choose coordinates

$$x^0 = t, x^1 = x, x^2 = y, x^3 = z$$

so that the metric becomes

$$g_{00} = 1 + 2U/c^2, g_{11} = g_{22} = g_{33} = -1/c^2, g_{ij} = 0, i \neq j$$

Then let the four potential be

$$(A^\mu) = (A^0, A^1, A^2, A^3) = (V/c^2, A_x, A_y, A_z)$$

We have

$$A_0 = (1 + 2U/c^2)V/c^2, A_1 = -A_x/c^2, A_2 = -A_y/c^2, A_3 = -A_z/c^2$$

and the field tensor components are

$$F_{01} = A_{1,0} - A_{0,1} = -\frac{1}{c^2}A_{x,t} - \frac{1}{c^2}((1 + 2U/c^2)V)_{,x}$$

$$= -\frac{1}{c^2}(A_{x,t} + V_{,x}) - \frac{2}{c^4}(UV)_{,x}$$

In general,

$$(F_{0r}, r = 1, 2, 3) = -\frac{1}{c^2}(A_{,t} + \nabla V) - \frac{2}{c^4}\nabla(UV)$$

$$F_{12} = A_{2,1} - A_{1,2} = -\frac{1}{c^2}(A_{y,x} - A_{x,y})$$

$$F_{23} = -\frac{1}{c^2}(A_{z,y} - A_{y,z})$$

$$F_{31} = -\frac{1}{c^2}(A_{x,z} - A_{z,x})$$

or

$$(F_{23}, F_{31}, F_{12}) = -\frac{1}{c^2}\nabla \times A$$

$$(F^{0r}) = (g^{00}g^{rr}F_{0r}) = -c^2(1 - 2U/c^2)(-\frac{1}{c^2}(A_{,t} + \nabla V) - \frac{2}{c^4}\nabla(UV))$$

$$= A_{,t} + \nabla V - \frac{2U}{c^2}(A_{,t} + \nabla V) + \frac{2}{c^2}\nabla(UV)$$

$$F^{rs} = g^{rr}g^{ss}F_{rs} = c^4 F_{rs}$$

so that

$$(F^{23}, F^{31}, F^{12}) = -c^2\nabla \times A$$

$$F^{0r}\sqrt{-g} = (1 + U/c^2)F^{0r} = A_{,t} + \nabla V - \frac{U}{c^2}(A_{,t} + \nabla V) + \frac{2}{c^2}\nabla(UV)$$

$$= A_{,t} + \nabla V - \delta E$$

where

$$\delta E = -(U/c^2)(A_{,t} + \nabla V) + (2/c^2)\nabla(UV)$$

$$F^{rs}\sqrt{-g} = -c^2(1 + U/c^2)\nabla \times A = -c^2\nabla \times A - c^2\delta B$$

where

$$\delta B = -U\nabla \times A/c^2$$

The equations

$$(F^{\mu\nu}\sqrt{-g})_{,\nu} = \mu_0 J^\mu$$

give for $\mu = 0$,

$$(\nabla, A_{,t} + \nabla V) - (\nabla, \delta E) = -\rho/\epsilon_0$$

or

$$(\nabla . A)_{,t} + \nabla^2 V - \nabla . \delta E = -\rho/\epsilon_0$$

and for $\mu = 1$,

$$(F^{12}\sqrt{-g})_{,2} + (F^{13}\sqrt{-g})_{,3} + (F^{10}\sqrt{-g})_{,0} = \mu_0 J^1$$

and likewise for $\mu = 2, 3$. These can be cast in vectorial form as

$$\nabla \times (\nabla \times A + \delta B) = \mu_0 J - (A_{,t} + \nabla V - \delta E)_{,t}/c^2$$

or

$$\nabla^2 A - \frac{1}{c^2}A_{,tt} = \nabla(\nabla . A) + \nabla \times \delta B + \nabla V_{,t}/c^2 - \delta E_{,t}/c^2 = -\mu_0 J$$

The Lorentz gauge condition reads

$$(A^\mu\sqrt{-g})_{,\mu} = 0$$

or

$$\frac{1}{c^2}(V(1+U/c^2))_{,t} + (\nabla, A(1+U/c^2)) = 0$$

or equivalently,

$$\frac{1}{c^2}V_{,t} + \nabla.A = -\frac{1}{c^4}(VU)_{,t} - \nabla.(UA)/c^2$$

Using this gauge condition, we get

$$\nabla^2 V - \frac{1}{c^2}V_{,tt} - \frac{1}{c^4}(VU)_{,tt} - \nabla.(UA)_{,t}/c^2 - \nabla.\delta E = -\rho/\epsilon_0$$

$$\nabla^2 A - \frac{1}{c^2}A_{,tt} = -\frac{1}{c^4}\nabla(VU)_{,t} - \nabla(\nabla.(UA))/c^2 + \nabla \times \delta B - \delta E_{,t}/c^2 = -\mu_0 J$$

Note that we have retained only terms linear in U since we are assuming $|U| << c^2$.

15. Torsion:

$$T(X,Y) = \nabla_X Y - \nabla_Y X - [X,Y]$$

Let e_i be a basis for the tangent space and e^i the dual basis. Then

$$\nabla_X Y = \nabla_X(Y^i e_i) = X(Y^i)e_i + Y^i \omega_i^j(X)e_j$$

so

$$T(X,Y) = (X(Y^i)e_i - Y(X^i) - [X,Y]^i)e_i + (Y^i\omega_i^j(X) - X^i\omega_i^j(Y))e_j$$

$$= (X(<e^i, X>) - Y(<e^i, X>) - e^i([X,Y]))e_i + (\omega_i^j(X)e^i(Y) - e^i(X)\omega_i^j(Y))e_j$$

or

$$T^i(X,Y) = e^i(T(X,Y)) = 2de^i(X,Y) + 2\omega_i^j \wedge e^i(X,Y)$$

or

$$\frac{1}{2}T^i = de^i + \omega_i^j \wedge e^i$$

This is Cartan's first equation of structure. In many problems, this equation can be used to compute the connection components ω_j^i from the metric. The second Cartan's equation of structure involves the Riemann curvature tensor. It is defined as

$$R(Z,X,Y) = \nabla_X\nabla_Y Z - \nabla_Y\nabla_X Z - \nabla_{[X,Y]}Z$$

We evaluate its components relative to a basis (e_i) for the tangent space.

$$X = X^i e_i, Y = Y^i e_i, Z = Z^i e_i$$

Then,

$$\nabla_Y Z = \nabla_Y(Z^i e_i) = Y(Z^i)e_i + Z^i\nabla_Y e_i = Y(Z^i)e_i + Z^i\omega_j^i(Y)e_j$$

$$= (Y(Z^i) + Z^j\omega_j^i(Y))e_i$$

$$\nabla_X\nabla_Y Z = X\{Y(Z^i) + Z^j\omega_j^i(Y)\}e_i + (Y(Z^i) + Z^j\omega_j^i(Y))\nabla_X e_i$$

$$= [XY(Z^i) + X(Z^j)\omega^i_j(Y) + Z^j X(\omega^i_j(Y))]e_i + [Y(Z^i) + Z^j\omega^i_j(Y)]\omega^k_i(X)e_k$$

$$= \{XY(Z^i) + X(Z^j)\omega^i_j(Y) + Z^j X(\omega^i_j(Y)) + Y(Z^k)\omega^i_k(X) + Z^j\omega^k_j(Y)\omega^i_k(X)\}e_i$$

Thus,

$$\nabla_X\nabla_Y Z - \nabla_Y\nabla_X Z$$

$$= \{[X,Y](Z^i) + Z^j(X(\omega^i_j(Y)) - Y(\omega^i_j(X)) + \omega^k_j(Y)\omega^i_k(X) - \omega^k_j(X)\omega^i_k(Y)\}e_i$$

so that

$$e^i(R(Z,X,Y)) =$$

$$e^i(\nabla_X\nabla_Y Z - \nabla_Y\nabla_X Z - \nabla_{[X,Y]}Z)$$

$$= Z^j(-\omega^i_j([X,Y]) + X(\omega^i_j(Y)) - Y(\omega^i_j(X)) + \omega^i_k(X)\omega^k_j(Y) - \omega^k_j(X)\omega^i_k(Y))$$

$$= 2Z^j(d\omega^i_j + \omega^i_k \wedge \omega^k_j)(X,Y)$$

We write

$$\omega(R(Z,X,Y)) = R^i_{jkm}\omega(e_i)e^j(Z)e^k \wedge e^m(X,Y)$$

Then we get from the above,

$$\frac{1}{2}R^i_{jkm}e_k \wedge e^m = d\omega^i_j + \omega^i_k \wedge \omega^k_j$$

This equation is called Cartan's second equation of structure. Once the connection forms ω^i_j are known, it can be used to determine the components of the Riemann curvature tensor R^i_{jkm}.

16.Orbits in the Schwarzchild space time: In the plane $\theta = \pi/2$, the metric is given by

$$d\tau^2 = \alpha(r)dt^2 - \alpha(r)^{-1}dr^2/c^2 - r^2d\phi^2/c^2$$

The Euler Lagrange equations give

$$\alpha(r)dt/d\tau = C_1, r^2d\phi/d\tau = C_2$$

so that

$$1 = \alpha(r)(dt/d\tau)^2 - c^{-2}\alpha(r)^{-1}(dr/d\tau)^2 - c^{-2}r^2(d\phi/d\tau)^2$$

$$= C_1^2/\alpha - c^{-2}\alpha^{-1}(dr/d\tau)^2 - c^{-2}r^{-2}C_2^2$$

$$= C_1^2/\alpha - c^{-2}\alpha^{-1}(dr/d\phi)^2C_2^2r^{-4} - c^{-2}r^{-2}C_2^2$$

so that

$$r^{-4}(dr/d\phi)^2 = c^2(C_1^2 - \alpha)/C_2^2 - \alpha r^{-2}$$

or letting $r = 1/u$,

$$(du/d\phi)^2 = c^2(E^2 - 1 + 2GMu/c^2)/L^2 - u^2(1 - 2GMu/c^2)$$

where

$$C_2 = E, C_1 = L$$

or

$$(du/d\phi)^2 = (E^2 - 1)c^2/L^2 + 2GMu^3 - u^2 + 2GMu/L^2$$

See the book "The mathematical theory of blackholes" by S.Chandrasekhar, page.98. This is the fundamental orbit equation for planar motion in the Schwarzchild metric. The term $2GMu^3$ is the general relativistic correction to the Newtonian orbit equation. Differentiating this equation with respect to ϕ gives us

$$d^2u/d\phi^2 + u = 3GMu^2 + GM/L^2$$

We can use this equation to evaluate first order corrections to the Newtonian orbit. Specifically, write

$$u = u_0 + u_1$$

where u_0 is the Newtonian orbit equation and u_1 is the first order correction produced by the general relativistic correction $3GMu^2$. We then get

$$u_0'' + u_0 = GM/L^2, u_1'' + u_1 = 6GMu_0u_1$$

The first equation has solution

$$u_0(\phi) = GM/L^2 + A.cos(\phi)$$

17.Maxwell equations in a small region near a Schwarzchild black-hole. Write

$$r = \sqrt{x^2 + y^2 + z^2}, x = X + u, y = Y + v, z = Z + w$$

where

$$|u| << |X|, |v| << |Y|, |w| << |Z|$$

$$\alpha(r) = 1 - 2m/r, m = GM/c^2$$

We let

$$\mathbf{R} = X\hat{x} + Y\hat{y} + Z\hat{z}, \rho = u\hat{x} + v\hat{y} + w\hat{z}$$

and retain terms upto $O(|\rho|)$ in all the computations.

$$1/r = (R^2 + 2(R,\rho))^{-1/2} = R^{-1}(1 - (R,\rho)/R^2) = 1/R - (R,\rho)/R^3$$

$$\alpha(r) = \alpha(R) + 2m(R,\rho)/R^3 = 1 - 2m/R + 2m(R,\rho)/R^3$$

$$\alpha(r)^{-1} = 1 + 2m/R - 2m(R,\rho)/R^3$$

$$dr = (xdx + ydy + zdz)/r$$

$$dr^2 = r^{-2}(x^2dx^2 + y^2dy^2 + z^2dz^2 + 2xydxdy + 2yzdydz + 2zxdzdx)$$

$$d\tau^2 = (1 - 2m/r)dt^2 - \frac{1}{c^2}(1 - 2m/r)^{-1}dr^2 - \frac{r^2}{c^2}(d\theta^2 + sin^2(\theta)d\phi^2)$$

$$= (1 - 2m/r)dt^2 - \frac{1}{c^2}((1 - 2m/r)^{-1} - 1)dr^2 - \frac{1}{c^2}(dx^2 + dy^2 + dz^2)$$

$$= (1 - 2m/r)dt^2 - \frac{1}{c^2}(2m/r)dr^2 - \frac{1}{c^2}(dx^2 + dy^2 + dz^2)$$

$$= (1 - 2m/R + 2m(R,\rho)/R^3)dt^2 - \frac{2m}{r^3c^2}(xdx + ydy + zdz)^2 - \frac{1}{c^2}(dx^2 + dy^2 + dz^2)$$

$$= (1 - 2m/R + 2m(R,\rho)/R^3)dt^2 - \frac{2m}{R^3c^2}(Xdu + Ydv + Zdw)^2 - \frac{1}{c^2}(du^2 + dv^2 + dw^2)$$

$$= (1 - 2m/R)dt^2 - \frac{2m}{R^3c^2}(X^2du^2 + Y^2dv^2 + Z^2dw^2 + 2XYdudv + 2YZdvdw + 2ZXdwdu) - \frac{1}{c^2}(du^2 + dv^2 + dw^2)$$

where m is of the first order of smallness and so are u, v, w. Second order quantities have been neglected. This approximate metric can be expressed as

$$d\tau^2 = (1 - 2m/R)dt^2 - \frac{1}{c^2}[(1 + 2mX^2/R^3c^2)du^2 + (1 + 2mY^2/R^3c^2)dv^2$$

$$+(1 + 2mZ^2/R^3c^2)dw^2 + 4mXYdudv/R^3c^2 + 4mYZR^3dvdw/R^3c^2 + 4mZXR^3dwdu/R^3c^2]$$

Note that R, X, Y, Z are constants. This metric has constant coefficients and hence corresponds to flat space-time. Formulation of the Maxwell equations is then elementary. To obtain general relativistic effects on the electromagnetic fields, we must go therefore to a higher order of approximation where the metric coefficients are no longer constants. The exact metric is given by

$$d\tau^2 = (1 - 2m/r)^2dt^2 - \frac{1}{c^2}((1 - 2m/r)^{-1} - 1)dr^2 - \frac{1}{c^2}(dr^2 + r^2(d\theta^2 + sin^2(\theta)d\phi^2))$$

$$= \alpha dt^2 - \frac{1}{r^2c^2}(\alpha^{-1} - 1)(xdx + ydy + zdz)^2 - \frac{1}{c^2}(dx^2 + dy^2 + dz^2)$$

$$= g_{tt}dt^2 + g_{xx}dx^2 + g_{yy}dy^2 + g_{zz}dz^2 + 2g_{xy}dxdy + 2g_{yz}dydz + 2g_{zx}dzdx$$

where

$$g_{tt} = \alpha(r),$$

$$g_{xx} = -\frac{1}{c^2}(1 + (\alpha^{-1} - 1))x^2/r^2), g_{yy} = -\frac{1}{c^2}(1 + (\alpha^{-1} - 1)y^2/r^2),$$

$$g_{zz} = -\frac{1}{c^2}(1 + (\alpha^{-1} - 1)z^2/r^2), g_{xy} = g_{yx} = -(\alpha^{-1} - 1)xy/r^2,$$

$$g_{yz} = g_{zy} = -(\alpha^{-1} - 1)yz/r^2, g_{xz} = g_{zx} = -(\alpha^{-1} - 1)zx/r^2$$

The metric coefficients are functions of x, y, z. Consider the electromagnetic four potential

$$A^\mu = (V/c^2, A_x, A_y, A_z)$$

In other words,

$$A^0 = V/c^2, A^1 = A_x, A^2 = A_y, A^3 = A_z$$

$$A_0 = g_{tt}A^0 = g_{tt}V/c^2, A_1 = g_{xx}A_x + g_{xy}A_y + g_{xz}A_z,$$

$$A_2 = g_{xy}A_x + g_{yy}A_y + g_{yz}A_z, A_3 = g_{xz}A_x + g_{yz}A_y + g_{zz}A_z$$

The covariant field tensor components are

$$F_{tx} = A_{1,t} - A_{0,x} = (g_{xx}A_x + g_{xy}A_y + g_{xz}A_z)_{,t} - (g_{tt}V/c^2)_{,x}$$

$$F_{ty} = A_{2,t} - A_{0,y} = (g_{xy}A_x + g_{yy}A_y + g_{yz}A_z)_{,t} - (g_{tt}V/c^2)_{,y}$$

$$F_{tz} = A_{3,t} - A_{0,z} = (g_{xz}A_x + g_{yz}A_y + g_{zz}A_z)_{,t} - (g_{tt}V/c^2)_{,z}$$

$$F_{xy} = A_{2,x} - A_{1,y} = (g_{xy}A_x + g_{yy}A_y + g_{yz}A_z)_{,x} - (g_{xx}A_x + g_{xy}A_y + g_{xz}A_z)_{,y}$$

$$F_{yz} = A_{3,y} - A_{2,z} = (g_{xz}A_x + g_{yz}A_y + g_{zz}A_z)_{,y} - (g_{xy}A_x + g_{yy}A_y + g_{yz}A_z)_{,z}$$

$$F_{zx} = A_{1,z} - A_{3,x} = (g_{xx}A_x + g_{xy}A_y + g_{xz}A_z)_{,z} - (g_{xz}A_x + g_{yz}A_y + g_{zz}A_z)_{,x}$$

We now compute the contravariant field tensor components. First note that the contravariant metric tensor $(g^{\mu\nu})$ has the form

$$(g^{\mu\nu}) = \begin{pmatrix} g^{tt} & 0 & 0 & 0 \\ 0 & g^{xx} & g^{xy} & g^{xz} \\ 0 & g^{xy} & g^{yy} & g^{yz} \\ 0 & g^{zx} & g^{yz} & g^{zz} \end{pmatrix}$$

In other words,

$$g^{tx} = g^{xt} = g^{ty} = g^{yt} = g^{tz} = g^{zt} = 0$$

The contravariant field tensor components are given by

$$F^{tx} = g^{tt}(g^{xx}F_{tx} + g^{xy}F_{ty} + g^{xz}F_{tz})$$

$$F^{ty} = g^{tt}(g^{xy}F_{tx} + g^{yy}F_{ty} + g^{yz}F_{tz})$$

$$F^{tz} = g^{tt}(g^{xz}F_{tx} + g^{yz}F_{ty} + g^{zz}F_{tz})$$

$$F^{xy} = -F^{yx} = g^{xx}(g^{yy}F_{xy} + g^{yz}F_{xz}) + g^{xy}(g^{yx}F_{yx} + g^{yz}F_{yz}) + g^{xz}(g^{yx}F_{zx} + g^{yy}F_{zy})$$

$$F^{yz} = -F^{zy} = g^{xy}(g^{yz}F_{xy} + g^{zz}F_{xz}) + g^{yy}(g^{xz}F_{yx} + g^{zz}F_{yz}) + g^{yz}(g^{zx}F_{zx} + g^{zy}F_{zy})$$

$$F^{zx} =$$

18. Energy-momentum tensor of the electromagnetic field:

$$L = F_{ij}F^{ij}\sqrt{-g}$$

$$\delta g = gg^{ij}\delta g_{ij}$$

so that

$$\delta\sqrt{-g} = \frac{1}{2}\sqrt{-g}g^{ij}\delta g_{ij}$$

giving the variation of L with respect to (g_{ij}) as

$$\delta L = \frac{1}{2}F_{ab}F^{ab}\sqrt{-g}g^{ij}\delta g_{ij} + \sqrt{-g}F_{ij}\delta F^{ij}$$

$$F^{ij} = g^{ia}g^{jb}F_{ab}$$

so that

$$\delta F^{ij} = F_{ab}(g^{ia}\delta g^{jb} + g^{jb}\delta g^{ia})$$
$$= -F_{ab}(g^{ia}g^{jc}g^{bd}\delta g_{cd} + g^{jb}g^{ic}g^{ad}\delta g_{cd})$$
$$F_{ij}\delta F^{ij} =$$
$$= (-F_{ab}F^{ac}g^{bd} - F_{ab}F^{cb}g^{ad})\delta g_{cd}$$

We define

$$F_b^a = g^{ac}F_{cb}$$

Then, the above becomes

$$F_{ij}\delta F^{ij} = (F_a^d F^{ac} - F_b^d F^{cb})\delta g_{cd} = 2F_a^d F^{ac}\delta g_{cd}$$

Thus,

$$\delta L = [2\sqrt{-g}F_a^d F^{ac} + \frac{1}{2}\sqrt{-g}F_{ab}F^{ab}g^{cd}]\delta g_{cd}$$

This means that the energy-momentum tensor of the electromagnetic field must be proportional to

$$S^{ij} = -F^{ir}F_r^j + \frac{1}{4}F_{pq}F^{pq}g^{ij}$$

Note that the Einstein field equation in space filled with radiation alone must be given by

$$R^{ij} - \frac{1}{2}Rg^{ij} = KS^{ij}$$

This equation can be used in the derivation of the Reissner-Nordstorm solution for the metric corresponding to a massive spherical body with spherically symmetric charge distribution. An axisymmetric stationary metric has the general form

$$ds^2 = exp(2\nu)dt^2 - exp(2\psi)(d\phi - \omega.dt)^2 - exp(2\mu_2)dr^2 - exp(2\mu_3)d\theta^2$$

where

$$\nu, \psi, \omega, \mu_2, \mu_3$$

are all functions of (r, θ) only. We define the one forms

$$\omega^0 = exp(\nu)dt, \omega^1 = exp(\psi)(d\phi - \omega dt), \omega^2 = exp(\mu_2)dr, \omega^3 = exp(\mu_3)d\theta$$

and then the metric can be expressed as

$$g = \omega^0 \otimes \omega^0 - \sum_{r=1}^{3} \omega^r \otimes \omega^r$$

The Riemann curvature tensor is computed using the Cartan equations of structure. (See the book by S.Chandrasekhar)

19. Tetrad formulation of the Maxwell equations for a space-time of sufficient generality: We set

$$(e_{0i}) = (exp(\nu),0,0,0),(e_{1i}) = (\omega.exp(\psi),-exp(\psi),q_2exp(\psi),q_3exp(\psi)),$$

$$(e_{2i}) = (0,0,-exp(\mu_2),0),(e_{3i}) = (0,0,0,-exp(\mu_3))$$

These are the covariant basis vectors for the orthonormal tetrad frame. Note that

$$\omega_0 = e_{0i}dx^i = exp(\nu)dt, \omega_1 = e_{1i}dx^i = -exp(\psi)(d\phi - \omega dt - q_2dx^2 - q_3dx^3)$$

$$\omega_2 = -exp(\mu_2)dx^2, \omega_3 = -exp(\mu_3)dx^3,$$

$$x^2 = r, x^3 = \theta, x^0 = t, x^1 = \phi$$

Then setting

$$(\eta^{ab}) = diag[1,-1,-1,-1]$$

gives us

$$g = \eta^{ab}\omega_a \otimes \omega_b = exp(2\nu)dt^2 - exp(2\psi)(d\phi - \omega dt - q_2dx^2 - q_3dx^3)^2 - exp(2\mu_2)(dx^2)^2 - exp(2\mu_3)(dx^3)^2$$

The corresponding contravariant tetrad basis is obtained from the equations

$$e_{ai}e_b^i = \eta_{ab}$$

It then follows that

$$(e_0^i) = (exp(-\nu),\omega.exp(-\nu),0,0),$$

$$(e_1^i) = (0,exp(-\psi),0,0),$$

$$(e_2^i) = (0,q_2exp(-\mu_2),exp(-\mu_2),0),(e_3^i) = (0,q_3exp(-\mu_3),0,exp(-\mu_3))$$

We have

$$e_0(f) = e_0^i\partial_i f = exp(-\nu)(f_{,0} + \omega f_{,1}) = exp(-\nu)f_{:0},$$

$$e_1(f) = e_1^i\partial_i f = exp(-\psi)f_{,1},$$

$$e_2(f) = e_2^i\partial_i f = exp(-\mu_2)(f_{,2} + q_2f_{,1}) = exp(-\mu_2)f_{:2}$$

$$e_3(f) = e_3^i\partial_i f = exp(-\mu_3)(f_{,3} + q_3f_{,1}) = exp(-\mu_3)f_{:3}$$

Recall that the spin coefficients are given by

$$\gamma_{abc} = e_a^i e_{bi:j}e_c^j$$

It follows that

$$\gamma_{abc} = -e_{bi}e^i_{a:j}e_c^j = -e_b^i e^i_{a:j}e_c^j = -\gamma_{bac}$$

ie, the spin coefficient is antisymmetric in its first two arguments. The Ricci rotation coefficients are given by

$$\lambda_{abc} = e_a^i(e_{bi,j} - e_{bj,i})e_c^j = e_a^i(e_{bi:j} - e_{bj:i})e_c^j$$

We have

$$\gamma_{abc} = \frac{1}{2}(\lambda_{abc} + \lambda_{cab} - \lambda_{bca})$$

Indeed, twice the right side evaluates to

$$e_a^i(e_{bi:j} - e_{bj:i})e_c^j + e_c^i(e_{ai:j} - e_{aj:i})e_b^j - e_b^i(e_{ci:j} - e_{cj:i})e_a^j$$

$$= \gamma_{abc} - \gamma_{cba} + \gamma_{cab} - \gamma_{bac} - \gamma_{bca} + \gamma_{acb}$$

$$= 2\gamma_{abc}$$

since

$$\gamma_{bac} = -\gamma_{abc}, \gamma_{cab} + \gamma_{acb} = 0, \gamma_{cba} + \gamma_{bca} = 0$$

This results in the above formula. Note that

$$\lambda_{abc} = e_a^i(e_{bi,j} - e_{bj,i})e_c^j = e_{bi,j}(e_a^i e_c^j - e_a^j e_c^i)$$

For the present metric,

$$\lambda_{100} = e_1^i(e_{0i,j} - e_{0j,i})e_0^j$$

Now, using

$$(e_0^i) = (exp(-\nu), \omega.exp(-\nu), 0, 0),$$

$$(e_1^i) = (0, exp(-\psi), 0, 0),$$

$$(e_2^i) = (0, q_2 exp(-\mu_2), exp(-\mu_2), 0), (e_3^i) = (0, q_3 exp(-\mu_3), 0, exp(-\mu_3))$$

$$(e_{0i}) = (exp(\nu), 0, 0, 0), (e_{1i}) = (\omega.exp(\psi), -exp(\psi), q_2 exp(\psi), q_3 exp(\psi)),$$

$$(e_{2i}) = (0, 0, -exp(\mu_2), 0), (e_{3i}) = (0, 0, 0, -exp(\mu_3))$$

we have

$$e_1^i e_0^j e_{0i,j} = 0$$

since the only nonvanishing component of e_0 is e_{00} and $e_1^0 = 0$.

$$e_1^i e_0^j e_{0j,i} = e_{00,1} e_1^1 e_0^0 = (exp(\nu))_{,1} exp(-\psi) exp(-\nu) = exp(-\psi)\nu_{,1}$$

Thus,

$$\lambda_{100} = -exp(-\psi)\nu_{,1}$$

$$\lambda_{101} = e_{0i,j}(e_1^i e_1^j - e_1^j e_1^i) = 0$$

$$\lambda_{110} = e_{1i,j})(e_1^i e_0^j - e_1^j e_0^i)$$

$$e_{1i,j} e_1^i e_0^j = e_1^1 e_0^j e_{11,j} = e_1^1(e_0^0 e_{11,0} + e_0^1 e_{11,1})$$

$$= exp(-\psi)(exp(-\nu)(-exp(\psi))_{,0} + \omega.exp(-\nu)(-exp(\psi))_{,1})$$

$$= exp(-\nu)(-\psi_{,0} - \omega\psi_{,1}) = -exp(-\nu)\psi_{:0}$$

Recall that by definition,

$$f_{:A} = f_{,A} + q_A f_{,1}$$

Next,

$$e_{1i,j}e_1^je_0^i = e_1^1e_0^ie_{1i,1} = exp(-\psi)(exp(-\nu)e_{10,1} + \omega.exp(-\nu)e_{11,1})$$

$$= exp(-\psi)(exp(-\nu)((\omega exp(\psi)),_1 - \omega.exp(-\nu)(exp(\psi)),_1)$$

$$= exp(-\nu)(\omega,_1 + \omega\psi,_1 - \omega\psi,_1) = exp(-\nu)\omega,_1$$

giving

$$\lambda_{110} = -exp(-\nu)(\psi_{:0} + \omega,_1) = -exp(-\nu)\Psi_0$$

where we define

$$\Psi_A = \psi_{:A} + q_{A,1}, q_0 = \omega$$

$$\lambda_{223} = e_{2i,j}(e_2^ie_3^j - e_2^je_3^i)$$

$$e_{2i,j}e_2^ie_3^j = e_{22,j}e_2^2e_3^j = e_2^2(e_3^1e_{22,1} + e_3^3e_{22,3})$$

$$= exp(-\mu_2)(q_3exp(-\mu_3)(-exp(\mu_2)),_1 + exp(-\mu_3)(-exp(\mu_2)),_3)$$

$$= -exp(-\mu_3)(q_3\mu_{2,1} + \mu_{2,3}) = -exp(-\mu_3)\mu_{2:3}$$

$$e_{2i,j}e_2^je_3^i = e_{22,j}e_3^2e_2^j = 0$$

Taking the difference, we get

$$\lambda_{223} = -exp(-\mu_3)\mu_{2:3}$$

$$\lambda_{200} = e_{0i,j}(e_2^ie_0^j - e_2^je_0^i)$$

$$e_{0i,j}e_2^ie_0^j = e_{00,j}e_2^0e_0^j = 0$$

since $e_2^0 = 0$. Also

$$e_{0i,j}e_2^je_0^i = e_{00,j}e_2^je_0^0 = e_0^0(e_2^1e_{00,1} + e_2^2e_{00,2})$$

$$= exp(-\nu)(q_2exp(-\mu_2)(exp(\nu)),_1 + exp(-\mu_2)(exp(\nu)),_2)$$

$$= exp(-\mu_2)(q_2\nu,_1 + \nu,_2) = exp(-\mu_2)\nu_{:2}$$

It follows that

$$\lambda_{200} = -exp(-\mu_2)\nu_{:2}$$

$$\lambda_{112} = e_{1i,j}(e_1^ie_2^j - e_1^je_2^i)$$

$$e_{1i,j}e_1^ie_2^j = e_{11,j}e_1^1e_2^j = e_1^1(e_{11,1}e_2^1 + e_{11,2}e_2^2)$$

$$= exp(-\psi)((-exp(\psi)),_1q_2exp(-\mu_2) + (-exp(\psi)),_2exp(-\mu_2))$$

$$= (-\psi,_1q_2 - \psi,_2)exp(-\mu_2) = -\psi_{:2}exp(-\mu_2)$$

$$e_{1i,j}e_1^je_2^i = e_{1i,1}e_1^1e_2^i = e_1^1(e_{11,1}e_2^1 + e_{12,1}e_2^2)$$

$$= exp(-\psi)((-exp(\psi)),_1q_2exp(-\mu_2) + (q_2exp(\psi)),_1exp(-\mu_2))$$

$$= exp(-\mu_2)(-\psi,_1q_2 + q_2\psi,_1 + q_{2,1}) = q_{2,1}exp(-\mu_2)$$

and hence, taking the difference,

$$\lambda_{112} = -(\psi_{:2} + q_{2,1})exp(-\mu_2) = -\Psi_2.exp(-\mu_2)$$

$$(e_0^i) = (exp(-\nu), \omega.exp(-\nu), 0, 0),$$

$$(e_1^i) = (0, exp(-\psi), 0, 0),$$

$$(e_2^i) = (0, q_2 exp(-\mu_2), exp(-\mu_2), 0), (e_3^i) = (0, q_3 exp(-\mu_3), 0, exp(-\mu_3))$$

$$(e_{0i}) = (exp(\nu), 0, 0, 0), (e_{1i}) = (\omega.exp(\psi), -exp(\psi), q_2 exp(\psi), q_3 exp(\psi)),$$

$$(e_{2i}) = (0, 0, -exp(\mu_2), 0), (e_{3i}) = (0, 0, 0, -exp(\mu_3))$$

20. Arthi's Ph.D thesis: Conclusions of the thesis and scope for further work: In this thesis, to start with we have formulated the equations governing the propagation of electromagnetic waves inside a waveguide having inhomogeneous permittivity. The waveguide cross section of the guide has been partitioned into sections and inside each section, the permittivity has been assumed to be constant. It has to start with been assumed that the fields satisfy the usual Helmholtz equation with the wave number being piecewise constant in accord with the standard relation between permittivity and wavelength. The method of moments has been applied to to obtain approximate solutions for the propagating modes. The test/basis functions have been chosen as sin functions that vanish on the boundary of the rectangular waveguide. The fields have been expanded as linear combinations of these test functions and substituted into the modified Helmholtz equation. Then inner products of this equation with the same test functions have been calculated resulting thereby in a system of linear equations for the coefficients. Some remarks on the nature of the resulting solutions have been made herein. The method of moments has thus been applied to the solution of the Helmholtz equation with spatially dependent wave vector. However, it has been noted that the in presence of inhomgeneous permittivity, the Maxwell equations do not yield the above mentioned form of the Helmholtz equation. The modification is more complicated and this has been analyzed with great detail in our work. The method of derivation involves first writing down the Maxwell curl equations for a given frequency taking into account the spatial dependence of the permittivity and then solving these to get the transverse components of the electric and magnetic fields in term of the partial derivatives of the longitudinal components and then substituting these into the last equation in the curl equations to arrive at the modified Helmholtz equation. Specifically, we have derived the explicit modified Helmholtz equation and have indicated the scheme by which numerical techniques for its solution can be developed. The next portion of the thesis deals with finite element solutions to the scalar wave propagation equation in general relativity. This model has been used as a prototype for describing a scalar wave field propagating inside a waveguide when there is a gravitating object present outside the guide. The first step involves using covariant differentiation to write down the scalar wave propagation equation in an arbitrary metric of space-time. This equation is also called the Laplace-Beltrami equation and can be derived from a variational principle. The variational principle can be approximated by a finite variable optimization problem by using the finite element method. This procedure leads to a generalized eigenvalue equation for the determination of the modes that propagate inside the waveguide in the presence of a gravitational field. In our simulations, we have assumed the gravitational field to be generated by a point mass located at a definite distance from the guide walls. The metric of space-time generated by such a point mass is taken to be the Newtonian metric. In order to implement the finite element method, we have partitioned the cross section of the guide into triangular elements and chosen

the test functions in each triangle as linear functions of x, y so that these test functions when evaluated at the triangle vertices yield the Kronecker δ symbols. This guarantees that the potential in each triangular element will coincide with the vertex potential when evaluated at that vertex. The vertex potentials are treated as the unknown variables. Several integrals involving the test functions and their derivatives as well as the gravitational potential in each element need to be evaluated. To carry out the numerical integration, we have changed the coordinate vectors in each triangle so that they fall along two sides of the triangle. The perturbation in the modes due to the gravitational field has been studied graphically. The next stage of the thesis involves the actual determination of the modes for an electromagnetic field inside a waveguide in the presence of a gravitational field. This involves starting with the covariant expression for the action of a free electromagnetic field in the presence of a metric of space-time and then specializing it to the case when the field is enclosed inside a waveguide. This specialization involves assuming sinusoidal dependence on time and exponential dependence on z, where $\hat{z}$ is the direction of the length of the guide. While computing the action we need to raise and lower the indices of tensors. This process leads to the presence of the gravitational potential in the action integral. Having done so, the finite element method has been applied to the mode determination. In each triangular element, the nodal potential is a four vector corresponding to four components of the electromagnetic potential. The action integral thus gives a quadratic function of the nodal potential and on optimizing this function, we get the generalized eigenvalue problem for the modes. The methods developed in this thesis admit a wider scope for research, namely, in obtaining finite element numerical solutions to the Einstein-Maxwell equations of gravitation and electricity. These field equations involve the Einstein tensor, built out of the Ricci tensor on the left hand side and the Maxwell stress tensor of energy momentum on the right side. The functions to be solved for are the ten gravitational potentials g_{ij}, ie, the metric tensor and the four electromagnetic potential A^i. If we know the metric tensor, then the electromagnetic four potential can be solved for using finite element methods for Maxwell's field equations in curved space time. This aspect has been dealt with in the thesis. Conversely, if the electromagnetic four potential is known, then the metric tensor can be solved for using the Einstein field equations. However if we adopt a finite element scheme for this, we would get very complicated nonlinear algebraic equations since the Ricci tensor depends in a highly nonlinear fashion on the metric tensor. It would be worth looking at quadratic and cubic approximations to the Einstein field tensor valid for weak gravitational fields and solving the resultant nonlinear equations that arise out of the finite element scheme using a gradient search approach.

21. Introduction to the special and general theories of relativity:salient features of the two theories: When electromagnetic theory was created by James Clerk Maxwell, it was thought that the electromagnetic waves needed an ether medium for propagation. Such an "ether wind" would cause the speed of the electromagnetic waves to change if the source moved relative to the observer. Such a change in speed could be detected by an interference experiment performed by Michelson and Morley. However no such change was observed and Lorentz concluded that the lengths of rods would have to contract when moved. This was however an artificial creation. The correct explanation was provided by Einstein who said that the Galileo-Newton postulates would have to be changed. Einstein in the process introduced the postulates of the special theory of relativity according to which the speed of light is the same for all observers moving relative to each other with uniform velocity and that the laws of physics remain invariant when one made a transition from one inertial frame to another. Einstein, using this postulate, arrived at the Lorentz transformations which significantly modify the Galilean transformations of space-time. The usual laws of addition of velocities does not hold good in this theory and the whole of Newtonian

mechanics needs to be modified. This leads to special-theory mechanics. The primary feature of the Lorentz transformations is that the Maxwell equations are covariant under this transformation while the Newtonian equations are not. Thus, if we agree to the postulates of the special theory of relativity, then we must retain the Maxwell equations as being correct and modify the Newtonian equations of motion. In the new mechanics, the mass of a particle increases with its velocity and the total energy of a body is proportional to the body's mass, in other words, the famous Einstein equation $E = mc^2$ becomes valid. In summary, Maxwell's equations describe correct laws of physics in the special theory of relativity. The laws of special relativity remain invariant under Loretnz transformations, ie under transformations between reference frames that are in motion relative to each other with a uniform velocity. However, we require that the laws of physics remain invariant under all coordinate transformations. This led Einstein to formulate his celebrated general theory of relativity. Einstein started developing this theory by first formulating his principle of equivalence. According to this if we have a region of space time where there is a uniform gravitational field, then by passing over to a freely falling frame, the gravitational field can be cancelled and in this freely falling frame, the laws of special relativity can be applied. This led Einstein to conclude that the metric of space time in a gravitational field will have non-constant coefficients. Einstein also guessed correctly that if the gravitational field is generated by a definite amount of matter, then the curvature tensor will be nonzero. If the gravitational field is uniform in a region of space-time, then it can be cancelled out over this region by passing over to a freely falling frame and such a field therefore has a flat space-time structure, ie, its curvature tensor vanishes. When the space time is curved, ie generated by a definite amount of matter, the coefficients of the metric which are ten in number may be termed the gravitational potentials. Einstein derived the field equations for such problems. In these field equations, the Einstein tensor constructed out of the curvature tensor appears on the left and on the right appears the energy momentum tensor of matter and electromagnetism. Thus Einstein's field equations, also called Einstein's law of gravitation, relate the curvature of space time to the distribution of matter and radiation. These field equations modify Newton's inverse square law of gravitation. Maxwell's equations in curved space time follow by replacing ordinary partial derivatives of the field quantities by covariant derivatives so as to yield tensor equations. If ordinary partial derivatives are used, the equations will not have tensor character, ie, they will not be invariant under all transformations of the space time manifold. Covariant differentiation takes into account the curvature of space-time induced by gravity and ensures that the equations will have tensor character.

22. The methods of this thesis can also be applied to the problem of studying gravitational radiation. We give a brief survey of this subject herein: First, we recall that the Einstein field equations in the presence of matter have the form

$$R_{\mu\nu} - \frac{1}{2}Rg_{\mu\nu} = KT_{\mu\nu}$$

where $T_{\mu\nu} = \rho.V_\mu V_\nu$ is the energy-momentum tensor of the matter distribution with ρ as the matter density and $V_\mu = dx_\mu/d\tau$ as the four velocity. The proper time $d\tau$ is defined by the metric: $d\tau^2 = g_{\mu\nu}dx^\mu dx^\nu$. These field equations can be contracted to give

$$R = -KT$$

where $R = g^{\mu\nu}R_{\mu\nu}$ is the scalar curvature of space-time and $T = g^{\mu\nu}T_{\mu\nu} = \rho$. The field equations can therefore be expressed in the alternative form as

$$R_{\mu\nu} = K(T_{\mu\nu} - \frac{1}{2}Tg_{\mu\nu}) = KS_{\mu\nu}$$

We note that

$$S_{00} = T_{00} - \frac{1}{2}Tg_{00} = \rho V_0^2 - \frac{1}{2}\rho g_{00} \approx = \rho/2$$

for weak gravitational fields and small matter velocities. For weak gravitational fields, we can write

$$g_{\mu\nu} = \eta_{\mu\nu} + h_{\mu\nu}$$

where

$$\eta_{00} = 1 = -\eta_{rr}, r = 1, 2, 3, \eta_{\mu\nu} = 0, \mu \neq \nu$$

and $|h_{\mu\nu}| << 1$. $h_{\mu\nu}$ represents the perturbation in the space time curvature caused by matter generating gravitational fields. Now, the Ricci tensor upto first order in $h_{\mu\nu}$ is given by

$$R_{\mu\nu} \approx \Gamma^{\alpha}_{\mu\alpha,\nu} - \Gamma^{\alpha}_{\mu\nu,\alpha}$$

and

$$\Gamma^{\alpha}_{\mu\alpha} \approx \eta^{\alpha\beta}(h_{\beta\mu,\alpha} + h_{\beta\alpha,\mu} - h_{\alpha\mu,\beta})/2$$

$$= \eta^{\alpha\beta}h_{\alpha\beta,\mu}/2 = h_{,\mu}/2$$

where h is the four dimensional trace of $(h_{\mu\nu})$. Also,

$$\Gamma^{\alpha}_{\mu\nu} \approx \eta^{\alpha\beta}(h_{\beta\mu,\nu} + h_{\beta\nu,\mu} - h_{\mu\nu,\beta})/2$$

giving the following approximate expression for the Ricci tensor:

$$R_{\mu\nu} \approx \frac{1}{2}(h_{,\mu} - h^{\alpha}_{\mu,\nu\alpha} - h^{\alpha}_{\nu,\mu\alpha} + \eta^{\alpha\beta}h_{\mu\nu,\alpha\beta})$$

We then pass over to harmonic coordinates. This involves assuming that in the given coordinate system,

$$g^{\mu\nu}\Gamma^{\alpha}_{\mu\nu} = 0$$

Let us write this condition in terms of the $h_{\mu\nu}$. It becomes

$$\eta^{\mu\nu}(h_{\alpha\mu,\nu} + h_{\alpha\nu,\mu} - h_{\mu\nu,\alpha}) = 0$$

or

$$h^{\mu}_{\alpha,\mu} = \frac{1}{2}h_{,\alpha}$$

We shall now show that a coordinate system can be chosen so that this condition is satisfied. Suppose we make an infinitesimal coordinate transformation

$$x^{\mu'} = x^{\mu} + \epsilon^{\mu}(x)$$

Then,

$$\eta_{\alpha\beta} + h_{\alpha\beta} = (\eta_{\mu\nu} + h'_{\mu\nu})(\delta^{\mu}_{\alpha} + \epsilon^{\mu}_{,\alpha})(\delta^{\nu}_{\beta} + \epsilon^{\nu}_{,\beta})$$
$$= \eta_{\alpha\beta} + h'_{\alpha\beta} + \epsilon_{\beta,\alpha} + \epsilon_{\alpha,\beta}$$

so that

$$h'_{\alpha\beta} = h_{\alpha\beta} - \epsilon_{\beta,\alpha} - \epsilon_{\alpha,\beta}$$

on retaining only terms of first order of smallness. It follows that

$$h'_{\alpha\beta,\rho} = h_{\alpha\beta,\rho} - \epsilon_{\beta,\alpha\rho} - \epsilon_{\alpha,\beta\rho}$$

Thus,

$$h'_{,\rho} = h_{,\rho} - 2\epsilon^{\alpha}_{,\rho\alpha}$$
$$\eta^{\beta\rho} h'_{\alpha\beta,\rho}$$
$$= h^{\rho'}_{\alpha,\rho} = h^{\rho}_{\alpha,\rho} - \epsilon^{\rho}_{,\alpha\rho} - \eta^{\beta\rho}\epsilon_{\beta,\alpha\rho}$$

Plug this into the gauge condition for $h'_{\alpha\beta}$:

$$h'_{,\rho} = 2h^{\mu'}_{\rho,\mu}$$

to get

$$h_{,\rho} - 2\epsilon^{\alpha}_{,\rho\alpha} = 2(h^{\alpha}_{\rho,\alpha} - \epsilon^{\alpha}_{,\alpha\rho} - \eta^{\alpha\beta}\epsilon_{\rho,\beta\alpha})$$

or

$$h_{,\rho} - 2h^{\alpha}_{,\rho\alpha} = -2\eta^{\alpha\beta}\epsilon_{\rho,\alpha\beta}$$

Suppose we choose ϵ_ρ so as to satisfy the pde

$$\eta^{\alpha\beta}\epsilon_{\rho,\alpha\beta} = h^{\alpha}_{\rho,\alpha} - \frac{1}{2}h_{,\rho}$$

Then, the transformed $h's$ will satisfy the required gauge condition. The gauge condition leads to a simplification of the equation satisfied by $h's$. The resulting equation has the form of a wave equation with source:

$$\eta^{\alpha\beta}h_{\mu\nu,\alpha\beta} = K.S_{\mu\nu}$$

the solution to which can be expressed in terms of the retarded potentials. Numerical integration packages for such an equation can be developed to study gravitational waves. Further, we can expand the Ricci tensor upto quadratic terms in the $h'_{\mu\nu}s$ and obtain the solution to a higher degree of accuracy using the method of moments. Specifically, such an equation will have the form

$$\eta^{\alpha\beta}h_{\mu\nu,\alpha\beta} + Q_{\mu\nu}(\{h_{\rho\sigma}) = KS_{\mu\nu}$$

where $Q_{\mu\nu}$ is a quadratic function of the $h'_{\mu\nu}s$. Such equations can be solved approximately using the method of moments in which $h_{\mu\nu}$ are expanded in terms of basis functions. The resulting algebraic equations will be a system of coupled quadratic equations for the coefficients and gradient search algorithms can be applied to this to obtain approximate solutions.

23.Numerical techniques for the evaluation of the energy momentum tensor of the gravitational field may also be developed. Suppose we expand the metric as $g_{\mu\nu} = \eta_{\mu\nu} + h_{\mu\nu}$ where $h_{\mu\nu}$ is a small perturbation. Let $R^{(1)}_{\mu\nu}$ and $R^{(2)}_{\mu\nu}$ denote the respectively the terms in $R_{\mu\nu}$ that are linear and quadratic in $h_{\mu\nu}$. Then, upto second order, we have

$$R = g^{\mu\nu}R_{\mu\nu} = (\eta^{\mu\nu} + h^{\mu\nu})(R^{(1)}_{\mu\nu} + R^{(2)}_{\mu\nu})$$
$$= \eta^{\mu\nu}R^{(1)}_{\mu\nu} + h^{\mu\nu}R^{(1)}_{\mu\nu} + \eta^{\mu\nu}R^{(2)}_{\mu\nu}$$

Thus,

$$Rg_{\alpha\beta} = \eta_{\alpha\beta}\eta^{\mu\nu}R^{(1)}_{\mu\nu} + \eta_{\alpha\beta}h^{\mu\nu}R^{(1)}_{\mu\nu} + \eta_{\alpha\beta}\eta^{(\mu\nu}R^{(2)}_{\mu\nu}$$
$$+\eta^{\mu\nu}R^{(1)}_{\mu\nu}h_{\alpha\beta}$$

upto second order. The linear component of the Einstein tensor $R_{\mu\nu} - \frac{1}{2}Rg_{\mu\nu}$ has vanishing ordinary divergence. It follows that the quadratic component of the Einstein tensor also has a vanishing divergence in a region where matter is absent provided that we treat third and higher order corrections as zero. It follows that the quadratic component should be regarded as the energy momentum tensor of the gravitational field and this can be evaluated once the $h_{\mu\nu}$ have been solved for using the moment method.

24. There is also scope for developing numerical packages based on the method of moments and the finite element method for solving gravitational problems using the post-Newtonian approximation. Herein we start with the geodesic equation giving the acceleration in terms of the velocities and the connection symbols. In the Newtonian approximation, the acceleration determined has order GM/r^2 which is of the order v^2/r. The post Newtonian approximation seeks to determine the acceleration upto order v^4. Using the geodesic equation, we then conclude that the various components of the connection must be determined to different powers of order v. Here, v is the characteristic velocity of the system of particles moving in the self generated gravitational field. It follows from basic principles that g_{00}, g_{ij} will have expansions in even powers of the velocity while g_{0i} will have expansions in odd powers of the velocity. These expansions are used to determine the different terms in the expansion of the connection coefficients in powers of the velocity. Having done so, the Ricci tensor is also expanded in powers of the velocity and so is the energy-momentum tensor of the matter distribution. The resulting field equations for gravitation are then formulated order by order in powers of the velocity and approximate solutions are developed. The solutions obtained using this post-Newtonian approximation can then be compared with that using the finite element method.

25.There is also scope for combining the moment method with the tetrad formalism. In the tetrad formalism, one starts with a basic set of four contravariant vector fields called a tetrad such that if the metric is contracted using these vector fields, the resulting metric will have constant coefficients. The Maxwell field tensor on contraction with this tetrad gives a four times four antisymmetric matrix of scalar fields. Maxwell's equations can be formulated in terms of these scalar fields. The resulting modification involves expressing contracted versions of the covariant derivatives of the field tensor with respect to the tetrad in terms of ordinary directional derivatives and the spin coefficients. The spin coefficients are independent of the Maxwell field. They depend only upon the tetrad basis chosen. In the tetrad formulation of the equations of general relativity, they play the role of the connection coefficients in the sense that just as the corrections in the covariant derivative involve the connection coefficients, the corrections to the contracted versions of the covariant derivative involve the spin coefficients. The detailed set of the Maxwell equations in a tetrad basis has been listed in Chandrasekhar's book, "The mathematical theory of blackholes". The tetrad formalism proves to be particularly useful in problems involving both gravitation and electromagnetism. It would be useful to develop moment methods for solving such problems in the tetrad formalism. One particular application of the tetrad formalism is the Newman-Penrose metric in which the problem is to solve for the metric of space time when a spherically symmetric distribution of matter acquires a uniform charge. Then, only certain tetrad components of the field tensor are nonvanishing by virtue of spherical symmetry and the problem can be solved exactly for the metric as well as the electromagnetic field. The problem is when this spherical charge distribution is perturbed slightly from uniform spherical. Then, the resulting tetrad equations can be linearized and transformed into a matrix problem using the moment method. The methods of this thesis will be applicable herein.

26. The work of this thesis also suggests that the method of moments and the finite element method can be applied to solve quantum mechanical problems. For example, consider the motion of a quantum mechanical particle in a potential $U(x,y,z)$. The stationary state Schrodinger equation for such a particle reads

$$\nabla^2\psi(x,y,z)+\frac{8\pi^2 m}{h^2}(E-U(x,y,z))\psi(x,y,z)=0$$

The problem is to determine the possible values of the energy eigenvalue E. This equation is in fact seen to be the modified Helmholtz equation with spatially dependent wave number $k(\mathbf{r})=\sqrt{2m(E-U(\mathbf{r}))/h^2}$. The analogous problem occurs in electromagnetic theory when the medium has an inhomogeneous permittivity.

27. Gravitation:

1.Spatial line element: Let A, B be two fixed points in space and let the spatial coordinates of A be x^α and those of B be $x^\alpha+dx^\alpha$. A light signal leaves B at time x^0+dx^{01} and arrives at A at time x^0. It then leaves A at time x^0 and arrives at B at time x^0+dx^{02}. We have with dx^0 denoting either dx^{01} or dx^{02},

$$g_{00}(dx^0)^2+2g_{0\alpha}dx^0dx^\alpha+g_{\alpha\beta}dx^\alpha dx^\beta=0$$

and solving this, we get

$$dx^0=-g_{0\alpha}dx^\alpha/g_{00}\pm\sqrt{(g_{0\alpha}g_{0\beta}-g_{00}g_{\alpha\beta})dx^\alpha dx^\beta}/g_{00}$$

The negative sign gives dx^{01} and the positive sign gives dx^{02}. The time delay in the light leaving B and arriving back at B is given by

$$dx^{02}-dx^{01}=\frac{2}{g_{00}}[(g_{0\alpha}g_{0\beta}-g_{\alpha\beta})dx^\alpha dx^\beta]^{1/2}$$

The corresponding interval of proper time is obtained by multiplying this delay by $\sqrt{g_{00}}$ and the spatial distance between the points A and B is obtained by multiplying this in turn by $1/2$ $(c=1)$. This spatial length element is given by

$$dl^2 = \gamma_{\alpha\beta} dx^\alpha dx^\beta$$

where

$$\gamma_{\alpha\beta} = g_{0\alpha} g_{0\beta}/g_{00} - g_{\alpha\beta}$$

We set

$$g_\alpha = -g_{0\alpha}/g_{00}$$

We claim that the inverse of the spatial metric $(\gamma_{\alpha\beta})_{1\leq\alpha,\beta\leq 3}$ is given by $-(g^{\alpha\beta})$. Indeed,

$$g^{ij} g_{km} = \delta^i_m$$

gives

$$g^{00} g_{0\alpha} + g^{0\beta} g_{\alpha\beta} = 0$$

$$g^{0\alpha} g_{00} + g^{\alpha\beta} g_{0\beta} = 0$$

$$g^{00} g_{00} + g^{0\alpha} g_{0\alpha} = 1$$

$$g^{0\alpha} g_{0\beta} + g^{\rho\alpha} g_{\rho\beta} = \delta^\alpha_\beta$$

The second equation gives

$$g^{0\alpha} = -g^{\alpha\beta} g_{0\beta}/g_{00}$$

and inserting this into the fourth equation, we get

$$g^{\alpha\beta} \gamma_{\beta\rho} = -\delta^\alpha_\beta$$

which is the stated claim.

2.Simultaneity: The time x^0 at which the light signal reaches A should be regarded as simultaneous with that time at B that is half way between the times of departure of the light signal at B and its rearrival at B, ie, the time $x^0 + (dx^{01} + dx^{02})/2$. This time is precisely given by (using the standard theory of quadratic equations)

$$x^0 - g_{0\alpha} dx^\alpha / g_{00}$$

This formula enables one to synchronize clocks along any curve. Specifically, if Γ is a curve with clocks placed all along it, then all these clocks will be synchronized if the time difference between two clocks placed at points P and Q on it is given by the line integral of $-g_{0\alpha} dx^\alpha / g_{00}$ along the curve from point P to point Q.

3.Maxwell's equations in Kerr metric(Ref:Chandrasekhar, "The mathematical theory of black-holes"). The Kerr metric (for a rotating blackhole) has the form

$$d\tau^2 = a(r,\theta)dt^2 - b(r,\theta)dr^2 - c(r,\theta)(d\phi - \omega(r,\theta)dt)^2 - f(r,\theta)d\theta^2$$

We may regard $dt, dr, d\phi, d\theta$ as one forms and define the one forms

$$\omega^0 = \sqrt{a}.dt, \omega^1 = \sqrt{b}dr, \omega^2 = \sqrt{f}d\theta, \omega^3 = \sqrt{c}(d\phi - \omega.dt)$$

so that the metric becomes

$$g = \omega^0 \otimes \omega^0 - \sum_{r=1}^{3} \omega^r \otimes \omega^r$$

Suppose that we have computed the Riemann curvature tensor R_{ijkm} relative to the coordinate chart (t, r, θ, ϕ). We then wish to compute the scalars $R_{(a)(b)(c)(d)}$ obtained by projecting R_{ijkm} onto the above one forms. Specifically, we write

$$\omega^a = e_i^a dx^i$$

so that

$$(e_i^0) = (\sqrt{a}, 0, 0, 0), (e_i^1) = (0, \sqrt{b}, 0, 0),$$
$$(e_i^2) = (0, 0, \sqrt{f}, 0), (e_i^3) = (-\omega\sqrt{c}, 0, 0, \sqrt{c})$$

We define

$$(\eta_{ab}) = diag[1, -1, -1, -1]$$

so that (η^{ab}), the inverse of (η_{ab}) is the same matrix. Then,

$$g = \eta_{ab}\omega^a \otimes \omega^b$$

We then have using

$$e_{ai} = \eta_{ab} e_i^b$$

that

$$e_{0i} = e_{0i}, e_{1i} = -e_i^1, e_{2i} = -e_i^2, e_{3i} = -e_i^3$$

The tetrad scalars $R_{(a)(b)(c)(d)}$ are obtained as

$$R_{(a)(b)(c)(d)} = R_{ijkm} e_a^i e_b^j e_c^k e_d^m$$

To obtain e_a^i from the e_{ai}, we must make use of the equations

$$e_{ai} e_b^i = \eta_{ab}$$

We recall how the curvature tensor components are defined. Take a covariant vector field A_i and consider

$$A_{i:j:k} - A_{i:k:j} = R_{ijk}^m A_m$$

This is the defining relation for the curvature tensor components R_{ijk}^m. Observe that

$$(A_i B_j)_{:k} = A_{i:k} B_j + A_i B_{j:k}$$

$$= (A_{i,k} - \Gamma_{ik}^m A_m) B_j + A_i (B_{j,k} - \Gamma_{jk}^m B_m)$$

$$= (A_i B_j)_{,k} - \Gamma_{ik}^m A_m B_j - \Gamma_{jk}^m A_i B_m$$

Thus, for any tensor T_{ij}, we have

$$T_{ij:k} = T_{ij,k} - \Gamma_{ik}^m T_{mj} - \Gamma_{jk}^m T_{im}$$

It follows that

$$A_{i:j:k} = A_{i:j,k} - \Gamma^m_{ik} A_{m:j} - \Gamma^m_{jk} A_{i:m}$$
$$= (A_{i,jk} - \Gamma^m_{ij} A_m)_{,k} - \Gamma^m_{ik}(A_{m,j} - \Gamma^n_{mj} A_n) - \Gamma^m_{jk}(A_{i,m} - \Gamma^n_{im} A_n)$$
$$= A_{i,jk} - \Gamma^m_{ij,k} A_m - \Gamma^m_{ij} A_{m,k} - \Gamma^m_{ik} A_{m,j} + \Gamma^m_{ik}\Gamma^n_{mj} A_n - \Gamma^m_{jk} A_{i,m}$$
$$+\Gamma^m_{jk}\Gamma^n_{im} A_n$$

so that

$$A_{i:j:k} - A_{i:k:j} = (-\Gamma^m_{ij,k} + \Gamma^m_{ik,j} + \Gamma^n_{ik}\Gamma^m_{nj} - \Gamma^n_{ij}\Gamma^m_{nk}) A_m$$

giving

$$R^m_{ijk} = \Gamma^m_{ik,j} - \Gamma^m_{ij,k} - \Gamma^n_{ij}\Gamma^m_{nk} + \Gamma^n_{ik}\Gamma^m_{nj}$$

Note that the Ricci tensor is given by

$$R_{ij} = R^m_{ijm}$$

ie, it is obtained by contracting the first and the last indices of the Riemann curvature tensor. Maxwell's equations in the Kerr metric background can be written down. We assume that the four vector potential A^i is independent of ϕ. Then, the field tensor F^{ij} is also independent of ϕ. The Maxwell equations are

$$F^{ij}_{:j} = 0$$

or

$$(F^{ij}\sqrt{-g})_{,j} = 0$$

We need to calculate g. Note that the covariant metric tensor (g_{ij}) obtained from the metric

$$d\tau^2 = a(r,\theta)dt^2 - b(r,\theta)dr^2 - c(r,\theta)(d\phi - \omega(r,\theta)dt)^2 - f(r,\theta)d\theta^2$$

is given by

$$g_{00} = a - c\omega^2, g_{11} = -b, g_{22} = -f, g_{33} = c, g_{03} = c\omega = g_{30}$$

with all the other metric coefficients zero. Thus,

$$g = det(g_{ij}) = g_{00}g_{11}g_{22}g_{33} - g_{03}^2 g_{11}g_{22} = g_{11}g_{22}(g_{00}g_{33} - g_{03}^2)$$
$$= (-b)(-f)((a - c\omega^2)c - (c\omega)^2) = bf((a - c\omega)^2 - c^2\omega^2) = bf(a^2 - 2ac\omega)$$

We write

$$A^0 = V, A^1 = A_r, A^2 = A_\theta, A^3 = A_\phi$$
$$A_0 = g_{00}V + g_{03}A_\phi = (a - c\omega^2).V + c\omega.A_\phi$$
$$A_1 = g_{11}A_r = -bA_r, A_2 = g_{22}A_\theta = -fA_\theta, A_3 = g_{33}A_\phi + g_{03}V = cA_\phi + c\omega V$$
$$F_{01} = A_{1,0} - A_{0,1} = (g_{11}A_r)_{,t} - (g_{00}V + g_{03}A_\phi)_{,r}$$
$$= g_{11}A_{r,t} - (g_{00}V)_{,r} - (g_{03}A_\phi)_{,r}$$
$$F_{02} = A_{2,0} - A_{0,2} = (g_{22}A_\theta)_{,0} - (g_{00}V + g_{03}A_\phi)_{,\theta}$$
$$= g_{22}A_{\theta,0} - (g_{00}V)_{,\theta} - (g_{03}A_\phi)_{,\theta}$$

$$F_{03} = A_{3,0} - A_{0,3} = (g_{03}V + g_{33}A_\phi)_{,0} - (g_{00}V + g_{03}A_\phi)_{,\phi}$$
$$= g_{03}V_{,t} + g_{33}A_{\phi,t} - g_{00}V_{,\phi} - g_{00}V_{,\phi} - g_{03}A_{\phi,\phi}$$
$$F_{12} = A_{2,1} - A_{1,2} = (g_{22}A_\theta)_{,r} - (g_{11}A_r)_{,\theta}$$
$$F_{23} = A_{3,2} - A_{2,3} = (g_{03}V + g_{33}A_\phi)_{,\theta} - (g_{22}A_\theta)_{,\phi}$$
$$= (g_{03}V)_{,\theta} + (g_{33}A_\phi)_{,\theta} - g_{22}A_{\theta,\phi}$$
$$F_{31} = A_{1,3} - A_{3,1} = (g_{11}A_r)_{,\phi} - (g_{03}V + g_{33}A_\phi)_{,r}$$
$$= g_{11}A_{r,\phi} - (g_{03}V)_{,r} - (g_{33}A_\phi)_{,r}$$

Now recall that

$$g^{00} = -g_{33}/(g_{03}^2 - g_{00}g_{33})$$

and

$$g^{03} = g_{03}/(g_{03}^2 - g_{00}g_{33})$$

Further,

$$x^3 = y_0/g_{03} - g_{00}(g^{00}y_0 + g^{03}y_3)/g_{03}$$

so that

$$g^{33} = -g_{00}g^{03}/g_{03} = -g_{00}/(g_{03}^2 - g_{00}g_{33})$$

Also

$$g^{11} = 1/g_{11}, g^{22} = 1/g_{22}$$

Written out explicitly, we have

$$g_{03}^2 - g_{00}g_{33} = c^2\omega^2 + c(a - c\omega^2)$$
$$= c^2\omega^2 + ca - c^2\omega^2 = ca$$

so that

$$g^{00} = c/ca = 1/a, g^{33} = (c\omega^2 - a)/ca = \omega^2/a - 1/c, g^{03} = c\omega/ca = \omega/a$$

This gives

$$F^{01} = g^{00}g^{11}F_{01} + g^{03}g^{11}F_{31}$$
$$= (1/a)(-1/b)F_{01} + (-1/b)(\omega/a)F_{31} = -F_{01}/ab - \omega.F_{31}/ab$$
$$F^{02} = g^{00}g^{22}F_{02} + g^{03}g^{22}F_{32} = (1/a)(-1/f)F_{02} + (\omega/a)(-1/f)F_{32}$$
$$= -F_{02}/af - \omega.F_{32}/af = -F_{02}/af + \omega F_{23}/af$$
$$F^{03} =$$

28.The following review comes under Chapter 5, Relativity.

1.S.Chandrasekhar, The mathematical theory of blackholes: The first chapter, titled mathematical preliminaries, provides an exhaustive survey of the mathematical tools of differential geometry needed in general relativity. The account is stressed on providing physical insight without sacrificing mathematical rigour. This chapter can be read easily by an advanced undergraduate student in mathematics or physics. The author starts this chatper by introducing differentiable manifolds, charts and coordinates. He then introduces the notion of a smooth vector field and a smooth one form. Then he defines the exterior derivative between two forms and uses this as a tool to introduce differential forms as skew symmetric multilinear functions on the tangent space to every point of the manifold. The calculus of such forms is discussed next wherein the author introduces Lie derivative of a vector field and more generally of a tensor field. The pushforward and pullback maps for tensor fields are discussed next.

Metric corresponding to a rotating blackhole:

$$d\tau^2 = f(t,r,\theta)dt^2 - g(t,r,\theta)dr^2 - h(t,r,\theta)(d\phi - \omega(t,r,\theta)dt)^2 - k(t,r,\theta)d\theta^2$$

The presence of the term $dtd\phi$ in the expansion of this metric indicates the presence of rotation about the z axis. Maxwell's equations need to be formulated in this space-time. Let us denote the components of the contravariant four vector potential by $A^t, A^r, A^\theta, A^\phi$. Then the components of the corresponding covariant potential can be computed once we know the metric tensor. We have

$$g_{tt} = f - h\omega^2, g_{rr} = -g, g_{\theta\theta} = -k, g_{\phi\phi} = -h, g_{t\phi} = g_{\phi t} = h\omega$$

29.The following review comes under Chapter 5, Relativity.

1.Steven Weinberg, Gravitation and cosmology:

Characteristic features of the book.

1.Notion of covariant derivatives of a vector field along a curve. Suppose that the curve is parameterized by a parameter τ. Take a vector field $A^\mu(x)$ and displace it parallely along the curve from $x^\mu(\tau)$ to $x^\mu(\tau+d\tau)$. The new vector is given by $A^\mu + \delta A^\mu$ where $\delta A^\mu = -\Gamma^\mu_{\alpha\beta}\delta x^\alpha A^\beta$. where $\delta x^\alpha = (dx^\alpha/d\tau)\delta\tau$. The covariant differential of the vector field along the curve is given by

$$A^\mu(x(\tau+\delta\tau)) - A^\mu - \delta A^\mu$$

The ratio of this differential to $d\tau$ gives us the covariant derivative of A^μ along the curve:

$$DA^\mu/D\tau = dA^\mu/d\tau + \Gamma^\mu_{\alpha\beta}A^\alpha dx^\beta/d\tau$$

The covariant derivative of a vector along the curve is again a vector on the curve, ie, it transforms like a vector field on the other hand, the ordinary derivative is not a vector.

2.Geodesic deviation: Let $x^\mu(\tau)$ be a geodesic and $x^\mu(\tau) + \delta x^\mu(\tau)$ a neighboring geodesic. Then,

$$\frac{d^2}{d\tau^2}(x^\mu + \delta x^\mu) + \Gamma^\mu_{\alpha\beta}(x+\delta x)\frac{d(x^\alpha + \delta x^\alpha)}{d\tau}\frac{d(x^\beta + \delta x^\beta)}{d\tau} = 0$$

$$\frac{d^2}{d\tau^2}(x^\mu) + \Gamma^\mu_{\alpha\beta}(x)(dx^\alpha/d\tau)(dx^\beta/d\tau) = 0$$

give on subtracting and linearizing,

$$\frac{d^2\delta x^\mu}{d\tau^2}+2\Gamma^\mu_{\alpha\beta}(x)\frac{dx^\alpha}{d\tau}\frac{d\delta x^\beta}{d\tau}$$

$$+\Gamma^\mu_{\alpha\beta,\nu}\delta x^\nu(dx^\alpha/d\tau)(dx^\beta/d\tau)=0$$

This is the equation for geodesic deviation. It can be cast in a well known form using covariant derivatives and the Riemann curvature tensor. The method has not been explained in full detail but it can be worked out as an exercise.

$$D\delta x^\mu/D\tau=d\delta x^\mu/d\tau+\Gamma^\mu_{\alpha\beta}(dx^\alpha/d\tau)\delta x^\beta$$

so that

$$D^2\delta x^\mu/D\tau^2=d/d\tau(d\delta x^\mu/d\tau+\Gamma^\mu_{\alpha\beta}(dx^\alpha/d\tau)\delta x^\beta)$$
$$+\Gamma^\mu_{\rho\sigma}(dx^\rho/d\tau)(d\delta x^\sigma/d\tau+\Gamma^\sigma_{\alpha\beta}(dx^\alpha/d\tau)\delta x^\beta)$$
$$=d^2\delta x^\mu/d\tau^2+\Gamma^\mu_{\alpha\beta}$$

3.Derivation of the Einstein field equations from very general considerations. In the weak field limit, the field equations must reduce to the Newtonian graviational field equation $\Delta\phi=4\pi G\rho$. Weinberg starts with an equation of the form $C_1R_{\mu\nu}+C_2Rg_{\mu\nu}=KT_{\mu\nu}$ and argues by taking the divergence and using the Bianchi identity, why $C_1=1, C_2=-1/2$.

4.Conditions for stellar equilibrium both in Newtonian theory of gravity and in general relativity have been derived. Chandrasekhar's limit for the mass of the white dwarf has then been obtained from these equations. The result is a classic combination of the Fermi-Dirac statistics and Newtonian gravitation. Let $M(r)$ denote the mass inside a sphere of radius r. Let $p(r)$ be the pressure at a radius r. Then for the pressure to balance the gravitational force, we must have

$$-p'(r)4\pi r^2dr=GM(r)4\pi r^2\rho.dr/r^2$$

or

$$-p'(r)=GM(r)\rho(r)/r^2, M(r)=\int_0^r 4\pi r^2\rho(r)dr$$

This is a integro-differential equation for $\rho(r)$ provided the equation of state $p(\rho)$ is known. Write $p(\rho)=F(\rho)$. Then we get $M'(r)=4\pi r^2\rho(r)$,

$$-F'(\rho(r))\rho'(r)=GM(r)\rho(r)/r^2$$

or

$$-F'(M'(r)/4\pi r^2)(M'/r^2)'=GMM'/r^4$$

For polytropes, $p=C\rho^\gamma$ and the resulting equation of state is called the Lane-Emden equation.

5.General relativistic equations for stellar structure: For static matter, the covariant four velocity components U_μ satisfy $U_r=U_\theta=U_\phi=0, U_t\neq 0$. U_t is determined from the condition

$$g^{\mu\nu}U_\mu U_\nu=-1$$

The metric tensor is given by

$$ds^2 = A(r)dr^2 + r^2 d\theta^2 + r^2 sin^2(\theta)d\phi^2 - B(r)dt^2$$

so that

$$g_{tt} = -B(r), g_{rr} = A(r), g_{\theta\theta} = r^2, g_{\phi\phi} = r^2 sin^2(\theta)$$

so that $g^{tt} = -1/B(r)$ and

$$g^{tt}U_t^2 = -1$$

gives

$$U_t = -\sqrt{B(r)}$$

The energy momentum tensor

$$T_{\mu\nu} = (p(r) + \rho(r))U_\mu U_\nu + p(r)g_{\mu\nu}$$

is calculated from this and inserted into the Einstein field equations

$$R_{\mu\nu} - \frac{1}{2}Rg_{\mu\nu} = -KT_{\mu\nu}$$

to arrive at the general relativistic equations for stellar structure. The Newtonian equation for stellar structure is obtained as a special case of this.

Remark: Consider cartesian coordinates with flat Minkowskian metric so that writing

$$ds^2 = dx^2 + dy^2 + dz^2 - dt^2$$

we have

$$g_{11} = g_{22} = g_{33} = 1, g_{44} = -1$$

then for static fields,

$$U_1 = U_2 = U_3 = 0, U_4 = 1$$

we have

$$T_{11} = T_{22} = T_{33} = p, T_{44} = (p + \rho) - p = \rho$$

as it should be. The fourth component is the energy density.

6.Gravitation: Assume that the field equations have the form

$$C_1 R^{\mu\nu} + C_2 R g^{\mu\nu} = kT^{\mu\nu}$$

where $T^{\mu\nu}$ is the energy-momentum tensor of the matter distribution and $R^{\mu\nu}$ is the Ricci tensor. We have using the Bianchi identity

$$R^{\mu\nu}_{:\nu} = \frac{1}{2}g^{\nu\rho}R_{,\rho}$$

that

$$(\frac{1}{2}C_1 + C_2)R_{,\nu} = kT^{\rho}_{\nu:\rho}$$

and since the right side is zero by matter conservation, it follows that $C_2 = -C_1/2$. Note that we cannot have $R_{,\mu} = 0$ as a field equation because it would imply that $T^{\nu}_{\nu,\rho} = 0$ which is certainly false.

30. P.A.M.Dirac, General relativity: The book starts by introducing the reader to curvilinear coordinates. A p dimensional curved surface immersed in n dimensional Euclidean space is specified by giving n smooth functions of p variables, ie, $y^n(x^1, ..., x^p), n = 1, 2, ..., N$. The metric in the n dimensional Euclidean space is standard $\sum_n (dy^n)^2$. This induces a metric on the p dimensional surface. Using standard parallel displacement in n dimensional Euclidean space combined with projections onto the tangent space, Dirac arrives at the general formula for parallel displacement on a p dimensional surface. The connection coefficients are derived in terms of the functions y^n. These formulae are expressed in terms of first and second order partial derivatives of the functions y^n with respect to its arguments. In order to derive the equations of infinitesimal parallel displacement, Dirac considers a vector tangential to surface expressed in terms of its curvilinear components and expressed the cartesian components of the vector in terms of the curvilinear components. He then uses the fact that the cartesian components do not undergo any change on infinitesimal parallel displacement and this fact is used to derive expressions for the curvilinear components of the vector at the neighbouring point. Specifically, the displaced vector at the neighbouring point is projected onto the tangent plane at this neighbouring point by using the fact that the normal component of the displaced vector on the surface is orthogonal to the tangent plane. In the final formula for the parallely displaced vector, the cartesian coordinates are eliminated and the final formula is thereby expressed entirely in terms of partial derivatives of the metric tensor.

31.Steven Weinberg, Gravitation and cosmology: A nice introduction to the equation of motion of point particles in curved space time has been given at the beginning. The author starts with the equation $d^2\xi^i/d\tau^2 = 0$ in an inertial system ξ^i and then transforms this equation into the curvilinear system. The connection coefficients arise as a natural consequence of this transformation. The Lorentz transformation of special relativity has been derived from general considerations without assuming linearity. Specifically, if $\eta_{\alpha\beta}$ is a constant metric, the proper time interval in the unprimed system is given by $d\tau^2 = \eta_{\alpha\beta} dx^\alpha.dx^\beta$. The condition that this time interval is preserved in in the primed system is that

$$\eta_{\rho\sigma} = \eta_{\alpha\beta} \frac{\partial x^{\alpha'}}{\partial x^{\rho'}} \frac{\partial x^\beta}{\partial x^\sigma}$$

From this condition, the Lorentz transformation equation namely that $\frac{\partial x^{\alpha'}}{\partial x^\rho} = \Lambda^\alpha_\rho$ are constants satisfying

$$\eta_{\rho\sigma} = \eta_{\alpha\beta} \Lambda^\alpha_\rho \Lambda^\beta_\sigma$$

has been derived by the author. The concept of time dilation has been explained very neatly using the Lorentz transformations. The author considers two observers, one at rest relative to a clock and the other in motion relative to the same clock. Suppose dt is the time measured by the first observer for a tick and dt' the time measured by the second observer for the tick. For the first observer, $dx = 0$ relative to the clock and hence the proper time measured by him for a tick is $(dt^2 - dx^2)^{1/2} = dt$. For the second observer, the proper time is $(dt'^2 - dx'^2)^{1/2} = dt'\sqrt{1 - v^2/c^2}$. Since both the observers are related by a Lorentz transformation, it follows that their proper times are the same and hence $dt' = dt/\sqrt{1 - v^2/c^2}$ where v is the velocity of the second observer relative to the clock. This is the phenomenon of time dilation.

32.The transformation law of the metric tensor from an cartesian to a curvilinear system has been explained and the author has provided us with the formula for Gaussian curvature for a two dimensional surface as an intrinsic property of the surface independent of the coordinate system used to specify the metric. The author mentions that this formula is nothing but a specialization of the Riemannian curvature scalar that is derived by contracting the Riemann curvature tensor.

33.Relativistic equations of motion are derived along the following lines. The author starts with the relativistic equations $d^2x^\alpha/d\tau^2 = f^\alpha$ and considers what this equation looks like in a frame in which the particle is instantaneously at rest. In such a frame, $dt = d\tau$ and if F^α is the four force in this frame, then F^α is related to f^α by a Lorentz transformation. This fact combined with $F^0 = 0$ is used by the author to derive the form of f^0 in terms of f^i. This relationship is the classic relationship between force and power. The author then introduces the notion of four momentum and shows that when the particle has velocities small compared with the speed of light, the zeroth component of the momentum is approximately the particle kinetic energy. Conservation of momentum in all inertial frames leads to conservation of energy. This fact has been proved elegantly. A neat discussion of the transformation laws of four vectors and tensors in special relativity has been given. A tensor for example $T^{\alpha\beta}$ transforms into

$$T^{\alpha\beta'} = \Lambda^\alpha_\rho \Lambda^\beta_\sigma T^{\rho\sigma}$$

For a system of point charges e_n moving along trajectories $x_n(t)$ respectively, the current density has been defined as

$$J^i = \sum_n e_n (dx^i_n(t)/dt)\delta^3(x - x_n(t))$$

and the charge density as

$$J^0 = \sum_n e_n \delta^3(x - x_n(t))$$

These four components form the four current density J^α and the author proves the charge conservation law for the four current density. Specifically,

$$\frac{\partial J^i}{\partial x^i} = -\sum_n e_n (dx^i_n/dt) \frac{\partial \delta^3(x - x_n(t))}{\partial x^i_n}$$

$$= -\sum_n e_n \frac{d}{dt}\delta^3(x - x_n(t)) = -\frac{\partial J^0}{\partial t}$$

which can be expressed in four vector notation as

$$\frac{\partial J^\alpha}{\partial x^\alpha} = 0$$

The author next introduces the energy momentum tensor for a system of point particles. Specifically if the nth particle is located at $x_n(t)$ at time t and moves with velocity $dx_n(t)/dt$, its three momentum and velocity are related by $dx^i_n/dt = p^i_n/E_n$. The energy density corresponding to this system is given by

$$T^{00} = \sum_n m_n \delta^3(x - x_n(t))$$

the flux of the ith component of the momentum across a unit area along the jth direction is given by

$$T^{ij} = \sum_n p_n^i (dx_n^j/dt)\delta^3(x - x_n(t)) = \sum_n (p_n^i p_n^j / E_n)\delta^3(x - x_n(t))$$

The corresponding flux of the energy density can be defined by replacing i with 0. The energy momentum tensor is therefore given by

$$T^{\alpha\beta} = \sum_n (p_n^\alpha p_n^\beta / E_n)\delta^3(x - x_n(t))$$

This shows that the energy momentum tensor is symmetric. Note that

$$\frac{\partial T^{\alpha 0}}{\partial t} = \frac{\partial}{\partial t}\sum_n p_n^\alpha \delta^3(x - x_n(t))$$

$$= \sum f_n^\alpha \delta^3(x - x_n(t)) - \sum_n p_n^\alpha (dx_n^i/dt)\frac{\partial}{\partial x^i}\delta^3(x - x_n(t))$$

$$= \sum f_n^\alpha \delta^3(x - x_n(t)) - \frac{\partial T^{\alpha i}}{\partial x^i}$$

where

$$f_n^\alpha = dp_n^\alpha / dt$$

is the four force acting on the nth particle. If no forces act, then we get energy momentum conservation:

$$\frac{\partial T^{\alpha\beta}}{\partial x^\beta} = 0$$

34.Maximally symmetric spaces have been discussed in considerable depth by Weinberg. Using this theory, Weinberg defines the notion of homogeneous and isotropic spaces which are later on used in cosmology. The first idea in this circle is the notion of form invariance of the metric. A metric is said to be form invariant or an isometry, if its form remains invariant under a coordinate transformation. This means that the metric $g_{\mu\nu}(x)$ satisfies the equation

$$g_{\mu\nu}(\bar{x}) = g_{\alpha\beta}(x)\frac{\partial x^\alpha}{\partial \bar{x}^\mu}\frac{\partial x^\beta}{\partial \bar{x}^\nu}$$

Under a given coordinate transformation $x \to \bar{x}$. The author considers infinitesimal coordinate transformations $\bar{x}^\mu = x^\mu + \xi^\mu(x)$ that leave the metric form invariant and thereby arrives at the Killing equation, ie

$$\xi_{\mu:\nu} + \xi_{\nu:\mu} = 0$$

ξ^μ satisfying this equation are called Killing vector fields. The author derives a second order partial differential equation satisfied by the Killing vector fields. This equation involves the Riemann Christoffel curvature tensor. The author then defines using Killing vector fields, the notions of homogeneous and isotropic spaces. He gives a typical example of a maximally symmetric space that forms the basis for all the modern theories of cosmology, namely the Robertson-Walker metric space. The metric for such a space has the form

$$d\tau^2 = dt^2 - R^2(t)dr^2/(1-kr^2) - r^2(d\theta^2 + sin^2(\theta)d\phi^2)$$

where k can assume the values $0, \pm 1$. $R(t)$ is called the scale factor of the universe. The author derives this metric as a special case of the metric

$$d\tau^2 = C_{\mu\nu}dx^\mu dx^\nu + dz^2$$

with z, x^μ satisfying the equation

$$C_{\mu\nu}x^\mu x^\nu + z^2 = 1$$

$C_{\mu\nu}$ are constants. The number of variables x^μ is N and hence the metric of this N dimensional curved space is obtained by restricting a metric on an $N+1$-dimensional space to an N-dimensional surface. The isometries that preserve the resulting metric are derived by the author to be generalized rotations and quasi-translations of space-time.

35.The basic equations of cosmology are derived by the author using the Robertson-Walker metric as the starting point. The author first shows that the Robertson Walker metric admits world lines in which the spatial coordinates r, θ, ϕ are constant. He therefore calls these coordinates as comoving coordinates. The galaxies move along world lines which means that for each galaxy, r, θ, ϕ are constant while t is a variable. The energy momentum tensor for the matter distribution inside the universe is given by

$$T_{\mu\nu} = (\rho(t) + p(t))U_\mu U_\nu + p(t)g_{\mu\nu}$$

where

$$g_{11} - R^2/(1-kr^2), g_{22} = r^2, g_{33} = r^2 sin^2(\theta), g_{44} = -1$$

The fact that the matter density and pressure ρ, p are functions of t alone is in accordance with the fact that the universe is homogeneous and isotropic. The Einstein field equations are formulated for this metric resulting thereby in the basic equations of cosmology.

Chapter 6

Fluid Dynamics

Some numerical problems in computational fluid dynamics:

Vorticity equation for two dimensional viscous fluid flow, computational methods. The velocity field for flow in two dimensions has the form

$$\mathbf{v(t},t) = v_x(x,y,t)\hat{x} + v_y(x,y,t)\hat{y}$$

Assume incompressibility, ie, the density ρ is a constant. Then the mass conservation equation reads

$$\frac{\partial v_x}{\partial x} + \frac{\partial v_y}{\partial y} = 0$$

which implies the existence of a function $\psi(x,y,t)$ (the stream function) such that

$$v_x = \frac{\partial \psi}{\partial y}, v_y = -\frac{\partial \psi}{\partial x}$$

The Navier-Stokes equation can be expressed as

$$\rho(\frac{\partial \mathbf{v}}{\partial t} + (\mathbf{v}, \nabla)\mathbf{v}) = -\nabla p + \nu \nabla^2 \mathbf{v}$$

where ν is the viscosity. This equation can be rearranged as

$$\frac{\partial \mathbf{v}}{\partial t} + \Omega \times \mathbf{v} + \frac{1}{2}\nabla v^2 = -\nabla p/\rho + \gamma . \nabla^2 \mathbf{v}$$

where

$$\gamma = \nu/\rho$$

is the ratio of the viscosity and density and

$$\Omega = \nabla \times \mathbf{v}$$

is the vorticity. Taking the curl of the above equation, we get

$$\frac{\partial \Omega}{\partial t} + \nabla \times (\Omega \times \mathbf{v}) = \gamma \nabla^2 \Omega$$

This equation must be expressed in terms of the stream function ψ. We have

$$\Omega = \nabla \times \mathbf{v} = (v_{y,x} - v_{x,y})\hat{z} = -\hat{z}\nabla^2\psi$$

so that

$$\Omega \times \mathbf{v} = -\nabla^2\psi\hat{z} \times (v_x\hat{x} + v_y\hat{y}) = -\nabla^2\psi(v_x\hat{y} - v_y\hat{x})$$

$$\nabla \times (\Omega \times \mathbf{v}) = [-(v_x\nabla^2\psi)_{,x} - (v_y\nabla^2\psi)_{,y}]\hat{z}$$

$$= -(v_x\nabla^2\psi_{,x} + v_y\nabla^2\psi_{,y})\hat{z}$$

since

$$v_{x,x} + v_{y,y} = 0$$

This gives

$$\nabla \times (\Omega \times v) = -\psi_{,y}\nabla^2\psi_{,x} + \psi_{,x}\nabla^2\psi_{,y}$$

and the vorticity equation reduces to

$$\nabla^2\psi_{,t} - \psi_{,x}\nabla^2\psi_{,y} + \psi_{,y}\nabla^2\psi_{,x} = \gamma\nabla^2\nabla^2\psi$$

This equation is for the stream function and can be cast in the following form: Let

$$s[\psi] = \psi_{,x}\nabla^2\psi_{,y} - \psi_{,y}\nabla^2\psi_{,x}$$

Then, the above equation can be expressed as

$$\nabla^2(\psi_{,t} - \gamma\nabla^2\psi) = s[\psi]$$

which on using the Greens function for the two dimensional Laplace operator becomes

$$\psi_{,t}(x, y, t) = \gamma\nabla^2\psi(x, y, t) + \int G(x - x', y - y')s[\psi(x', y', t)]dx'dy'$$

Note that the Green's function for the two dimensional Laplace operator is given by the formula

$$G(x, y|x', y') = -\frac{1}{4\pi}\int_{-\infty}^{\infty}\left(\frac{1}{(x - x')^2 + (y - y')^2 + z^2} - |z|^{-1}\right)dz$$

Writing

$$\rho^2 = (x - x')^2 + (y - y')^2$$

we have

$$-4\pi G = \int_{-\infty}^{\infty}\left(\frac{1}{\sqrt{\rho^2 + z^2}} - |z|^{-1}\right)dz$$

The second term $|z|^{-1}$ in the integrand has been subtracted in order to guarantee convergence of the integral as $|z| \to \infty$. In order to solve this integro differential equation approximately, we can make use of the method of moments. The velocity field is specified at $t = 0$. Assume that the fluid flow takes place inside the square $[0, L] \times [0, L]$. We have specified the velocity field at time $t = 0$. This amounts to specifying the stream function $\psi(x,y,0)$ at $t=0$. We choose an appropriate set of basis functions $\phi_k(x,y), k=1,2,...,N$ and write

$$\psi(x,y,t)=\sum_{k=1}^{N} c_k(t)\phi_k(x,y)$$

Then,

$$s[\psi](x,y,t)=\sum_{k,m} c_k(t)c_m(t)(\phi_{k,x}\nabla^2\phi_{m,y}-\phi_{k,y}\nabla^2\phi_{m,x})(x,y)$$

Then the integral equation can be expressed as

$$\sum_k c_k'(t)\phi_k(x,y)=\gamma\sum_k c_k(t)\nabla^2\phi_k(x,y)+\sum_{k,m} c_k(t)c_m(t)$$

$$\int G(x-x',y-y')(\phi_{k,x}\nabla^2\phi_{m,y}-\phi_{k,y}\nabla^2\phi_{m,x})(x',y')dx'dy'$$

from which, we derive the equations

$$\sum_k c_k'(t)\int \phi_k\phi_p dxdy=\gamma\sum_k c_k(t)\int \phi_p\nabla^2\phi_k dxdy+\sum_{k,m} c_k(t)c_m(t)$$

$$\int G(x-x',y-y')\phi_p(x,y)(\phi_{k,x}(x',y')\nabla^2\phi_{m,y}(x',y')-\phi_{k,y}(x',y')\nabla^2\phi_{m,x}(x',y'))dx'dy'dxdy$$

Define the real numbers

$$A(p,k)=\int \phi_p(x,y)\phi_k(x,y)dxdy,$$

$$B(p,k)=\gamma\int \phi_p(x,y)\nabla^2\phi_k(x,y)dxdy,$$

$$C(p,k,m)=\int G(x-x',y-y')\phi_p(x,y)(\phi_{k,x}(x',y')\nabla^2\phi_{m,y}(x',y')$$
$$-\phi_{k,y}(x',y')\nabla^2\phi_{m,x}(x',y'))dx'dy'dxdy$$

Then the dynamical equations for $\{c_k(t)\}$ can be cast as

$$\sum_{k=1}^{N} A(p,k)c_k'(t)=\sum_{k=1}^{N} B(p,k)c_k(t)+\sum_{k,m=1}^{N} C(p,k,m)c_k(t)c_m(t), p=1,2,...,N$$

Let $\mathbf{A}$ denote the $N\times N$ matrix $((A(p,k))$ and let $\mathbf{P}=\mathbf{A}^{-1}$. Let

$$Q(p,k)=\sum_{q=1}^{N} P(p,q)B(q,k), S(p,k,m)=\sum_{q=1}^{N} P(p,q)C(q,k,m)$$

Then the above dynamical system can be expressed as

$$c_p'(t)=\sum_k Q(p,k)c_k(t)+\sum_{k,m} S(p,k,m)c_k(t)c_m(t)$$

and this can be solved by iterative techniques.

2.Motion of a fluid inside a straight pipe having arbitrary cross section (Ref:V.K.Kapoor and Neetu, Analytical study of Carotid artery in different stages of atherosclerosis with MATLAB): Choose cylindrical coordinates and let the z axis be directed along the length of the pipe. The cross section of the pipe is the $\rho - \phi$ plane and the boundary of the pipe has the equation $\rho = f(\phi)$. The velocity field is $v_z(\rho, \phi)$ and let $p(\rho, \phi, z)$ be the pressure of the fluid inside the pipe. The mass conservation equation $\frac{\partial v_z}{\partial z} = 0$ is automatically satisfied since v_z is assumed to be independent of z. We have

$$v.\nabla v = v_z v_{z,z}\hat{z} = 0$$

so that Navier-Stokes equation gives

$$\gamma \nabla^2 v_z = dp/dz$$

The left side is independent of z and hence so must be the right side. Denoting $dp/dz = \eta(\rho, \phi)$, we get Poisson's equation

$$\nabla^2 v_z(\rho, \phi) = \eta(\rho, \phi)/\gamma$$

or in full notation

$$\frac{1}{\rho}\frac{\partial}{\partial \rho}(\rho \frac{\partial v_z(\rho, \phi)}{\partial \rho}) + \frac{1}{\rho^2}\frac{\partial^2 v_z(\rho, \phi)}{\partial \phi^2}$$
$$= \eta(\rho, \phi)/\gamma$$

3.Motion of a fluid in a pipe whose cross sectional diameter varies with z, t. Choose cylindrical coordinates so that the z axis is along the length of the pipe and the coordinates are (ρ, ϕ, z). The velocity field has the form

$$\mathbf{u} = u_\rho(\rho, z, t)\hat{\rho} + u_z(\rho, z, t)\hat{z}$$

Then,

$$\mathbf{u}.\nabla = u_\rho \frac{\partial}{\partial \rho} + u_z \frac{\partial}{\partial z}$$

$$\mathbf{u}.\nabla\mathbf{u} = (u_\rho \frac{\partial}{\partial \rho} + u_z \frac{\partial}{\partial z})(u_\rho \hat{\rho} + u_z \hat{z})$$

$$= (u_\rho u_{\rho,\rho} + u_z u_{\rho,z})\hat{\rho} + (u_\rho u_{z,\rho} + u_z u_{z,z})\hat{z}$$

Thus, the Navier-Stokes equations are with σ denoting the constant fluid density,

$$\sigma(u_\rho u_{\rho,\rho} + u_z u_{\rho,z} + u_{\rho,t}) = -\frac{\partial p}{\partial \rho} + (\eta \nabla^2 u)_\rho$$

$$\sigma(u_\rho u_{z,\rho} + u_z u_{z,z}) + u_{z,t}) = -\frac{\partial p}{\partial z} + \eta(\nabla^2 u)_z$$

To complete this set of equations, we need to express $(\nabla^2 u)_\rho$ and $(\nabla^2 u)_z$. We have

$$\nabla^2 \mathbf{u} = \rho^{-1} u_{,\rho} + u_{,\rho\rho} + \frac{1}{\rho^2} u_{,\phi\phi} + u_{,zz}$$

Now observe that

$$\rho^{-1}u_{,\rho} + u_{,\rho\rho} = (\rho^{-1}u_{\rho,\rho} + u_{\rho,\rho\rho})\hat{\rho} + (\rho^{-1}u_{z,\rho} + u_{z,\rho\rho})\hat{z}$$

$$u_{,\phi\phi} = (u_\rho\hat{\rho} + u_z\hat{z})_{,\phi\phi}$$
$$= -u_\rho\hat{\rho}$$

since $u_\rho, u_z, \hat{z}$ are independent of ϕ while

$$(\hat{\rho})_{,\phi} = \hat{\phi}, (\hat{\rho})_{,\phi\phi} = -\hat{\rho}$$

Thus,

$$\nabla^2\mathbf{u} = \hat{\rho}(\rho^{-1}u_{\rho,\rho} + u_{\rho,\rho\rho} - u_\rho/\rho^2 + u_{\rho,zz}) + \hat{z}(u_{z,\rho\rho}/\rho + u_{z,zz})$$

so that the full set of Navier-Stokes equation reads

$$u_\rho u_{\rho,\rho} + u_z u_{\rho,z} + u_{\rho,t} = -\sigma^{-1}p_{,\rho} + \nu(\rho^{-1}u_{\rho,\rho} + u_{\rho,\rho\rho} - u_\rho/\rho^2 + u_{\rho,zz})$$

$$u_\rho u_{z,\rho} + u_z u_{z,z} + u_{z,t} = -\sigma^{-1}p_{,z} + \nu(\rho^{-1}u_{z,\rho} + u_{z,\rho\rho} + u_{z,zz})$$

Approximation: Suppose that the velocities are small. Then the nonlinear terms may be neglected and the resulting linearized equations read

$$u_{\rho,t} = -\sigma^{-1}p_{,\rho} + \nu(\rho^{-1}u_{\rho,\rho} + u_{\rho,\rho\rho} - u_\rho/\rho^2 + u_{\rho,zz})$$
$$u_{z,t} = -\sigma^{-1}p_{,z} + \nu(\rho^{-1}u_{z,\rho} + u_{z,\rho\rho} + u_{z,zz})$$

The mass conservation equation reads

$$\rho^{-1}(\rho.u_\rho)_{,\rho} + u_{z,z} = 0$$

or

$$u_{\rho,\rho} + \rho^{-1}u_\rho + u_{z,z} = 0$$

We assume a solution of the form

$$u_\rho(\rho, z, t) = F(\rho)exp(i(\omega t - kz)), u_z(\rho, z, t) = G(\rho)exp(i(\omega t - kz)), p(\rho, z, t) = H(\rho)exp(i(\omega t - kz))$$

This solution corresponds to wave like motion propagating in time along the axis of the pipe. It is particularly applicable to the motion of blood in an artery whose cross section varies with t, z in a sinusoidal fashion. Then,

$$i\omega.F(\rho) = -\sigma^{-1}H'(\rho) + \nu(\rho^{-1}F'(\rho) + F''(\rho) - F(\rho)/\rho^2 - k^2F(\rho))$$

$$i\omega.G(\rho) = i\sigma^{-1}H(\rho) + \nu(\rho^{-1}G'(\rho) + G''(\rho) - k^2G(\rho))$$
$$F'(\rho) + \rho^{-1}F(\rho) - ikG(\rho) = 0$$

The solution to these coupled equations determine the motion.

4.One dimensional flow. Mass-conservation equation:

$$\frac{\partial \rho}{\partial t}+\frac{\partial(\rho u)}{\partial x}=0$$

Navier-Stokes equation:

$$\frac{\partial u}{\partial t}+u\frac{\partial u}{\partial x}=-\rho^{-1}\frac{\partial p}{\partial x}$$

Equation of state:

$$p=f(\rho)$$

Thus, the Navier equation can be rewritten as

$$\frac{\partial u}{\partial t}+\frac{\partial(u^2/2)}{\partial x}=-\rho^{-1}f'(\rho)\frac{\partial \rho}{\partial x}$$

Define the function

$$g(\rho)=\int_0^{\rho}\rho^{-1}f'(\rho)d\rho$$

Then the Navier equation can be expressed as

$$\frac{\partial u}{\partial t}+\frac{\partial}{\partial x}(u^2/2+g(\rho))=0$$

Define the two component field

$$U=(\rho,u),F(U)=(\rho.u,u^2/2+g(\rho))^T$$

Then the mass conservation and Navier equations can be clubbed into a single equation

$$\frac{\partial U}{\partial t}+\frac{\partial F(U)}{\partial x}=0$$

This equation is particularly suitable for finite difference implementations.

5.Ph.D thesis of Dr.Neetu: Gas flow in one dimension (Ph.d problem of Dr.Neetu): $\rho(t,x)$ is the fluid density, $U(t,x)$ is the fluid velocity and $p(t,x)$ is the fluid pressure. The mass conservation equation reads

$$\rho_{,t}+(\rho.U)_{,x}=0,$$

The Navier-Stokes equation reads

$$\rho(UU_{,x}+U_{,t})=-p_{,x}$$

This can also be expressed as

$$\rho(\frac{1}{2}(U^2)_{,x}+U_{,t})=-p_{,x}$$

Assume that the equation of state of the gas is given by

$$p=p(\rho)$$

Then define a function

$$g(\rho) = \int^{\rho} dp(\rho)/\rho$$

Then the second equation can be expressed as

$$\frac{1}{2}(U^2)_{,x} + U_{,t} + g(\rho)_{,x} = 0$$

These two equations can be rearranged as

$$\rho_{,t} = -(\rho U)_{,x},$$

$$U_{,t} = -\frac{1}{2}(U^2)_{,x} - g(\rho)_{,x}$$

Define the state vector

$$X = (\rho, U)$$

and the nonlinearly transformed state vector

$$F(X) = F(\rho, U) = [-\rho U, -\frac{1}{2}U^2 - g(\rho)]$$

Then the equations of mass and momentum conservation can be expressed in compact state variable form as

$$\frac{\partial X}{\partial t} = \frac{\partial F(X)}{\partial x}$$

This equation can be solved by a finite difference scheme. Specifically, assume that $X(t, x)$ is defined for $t \geq 0, 0 \leq x \leq L$. Divide the interval $[0, L]$ into N equal parts, each of length Δ so that $\Delta = L/N$. Also divide the time axis into equal intervals of length τ. Define $\rho[n, k] = \rho(n\tau, k\Delta)$ and $U[n, k] = U(n\tau, k\Delta)$. Thus,

$$X[n, k] = (\rho[n, k], U[n, k])$$

and the discretized equations of motion are

$$\tau^{-1}(X[n+1, k] - X[n, k]) = \Delta^{-1}(F(X[n, k+1]) - F(X[n, k]))$$

or

$$X[n+1, k] = X[n, k] + \frac{\tau}{\Delta}(F(X[n, k+1]) - F(X[n, k]))$$

We start with the state vector at time $n = 0$, ie,

$$X[0, k], 0 \leq k \leq N-1, X[0, N] = 0$$

and for each $n = 1, 2, ...$ and $k = 0, 1, 2, ..., N-1$ compute $X[n, k]$ and set $X[n, N] = 0$. In this way, the evolution of the state with time is completely determined by the above difference scheme. As typical examples of the computation of $g(\rho)$, we consider first an isothermal gas for which the equation of state is $p = C\rho$ where C is a constant. Then, $g(\rho) = \int C d\rho/\rho = C.log(\rho)$ and thus

$$F(\rho, U) = [-\rho U, -\frac{1}{2}U^2 - g(\rho)]$$

The other example is an adiabatic gas for which $p = C\rho^\gamma$ where γ is the ratio of the specific heat at constant pressure to the specfic heat at constant volume. Then,

$$g(\rho) = \int dp/\rho = \int C\gamma\rho^{\gamma-2}d\rho = \frac{C\gamma}{\gamma-1}\rho^{\gamma-1}$$

Thus,

$$F(\rho, U) = [-\rho U, -U^2/2 - (C\gamma/(\gamma-1))\rho^{\gamma-1}]$$

We next consider one dimensional viscous gaseous flow. If ν denotes the viscosity, the mass conservation equation is unchanged while the Navier-Stokes equation becomes

$$\rho(UU_{,x} + U_{,t}) = -p_{,x} - \nu U_{,xx}$$

This can be expressed as

$$U_{,t} = -(U^2/2)_{,x} - g(\rho)_{,x} - \nu\rho^{-1}U_{,xx}$$

The iteration for the mass conservation equation then reads

$$\rho[n+1, k] = \rho[n, k] + (\tau/\Delta)(\rho[n, k+1]U[n, k+1] - \rho[n, k]U[n, k])$$

and for momentum,

$$U[n+1, k] - U[n, k] = -\frac{\tau}{2\Delta}(U^2[n, k+1] - U^2[n, k]) - \frac{\tau}{\Delta}(g(\rho[n, k+1]) - g(\rho[n, k]))$$

$$-\frac{\nu\tau}{\Delta^2}\rho[n, k]^{-1}(U[n, k+1] - U[n, k] + U[n, k-1])$$

We now look at the two ideal dimensional gas flow problem. The density field is given by

$$\rho(t, x, y)$$

and the velocity components are

$$(\mathbf{U}(t, x, y) = (U_x(t, x, y), U_y(t, x, y))$$

the mass conservation equation reads

$$\rho_{,t} + (\rho U_x)_{,x} + (\rho U_y)_{,y} = 0$$

and the two Navier-Stokes equations are

$$\rho(U_x U_{x,x} + U_y U_{x,y} + U_{x,t}) = -p_{,x}$$

$$\rho(U_x U_{y,x} + U_y U_{y,y} + U_{y,t}) = -p_{,y}$$

We have

$$(\rho U_x)_{,t} + (\rho U_x^2)_{,x} + (\rho U_x U_y)_{,y}$$

$$= \rho(U_{x,t} + U_x U_{x,x} + U_y U_{x,y}) + U_x(\rho_{,t} + (\rho U_x)_{,x} + (\rho U_y)_{,y})$$

$$= \rho(U_{x,t} + U_x U_{x,x} + U_y U_{x,y})$$

in view of the mass conservation equation. This means that the two Navier-Stokes equations can be written as

$$(\rho U_x)_{,t} + (\rho U_x^2)_{,x} + (\rho U_x U_y)_{,y} = -p_{,x},$$
$$(\rho U_y)_{,t} + (\rho U_y U_x)_{,x} + (\rho U_y^2)_{,y} = -p_{,y}$$

Define the state vector

$$X = (\rho, \rho U_x, \rho U_y)$$

and also

$$F_x(X) = -(\rho U_x, \rho U_x^2 + p(\rho), \rho U_y U_x),$$
$$F_y(X) = -(\rho U_y, \rho U_x U_y, \rho U_y^2 + p(\rho)),$$

Then, the mass conservation and Navier-Stokes equations can be expressed as

$$\frac{\partial X}{\partial t} = \frac{\partial F_x(X)}{\partial x} + \frac{\partial F_y(X)}{\partial y}$$

This equation can be simulated using the finite difference scheme. The assumption is that the equation of state $p(\rho)$ (pressure as a function of density is known). The extension of this formulation to three dimensions is also straightforward. We shall formulate this for an n dimensional fluid. The equation of mass conservation in n dimensions is

$$(\rho U_i)_{,i} + \rho_{,t} = 0$$

where the summation is over $i = 1, 2, ..., n$. There are n spatial indices and one time index. The notation $(\rho U_i)_{,i}$ stands for $\sum_{i=1}^{n} (\rho U_i)_{,i}$. The Navier-Stokes equation reads

$$\rho(U_j U_{i,j} + U_{i,t}() = -p_{,i}, i = 1, 2, ..., n$$

and finally the equation of state $p = p(\rho)$ determines the pressure as a function of the density. Now by matter conservation, we have

$$(\rho U_i U_j)_{,j} + (\rho U_i)_{,t} =$$
$$(\rho U_j)_{,j} U_i + \rho U_j U_{i,j} + \rho_{,t} U_i + \rho U_{i,t}$$
$$= U_i(\rho_{,t} + (\rho U_j)_{,j}) + \rho(U_j U_{i,j} + U_{i,t})$$
$$= \rho(U_j U_{i,j} + U_{i,t})$$

which is the left side of the Navier-Stokes equation. Thus, we can express the Navier-Stokes equation in the form

$$(\rho U_i U_j)_{,j} + (\rho U_i)_{,t} = -p_{,i}$$

Define the state vector

$$\mathbf{X} = (\rho, \rho U_1, ..., \rho U_n) = (\rho, \rho U_i, i = 1, 2, ..., n)$$

and its nonlinearly transformed versions

$$F_i(X) = (\rho U_i, \rho U_i U_j + p\delta_{ij}, j = 1, 2, ..., n)$$

or in expanded form,

$$F_i(X) = (\rho U_i, \rho U_i U_1, ..., \rho U_i U_{i-1}, \rho U_i^2 + p, \rho U_i U_{i+1}, ..., \rho U_i U_n)$$

The equation of mass conservation along with the Navier-Stokes equations are then

$$\frac{\partial X}{\partial t} + \sum_{i=1}^{n} \frac{\partial F_i(X)}{\partial x_i} = 0$$

or using the abbreviated notation for partial derivatives,

$$X_{,t} = -\sum_{i=1}^{n} F_{i,i}(X)$$

This can be solved using an iterative method.

n dimensional fluid moving in a curved space-time manifold. The energy-momentum tensor of such a fluid is given by

$$T^{ij} = (\rho + p)U^i U^j + pg^{ij}$$

and the equations of matter and momentum conservation are

$$T^{ij}_{:j} = 0$$

For a symmetric tensor the covariant divergence can be expressed in terms of ordinary derivatives as follows. Let $T^{ij} = A^i B^j$ where A^i, B^i are two contravariant vectors.

Chapter 7

Group Representations

1.Induced representations:Let G be a group and H a subgroup of G. Assume that π is a representation of H in the vector space V. Then let $Ind(V)_H^G$ be the space of all functions $f: G \to V$ with the property $f(gh^{-1}) = \pi(h)f(g)$ for all $h \in H, g \in G$. The set of all such functions forms a vector space. To see what this space looks like, we suppose that G is a finite group. Let the distinct cosets of H in G be denoted by $g_iH, i = 1, 2, ..., n$. Of course one of them will be obtained by taking $g = e$. Given any $g \in G$, we have $g = g_ih$ for some i and some $h \in H$. Then if $f \in Ind(V)_H^G$, it follows that $f(g) = f(g_ih) = \pi(h)^{-1}f(g_i)$. We can assign $f(g_i)$ arbitrary values in V. Thus, choose $m_1, ..., m_n \in V$ and set $f(g_i) = m_i, i = 1, 2, ..., n$. It follows then that

$$f(g_ih) = \pi(h)^{-1}m_i, i = 1, 2, ..., N$$

It follows from this that

$$Ind(V)_H^G = \bigoplus_{i=1}^{n} V_i, V_i = V$$

Assume that for f in this space,

$$\rho(g)f(x) = f(g^{-1}x)$$

Then, clearly ρ is a representation of G in $Ind(V)_H^G$. Note that if $f \in Ind(V)_H^G$, then $\rho(g)f = fog^{-1} \in Ind(V)_H^G$ as can be seen from

$$(\rho(g)f)(xh) = f(g^{-1}xh) = \pi(h)^{-1}f(g^{-1}x) = \pi(h)^{-1}(\rho(g)f)(x)$$

Denote the right translation operator of a function f by $h \in H$ by R_h. Then the above can be expressed as

$$R_h\rho(g)f(x) = \pi(h)^{-1}(\rho(g)f)(x)$$

which is the desired property. Let $f \in Ind(V)_H^G$ be given. Let $f_i : G \to V$ be defined by $f_i(g_ih) = \pi(h)^{-1}m_i$ where $m_i \in V$ is defined by $f(g_i) = m_i$ and $f_i(g_jh) = 0$ for all $j \neq i$ and $h \in H$. This construction shows that we can write $Ind(V)_H^G = \bigoplus_{i=1}^{N} g_iV$. More precisely, given a group G and a subgroup H of G, we consider the set $M = G/H$ of all left cosets of H. G acts transitively on G/H and for each coset $g_iH \in G/H$ we attach a vector space $V_i = V$ on it. Thereby we obtain a vector bundle

E consisting of the unions $\bigcup_i g_i V$. An element of $g_i V$ is denoted by (x, v) where $x \in G$ and $v \in V$. We identify the element (xh, v) with $(x, \pi(h)v)$. The action of $g \in G$ on (x, v) is given by

$$g.(x, v) = (g.x, v)$$

This is consistent with the fact that for $h \in H$,

$$g.(xh, \pi(h)^{-1}v) = (gxh, \pi(h)^{-1}v) = (gx, v)$$

We can express an element $f \in Ind_H^G(V)$ by $f(x) = (x, v(x))$ $x \in G$. Then for all $h \in H$, we must identify $(x, v(x))$ with $(xh, v(xh))$. But this is in turn to be identified with $(x, \pi(h)v(xh))$ and hence we must have that $v(xh) = \pi(h)^{-1}v(x)$. This is the original starting point for defining the induced representation. Another way to see all this is as follows. Let $V_i = V, i = 1, 2, ..., N$ and consider the map $\Phi : Ind_H^G(V) \to \bigoplus_{i=1}^N V_i$ defined by $\Phi(f) = (m_1, m_2, ..., m_N)$ where $m_i = f(g_i), i = 1, 2, ..., N$. Then obviously, Φ is linear, one one and onto. Hence, as a vector space, $Ind_H^G(V)$ is isomorphic to $\bigoplus_{i=1}^N V_i$. Suppose $g \in G$. Then for each i, there exists a j such that $g^{-1}g_i \in g_j H$, say $g^{-1}g_i = g_j h_{ij}$ for some $h_{ij} \in H$. Then,

$$f(g^{-1}g_i) = f(g_j h_{ij}) = \pi(h_{ij})^{-1} f(g_j) = \pi(h_{ij})^{-1} m_j$$

Conversely, given $m_i \in V_i, i = 1, 2, ..., N$, we may define $f : G \to V$ by $f(g_i h) = \pi(h)^{-1} m_i, h \in H$ and it then follows that $f \in Ind_H^G(V)$. This proves the isomorphism of $Ind_H^G(V)$ with $\bigoplus V_i = \bigoplus_{i=1}^N g_i V$.

Character of the induced representation(George Mackey): Let G be a finite group and H a subgroup of G. Assume that π is a representation of H in a finite dimensional vector space V. We want to calculate the character of the representation $Ind_H^G(\pi)$ induced by π. We note that the space $Ind_H^G(V)$ is a direct sum $\bigoplus g_i V$ where $g_i H, i = 1, 2, ..., N$ are the distinct cosets of H in G. Now let $g \in G$. Then $g(g_i V) = g_i V$ iff $gg_i \in g_i H$ iff $g_i^{-1}gg_i \in H$. In this case, the action of g on $g_i V$ is equivalent to the action of $g_i^{-1}gg_i \in H$ on V. Let $E(g)$ denote the set of all $i = 1, 2, ..., N$ for which $g_i^{-1}gg_i \in H$. Then, it follows that the character of $Ind_H^G(\pi)$ is given by

$$\chi(g) = \sum_{i \in E(g)} \chi_\pi(g_i^{-1}gg_i)$$

This formula can be simplified considerably. The details can be found in the book by Procesi, "Lie groups, an approach through invariants and representations".

2.Let U be a vector space and let f be a non-degenerate bilinear form on U. Then f induces a map $\phi :\to U^*$ by the equation $\phi(u)(v) = f(u, v)$. The map ϕ is an isomorphism. Let V be an isotropic subspace of V, ie, the restriction of the non-degenerate bilinear form f to V is identically zero. Then we wish to show that $dimV \leq dimU/2$. Indeed, we have $f(u, v) = 0$ for all $u, v \in V$. Define

$$V^\perp = \{v \in U : f(u, v) = 0 \forall u \in V\}$$

Then obviously $dimV^\perp = dimU - dimV$. Assume that $dimU = n$, $dimV = m$. Choose a basis $\{e_1, ..., e_m\}$ for V. Choose a basis $\{f_1, ..., f_{n-m}\}$ for $V^\perp$. The forms $\phi(e_i), i = 1, 2, ..., m$ are linearly independent in V^*. Also $\phi(e_i)(e_j) = 0, j = 1, 2, ..., m, \phi(e_i)(f_j) = 0, j = 1, 2, ..., n - m$. It follows that the dimension of the space spanned by the vectors $\{e_1, ..., e_m, f_1, ..., f_{n-m}\}$ cannot be more than $n - m$. In particular, $m \leq n - m$, so that $m \leq n/2$.

Problem: If f is a non-degenerate bilinear form on U and V is a subspace of U having dimension m, then prove that $dimV^\perp = n - m$. Indeed, choose a basis $\{e_1, ..., e_m\}$ for V. Then $\phi(e_i) = f(e_i, .), i = 1, 2, ..., m$ are linearly independent in U^* and $u \in V^\perp$ iff $\phi(e_i)(u) = 0, i = 1, 2, ..., m$. In a basis for V, f is represented by a nonsingular matrix A of size $n \times n$ and e_i by column vectors of size $n \times 1$ so that the above condition boils down to $e_i^T Au = 0, i = 1, 2, ..., m$. Noting that the matrix $[A^T e_1, ..., A^T e_m]$ is of size $n \times m$ and has rank m, it follows that a basis for solutions u has dimension $n - m$ (use the rank-nullity theorem) and hence the proof is complete.

3.Problem:Let A be an $n \times n$ semisimple matrix with eigenvalues $c_1, ..., c_n$ and corresponding eigenvectors $e_1, ..., e_n$. Set

$$A = UDU^{-1}, U = [e_1, ..., e_n], D = diag[c_1, ..., c_n]$$

Let f_i denote the $n \times 1$ vector with a one in the ith position and zeros at other positions. Then we have

$$U^{-1}U = I$$

and hence

$$[U^{-1}e_1, ..., U^{-1}e_n] = [f_1, ..., f_n]$$

or equivalently,

$$U^{-1}e_i = f_i, i = 1, 2, ..., n$$

equivalently,

$$e_i = Uf_i, i = 1, 2, ..., n$$

Also the equation

$$U^{-1}U = I$$

implies

$$g_i^T U = f_i^T$$

where $g_i^T, i = 1, 2, ..., n$ are the rows of U^{-1}. This means that

$$g_i^T = f_i^T U^{-1}, i = 1, 2, ..., n$$

Thus,

$$E_{ij} = e_i g_j^T = Uf_i f_j^T U^{-1}$$

Then,

$$AE_{ij} = UDU^{-1}Uf_i f_j^T U^{-1} = UDf_i f_j^T U^{-1} = c_i Uf_i f_j^T U^{-1} = c_i E_{ij},$$
$$E_{ij}A = Uf_i f_j^T U^{-1}UDU^{-1} = Uf_i f_j^T DU^{-1} = c_j Uf_i f_j^T U^{-1} = c_j E_{ij}$$

Thus,

$$ad(A)(E_{ij}) = AE_{ij} - E_{ij}A = (c_i - c_j)E_{ij}$$

which means that $\{E_{ij} : i, j = 1, 2, ..., n\}$ is a basis for $\mathbb{C}^{n \times n}$ relative to which $ad(A)$ is diagonal with eigenvalues $c_i - c_j$.

4.Symmetric and antisymmetric polynomials in n variables $x_1, ..., x_n$. Let $f(x_1, ..., x_n)$ be an antisymmetric polynomial. Then f vanishes when $x_i = x_j$ for any i, j with $i \neq j$. It follows that

$$f(x_1, ..., x_n) = [\Pi_{i<j}(x_i - x_j)]\psi(x_1, ..., x_n)$$

where ψ is a symmetric polynomial. Note that by antisymmetric we mean that

$$f(x_{\sigma 1}, ..., x_{\sigma n}) = \epsilon(\sigma) f(x_1, ..., x_n)$$

and by symmetric we mean the same without the factor $\epsilon(\sigma)$. Let $p_i, i = 1, 2, ..., n$ be non-negative integers and define $\lambda = (p_1, ..., p_n)$. λ can be viewed as a partition of the integer $m = \sum_{i=1}^n p_i$. Without loss of generality, assume that $p_1 \geq p_2 \geq ... \geq p_n$. Define $\rho = (n-1, n-2, ..., 2, 1)$ so that

$$\lambda + \rho = (p_1 + n - 1, p_2 + n - 2, ..., p_{n-1} + 2, p_n + 1)$$

The map $\lambda \to \lambda + \rho$ maps decreasing sequences to strictly decreasing sequences. Let J be the set of all strictly decreasing n-length sequences of positive integers, ie,

$$J = \{(q_1, ..., q_n) : q_1 > q_2 > ... > q_n > 0\}$$

Then for $q \in J$, we define the polynomial

$$A_q(x) = \sum_{\sigma \in S_n} \epsilon(\sigma) x_{\sigma 1}^{q_1} ... x_{\sigma n}^{q_n}$$

It is clear that $\{A_q(x) : q \in J\}$ forms a basis for the set of all antisymmetric polynomials in n variables. Let $\mathcal{P}_m$ denote all partitions of m of length n, ie, all sequences $(p_1, ..., p_n)$ with $p_1 \geq p_2 \geq ... \geq p_n \geq 0$. Then $\lambda + \rho \in J$ for $\lambda \in \mathcal{P}$ and we have that the set

$$S_\lambda = A_{\lambda+\rho}/V, \lambda \in \mathcal{P} = \bigcup_{m \geq 0} P_m$$

is a basis for all symmetric polynomials in n variables. In the theory of irreducible representations of the permutation group S_n, each partition λ of n determines an irreducible representation. Let c_λ denote the associated character. Let s be any permutation expressed as a product of cycles, ie s is product of m cycles of lengths $p_1 \geq p_2 \geq ... \geq p_m > 0$ with $\sum_{i=1}^m p_i = n$. Let $\mu = (p_1, ..., p_m)$. Then $c_\lambda(s) = c_\lambda(\mu)$. It depends only on the partition μ and not on the permutation s. Specifically, if s' is conjugate to s, then s' is also a product of cycles of lengths $p_1, ..., p_m$ and $c_\lambda(s) = c_\lambda(s')$. Conversely, if s' is any permutation expressible as a product of p_i i-cycles, $i = 1, 2, ..., m$, then s' is conjugate to s and hence $c_\lambda(s) = c_\lambda(s')$. Associated to each partition μ, we define a symmetric function ψ_μ. Then it can be shown (Procesi) that

$$\psi_\mu = \sum c_\lambda(\mu) S_\lambda$$

and the characters of the irreducible representations of S_n can thence be determined.

5.Let G be a linearly reductive group and let U be a finite dimensional module. Then by definition of linearly reductive, we have the direct sum decomposition

$$U = U^G \oplus U_G$$

where U^G is the set of invariants (ie a direct sum of one dimensional representations) and U_G is the direct sum of non-trivial irreducible representations. Let U_1, U_2 be two finite dimensional modules and let $f : U_1 \to U_2$ be a G equivariant map, ie, $g.f(u) = f(g.u)$ for all $g \in G, u \in U_1$, or equivalently, $g.f = f.g$ for all $g \in G$. Let $u \in U_1$, so that $f(u_1) \in U_2$. π_2 denotes the projection of U_2 onto U_2^G. Thus, $\pi_2(f(u_1)) \in U_2^G$. Let π_1 denote the projection of U_1 onto U_1^G. Thus, $\pi_1(u_1) \in U_1^G$ and $f(\pi_1(u_1)) \in U_2^G$. We want to show that $\pi_2(f(u_1)) = f(\pi_1(u_1))$. Since $\pi_1(u_1) \in U_1^G$, it follows that $g.\pi_1(u_1) = \pi_1(u_1)$ for all $g \in G$ and hence by definition of equivariance,

$$f(\pi_1(u_1)) = f(g.\pi_1(u_1)) = g.f(\pi_1(u_1)), g \in G$$

which implies

$$f(\pi_1(u_1)) \in U_2^G$$

whence

$$f(\pi_1(u_1)) = \pi_2 o f o \pi_1(u_1), u_1 \in U_1$$

or

$$f o \pi_1 = \pi_2 o f o \pi_1$$

We wish to show that

$$f o \pi_1 = \pi_2 o f$$

It suffices in view of the above to show that if π_1' is the projection of U_1 onto $U_{1,G}$, then, $\pi_2 o f o \pi_1' = 0$. In other words, it suffices to show that for $u \in U_{1,G}$, we have $\pi_2(f(u)) = 0$. Now $u \in U_{1,G}$ implies $g.u \neq u$ for some $g \in G$. This implies $f(g.u) \neq f(u)$ assuming f to be one-one. This implies $g.f(u) \neq f(u)$ by equivariance of f. This implies $f(u) \in U_{2,G}$ and hence $\pi_2(f(u)) = 0$. Suppose that we do not make the assumption that f is one-one. Choose $0 \neq u \in U_{1,G}$. Then $g.u, g \in G$ spans an irreducible submodule U_{10} of $U_{1,G}$. Likewise, assuming $f(u) \neq 0$, it follows that $g.f(u), g \in G$ spans an irreducible submodule U_{20} of $U_{2,G}$. Since $f(g.u) = g.f(u), g \in G$, it follows that the restriction map $f : U_{10} \to U_{20}$ is a well defined linear map that intertwines the G action. It follows that this restriction map is an isomorphism since it intertwines two irreps. Now $u \in U_{1,G}$ implies that $g.u \neq u$ for some $g \in G$ and by what we just saw, we have $g.f(u) = f(g.u) \neq f(u)$ whence $f(u) \in U_{2,G}$ proving that $\pi_2(f(u)) = 0$ and we are through.

6.Let U_1, U_2 be two subspaces of a vector space V so that $V = U_1 \oplus U_2$. Let $\otimes_a$ denote the antisymmetric tensor product. We have for any two vectors $x, y \in V$, $x = x_1 + x_2, y = y_1 + y_2$ with $x_1, y_1 \in U_1, x_2, y_2 \in U_2$. Then

$$x \otimes y = (x_1 + x_2) \otimes (y_1 + y_2) = x_1 \otimes y_1 + x_1 \otimes y_2 + x_2 \otimes y_1 + x_2 \otimes y_2$$

$$y \otimes x = (y_1 + y_2) \otimes (x_1 + x_2) = y_1 \otimes x_1 + y_1 \otimes x_2 + y_2 \otimes x_1 + y_2 \otimes x_2$$

so that

$$x \otimes y - y \otimes x = (x_1 \otimes y_1 - y_1 \otimes x_1) + (x_2 \otimes y_2 - y_2 \otimes x_2) + (x_1 \otimes y_2 - y_2 \otimes x_1)$$
$$+(x_2 \otimes y_1 - y_2 \otimes x_1)$$

This is an element of

$$U_1 \wedge U_1 + U_2 \wedge U_2 + U_1 \wedge U_2 + U_2 \wedge U_1$$

Note that $U_1 \wedge U_2 + U_2 \wedge U_1$ is isomorphic to $U_1 \otimes U_2$. Indeed, if (e_i) is a basis for U_1 and (f_i) a basis for U_2, then $e_i \otimes f_j$ is a basis for $U_1 \otimes U_2$ which has dimension $dimU_1.dimU_2$. On the other hand, $e_i \wedge f_j$ is a basis for $U_1 \wedge U_2$ having dimension nm while $f_i \wedge e_j$ is a basis for $U_2 \wedge U_1$ and since $f_i \wedge e_j = -e_j \wedge f_i$, it follows that $U_2 \wedge U_1 = U_1 \wedge U_2$ proving the claim. We have the more general formula that

$$\wedge^p V = \bigoplus_{i=0}^{n} \wedge^i(U_1) \otimes \wedge^{p-i}(U_2)$$

Indeed, let $x_i \in V$ so that $x_i = x_{i1} + x_{i2}, i = 1, 2, ..., p$ with $x_{i1} \in U_1, x_{i2} \in U_2$. Then we compute

$$x_1 \wedge x_2 \wedge ... \wedge x_p = \wedge_{i=1}^{p}(x_{i1} + x_{i2})$$

The right side can be written as a sum, a general term in this sum will have the component from U_1 selected from r places and the component from U_2 selected from $p - r$ places. It follows then that

$$\wedge^p V \subset \bigoplus_{r=0}^{p} (\wedge^r(U_1) \otimes \wedge^{p-r}(U_2))$$

Now the dimension of the left side is $\binom{n+m}{p}$ while the dimension of the right side is

$$\sum_{r=0}^{p} \binom{n}{r}\binom{m}{p-r} = \binom{n+m}{p}$$

and we are through.

7.Result: Let U be a rational representation of an algebraic group G. Then, this representation can be embedded in an equivariant way in the direct sum of finitely many copies of the coordinate ring $k(G)$ under right or left action. For take a basis $u^i, i = 1, 2, ..., n$ for U^*. Define $j : U \to k(G)^n$ by

$$j(u)(g) = (< u^1, gu >, ..., < u^n, gu >)$$

Then

$$j(g_1 u)(g) = (< u^1, gg_1 u >, ..., < u^n, gg_1 u >) = j(u)(gg_1) = (g_1.j(u))(g)$$

or equivalently, under right action,

$$jog_1 = g_1.j$$

proving the claim.

Let V be a vector space and $G \subset GL(V)$ be a linear algebraic group. Let d denote the determinant function as a one dimensional representation. Let A, B be the coordinate rings of $End(V)$ and $GL(V)$ respectively. Let I be the ideal in $B = k(GL(V))$ that generates G, ie, I is a set of elements of B (polynomials in the matrix elements and reciprocal of the determinant of $g \in GL(V)$) such that

$$G = \{g \in GL(V) : f(g) = 0 \forall f \in I\}$$

Note that conversely, if we are given an algebraic subgroup G of $GL(V)$ and we define

$$I = \{f \in B : f(g) = 0 \forall g \in G\}$$

then we have that I is an ideal in B and this ideal is said to generate G. Suppose that we are given elements $u_1, ..., u_r \in B$ such that

$$G = \{g \in GL(V) : u_i(g) = 0, i = 1, 2, ..., r\}$$

and if I is the ideal generated by the $u_i's$, ie,

$$I = \sum_{i=1}^{r} Bu_i$$

then I generates G. It is easy to see that

$$k(G) = B/I$$

For suppose $f \in k(G)$. Then for any $h \in I$ and $g \in G$, we have $(f + h)(g) = f(g) + h(g) = f(g)$ implying that f can be identified with $f + I$. In fact consider the map

$$\Phi : B/I \to k(G)$$

defined by

$$\Phi(f + I)(g) = f_G$$

where f_G denotes restriction of f to G. This map is one one since $f \in I$ implies $f_G = 0$. It is clearly onto and further if $f_1, f_2 \in B$, then

$$(f_1 + I)(f_2 + I) = f_1 f_2 + I$$

and hence

$$\Phi((f_1 + I)(f_2 + I)) = (f_1 f_2)_G = f_{1G}.f_{2G} = \Phi(f_1 + I).\Phi(f_2 + I)$$

and this completes the proof. Let U be a rational representation of G. Then U embeds in a direct sum of $k(G)$, ie, there exists a G-equivariant map $j : U \to k(G)^p$ under right multiplication. By definition of coordinate ring, we have

$$B = \bigcup_{i=0}^{\infty} d^{-i} A$$

A digression: Let V be a vector space and W a subspace of V. Then W^p is a subspace of V^p. We claim that V^p/W^p is isomorphic to $(V/W)^p$ where raising to the power p means taking p fold direct sum. Indeed, define the map

$$\psi : V^p/W^p \to (V/W)^p$$

by

$$\psi((v_1, ..., v_p) + W^p) = (v_1 + W, ..., v_p + W)$$

Then this is clearly the required isomorphism. It follows that

$$U \subset k(G)^p = B^p/I^p = (\bigcup_i d^{-i} A)^p/I^p$$

Let $u \in U$. Then from the above, we have that

$$u = (d^{-i_1} a_1, ..., d^{-i_p} a_p) + I$$

for some $a_1, ..., a_p \in A$ and non-negative integers $i_1, ..., i_p$. Letting $i = max_j i_j$, it follows that

$$u \in d^{-i} A^p / I$$

By letting u vary over a basis (which by assumption is finite), we get

$$U \subset d^{-r} A^p / I$$

for some non-negative integer r. Note that $d^{-i}A \subset d^{-j}A$ if $j > i$. In other words, we can construct a linear map $\psi : A^p \to k(G)^p$ such that U is in the image of ψ.

8.Roots: Let L be a semisimple Lie algebra. A toral subalgebra of L is a subalgebra made only of semisimple elements. A toral subalgebra must necessarily be Abelian. For suppose $\mathbb{T}$ is a toral subalgebra of L that is not Abelian. Then we can find an $x \in \mathbb{T}$ with $ad(x)(\mathbb{T}) \neq 0$. It follows that there exists a nonzero complex number a and a nonzero $y \in L$ such that $ad(x)(y) = [x, y] = ay$. (a is an eigenvalue for the restriction of $ad(x)$ to $\mathbb{T}$. Consider the space spanned by $\{x, y\}$. We have $ad(y)(x) = -ay, ad(y)(y) = 0$. Thus, $ad(y)$ restricted to this space is a nonzero nilpotent matrix and this implies that $ad(y)$ is not semisimple, a contradiction. It follows that if $\mathbb{T}$ is a toral subalgebra, then the operators $ad(x), x \in \mathbb{T}$ are all simultaneously diagonable. A maximal toral subalgebra is called a cartan subalgebra. Let $\mathbb{T}$ be a Cartan subalgebra of L. Suppose that For each $h \in \mathbb{T}$, $\alpha(h)$ is an eigenvalue of $ad(x)$ on L. The common eigenspace is then given by

$$L_\alpha = \{x \in L : ad(h)(x) = \alpha(h)x \forall h \in \mathbb{T}\}$$

For $h, h' \in \mathbb{T}$, we then have

$$\alpha(h + h')x = ad(h + h')(x) = ad(h)(x) + ad(h')(x) = \alpha(h)x + \alpha(h')x$$

where x is any nonzero element of L_α. It follows that $\alpha \in \mathbb{T}^*$.

9.Every permutation is a product of its cycles. Consider the symmetric group S_n acting on $\{1, 2, ..., n\}$. Any $\sigma \in S_n$ is represented by the ordered sequence $\{\sigma 1, ..., \sigma n\}$. Start with $i_1 = 1$ and keep applying σ to it. Then since the set is finite, we must have $\sigma^k = 1$ for some $k > 0$. Let k_1 be the smallest value of k for which this happens. Then, the cycle generated by the element 1 for the permutation σ is represented as

$$(1, \sigma(1), \sigma^2(1), ..., \sigma^{k_1 - 1}(1))$$

By the definition of k_1, the numbers $\sigma^r(1), r = 0, 1, ..., k_1 - 1$ are all distinct. We then choose a number i_2 and repeat the same with the permutation σ. Then we obtain another disjoint cycle

$$(i_2, \sigma(i_2), \sigma^2(i_2), ..., \sigma^{k_2 - 1}(i_2))$$

It is easily seen that the set of element in the cycle generated by 1 is disjoint from the set of elements in the cycle generated by i_2. Keep repeating until we get p cycles whose union is the whole set $\{1, 2, ..., n\}$. In this way, the permutation σ can be represented as a product of cycles

$$\sigma = \Pi_{m=1}^{p}(i_m, \sigma(i_m), \sigma^2(i_m), ..., \sigma^{k_m - 1}(i_m))$$

Given any element $j = 1, 2, ..., n$, it follows that $j = \sigma^r(i_r)$ for some r and then the action of σ on this element is given by

$$\sigma(j) = \sigma^{r+1}(i_r)$$

10. Invariant polynomials: Let V be a vector space and V^* its dual. Assume that $dimV = n < \infty$. Consider $G = GL(V)$, the general linear group acting on V. Any element $v \in V$ under the action of $g \in G$ transforms to $g.v \in V$. The action of $GL(V)$ on V^* is given by $f \rightarrow = g.f = fg^{-1}$. Thus, if $f \in V^*$ and $v \in V$, we have that $f(v)$ is invariant. more generally, we can consider a polynomial P on $V^p \times V^{*q}$ defined by

$$P(v_1, ..., v_p, \alpha_1, ..., \alpha_p) = \Pi_{j=1}^n \alpha_{\sigma j}(v_j)$$

This polynomial is invariant under G action in the sense that if $g \in G$, then

$$P(g.v_1, ..., g.v_p, g.\alpha_1, ..., g.\alpha_p) = \Pi_{j=1}^n \alpha_{\sigma j} o g^{-1}(gv_j) = \Pi_j \alpha_{\sigma j}(v_j) = P(v_1, ..., v_p, \alpha_1, ..., \alpha_p)$$

11.Permutation representation: Let G be a finite group and let $\mathbb{F}$ be a field. The group algebra of G consists of all functions $a : G \rightarrow \mathbb{F}$. We denote the group algebra by $F(G)$. Any $a \in F(G)$ can be expressed as

$$a = \sum_{g \in G} a(g)g$$

formally. Given any $g \in G$ and $a \in F(G)$, the action of g on a is given by

$$g.a = \sum_{h \in G} a(h)g.h = \sum_h a(g^{-1}h)h$$

We may regard the function as a column vector. For instance if $G = \{g_1, ..., g_n\}$ and $a \in F(G)$, we set $a_i = a(g_i)$. Then, a can be represented by the column vector

$$a = [a_1, a_2, ..., a_n]^T$$

Now, for a given $g \in G$, there exists a permutation σ of $\{1, 2, ..., n\}$ such that $gg_i = g_{\sigma(i)}, i = 1, 2, ..., n$. It then follows that $g.a$ is is given by

$$g.a = \sum_h a(h)g.h = \sum_i a(g_i)g.g_i = \sum_i a_i g_{\sigma i} = \sum_i a_{\sigma^{-1}i} g_i$$

and this is represented by the column vector

$$g.a = [a_{\sigma^{-1}1}, a_{\sigma^{-1}2}, ..., a_{\sigma^{-1}n}]^T$$

The action of G on $F(G)$ is a representation of the group G and for the above reason, it is called the permutation representation.

12.Young's Tableaux for the permutation group: Let n be a positive integer and consider a tableaux T consisting of h_1 boxes in the first row, h_2 boxes in the second row and so on upto h_r boxes in the last row. The rows are arranged one below the other so that the first box in each row is aligned. We assume that $h_1 \geq h_2 \geq ... \geq h_r$ is valid and $\sum_{i=1}^{r} h_i = n$. In other words, $\{h_i : i = 1, 2, ..., r\}$ is partition of the integer n. For example suppose $n = 6$ and the tableaux is specified by $(3, 2, 1)$. Then the first row has three boxes, the second has two boxes and finally the third row has one box. The first column therefore has three boxes, the second column has two boxes and the third has one. We can specify the coordinates of the boxes as $(1,1), (1,2), (1,3), (2,1), (2,2), (3,1)$. Thus the three boxes of the first row have positions $(1, i), i = 1, 2, 3$, the two boxes of the second row have positions $(2, i), i = 1, 2, 3$ and finally the only box in the third row has coordinates $(3, 1)$. More precisely, we call the arrangement of boxes corresponding to a partition of n as a Young diagram and after filling these boxes with numbers $1, 2, ..., n$, the Young diagram becomes a Tableaux or more precisely a Young Tableaux. Thus, given a diagram D (which is a set of boxes specified by coordinates as explained above), we can construct a bijective map $f : D \to \{1, 2, ..., n\}$. This map assigns a number i to each box. Such a map is called a Tableaux. For a given diagram, there are $n!$ Tableaux's. A bijective map of the above kind can be thought of as filling the Young diagram with numbers from one to n. Let σ be a permutation of $X = \{1, 2, ..., n\}$ and D a Young diagram and $f : D \to X$ a Young tableaux. Then, $\sigma o f : D \to X$ is another tableaux. if the box having coordinates (i, j) in the Young tableaux has the number r, ie, $f(i, j) = r$. Then the same box in the tableaux $\sigma o f$ will have the number

$$\sigma o f(i, j) = \sigma(r)$$

Let T be a Young diagram (a Tableaux filled with numbers) and let us denote the set of numbers in the ith row by R_i, where $i = 1, 2, ..., k$. Then $X = \bigcup_{i=1}^{k} R_i$ is a partition of X. Let G_i be the set of all the elements of S_n that permute only the ith row R_i of the tableaux T. Then G_i is a subgroup of S_n. It is clear that $\Pi_{i=1}^{k} G_i$ is also a subgroup of S_n. It consists of all the permutations which permute each row of T, there is no permutation of the columns. We denote this subgroup by R_T. The elements of R_T permute the rows of T amongst themselves. Likewise C_T is the subgroup of permutations of the column elements. We claim that if s is any permutation, ie, $s \in S_n$, then

$$R_{sT} = sR_T s^{-1}, C_{sT} = sC_T s^{-1}$$

To see this we consider for example the first row of T. Let it be given by the elements $(i_1, i_2, ..., i_m)$ so that $i_1, ..., i_m$ are distinct elements of $\{1, 2, ..., n\}$. Then, the first row of sT is given by $(s(i_1), s(i_2), ..., s(i_m))$. Let $\sigma \in R_T$ so that

$$(\sigma(i_1), \sigma(i_2), ..., \sigma(i_m)) = (i_{\rho(1)}, ..., i_{\rho(m)})$$

where ρ is a permutation of $(1, 2, ..., m)$. Consider now the action of the permutation $s\sigma s^{-1}$ on the first row of R_{sT}. It is given by

$$s.\sigma.s^{-1}(s(i_1), s(i_2), ..., s(i_m)) = s(\sigma(i_1), \sigma(i_2), ..., \sigma(i_m))$$

$$= s(i_{\rho(1)}, i_{\rho(2)}, ..., i_{\rho(m)}) = (s(i_{\rho(1)}), ..., s(i_{\rho(m)}))$$

which is evidently a permutation of the first row of sT. The other rows are handled analogously and likewise the columns. S_n is the symmetric group of all permutations of n elements and $Q(S_n)$ is the set of all functions on S_n with values in the rational field Q. Let $R = Q(S_n)$. This is the group algebra of S_n. Let T be a tableaux and define

$$s_T = \sum_{\sigma \in R_T} \sigma, a_T = \sum_{\sigma \in C_T} sgn(\sigma)\sigma$$

s_T symmetrizes the rows of the tableaux T while a_T symmetrizes the columns of the tableaux. T. For example, suppose $n = 6$ and the tableaux has rows of length 3, 2, 1. Set

$$f(1,1) = 4, f(1,2) = 3, f(1,3) = 1,$$

$$f(2,1) = 2, f(2,2) = 6,$$

$$f(3,1) = 5$$

Let G_1 denote the subgroup of S_6 consisting of all the permutations of $(4,3,1)$, G_2 the subgroup of all permutations of $(2,6)$, G_3 the subgroup consisting only of the identity. Specifically, the elements of G_1 are (a) $\sigma_{11} : (4,3,1) \to (3,4,1)$, (b)$\sigma_{12} : (4,3,1) \to (4,1,3)$, (c) $\sigma_{13} : (4,3,1) \to (1,3,4)$, (d) $\sigma_{14} : (4,3,1) \to (1,4,3)$, (e) $\sigma_{15} : (4,3,1) \to (3,1,4)$ and (f) and the identity σ_{16}. Likewise, G_2 consists of the elements $\sigma_{21} = (26)$ and $\sigma_{22} = e$ and finally $G_3 = \{e = \sigma_{31}\}$. Then

$$s_T = \sum_{j_1,j_2,j_3} \sigma_{1,j_1}\sigma_{2,j_2}\sigma_{3,j_3} =$$

s_T is the operator inducing symmetrizers on the rows of T. We can repeat the process for the columns. Specifically, H_1 will denote all permutations of $(4,2,5)$, H_2 will denote all permutations of $(3,6)$ and finally, H_3 will denote the identity. Then

$$a_T = \sum$$

Let G_i denote the subgroup of permutations of the ith row. Then

$$s_T = \sum_i \sum_{\sigma \in G_i} \sigma$$

$$s_T^2 = \sum_{i,j} \sum_{\sigma \in G_i, \rho \in G_j} \sigma\rho$$

We clearly have

$$\sum_{\sigma,\tau \in G_i} \sigma\tau = |G_i|! \sum_{\sigma \in G_i} \sigma$$

since for $\sigma \in G_i$, we have

$$\sigma. \sum_{\tau \in G_i} \tau = \sum_{\tau \in G_i} \sigma\tau = \sum_{\rho \in G_i} \rho$$

where we have set $\rho = \sigma\tau$. The group property of G_i has been used. Also if $i \neq j$, then for $\sigma \in G_i$ and $\tau \in G_j$, we have

13.Semisimple ring: Let R be a ring. Then R is a left module over itself. If R is semisimple, then by definition R as a vector space over itself can be decomposed into irreducible modules, ie, we can write

$R = \bigoplus R_i$ where (a) R_i is a submodule over R, (b) R_i is irreducible, ie, it has no nontrivial submodules over R. Let G be a group and let U be the left regular representation of G. Then, assuming G to be compact, we can decompose U into a direct sum of irreducible representations U_i with U_i appearing $dimU_i$ times in the decomposition. We can thus write

$$U = \bigoplus_i U_i \otimes \mathbb{C}^{d_i}, d_i = dimU_i$$

Let G be a group and $F(G)$ denote its group algebra. Let T be a representation of G in a vector space U. Then U can be regarded as an $F(G)$-module so that if $a \in F(G)$ and $x \in U$, then

$$a.x = \sum_{g \in G} a(g)T(g)x$$

We have for any two elements $a, b \in F(G)$,

$$b.(a.x) = \sum_g b(g)T(g)\sum_h a(h)T(h)x = \sum_{g,h} b(g)a(h)T(gh)x = \sum_k (\sum_{g.h=k} b(g)a(h))T(k)x = (b * a).x$$

This confirms that U is indeed a module over $F(G)$ with the product in $F(G)$ being defined by the convolution operation. Suppose R is a ring. Then, R considered a module over itself is called the regular representation of R. In this representation, we view the elements of R as linear operators acting on the vector space R so that $(r, x) \to r.x$. Suppose T is an endomorphism of the regular representation, ie, $T : R \to R$ is R linear. We write $T \in End_R(R)$. This means that for any $x, y \in R$, we have $T(xy) = xT(y)$. In particular, $T(x) = T(x.1) = xT(1)$. Thus, T can be identified with $T(1)$ acting on R via right multiplication. Suppose $T, S \in End_R(R)$. Then for any $x \in R$, we have

$$TS(x) = T(xS(1)) = x.S(1)T(1) = x.TS(1)$$

Thus, TS is identified with right multiplication with $S(1)T(1)$. R^0 is the ring R with product taken in reverse order. In other words, if $x, y \in R^0$, then $x, y \in R$ and composition is defined by $\phi(x, y) = y.x$. Then, it is clear that if we define the mapping $\Phi : End_R(R) \to R^0$ by $\Phi(T) = T(1)$, we get

$$\Phi(TS) = TS(1) = S(1)T(1) = \phi(T(1), S(1)) = \phi(\Phi(T), \Phi(S))$$

In this way, $End_R(R)$ is isomorphic as a ring to R^0.

Consider now the action of $R \otimes R^0$ on R given by $(a \otimes b)(c) = acb$. We denote this action by Φ. Thus,

$$\Phi(a \otimes b, c) = acb$$

We have

$$\Phi(a \otimes b, \Phi(a' \otimes b', c)) = \Phi(a \otimes b, a'cb') = aa'cb'b$$

on the one hand and on the other

$$(a \otimes b)(a' \otimes b') = aa' \otimes b'b$$

since multiplication in R^0 is defined by reversing the order of multiplication in R. It follows that

$$\Phi((a \otimes b)(a' \otimes b'), c) = aa'cb'b$$

and hence Φ defines a valid action. Let R be a ring and U a module over R. If U is a cyclic module, then there exists an element $x \in U$ such that $R.x = U$. In this case, suppose $y \in U$ is arbitrary. Then there exists an $r \in R$ such that $y = r.x$ and we get $R.y = R.r.x$. This need not be equal to $R.x$ since $R.r$ need not equal R. In other words, if a module is cyclic, it does not mean that it is irreducible, it can still have submodules. For example, let R denote the ring of all polynomials in one variable and let U be the R module consisting of all polynomials divisible by a non-constant polynomial $p(z)$. Then $p(z)$ is a cyclic element of U and hence U is a cyclic module. However, if $q(z)$ is a polynomial divisible by $p(z)$ and $degq > degp$, then all $Rq \neq U$ and hence Rq is a non-trivial submodule of U. Suppose U is a cyclic module and let x generate U. Define $\phi : R \to U$ by $\phi(r) = r.x$. ϕ is surjective and $ker\phi$ is a left ideal in R. Indeed, $\phi(r) = 0$ implies that for any $s \in R$, we have $\phi(sr) = (sr).x = s.(r.x) = s.0 = 0$. It follows that U is isomorphic as a module to $R/ker\phi$. On the other hand, suppose J is any ideal in R. We then claim that R/J is a cyclic module over R. This follows from the fact that $1 + J$ generates R/J. (We are assuming that the ring R has an unit element). This proves that a module is cyclic iff it is a quotient of the regular representation.

14.Let F be a field and $R = M_n(F)$ the ring of all $n \times n$ matrices with entries in F. F^n is a module over R. F^n is clearly a cyclic module. Let e_1 be the element of F^n having first entry equal to one and other entries zero. Then the annihilator I_1 is the left ideal of all matrices having first column zero. Let J_1 be the ideal of R having all columns except the first zero. Then we have $R = I_1 \oplus J_1$ and in particular, $J_1 = R/I_1$. J_1 is isomorphic to F^n as a module over R.

15.Vector fields on a manifold and flows generated by them: Let M be a C^∞-manifold and X a vector field on it. For $x \in M$, assume that $X(x)$ in a local coordinate system has components

$$(X^i(x) : i = 1, 2, ..., n)$$

ie,

$$X(x) = \sum_{i=1}^{n} X_i(x) \frac{\partial}{\partial x^i}$$

Let $F(t, .)$ be the flow generated by this vector field. Thus, in terms of components, we have

$$\frac{dF^i(t, x)}{dt} = X^i(F(t, x)), t \geq 0, F(0, x) = x$$

We can consider the vector field to act on functions. Assume that this action is given by the operator $\phi(t)$. Thus,

$$\phi(t)f(x) = f(F(-t, x))$$

and we get

$$\frac{d}{dt}\phi(t)f(x)|_{t=0} = -\sum_i X_i(x)\frac{\partial f(x)}{\partial x^i}$$

We denote the differential operator corresponding to X by the same symbol. Then, from the multivariate Taylor expansion, we have the operator equation

$$\phi(t) = exp(-tX) = \sum_{n=0}^{\infty} (-t)^n X^n / n!$$

(Procesi, Lie group, an approach through invariants)

16. Let X, Y be vector fields on a manifold and $\phi(t)$ the one parameter group of operators generated by X. We regard the vector fields as operators. Then define

$$Y(t) = \phi(t).Y.\phi(t)^{-1}$$

Then, it is easy to see that for any smooth function f on the manifold, we have

$$\frac{d}{dt}Y(t)f = -XY(t)f + Y(t)Xf = [Y(t), X]f$$

This follows at once from the operator equation

$$\frac{d\phi(t)}{dt} = -X\phi(t) = -\phi(t)X$$

Remark:

$$\phi(t) = exp(-tX) = exp(-t\sum_i X_i(x)\frac{\partial}{\partial x^i})$$

Also

$$\phi(t)f(x) = f(F(-t, x))$$

so that

$$\frac{d}{dt}\phi(t)f(x) = -\sum_i \frac{dF^i(-t,x)}{dt}\frac{\partial}{\partial z^i}f(z)|_{z=F(-t,x)}$$

$$= -\sum_i X^i(z)\frac{\partial f(z)}{\partial z^i}|_{z=F(-t,x)} = -(\phi(t)Xf)(x)$$

or in terms of operators,

$$\frac{d\phi(t)}{dt} = -\phi(t)X$$

That the operators $\phi(t)$ and X commute so that $\phi(t)X = X\phi(t)$ is an easy consequence of the operator identity $\phi(t) = exp(-tX)$.

17. Assume that G is a Lie group acting on a manifold M by $\rho : G \times M \to M$ so that $(g, m) \to \rho(g, m)$. Let $\phi : \mathbb{R} \to G$ be a one parameter subgroup of G, ie, $\phi(s+t) = \phi(s)\phi(t)$. Let L_g denote left action on G by g and R_g right action on G by G. Thus, $L_g(h) = gh, R_g(h) = hg^{-1}$. Let $T : G \to G$ be a transformation that commutes with left action, ie, $TL_g = L_gT$. Then, we have

$$T(g) = TL_g(e) = L_gT(e)$$

so that T is the right action by $T(e)^{-1}$, ie,

$$T = R_{T(e)^{-1}}$$

Assume that a group G acts on a manifold M. Take an element $g \in G$ and points $p, q \in M$ so that $p = g.q$. Let X be a vector field on M. For any smooth function f on M, we have

$$dg_q(X_q)(f) = X_q(fog) = X(f(gq)) = Xg^{-1}f(q) = (Xg^{-1}f)(g^{-1}p) = (gXg^{-1}f)(p)$$

so that

$$dg_q(X_q) = (gXg^{-1})_p$$

Note: An element $g \in G$ acts on functions f on M in accord with the rule

$$(g.f)(x) = f(g^{-1}.x)$$

If a vector field X on G satisfies

$$gXg^{-1} = X, g \in G$$

then we say that X is a left invariant vector field. Let $X(e) = a \in TG_e$. Then the condition for X to be left invariant is that

$$X(g) = gXg^{-1}(g) = (dg)_e(X_e) = dg_e(a)$$

18.Let M be a C^∞-manifold and take a vector field X on M. Let $\phi(t)$ denote the one parameter group of diffeomorphisms of M generated by X (obtained as solutions to a system of first order differential equations). Let $g : M \to M$ be a diffeomorphism. For any smooth function f on M, we compute with $q = g.p$,

$$\frac{d}{dt}f(g.\phi(t).g^{-1}p)|_{t=0} = \frac{d}{dt}(fog)(\phi(t)q)|_{t=0}$$
$$= X_q(fog) = dg_q(X_q)(f) = (gXg^{-1})_p(f)$$

This means that $t \to g.\phi(t).g^{-1}$ is a one parameter group of diffeomorphisms of M generated by the vector field gXg^{-1}. Suppose G is a Lie group and $a \in TG_e$. Let X be the left invariant vector field on G generated by a and let $\phi(t)$ be the one parameter group of diffeomorphisms of G generated by X_a. Then, we have $gXg^{-1} = X$ and hence $g\phi(t)g^{-1}$ is the one parameter group of diffeomorphisms of G generated by X, ie,

$$g\phi(t)g^{-1} = \phi(t), t \in \mathbb{R}$$

or equivalently,

$$g.\phi(t) = \phi(t).g, t \in \mathbb{R}$$

Consider

$$\psi(t) = \phi(t)(e) \in G$$

We have

$$\phi(t)(g) = g.\phi(t)(e) = g\psi(t)$$

and hence

$$\psi(t+s) = \phi(t+s)(e) = \phi(t)(\phi(s)(e)) = \phi(t)(\psi(s)) = \psi(s).\psi(t)$$

proving that $\{\psi(t)\}$ is a one parameter subgroup of G. Note that since $\phi(t)$ commutes with left translations, it follows that $\phi(t)$ is a right translation and this right translation is precisely $R_{\psi(t)^{-1}}$.

19.Roots:} is a finite dimensional semisimple Lie algebra and ⟨ a maximal Abelian subalgebra. For each $H \in \langle$, $ad(H)$ is a linear operator on }. The Cartan-Killing form on } is defined as

$$< X, Y >= Tr(ad(X).ad(Y)), X, Y \in \}$$

Since $\}$ is semisimple, Cartan's theorem implies that $< .,. >$ is nonsingular on $\} \times \}$. Since the operators $ad(H), H \in \langle$ all commute:$[ad(H), ad(H')] = ad([H, H']) = 0$ for all $H, H' \in \langle$, it follows that these operators can be diagonalized simultaneously. In other words there exist a finite set $\Delta \subset \langle^*$, such that if

$$\}_\alpha = \{X \in \} : ad(H)(X) = \alpha(H)X \forall H \in \langle\}$$

then

$$\} = \langle \oplus \oplus_{\alpha \in \Delta} \}_\alpha$$

In this decomposition, we have used the fact that $ad(H)$ for each $H \in \langle$ is a semisimple operator on $\langle$ and hence the set $ad(H), H \in \langle$ of operators in $\}$ can be diagonalized simultaneously. Since $\langle$ is maximal Abelian, it follows that $\alpha \neq 0$ for all $\alpha \in \Delta$. It is clear that $< .,. >$ is non-singular on $\langle \times \langle$. For suppose that for some $H \in \langle$, we have $< H, H' >= 0$ for all $H' \in \langle$. Then consider any $X \in \}_\alpha$. We have for $Y \in \}_\beta$ that $[X, Y] \in \}_{\alpha+\beta}$. In this formula, $\}_{\alpha+\beta}$ should be interpreted as the zero subspace if $\alpha+\beta \notin \Delta$. The set Δ is called the set of roots and an element $\alpha \in \Delta$ is called a root. To see the validity of this, we note that for $X \in \}_\alpha, Y \in \}_\beta, H \in \langle$, we have

$$ad(H)([X, Y]) = [H, [X, Y]] = -([X, [Y, H]] + [Y, [H, X]]) = [X, \beta(H)Y] - [Y, \alpha(H)X]$$

$$= (\alpha(H) + \beta(H))[X, Y]$$

so that

$$[X, Y] \in \}_{\alpha+\beta}$$

Note that if $\}_{\alpha+\beta} = 0$, ie, $\alpha + \beta \notin \Delta$, then $[X, Y] = 0$. It follows that if $X \in \}_\alpha$ for some $\alpha \in \Delta$, then for every $\beta \in \Delta$,

$$ad(H)ad(X)(\}_\beta) \subset \}_{\alpha+\beta}$$

We write $\}_0 = \langle$. Then the above formula is also valid for $\beta = 0$. It follows that the matrix of $ad(H)ad(X)$ relative to a basis corresponding to the direct sum decomposition of $\}$ into root spaces cannot have any diagonal element and hence

$$< H, X >= Tr(ad(H).ad(X)) = 0$$

In other words, $H \perp \}_\alpha$ for all $\alpha \in \Delta$. Since $H \perp \langle$, by hypothesis, it follows that $H \perp \}$ implying that $< .,. >$ is singular on $\}$ which is a contradiction. If $\alpha, \beta \in \Delta$, $\alpha + \beta \neq 0$, then it is clear that $\}_\alpha \perp \}_\beta$. This formula remains valid if either α or β is zero with the interpretation $\}_0 = \langle$. To see this, let $X \in \}_\alpha, Y \in \}_\beta$. Then $ad(X)ad(Y)$ maps $\}_\nu$ to $\}_{\nu+\beta+\alpha}$ and hence if $\alpha + \beta \neq 0$, $ad(X).ad(Y)$ will have no diagonal element in the basis corresponding to the direct sum decomposition of $\}$ implying that $< X, Y >= Tr(ad(X).ad(Y)) = 0$. We now claim that $< .,. >$ is non-singular on $\}_\alpha \times \}_{-\alpha}$. For suppose $X \in \}_\alpha$ is such that $X \perp \}_{-\alpha}$. Then since we have just proved that $X \perp \}_\nu$ for all $\nu \in \Delta \cup \{0\} - \{-\alpha\}$, it follows that $X \perp \}$ a contradiction unless $X = 0$. We now want to show that the roots Δ span $\langle^*$. Indeed, let $H, H' \in \langle$. Then $\langle$ is an eigensubspace of $ad(H)ad(H')$ with zero eigenvalue while $\}_\alpha$ is an eigensubspace of the same with eigenvalue $\alpha(H)\alpha(H')$. It follows that

$$< H, H' >= \sum_{\alpha \in \Delta} dim(\}_\alpha)\alpha(H)\alpha(H')$$

Suppose that Δ does not span $\langle^*$. Then there will exist an $H \in \langle$ such that $\alpha(H) = 0$ for all $\alpha \in \Delta$ from which it would follow that $< H, H' >= 0$ for all $H' \in \langle$, a contradiction. Since the Cartan-Killing form

$<.,.>$ on $\langle \times \langle$ is non-singular, and since the roots Δ span $\langle^*$, there is a unique isomorphism from $\langle \to \langle^*$ such that $H \to \phi(H)$ satisfies

$$\phi(H)(H') =< H, H' >$$

or equivalently,

$$\phi(H) =< H, . >$$

We define $H_\alpha \in \langle$ by the equation

$$\phi(H_\alpha) = \alpha$$

This means that

$$\alpha(H') =< H_\alpha, H' >, H' \in \langle$$

We may more generally define for any $\lambda \in \langle^*$, $H_\lambda \in \langle$ by the equation

$$< H_\lambda, H' >= \lambda(H'), H' \in \langle$$

or equivalently,

$$\phi(H_\lambda) = \lambda$$

We now show that if $X \in \}_\alpha, Y \in \}_{-\alpha}$, then $[X, Y] =< X, Y > H_\alpha$. We start with the fact that $[X, Y] \in \langle$ for if $H \in \langle$, then

$$[H, [X, Y]] = -([X, [Y, H]] + [Y, [H, X]]) = -[X, \alpha(H)Y] - [Y, \alpha(H)X] = 0$$

and hence $[X, Y]$ commutes with $\langle$. Since however $\langle$ is maximal Abelian, it follows that $[X, Y] \in \langle$. Thus,

$$< H, [X, Y] >=< [H, X], Y >= \alpha(H) < X, Y >=< H, H_\alpha >< X, Y >=< H, < X, Y > H_\alpha >$$

Since $[X, Y], H_\alpha \in \langle$ and since $< .,. >$ is non-singular, it must follows that

$$[X, Y] =< X, Y > H_\alpha, X \in \}_\alpha, Y \in \}_{-\alpha}$$

Since $< .,. >$ is non-singular on $\}_\alpha \times \}_{-\alpha}$, it follows that for some $(X, Y) \in \}_\alpha \times \}_{-\alpha}$, we have $< X, Y > \neq 0$ and hence

$$[g_\alpha, \}_{-\alpha}] = \mathbb{C}.H_\alpha$$

(Ref:V.S.Varadarajan, Lie groups, Lie algebras and their representations).

20.Let L be a finite dimensional Lie algebra and let A be a subalgebra of L such that $x \in A$ implies $ad(x)$ is semisimple. Then A must be Abelian. For suppose $x, y \in A$ are such that $[x, y] \neq 0$. Then, $ad(x)$ is a nonzero linear operator in A and hence has an eigenvector, say z. We have $[x, z] = c.z$ with c nonzero and $z \in A$ nonzero. Then $ad(z)(z) = 0, ad(z)(x) = -cz$. Thus the two dimensional space spanned by x, z is invariant under $ad(z)$ and the matrix of the restriction of $ad(z)$ to $span\{x, z\}$ relative to the ordered basis $\{z, x\}$ is nonzero nilpotent and hence $ad(z)$ cannot be semisimple, a contradiction.

21.Direct sum of representations of a Lie algebra. Let L be a Lie algebra and let π_1 and π_2 be two representations of L respectively in the vector spaces V_1, V_2. Then the representation π of L in the space $V_1 \otimes V_2$ defined by

$$\pi(X)(u \otimes v) = \pi_1(X)u \otimes v + u \otimes \pi_2(X)v$$

is called the direct product of the two representations, and is denoted by

$$\pi = \pi_1 \otimes \pi_2$$

22.Let $V = \mathbb{C}^2$. Then, $S^k(V)$, the k-fold symmetric tensor product of V with itself can be identified with all homogeneous polynomials in two variables (x, y) having degree k. A basis for ths space is given by the polynomials

$$p_i(x, y) = \binom{k}{i} x^{k-i} y^i, i = 0, 1, 2, ..., k$$

We shall compute the action of X, Y, H on these polynomials and show that this defines an irreducible representation of $sl(2, \mathbb{C})$ of dimension $k+1$. Let g be any element of the group $SL(2, \mathbb{C})$. Then the action of g on p_i is given by $p_i(x, y) \to p_i(g^{-1}(x, y))$. It follows that if $g = exp(tZ)$ so that $Z \in sl(2, \mathbb{C})$, then the action of Z on $p_i(x, y)$ is given by

$$p_i(x, y) \to \frac{d}{dt} p_i(exp(-tZ)(x, y))|_{t=0}$$

Remark: Let us take a vector space V on which a group of matrices G acts. Then, G acts on $V \otimes V$ in the canonical way. Specifically, $u \otimes v$ gets mapped under g to $gu \otimes gv$. G also acts on the symmetric space $S^2(V)$. Specifically, we choose a basis $e_1, ..., e_n$ for V and then obtain a basis $e_i.e_i, e_i.e_j + e_je_i, i < j$ for $S^2(V)$. We identify the vectors $e_i.e_i$ with the polynomials x_i^2 and $e_i.e_j + e_j.e_i$ with polynomials $2x_ix_j$ in n variables $x_1, ..., x_n$. Then, let $g = exp(tX)$, so that X belongs to the Lie algebra of G. The action of X on $e_i^2 = e_i.e_i$ is given by

$$e_i^2 \to \frac{d}{dt}(exp(tX)e_i).(exp(tX)e_i)|_{t=0} = (Xe_i).e_i + e_i.Xe_i$$

Let X in the basis $B = e_1, ..., e_n$ have matrix elements X_{ij}. Then,

$$Xe_i^2 = X_{ji}e_j.e_i + X_{ji}e_i.e_j = X_{ji}(e_j.e_i + e_i.e_j)$$

summation over the index j being assumed. Also for $i < j$, we have

$$X(e_ie_j + e_je_i) = (Xe_i).e_j + e_i.Xe_j + (Xe_j).e_i + e_j.Xe_i$$

$$= X_{ki}e_k.e_j + X_{kj}e_i.e_k + X_{kj}e_k.e_i + X_{ki}e_j.e_k$$

$$= X_{ki}(e_k.e_j + e_j.e_k) + X_{kj}(e_i.e_k + e_k.e_i)$$

Let us now compute the action of X on the polynomials x_i^2, x_ix_j. First observe that if $g = exp(tX)$, then the action is defined in a dual fashion, ie, as $p(x) \to p(g^Tx)$. Also

$$(g^Tx)_i^2 = (\sum_j g_{ji}x_j)^2$$

and its derivative with respect to t at $t = 0$ equals

$$2X_{ji}x_jx_i$$

which corresponds to the vector $X_{ji}(e_j.e_i + e_i.e_j)$. Also the action of g on $2x_ix_j$ is defined by $2x_ix_j \to 2(g^Tx)_i(g^Tx)_j$ and its derivative with respect to t at $t = 0$ is given by

$$2(X_{ki}x_kx_j + X_{kj}x_ix_k)$$

This polynomial corresponds to the vector

$$X_{ki}(e_kx_j + e_je_k) + X_{kj}(e_ie_k + e_ke_i)$$

which is the formula given above. More generally, Suppose V is $\mathbb{C}^2$. We choose a basis e_1, e_2 of vectors for this space and let $e_1^ie_2^{k-i}$ denote the vector obtained by adding all the **distinct permutations** of $e_1...e_1.e_2...e_2$ in the tensor product $V^{\otimes_s k}$. Note that there will in all be $\binom{k}{i}$ distinct permutations. This symmetric vector corresponds to the symmetric polynomial x^iy^{k-i} in two variables. For example, let us take $e_1.e_1.e_2$.. Then,

$$e_1^2.e_2 = e_1.e_1.e_2 + e_1.e_2.e_1 + e_2.e_1.e_1$$

$$X(e_1^2.e_2) = (Xe_1).e_1.e_2 + e_1.Xe_1.e_2 + e_1.e_1.Xe_2 + (Xe_1).e_2.e_1 + e_1.Xe_2.e_1 + e_1.e_2.Xe_1$$
$$+Xe_2.e_1.e_1 + e_2.Xe_1.e_1 + e_2.e_1.Xe_1$$
$$= (X_{11}e_1 + X_{21}e_2)e_1.e_2 + e_1.(X_{11}e_1 + X_{21}e_2)e_2 + ...$$

It is equivalent to the dual action of X on the symmetric polynomial x^iy^{k-i}. Note that $X^Tx = X_{11}x+X_{21}y$ which for the X of $sl(2,\mathbb{C})$ is zero. $X^Ty = X_{12}x + X_{22}y = x$. Thus,

$$X(x^iy^{k-i}) = (k-i)x^{i+1}y^{k-i-1}$$

Writing

$$v_i = \binom{k}{i}x^{k-i}y^i$$

we get

$$Xv_i = \binom{k}{i}ix^{k-i+1}y^{i-1} = \frac{\binom{k}{i}i}{\binom{k}{i-1}}v_{i-1} = (k-i+1)v_{i-1}$$

Likewise, for Yv_i, Hv_i. This means that the tensor representation $S^k(V)$ is an irreducible representation of $sl(2,\mathbb{C})$ of dimension $k+1$.

23.Let L be a Lie algebra with $L = [L, L]$. Then obviously any one dimensional representation of L is trivial. We denote this trivial representation (Module) by $\mathbb{C}$. The L action is given by $X.x = 0 \forall x \in \mathbb{C}, X \in L$.

Assume that if M is a finite dimensional representation of L and $N \subset M$ is an irreducible submodule with M/N one dimensional, then $M = N \oplus \mathbb{C}$ as submodules. Then we want to prove that if the irreducibility condition on N is dropped, the same result holds. If N is irreducible, there is nothing to prove. Suppose N is not irreducible. Then N contains a nonzero irreducible submodule P. Define $M' = M/P$, $N' = N/P$. Then M'/N' is one dimensional. Indeed, let $u \in M - N$. Then, $u + P \in M/P = M'$ but $u + P \notin N/P = N'$. Further, suppose $X \in M'/N'$ is nonzero. Then $X = A + N'$ where $A \in M' - N'$. $A \in M'$ implies $A = x + P$ for some $x \in M$. $A \notin N'$ implies $x + P \notin N/P$ implies $x \notin N$. Thus, $x \in M - N$ and hence x is a scalar multiple of u (M/N is one dimensional). By induction, there is a complement $\mathbb{C}$ to N' in M'. Define the submodule Q of M so that $Q/P = \mathbb{C}$. Note that $M' = M/P = N' \oplus \mathbb{C} = N/P \oplus \mathbb{C}$, so we can choose an element $a + P$ in M' ($a \in M$) such that $a + P$ spans $\mathbb{C}$. Obviously $a \notin N$. Then we take Q to be the module spanned by P and a. By hypothesis, P has a one dimensional complement in Q, ie, we can write $Q = P \oplus R$ where R is isomorphic to $\mathbb{C}$. Then we claim that R is a complement of N in M. Suppose $0 \neq x \in R$. Then, $x \notin P$ and hence $x + P$ is a nonzero element of $Q/P = \mathbb{C}$. Then we have from the equation $M' = N' \oplus (Q/P)$ that $x + P \notin N' = N/P$. This means that $x \notin N$ and hence we are through.

Suppose now that if M is a finite dimensional representation of L and N a submodule of M with M/N one dimensional then $M = N \oplus \mathbb{C}$ as submodules. Then, we wish to show that every finite dimensional representation of L is completely reducible. For let M be a finite dimensional module and N a submodule. We wish to produce a submodule P such that $M = N \oplus P$. $Hom_L(M, N)$ denotes the space of all maps $T : M \to N$ with the property that $T(lx + y) = lT(x) + T(y)$ for all $l \in L, x, y \in M$, ie, $Hom_L(M, N)$ is the space of all L-linear maps from M into N. $Hom(M, N)$ on the other hand, denotes the space of all linear maps $T : M \to N$. Note that $Hom(M, N)$ is an L module, once we introduce the multiplication

$$(lT)(x) = lT(x) - Tl(x), x \in M, l \in L$$

24. Let V be a vector space and let $T \in End(V)$. Define $f_T \in (V \otimes V^*)^*$ so that

$$f_T(u \otimes \phi) = \phi(Tu), u \in V, \phi \in V^*$$

This map is well defined. Note that we can equivalently view f_T as a bilinear functional on $V \times V^*$ in accord with

$$f_T(u, \phi) = \phi(Tu)$$

In fact if ψ is any bilinear functional on $V \times V^*$, then there exists a unique element θ in $(V \otimes V^*)^*$ so that $\theta(u \otimes \phi) = \psi(u, \phi)$ by the definition of tensor product. Now suppose A group G acts linearly on V. This group G is a subgroup of $GL(V)$. Suppose $T \in End(V)$ is G invariant, ie, $gTg^{-1} = T$ for all $g \in G$. Then noting that G acts on V^* as $(g.\phi)(u) = \phi(g^{-1}u)$, it follows that

$$f_T(g.u, g.\phi) = (g.phi)(Tg.u) = \phi(g^{-1}Tgu) = \phi(u) = f_T(u, \phi)$$

In other words, f_T is a G-invariant element of $(V \otimes V^*)^*$. In other words, we have a one-one correspondence between G-invariant elements of $End(V)$ and G-invariant elements of $(V \otimes V^*)^*$. More generally, we can consider the following situation. Let A be a vector space and let $\otimes$ be a tensor product. We can identify $End(A^{\otimes n})$ with $End(A)^{\otimes n}$ so that if $T_1, ..., T_n \in End(A)$, then the element $T_1 \otimes ... \otimes T_n \in End(A)^{\otimes n}$ is identified with an element of $End(A^{\otimes n})$ by the rule

$$(T_1 \otimes ... \otimes T_n)(u_1 \otimes ... \otimes u_n) = T_1 u_1 \otimes T_2 u_2 \otimes ... \otimes T_n u_n$$

We can identify $End(A) \otimes n$ with $(A^{*\otimes n} \times A^{\otimes n})^*$ so that if $T_i \in End(A), i = 1, 2, ..., n$, and $\phi_i \in A^*, u_i \in A, i = 1, 2, ..., n$, then with $T = T_1 \otimes ... \otimes T_n$,

$$f_T(\phi_1 \otimes ... \otimes \phi_n, u_1 \otimes ... \otimes u_n) = \Pi_{i=1}^n \phi_i(T_i u_i)$$

Suppose G is a linear group acting on A. Then this action induces an action on A^*. We also have a standard action of G induced on $A^{\otimes n}$ and on $A^{*\otimes n}$ and we get

$$f_T(g.(\phi_1 \otimes ... \otimes \phi_n), g.(u_1 \otimes ... \otimes u_n)) = f_T(g.\phi_1 \otimes ... \otimes g.\phi_n, g.u_1 \otimes ... \otimes g.u_n)$$

$$= \Pi_{i=1}^n (g.\phi_i)(T_i g u_i) = \Pi_{i=1}^n \phi_i(g^{-1} T_i g u_i)$$

and it follows that f_T is G-invariant iff

$$gTg^{-1} = T, g \in G$$

or equivalently, iff

$$g.T = T.g, g \in G$$

if iff $T \in End(A)^{\otimes n}$ is G-invariant. In this way, we obtain a linear one-one correspondence with G invariant elements of $End(A)^{\otimes n}$ and G-invariant elements of $(A^{*\otimes n} \otimes A^{\otimes n})^*$, or equivalently G-invariant multilinear functionals on $A^{*n} \times A^n$.

25. Let V be a vector space and let $\{e_1, e_2, ..., e_n\}$ be a basis for V. Then a basis for $V^{\otimes k}$ is given by $e_{i_1} \otimes ... \otimes e_{i_k}$ where $1 \leq i_1, ..., i_k \leq n$. Let σ be an element of S_k, ie, a permutation of the set $\{1, 2, ..., k\}$. Then fix $i_1, ..., i_k$ and assume that j appears h_j times in the sequence $\{i_1, ..., i_k\}$. Define the action of S_k on $V^{\otimes k}$ so that

$$\sigma(v_1 \otimes ... \otimes v_k) = v_{\sigma^{-1}1} \otimes ... \otimes v_{\sigma^{-1}k}$$

Consider the vector

$$v = \sum_{j=1}^n x_j e_j$$

We want to compute v^k. For $n = 2, k = 3$, this evaluates to

$$v^3 = (x_1e_1 + x_2e_2)^3 = x_1^3 e_1.e_1.e_1 + x_1^2 x_2(e_1.e_1.e_2 + e_1.e_2.e_1 + e_2.e_1.e_1) + x_1 x_2^2(e_1.e_2.e_2 + e_2.e_1.e_2 + e_2.e_2.e_1)$$
$$+ x_2^3(e_2.e_2.e_2)$$

We can write this as

$$\sum_{h_1+h_2=3} x_1^{h_1} x_2^{h_2} e(h_1, h_2)$$

where

$$e(h_1, h_2)$$

is the sum of all distinct tensor products of e_1, e_2 in which e_1 appears h_1 times and e_2 h_2 times. In general, we have

$$v^{\otimes n} = \sum_{h_1+...+h_n=k} x_1^{h_1} ... x_n^{h_n} e(h_1, ..., h_n)$$

We want to compute the symmetric vector

$$e_h = \sum_{\sigma \in S_k} \sigma.(e_{i_1} \otimes ... \otimes e_{i_k}), h = (h_1, ..., h_n), \sum_{j=1}^{n} h_j = k$$

Let $k = 3, n = 2, h_1 = 2, h_2 = 1$ and consider $e_1.e_1.e_2$ where dot denotes tensor product. The above symmetrizing operation gives

$$e_h = 2e_1.e_1.e_2 + 2e_1.e_2.e_1 + 2e_2.e_1.e_1 = h_1!h_2!e(h_1, h_2)$$

More generally, we have

$$e_h == h_1!...h_n!e(h_1, ..., h_n)$$

We define

$$e_1^{h_1}...e_n^{h_n} = e_h/k!$$

ie,

$$e_1^{h_1}...e_n^{h_n} = \frac{1}{k!} \sum_{\sigma \in S_k} e_{i_{\sigma 1}}...e_{i_{\sigma k}}$$

Then, we get

$$e(h_1, ..., h_n) = \frac{k!}{h_1!...h_n!} e_1^{h_1}...e_n^{h_n}$$

26.Tensor product: Let U, V, W be vector spaces all over the same field and $\phi : U \times V \to W$ be a bilinear map such that (1) $dimW = dimU.dimV$ and (2) the range of ϕ spans the whole of W. Then ϕ is called a tensor product. We claim that if this is the case and if S is any vector space and $h : U \times V \to S$ is any bilinear map, then there exists a unique linear map $T : W \to S$ such that $h = To\phi$. For let h be as above and let $e_1, ..., e_n$ be a basis for U and $f_1, ..., f_m$ a basis for V. Then we claim that $\{\phi(e_i, f_j)\}$ is a basis for W. Indeed, since $dimW = mn$ and since $\{\phi(e_i, f_j)\}$ span W, the claim is obvious. Given $w \in W$, there exists a unique set of scalars c_{ij} such that

$$w = \sum_{i,j} c_{ij}\phi(e_i, f_j)$$

and then define

$$T(w) = \sum c_{ij}h(e_i, f_j)$$

Then $T : W \to S$ is a well defined linear map. That T is well defined follows from the fact that $c_{ij}'s$ are uniquely determined by w. Linearity follows from the argument that if $w' \in W$ is another vector with

$$w' = \sum c'_{ij}\phi(e_i, f_j)$$

then

$$aw + bw' = \sum_{i,j} (ac_{ij} + bc'_{ij})\phi(e_i, f_j)$$

giving

$$T(aw+bw') = \sum(ac_{ij}+bc'_{ij})h(e_i,f_j) = a\sum c_{ij}h(e_i,f_j) + b\sum c'_{ij}h(e_i,f_j)$$
$$= aT(w) + bT(w')$$

Finally, let

$$u = \sum u_i e_i \in U, v = \sum v_i f_i \in W$$

Then,

$$To\phi(u,v) = T(\sum u_i v_j \phi(e_i,f_j)) = \sum u_i v_j h(e_i,f_j) = h(\sum u_i e_i, \sum v_j f_j) = h(u,v)$$

completing the proof.

27. Let U be a vector space and let $T(U)$ be the tensor algebra on U. In other words, $T(U) = \bigoplus_{n=0}^{\infty} T_n(U)$ where $T_0(U) = \mathbb{F}$ and $T_n(U)$ is the vector space spanned by all the homogeneous elements of degree n, ie, by elements of the form $u_1 \otimes ... \otimes u_n$ with $u_i \in U, i = 1,2,...,n$. We can multiply two elements of $T(U)$ in the obvious fashion, use distributivity of multiplication over addition combined with

$$(u_1 \otimes ... \otimes u_n) \otimes (v_1 \otimes ... \otimes v_m) = u_1 \otimes ... \otimes u_n \otimes v_1 \otimes ... \otimes v_m$$

Then we have the obvious equation

$$T_n(U)T_m(U) \subset T_{n+m}(U)$$

or in other words, $T(U)$ is a graded algebra. Suppose A is any algebra and let I denote the two sided ideal generated by all elements of the form $ab - ba$ with $a, b \in A$. This ideal consists of linear combinations of all elements of the form $c(ab - ba)d$ with a, b, c, d ranging over A. Consider the quotient algebra A/I. Any element here is of the form $a + I$ with $a \in A$. The product of two elements $a + I$ and $b + I$ is defined by $ab + I$. This is well defined since if $a' - a, b' - b \in I$, the $a' = a + e_1, b' = b + e_2$ for some $e_1, e_2 \in I$ and since I is a two sided ideal, it follows that $ae_2, e_1 b \in I$ and hence,

$$a'b' + I = (a + e_1)(b + e_2) + I = ab + ae_2 + e_1 b + e_1 e_2 + I = ab + I$$

Further, we claim that A/I is an Abelian algebra. For if $a, b \in A$, then

$$(a + I)(b + I) - (b + I)(a + I) = (ab - ba) + I = I$$

since $ab - ba \in I$. This tells us how to derive a commutative algebra from a non-commutative algebra by forming quotients.

28. Let U be a vector space and $T(U)$ the tensor algebra. Thus, $T(U) = \bigoplus_n T_n(U)$ where $T_n(U)$ is the vector space of homogeneous tensors of degree n. Let I be the two sided ideal in $T(U)$ generated by elements of the form $u \otimes v - v \otimes u$ with $u, v \in U$. Thus, any element in I can be expressed as linear combinations of elements of the form $t \otimes (u \otimes v - v \otimes u) \otimes t'$ where $t \in T_k(U), t' \in T_l(U), k, l = 0, 1, 2, ...$ Note that $T_0(U) = \mathbb{F}$. I is a homogeneous ideal since it can be expressed as $I = \bigoplus I \cap T_n(U)$. Note that $I \cap T_0(U) = 0$ and $I \cap T_1(U) = 0$. Set $I_n = I \cap T_n(U)$. Then we claim that $T(U)/I$ is isomoprhic as a vector space to $\bigoplus T_n(U)/I_n$. More generally, let V be a vector space and let $V = \bigoplus V_n$. Let W

be a vector space and $W = \bigoplus W_n$ with $W_n \subset V_n$. Then, we claim that V/W is naturally isomorphic to $\bigoplus_n(V_n/W_n)$. Let $x \in V$ and let $x = \sum x_n$ with $x_n \in V_n$. Define $T(x+W) = \sum_n(x_n + W_n)$. Then this defines the necessary isomorphism. Suppose $y - x \in W$. Then we can write $y - x = \sum z_n$ with $z_n \in W_n$ and $y = \sum_n(x_n + z_n)$ with $y_n = x_n + z_n \in V_n$. Thus, $T(y+W) = \sum_n(y_n + W_n) = \sum_n(x_n + z_n + W_n) = \sum_n(x_n + W_n)$ showing that T is well defined. T is oneone onto and linear, can be proved easily. We have thus, $T(U)/I = \bigoplus_n(T_n(U)/I_n)$. I_n is a subspace of homogeneous tensors of degree n and is also a subset of the ideal generated by commutators in $T(U)$. Thus, $T_n(U)/I_n$ can be identified with the space of symmetric tensors of degree n. Suppose for example $n = 2$. Take $u, v \in T_2(U)$. Then $(u + I_2)(v + I_2) = u \otimes v + I_2 = v \otimes u + I_2$ justifying our claim.

29.Let G be a group and let $f : G \to \mathbb{R}$ be a function such that the vector space spanned by all left translates of f is a finite dimensional vector space V. Choose a basis $u_1, ..., u_n$ for this space. Then for each $g, x \in G$, we have

$$f(g^{-1}x) = \sum_{i=1}^{n} c_i(g)u_i(x)$$

for some functions $c_1, ..., c_n$ on G. We claim that the subspace V of all functions on G is invariant under all left translates. To prove this, we need only show that for each $g, x \in G$, we can write

$$u_i(g^{-1}x) = \sum_{j=1}^{n} c_{ji}(g)u_j(x)$$

and hence corresponding to each f satisfying the above property, we obtain a finite dimensional representation of G. To prove the above statement, observe that the functions c_i can be chosen to be linearly independent functions for other wise, we may redefine the $u_i's$ in terms of appropriate linear combinations to get the desired linear independence. For example, suppose $n = 2$ and

$$f(g^{-1}x) = c_1(g)u_1(x) + c_2(g)u_2(x)$$

suppose the functions c_1, c_2 are linearly dependent, say $c_2(g) = \alpha.c_1(g)$. Then,

$$f(g^{-1}x) = c_1(g)(u_1(x) + \alpha.u_2(x))$$

and we can take $n = 1$ and replace u_1 by $u_1 + \alpha.u_2$. We now prove that V is invariant under left translates. Indeed, we have

$$f(g^{-1}h^{-1}x) = \sum_i c_i(hg)u_i(x)$$

on the one hand and

$$f(g^{-1}h^{-1}x) = \sum_i c_i(g)u_i(h^{-1}x)$$

so that

$$\sum_i c_i(g)u_i(h^{-1}x) = \sum_i c_i(hg)u_i(x)$$

since the functions c_i are linearly independent, we can choose $g_1, ..., g_n \in G$ such that the matrix $((c_i(g_j)))$ is non-singular. Hence the system of equations

$$\sum_i c_i(g_j)u_i(h^{-1}x) = \sum_i c_i(hg_j)u_i(x)$$

can be solved to express the functions $u_i(h^{-1}x)$ as linear combinations of the $u_i's$ and this proves that V is a left invariant vector space.

30.Lie algebras:} is a vector space with a bilinear map $[.,.] : \} \times \} \rightarrow \}$ such that $[Y, X] = -[X, Y]$,

$$[X, [Y, Z]] + [Y, [Z, X]] + [Z, [X, Y]] = 0$$

Let $\{X_1, ..., X_n\}$ be an ordered basis for }. Then there exist n^3 constants $c(m, k, p), m, k, p = 1, 2, ..., n$ such that

$$[X_m, X_k] = \sum_p c(m, k, p)X_p$$

We have

$$\sum_p c(m, k, p)X_p = [X_m, X_k] = -[X_k, X_m] = -\sum_p c(k, m, p)X_p$$

and hence

$$c(m, k, p) = -c(k, m, p)$$

Further the Jacobi identity gives

$$[[X_m, X_k], X_p] + [X_k, X_p], X_m] + [[X_p, X_m], X_k] = 0$$

so that

$$\sum_r [c(m, k, r)[X_r, X_p] + c(k, p, r)[X_r, X_m] + c(p, m, r)[X_r, X_k]] = 0$$

or

$$\sum_r [c(m, k, r)c(r, p, q) + c(k, p, r)c(r, m, q) + c(p, m, r)c(r, k, q)] = 0$$

The scalars $\{c(m, k, p)\}$ are called the structure constants of the Lie algebra }. The skew symmetry of the Lie bracket and the Jacobi identity show that these constants are not all independent. There exist relations amongst them.

31.Basic results on roots and root spaces:(Rank of a Lie algebra, regular and singular elements) (Ref.V.S.Varadarajan, Lie groups, Lie algebras and their Representations): Let } be a semisimple Lie algebra of dimension m and ⟨ a Cartan sub-algebra. We shall introduce the notion of a Cartan subalgebra first. Let z be a complex number and define

$$F(X, t) = det(t.I - ad(X)) = \sum_{i=0}^{m} (-1)^{m-i} p_i(X)t^i$$

so that $p_i's$ are all polynomials on $\}$

Remark:Let $X_1, ..., X_m$ be a basis for $\}$. Fix this basis. For $X \in \}$, set

$$X = t_1X_1 + t_2X_2 + ... + t_mX_m, t_1, ..., t_m \in \mathbb{C}$$

Then,

$$F(X,t) = \sum_{i=0}^{m-1} (-1)^{m-i} p_i(X) t^i = det(t.I - \sum_{k=1}^{m} t_k X_k)$$

The right hand side is a polynomial in the complex variables $t, t_1, ..., t_k$ and equating coefficients of same powers of t shows that $p_i(X)$ is a polynomial in the variables $\{t_1, ..., t_m\}$. This justifies saying that $p_i(X)$ is a polynomial in X. Putting $t = 0$ gives

$$(-1)^m detad(X) = (-1)^m p_0(X)$$

Now observe that $ad(X)(X) = 0$, ie, $ad(X)$ has zero as an eigenvalue and hence $p_0(X) = 0$, ie, $detad(X) = 0$ (Determinant of a linear operator is the product of its eigenvalues). Let $l \geq 0$ be the smallest integer r such that $p_r \neq 0$. r is called the rank of the Lie algebra $\}$. Thus, we can write

$$F(X,t) = \sum_{i=l}^{m} (-1)^{m-i} p_i(X) t^i$$

We write

$$l = rk(\})$$

Since $p_0 = 0$, it follows that $l \geq 1$. We claim that $l = m$ iff $\}$ nilpotent. To see this, first assume that $\}$ is nilpotent. Then, $ad(X)$ is nilpotent for all $X \in \}$ and hence all the eigenvalues of $ad(X)$ are zero This means that

$$F(X,t) = t^m$$

and hence $l = m$. Conversely, if $l = m$, then $F(X,t) = t^m$ and since every linear operator satisfies its characteristic equation, it follows that $ad(X)^m = 0$, proving that $ad(X)$ is nilpotent. Define $\eta = p_l$. We say that $X \in \}$ is regular if if $\eta(X) \neq 0$ and that X is singular if $\eta(X) = 0$. Let $\}'$ be the set of all regular elements of $\}$. Let α be any automorphism of $\}$. Then for any $X \in \}$ and $Y \in \}$, we have

$$[\alpha(X), \alpha(Y)] = \alpha([X,Y])$$

or

$$\alpha o ad(X) = ad(\alpha(X)) o \alpha$$

or

$$ad\alpha(X) = \alpha o ad(X) o \alpha^{-1}$$

It follows that since the determinant is invariant under a similarity transformation,

$$F(\alpha(X), t) = F(X,t)$$

and hence

$$p_i(\alpha(X)) = p_i(X), 0 \leq i \leq m$$

which states that the polynomials $\{p_i\}$ are all invariant under automorphisms of $\}$. $\}'$, the set of regular elements of $\}$ is an open set for the following reason. Suppose $l = rk(\}))$, so that $\eta = p_l$. Let $X \in \}'$. Then, $\eta(X) \neq 0$. Since η is a polynomial, it is continuous and hence there exists a nhood U of X in $\}$ such that $\eta(Y) \neq 0$ for all $Y \in U$. $\}'$ is invariant under the adjoint group of $\}$. Indeed, let $g \in G$ where G is the group associated to the Lie algebra $\}$. This means that G consists of all elements of the form $exp(X)$ with X varying over $\}$. The adjoint group of $\}$ is defined by the set of all transformations $Ad(g)$ as $g \in G$. Equivalently, the adjoint group of $\}$ consists of all linear transformations of $\}$ having the form $exp(ad(X))$ with X varying over $\}$. For any $X \in \}$, $exp(ad(X))$ is an automorphism of $\}$. If $\}$ is a subspace of $End(V)$ where V is a vector space, then

$$exp(ad(X))(Y) = exp(X)Yexp(-X)$$

Indeed, set

$$g(t) = exp(t.ad(X))$$

Then, on the one hand,

$$\frac{d}{dt}g(t)Y = ad(X)(g(t)Y) = [X, g(t)Y]$$

while on the other hand, if

$$F(t) = exp(tX)Yexp(-tX)$$

then

$$\frac{d}{dt}F(t) = XF(t) - F(t)X = [X, F(t)]$$

which porves that $F(t)$ and $g(t)Y$, both satisfy the same differential equation with same initial conditions and are hence equal. We want to prove that $exp(ad(X))$ is an automorphism of $\}$. Indeed, write

$$W(t) = [exp(t.ad(X))(Y), exp(t.ad(X))(Z)], Q(t) = exp(t.ad(X))[Y, Z]$$

Then,

$$Q'(t) = ad(X)(Q(t)), Q(0) = [Y, Z]$$

$$\frac{dW(t)}{dt} = \frac{d}{dt}[exp(t.ad(X))(Y), exp(t.ad(X))(Z)]$$

$$= [[ad(X)(exp(t.ad(X))(Y)), exp(t.ad(X))(Z)] + [exp(t.ad(X))(Y), ad(X)(exp(t.ad(X))(Z))]$$

$$= ad(X)(W(t)), W(0) = [Y, Z]$$

and this confirms that $W(t) = Q(t)$ and hence that $exp(ad(X)) : \} \to \}$ is a Lie algebra automorphism. For $X \in \}$, let $\nu(X)$ denote the multiplicity of zero as a root of $F(X,t)$. It is clear that $\nu(X) \geq l$. Indeed, this follows from

$$F(X,t) = \sum_{i=l}^{m} (-1)^{m-i} p_i(X) t^i$$

so that the multiplicity of zero as a root of $F(X,t)$ is $\geq l$. If $p_l(X) \neq 0$, ie, X is regular, then $\nu(X) = l = rk(\})$.

Example:Let V be a vector space over the complex field $\mathbb{C}$ such that $dimV = l+1$ and let $\} = sl(V)$, the space of all endomorphisms of V having trace zero. For $X \in \}$, let $\tilde{X} \in End(V^*)$ be defined by the equation

$$(\tilde{X}f)(v) = -f(Xv)$$

Consider the linear operator

$$X_{1,1} = X \otimes I + I \otimes \tilde{X}$$

acting on

$$V \otimes V^* = V_{1,1}$$

Let $\{c_1, ..., c_r\}$ be the set of distinct eigenvalues of X with multiplicities $\nu_1, ..., \nu_r$. Then the eigenvalues of $\tilde{X}$ (which is X^T) are $c_1, ..., c_r$ with multiplicities $\nu_1, ..., \nu_r$. Suppose we write the eigenvalues of X as $\lambda_i, i = 1, 2, ..., l+1$ multiplicities inclusive. Then the eigenvalues of $\tilde{X}$ are $-\lambda_1, ..., -\lambda_{l+1}$ multiplicities inclusive. It follows that the eigenvalues of $X_{1,1}$ are $\lambda_i - \lambda_j, 1 \leq i, j \leq l+1$ multiplicities inclusive. The zero eigenvalue occurs iff $\lambda_i = c_k = \lambda_j$ for some k. For a fixed k, this can happen in ν_k^2 number of ways. This implies that zero is an eigenvalue of $X_{1,1}$ of multiplicity $\sum_{k=1}^r \nu_k^2$.

Let $T \in gl(V)$. Then, the formula $T = T - tr(T)I + tr(T)I$ implies that $gl(V) = sl(V) \oplus \mathbb{C}.I$. Note that $gl(V)$ consists of all endomorphisms of V while $sl(V)$ consists of all endomorphisms of V having trace zero. Let $X \in \} = sl(V)$. Then $ad(X)$ leaves invariant $sl(V)$ as well as $\mathbb{C}.I$. We have $ad(X)(I) = 0$ and hence the multiplicity of zero as an eigenvalue of $ad(X)$ on $sl(V)$ is $\nu(X) = \sum_{k=1}^r \nu_k^2 - 1$. Now

$$\sum \nu_r^2 - 1 \geq \sum \nu_r - 1 = l$$

and hence

$$\nu(X) \geq l$$

with equality iff $\nu_k = 1$ for all k, ie, iff X has all eigenvalues distinct. It is easy to see that $l = rk(\})$. Indeed, suppose $\lambda_i, i = 1, 2, ..., l+1$ are all distinct eigenvalues of X. Then, the multiplicity of zero as an eigenvalue of $ad(X)$ on $gl(V)$ $l+1$ and hence the multiplicity of zero as an eigenvalue of $ad(X)$ on $\} = sl(V)$ is l. This proves that

$$F(X,t) = det(t.I - ad(X)) = t^l \Pi_{i \neq j}(t - (\lambda_i - \lambda_j))$$

and hence,

$$\eta(X) = p_l(X) = \Pi_{i \neq j}(\lambda_i - \lambda_j)$$

Remark: Let $X \in End(V)$. Then $X_{1,1} = X \otimes I + I \otimes \tilde{X}$ in $End(V \otimes V^*)$ is related to $ad(X)$ in $EndEnd(V)$ by a similarity transformation. Indeed, define a linear map

$$\beta : V \otimes V^* \rightarrow End(V)$$

so that

$$\beta(u \otimes f)(v) = f(v)u$$

β is then a linear map and in fact an isomorphism of vector spaces. We have

$$\beta(X_{1,1}(u \otimes f))(v) = \beta(Xu \otimes f + u \otimes \tilde{X}f)(v)$$

$$= f(v)Xu - f(Xv)u$$

On the other hand,

$$ad(X)(\beta(u \otimes f))v = X\beta(u \otimes f)(v) - \beta(u \otimes f)Xv$$

$$= X(f(v)u) - f(Xv)u = f(v)Xu - f(Xv)u$$

This shows that

$$\beta.X_{1,1}(u \otimes f) = ad(X)\beta(u \otimes f)$$

so that

$$ad(X) = \beta.X_{1,1}\beta^{-1}$$

proving that $X_{1,1}$ is related to $ad(X)$ via a similarity transformation.

32.Submersions and immersions: Let M, N be C^∞-manifolds with $dimM = m, dimN = n$ and let $\pi : M \to N$ be a C^∞ map such that $(d\pi)_x$ is surjective. Let $y = \pi(x)$. Let $(x_1, ..., x_m)$ be coordinates around x and $(y_1, ..., y_n)$ coordinates around y. Without loss of generality, we may assume that $x_i(x) = 0, y_i(y) = 0$ for all x, y. Choose C^∞ functions $F_1, ..., F_n$ around 0_m such that

$$y_j o\pi = F_j(x_1, ..., x_m), j = 1, 2, ..., n$$

$(d\pi)_x$ is surjective implies that $((\frac{\partial F_i}{\partial x_j}))$ is a matrix of size $n \times m$ having rank n at 0_m. It follows that this matrix has n linearly independent columns and without loss of generality, we may assume that the first n columns of this matrix are linearly independent so that $((\frac{\partial F_i}{\partial x_j}))_{1 \le i,j \le n}$ is a non-singular square matrix. Given numbers $y_1, ..., y_n$ in the vicinity of $y_i o\pi(x), i = 1, 2, ..., n$ and $x_{m+1}, ..., x_n$, since this matrix is nonsingular, it follows that we can invert the system of equations

$$y_i = F_i(x_1, ..., x_n, x_{n+1}, ..., x_m), i = 1, 2, ..., n$$

and hence obtain $(x_1, ..., x_n)$ as a function of $y_1, ..., y_n, x_{n+1}, ..., x_m$. It follows that since $x_1, ..., x_m$ is a coordinate system around x that $y_1 o\pi, y_2 o\pi, ..., y_n o\pi, x_{n+1}, ..., x_m$ is a system of coordinates around x in M. In particular, we have $m \ge n$ and that with respect to these local coordinates, π is a projection $(t_1, ..., t_m) \to (t_1, ..., t_m)$. In fact, we have, setting

$$t_i = y_i o\pi(x), i = 1, 2, ..., n, t_i = x_i(x), i = n + 1, ..., m$$

that

$$y_i = t_i, i = 1, 2, ..., n$$

proving the projection property. In particular, we have shown that if $(d\pi)_x$ is surjective, then $(d\pi)_{x'}$ is surjective for all x' in a nhood of x. That is,

$$M_1 = \{z \in M : (d\pi)_z surjective\}$$

is an open subset of M. Moreover, $\pi(M_1)$ is open in N. This follows from the following observation. Suppose $X_1, X_2, ..., X_r$ are topological spaces and $\pi_i : X_1 \times X_2 \times ... \times X_r \to X_i$ is the canonical projection.

Then by definition of product topology, $U \subset X_1 \times ... \times X_r$ is open iff $\pi_i(U)$ is open in X_i for every i. Another way to see this is that an open subset U of $X_1 \times .... \times X_r$ can be expressed as

$$U = \bigcup_{\alpha\in\Gamma} U_{1\alpha} \times U_{2\alpha} \times ... \times U_{r\alpha}$$

where $U_{i\alpha}$ is open in X_i for every i. It follows that

$$\pi_i(U) = \bigcup_{\alpha\in\Gamma} U_{i\alpha}$$

is open in X_i.

33.Let M, N be C^∞ manifolds and $\pi : M \to N$ a C^∞ map. Assume that for some positive integer p, $dim(d\pi)_x(TM_x) = p$ for all $x \in M$. Then $x \to ker(d\pi)_x$ is an involutive analytic system of tangent spaces on M. Further, for each $y \in M$, we can find a connected open set U containing y such that $\pi(U)$ is a regular p dimensional submanifold of N and that π is a submersion of U onto $\pi(U)$. Without loss of generality, we may assume that M, N are open nhoods of the origin in $\mathbb{R}^m$ and $\mathbb{R}^n$ respectively. Let π be the map

$$(t_1, ..., t_m) \to (u_1, ..., u_n)$$

with $u_i's$ analytic functions of $t_j's$. We may assume that $u_j(0_m) = 0_n$ for all j. The condition $dim(d\pi)_x$ $(TM_x) = p$ translates to the condition that the $n \times m$ matrix $((\frac{\partial u_i}{\partial x_j}))$ has rank p on M. We may assume without loss of generality that the submatrix $((\frac{\partial u_i}{\partial x_j}))_{1\le i\le p, 1\le j\le m}$ also has rank p. Let π_1 be the projection

$$(x_1, ..., x_n) \to (x_1, ..., x_p)$$

Then, set

$$\pi' = \pi_1 o \pi$$

This mapping is represented by

$$(x_1, ..., x_n) \to (u_1, ..., u_p)$$

and by the above remark, the Jacobian of this matrix has rank p for all $x \in M$, ie, $(d\pi')_x$ is surjective for all $x \in M$. Thus, $\pi'(M) = P$ is open in $\mathbb{R}^p$ and contains the origin. From the rank-nullity theorem, $Ker(d\pi')_x$ has dimension $m - p$ and moreover if X, Y are elements of $Ker(d\pi')_x$, then $(d\pi')_x(X) = (d\pi')_x(Y) = 0$ whence, $(d\pi')_x([X, Y]) = [(d\pi')_x(X), (d\pi')_x(Y)] = 0$, ie, $Ker(d\pi')_x, x \in M$ is an involutive analytic system of tangent spaces of rank $m - p$.

34.Definition of connected topological space: X is a topological space. $A \subset X$. A is said to be disconnected if there exist open sets U, V in X such that $A \cap U \cap V = \phi$ and $A \subset U \cup V$ but $A \cap U \neq \phi, A \cap V \neq \phi$.

35. **SL(2,C)**

Consider the Lie algebra $sl(2,\mathbb{C})$. Define the matrices

$$e = \begin{pmatrix} 0 & 1 \\ 0 & 0 \end{pmatrix}, f = \begin{pmatrix} 0 & 0 \\ 1 & 0 \end{pmatrix},$$

$$h = \begin{pmatrix} 1 & 0 \\ 0 & -1 \end{pmatrix}$$

Reflection: Let $\alpha \in V$ where V is a real inner product space. Define an operator $s_\alpha \in L(V)$ via the equation

$$s_\alpha(x) = x - 2(\alpha, x)\alpha/(\alpha, \alpha)$$

We have

$$s_\alpha(\alpha) = -\alpha$$

and if $(\alpha, x) = 0$, then

$$s_\alpha(x) = x$$

s_α is therefore the reflection about the hypeplane $\{x \in V : (\alpha, x) = 0\}$.

36.Problem: Define a semisimple and nilpotent operator on a finite dimensional vector space and give examples. Prove the Jordan decomposition theorem that gives a finite dimensional vector space V and a $T \in \mathcal{L}(V)$, there exist unique linear operators S, N in V such that $SN - NS = 0$ and $T = S + N$. By making use of the primary decomposition theorem, derive explicit formulae for S, N given a $T \in \mathcal{L}(V)$.

37.Problem: Let V be a finite dimensional vector space and $\mathcal{L}(V)$, the algebra of all linear transformations on V. Given a $T \in \mathcal{L}(V)$, let $T = S + N$ be its Jordan decomposition. Define the operator $ad(T) \in \mathcal{L}(\mathcal{L}(V))$ so that $ad(T)(A) = TA - AT = [T, A]$. Show that $ad(T) = ad(S) + ad(N)$ is the Jordan decomposition of $ad(T)$. Give an example.

38.Problem:Let L be a Lie algebra and for each $x \in L$, define the linear map $ad(x) : L \to L$ by the formula $ad(x)(y) = [x, y], y \in L$. Show that $x \to ad(x)$ is a Lie algebra representation. Assuming appropriate structure constants for a basis of L, determine the matrix of the operator $ad(x)$ relative to this basis.

39.Problem:Define a semisimple Lie algebra as a Lie algebra having no non-trivial solvable ideals. By saying that a Lie algebra is solvable means that there exists a finite integer n such that $D^n L = 0$ where

$$DL = [L, L], D^2 L = [DL, DL], ...$$

with $[L, L]$ denoting the vector space spanned by all elements in L of the form $[X, Y]$ where $X, Y \in L$.

Note: Suppose L is a solvable Lie algebra. Then we claim that $L = [L, L]$. Conversely if $[L, L] = L$, then L is solvable. Indeed, let m be the smallest positive integer for which $D^m L = 0$. Then $D^{m-1}DL = 0$ But if $DL = L$, then we get $D^{m-1}L = 0$, a contradiction to minimality of m. Hence DL is a proper subset of L. It follows that if $DL = L$, then L is not solvable. Conversely, suppose L is not solvable.

Then there exists an integer p such that $D^{p+1}L = D^pL \neq 0$. It follows that $D^pDL = D^{p-1}DL$. Suppose DL is a proper subspace of L. Then by the induction hypothesis on the dimension of L, it follows that $D^2L = DL$.

Remark: Let L_1, L_2 be a two Lie algebras and let $\pi : L_1 \to L_2$ be a homomorphism. If L_1 is solvable, then it is clear that $\pi(L_1)$ is solvable. Suppose I is a solvable ideal of L_1. Then, $\pi(I)$ is a solvable ideal of $\pi(L_1)$. Let L be a Lie algebra and let I be an ideal of L. Then, L/I is a Lie algebra with bracket defined by

$$[X+I, Y+I] = [X,Y] + I, X, Y \in L$$

That this operation is well defined follows from the fact that if $X_2 - X_1, Y_2 - Y_1 \in I$, then

$$[X_2, Y_2] + I = [X_1, Y_1] + I$$

since

$$[X_2, Y_1], [X_1, Y_2] \in I$$

I being an ideal.

40.Suppose L is a Lie algebra with $[L, L] = L$. Then a one dimensional representation is necessarily trivial. For if π is a one dimensional representation, then for any $X \in L$, we choose $Y, Z \in L$ with $X = [Y, Z]$. Then, $\pi(X) = [\pi(Y), \pi(Z)] = 0$ since $\pi(Y), \pi(Z)$ are scalars proving the claim.

41.Suppose L is a Lie algebra with $[L, L]$ a proper subspace of L. Then consider the Lie algebra $L' = L/[L, L]$. L' is an Abelian Lie algebra. Let $\pi : L \to L'$ denote the canonical projection. We choose a basis $X_1, ..., X_k$ for L' and define a representation ρ of L' such that $\rho(X_1) = T, \rho(X_i) = 0, i = 2, 3, ..., k$ where T is not a semisimple operator. Then ρ is not a semisimple representation of L', ie it is not completely reducible. Define a representation ν of L such that

$$\nu(X) = \rho(\pi(X))$$

Then ν is not semisimple, ie, it is not completely reducible. For suppose W is a subspace invariant under ρ with no complementary invariant subspace. The subspace $\pi^{-1}(W)$ is then invariant under ν. Suppose W' is a complementary invariant subspace under ν. Then it is clear that $\pi(W')$ is complementary to W and invariant under ρ, a contradiction. We are done.

42.Root systems:Let E be a finite dimensional real vector space with a scalar product (u, v) and let Φ be a set of non-zero vectors in E. We say that Φ is a reduced root system if (a) $span(\Phi) = E$, (b) $\alpha \in \Phi$ and $c \in \mathbb{R}$ with $c\alpha \in \Phi$ implies that $c = \pm 1$, (c) $2(\alpha, \beta)/(\beta, \beta)$ are integers for all $\alpha, \beta \in \Phi$.

43.Problem: Let the equation of a plane in $\mathbb{R}^n$ be given by $\alpha(x) = 0$ where α is a linear functional on $\mathbb{R}^n$. Determine an expression for the reflection operator about this plane.

44.Group theory, DSP implementation

1.$sl(2, \mathbb{C})$. The group is $SL(2, \mathbb{C})$ which consists of all 2×2 complex matrices having unit determinant, ie, all matrices of the form

$$\begin{pmatrix} a & b \\ c & d \end{pmatrix}, ad - bc = 1$$

The Lie algebra is given by $sl(2,\mathbb{C})$ and consists of all 2×2 complex matrices having trace zero. Any element of $sl(2,\mathbb{C})$ can be expressed as

$$\mathbf{A} = \begin{pmatrix} a & b \\ c & -a \end{pmatrix}, a, b, c \in \mathbb{C}$$

Note that we can write,

$$\mathbf{A} = a\mathbf{H} + b\mathbf{X} + c\mathbf{Y}$$

where

$$\mathbf{H} = \begin{pmatrix} 1 & 0 \\ 0 & -1 \end{pmatrix}, \mathbf{X} = \begin{pmatrix} 0 & 1 \\ 0 & 0 \end{pmatrix},$$

$$\mathbf{Y} = \begin{pmatrix} 0 & 0 \\ 1 & 0 \end{pmatrix}$$

The commutation relations can be written down:

$$[H, X] = 2X, [H, Y] = -2Y, [X, Y] = H$$

Suppose $\pi : sl(2,\mathbb{C}) \to M(V)$ is an irreducible representation of $sl(2,\mathbb{C})$ in a finite dimensional vector space V. We then have the commutation relations

$$[\pi(H), \pi(X)] = \pi([H, X]) = 2\pi(X),$$

$$[\pi(H), \pi(Y)] = \pi([H, Y]) = -2\pi(Y),$$

$$[\pi(X), \pi(Y)] = \pi([X, Y]) = \pi(H)$$

Let V be finite dimensional. Then $\pi(H)$ has an eigenvector in V, say k. We write

$$\pi(H).v = k.v$$

Then,

$$\pi(H)\pi(X)v = \pi([H, X])v + \pi(X)\pi(H)v = \pi(2X)v + k\pi(X)v = (k+2)\pi(X)v$$

$$\pi(H)\pi(Y)v = \pi([H, Y])v + \pi(Y)\pi(H)v = -2\pi(Y)v + \pi(Y)k.v = (k-2)\pi(Y)v$$

This implies that $\pi(X)$ raises the eigenvalue of an eigenvector of $\pi(H)$ by two while $\pi(Y)$ lowers the eigenvalue of an eigenvector of $\pi(H)$ by two. In particular, it follows that since V is finite dimensional, say $dim V = n$, there exist finite positive integers a, b such that $\pi(X)^a v = 0 = \pi(Y)^b v$. Let $w = \pi(X)^{a-1}v$. Then, w is an eigenvector of $\pi(H)$ having maximum eigenvalue, say p. It follows that $\pi(Y)^r w$ for $r = 1, 2, ...$, if nonzero, are eigenvectors of $\pi(H)$ with eigenvalues $k - 2r, r = 1, 2, ...$ respectively. There is a smallest integer r such that $\pi(Y)^{r+1}w = 0$. Then, $\{\pi(Y)^m w, m = 0, 1, 2, ..., r\}$ forms a basis for V by irreducibility. It follows that $n = r + 1$. It is now easy to see that $p = n - 1$. Suppose for example, $n = 2$. Then, $\pi(H)w = pw$, $\pi(H)\pi(Y)w = (p-2)\pi(Y)w$,

$$\pi(Y)^2 w = 0$$

implies

$$\pi(X)\pi(Y)^2 w = 0$$

implies

$$(\pi([X,Y])\pi(Y)+\pi(Y)\pi([X,Y]))w=0$$

since $\pi(X)w=0$, giving

$$\pi(H)\pi(Y)w+\pi(Y)\pi(H)w=0$$

or

$$(2p-2)\pi(Y)w=0$$

so that $p=1$. Suppose now that $n=2$. Then a basis for V is given by $\{w,\pi(Y)w,\pi(Y)^2w\}$. Now, $\pi(Y)^3w=0,\pi(X)w=0$ so that

$$[\pi(X),\pi(Y)^3]w=0$$

or

$$(\pi([X,Y])\pi(Y)^2+\pi(Y)\pi([X,Y])\pi(Y)+\pi(Y)^2\pi([X,Y]))w=0$$

or

$$\pi(H)\pi(Y)^2w+\pi(Y)\pi(H)\pi(Y)w+\pi(Y)^2\pi(H)w=0$$

giving

$$(p-4)\pi(Y)^2w+(p-2)\pi(Y)^2w+p\pi(Y)^2w=0$$

or

$$(3p-6)\pi(Y)^2w=0$$

giving $p=2$. The argument for general p is now straightforward. We have a basis $\{\pi(Y)^m w, m=0,1,2,...,n-1\}$ for V with $\pi(H)w=pw$ and more generally,

$$\pi(H)\pi(Y)^m w=(p-2m)\pi(Y)^m w, m=0,1,...,n-1$$

Then,

$$[\pi(X),\pi(Y)^n]w=0$$

giving

$$([\pi(X),\pi(Y)]\pi(Y)^{n-1}+\pi(Y)[\pi(X),\pi(Y)]\pi(Y)^{n-2}+...+\pi(Y)^{n-1}[\pi(X),\pi(Y)])w=0$$

so that

$$p-2(n-1)+p-2(n-2)+...+p=0$$

or

$$np-n(n-1)=0$$

or

$$p=n-1$$

proving the claim. This result suggests to us how to construct irreducible representations of $sl(2,\mathbb{C})$

45.Invariant subspaces: Let V be a vector space over the complex field $\mathbb{C}$ and assume that W_1, W_+2 are two subspaces of V such that $V=W_1\oplus W_2$. This means that any vector $x\in V$ can be uniquely expressed as $x=x_1+x_2$ with $x_i\in W_i, i=1,2$. An equivalent condition is that $V=W_1+W_2$ and

$W_1 \cap W_2 = (0)$, ie, any vector $x \in V$ can be expressed as $x = x_1 + x_2$ with $x_i \in W_i, i = 1, 2$ and $W_1 \cap W_2 = 0$. Let $T : V \to V$ be a linear transformation that leaves both the $W_i's$ invariant, ie, $T(W_i) \subset W_i, i = 1, 2$. Then choose a basis B_i for $W_i, i = 1, 2$. It follows that $B = B_1 \cup B_2$ is a basis for V and that if $A_i = [T_i|_{W_i}]_{B_i}, i = 1, 2$, then we have

$$[T]_B = \begin{pmatrix} \mathbf{A}_1 & \mathbf{0} \\ \mathbf{0} & \mathbf{A}_2 \end{pmatrix}$$

46.Module over a ring: Let R be a ring, ie, R is a set equipped with two binary operations, addition and multiplication. R has a zero element as well as a unity element. Under addition, R is an Abelian group with zero as the identity. Under multiplication, R is a semigroup and multiplication may be non-commutative. For example, let R be the set of all $n \times n$ matrices over a field $\mathbb{F}$. Then, R is a non-commutative ring. Note that in R, the non-singular matrices are invertible while the singular matrices are not invertible. The identity matrix is the identity element of R as a semigroup and the zero matrix is the identity element of R as an additive group. Another example of a ring is the set of all polynomials in n complex variables. The zero element of the additive group here is the zero polynomial while the unit polynomial is the multiplicative identity element. Let R be a ring and V a set which we call a module such that there exists a binary operation on V called addition under which V is closed and there exists a mapping $R \times V \to V$ such that $r(v_1 + v_2) = rv_1 + rv_2$ for all $r \in R, v_i \in V, i = 1, 2$, $0.v = 0$ and $(r_1 + r_2)v = r_1 v + r_2 v$, $r_1(r_2 v) = (r_1 r_2)v, 1.v = v$. Then, V is said to be a module over the ring R. For example, let V be a vector space over a field $\mathbb{F}$ and R the set of all linear transformations acting on V. Then V becomes a module over R. Let V be a module over the ring R and let $T : V \to V$ be an R-linear transformation, ie $T(r_1 v_1 + r_2 v_2) = r_1 T(v_1) + r_2 T(v_2)$ for all $r_i \in R$ and $v_i \in V$ $i = 1, 2$. For example, if R is an algebra of linear transformations acting on a vector space V, then a linear map $T : V \to V$ is an R-module iff T commutes with every element of R. Let V be a module over a ring R and let $W \subset V$. We say that W is a submodule of V if $rW \subset W$ for all $r \in R$. We say that the module (R, V) is irreducible if it has no trivial proper submodule, ie, if W is a submodule of V, then either $W = (0)$ or else $W = V$. Suppose (R, V) is an irreducible module. Then, we claim that V is cyclic. Specifically, if v is any non-zero element of V, then $\{r.v : r \in R\} = V$. Indeed the set $W = \{r.v : r \in R\}$ is a nonzero submodule of V. Nonzero because $1.v = v$ and submodule because if $s, r \in R$ are arbitrary, then $s.(r.v) = (s.r).v$ for all $s, r \in R$. This implies $s.W \subset W$ for all $s \in R$. Suppose G is a group and T is a representation of G in a vector space V, ie, $T : G \to GL(V)$ is a homomorphism where $GL(V)$ is the general linear group (all nonsingular linear transformations acting on V). By homomorphism, we mean that $T(g_2 g_1) = T(g_2)T(g_1)$ for all $g_1, g_2 \in G$ and since $T(g)$ is non-singular for all g, it follows that $T(e) = I$. Let $a : G \to \mathbb{C}$ be any function on the group such that a vanishes at all but a finite number of points. Then we can form the element

$$T(a) = \sum_g a(g)T(g)$$

This is an element of $\mathcal{L}(V)$, the algebra of all linear transformations on V. As a ranges over all functions on G that vanish at all but a finite number of points, $T(a)$ ranges over a subalgebra of $\mathcal{L}(V)$. Indeed, can see that if a, b are any two such functions, then

$$T(a) + T(b) = \sum_g a(g)T(g) + \sum_g b(g)T(g) = \sum_g (a(g) + b(g))T(g) = T(a + b)$$

if c is a scalar, then

$$T(c.a) = \sum_g ca(g)T(g) = c\sum_g a(g)T = c.T(a)$$

and finally, if a, b are two such functions, then

$$T(a)T(b) = \sum_g a(g)T(g)\sum_h b(h)T(h) = \sum_{g,h} a(g)b(h)T(gh) = \sum_g (a*b)(g)T(g) = T(a*b)$$

where

$$a*b(g) = \sum_{h_1h_2=g} a(h_1)b(h_2) = \sum_h a(h)b(h^{-1}g) = \sum_h a(gh^{-1})b(h)$$

is the convolution of the functions a, b. The algebra of all operators in V of the form $\sum_g a(g)T(g)$ is denoted by $\mathcal{A}(G)$. We use the notation

$$T(a) = \sum_g a(g).g$$

The set of all such transformations in V, namely $\mathcal{A}(G)$ is called the group algebra of G. Note the notation used. For an element $g \in G$ and $v \in V$, $T(g)v$ is written as $g.v$ and $T(a)v$ is written as $a.v$ The group algebra $\mathcal{A}$ acts on V and under this action, V becomes a module over $\mathcal{A}(G)$. Suppose T is an irreducible representation of G. Then $(\mathcal{A}, V)$ becomes an irreducible module. Indeed, suppose W is a submodule of V. Then $a.W \subset W$ for all $a \in \mathcal{A}$. It follows on taking $a = \delta_g$, ie, $a(h) - \delta_{g,h}$ that $T(a) = T(g)$ and hence $T(g)(W) \subset W$ for all $g \in G$ which means that W is invariant under the representation T of G. If T irreducible, then $W = (0)$ or else $W = V$ and this proves the claim. Conversely, if $(\mathcal{A}, V)$ is an irreducible module, then T is an irreducible representation of G since if W is $T(g)$ invariant for all $g \in G$, then for any $a : G \to \mathbb{C}$ that vanishes at all but a finite number of elements, we have $T(a)W = \sum_g a(g)T(g)(W) \subset W$.

47.Consider a semisimple Lie algebra } and assume ⟨ is its Cartan subalgebra. Let Δ denote the set of roots. Thus ⟨ is a maximal Abelian algebra and Δ is a finite subset of ⟨* with the property that if $X \in g_\alpha$ iff

$$[H, X] = \alpha(H)X, H \in \langle$$

or equivalently,

$$ad(H)(X) = \alpha(H)X, H \in \langle$$

$\}_\alpha$ is called the root space corresponding the the root α. If $\alpha \in \Delta$, then it can be shown that $dim\}_\alpha = 1$ while if $\alpha \notin \Delta$, then obviously $\}_\alpha = \{0\}$. We have the direct sum decomposition,

$$\} = \langle \oplus \bigoplus_{\alpha\in\Delta} \}_\alpha$$

It can be shown that $dim\}_\alpha = 1$ for all $\alpha \in \Delta$. In fact, let $H, H' \in \langle$. Then $ad(H)ad(H')(X) = \alpha(H)\alpha(H')X$ for all $X \in \}_\alpha$ while $ad(H)ad(H')(\langle) = 0$. It follows that

$$Tr(ad(H)ad(H')) = \sum_{\alpha\in\Delta} dim(\}_\alpha)\alpha(H)\alpha(H')$$

Now let $\alpha \in \Delta$ and consider the subspace

$$V = \mathbb{C}.Y + \langle + \bigoplus_{k\geq 1} \}_{k\alpha}$$

of $\}$. We have seen that if $X \in \}_\alpha$ and $Y \in \}_{-\alpha}$, then $[X,Y] =< X,Y > H_\alpha$ where $H_\alpha \in \langle$ is the unique element for which $< H_\alpha, H >= \alpha(H)$ for all $H \in \langle$. Since $< .,. >$ is nonsingular on $\}_\alpha \times \}_{-\alpha}$, it follows that X,Y can be chosen so that $[X,Y] = H_\alpha$. It follows that $ad(H_\alpha) = [ad(X), ad(Y)]$ and hence if W is any subspace of $\}$ that is invariant under both $ad(X)$ and $ad(Y)$, then it is invariant under also $ad(H_\alpha)$ and this implies that

$$Tr(ad(H_\alpha)|W) = Tr([ad(X), ad(Y)]|_W) = 0$$

But clearly V is invariant under both $ad(X), ad(Y)$. Indeed,

$$ad(X)(H) = -\alpha(H)X \in \}_\alpha, H \in \langle,$$

$$ad(X)(Y) = [X,Y] = H_\alpha \in \langle,$$

$$ad(X)(Z) = [X,Z] \in \}_{(k+1)\alpha} \forall Z \in \}_{k\alpha}$$

These three identities prove that V is invariant under $ad(X)$. likewise,

$$ad(Y)(H) = [Y,H] = \alpha(H)Y,$$

$$ad(Y)(Y) = 0,$$

$$ad(Y)(Z) = [Y,Z] \in \}_{(k-1)\alpha}, Z \in \}_{k\alpha}, k \geq 1$$

where $\}_0 = \langle$. Thus,

$$Tr[ad(H_\alpha)|_V] = 0$$

On the other hand,

$$ad(H_\alpha)(H) = 0, ad(H_\alpha)(Y) = -\alpha(H)Y, ad(H_\alpha)(Z) = k\alpha(H_\alpha)Z, Z \in \}_{k\alpha}, H \in \langle$$

It follows that

$$0 = Tr(ad(H_\alpha)|_V) = -\alpha(H_\alpha) + \sum_{k=1}^{\infty} k\alpha(H_\alpha)dim(\}_{k\alpha})$$

and since $\alpha(H_\alpha) \neq 0$, it follows that

$$dim\}_{k\alpha} = 0, k > 1, dim\}_\alpha = 1$$

48.Some important theorems in Lie algebra theory:

1.Let $\}$ be a Lie algebra, ρ a representation of $\}$ in a finite dimensional vector space V and for any $X,Y \in \}$, let $B(X,Y) = Tr(\rho(X).\rho(Y))$. Define

$$\surd = \{X \in \} : B(X,Y) = 0 \forall Y \in \}\}$$

Then $\surd$ is an ideal of $\}$. Indeed, let $X \in \surd$ and $Z, Y \in \}$. Then we must prove that $B([Z, X], Y) = 0$. But,

$$B([Z,X],Y) = Tr(\rho([Z,X])\rho(Y)) = Tr((\rho(Z)\rho(X) - \rho(X)\rho(Z))\rho(Y))$$

$$= Tr(\rho(X)(\rho(Y)\rho(Z) - \rho(Z)\rho(Y))) = Tr(\rho(X).\rho([Y,Z]) = B(X,[Y,Z]) = 0$$

implying that $[Z, X] \in \surd$. We now claim that $\rho(X)$ is nilpotent for all $X \in [\surd, \}]$. Let $X \in [\surd, \}]$. To prove that $\rho(X)$ is nilpotent, it suffices to show that for each replica R of $\rho(X)$, we have $Tr(\rho(X).R) = 0$ (V.S.Varadarajan, Lie groups, Lie algebras and their representations). Let R be a replica of $\rho(X)$. Then, adR is a replica of $ad\rho(X)$. Hence, adR is a polynomial in $ad\rho(X)$. Since $ad\rho(X)$ leaves $\rho[\}]$ invariant, it follows that adR leaves $\rho[\}]$ invariant. Let $X \in [\surd, \}]$. Then,

$$X = \sum_i [Y_i, Z_i], Y_i \in \surd, Z_i \in \}$$

Since adR leaves $\rho[\}]$ invariant, it follows that there exist $U_i \in \}$ such that

$$[\rho(Z_i), R] = \rho(U_i)$$

Then,

$$Tr(\rho(X).R) = \sum_i Tr(\rho([Y_i, Z_i])R) = \sum_i Tr((\rho(Y_i)\rho(Z_i) - \rho(Z_i)\rho(Y_i))R)$$

$$= \sum_i Tr(\rho(Y_i)[\rho(Z_i), R]) = \sum_i Tr(\rho(Y_i)\rho(U_i)) = \sum_i B(Y_i, U_i) = 0$$

proving that $\rho(X)$ is nilpotent.

2.Let $\}$ be a semisimple Lie algebra. If $\langle$ is an ideal of $\}$, then $\langle^\perp$ is also an ideal, $[\langle, \langle^\perp] = 0$ and $\} = \langle \oplus \langle^\perp$

Let $X \in \}, Y \in \langle^\perp$. We want to show that $[X, Y] \in \langle^\perp$ or equivalently that $< [X, Y], Z >= 0$ for all $Z \in \langle$. By associativity of the Cartan-Killing form, we have

$$< [X, Y], Z >= - < [Y, X], Z >= - < Y, [X, Z] >= 0$$

since $[X, Z] \in \langle$ by definition of ideal. Suppose $\langle \cap \langle^\perp \neq 0$. Then, choose any two non-zero elements X, X' in this intersection and we then have

$$< X, X' >= 0$$

implying that $\langle \cap \langle^\perp$ is solvable. (Cartan's criterion for solvability of a Lie algebra). This contradicts the semisimplicity of $\}$. ($\langle \cap \langle^\perp$ is an ideal, it being the intersection of two ideals and the definition of semisimplicity of a Lie algebra means that it does not have any solvable ideal). Since $< .,. >$ is a non-degenerate bilinear form, it follows that

$$dim\} = dim\langle + dim\langle^\perp$$

In fact, if $dim\langle = m$ and $dim\} = n$, then we choose a basis $\{H_1, ..., H_m\}$ for $\langle$. It follows then that the elements $< H_i, . >, i = 1, 2, ..., m$ of $\langle^*$ are linearly independent because of non-degeneracy of the

Cartan-Killing form and hence the dimension of the space $\{H \in \langle :< H_i, H >= 0, i = 1, 2, ..., m\}$ is exactly $n - m$ (use the rank-nullity theorem). It follows that This space is precisely $\langle^{\perp}$ and hence we are through. The above dimension equation in conjunction with the equation $\langle \cap \langle^{\perp}$ imply that

$$\} = \langle \oplus \langle^{\perp}$$

Since now both $\langle$ and $\langle^{\perp}$ are ideals, it follows that

$$[\langle, \langle^{\perp}] \subset \langle \cap \langle^{\perp} = 0$$

$\langle, \langle^{\perp}$ are Lie algebras (they being ideals) and hence $\langle \times \langle^{\perp}$ is also a Lie algebra. The Lie bracket on this algebra is defined as usual

$$[(X, X'), (Y, Y')] = ([X, Y], [X', Y'])$$

Define the map

$$\Phi : \langle \times \langle^{\perp} \rightarrow \}$$

so that

$$\Phi(X, X') = X + X', X \in \langle, X' \in \langle^{\perp}$$

Then, Φ is a Lie algebra isomorphism. That Φ is linear and one-one is immediate. One-one follows from $\langle \cap \langle^{\perp} = 0$. That Φ is onto follows from $\} = \langle + \langle^{\perp}$. That Φ preserves the Lie bracket follows from

$$\Phi([(X, X'), (Y, Y')]) = \Phi(([X, Y], [X', Y'])) = [X, Y] + [X', Y']$$

on the one hand, while on the other,

$$[\Phi(X, X'), \Phi(Y, Y')] = [X + X', Y + Y'] = [X, Y] + [X', Y']$$

since $[\langle, \langle^{\perp}] = 0$. We now claim that $\langle$ is semisimple. We shall show that $rad\langle$ is an ideal of $\}$. Let D be a derivation of $\}$. $exp(tD)$ is then an automorphism of $\}$. $exp(tD)$ therefore leaves $rad\}$ invariant. (If α is an automorphism of $\}$, then $\alpha(rad\})$ is also a solvable ideal of $\}$. It follows then that $\alpha(rad\}) \subset rad\}$. Replacing α by α^{-1} gives the reverse containment. Thus, $\alpha(rad\}) = rad\}$. In particular, taking $D = adX$, we see that $exp(t.ad(X))$ leaves $rad\langle$ invariant for every $t \in \mathbb{R}$ and every $X \in \}$. (adX is a derivation of $\langle$ since $\langle$ is an ideal in $\}$.) Thus, $rad\langle$ is an ideal of $\}$. But, $\}$ is semisimple. Hence, $rad\langle = 0$, ie $\langle$ is semisimple. Likewise, $\langle^{\perp}$ is also semisimple. It follows that since $\}/\langle$ is isomorphic to $\langle^{\perp}$, $\}/\langle$ is also semisimple.

49.Representation of the universal enveloping algebra of a Lie algebra: Let $\}$ be a Lie algebra and $U(\})$ its universal enveloping algebra. If $X, Y \in \}$, then viewed as elements of $U(\})$, we have that

$$X \otimes Y - Y \otimes X = XY - YX$$

Let π be a representation of $\}$ in a vector space $\}$. Then, π extends to a representation of $U(\})$ in the natural way: π is linear and if $X_1, ..., X_k \in \}$, then $X_1 ... X_k \in U(\})$ and

$$\pi(X_1 ... X_k) = \pi(X_1) ... \pi(X_k)$$

as a linear operator in V. We have the obvious equation $\pi(ab) = \pi(a)\pi(b), a, b \in U(\})$. Suppose π is a representation of $U(\})$ in V and that $v \in V$ is a cyclic vector for π, ie,

$$V = \{\pi(a)v : a \in U(\})\}$$

Define

$$\mathcal{M} = \{a \in U(\}) : \pi(a)v = 0\}$$

That $\mathcal{M}$ is an ideal follows from the fact that if $a \in \mathcal{M}, b \in U(\})$, then

$$\pi(ba)v = \pi(b)\pi(a)v = 0$$

so that $ba \in \mathcal{M}$, ie,

$$U(\}).\mathcal{M} \subset \mathcal{M}$$

Consider the quotient $Q = U(\})/\mathcal{M}$ as a vector space and define a map

$$\Phi : Q \to V$$

so that

$$\Phi(x + \mathcal{M}) = \pi(x)v, x \in U(\})$$

Then Φ is a well defined linear map since $\pi(x)v = 0$ implies $x \in \mathcal{M}$ implies $x + \mathcal{M} = \mathcal{M}$. Φ is surjective since v is π-cyclic. Φ is obviously injective. In other words, Φ is a vector space isomorphism. Define a representation ρ of $U(\})$ in Q by the equation

$$\rho(x)(y + \mathcal{M}) = xy + \mathcal{M}$$

Then we have,

$$\Phi o\rho(x)(y + \mathcal{M}) = \pi(xy)v = \pi(x)\pi(y)v = \pi(x)\Phi(y + \mathcal{M})$$

Thus,

$$\Phi\rho\Phi^{-1} = \pi$$

proving that the representation π is equivalent to ρ. Let now π be an irreducible representation of $U(\})$ in a vector space V. Choose a $v \in V$. Then, v is cyclic. Define

$$\mathcal{M} = \{a \in U(\}) : \pi(a)v = 0\}$$

We've seen that $\mathcal{M}$ is an ideal in $U(\})$ and that the natural representation of $U(\})$ in $U(\})/\mathcal{M}$ is equivalent to π. Suppose I is an ideal containing $\mathcal{M}$ and suppose $I \neq \mathcal{M}$. Then choose $a \in I - \mathcal{M}$. Define

$$U = \{\pi(a)v : a \in I\}$$

Then U is π-invariant and hence $U = V$. Let $a \in U(\})$ be arbitrary. Then, $\pi(a)v \in V = U$ and hence there exists $b \in I$ such that $\pi(a)v = \pi(b)v$, ie, $\pi(a - b)v = 0$, or $a - b \in \mathcal{M} \subset I$. Thus, $a \in I + b = I$ and hence $U(\}) = I$. In other words, $\mathcal{M}$ is a maximal ideal in $U(\})$. This argument shows that corresponding to each irreducible representation of $U(\})$ is a maximal ideal and that the representation can be realized as the quotient of the maximal ideal. Conversely, let $\mathcal{M}$ be a maximal ideal in $U(\})$. Consider the natural representation ρ of $U(\})$ in $U(\})/\mathcal{M}$. We claim that this representation is irreducible. Indeed, suppose

W is a nonzero invariant subspace of $U(\})/\mathcal{M}$. Then choose any nonzero element in W. It has the form $a+\mathcal{M}$ for some $a \notin \mathcal{M}$. Let $\mathcal{N}$ be the ideal generated by a and $\mathcal{M}$. Then $\mathcal{N}/\mathcal{M}$ is a subspace of W. Since $\mathcal{N}$ is an ideal properly containing $\mathcal{M}$, maximality of $\mathcal{M}$ implies that $\mathcal{N} = U(\})$ and hence that $W = U(\})/\mathcal{M}$.

Let π now be a representation of $U(\})$ in a vector space V and let v be a cyclic vector. Let U be a π-invariant subspace of V that does not contain v. Define

$$\mathcal{M}(U) = \{a \in U(\}) : \pi(a)v \in U\}$$

Then $\mathcal{M}(U)$ is an ideal. Since v is a cyclic vector, it follows that

$$U = \{\pi(a)v : a \in \mathcal{M}(U)\}$$

Let

$$\mathcal{M}_0 = \{a \in U(\}) : \pi(a)v = 0\}$$

Then $\mathcal{M}_0$ is an ideal and $\mathcal{M}_0 \subset \mathcal{M}(U)$. Since $v \notin U$, it follows that $1 \notin \mathcal{M}(U)$ and hence $\mathcal{M}(U)$ is a proper ideal of $U(\})$. Conversely, let $\mathcal{M}$ be a proper ideal of $U(\})$ and define

$$U = \{\pi(a)v : a \in \mathcal{M}\}$$

Then U is a proper π-invariant subspace of V. It is proper since $1 \notin \mathcal{M}$ and hence $v \notin U$. Further, $\mathcal{M} = \mathcal{M}(U)$. For let $a \in \mathcal{M}$. Then, $\pi(a)v \in U$ and hence $a \in \mathcal{M}(U)$. Conversely, let $a \in \mathcal{M}(U)$. Then, $\pi(a)v \in U$ and hence $\pi(a)v = \pi(b)v$ for some $b \in \mathcal{M}$. It follows that $a - b \in \mathcal{M}_0 \subset \mathcal{M}$ and hence $a \in \mathcal{M}$. The map $U \to M(U)$ is a bijection between all π-invariant subspaces of V not containing v and the set of all proper ideals of $U(\})$ containing $\mathcal{M}_0$.

50.Let X, Y be vector fields on a manifold. f is a function defined on the manifold. Then the Lie derivative $L_X Y$ of Y along the direction of X is defined so that

$$(L_X Y)(f) + Y(L_X(f)) = L_X(Y(f))$$

ie something like the Leibniz rule is satisfied. For functions, we have $L_X(f) = X(f)$. It follows therefore that

$$(L_X Y)(f) = (XY - YX)(f) = [X,Y](f)$$

51.Casimir element of a Lie algebra:$\}$ is a semisimple Lie algebra and $\}^*$ is the space of all linear functionals on $\}$. $< .,. >$ denotes the Cartan-Killing form on $\}$. Thus,

$$< X,Y >= Tr(ad(X).ad(Y))$$

This form is nonsingular since $\}$ is semisimple. For $f \in \}^*$, define $\tilde{f} \in \}$ so that

$$< \tilde{f}, X >= f(X), X \in \}$$

This is a valid definition since the Cartan-Killing form is non-singular. For a given endomorphism L of }, define the endomorphism $\tilde{L}$ of }* by the equation

$$(\tilde{L}f)(X) = -f(LX)$$

In other words,

$$\tilde{L} = -L^T$$

Let $f \in \}^*$. Let $X \in \}$ and set $D_X = ad(X)$. Then we have

$$< D_X\tilde{f}, Y >=< ad(X)\tilde{f}, Y >=< [X, \tilde{f}], Y >= - < [\tilde{f}, X], Y >= - < \tilde{f}, [X, Y] >= -f([X, Y])$$

on the one hand and on the other,

$$< \tilde{(\tilde{D}_X f)}, Y >= (\tilde{D}_X f)(Y) = -f(D_X Y) = -f([X, Y])$$

so we have

$$D_X\tilde{f} = \tilde{(\tilde{D}_X f)}, f \in \}^*$$

Suppose we extend the domain of D_X to include all elements of the symmetric algebra on }. We have

$$\tilde{D}_X(f^2) = (\tilde{D}_X f).f + f.\tilde{D}_X f$$

so that

$$\tilde{(\tilde{D}_X(f^2))} = \tilde{(\tilde{D}_X f)} \otimes f + f \otimes \tilde{(\tilde{D}_X f)}$$

On the other hand,

$$D_X(\tilde{f} \otimes \tilde{f}) = D_X\tilde{f} \otimes \tilde{f} + \tilde{f} \otimes D_X\tilde{f}$$

The isomorphism *tilde* : $\}^* \to \}$ extends naturally to an isomorphism *eeeetilde* of the polynomial algebra $\mathcal{P}$ on } onto the symmetric algebra $\mathcal{S}$ on }. For example, we have

$$f^2 \to \tilde{f} \otimes \tilde{f}$$

If $f, g \in \}^*$, then fg is in $\mathcal{P}$ and is defined by the equation

$$fg(X) = f(X)g(X)$$

we have

$$fg = ((f+g)^2 - f^2 - g^2)/2 \to ((\tilde{f} + \tilde{g}) \otimes (\tilde{f} + \tilde{g})/2 - tildef \otimes \tilde{f} - \tilde{g} \otimes \tilde{g})/2$$
$$= \frac{1}{2}(\tilde{f} \otimes \tilde{g} + \tilde{g} \otimes \tilde{f})$$

and it is easy to see that *tilde* is an algebra isomorphism. Define

$$\xi(X) = tr((adX)^2)$$

ξ is called the Casimir polynomial on }. We have in view of what we just proved,

$$D_X\tilde{\xi} = \tilde{(\tilde{D}_X \xi)}$$

But,

$$\tilde{D}_X\xi(Y) = -2.tr(ad([X,Y]).(adY)) = -2.tr((ad(X)ad(Y) - ad(Y)ad(X)).ad(Y)) = 0$$

Thus,

$$D_X\tilde{\xi} = 0$$

λ denotes the symmetrizer map of $\mathcal{S}$ onto the universal enveloping algebra $U(\})$ of $\}$. Note that λ is an isomorphism. For example, suppose $X, Y \in \}$. The element $X.Y$ of $U(\})$ can be expressed as $X.Y = \frac{1}{2}(XY - YX) + \frac{1}{2}(XY + YX)$. The second term is an element of $\mathcal{S}$ while the first term is the same as $[X, Y]$ which being in $\}$, is also an element of $\mathcal{S}$. The element $\omega = \lambda(\tilde{\xi})$ is called the Casimir element of $U(\})$. We want to show that ω belongs to the centre of $U(\})$. Let $\{X_1, ..., X_n\}$ be a basis for $\}$ and $\{X^1, ..., X^n\}$ the dual basis relative to the Cartan-Killing form, ie,

$$< X_i, X^j >= \delta_{ij}$$

Then, we claim that

$$\omega = \sum_{i=1}^{n} X_i X^i$$

The equation $D_X\tilde{\xi} = 0$ implies that ω belongs to the centre of $U(\})$. For any $X \in \}$, we have

$$X = \sum_j < X, X^j > X_j$$

and hence

$$\xi(X) = tr(adX)^2 = \sum_{j,k} < X, X^j >< X, X^k >< X_j, X_k >$$

Consequently,

$$\tilde{\xi} = \sum_{j,k} < X_j, X_k > X^j X^k$$

Thus, viewing this as an element of $U(\})$, we have

$$\omega = \sum_{j,k} < X_j, X_k > X^j X^k = \sum_k (\sum_j < X_j, X_k > X^j) X^k = \sum_k X_k X^k$$

(Taken from V.S.Varadarajan, "Lie groups, Lie algebras and their representations").

52.Problem: Let V be a finite dimensional vector space T a linear operator in V. Let W be a T-invariant subspace of V. Prove the existence of a linear map T' in V/W induced by T. Suppose $V = W \oplus W'$ with W' also T-invariant. Then show that T' restricted to V/W is non-singular iff T' restricted to W' is non-singular.

53.Problem:Suppose V is a vector space of finite dimension and V^* is its dual. Let $L \in End(V)$, all endomorphisms on V. Define $L_{1,1} \in End(V \otimes V^*)$ via the equation

$$L_{1,1}(x \otimes y) = Lx \otimes y - x \otimes L^T y$$

Also define $ad(L) \in End(End(V))$ via the equation

$$adL(M) = LM - ML$$

Then show that $ad(L)$ is related to $L_{1,1}$ via a similarity transformation. Specifically define a map $\zeta : V \otimes V^* \to End(V)$ via the equation

$$\zeta(x \otimes y)(z) = y(z)x$$

Then note that

$$\zeta.L_{1,1}(x \otimes y)(z) = \zeta(Lx \otimes y - x \otimes L^T y)(z) = (Lx)y(z) - xy(Lz)$$

On the other hand,

$$(ad(L)\zeta(x \otimes y))(z) = L\zeta(x \otimes y)z - \zeta(x \otimes y)Lz$$

$$= L(y(z)x) - y(Lz)x) = y(z)Lx - y(Lz)x$$

It follows that

$$\zeta.L_{1,1}.(x \otimes y) = ad(L)\zeta$$

or

$$ad(L) = \zeta.L_{1,1}.\zeta^{-1}$$

Note that ζ as defined can be extended to an isomorphism between the vector spaces $V \otimes V^*$ and $End(V)$.

54.Problem:Let $\mathcal{M} = GL_n(\mathbb{R})$. This is a Lie group. Its tangent space at the origin is $M_n(\mathbb{R})$, the space of all $n \times n$ matrices. Let $X, Y \in M_n(\mathbb{R})$ and $x \in GL_n(\mathbb{R})$. Then the vector field at x corresponding to X is given by $x.X$ and that corresponding to Y by $x.Y$. We have in terms of differential operators,

$$X_x(f) = \frac{d}{dt} f(x.exp(tX)) \frac{d}{dt} f(x + txX)|_{t=0} = f(x; X)$$

in the language of Varadarajan (Lie groups, Lie algebras and their representations). We have

$$Y_x X_x(f) = \frac{d}{ds} X_x(f)(x + sxY)|_{s=0} = \frac{\partial^2}{\partial t \partial s} f(x + sxY + t(x + sxY)X)|_{t=s=0}$$

$$= \frac{\partial^2}{\partial t \partial s} f(x + sxY + txX + tsxYX)|_{t=s=0}$$

$$= \frac{d}{dt}(xY + txYX)_{ij} f_{,ij}(x + txX)|_{t=0} = (xYX)_{ij} f_{,ij}(x) + (xY)_{ij}(xX)_{km} f_{,ij,km}$$

Interchanging X, Y gives us

$$X_x Y_x(f) = (xXY)_{ij} f_{,ij}(x) + (xX)_{km}(xY)_{ij} f_{,ij,km}$$

so that

$$(X_x Y_x - Y_x X_x)(f) = (x(XY - YX))_{ij} f_{,ij}(x)$$

This means that

$$X_x Y_x - Y_x X_x = (XY - YX)_x$$

55.Problem: Let $\}$ be a Lie algebra and $\mathcal{T}$ the corresponding tensor algebra. Let $\dashv$ be a homogeneous ideal in $\mathcal{T}$, ie, if we write

$$\mathcal{T} = \sum_{r=0}^{\infty} \mathcal{T}_r$$

where $\mathcal{T}_r$ is the space spanned by homogeneous tensors of degree r, ie, $\mathcal{T}_0 = \mathbb{F}$ and

$$\mathcal{T}_r = span\{u_1 \otimes ... \otimes u_r : u_i \in \}, i = 1, 2, ..., r\}$$

then

$$\dashv = \sum_r \mathcal{T}_r \cap \dashv$$

An example where this is not valid is the following: Suppose $\dashv$ is generated by the element $u_1 + u_2 \otimes u_3$, ie,

$$\dashv = \sum_{t,t' \in \mathcal{T}} t \otimes (u_1 + u_2 \otimes u_3) \otimes t'$$

Then it is clear that unless $u_i = 0$ for some i, $\dashv$ is not a homogeneous ideal. Indeed, $\mathcal{T}_r \cap \dashv = 0$ for all r.

56.Problem: Let $X_1, ..., X_p$ be smooth vector fields on a manifold M such that they are linearly independent at each point and further, $[X_i, X_j] = 0$ for all i, j on M. Then there exists a coordinate system $x_1, ..., x_m$ such that

$$X_j = \frac{\partial}{\partial x_j} + \sum_{s<j} a_{js} \frac{\partial}{\partial x_s}, j = 1, 2, ..., p$$

where $a_{js}(x)$ are smooth functions on M. If $p = 1$, there is nothing to prove. Let $p > 1$ and assume the result for $X_1, ..., X_{p-1}$. Then the induction hypothesis states that there exist coordinates $u_1, ..., u_m$ such that

$$X_j = \frac{\partial}{\partial u_j} + \sum_{1 \le s < j} b_{js} \frac{\partial}{\partial u_s}, j = 1, 2, ..., p-1$$

where b_{js} are smooth functions. Set

$$X_p = \sum_{1 \le s \le m} g_s \frac{\partial}{\partial u_s}, X_p' = \sum_{p \le s \le m} g_s \frac{\partial}{\partial u_s}$$

Suppose $X_p' = 0$ at some point y. Then it would follow that at this point,

$$X_p = \sum_{1 \le s < p} g_s \frac{\partial}{\partial u_s}$$

which would imply that $X_1, ..., X_p$ are linearly dependent at y, a contradiction. Hence, $X'_p \neq 0$ at any point y. The conditions $[X_p, X_j] = 0$ give

$$[\sum_{1\leq s\leq m} g_s \frac{\partial}{\partial u_s}, X_j] = 0$$

or

$$\sum_{1\leq s\leq m} g_s[\frac{\partial}{\partial u_s}, X_j] - \sum_{1\leq s\leq m} X_j(g_s)\frac{\partial}{\partial u_s}, j = 1, 2, ..., p-1$$

For $j = 1, 2, ..., p-1$ and $s = 1, 2, ..., m$, we have

$$[\frac{\partial}{\partial u_s}, X_j] = [\frac{\partial}{\partial u_s}, \frac{\partial}{\partial u_j} + \sum_{1\leq s<j} b_{js}\frac{\partial}{\partial u_s}]$$

and this is clearly a linear combination of $\frac{\partial}{\partial u_t}, t = 1, 2, ..., p-1$. It follows from the above that $X_j(g_s) = 0$ for $j = 1, 2, ..., p-1, s = p, p+1, ..., m$. It follows by recursion that $\frac{\partial g_s}{\partial u_j} = 0$ for $1 \leq j \leq p-1$ and $p \leq s \leq m$. This implies that for each $s = p, p+1, ..., m$, g_s is a function of $u_p, u_{p+1}, ..., u_m$ only. From this it immediately follows that $u_p, ..., u_m$ can be replaced by $v_p, ..., .v_m$ such that $X'_p = \frac{\partial}{\partial v_p}$ from which the induction carries forward.

57.Group theory:Let $GL_n(\mathbb{R})$ denote the multiplicative group of all non-singular linear transformations on $\mathbb{R}^n$. $GL_n(\mathbb{R})$ can thus be viewed as the group of all $n \times n$ non-singular matrices. $O_n(\mathbb{R})$ is the subroup of all orthogonal matrices. We define a map $\pi_{ij} : GL_n(\mathbb{R}) \to \mathbb{R}$ so that

$$\pi_{ij}(A) = \sum_{k=1}^{n} a_{ik}a_{jk} - \delta_{ij}, 1 \leq i \leq j \leq n$$

Then

$$O_n(\mathbb{R}) = \{A \in GL_n(\mathbb{R}) : \pi_{ij}(A) = 0, 1 \leq i \leq j \leq n\}$$

Also for any $X \in M_n(\mathbb{R})$, (the Lie algebra of $GL_n(\mathbb{R})$, namely the space of all $n \times n$ real matrices),

$$d\pi_{ij}(A)(X) = d(\sum_k a_{ik}a_{jk} - \delta_{ij}) = \sum_k (a_{jk}da_{ik}(X) + a_{ik}da_{jk}(X))$$

$$= \sum_k (a_{jk}x_{ik} + a_{ik}x_{jk}) = (XA^T + AX^T)_{ij}$$

It follows that $kerd\pi_I$ is the subset of $M_n(\mathbb{R})$ consisting of all $n \times n$ real valued skew-symmetric matrices. Thus, $ker\pi$ is precisely $O_n(\mathbb{R})$ and $kerd\pi$ is precisely the Lie algebra $o_n(\mathbb{R})$ of $O_n(\mathbb{R})$.

58.Problem: Let G_1 be a Lie group and let TG_{1e} be its Lie algebra. Let G_2 be another Lie group and TG_{2e} its Lie algebra. Let $f : G_1 \to G_2$ be a smooth homomorphism. Then for all $x, y \in G_1$, we have $f(xy) = f(x)f(y)$. We introduce local coordinates in G_1 and G_2 so that 0 is the identity element of both

G_1 and G_2. Let Φ_1 be the composition operation in G_1 and Φ_2 that in G_2. Thus the homomorphism property of f can be expressed as

$$\Phi_2(f(x), f(y)) = f(\Phi_1(x, y))$$

Let $\xi \in TG_{1e}$ and let X_ξ be the left invariant vector field on G_1 generated by ξ, ie,

$$X(x) = (Tf)_x(\xi)$$

Let $exp(tX_\xi)$ denote the one parameter group of diffeomorphisms of G_1 corresponding to the vector field X_ξ. In other words, $exp(tX_\xi) : G_1 \to G_1$ satisfies the equation

$$\frac{d}{dt}exp(tX_\xi)(x)|_{t=0} = X_\xi(x), x \in G_1$$

It follows that

$$\Phi_2(f(x), f(exp(tX_\xi)(0))) = f(\Phi_1(x, exp(tX_\xi)(0)))$$

Let $\Phi_{1,1}$ denote the Jacobian matrix of Φ_1 with respect to its first argument and $\Phi_{1,2}$ the Jacobian with respect to the second argument and likewise for Φ_2. We then get on differentiating the above expression with respect to t at $t = 0$ that

$$\Phi_{2,2}(f(x), 0)f'(0)X_\xi(0) = f'(x)\Phi_{1,2}(x, 0)X_\xi(0)$$

or equivalently,

$$\Phi_{2,2}(f(x), 0)f'(0)\xi = f'(x)\Phi_{1,2}(x, 0)\xi$$

Note that $f'(x)$ is the Jacobian matrix of f at x. Now,

$$f'(0)\xi = (Tf)_e(\xi)$$

and hence, by left invariance, it follows that

$$\Phi_{2,2}(f(x), 0)f'(0)\xi = \Phi_{2,2}(f(x), 0)(Tf)_e(\xi) = X_{(Tf)_e(\xi)}(f(x))$$

Likewise,

$$\Phi_{1,2}(x, 0)\xi = X_\xi(x)$$

and

$$f'(x)\Phi_{1,2}(x, 0)\xi = Tf(X_\xi)(f(x))$$

This gives

$$X_{(Tf)_e(\xi)} = Tf(X_\xi)$$

(See the proof in Abraham and Marsden. There is a small error in their proof. They have assumed left invariance of the vector fields but have based their proof on right invariance.)

59.Some general remarks on rings, modules, algebras and representations.

Let R be a ring. and let $f : R \to R$ be a R-linear endomorphism. Then $T(rx) = rT(x)$. In particular, $T(r) = T(r.1) = r.T(1)$. Thus, we can consider T acting on R as being the same as right multiplication in R by $T(1)$. (Ref:Procesi, Lie groups, an approach through invariants).

Let R be a ring and M an R module. Let $m \in M$ be a cyclic element. In other words, we have $R.m = M$. Consider the map $\phi : R \to M$ defined by $\phi(r) = r.m$. By cyclicity of m, it follows that ϕ is a surjective R linear map. Let $J = ker\phi$. Then, J is a left ideal of R. Indeed, $x \in J$ implies $\phi(x) = 0$ implies $\phi(r.x) = rxm = r\phi(x) = 0$ implies $r.x \in J$, ie, $r.J \subset J$. We claim that R/J is isomorphic to M as an R module. Indeed, Define $\psi : R/J \to M$ so that $\psi(r + J) = \phi(r) = rm$. Then ψ is well defined since $r_1 - r_2 \in J$ implies $\phi(r_1) - \phi(r_2) = \phi(r_1 - r_2) = 0$. It is clear that ψ is onto. That ψ is injective follows from the fact that if $\phi(r_1) = \phi(r_2)$, then $\phi(r_1 - r_2) = 0$ and this implies that $r_1 - r_2 \in J$ so that $r_1 + J = r_2 + J$. Conversely, if J is any left ideal of R, then R/J is an R-cyclic module. In other words, we have proved that every cyclic R-module is isomorphic to the quotient of R with a left ideal.

Semisimple ring:A ring R is semisimple if it is semisimple as a module over itself. Recall that a module M over R is semisimple if M is expressible as a direct sum of irreducible R-modules. Suppose R is a semisimple ring. Then there is a family of R-modules $\{M_i : i \in I\}$ such that M_i is irreducible for every i, every element x of R is expressible as $x = \sum_{k=1}^{n} x_k$ for some finite integer n and with $x_k \in M_{i_k}, k = 1, 2, ..., n$ for some distinct $i_k, k = 1, 2, ..., n$. We consider $1 \in R$. Then $1 = \sum_{k=1}^{n} x_k$ for some $x_k \in M_{i_k}$. Then for any $x \in R$, we have $x = \sum_{k=1}^{n} xx_k$ with $xx_k \in M_{i_k}$. This shows that $R = \bigoplus_{k=1}^{n} M_{i_k}$, or in other words, every semisimple ring considered a module over itself is a finite direct sum of irreducible modules.

60.Theorem on Nilpotent Lie algebra(Engel's theorem): Let V be a vector space and $End(V)$ the Lie algebra of all endomorphisms of V. Let $L \subset End(V)$ be a Lie algebra all of whose elements are nilpotent. Then there is a nonzero vector $u \in V$ such that $au = 0$ for all $a \in L$, ie, $Lu = 0$. To see this, let A denote any proper maximal Lie subalgebra of L. Then $ad(A)$ consists of nilpotent operators acting on L/A and hence by the induction hypothesis, there exists a $u \in L$ such that $ad(A)u \in A$, ie, $[A, u] \subset A$. It follows that $A + \mathbb{F}.u$ is a Lie subalgebra of L and is strictly larger than A. Maximality of A then implies that $A + \mathbb{F}.u = L$. Let $W = \{x \in V : Ax = 0\}$. Then for $w \in W$ and $a \in A$, we have $auw = [a, u]w + uaw = 0$ since $[a, u] \in A$. It follows that $uw \in W$, ie, W is u-stable. Thus, $L = A + \mathbb{F}.u$. Hence, L is a nilpotent family of operators acting on W and hence by the induction hypothesis on $dimV$, there exists a $w \in W$ such that $Lw = 0$.

61.Symmetric algebra and Antisymmetric algebra:Let A be an algebra and let I be the ideal generated by all the commutators $ab - ba, a, b \in A$. Consider the quotient algebra A/I. This algebra is Abelian. Indeed, if $a, b \in A$, then $(a + I)(b + I) - (b + I)(a + I) = ab - ba + I = I$. A/I is a symmetric algebra for if $X, Y \in A/I$, then $XY = YX$. This operation can be carried out for the tensor algebra on a vector space. Suppose U is a vector space and $\otimes$ a tensor product on U. Let $T(U)$ denote the tensor algebra on U, ie, $T(U)$ is a linear combination of elements of the form $u_1 \otimes ... \otimes u_n$ with $u_1, ..., u_n \in U, n = 1, ...,$. Let I be the ideal in $T(U)$ generated by elements of the form $u \otimes v - v \otimes u$ with $u \in U$. Equivalently, I is a linear combination of elements of the form $t \otimes (u \otimes v - v \otimes u) \otimes t'$ with $t, t' \in T(U)$. Then $T(U)/I$ is isomorphic to the symmetric algebra $S(U)$. We may map the element $u \otimes v + v \otimes u$ in $S(U)$ to the element $u \otimes v + I = v \otimes u + I$ and extend this map to the whole of $S(U)$. It follows that if $u_1, ..., u_k \in U$, then $\sum_{\sigma \in S_n} u_{\sigma 1} \otimes ... \otimes u_{\sigma n}$ gets mapped to the element $u_1 \otimes ... \otimes u_n + I$. Likewise, suppose A is an algebra and I the ideal generated by elements of the form a^2 with $a \in A$. It follows from the relation $(a+b)^2 - a^2 - b^2 = ab + ba, a, b \in A$ that I is generated by the anticommutators $ab + ba$ with $a, b \in A$. Then, A/I is the antisymmetric algebra, ie, for $X, Y \in A/I$, one has $XY = -YX$. Suppose B is an algebra and $j : A \to B$ is an algebra homomorphism such that $j(a)^2 = 0$ for all $a \in A$. Then define a map $h : A/I \to B$ so that $h(a+I) = j(a), a \in A$. It follows that h is well defined and linear. That h is well defined follows from the following observation. Let $a - b \in I$. Then, $a - b$ is a linear combination of elements of the form $t \otimes u^2 \otimes t'$ with $u \in A$. Then,

$$j(a) - j(b) = j(a-b) = j(t)j(u)^2 j(t') = 0$$

In other words, let $F : A \to A/I$ be the canonical map, ie, $F(a) = a + I$. Then, if $j : A \to B$ is a homomorphism such that $j(a)^2 = 0$ for all $a \in A$, there exists a unique homomorphism $h : A/I \to B$ such that $j = hoF$.

62.Let M, N be irreducible R modules where R is a ring and let $f : M \to N$ be a nonzero R-homomorphism, ie f is R-linear. We then claim that f is an isomorphism. Indeed, $range(f)$ is a submodule of N invariant under R and $ker(f)$ is a submodule of M also invariant under R. These follow since $f(rm) = rf(m)$ for all $r \in R, m \in M$. Since N, M are irreducible and f is nonzero, it must follow that $ker(f) = 0$ and $range(f) = M$, ie, f is an R-isomorphism.

63.Let M be an irreducible R-module where R is a ring. The centralizer Δ of M consists of all R-homomorphisms $T : M \to M$, ie, $T(r_1 x + r_2 y) = r_1 T(x) + r_2 T(y)$ with $r_1, r_2 \in R$ and $x, y \in M$. $T \in \Delta$ implies in particular that $rT = Tr$ for all $r \in R$, ie T commutes with R which is the reason why Δ is called the centralizer of M. We claim that Δ is a division ring. Indeed, we need only prove invertibility of every $T \in \Delta$ and this follows from the previous remark.

64.Structure theorem: Let R be a semisimple ring. Then by definition of semisimplicity, we can write $R = \bigoplus_i N_i^{m_i}$ as a left module where $N_i^{m_i}$ stands for direct sum of m_i copies of the irreducible module N_i. Let $T : R \to R$ be a homomorphism, ie, $T \in End_R(R)$. Then, we have $T(r) = rT(1)$ and identifying T with $T(1)$, it follows that $End_R(R) = R^o$ where R^o stands for "opposite ring", ie, for $r, s \in R^o$, multiplication is defined by $\phi(r, s) = sr$. Note that if $T : N_i^{m_i} \to N_j^{m_j}$ is an R-homomorphism and $i \neq j$, then $T = 0$ since N_i, N_j are inequivalent representations while if $T : N_i^{m_i} \to N_i^{m_i}$ is an R-homomorphism, then T is an $m_i \times m_i$ matrix with elements from $\Delta_i = End_R(N_i)$. Note that Δ_i is a division ring by the previous remark and hence we get the structure theorem,

$$R^o = End_R(R) = \bigoplus_i M_{m_i}(\Delta_i)$$

This gives

$$R = \bigoplus_i M_{m_i}(\Delta_i^o)$$

ie R is expressible as a direct sum of matrix rings over division rings.

65. Collected papers of Harish-Chandra: The initial part of Harish-Chandra's work focussed on the Dirac equation like the algebra of the Dirac matrices and relativistic wave equations. The Dirac matrices $\gamma^\mu, \mu = 0, 1, 2, 3$ satisfy the anticommutation relations

$$\gamma^a\gamma^b + \gamma^b\gamma^a = 2g^{ab}$$

where

$$g^{00} = 1, g^{11} = g^{22} = g^{33} = -1$$

The Dirac equation reads

$$(\gamma^a\partial_a - m)\psi = 0$$

It follows from this equation that

$$(\gamma^b\partial_b + m)(\gamma^a\partial_a - m)\psi = 0$$

and noting that

$$\gamma^b\gamma^a\partial_a\partial_b = g^{ab}\partial_a\partial_b$$

we get

$$(g^{ab}\partial_a\partial_b - m^2)\psi = 0$$

which is the Klein-Gordon equation. The set of sixteen matrices $\{1, \gamma^a, \gamma^a\gamma^b, \gamma^a\gamma^b\gamma^c, \gamma^0\gamma^1\gamma^2\gamma^3\}$ $(1 + 4 + 6 + 4 + 1 = 16)$ is closed under multiplication except for scalar multiplicative factors and hence defines a sixteen dimensional algebra. Upto equivalence, there is only one four dimensional representation of this algebra and this representation leads to the standard four component Dirac wave equation. This representation is irreducible. Harish-Chandra discusses the problem of Lorentz invariance of generalized Dirac equations. For example, under a Lorentz transformation of the coordinates, the Dirac matrices γ^a get transformed to $t^a_b\gamma^b$ where (t^a_b) is a Lorentz transformation matrix. Suppose the wave function ψ gets transformed under such a Lorentz transformation to $T\psi$ where T is a linear transformation on the space of wave functions. Suppose further that

$$t^a_b\gamma^b = T\gamma^a T^{-1}$$

Then, the Dirac equation remains invariant under the Lorentz transformation as can be seen from

$$(t^a_b\gamma^b\partial_a - m)T\psi = T(\gamma^a\partial_a - m)T^{-1}T\psi = T(\gamma^a\partial_a - m)\psi = 0$$

Harish-Chandra has studied properties of the T matrix.

One of the papers of Harish-Chandra deals with a wave equation of the form

$$(\partial^\mu\partial_\mu + \chi^2)\psi = \delta(x), x \in \mathbb{R}^4$$

This equation can be solved using the Fourier transform. Taking the four dimensional Fourier transform gives us

$$\hat{\psi}(\omega, \mathbf{k}) = (-\omega^2 + k^2 + \chi^2)^{-1}$$

Thus,

$$\psi(t, \mathbf{r}) = \int d^4k(-\omega^2 + k^2 + \chi^2)^{-1}exp(i(\omega t - \mathbf{k.r}))d\omega d^3k$$

The ω-integral is carried out using contour integration. Assuming that the contour encloses just one of the poles $\pm\sqrt{k^2+\chi^2}$, we get

$$\psi(t,\mathbf{r}) = \int d^3k.exp(i(\sqrt{k^2+\chi^2}t - \mathbf{k}.\mathbf{r}))/\sqrt{k^2+\chi^2}$$

This integral can be evaluated using spherical polar coordinates:

$$\psi(t,\mathbf{r}) = 2\pi\int_0^\infty\int_0^\pi k^2 sin(\theta)dk.d\theta.exp(i(\sqrt{k^2+\chi^2}t - kr.cos(\theta)))$$

The integral with respect to θ can be carried out easily giving

$$\psi(t,\mathbf{r}) = ir^{-1}\int_0^\infty k.dk.exp(i\sqrt{k^2+\chi^2}t).sin(kr)$$

The evaluation of this integral is not easy. Harish Chandra has considered radiation emitted by a particle subject to the above wave equation for a particle moving along the world line $z^\mu(\tau)$ so that $z^\mu(\tau) = (t,\mathbf{R}(\tau))$. Let x^μ denote the coordinates of an event in space-time. The retarded time $\tau_r(x)$ associated with the particle's motion satisfies the equation

$$(x^\mu - z^\mu(\tau))(x_\mu - z_\mu(\tau)) = 0$$

or expanded in full,

$$(t-\tau)^2 - (\mathbf{r} - \mathbf{R}(\tau))^2 = 0$$

so that

$$\tau_r = t - |\mathbf{r} - \mathbf{R}(\tau_r)|$$

We get some identities(discussed in HarishChandra's paper) for the partial derivatives of the retarded time with respect to the space-time coordinates:

$$0 = \partial_\nu[(x^\mu - z^\mu)(x_\mu - z_\mu)] = 2(\delta^\mu_\nu - v^\mu\partial_\nu\tau)(x_\mu - z_\mu)$$

so that

$$x_\nu - z_\nu - \partial_\nu\tau v.(x-z), v^\mu = dz^\mu(\tau)/d\tau$$

or

$$\partial_\nu\tau = (x_\nu - z_\nu)/[v.(x-z)]$$

where

$$v.(x-z) = v^\mu(x_\mu - z_\mu)$$

We show that a solution to the wave equation when $\chi = 0$ is given by

$$S(\tau)/[(x^\mu - z^\mu(\tau))v_\mu(\tau)]$$

where S is any function. To see this, we observe first that with the notation $(x-z)^2 = (x^\mu - z^\mu)(x_\mu - z_\mu)$, we have

$$\delta((x-z(\tau))^2) = \delta(\tau - \tau_r)/v.(x - z(\tau_r))$$

Thus, it is sufficient to show that $\Delta(x) = \delta(x^2)$ satisfies the wave equation with $\chi = 0$. Note that

$$\partial_\nu \Delta(x) = 2x_\nu \delta'(x^2)$$

and

$$\partial^\nu \partial_\nu \Delta(x) = 8\delta'(x^2) + 4x^2\delta''(x^2)$$

Consider now

$$\int f(t)\delta'(t)dt = -f'(0),$$

$$\int tf(t)\delta''(t)dt = \int (tf(t))''\delta(t)dt = 2f'(0)$$

From these two, it follows that

$$\int (8\delta'(t) + 4t\delta''(t))f(t)dt = -8f'(0) + 8f'(0) = 0$$

proving the claim. Thus, $\Delta(x - z(\tau))$ also satisfies the wave equation for all τ and hence so does $\int \Delta(x - z(\tau))S(\tau)d\tau$. This proves the claim made.

Lie ideal: Let $x_1, ..., x_n$ be indeterminates and let $\mathcal{U}$ denote the free associative algebra generated by these elements. Let $\mathcal{B}$ be an ideal in $\mathcal{U}$ such that for any two linear elements $x, y \in \mathcal{U}$, there exists another linear element $z \in \mathcal{B}$ such that $[x, y] = z(mod\mathcal{B})$. Then $\mathcal{B}$ is called a Lie ideal of $\mathcal{U}$. Consider an element $x_1x_2x_3 \in \mathcal{U}$. Let $\mathcal{B}$ be a Lie ideal in $\mathcal{U}$. We have $x_1x_2x_3 = x_2x_1x_3 + zx_3 mod\mathcal{B}$ where $x_1x_2 - x_2x_1 = z mod\mathcal{B}$. It follows using this fact and induction that

$$x_1x_2x_3 = \frac{1}{6}\sum_{\sigma \in S_3} x_{\sigma 1}x_{\sigma 2}x_{\sigma 3} + w mod\mathcal{B}$$

where w is a canonical element. A canonical element of $\mathcal{U}$ has the form

$$\sum_r \sum_{i_1,...,i_r} a_r(i_1, ..., i_r)X_{i_1}...X_{i_r}$$

where $a_r(j_1, ..., j_r) = a_r(i_1, ..., i_r)$ for any permutation $(j_1, ..., j_r)$ of $(i_1, ..., i_r)$. Let $\mathcal{L}$ be a Lie algebra and let $X_1, ..., X_n$ be a basis for this Lie algebra. Define a mapping $\omega(X_i) = x_i, i = 1, 2, ..., n$ and extend it linearly to the whole of $\mathcal{L}$. Let $\mathcal{U}(\mathcal{L})$ be the universal enveloping algebra of $\mathcal{L}$. Consider the map $\pi : \mathcal{U} \to \mathcal{U}(\mathcal{L})$ so that $\pi(x_{i_1}...x_{i_r}) = X_{i_1}...X_{i_r}$ and extend the domain of π to the whole of $\mathcal{U}$. Let $\mathcal{B}$ denote the kernel of π. We want to show that $\mathcal{B}$ is a Lie ideal in $\mathcal{U}$. Indeed, let i, j be two indices. Then, $[X_i, X_j] = \sum_k c(kij)X_k$ so that

$$\pi(x_ix_j - x_jx_i - \sum_k c(kij)x_k) = X_iX_j - X_jX_i - \sum_k c(kij)X_k = 0$$

so that

$$x_ix_j - x_jx_i = \sum_k c(kij)x_k mod\mathcal{B}$$

proving the claim. Note that $\mathcal{B}$ cannot have any canonical non-zero element.

On some applications of the universal enveloping algebra of a semsimple Lie algebra: Harish Chandra starts with a set of $3l$ generators $H_i, X_i, Y_i, i = 1, 2, ..., l$ of a Lie algebra satisfying the commutation relations $[H_i, H_j] = 0, [H_i, X_j] = a_{ij}X_j, [H_i, Y_j] = -a_{ij}Y_j, [X_i, Y_j] = \delta_{ij}H_i$ where (a_{ij}) is a Cartan matrix. More precisely, he shows how a universal enveloping algebra can be constructed satisfying these commutation relations. The way is to start with 3l generators $H_i, X_i, Y_i, i = 1, 2, ..., l$ of a free associative algebra $\mathcal{U}$ and to consider the ideal $\mathcal{L}$ generated by elements of the form $[H_i, X_j] - a_{ij}X_j, [H_i, Y_j] + a_{ij}Y_j, [X_i, Y_j] - \delta_{ij}H_i$. Then consider the quotient algebra $\mathcal{U}' = \mathcal{U}/\mathcal{L}$. Let H_i', X_i', Y_i' denote the images of H_i, X_i, Y_i in this quotient algebra, ie,

$$H_i' = H_i + \mathcal{L}, X_i' = X_i + \mathcal{L}, Y_i' = Y_i + \mathcal{L}, i = 1, 2, ..., l$$

Then,

$$[H_i', X_j'] - a_{ij}X_j' = [H_i, X_j] - a_{ij}X_j + \mathcal{L} = \mathcal{L}$$

and likewise for the others. In other words, $\mathcal{U}'$ is an associative algebra in which the H_i', X_i', Y_i' satisfy the desired commutation relations. Using the theory of roots, Harish Chandra then explains how to construct finite dimensional irreducible representations of the algebra $\mathcal{U}'$. Suppose π is a representation of $\mathcal{U}'$ in a vector space V. Let $\langle = span\{H_1', ..., H_l'\}$. Let $\lambda \in \langle^*$. A nonzero vector $v \in V$ is said to be a weight vector of π with weight λ if $\pi(H)v = \lambda(H)v$ for all $H \in \langle$. In this case, we say that λ is a weight of the representation π. Let λ be a weight and let v be a weight vector with weight λ. Then,

$$\pi(H_j)\pi(X_i)v = \pi(X_i)\pi(H_j)v + \pi([H_j, X_i])v = \pi(X_i)\lambda(H_j)v + a_{ji}\pi(X_i)v = (\lambda(H_j) + a_{ji})\pi(X_i)v$$

so that if we define an $\alpha_i \in \langle^*$ so that

$$\alpha_i(H_j) = a_{ji}$$

then,

$$\pi(H_j)\pi(X_i)v = (\lambda + \alpha_i)(H_j)\pi(X_i)v$$

and hence for every $H \in \langle^*$, we have

$$\pi(H)\pi(X_i)v = (\lambda + \alpha_i)(H)\pi(X_i)v$$

implying that $\pi(X_i)v$ is a weight vector with weight $\lambda + \alpha_i$. Likewise, $\pi(Y_i)v$ is a weight vector with weight $\lambda - \alpha_i$. Suppose v is a weight vector with weight λ. Consider the vector $w(i_1, ..., i_k) = \pi(X_{i_1}X_{i_2}...X_{i_k})v$. It is clear by induction that $w(i_1, ..., i_k)$ is a weight vector with weight $\lambda + \sum_{m=1}^{k} \alpha_{i_m}$. Consider the adjoint representation of $\mathcal{U}'$ in $\mathcal{U}$. This representation is defined by

$$\pi(u)x = [u, x], x \in \mathcal{U}, u = H_i, X_i, Y_i$$

Note that

$$\pi(uv)x = \pi(u)\pi(v)x = [u, [v, x]]$$

Suppose $f: G \to \mathbb{C}$ is an invariant function on G, ie, $f(xyx^{-1}) = f(x)$ for all $x, y \in G$. Then let π be an irreducible unitary representation of G with character $\chi_\pi(x) = Tr(\pi(x))$. Consider the operator

$$A = \int_G f(x)\pi(x)dx$$

We have for all $y \in G$

$$A = \int_G f(yxy^{-1})\pi(x)dx = \int_G f(x)\pi(y^{-1}xy)dx = \pi(y)^* A\pi(y)$$

It follows that

$$\pi(y)A = A\pi(y), y \in G$$

and hence by Schur's lemma, $A = c.I$ for some constant c. We write

$$c.I = \int_G f(x)\pi(x)dx$$

and taking trace gives us

$$c.d_\pi = \int f(x)\chi_\pi(x)dx$$

so that

$$\int_G f(x)\pi(x)dx = d_\pi^{-1}(\int f(x)\chi_\pi(x)dx).I$$

Now let $h(x)$ be any smooth function on G and define the invariant function

$$f(x) = \int_G h(yxy^{-1})dy$$

Then, we get from the above,

$$d_\pi^{-1}(\int h(yxy^{-1})\chi_\pi(x)dxdy)I = \int h(yxy^{-1})\pi(x)dxdy$$

Now,

$$\int h(yxy^{-1})\chi_\pi(x)dxdy = \int h(z)\chi_\pi(z)dydz = \int h(z)\chi_\pi(z)dz$$

giving the formula

$$d_\pi^{-1}\int h(z)\chi_\pi(z)dz = \int h(yxy^{-1})(\phi, \pi(x)\phi)dxdy$$

for any unit vector ϕ. This formula is the starting point for Harish Chandra's investigation of the distribution characters for complex semisimple Lie groups and Lie algebras.

Free Lie algebra with n generators. Let $X_1, ..., X_n$ be n symbols. The free Lie algebra generated by these elements is the smallest Lie algebra containing $X_1, ..., X_n$. Note that such a Lie algebra will necessarily contain elements of the form $X_iX_j - X_jX_i$ and their formal linear combinations.

Let $\mathcal{G}$ be the free associative algebra over a field K with generators $e_1, ..., e_l$. Define

$$\pi(Y_i)e_{i_1}...e_{i_r} = e_i e_{i_1}...e_{i_r},$$

$$\pi(H) = (\mu(H) - \alpha_{i_1}(H) - ... - \alpha_{i_r}(H))e_{i_1}...e_{i_r}$$

$$\pi(X_i) = 0$$

$$\pi(X_i)e_{i_1}...e_{i_r} = \delta_{i,j_1}\pi(H_i)e_{i_2}...e_{-i_r} + e_{i_1}\pi(X_i)e_{i_2}...e_{i_r}$$

Here μ is a linear function on $span\{H_1, ..., H_l\}$. Extend the domain of π to the whole of $\mathcal{U}$, the free associative algebra generated by $H_i, X_i, Y_i, i = 1, 2, ..., l$. We then have

$$\pi([H_i, H_j])e_{i_1}...e_{i_r} = (\pi(H_i)\pi(H_j) - \pi(H_j)\pi(H_i))e_{i_1}...e_{i_r}$$

Now,

$$\pi(H_j)e_{i_1}...e_{i_r} = (\mu(H_j) - \alpha_{i_1}(H_j) - ... - \alpha_{i_r}(H_j))e_{i_1}...e_{i_r}$$

$$\pi(H_i)\pi(H_j)e_{i_1}...e_{i_r} = (\mu(H_j) - \sum_{k=1}^{r}\alpha_{i_k}(H_j))(\mu(H_i) - \sum_{k=1}^{r}\alpha_{i_k}(H_i))e_{i_1}...e_{i_r}$$

Interchanging i and j and subtracting gives us

$$\pi([H_i, H_j])e_{i_1}...e_{i_r} = 0$$

so that

$$\pi([H_i, H_j]) = 0, i, j = 1, 2, ..., l$$

Again,

$$\pi(X_i)\pi(Y_i)e_{i_1}...e_{i_r} = \pi(X_i)e_ie_{i_1}...e_{i_r} = \pi(H_i)e_{i_1}...e_{i_r} + e_i\pi(X_i)e_{i_1}...e_{i_r}$$

$$\pi(Y_i)\pi(X_i)e_{i_1}...e_{i_r} = e_i\pi(X_i)e_{i_1}...e_{i_r}$$

and hence, we get

$$\pi([X_i, Y_i])e_{i_1}...e_{i_r} = \pi(H_i)e_{i_1}...e_{i_r}$$

Thus,

$$\pi([X_i, Y_i] - H_i) = 0$$

$$\pi(H_i)\pi(X_j)e_{i_1}...e_{i_r} = \pi(H_i)[\delta_{j,i_1}\pi(H_j)e_{i_2}...e_{i_r} + e_{i_1}\pi(X_j)e_{i_2}...e_{i_r}]$$

$$\pi(X_j)\pi(H_i)e_{i_1}...e_{i_r} = \pi(X_j)[\mu(H_i) - \sum\alpha_{i_k}(H_i)]e_{i_1}...e_{i_r}$$

$$= [\mu(H_i) - \sum\alpha_{i_k}(H_i)]\pi(X_j)e_{i_1}...e_{i_r}$$

Difficult to compute. Let us compute for $r = 2$.

$$\pi(H_i)\pi(X_j)e_{i_1}e_{i_2} = \pi(H_i)[\delta_{j,i_1}\pi(H_j)e_{i_2} + e_{i_1}\pi(X_j)e_{i_2}]$$

$$= \pi(H_i)[\delta_{j,i_1}(\mu(H_j) - \alpha_{i_2}(H_j))e_{i_2} + e_{i_1}\delta_{j,i_2}\mu(H_j)]$$

$$= \delta_{j,i_1}(\mu(H_j) - \alpha_{i_2}(H_j))\mu(H_i)e_{i_2} + \delta_{j,i_2}\mu(H_j)(\mu(H_i) - \alpha_{i_1}(H_i))e_{i_1}$$

on the other hand,

$$\pi(X_j)\pi(H_i)e_{i_1}e_{i_2} = \pi(X_j)(\mu(H_i) - \alpha_{i_1}(H_i) - \alpha_{i_2}(H_i))e_{i_1}e_{i_2}$$
$$= (\mu(H_i) - \alpha_{i_1}(H_i) - \alpha_{i_2}(H_i))(\delta_{j,i_1}\pi(H_i)e_{i_2} + e_{i_1}\pi(X_j)e_{i_2})$$

In order to evaluate all these, we need the full expansion for $\pi(X_i)e_{i_1}...e_{i_r}$. Note that

$$\pi(X_i)e_{i_1}...e_{i_r} =$$
$$\delta(i,i_1)(\mu(H_i) - \sum_{k=2}^{r}\alpha_{i_k}(H_i))e_{i_2}...e_{i_r} + e_{i_1}\pi(X_i)e_{i_2}...e_{i_r}$$

$$= \delta(i,i_1)(\mu(H_i) - \sum_{k\geq 2}\alpha_{i_k}(H_i))e_{i_2}...e_{i_r} +$$

$$\delta(i,i_2)(\mu(H_i) - \sum_{k\geq 3}\alpha_{i_k}(H_i))e_{i_1}e_{i_3}...e_{i_r} + e_{i_1}e_{i_2}\pi(X_i)e_{i_3}...e_{i_r}$$

$$+\delta(i,i_3)(\mu(H_i) - \sum_{k\geq 4}\alpha_{i_k}(H_i))e_{i_1}e_{i_2}e_{i_4}...e_{i_r} + ...$$

This can be abbreviated to

$$\pi(X_i)e_{i_1}...e_{i_r} =$$
$$\sum_m \delta(i,i_m)(\mu(H_i) - \sum_{k>m}\alpha_{i_k}(H_i)e_{i_1}...e'_{i_m}...e_{i_r}$$

where the prime over a symbol implies the omission of that symbol. Thus,

$$\pi(H_j)\pi(X_i)e_{i_1}...e_{i_r} = \sum_m \delta(i,i_m)(\mu(H_i) - \sum_{k>m}\alpha_{i_k}(H_i))(\mu(H_j) - \sum_{p\neq m}\alpha_{i_p}(H_j))e_{i_1}...e'_{i_m}...e_{i_r}$$

Next,

$$\pi(X_i)\pi(H_j)e_{i_1}...e_{i_r} = (\mu(H_j) - \sum \alpha_{i_k}(H_j))\pi(X_i)e_{i_1}...e_{i_r}$$
$$= (\mu(H_j) - \sum \alpha_{i_k}(H_j))\sum_m \delta(i,i_m)(\mu(H_i) - \sum_{k>m}\alpha_{i_k}(H_i))e_{i_1}..e'_{i_m}.e_{i_r}$$

It follows that the coefficient of $e_{i_1}...e'_{i_m}...e_{i_r}$ in $\pi([H_j,X_i])e_{i_1}...e_{i_r}$ is given by

$$\delta(i,i_m)(\mu(H_i) - \sum_{k>m}\alpha_{i_k}(H_i))(\mu(H_j) - \sum_{p\neq m}\alpha_{i_p}(H_j))$$
$$-\delta(i,i_m)(\mu(H_j) - \sum \alpha_{i_k}(H_j))(\mu(H_i) - \sum_{k>m}\alpha_{i_k}(H_i))$$
$$= \delta(i,i_m)\alpha_{i_m}(H_j)(\mu(H_i) - \sum_{k>m}\alpha_{i_k}(H_i))$$
$$= \delta(i,i_m)\alpha_i(H_j)(\mu(H_i) - \sum_{k>m}\alpha_{i_k}(H_i))$$

which is also equal to the coefficient of the same monomial in the expansion of $\alpha_i(H_j)\pi(X_i)e_{i_1}...e_{i_r}$. This shows that

$$\pi([H_j, X_i] - \alpha_i(H_j)X_i) = 0$$

Problem: Discuss Harish Chandra's work of constructing all the finite dimensional irreducible representations of a semisimple Lie algebra. Let $H_i, X_i, Y_i, i = 1, 2, ..., l$ be the generators of the Lie algebra and let $\mathcal{U}$ be the universal enveloping algebra of this Lie algebra. Let λ be a linear functional on $\langle = sp\{H_1, ..., H_l\}$. Consider the left ideal

$$\mathcal{U}_\lambda = \sum_{i=1}^{l} \mathcal{U}X_i + \sum_{i=1}^{l} \mathcal{U}(H_i - \lambda(H_i))$$

Let We have

$$H_i.1 = H_i - \lambda(H_i) + \lambda(H_i)1 = \lambda(H_i) mod \mathcal{U}_\lambda,$$

$$X_i.1 = 0 mod \mathcal{U}_\lambda$$

Suppose $\mathcal{M}_\lambda$ is a maximal ideal containing $\mathcal{U}_\lambda$. Then Harish Chandra proves that this maximal ideal is unique and that if $\lambda(H_i), i = 1, 2, ..., l$ are positive integers, the representation of $\mathcal{U}$ in $\mathcal{U}/M_\lambda$ is irreducible. The following results are used by Harish Chandra. Suppose π is an irreducible representation of $\mathcal{U}$ in a vector space V. Take a nonzero vector $v \in V$ and define

$$\mathcal{M} = \{a \in \mathcal{U} : \pi(a)v = 0\}$$

Then, $\mathcal{M}$ is a left ideal in $\mathcal{U}$ since $\pi(a)v = 0$ implies $\pi(ba)v = \pi(b)\pi(a)v = 0$ so that $a \in \mathcal{M}$ implies $ba \in \mathcal{M}$. Consider the mapping $a \rightarrow \pi(a)v$ from $\mathcal{U}$ into V. Denote this mapping by $T : \mathcal{U} \rightarrow V$. This mapping is linear. The kernel of this mapping is $\mathcal{M}$. It follows that the induced mapping $\tilde{T} : \mathcal{U}/\mathcal{M} \rightarrow V$ defined by $\tilde{T}(x + \mathcal{M}) = \pi(x)v$ is an isomorphism. This mapping intertwines the action of $\mathcal{U}$ on $\mathcal{U}/\mathcal{M}$ with the action of π on V. It follows that the representation of $\mathcal{U}$ in $\mathcal{U}/\mathcal{M}$ is an irreducible representation. Thus, $\mathcal{M}$ is a maximal left ideal of $\mathcal{U}$.

66.Problem: Prove that a family of commuting matrices, each of which is diagonable, is simultaneously diagonable.

Remark:Cartan's theory of roots leading to the classification of semisimple Lie algebras is based on this fact. Given a Lie algebra $\}$, $\langle$ denotes a maximal Abelian subalgebra and $\{ad(H) : H \in \langle\}$ is a commuting family of semisimple linear operators acting on $\}$. It follows that there exist linear functionals $\lambda_i, i = 1, 2, ..., M$ on $\langle$ such that if

$$\}_\lambda = \{X \in \} : [H, X] = \lambda(H)X, \forall H \in \langle\}$$

then

$$\} = \bigoplus_{i=1}^{M} \}_{\lambda_i}$$

67.Let $\}$ be a Lie algebra of a Lie group G and consider the exponential map $exp : \} \to \mathcal{G}$. Let $X, Y \in \}$ and assume that $ad(X), ad(Y)$ are semisimple linear transformations in $\}$. The equation $exp(X) = exp(Y)$ implies that for any $g \in G$,

$$exp(ad(X)).g = exp(X).g.exp(-X) = exp(Y).g.exp(-Y) = exp(ad(Y))g, g \in G$$

so that

$$exp(ad(X)) = exp(ad(Y))$$

From this it can be concluded that if $ad(X)$ and $ad(Y)$ are such that the imaginary part of their eigenvalues all fall in the range $[-\pi, \pi]$, then $ad(X) = ad(Y)$ implying that $ad(X-Y) = 0$, or equivalently, $X - Y \in \ddagger$, where $\ddagger$ is the centre of the Lie algebra $\}$. Set $X - Y = U$. Then $X = Y + U$ and since $U \in \ddagger$, it follows that

$$exp(X) = exp(Y)exp(U)$$

whence $exp(U) = 1$. It follows that if we define $\lfloor$ to be the set of all $X \in \}$ such that the imaginary parts of all the eigenvalues of $ad(X)$ fall in the range $[-\pi, \pi]$, then $exp(X) = exp(Y)$ for $X, Y \in \lfloor$ implies that $X - Y \in D$ where $D = \{U \in \ddagger : exp(U) = 1\}$. Thus, $exp : \lfloor : exp(\lfloor) \subset G^0$ is onto and its kernel is D. It follows that $exp : \lfloor/D \to exp(\lfloor)$ is a diffeomorphism. Here, G^0 is the connected component of G containing e.

68.Let $\mathcal{U}$ be an associative algebra, for example, $\mathcal{U}$ may be taken as the universal enveloping algebra of a Lie algebra and let ρ be a representation of $\mathcal{U}$ in a vector space V. Let $v \in V$ be a nonzero cyclic vector for ρ, ie $V = \rho(\mathcal{U})v$. Define

$$\mathcal{M}_v = \{a \in \mathcal{U} : \rho(a)v = 0\}$$

Then, $\mathcal{M}_v$ is a left ideal of $\mathcal{U}$. Define the map

$$\xi : \mathcal{U}/\mathcal{M}_v \to V$$

so that

$$\xi(a + \mathcal{M}_v) = \rho(a)v$$

Then, ξ is an isomorphism and intertwines the representations ρ_v of $\mathcal{U}$ in $\mathcal{U}/\mathcal{M}_v$ and ρ where

$$\rho_v(a)(b + \mathcal{M}_v) = ab + \mathcal{M}_v$$

The intertwining property follows from

$$\xi.\rho_v(a)(b + \mathcal{M}_v) = \xi(ab + \mathcal{M}_v) = \rho(ab)v = \rho(a)\rho(b)v$$

on the one hand and on the other

$$\rho(a)\xi(b + \mathcal{M}_v) = \rho(a)\rho(b)v$$

so that

$$\xi\rho_v(a) = \rho(a)\xi$$

or

$$\rho_v(a) = \xi^{-1}\rho(a)\xi$$

Let W be any ρ-invariant subspace of V. Consider the set

$$\mathcal{M}_W = \{a \in \mathcal{U} : \rho(a)v \in W\}$$

Then $\mathcal{M}_W$ is a left ideal of $\mathcal{U}$ containing $\mathcal{M}_v$, Conversely, let $\mathcal{M}$ be a left ideal of $\mathcal{U}$ containing $\mathcal{M}_v$. Define

$$W = \{\rho(a)v : a \in \mathcal{M}\}$$

Then, W is a ρ invariant subspace of V and we claim that $\mathcal{M}_W = \mathcal{M}$. Indeed, let $a \in \mathcal{M}_W$. Then, $\rho(a)v \in W$ and hence there exists a $b \in \mathcal{M}$ such that $\rho(a)v = \rho(b)v$, ie, $b - a \in \mathcal{M}_v \subset M$. It follows that $a \in \mathcal{M}$ and this proves that $\mathcal{M}_W \subset \mathcal{M}$. Conversely, let $a \in \mathcal{M}$. Then, $\rho(a)v \in W$, whence, $a \in \mathcal{M}_W$. This proves the claim.

69.Let } be a semisimple Lie algebra and let ⟨ be a Cartan subalgebra of }. ⟨ is a maximal Abelian subalgebra of } and let G_0 denote the normalizer of ⟨, ie, G_0 is the set of all $g \in G$ for which $Ad(g)$⟨ =⊂ ⟨. Then if H is the subgroup of G corresponding to ⟨, $W = G_0/H$ is the Weyl group. It is generated by reflections. Let $\alpha_1, ..., \alpha_r$ denote the positive roots in }. For any root α, s_α is the linear transformation on ⟨ defined by

$$s_\alpha(H) = H - \alpha(H)H_\alpha$$

where $H_\alpha \in$ ⟨ is defined by

$$B(H_\alpha, H) = \alpha(H), H \in \langle$$

where $B(X,Y) = tr(ad(X).ad(Y))$. Note that $\alpha(H_\alpha) = 2$. Note that

$$s_\alpha(H_\alpha) = -H_\alpha$$

Let S(}) denote the symmetric algebra on }. Then S(}) can be identified with the algebra of all polynomials on } using the duality imposed by the Cartan-Killing form. For example, if f is a linear functional on }, then f is identified with the element $X_f \in$ } defined by

$$B(X_f, Y) = f(Y), Y \in \}$$

Note that we are here making use of Cartan's theorem that the Cartan Killing form for a semisimple Lie algebra is non-singular. Suppose for example, f_1, f_2 are linear functionals on }. Then, $X \to f_1(X)f_2(X)$ is a quadratic polynomial on } and the corresponding element in S(}) is given by

$$\frac{1}{2}(X_{f_1} \otimes X_{f_2} + X_{f_2} \otimes X_{f_1}).$$

If $p \in S$(}), then we can define a differential operator $\partial(p)$ acting on smooth functions defined on }. For example if $X \in$ }, then $\partial(X)$ is the differential operator defined by

$$f(Y; \partial(X)) = \frac{d}{dt}f(Y + tX)|_{t=0} = Xf(Y)$$

If $p = \frac{1}{2}(X_1 \otimes X_2 + X_2 \otimes X_1)$, then $\partial(p)$ is the differential operator defined by

$$f(Y, \partial(p)) = X_1 X_2 f(Y) = X_2 X_1 f(Y) = \frac{\partial^2}{\partial t_1 \partial t_2} f(Y + t_1 X_1 + t_2 X_2)|_{t_1 = t_2 = 0}$$

Let now f be a smooth function and define for $x \in G$, $f^x(Y) = f(x^{-1}Y)$, ie $f^x = fox^{-1}$, where xY means $Ad(x)Y = xYx^{-1}$. Let $X \in \}$ and define $\partial(X)^x = \partial(xX)$. We have

$$\partial(X)^x f(Y) = \frac{d}{dt} f(Y + txX)|_{t=0} = \frac{d}{dt} f(x(x^{-1}Y + tX))|_{t=0}$$

$$= X(fox)(x^{-1}Y) = X(f^{x^{-1}})^x(Y)$$

In other words,

$$\partial(X)^x(f) = X(f^{x^{-1}})^x$$

Suppose p, q are elements of $S(\})$. Then, p can be identified with a polynomial in $\}$ while q can be identified with a differential operator $\partial(p)$ acting on smooth functions defined on $\}$. We can now consider the differential operator $p.\partial(q)$. For example, suppose $p = X \in \}$ and $q = Y \in \}$. Then $p\partial(q) = X.\partial(Y)$ is the differential operator defined as follows.

$$p\partial(q) f(Z) = B(X, Z) \frac{d}{dt} f(Z + tY)|_{t=0}$$

Likewise, we can consider the differential operator $\partial(q)p$. This operator is defined so that

$$\partial(q) p f(Z) = \frac{d}{dt} B(Z + tY, X) f(Z + tY)|_{t=0}$$

$$= B(Y, X) f(Z) + B(Z, X) f(Z; Y)$$

70.Let V be a vector space and for $x, y \in V$, define $[x, y] = 0$. Then V becomes an Abelian Lie algebra. Let $\mathcal{G}$ be its universal enveloping algebra. Then for any $x, y \in V$, we have $xy = yxin\mathcal{G}$ since $x \otimes y - y \otimes x - [x, y] = 0$ in $\mathcal{G}$ and $[x, y] = 0$. Thus $\mathcal{G}$ can be identified with the symmetric algebra $S(V)$ on V. Let τ denote the tensor algebra on V and let $\gamma : \tau \to S(V)$ be the natural homomorphism. Note that

$$\gamma(x) = x, \gamma(x \otimes y) = (x \otimes y + y \otimes x)/2,$$

and more generally,

$$\gamma(x_1 \otimes ... \otimes x_n) = \frac{1}{n!} \sum_{\sigma \in S_n} x_{\sigma 1} \otimes ... \otimes x_{\sigma n}$$

$$\gamma(x \otimes y)\gamma(z) = \gamma(x \otimes y \otimes z) = \frac{1}{3!}(xyz + xzy + yxz + yzx + zxy + zyx)$$

Let $\mathcal{M}$ be the two sided ideal in τ generated by elements of the form $x \otimes y - y \otimes x$. Then, it is clear that τ is the direct sum of $S(V)$ and $\mathcal{M}$. The proof is elementary. Let s denote a transposition of two successive elements in $x_1 x_2 ... x_r$. Then it is clear that

$$x_1 x_2 ... x_r - s(x_1 x_2 ... x_r) \in \mathcal{M}$$

Now any permutation σ of $1,2,...r$ is a product of successive transpositions, say $\sigma = s_m s_{m-1}...s_1$. It follows that

$$s_{i+1}s_i...s_1(x_1x_2...x_r) - s_is_{i-1}...s_1(x_1x_2...x_r) \in \mathcal{M}, i = 0,1,...,m-1$$

and hence summing over $i = 0,1,...,m-1$ gives us

$$s_ms_{m-1}...s_1(x_1x_2...x_r) - x_1x_2...x_r \in \mathcal{M}$$

or

$$x_1x_2...x_r - \sigma(x_1x_2...x_r) \in \mathcal{M}, \sigma \in S_r$$

summing over all permutations gives us

$$x_1x_2...x_r - \frac{1}{r!}\sum_{\sigma\in S_r}\sigma(x_1x_2...x_r) \in \mathcal{M}$$

and since

$$\frac{1}{r!}\sum_{\sigma\in S_r}\sigma(x_1x_2...x_r) \in S(V)$$

the claim follows.

71.Problem in groups: Let } be a semisimple Lie algebra and let $B(X,Y) = Tr(ad(X)ad(Y))$ be the Cartan-Killing form on this Lie algebra. Then, $X \to Tr(ad(X)^2)$ is a polynomial. What is the element in the symmetric algebra $S(\})$ associated with this element ?.

72.Another problem in groups: Let $e_1,...,e_n$ be vectors in a vector space V. Consider the differential operator $D = \sum_{i=1}^n \partial(e_i)^2$. Let $B(.,.)$ be a symmetric non-degenerate bilinear form on V. Let $X \in V$ and consider the linear functional $f = B(X,.)$ on V. Determine $[D,f]$.

Ans:

$$[D,f]\psi(Z) = [\sum_{i=1}^n \partial(e_i)^2, B(X,Z)]\psi(Z)$$

$$= \sum_i(\partial(e_i)[\partial(e_i), B(X,Z)]\psi(Z) + [\partial(e_i), B(X,Z)]\partial(e_i)\psi(Z))$$

$$= 2\sum_i B(X,e_i)\partial(e_i)\psi(Z)$$

On the other hand, we note that if $e_i's$ form an onb with respect to $B(.,.)$, then

$$X = \sum_i B(X,e_i)e_i$$

and hence

$$\partial(X) = \sum_i B(X,e_i)\partial(e_i)$$

Thus we get the formula,

$$[D, B(X,.)] = 2\partial(X)$$

73. Groups: $\} = SL(2,\mathbb{C})$ is the Lie group of all 2×2 complex matrices having unit determinant. Its Lie algebra is $sl(2,\mathbb{C})$ consisting of all 2×2 complex matrices having zero trace. $\}_0 = sl(2,\mathbb{R})$ is a real form of $sl(2,\mathbb{C})$. $\mathcal{U}$ consists of all 2×2 orthogonal real matrices having unit determinant, ie, $\mathcal{U} = SO(2)$. $\sqcap$ is a compact real form of $\}$. Here, $\sqcap$ is the Lie algebra of $\mathcal{U}$. We make things more precise here. Let $\sqcap$ consist of all 2×2 complex skew Hermitian matrices of trace zero. Then, $\sqcap$ is a real vector space. Any element $X \in \sqcap$ can be expressed as

$$X = i\begin{pmatrix} a & b+ic \\ b-ic & -a \end{pmatrix}, a, b, c \in \mathbb{R}$$

Let $\surd$ consist of all 2×2 complex Hermitian matrices of trace zero. Then clearly, $\surd = i\sqcap$. $\surd$ is also a real vector space. Suppose X is any matrix in $\}$ consider the linear transformation $ad(X)$ on $\}$. Consider the inner product $(A,B) = Tr(AB^*)$ on $\}$. We then have

$$(ad(X)(A),B) = (XA-AX,B) = Tr((XA-AX)B^*) = Tr(A(B^*X - XB^*))$$

$$= Tr(A(X^*B - BX^*)^*) = (A,[X^*,B]) = (A, ad(X^*)B)$$

It follows that $ad(X)^* = ad(X^*)$. Suppose X is Hermitian and Y is skew-Hermitian in $sl(n,\mathbb{R})$. Then X, Y are real matrices and the transformations $ad(X), ad(Y)$ are real transformations in $sl(n,\mathbb{R})$. Thus, $Tr(ad(X)ad(Y))$ is real, where the trace can be regarded as being in $sl(n,\mathbb{C})$ (We choose a basis for $sl(n,\mathbb{C})$ consisting of vectors in $sl(n,\mathbb{R})$). On the other hand,

$$B(X,Y)^* = Tr((ad(X).ad(Y))^*) = Tr(ad(Y)^*ad(X)^*)$$

$$= Tr(ad(Y^*)ad(X^*)) = -Tr(ad(X)ad(Y)) = -B(X,Y)$$

It follows that $B(X,Y) = 0$. In other words, we have proved the following:Let

$$\sqcap \cap \}_0 = \sqcup_0, i\sqcap \cap \}_0 = \surd_0$$

Then

$$\}_0 = \sqcup_0 \oplus \surd_0, \sqcup_0 \perp \surd_0$$

In our case $n = 2$, we have that $\approx_0$ is spanned by the single element $\begin{pmatrix} 0 & 1 \\ -1 & 0 \end{pmatrix}$ and $\surd_0$ is spanned by the elements $\begin{pmatrix} 1 & 0 \\ 0 & -1 \end{pmatrix}$ and $\begin{pmatrix} 0 & 1 \\ 1 & 0 \end{pmatrix}$. Suppose $\sqcap$ is a compact real form of a semisimple Lie algebra $\}$. Let $\}_0$ be a real form of $\}$ and define $\approx_0 = \sqcap \cap \}_0$ and $\surd_0 = i\sqcap \cap \}_0$. We are assuming here that η is a conjugation of $\}$ with respect to $\}_0$ and that $\eta(\sqcap) = \sqcap$. This means that for each $X \in \}_0$, we have $\eta(X) = X$. Let $\approx$ denote the complexification of $\approx_0$ and $\surd$ the complexification of $\surd_0$. Then we have $\}_0 = \sqcup_0 + \surd_0$ and $\} = \sqcup + \surd$ and both the sums are direct. Consider the conjugation $\theta(X+Y) = X - Y$

where $X \in \sqcup$ and $Y \in \surd$. This has been used as the starting point in Harish Chandra's investigations into the problem of determining the characters of a semisimple Lie group.

74.Let G be a Lie group with Lie algebra $\}$ and let $f : G \to \mathbb{C}$ be a smooth function. Consider the function $\phi : G \times G \to \mathbb{R}$ defined by $\phi(x, m) = f(xmx^{-1})$. Let $X, Y \in \}$ and regard X, Y as differential operators consider

$$\phi(x, X, m, Y) = \frac{\partial^2}{\partial t \partial s}\phi(x.exp(tX), m.exp(sY))|_{t=s=0}$$

Note that

$$\phi(x.exp(tX), m.exp(sY)) = \phi(x.exp(tX).m.exp(sY).exp(-tX)x^{-1})$$

Now observe that

$$\frac{\partial}{\partial t}x.exp(tX).m.exp(sY).exp(-tX)x^{-1}|_{t=0} = x.X.m.exp(sY).x^{-1} - x.m.exp(sY)Xx^{-1}$$

$$= xmm^{-1}Xmexp(sY)x^{-1} - xmx^{-1}xexp(sY)Xx^{-1}$$

$$= xmx^{-1}xAd(m^{-1})(X)exp(sY)x^{-1} - xmx^{-1}xexp(sY)Xx^{-1}$$

$$= xmx^{-1}(Ad(x)Ad(m^{-1})(X))exp(sAd(x)(Y)) - xmx^{-1}exp(s.Ad(x)Y)Ad(x)X$$

A further differentiation with respect to s at $s = 0$ gives

$$\frac{\partial^2}{\partial t \partial s}x.exp(tX)m.exp(sY)exp(-tX)x^{-1}|_{t=s=0}$$

$$= xmx^{-1}Ad(x)Ad(m^{-1}X)Ad(x)(Y) - xmx^{-1}Ad(x)(Y)Ad(x)(X)$$

$$= xmx^{-1}Ad(x)(L_{Ad(m^{-1})X}Y) - xmx^{-1}Ad(x)R_XY$$

so that

$$\phi(x, X, m, Y) = f(xmx^{-1}, Ad(x)W)$$

where

$$W = (L_{Ad(m^{-1})X} - R_X)Y$$

This formula has been used extensively by Harish Chandra in deriving the characters of semisimple Lie groups.

75.Harmonic analysis on semisimple Lie groups by V.S.Varadarajan. This book is mostly about the work of the great Indian mathematician Harish Chandra. The author begins by giving a brief but complete survey of the history of harmonic analysis until the point of Harish Chandra's work. The classical Fourier series analysis has been explained from the standpoint of representation theory. Herein, the group is $SO(2)$ and any smooth function $f : SO(2) \to \mathbb{C}$ can be viewed as a function $f(\theta)$ of a single variable θ having period 2π. Such a function can be developed into a Fourier series

$$f(\theta) = \sum_n c_n exp(in\theta)$$

and this corresponds to a decomposition

$$L^2(SO(2)) = \bigoplus_n \mathcal{H}_n$$

where $\mathcal{H}_n =< exp(in\theta) >$. The regular representation in $L^2(SO(2))$ is thus decomposed into a direct sum of one dimensional representations. Specifically, given an $f \in H_n$, we have

$$f(\theta - \alpha) = exp(-in\alpha)f(\theta)$$

so that the regular representation $U(\alpha)$ when restricted to H_n is simply multiplication by $exp(-in\alpha)$. The author gives a brief summary of the Peter Weyl theorem, the first major result in group representation theory due to Weyl. The idea in brief is as follows. Let G be a compact group and consider continuous functions $f : G \to \mathbb{C}$. This space of functions is denoted by $C(G)$ and we choose an element $a \in C(G)$. The resulting kernel $K_a(g,h) = a(h^{-1}g)$ defines an integral operator $T_a : C(G) \to C(G)$ so that

$$T_a(f)(g) = \int_G K_a(g,h)f(h)d\mu(h)$$

where μ is the Haar measure on G. The operator T_a commutes with the left regular representation. Indeed,

$$(T_aL_hf)(g) = \int K_a(g,x)f(h^{-1}x)d\mu(x) = \int K_a(hg,hx)f(x)d\mu(x) = \int K_a(g,x)f(x)d\mu(x)$$

since

$$K_a(hg,hx) = a((hx)^{-1}hg) = a(x^{-1}g) = K_a(g,x)$$

T_a is moreover a compact operator in $L^2(G)$. it follows that every eigensubspace of T_a is invariant under the left regular representation. From the theory of compact operators in a Hilbert space, it follows that the eigensubspaces of T_a corresponding to nonzero eigenvalues are finite dimensional. The restriction of the left regular representation to each such eigensubspace therefore yields a finite dimensional representation. These facts are then used to prove that the matrix elements of irreducible representations of G span a dense linear manifold in $L^2(G)$ and thereby yield a decomposition of the left regular representation into irreducibles. The author is able to explain these facts with effortless ease, brevity preserving mathematical accuracy. Specifically, suppose f is orthogonal to all the finite dimensional ρ-invariant subspaces of $L^2(G)$ where ρ is the left regular representation. Then, f is orthogonal to all the eigenspaces of T_a corresponding to its nonzero eigenvalues and hence $T_a(f) = 0$. By choosing a sequence $a_n \in C(G)$ such that $a_n * g \to g$ for all $g \in L^2(G)$, it follows that $0 = lim a_n * f = f$ proving the desired completeness property. The author next gives a complete picture of Weyl's character formula for $U(n)$ and states that by using theory of roots, this formula can be generalized to all compact semisimple groups. The key to the Weyl character formula is the Weyl integration formula in which the group $G = U(n)$ acts on itself by conjugation, ie, $x \to gxg^{-1}$. Let D be the subgroup of diagonal matrices. Then since every unitary operator is diagonable, it follows that in every orbit of this action of G on itself, there is at least one element in D. Given any $x \in G$, we have the decomposition $x = gtg^{-1}$ for some $g \in G$. We can look at the transformation $\psi : G/D \times D \to G$ given by $\psi(gD,t) = gtg^{-1}$ and this mapping is a bijection. The coset $gD \in G/D$ can be looked upon as the radial coordinates of gtg^{-1} and $t \in D$ as the angular coordinates of gtg^{-1}. Let

$x \in G$ and let $t_1, ..., t_n$ be the eigenvalues of x. Assume that the distinct eigenvalues of x are $c_1, ..., c_r$ with c_k occurring m_k times. Without loss of generality, assume that

$$x = diag(t_1, ..., t_n) = diag(c_1, ..., c_1, c_2, ..., c_2, ..., c_r, ..., c_r)$$

Let G_x be the stability subgroup of x, ie, $G_x = \{g \in G : gxg^{-1} = x\}$. Then it is clear that $dimG_x = m_1^2 + ... + m_r^2$. Specifically, any element $g \in G_x$ iff $g = diag[A_1, ..., A_r]$ where A_j is any $n_j \times n_j$ matrix. It follows that $dim(G/G_x)$ is a maximum iff $\sum m_j^2$ is a minimum, namely n and this happens iff all the eigenvalues of x are distinct. The subset G' of G consisting of elements x having distinct eigenvalues is dense and open in G and if D' is the corresponding subset of D, we see at once that $D' = G' \cap D$. The author considers the map $\psi : G/D \times D' \to G'$. This map has a bijective differential and its covering map has exactly $n!$ fibres. This can be understood from the fact that if $t_1, ..., t_n$ are n distinct complex numbers, all having unit modulus, then for any permutation $\sigma \in S_n$, we have that $diag(t_{\sigma 1}, ..., t_{\sigma n})$ are all mutually conjugate. This means that if $x \in G'$, then there are exactly $n!$ elements in D' that are conjugate to x. Let $d\bar{x}$ denote the invariant measure on G/D and dt the normalized Haar measure on D. Let dx denote the Haar measure on G. We can view these measures as exterior forms. We are using the usual pullback formula

$$\psi^*(dx) = \Delta\bar{\Delta}d\bar{x}dt$$

where Δ is a function on D which we shall shortly compute. In fact, writing $\psi^*(dx) = \omega(\bar{x}, t)d\bar{x}dt$, it follows that $\omega(\bar{x}, t)$ is the Jacobian of ψ at $(\bar{x}, t)$. Now for any $g \in G$,

$$\psi^*(dx) = \psi^*(d(gx)) = \omega(g\bar{x}, t)d(g\bar{x})dt = \omega(g\bar{x}, t)d\bar{x}dt$$

since the measures dx on G and $d\bar{x}$ on G/D are G-invariant. It follows that $\omega(\bar{x}, t)$ is independent of the choice of the coset $\bar{x}$. Thus, we can write $\omega(\bar{x}, t) = \omega(t)$. To evaluate ω, we need the differential $d\psi$ of ψ at $(1, t)$. We have chosen 1 for the first component because the Jacobian is independent of the first coordinate as we just saw. Now let Z be an element in }, the Lie algebra of G. Then for any real s, $exp(sZ) \in G$ and if $t \in D'$, we have

$$\psi(exp(sZ)D, t) = exp(sZ).t.exp(-sZ) = t.t^{-1}.exp(sZ).t.exp(-sZ) = t.exp(st^{-1}Zt)exp(-sZ)$$

Thus,

$$d\psi_{(1,t)}(Z, H) = H + t^{-1}Zt - Z$$

We need to calculate the determinant of $d\psi_{(1,t)}$. Consider the decomposition $\} = V_1 \oplus V_2$ where V_1 is the space of diagonal matrices having trace zero and V_2 is the space of matrices in } all of whose diagonal elements are zero. The pair (Z, H) assumes values in $\} \times \mathcal{V}_1$. It is clear that $t^{-1}Zt - Z \in V_2$ and hence we get taking a basis $E_{rs}, r \neq s$ for V_2,

$$\omega(t) = det(d\psi_{(1,t)}) = \Pi_{r \neq s}(t_r/t_s - 1) = \Delta(t)\bar{\Delta}(t)$$

where

$$\Delta(t) = \Pi_{r<s}(t_r - t_s)$$

This immediately leads to the Weyl integration formula

$$\int_G f dx = \frac{1}{n!}\int_{G \times D} \Delta(t)\bar{\Delta}(t)f(xtx^{-1})dxdt$$

This is precisely Weyl's integration formula. Suppose $m_1 > m_2 > ... > m_n > 0$ and let $\chi_{m_1...m_n}$ denote the character of D defined by

$$\chi_{m_1...m_n}(t_1, ..., t_n) = t_1^{m_1}...t_n^{m_n}$$

Let $w \in S_n$ and define the character of D $\chi^w_{m_1...m_n}$ by the formula

$$\chi^w_{m_1...m_n}(t_1, ..., t_n) = \chi_{m_1...m_n}(t_{\sigma 1}, ..., t_{\sigma n})$$

Then set

$$\theta_{m_1...m_n} = \Delta^{-1} \sum_{w \in S_n} sgn(w)\chi^w_{m_1...,m_n}$$

where

$$\Delta(t_1, ..., t_n) = \Pi_{1 \le i < j \le n}(t_i - t_j)$$

Note that $\theta_{m_1...m_n}$ is defined on D but its domain can be extended to the whole of G by setting its value at $x \in G$ to be equal to its value at any one of the elements of D conjugate to it. Note that this is consistent since if $\sigma \in S_n$ is arbitrary, then $\theta_{m_1...m_n}$ is symmetric function, ie,

$$\theta_{m_1...m_n}(t_{\sigma 1}, ..., t_{\sigma n}) = \theta_{m_1...m_n}(t_1, ..., t_n)$$

as follows from the antisymmetry of Δ. For example, suppose $n = 3$. Then

$$\Delta(t_1, t_2, t_3) = (t_1 - t_2)(t_2 - t_3)(t_1 - t_3)$$

and this is antisymmetric with respect to interchange of any two indices, ie, under a transposition. Since every permutation is a product of transpositions, the desired skew symmetry property of $\Delta(t)$ follows, ie,

$$\Delta(t_{\sigma 1}, t_{\sigma 2}, ..., t_{\sigma n}) = sgn(\sigma).\Delta(t_1, ..., t_n)$$

Representation theory of $SL(2, \mathbb{C})$ and $SL(2, \mathbb{R})$ has been discussed in considerable detail.

The notion of induced representations has been dealt with with clarity but in brief. The idea was originally introduced into quantum mechanics by G.W.Mackey. Suppose G is a group and H a subgroup of G. Let σ be a representation of H in a Hilbert space X. Consider the set of all functions $f \in L^2(G : X)$ where the measure μ on G is quasi invariant, ie, for every $g \in G$, $\mu o g^{-1}$ is equivalent to μ. Note that two measures μ, ν are said to be equivalent iff given any measurable set E, we have $\mu(E) = 0$ iff $\nu(E) = 0$. A function f is said to be in $L^2(G : X)$ iff $f : G \to X$ is a measurable map and $\int_X \| f(x) \|^2 d\mu(x) < \infty$. Let

$$Y = \{f \in L^2(G : X) : f(gh) = \sigma(h)^{-1} f(g) \forall h \in H\}$$

Suppose $f \in Y$ and $h_1, h_2 \in H$. Then,

$$f(gh_1h_2) = \sigma(h_1h_2)^{-1} f(g) = (\sigma(h_1)\sigma(h_2))^{-1} f(g) = \sigma(h_2)^{-1}\sigma(h_1)^{-1} f(g)$$
$$= \sigma(h_2)^{-1} f(gh_1) = f(gh_1h_2)$$

which shows that the definition is consistent. Then define a representation π_σ of G in Y, so that for $f \in Y$, we have

$$\pi_\sigma(g) f(x) = f(g^{-1}x)$$

Note that $f \in Y$ implies $fog^{-1} \in Y$ from definition. Indeed $f \in Y$ implies $f(xh) = \sigma(h)^{-1}f(x)$ for all $x \in G$ and $h \in H$ and this in turn implies replacing x by $g^{-1}x$ that $f(g^{-1}xh) = \sigma(h)^{-1}f(g^{-1}x)$ for all $g, x \in G, h \in H$ proving that $fog^{-1} \in Y$.

$G = SL(2,\mathbb{R})$:Standard basis:

$$H = \begin{pmatrix} 1 & 0 \\ 0 & -1 \end{pmatrix},$$

$$X = \begin{pmatrix} 0 & 1 \\ 0 & 0 \end{pmatrix},$$

$$Y = \begin{pmatrix} 0 & 0 \\ 1 & 0 \end{pmatrix}$$

Iwasawa decomposition:

$$G = KAN$$

where K is the group of 2×2 real unitary matrices, A is the group of real diagonal matrices with positive diagonal entries and finally N is the group of 2×2 upper triangular matrices with ones on the diagonal. Note that K, A, N are all subgroups of G. Note that this decomposition is also valid for $SL(n,\mathbb{R})$. For $x \in G$, we define

$$u_\theta = \begin{pmatrix} cos(\theta) & -sin(\theta) \\ sin(\theta) & cos(\theta) \end{pmatrix} \in K,$$

$$a_t = \begin{pmatrix} exp(t) & 0 \\ 0 & exp(-t) \end{pmatrix} \in A,$$

$$n_s = \begin{pmatrix} 1 & s \\ 0 & 1 \end{pmatrix} \in N$$

so that

$$x = u_\theta a_t n_s$$

and the real numbers $\theta(mod 2\pi), t, s$ are uniquely determined by $x \in G$. The commutation relations are

$$[H, X] = 2X, [H, Y] = -2Y, [X, Y] = H$$

$SL(2,\mathbb{R})$ is a subgroup of $SL(2,\mathbb{C})$ and we can regard H, X, Y as a basis for $SL(2,\mathbb{C})$ by taking into account complex multiplication factors. Define

$$H' = -i(X - Y), X' = (H + i(X + Y))/2, Y' = (H - i(X + Y))/2$$

Then we claim that the transformation $(H, X, Y) \rightarrow (H', X', Y')$ defines an automorphism of the Lie algebra $\mathit{fl}(2,\mathbb{C})$. To prove this, we evaluate the commutators:

$$[H', X'] = [-i(X - Y), (H + i(X + Y))/2] = -i[X - Y, H]/2 + [X - Y, X + Y]/2$$

$$= -i([X, H] - [Y, H])/2 + [X, Y]/2 - [Y, X]/2 = -i(-X - Y) + H) = H + i(X + Y) = 2X',$$

and likewise,

$$[H',Y'] = -2Y', [X',Y'] = H'$$

Note that

$$[H',Y'] = [-i(X-Y),(H-i(X+Y))/2] = -i[X-Y,H]/2 - [X-Y,X+Y]/2$$

$$= -i(-2X-2Y)/2 - H/2 - H/2 = -H + i(X+Y) = -2Y'$$

Consider the Iwasawa decomposition $G = K.A.N$. If $G = SL(n,\mathbb{C})$, then $K = SU(n)$, A is the subgroup of G consisting of diagonal matrices with positive diagonal entries and N is the subgroup of upper triangular matrices with ones on the diagonal. $B = M.A.N$ is the upper triangular group with M as the diagonal subgroup of $SU(n)$. If $G = SL(n,\mathbb{R})$, then $K = SO(n)$ and M is diagonal with diagonal entries ± 1, ie, M becomes the diagonal subgroup of $SO(n)$. The mapping $(k,a,n) \to kan$ from $K \times A \times N \to G$ is an isomorphism, or more precisely, a diffeomorphism of manifolds. The mapping $K \times B \to G$ defined by $(k,b) \to kb$ is onto but not injective. However the mapping $K/M \times B \to G$ defined by $(kM,b) \to kb$ is a diffeomorphism of manifolds. This follows from the fact that $k_1 b_1 = k_2 b_2$ implies taking $b_1 = m_1 a_1 n_1$ and $b_2 = m_2 a_2 n_2$ that $k_1 m_1 a_1 n_1 = k_1 m_2 a_2 n_2$, then, $k_1 m_1 = k_2 m_2$ and this gives $k_1 M = k_2 M$. Suppose for $x \in G$, we define $k(x) \in K, a(x) \in A, n(x) \in N$ so that $x = k(x).a(x).n(x)$. We then have (a) $k(xy) = k(xk(y))$. This follows since

$$xy = xk(y)a(y)n(y) = k(xk(y))a(xk(y))n(xk(y))a(y)n(y)$$

$$= k(x(k(y))a(xk(y))a(y)a(y)^{-1}n(xk(y))a(y)n(y)$$

and $a(y)^{-1}n(xk(y))a(y) \in N$. Thus,

$$k(xy) = k(xk(y))$$

We next verify that $a(xy) = a(xk(y))a(y)$. This follows immediately from the above formula. Finally, we obtain from the above formula that

$$n(xy) = a(y)^{-1}n(xk(y))a(y)n(y)$$

Put $a(x:u) = a(xu)$ for $x \in G, u \in K$. Then we have obviously $a(1:u) = a(u) = 1$. We can view K as a G-space by the following action $(x,u) \to x[u] = k(xu)$ for $x \in G, u \in K$. The action property follows from

$$x_2 x_1[u] = k(x_2 x_1 u) = k(x_2 k(x_1 u)) = x_2[x_1[u]], x_1, x_2 \in G, u \in K$$

We verify the cocycle identity:

$$a(x_1 x_2 : u) = a(x_1 : x_2[u])a(x_2 : u)$$

Indeed, the left side equals

$$a(x_1 x_2 u) = a(x_1 k(x_2 u))a(x_2 u)$$

from the formula proved above and noting that $x_2[u] = k(x_2 u)$, the proof is complete. To derive the principal series representations, some background in induced representations is needed. Suppose H is a subgroup of G and we have a unitary representation σ of H in a Hilbert space $\mathcal{K}$. Let μ be a quasi invariant measure on G/H. Let $\mathcal{H}$ denote the collection of all functions $f : G \to \mathcal{K}$ for which (a) $f(x\xi) = \sigma(\xi)^{-1}f(x)$ for all $x \in G, \xi inH$ and (b) $\int_{G/H} \| f(x) \|^2 d\mu(x) < \infty$ The second integral is well defined since from the first condition, $\| f(x\xi) \| = \| f(x) \|$ for all $x \in G, \xi \in H$, ie, the function $x \to \| f(x) \|$ is a constant on each H-coset. Then, define the action $\pi(.)$ of G on $\mathcal{H}$ so that

$$\pi(g)f(g') = \rho(g, g^{-1}[g'])^{1/2} f(g^{-1}g'), g' \in G, g \in G$$

where

$$\rho(g, x) = \frac{d\mu}{d\mu o g}(x), x \in G/H$$

We need to check that π is a unitary representation of G. Note first that $\| \pi(g)f(g') \|$ for a fixed g depends only on the coset $g'H$. This follows from the equation

$$(\pi(g)f)(g'h) = \rho(g, g^{-1}[g'])^{1/2} f(g^{-1}g'h) = \rho(g, g^{-1}g')\sigma(h)^{-1}f(g^{-1}g')$$

This means that we can write for $x \in G/H$,

$$\| (\pi(g)f)(x) \|^2 = |\rho(g, g^{-1}x)| \, \| f(g^{-1}x) \|^2$$

Then,

$$\int_{G/H} \| \pi(g)f(x) \|^2 d\mu(x) = \int_{G/H} (d\mu o g^{-1}(x)/d\mu(x)) \| f(g^{-1}x) \|^2 d\mu(x)$$

$$= \int_{G/H} \| f(g^{-1}x) d\mu o g^{-1}(x) = \int_{G/H} \| f(x) \|^2 d\mu(x)$$

proving unitarity of the operator $\pi(g)$. The representation property follows immediately using

$$(\pi(g_2)\pi(g_1)f)(g') = \rho(g_2, g_2^{-1}[g'])^{1/2}(\pi(g_1)f)(g_2^{-1}g')$$

$$= \rho(g_2, g_2^{-1}[g'])^{1/2}\rho(g_1, g_1^{-1}g_2^{-1}[g'])^{1/2} f(g_1^{-1}g_2^{-1}g')$$

Now observe that

$$\rho(g_2, g_2^{-1}[g'])\rho(g_1, g_1^{-1}g_2^{-1}[g'])$$

$$= \rho(g_2g_1, g_1^{-1}g_2^{-1}[g'])$$

In fact, we set $x = g_1^{-1}g_2^{-1}[g']$ and then note that

$$\rho(g_2, g_2^{-1}[g'])\rho(g_1, g_1^{-1}g_2^{-1}[g']) = \rho(g_2, g_1x)\rho(g_1, x)$$

$$= (d\mu(g_1x)/d\mu(g_2g_1x))(d\mu(x)/d\mu(g_1x)) = d\mu(x)/d\mu(g_2g_1x) = \rho(g_2g_1, x)$$

proving the claim. Suppose $(g, x) \to \gamma(g, x)$ is a cocycle. Here, H is a subgroup of G and $X = G/H$. Define

$$\pi(g)f(x) = \rho(g, g^{-1}[x])^{1/2}\gamma(g, g^{-1}[x])f(g^{-1}[x])$$

where $f \in L^2(G/H : K, \mu)$ We want to show that π is equivalent to the induced representation by σ for an appropriate choice of the cocycle γ, ie, if γ is $U(K)$ valued and $\gamma(h, H) = \sigma(h), h \in H$. Let $s : G/H \to G$ be such that $s(x)H = x$ for all $x \in G/H$. Define

$$\xi(g : x) = s(g[x])^{-1}gs(x)$$

Then

$$\xi(g : x)H = s(g[x])^{-1}gx$$

and

$$s(g[x])H = g[x] = gx$$

so that

$$s(g[x])^{-1}gx \in H$$

ie,

$$\xi(g : x) \in H$$

Define

$$\gamma_s(g : x) = \sigma(\xi(g : x))$$

Note that σ is a unitary representation of H in the Hilbert space K. We want to show first that γ_s is a cocycle. Indeed,

$$\gamma_s(g_2, g_1x)\gamma_s(g_1, x) = \sigma(\xi(g_2, g_1x)\xi(g_1, x))$$

and

$$\begin{aligned}\xi(g_2, g_1x)\xi(g_1, x) &= s(g_2g_1[x])^{-1}g_2s(g_1[x])s(g_1[x])^{-1}g_1s(x)\\ &= s(g_2g_1[x])^{-1}g_2g_1s(x) = \xi_s(g_2g_1, x)\end{aligned}$$

proving the cocycle property. Let γ be any σ-cocycle. Then,

$$\gamma(g, x) = \gamma(g, s[x]H)$$

But,

$$\gamma(g, s[x]H)\gamma(s[x], H) = \gamma(gs[x], H)$$

so that

$$\gamma(g, x) = \gamma(gs[x], H)\gamma(s[x], H)^{-1}$$

Next,

$$s(g[x])^{-1}gs(x) = \xi(g, x)$$

so that

$$gs(x) = s(g[x])\xi(g, x)$$

and we get

$$\begin{aligned}\gamma(g, x) &= \gamma(s(g[x])\xi(g, x), H)\gamma(s[x], H)^{-1}\\ &= \sigma(s(g[x])\xi(g, x))\gamma(s[x], H)^{-1}\\ &= \sigma(s(g[x]))\sigma(\xi(g, x))\gamma(s[x], H)^{-1}\end{aligned}$$

$$= \gamma(s(g[x]), H)\gamma_s(g, x)\gamma(s[x], H)^{-1}$$

Define

$$a(x) = \gamma(s[x], H) = \sigma(s[x])$$

Then,

$$\gamma(g, x) = a(g[x])\gamma_s(g, x)a(x)^{-1}$$

Define the map $f \to af$ in $L^2(G/H, K, \mu)$. Call this map T. Clearly T is a unitary operator. We have Let π denote the representation corresponding to the cocycle γ and π_s that corresponding to γ_s. Then,

$$T^{-1}\pi(g)f(x) = a(x)^{-1}\rho(g, g^{-1}[x])^{1/2}\gamma(g, g^{-1}[x])f(g^{-1}[x])$$

$$= \rho(g, g^{-1}[x])^{1/2}a(x)^{-1}a(x)\gamma_s(g, g^{-1}[x])a(g^{-1}[x])^{-1}f(g^{-1}[x])$$

$$= \rho(g, g^{-1}[x])\gamma_s(g, g^{-1}[x])(T^{-1}f)(g^{-1}[x]) = \pi_s(g)T^{-1}f(x)$$

so that

$$T^{-1}\pi(g) = \pi_s(g)T^{-1}$$

or

$$\pi(g) = T\pi_s(g)T^{-1}$$

showing that π and π_s are unitarily equivalent. Suppose now that we define $Uf : G \to K$ so that $Uf(g) = \sigma(\xi(g, H))^{-1}f(gH)$. We then have

$$\int_{G/H} \| Uf(g) \|^2 d\mu(g) = \int_{G/H} \| f(x) \|^2 d\mu(x)$$

and further,

$$Uf(gh) = \sigma(\xi(gh, H))^{-1}f(gH)$$

for $h \in H$. But,

$$\sigma(h)^{-1}Uf(g) = \sigma(h)^{-1}\sigma(\xi(g, H))^{-1}f(gH) = \sigma(\xi(g, H)h)^{-1}f(gH)$$

and

$$\xi(g, H)h = s(gH)^{-1}gh = s(ghH)^{-1}gh = \xi(gh, H)$$

showing that

$$Uf(gh) = \sigma(h)^{-1}Uf(g)$$

For general compact semisimple Lie groups, using Elie Cartan's theory of roots, a brief presentation of the Weyl character formula has been made. Let $G = U(n)$ and let D be the diagonal subgroup of $U(n)$, ie, the set of all t such that

$$t = diag[t_1, ..., t_n], |t_i| = 1, \Pi_{i=1}^n t_i = 1$$

Suppose $g \in SU(n)$. Then by the spectral theorem, there exists a $W \in U(n)$ such that

$$g = WtW^{-1},$$

Since W is unitary, $WW^* = I$ and so $|detW| = 1$ giving $detW = exp(i\alpha)$ for some $\alpha \in \mathbb{R}$. It follows that on defining $V = exp(-i\alpha/n)W$ that $V \in SU(n)$ and $g = VtV^{-1}$. This means that every element of G is conjugate to at least one element of the torus. Let G' denote the subset of all $g \in G$ having distinct eigenvalues. Then G' is conjugate to D' where D' consists of precisely those elements of D having distinct diagonal entries. Every element of G' is conjugate to exactly $n!$ elements of D' each one being related to the other via a permutation. Weyl's character formula states that suppose θ is the character of an irreducible representation π of G, ie, $\theta(x) = tr\pi(x)$. Then there exist non-negative integers $m_1 > m_2 > ... > m_n$ such that if $\chi_{m_1,...,m_n}(t) = \Pi t_i^{m_i}$, then

$$\theta(t) = \Delta(t)^{-1} \sum_{w \in W} sgn(w)\chi^w_{m_1,...,m_n}(t)$$

where $W = S_n$ is the permutation group, also called in general terms, the Weyl group. Here, $\Delta(t) = \Pi_{i<j}(t_i - t_j)$. The justification for this formula is seen as follows. Suppose we restrict π to D. Call this representation π_D. Then π_D is a direct sum of irreducible characters of D. It follows on taking trace that $\theta(t)$ is a finite Fourier series with positive integral coefficients. We must therefore check that the above formula is indeed a finite Fourier series with positive integer coefficients. Take first the case $n = 2$. We have

$$\theta(t) = (t_1^{m_1}t_2^{m_2} - t_1^{m_2}t_2^{m_1})/(t_1 - t_2) = t_1^{m_2}t_2^{m_2}(t_1^{m_1-m_2} - t_2^{m_1-m_2})/(t_1 - t_2)$$

which is evidently a finite Fourier series with positive integer coefficients. Let us now work out the case $n = 3$.

$$\theta(t) = t_1^{m_1}(t_2^{m_2}t_3^{m_3} - t_2^{m_3}t_3^{m_2}) + t_2^{m_1}($$

Let G be a matrix Lie group. Its Lie algebra $\}$ may be defined as the set of all matrices X such that $exp(tX) \in G\ \forall t \in \mathbb{R}$. We denote the Lie algebra of G by $\}$. Then, $X, Y \in \}$ implies $tX + Y \in \}, [X, Y] = XY - YX \in \}$. These can be proved by using elementary properties of groups and properties of the exponential map. For example, $X, Y \in \}$ implies for any integer n that $exp(X/n), exp(Y/n) \in G$ and these imply

$$(exp(X/n).exp(Y/n))^n \in G$$

Taking $\lim n \to \infty$ gives us $exp(X+Y) \in G$ and this holds if we replace X by tX and Y by tY for any real number t. Hence, $X+Y \in \}$. Likewise, suppose $X, Y \in \}$. We want to show that $[X, Y] = XY - YX \in \}$. This follows from the fact that $exp(X/n)exp(Y/n)exp(-X/n)exp(-Y/n) \in \mathcal{G}$ for all $n = 1, 2, ...$ and hence that

$$(exp(X/n)exp(Y/n)exp(-X/n)exp(-Y/n))^{n^2} \in G, n = 1, 2, ...$$

Now,

$$exp(X/n)exp(Y/n) = (1 + X/n + X^2/2n^2 + O(1/n^3))(1 + Y/n + Y^2/2n^2 + O(1/n^3))$$

$$= 1 + X/n + Y/n + (2XY + X^2 + Y^2)/n^2 + O(1/n^3)$$

$$exp(-X/n)exp(-Y/n) = (1 - X/n + X^2/2n^2 + O(1/n^3))(1 - Y/n + Y^2/2n^2 + O(1/n^3))$$
$$= 1 - (X+Y)/n + (2XY + X^2 + Y^2)/2n^2 + O(1/n^3)$$

giving

$$exp(X/n)exp(Y/n)exp(-X/n)exp(-Y/n)$$
$$= 1 + \frac{1}{n^2}(2XY + X^2 + Y^2 - (X+Y)^2) + O(1/n^3) = 1 + \frac{1}{n^2}(XY - YX)$$

from which it follows that

$$(exp(X/n)exp(Y/n)exp(-X/n)exp(-Y/n))^{n^2} \to exp([X,Y]), n \to \infty$$

Replacing X by tX shows that $exp(t[X,Y]) \in G$ for all $t \in \mathbb{R}$ which proves that $[X,Y] \in \}$.

Differential of a representation of a Lie group. Suppose G is a Lie group and $\}$ is its Lie algebra. Let $X \in \}$. Then, X is a left invariant vector field on G. Define $g(t) = exp(tX)(e)$. Thus, $g(0) = e$ and $g'(t) = X(g(t))$. We want to show that $g(t+s) = g(t)g(s)$. To do so, it suffices to show that $\frac{d}{ds}g(t)g(s) = X(g(t)g(s))$. In terms of local coordinates, we write

$$g(t)g(s) = \phi(g(t), g(s))$$

so that

$$\frac{d}{ds}g(t)g(s) = \phi_y(g(t), g(s))X(g(s)) = dL_{g(t)}(X(g(s)) = X(g(t)g(s))$$

by left invariance. This proves the claim.

Representations of $sl(2,\mathbb{C})$. $G = SU(2)$ is the Lie group of 2×2 unitary matrices having unit determinant. Its Lie algebra is $su(2)$ and consists of 2×2 skew Hermitian matrices. Denote this Lie algebra by $\}$. Its complexification is $\}_C$ and it consists of 2×2 complex matrices having zero trace. Indeed, a typical element of $\}_C$ has the form $X + iY$ where $X, Y \in \}$ and any complex matrix A having zero trace is of the form $X + iY$ where X, Y are skew Hermitian matrices. In other words, the complexification $\}_C = sl(2,\mathbb{C})$ and the corresponding Lie group is $G_C = SL(2,\mathbb{C})$. The commutation $sl(2,\mathbb{C})$ is generated by the elements H, X, Y that satisfy the commutation relations

$$[H,X] = 2X, [H,Y] = -2Y, [X,Y] = H$$

Let τ be an irreducible representation of $sl(2,\mathbb{C})$ in a finite dimensional complex vector space. Let λ be an eigenvalue of $\tau(H)$ for which $\lambda + 2$ is not an eigenvalue. Let v be an eigenvector of $\tau(H)$ with eigenvalue μ. Then,

$$\tau(H)\tau(X) - \tau(X)\tau(H) = 2\tau(X)$$

implies

$$\tau(H)\tau(X)v = (\mu + 2)\tau(X)v$$

Thus, $\tau(X)v$ is an eigenvector of $\tau(H)$ with eigenvalue $\mu + 2$. Similarly, $\tau(Y)v$ is an eigenvector of $\tau(H)$ with eigenvalue $\mu - 2$. Let S be the set of eigenvalues of $\tau(H)$ and for any $\mu \in S$, let V_μ denote the corresponding eigenspace, ie,

$$V_\mu = \{v \in V : \tau(H)v = \mu v\}$$

Define

$$V' = \bigoplus_{\mu \in S} V_\mu$$

Then, V' is π-stable. Indeed, if $v \in V'$, then $v \in V_\mu$ for some $\mu \in S$. It follows that $\pi(H)v = \mu v$ and by the above argument, $\pi(H)\pi(X)v = (\mu+2)\pi(X)v$ and $\pi(H)\pi(Y)v = (\mu-2)\pi(Y)v$ so that if $\pi(X)v \neq 0$, then $\mu+2 \in S$ and $\pi(X)v \in V_{\mu+2}$ and likewise, if $\pi(Y)v \neq 0$, then $\mu-2 \in S$ and $\pi(Y)v \in V_{\mu-2}$. Thus, the operators $\pi(X), \pi(Y), \pi(H)$ leave V' invariant and hence, by irreducibility of V, we have $V = V'$, ie, V is a direct sum of weight spaces. We start with a $v \in V_\mu$ and then successively apply the operators $\pi(X)$ to v to get a sequence $\pi(X)^m v, m = 0, 1, 2, ...$ with $\pi(X)^m v$ being an eigenvalue of $\pi(H)$ with eigenvalue $\mu + 2m$. Since the representation is finite dimensional, we must have for some m_0, $\mu(X)^{m_0} v = 0$. Likewise for some positive integer m_1, we must have $\pi(Y)^{m_1} v = 0$. These facts can be used to construct the irreducible representations of $sl(2, \mathbb{C})$. The following commutation relations are used in the derivation. Suppose $Xv = 0$, $Hv = \mu v$ and $v \neq 0$. Let m be the smallest positive integer for which $Y^m v = 0$. For simplicity suppose $Y^2 v = 0$ and $Yv \neq 0$. Then,

$$HYv = (HY - YH + YH)v = (-2Y + YH)v = (\mu - 2)Yv$$

$Y^2 v = 0$ gives

$$XY^2 v = 0$$

so that

$$(XY - YX + YX)Yv = 0$$

or

$$(H + YX)Yv = 0$$

or

$$HYv + Y(XY - YX + YX)v = 0$$

or

$$(\mu - 2)Yv + \mu Yv = 0$$

giving

$$\mu = 1$$

Suppose more generally that $Xv = 0, Hv = \mu v$ and $Y^m v = 0$ but $Y^{m-1} v \neq 0$. Then we have

$$XY^m v = 0$$

from which we can deduce that $\mu = m - 1$. For deducing this, we need a commutation relation between X and Y^m. We have

$$[X, Y^m] = [X, Y]Y^{m-1} + Y[X, Y]Y^{m-2} + ... + Y^{m-1}[X, Y]$$

$$= HY^{m-1} + YHY^{m-2} + ... + Y^{m-1}H$$

from which it follows that

$$0 = XY^m v = [(\mu - 2(m-1)) + (\mu - 2(m-2)) + ... + \mu]Y^{m-1} v = 0$$

giving

$$m\mu = 2(1+2+...+m-1) = m(m-1)$$

or

$$\mu = m-1$$

A commutation relation between X and Y^m:

$$[X,Y^2] = HY + YH = HY + (YH - HY + HY) = HY + 2Y + HY = 2HY + 2Y$$

$$[X,Y^3] = [X,Y]Y^2 + Y[X,Y]Y + Y^2[X,Y] = HY^2 + YHY + Y^2H$$

$$= HY^2 + (YH - HY + HY)Y + Y^2H = HY^2 + (2Y + HY)Y + Y^2H = HY^2 = 2HY^2 + 2Y^2 + Y^2H$$

We want to shift all the $Y's$ to the right. We have

$$Y^2H = HY^2 + [Y^2,H] = HY^2 + Y[Y,H] + [Y,H]Y = HY^2 + 4Y^2$$

so that

$$[X,Y^3] = 2HY^2 + 2Y^2 + HY^2 + 4Y^2 = 3HY^2 + 6Y^2$$

Note that

$$[H,Y^m] = [H,Y]Y^{m-1} + Y[H,Y]Y^{m-2} + ... + Y^{m-1}[H,Y]$$

$$= -2mY^m$$

Thus,

$$[X,Y^m] = [X,Y]Y^{m-1} + Y[X,Y]Y^{m-2} + ... + Y^{m-1}[X,Y]$$

$$= HY^{m-1} + YHY^{m-2} + ... + Y^{m-1}H = \sum_{k=0}^{m-1} Y^k HY^{m-1-k}$$

$$= \sum_{k=0}^{m-1}([Y^k,H]Y^{m-1-k} + HY^{m-1}) = \sum_{k=0}^{m-1} 2kY^{m-1} + mHY^{m-1}$$

$$= m(m-1)Y^{m-1} + mHY^{m-1}$$

We have the alternate formula

$$[X,Y^m] = \sum_{k=0}^{m-1} Y^k HY^{m-1-k} = \sum_{k=0}^{m-1} Y^k([H,Y^{m-1-k}] + Y^{m-1-k}H)$$

$$= \sum_{k=0}^{m-1} Y^k(-2(m-1-k)Y^{m-1-k} + Y^{m-1-k}H) = -2[\sum_{k=0}^{m-1}(m-1-k)]Y^m + mY^mH$$

$$= -m(m-1)Y^m + mY^mH$$

From this relation, we can deduce that the finite dimensional irreducible representations of $sl(2,\mathbb{C})$ have dimensions $m, m = 1,2,....$ (Varadarajan, Harmonic analysis on semisimple Lie groups) such that the highest weight associated with the dimension m subspace is $m-1$. Note that μ is by definition a highest weight if there is a nonzero vector v such that $\pi(H)v = \mu v$ and $\pi(X)v = 0$.

The Iwasawa decomposition for $SL(n,\mathbb{R})$: Let $G = SL(n,\mathbb{R})$. Let $g \in G$. Then by the Gram-Schmidt process, we can write $g = kr$ where k is a real orthogonal matrix and r is upper triangular and real. We can write $r = man$ where m is diagonal with diagonal entries ± 1, a is diagonal with positive diagonal entries and n is upper triangular with ones on its diagonal. Then we have the decomposition $G = KMAN$ or $G = KB$ with K compact and B the upper triangular subgroup of G. Let us take an example illustrating this decomposition. $n = 2$ which means that $G = Sl(2,\mathbb{R})$. The Lie algebra is $sl(2,\mathbb{R})$ and a basis for this is

$$X = \begin{pmatrix} 0 & 1 \\ 0 & 0- \end{pmatrix}, Y = \begin{pmatrix} 0 & 0 \\ 1 & 0 \end{pmatrix},$$

$$H = \begin{pmatrix} 1 & 0 \\ 0 & -1 \end{pmatrix}$$

The subgroup K is $SO(2)$, ie, K consists precisely of the matrices

$$u_\theta = \begin{pmatrix} cos(\theta) & sin(\theta) \\ -sin(\theta) & cos(\theta) \end{pmatrix}$$

K is a one dimensional Lie group with generator $X - Y$. In fact, we have

$$u_\theta = exp(\theta(X - Y))$$

K^G is called the elliptic subgroup of G. Note that

$$K^G = \{gkg^{-1} : g \in G\}$$

Also,

$$exp(tH) = \begin{pmatrix} e^t & 0 \\ 0 & e^{-t} \end{pmatrix}$$

We set

$$A = \{exp(tH) : t \in \mathbb{R}\}$$
$$A^G = \{gag^{-1} : g \in G\}$$

is called the hyperbolic subgroup of G. Note that K is a maximal Abelian subgroup of G. Indeed suppose Z is a 2×2 matrix that commutes with A. Choose a basis for $\mathbb{C}^2$ such that relative to this basis

$$u_\theta = diag[e^{i\theta}, e^{-i\theta}], \theta \in [0, 2\pi)$$

Note that this is possible since $u_\theta, \theta \in [0, 2\pi)$ are semisimple and commute with each other. Then relative to this basis, let

$$Z = \begin{pmatrix} z_{11} & z_{12} \\ z_{21} & z_{22} \end{pmatrix}$$

The equation

$$u_\theta Z = Z u_\theta$$

gives

$$exp(i\theta)z_{11} = z_{11}exp(i\theta), exp(i\theta)z_{12} = z_{12}exp(-i\theta),$$

$$exp(-i\theta)z_{21} = z_{21}exp(i\theta), exp(-i\theta)z_{22} = z_{22}exp(-i\theta)$$

These hold good for all $\theta \in [0, 2\pi)$ and hence

$$z_{12} = z_{21} = 0$$

The condition $det(Z) = 1$ gives $z_{22}^{-1} = z_{11} = z$. Now the basis for $\mathbb{C}^2$ chosen can be seen to be $(1, i)/\sqrt{2}, (1, -i)/\sqrt{2}$ and hence relative to the original basis, we have

$$Z = \frac{z}{2}\begin{pmatrix} 1 & -i \\ i & 1 \end{pmatrix} + \frac{z^{-1}}{2}\begin{pmatrix} 1 & i \\ -i & 1 \end{pmatrix}$$

$$= \frac{1}{2}\begin{pmatrix} z + z^{-1} & -i(z - z^{-1}) \\ i(z - z^{-1}) & z + z^{-1} \end{pmatrix}$$

This must be a real matrix since it belongs to G. It follows that $z^{-1} = \bar{z}$ or $|z|^2 = 1$ proving that $Z \in K$. Likewise, we may easily show that MA is maximal where $M = \{1, \gamma\}$ where $\gamma = -1$. MA is maximal Abelian. Define

$$L = MA = A \cup \gamma A$$

Then, L is maximal Abelian. Note that $G = KB = KMAN$. Explicitly

$$x = u_\theta.a_t.n_s$$

or

$$x = u_\theta.\gamma.a_t.n_s$$

where

$$a_t = exp(tH) = diag[exp(t), exp(-t)], n_s = \begin{pmatrix} 1 & s \\ 0 & 1 \end{pmatrix}$$

Problem: Compute the left and right invariant measures on G. Is G unimodular ? Suppose H is any Lie group and K, B are subgroups of H such that $H = KB$ meaning that the map $(k, b) \to kb$ from $K \times B \to G$ is a diffeomorphism of manifolds. Suppose H is unimodular. H. Then, we can write $dx = j(k, b)d_l k d_r b, x = kb$ where $j : K \times B \to [0, \infty)$ is a differentiable map and $d_l k$ is the left invariant measure on K while $d_r b$ is the right invariant measure on B. Left invariance of dx under K implies that j is independent of k while right invariance of dx under B implies that j is independent of b. Thus, $dx = d_l k d_r b$. (Ref.V.S.Varadarajan, Harmonic analysis on semisimple Lie groups, Cambridge University Press).

Haar measure on $G = SL(2, \mathbb{R})$. Any $x \in G$ can be expressed as $x = u_\theta a_t n_s$ or $x = u_\theta \gamma a_t n_s$. We look at the first connected component. $u_\theta = exp(\theta(X - Y)), a_t = exp(tH), n_s = I + sX = exp(sX)$. We have

$$\frac{\partial x}{\partial s} = u_\theta a_t X = u_\theta a_t n_s X = xX$$

so that

$$\frac{\partial}{\partial s} \to X$$

$$\frac{\partial x}{\partial t} = u_\theta a_t H n_s = u_\theta a_t n_s n_{-s} H n_s = x n_{-s} H n_s$$

so that

$$\frac{\partial}{\partial t} \to n_{-s} H n_s = (I - sX)H(I + sX) = H + s(HX - XH) = H + 2sX$$

Finally,

$$\frac{\partial x}{\partial \theta} = u_\theta (X - Y) a_t n_s = u_\theta a_t n_s n_{-s} a_{-t} (X - Y) a_t n_s$$
$$= x.n_{-s} a_{-t} (X - Y) a_t n_s$$

so that

$$\frac{\partial}{\partial \theta} \to n_{-s} a_{-t} (X - Y) a_t n_s$$

We compute this below: Note that $n_s = I + sX$,

$$\frac{\partial}{\partial \theta} \to n_{-s} a_{-t} (X - Y) a_t n_s =$$

$$= \begin{pmatrix} 1 & -s \\ 0 & 1 \end{pmatrix} \cdot \begin{pmatrix} e^{-t} & 0 \\ 0 & e^t \end{pmatrix} \begin{pmatrix} 0 & 1 \\ -1 & 0 \end{pmatrix}$$
$$\times \begin{pmatrix} e^t & 0 \\ 0 & e^{-t} \end{pmatrix} \cdot \begin{pmatrix} 1 & s \\ 0 & 1 \end{pmatrix}$$
$$= \begin{pmatrix} se^{2t} & s^2 e^{2t} + e^{-2t} \\ -e^{2t} & -se^{2t} \end{pmatrix}$$
$$= se^{2t} H + (s^2 e^{2t} + e^{-2t}) X - e^{2t} Y$$

By inverting these equations, we get

$$X \to \frac{\partial}{\partial s},$$

$$H \to \frac{\partial}{\partial t} - 2s\frac{\partial}{\partial s},$$

$$Y \to s\frac{\partial}{\partial t} + (s^2(1 + exp(-4t)) - 2s^2)\frac{\partial}{\partial s} - e^{-2t}\frac{\partial}{\partial \theta}$$

The exterior product of the three left invariant vector fields X, H, Y therefore evaluates to $exp(-2t)$ apart from a sign factor and hence the Haar measure on G_1 is given by $exp(2t)d\theta dt ds$. Here, G_1 is the connected component of G consisting of matrices having the form $u_\theta a_t n_s$.

Consider $B = AN$ where A is the diagonal subgroup and N is the upper triangular subgroup of $SL(n, \mathbb{R})$. It is clear that for every $a \in, n \in N$ that $ana^{-1} \in N$. Let dn denote the left invariant measure on N. Then it easily follows that $d(ana^{-1}) = \delta(a) dn$ where $a \to \delta(a)$ is a homomorphism from A into $\mathbb{R}$.

Determine the homomorphism $\delta(a)$. Let us do the sample computation for $SL(3,\mathbb{R})$. First we compute dn. Note that $n \in N$ has the form

$$n = \begin{pmatrix} 1 & n_{12} & n_{13} \\ 0 & 1 & n_{23} \\ 0 & 0 & 1 \end{pmatrix}$$

We claim that

$$dn = dn_{12}dn_{13}dn_{23}$$

Let $p \in N$. Then,

$$q = pn = \begin{pmatrix} 1 & p_{12}+n_{12} & p_{13}+p_{12}n_{23} \\ 0 & 1 & n_{23}+p_{23} \\ 0 & 0 & 1 \end{pmatrix}$$

It follows that

$$\frac{\partial(q_{12}, q_{13}, q_{23})}{\partial(n_{12}, n_{13}, n_{23})} = 1$$

This is result that is true for all dimensions. In other words,

$$dn = \Pi_{1\leq i<j\leq n} dn_{ij}$$

We now have for $a = diag[a_1, a_2, a_3]$ the result

$$ana^{-1} = p$$

where

$$p_{ij} = a_i n_{ij}/a_j, i < j, p_{ii} = 1, p_{ij} = 0, i > j$$

This gives

$$\Pi_{i<j} dp_{ij} = [\Pi_{i<j}(a_i/a_j)]\Pi_{i<j} dn_{ij}$$

so that

$$\delta(a) = \Pi_{i<j}(a_i/a_j)$$

Principal series representation:Consider the decomposition $G = KB$ where K is compact and $B = MAN$ with M diagonal having diagonal entries ± 1 A the diagonal subgroup with positive diagonal entries and N upper triangular having ones on its diagonal. Let $\chi = (\xi, \eta)$ be a character of MA so that $\chi(ma) = \eta(a)\xi(m)$ where $m \in M, a \in A$. Then χ can be regarded as a character of B so that $\chi(man) = \eta(a)\xi(m)$. This character is a one dimensional representation of B. It induces a representation of G and we denote this representation by L_χ. Thus, $L_\chi = Ind_B^G \chi$. This is precisely the principal series introduced by Gelfand and Naimark.

Consider $H = KB$. We can view K as an H-space by identifying K with H/B. The action of $x \in H$ on $u \in K$ is given by $u \to x[u]$. Specifically, we have the decomposition $x = k(x)b(x)$ for any $x \in H$ with $k(x) \in K, b(x) \in B$ and $x[u] = k(x.u)$. We have with $B = AN$, $d(ana^{-1}) = \delta(a)^2 dn$ and moreover if μ_B denotes the modular function on B, then for $x = kb$ we have $dx = d_l k d_r b = \mu_B(b) d_l k d_l b$.

For fixed $a \in A$, the map $n \to ana^{-1}$ is an automorphism of N and so $d(ana^{-1}) = c(a)dn$ for some $c(a) > 0$. now for $a, b \in A$, we have

$$d(abn(ab)^{-1}) = c(ab)dn$$

and

$$d(abn(ab)^{-1}) = d(abnb^{-1}a^{-1}) = c(a)d(bnb^{-1}) = c(a)c(b)dn$$

proving that

$$c(ab) = c(a)c(b)$$

Setting $\delta(a) = c(a)^{1/2}$ gives us $d(ana^{-1}) = \delta(a)^2 dn$ where $a \to \delta(a)$ from $A \to \mathbb{R}_+$ is an analytic homomorphism. We have for $a_0 \in A, n_0 \in N$,

$$\int f(a_0 an)dadn = \int f(an)dadn$$

$$\int f(n_0 an)dadn = \int f(aa^{-1}n_0 an)dadn = \int f(an)dadn$$

since $a^{-1}n_0 a \in N$ and dn is left invariant. This shows that in the decomposition $B = AN$, the measure $da.dn$ on B is left invariant under both A and N and is therefore left invariant under B. Thus, $d_l b = dadn$. We next wish to show that the modular function on B is given by $\mu_B(an) = \delta(a)^2$, ie, $d_r b = \delta(a)^2 d_l b$. This follows from the following chain of formulae. First note that N is unimodular and so is A.

$$\int f(ann_0)\delta(a)^2 dadn = \int f(an)\delta(a)^2 dadn$$

$$\int f(ana_0)\delta(a)^2 dadn = \int \delta(a)^2 da f(aa_0 a_0^{-1} na_0)dn = \int \delta(a)^2 da.f(aa_0 n)d(a_0 na_0^{-1})$$

$$= \int \delta(a)^2 da.f(aa_0 n)\delta(a_0)^2 dn = \int \delta(aa_0)^2 f(aa_0 n)dadn = \int \delta(a)^2 f(an)dadn$$

showing that the measure $\delta(a)^2 dadn$ on $B = AN$ is invariant under right translations by both A and N. This proves that $d_r b = \delta(a)^2 dadn$.

Consider the action of H on K. We want to show that for this action,

$$d(x^{-1}[u]) = \mu_B(b(x^{-1}[u]))du$$

To this end, consider a function $g(ub) = f(u)E(b)$ where $u \in K, b \in B$ and $f \in C(K), E \in C(B)$. Thus, $g \in C(B)$. We have Assume that $\int_B E(b)d_r b = 1$. We have

$$\int_K f(u)du = \int f(u)E(b)dud_r b = \int_H g(y)dy = \int_H g(x^{-1}y)dy = \int g(x^{-1}ub)dud_r b$$

$$= \int g(x^{-1}ub)\mu_B(b)dud_l b$$

But,

$$x^{-1}ub = k(x^{-1}u)b(x^{-1}u)b$$

so that

$$\int \mu_B(b)g(x^{-1}ub)dud_lb = \int g(k(x^{-1}u)b)\mu_B(b(x^{-1}u)^{-1}b)dud_lb$$

$$= \int f(x^{-1}[u])E(b)\mu_B(b(x^{-1}u)^{-1}b)dud_lb$$

$$= \int f(x^{-1}[u])E(b)\mu_B(b(x^{-1}u))\mu_B(b)dud_l = \int f(x^{-1}[u])\mu_B(b(x^{-1}u))^{-1}du$$

Thus,

$$d(x^{-1}[u]) = \mu_B(b(x^{-1}u))^{-1}du = \delta(a(x^{-1}u))^{-2}du$$

since as shown above,

$$\mu_B(an) = \delta(a)^2, a \in A, n \in N, B = A.N$$

This formula for the Radon-Nikodym derivative of H action on K is crucial in constructing the principal series representation of $SL(2,\mathbb{C})$ as per Gelfand and Naimark.

Remark:Let $G = SL(2,\mathbb{R})$ so that $G = KAN$ where $K = SO(2)$, $A = diag[e^t, e^{-t}]$ and N consists of matrices of the form $n = \begin{pmatrix} 1 & s \\ 0 & 1 \end{pmatrix}$. We write $x = k(x)a(x)n(x)$ in this Iwasawa decomposition. A normalizes N. The invariant measure on N is $dn = ds$, the Lebesgue measure. We have $d(ana^{-1}) = \delta(a)^2 dn$ where $\delta(a)^2 = a_1/a_2 = a_1^2$ for $a = diag[a_1, a_2]$. Writing $a_1 = exp(t), a_2 = exp(-t)$, we get $\delta(a) = exp(t)$. G acts on K in accord with $(x, u) \to x[u] = k(xu)$ where $x \in G$ and $u \in K$. We define $a(x, u) = a(xu)$ for $x \in G$ and $u \in K$. Then $a : G \times K \to A$ is a cocycle, ie,

$$a(x_1, x_2[u])a(x_2, u) = a(x_1x_2, u)$$

This follows since

$$x_1x_2u = x_1k(x_2u)a(x_2u)n(x_2u) = k(x_1k(x_2u))a(x_1k(x_2u))n(x_1k(x_2u))a(x_2u)n(x_2u)$$

$$= k(x_1k(x_2u))a(x_1k(x_2u))a(x_2u)a(x_2u)^{-1}n(x_1k(x_2u))a(x_2u)n(x_2u)$$

implying that

$$a(x_1x_2u) = a(x_1k(x_2u))a(x_2u) = a(x_1, x_2[u])a(x_2, u)$$

settling the cocycle property. We've seen above that $dx^{-1}[u]/du = \delta(a(x^{-1}u))^{-2}$ and hence the representation π_η of G induced by a quasicharacter η on A is given by

$$\pi_\eta(x)f(u) = (\eta\delta)(a(x^{-1}u)^{-1}f(x^{-1}[u])$$

where the representation space now consists of all square integrable functions f on K. Note that the factor $\delta(a(x^{-1}[u]) = \rho(x, x^{-1}[u])^{1/2} = (dx^{-1}[u]/du)^{1/2}$ arises from the definition of the induced representation. Actually in the definition of the principal series representation, the representation space is all functions $g(an)$ on $G = KB = KAN$ such that $g(xan) = \eta(a)^{-1}g(x)$. In other words, the function g is defined completely by its values on G/B which is isomorphic to K. Define the map $U : g \to Ug$ where g is defined on G in the above fashion and Ug is defined on K so that $g(uan) = \eta(a)^{-1}Ug(u)$, ie, $Ug(u) = \eta(a)g(uan) = g(u), u \in K$ Then,

$$\rho(x, x^{-1}[u])^{1/2}g(x^{-1}y) = \delta(a(x^{-1}u))^{-1}g(k(x^{-1}k(y))a(x^{-1}y)n(x^{-1}y))$$

$$= \delta(a(x^{-1}u))^{-1}\eta(a(x^{-1}u))^{-1}f(x^{-1}[k(y)])$$

where $f = Ug$ and this shows that π_η is equivalent to the induced representation.

76. V.S.Varadarajan, Lie groups, Lie algebras and their representations: Problem in Lie groups: Let G be a Lie group with Lie algebra $\}$. Choose a basis $B = \{X_1, ..., X_m\}$ for $\}$ and introduce corresponding coordinates $\{x_1, ..., x_m\}$ in $\}$. Thus, if $X \in \}$, then the coordinates of x are $dx_i(X) = \alpha_i, i = 1, 2, ..., m$ where $X = \sum \alpha_i X_i$. Let $\{\omega_1, ..., \omega_m\}$ be the basis for $\}^*$ that is dual to the basis B. Thus, $\omega_i(X_j) = \delta_{ij}$. Let $\bar{\omega}_i, i = 1, 2, ..., m$ be the left invariant one forms on G that correspond to $\omega_i, i = 1, 2, ..., m$ respectively at e. Let $\bar{\omega}^*$ be the one form on $\}$ that corresponds to the form $\bar{\omega}$ on G under the exponential mapping. Thus, $\bar{\omega}^* = exp^*\bar{\omega}$. Let $c(mpq)$ be the structure constants of the Lie algebra $\}$ corresponding to the basis B. Thus,

$$[X_i, X_j] = \sum_p c(ijp)X_p$$

Use the abbreviation ω for $\bar{\omega}$. (Suppose X and ω are respectively left invariant vector fields and one forms with components X_i, ω_i relative to a coordinate basis. We have for $y = gx$, with $y_{i,j} = \partial y_i/\partial x_j$,

$$X_i(y) = X_j(x)\partial y_i/\partial x_j, \omega_i(x) = \omega_j(y)\partial y_j/\partial x_i$$

so that

$$\omega_i(y) = \omega_j(x)\frac{\partial x_j}{\partial y_i}$$

implying that

$$\omega_i(y)X_i(y) = \omega_j(x)(\partial x_j/\partial y_i)X_k(x)\partial y_i/\partial x_k = \omega_j(x)X_k(x)\delta_{jk} = \omega_j(x)X_k(x)$$

) This means that $\omega(X)(x)$ is a function independent of x. Now consider

$$\omega_i(X_j) = \delta_{ij}$$

which in terms of components, reads

$$\omega_{im}X_{jm} = \delta_{ij}$$

Differentiate both sides to get

$$\omega_{im,r}X_{jm} + \omega_{im}X_{jm,r} = 0$$

Now,

$$\omega_i = \omega_{im}dx^m$$

$$d\omega_i = \omega_{im,r}dx^r \wedge dx^m$$

so that

$$d\omega_i(X_r, X_s) = \omega_{im,p}(X_{rp}X_{sm} - X_{rm}X_{xp})/2$$

On the other hand, from the above equation, we have

$$\omega_{im,r}X_{jm} = -\omega_{im}X_{jm,r}$$

so that

$$\omega_{im,r}X_{jm}X_{kr} = -\omega_{im}X_{jm,r}X_{kr}$$

and hence,

$$\omega_{im,r}(X_{jm}X_{kr} - X_{km}X_{jr}) = -\omega_{im}(X_{jm,r}X_{kr} - X_{km,r}X_{jr})$$

$$= -\omega_{im}[X_k, X_j]_m = -\omega_i([X_k, X_j]) = -\omega_i(\sum c(kjp)X_p) = -c(kji)$$

From the above two equations, we deduce that

$$d\omega_i(X_r, X_s) = -c(rsi)/2$$

so that

$$d\omega_i = -\frac{1}{2}\sum_{r,s} c(rsi)\omega_r \wedge \omega_s$$

Write

$$\bar{\omega}_i^* = \sum_j a_{ij}(x_1, ..., x_m)dx_j$$

Evaluate both sides at

$$X = t\sum_{j=1}^{m} \alpha_j X_j$$

The right side gives

$$\sum_j a_{ij}(t\alpha_1, ..., t\alpha_m)t\alpha_j$$

The left side gives

$$\bar{\omega}_i^*(X) = (\bar{\omega}_i)|_{expX}(\frac{d}{dt}exptX)|_{t=1})$$

$$= (\bar{\omega}_i)_{expX}(exp(X)X) = \omega_i(X) = t\alpha_i$$

Thus we get the equation

$$\sum_j a_{ij}(t\alpha_1, ..., t\alpha_m)\alpha_j = \alpha_i, i = 1, 2, ..., m$$

Now consider the mapping

$$(t, x_1, ..., x_m) \rightarrow (tx_1, ..., tx_m)$$

from $\mathbb{R}^{m+1}$ into $\mathbb{R}^m$. Let ω_i' denote the pullback of $\bar{\omega}_i^*$ under this map. Then, writing $y_i = tx_i$, we have

$$\omega_i' = \sum_j a_{ij}(y)(\partial y_j/\partial t)dt + \sum_{j,k} a_{ij}(y)(\partial y_j/\partial x_k)dx_k$$

$$= \sum_j a_{ij}(y)x_j dt + \sum_j a_{ij}(y)tdx_j$$

$$= x_i dt + t \sum_j a_{ij}(tx) dx_k$$

where $tx = y$ is the abbreviation for $(tx_1, ..., tx_m) = (y_1, ..., y_m)$. Now we have using properties of the pullback and the identity for $d\omega_i$ that

$$d\omega'_i = -\sum_{r,s} c(rsi)\omega'_r \wedge \omega'_s$$

(If $\phi : \mathcal{M} \to \mathcal{N}$ is a smooth map and $\omega_1, ..., \omega_k$ are forms on $\mathcal{N}$, then $\phi^*(\omega_1 \wedge ... \wedge \omega_k) = \phi^*\omega_1 \wedge ... \wedge \phi^*\omega_k$ and $\phi^* d\omega = d\phi^*\omega$. These are properties of the pullback map that are proven easily.) Now on the one hand,

$$d\omega'_i = d(x_i dt + t \sum_j a_{ij}(tx) dx_j)$$

$$= dx_i \wedge dt + \sum_j (\partial(ta_{ij}(tx))/\partial t) dt \wedge dx_j + t^2 \sum_j (\partial a_{ij}(tx)/\partial y_r) dx_r \wedge dx_j$$

On the other, using the above formula,

$$d\omega'_i = -\sum_{r,s} c(rsi)(x_r dt + t \sum_j a_{rj}(tx) dx_j) \wedge (x_s dt + t \sum_k a_{sk}(tx) dx_k)$$

$$- -\sum tc(rsi) x_r a_{sk}(tx) dt \wedge dx_k - \sum ta_{rj}(tx) c(rsi) x_s dx_j \wedge dt - t^2 \sum c(rsi) a_{rj}(tx) a_{sk}(tx) dx_j \wedge dx_k$$

Comparing coefficients of $dx_j \otimes dt$ gives

$$\delta_{ij} - \partial(ta_{ij}(tx)/\partial t) = t \sum c(rsi) x_r a_{sj}(tx)$$

Defining the matrices $\mathbf{A}(tx) = ((a_{ij}(tx)))$ and $\mathbf{L}(x) = ((l_{is})) = ((\sum_r x_r c(rsi)))$, we can write this equation as

$$\mathbf{I} - \frac{\partial t\mathbf{A}(tx)}{\partial t} = t\mathbf{L}(x)\mathbf{A}(tx)$$

Let $\}$ be a nilpotent Lie algebra, ie if $\}^0 = \}$ and $\}^{i+1} = [\}, \}^i], i = 1, 2, ...$ then for sufficiently large i, we have $\}^i = 0$. This is the same as requiring that there exists a positive integer n such that for all $X_i \in \}, i = 1, 2, ..., n$, we have

$$ad(X_1).ad(X_2)...ad(X_n) = 0$$

We are assuming $\}$ to be a finite dimensional Lie algebra. Let ρ be a representation of $\}$ in a finite dimensional vector space V. For $\lambda \in \}^*$, denote by $V_{\rho,\lambda}$ the vector space of all elements $v \in V$ for which there is a finite positive integer n with the property

$$(\rho(X) - \lambda(X)I)^n v = 0$$

In other words,

$$V_{\rho,\lambda} = \bigcup_{n=0}^{\infty} \mathcal{N}((\rho(X) - \lambda(X)I)^n)$$

We then claim that there exist $\lambda_1, ..., \lambda_k \in \}^*$ such that

$$V = \bigoplus_{i=1}^{k} V_{\rho,\lambda_i}$$

The proof is by induction. We first choose a nonzero element $X_0 \in \}$ and apply the primary decomposition theorem to $\rho(X_0)$., We then get that if $\mu_1, ..., \mu_m$ are the distinct eigenvalues of $\rho(X_0)$, then

$$V = \bigoplus_{i=1}^{m} V_{\rho(X_0),\mu_i}$$

where $dimV_{\rho(X_0),\mu_i} = n_i < dimV, i = 1, 2, ..., m$. We now use induction on the dimension of V. If $dimV = 1$, there is nothing to prove, so suppose that $dimV > 1$ and assume the result to be true when the dimension of V is $N - 1$. Let $dimV = N$ and let $X \in \}$. Then we have

$$ad(X_0)^p X = 0$$

for sufficiently large p since $\}$ is nilpotent. It follows that since ρ is a representation,

$$ad(\rho(X_0))^p \rho(X) = 0$$

and hence $V_{\rho(X_0),\mu_i}$ is invariant under $\rho(X), X \in \}$ (see the result below). By replacing V with $V_{\rho(X_0),\mu_i}$, the required decomposition follows using induction.

Remark: Let V be a vector space and let $T, S \in L(V)$. Let $c_1, ..., c_r$ be the distinct eigenvalues of T. Consider the primary decomposition

$$V = \bigoplus_{i=1}^{r} W_i, W_i = \bigcup_{m \geq 1} \mathcal{N}((T - c_i)^m)$$

Let

$$ad(T)^M S = 0$$

for some positive integer M. Then, we claim that each subspace W_i is invariant under S. Use induction on M. If $M = 1$, then $[T, S] = 0$ and the proof is trivial. So suppose the result is true if M is replaced by $M - 1$. Let E_i be the projection onto W_i. Then, E_i is a polynomial in T. We have that W_i is invariant under $ad(T)^{M-1}S$ since $[T, ad(T)^{M-1}S] = 0$. It follows that

$$E_i ad(T)^{M-1}(S)E_j = 0, i \neq j$$

and hence since $[T, E_i] = 0$ for all i,

$$ad(T)^{M-1}(E_i S E_j) = 0, i \neq j$$

The induction hypothesis now implies that W_j is invariant under $E_i S E_j$ for $i \neq j$ and hence $E_i S E_j = 0$ if $i \neq j$. Thus,

$$S = \sum_i E_i S E_i$$

proving that W_i is invariant under S. Another simpler proof can be given. Since $ad(T)^M S = 0$ with $M \geq 1$, it follows that $ad(T - \mu_i)^M S = 0$ for each i. Write $T_i = T - \mu_i$. Then, $ad(T_i)^M S = 0$. Let $l_i S = T_i S, r_i S = ST_i$. Then $ad(T_i) = l_i - r_i$ and l_i, r_i commute. Thus, r_i and $ad(T_i)$ commute and we have

$$l_i^k S = (r_i + ad(T_i))^k S = \sum_{m=0}^{k} \binom{k}{m} (ad(T_i)^m S) T_i^{k-m}$$

Let $v \in W_i$. Then, $T_i^{k-m} v = 0$ if $k - m$ is sufficiently large and $ad(T_i)^m S = 0$ if m is sufficiently large. it follows that $l_i^k Sv = T_i^k Sv = 0$ if k is sufficiently large and hence $Sv \in W_i$ proving the claim.

Lie groups:Let D be a differential operator acting on functions defined on a manifold M. Let α be a diffeomorphism of M. For a smooth function $u : M \to \mathbb{R}$, define $u^\alpha : M \to \mathbb{R}$ so that

$$\tau_\alpha(u) = u^\alpha = uo\alpha^{-1}$$

Suppose β is another diffeomorphism of M. Then consider

$$\tau_\beta \tau_\alpha(u) = uo\alpha^{-1} o\beta^{-1} = uo(\beta o\alpha)^{-1} = \tau_{\beta\alpha}(u)$$

Thus $\alpha \to \tau_\alpha$ is a homomorphism from the group of diffeomorphisms of M into the group of automorphisms of the algebra $\mathcal{F}(M)$ of smooth functions on M. Note that

$$\tau_\alpha(u+v) = (u+v)o\alpha^{-1} = uo\alpha^{-1} + vo\beta^{-1} = \tau_\alpha(u) + \tau_\alpha(v)$$

and

$$\tau_\alpha(uv) = (uv)o\alpha^{-1} = (uo\alpha^{-1})(vo\alpha^{-1}) = \tau_\alpha(u)\tau_\alpha(v)$$

Now define the action of $\alpha \in Diff(M)$ on differential opertators on $\mathcal{F}(M)$. Let D be any differential operator acting on $\mathcal{F}(M)$. Then we define

$$\tau_\alpha(D) = \tau_\alpha o D o \tau_\alpha^{-1}$$

We see that $\tau_\alpha(D)$ is a differential operator and that if $\alpha, \beta \in Diff(M)$, then

$$\tau_\beta(\tau_\alpha(D)) = \tau_\beta \tau_\alpha D \tau_\alpha^{-1} \tau_\beta^{-1}$$

$$= \tau_{\beta\alpha} o D o \tau_{\beta\alpha}^{-1} = \tau_{\beta\alpha}(D)$$

Differential of a one form on a manifold: Let ω be a one form. In terms of local coordinates, $\omega(X) = \omega_a X^a$ where X is a vector field. Then,

$$\omega = \omega_a dx^a$$

so that

$$d\omega = \omega_{a,b} dx^b \wedge dx^a$$

and

$$d\omega(X,Y) = \omega_{a,b}(X^b Y^a - X^a Y^b)$$

On the other hand,

$$X(\omega(Y)) = X^a(\omega_b Y^b)_{,a} = \omega_{b,a} X^a Y^b + \omega_b X^a Y^b_{,a}$$
$$Y((\omega(X)) = \omega_{b,a} Y^a X^b + \omega_b Y^a X^b_{,a}$$

giving

$$Y(\omega(X)) - X(\omega(Y)) = \omega_{b,a}(X^b Y^a - Y^b X^a) + \omega_b [Y,X]^b$$

It follows that

$$d\omega(X,Y) = Y(\omega(X)) - X(\omega(Y)) + \omega([X,Y])$$

(Varadarajan, Lie groups, Lie algebras and their representations and Abraham and Marsden, Foundations of mechanics).

Definition of a left invariant one form on a Lie group. Let G be a Lie group with Lie algebra }. } is identified with TG_e, the tangent space to G at e. Equivalently, } may be identified with the Lie algebra of all left invariant vector fields on G. If X is a left invariant vector field, then we have

$$X_a = dl_a(X_e)$$

where $l_a : x \to ax$ for any $a, x \in G$ is left translation by a. Let $\phi \in TG^*$. For each $a \in G$, then $\phi_a \in TG^*_a$. l^*_a is pullback. l^*_a maps TG^*_{ax} onto TG^*_x for each $x \in G$. If $l^*_a \phi_{ax} = \phi_x$ for each $a, x \in G$, we say that ϕ is a left invariant one form on G. Let $f : G \to G$ be a diffeomorphism and suppose $f^*\phi_{f(x)} = \phi_x$ for all $x \in G$ and for every left invariant one form ϕ on G. Then, we claim that $f = l_a$ for some $a \in G$. Let $a = f(e)$. Then, $l^*_a \phi_a = \phi_e$ and more generally, $l^*_a \phi_{ax} = \phi_x$ for all $x \in G$. It follows that $l^*_a \phi_{ax} = f^*\phi_{f(x)}$ for all $x \in G$. Now, $\phi_y = l^*_{xy^{-1}} \phi_x$ for all $x, y \in G$ so that, $\phi_{f(x)} = l^*_{xf(x)^{-1}} \phi_x$. It therefore follows that

$$\phi_x = l^*_a \phi_{ax} = f^*\phi_{f(x)} = f^* l^*_{xf(x)^{-1}} \phi_x$$

The components of ϕ_{ax} may be chosen arbitrarily and hence, since l^*_a maps TG^*_{ax} isomorphically onto TG_x, it follows that

$$f^* l^*_{xf(x)^{-1}} = Id$$

when the left side is viewed as a map on TG^*_x. It follows that

$$df^{-1}(x) = dl_{xf(x)^{-1}}$$

when both sides are viewed as linear maps from $TG_{f(x)}$ onto TG_x. Note that f^{-1} maps $f(x)$ to x and so does $l_{xf(x)^{-1}}$. It follows therefore that

$$f^{-1}(y) = l_{xf(x)^{-1}} y$$

for all y in a small nhood of $f(x)$. In terms of local coordinates, writing $y = f(x)\delta$ where δ is close to the identity, we have

$$f^{-1}(f(x)\delta) = x\delta$$

or

$$f(x)\delta = f(x\delta)$$

This is valid upto first order smallness in $\delta - e$. Iterating this, we get

$$f(x)\delta^2 = f(x\delta)\delta = f(x\delta^2)$$

and in general,

$$f(x)\delta^n = f(x\delta^n), n = 1, 2, ...$$

so that

$$f(x)y = f(xy), x, y, \in G$$

from which, we deduce that

$$f(y) = f(e)y$$

which means that f is simply left translation by $f(e)$.

Let X, Y be two left invariant vector fields on a Lie group G. We denote the Lie algebra of G by }. Consider the map $\psi : (s, t) \to G$ defined by

$$\psi(s, t) = exp(sX).exp(tY)$$

Then we have

$$\frac{\partial\psi(s,t)}{\partial s} = exp(sX).X.exp(tY) = exp(sX)exp(tY)exp(-tY)Xexp(tY)$$

$$= \psi(s, t)Ad(exp(-tY))X$$

This means that

$$d\psi(\frac{\partial}{\partial s}) = Ad(exp(-tY))X$$

More generally, let $X_1, ..., X_m$ be left invariant vector fields and define the map $\psi : (t_1, ..., tm) \to exp(t_1X_1)...exp(t_mX_m)$. Then,

$$\frac{\partial\psi}{\partial t_i} = exp(t_1X_1)...exp(t_{i-1}X_{i-1}).X_i.exp(t_{i+1}X_{i+1})...exp(t_nX_n)$$

$$= \psi(t_1, ..., t_n)Ad(y^{-1})X_i$$

where

$$y = exp(t_{i+1}X_{i+1})...exp(t_nX_n)$$

77.Lie groups, an approach through invariants by Procesi:

Salient features:

Given a group G acting on a set X, we consider functions $f : X \to \mathbb{Y}$ where Y is any set. The space of all such functions is denoted by Y^X. The group G acts on this space by left translation, ie,

$g.f(x) = f(g^{-1}x)$. A function f is an invariant iff it is constant on all the G orbits, ie, f takes the same value on all points of $G.x$ for every $x \in X$. Such an invariant function has the form $c.\sum_{x\in O} x$ where O is any G-orbit. x stands for the indicator of the one point set $\{x\}$. This function assumes the value $c \in Y$ at each $x \in O$. Given a group G, $G \times G$ is another group. Its action on G is given by $(g,k)x = gxk^{-1}$. The stability subgroup of $e \in G$ is given by $\Delta = \{(g,g) : g \in G\}$. It follows that the $G \times G$ space G is isomorphic to $G \times G/\Delta$.

The notion of irreducibility has been explained in considerable generality from the stand-point of a ring acting on a module. Given a group G and a vector space V on which the group G acts as a representation, we can consider the group algebra as consisting of all linear operators that can be obtained as linear combinations of the group representation operators, ie, if we denote the group representation element corresponding to the group element g by g itself, then the group algebra elements have the form $\sum_{g\in G} a(g)g$ where $a : G \to \mathbb{C}$ are functions that vanish on all but a finite number of elements of G. The group algebra is denoted by $\mathbb{C}(G)$. It acts on the representation space V just as a ring acts on a module. A ring is a module over itself. If it is semisimple, then this ring as a module can be decomposed into the direct sum of irreducible modules N_i. Each N_i is therefore invariant under left action by R (by definition of module). N_i will in general appear with a certain multiplicity m_i, so that as a module, we can write $R = \bigoplus_i N_i^{m_i}$. Suppose $T \in End_R(R)$. Then, T is in the centralizer of R. It follows from Schur's lemma that the component of T that maps $N_i^{m_i}$ into $N_j^{m_j}$ will be zero if $i \neq j$ and will in general be nonzero when $j = i$. Writing $\Delta_i = End_R(N_i)$, we find that Δ_i is a division algebra and that the $End_R(R) = \bigoplus M_{m_i}(\Delta_i)$. This decomposition has been used with great power in Procesi's book.

Suppose R is a finite dimensional semisimple algebra. We regard R as a module over itself and hence decompose it as $R = \bigoplus R_i$ where $R_i = N_i^{m_i}$ and N_i is an irreducible R-module. It follows from Schur's lemma that

$$End_R(R) = R^o = \bigoplus M_{m_i}(\Delta_i)$$

where $\Delta_i = End_R(N_i)$ is a division ring. Note that we have identified $End_R(R)$ with R^o. Again taking the o operation gives us

$$R = \bigoplus M_{m_i}(\Delta_i^o)$$

Assume that R is a semisimple algebra and has the decomposition

$$R = \bigoplus M_{m_i}(\Delta_i)$$

where Δ_i are division rings. Let N be an irreducible R module. Choose any nonzero $n \in N$ and observe that $Rn = N$. Define the map $T : R \to N$ so that $T(r) = r.n$. Then T is surjective. If $R = \bigoplus N_i^{m_i}$ is the decomposition of R so that N_i is irreducible, and we assume that $n \in N_i$, then Rn is an R invariant subspace of N_i and hence $Rn = N_i$, ie, $N = N_i$. This means that if M is any semisimple module over R, it has the decomposition

$$M = \bigoplus N_i^{p_i}, N_i = \Delta_i^{m_i}$$

Note that R as a module over itself has the decomposition $R = \bigoplus N_i^{m_i}$ since $M_{m_i}(\Delta_i)$ acts irreducibly on $\Delta_i^{m_i}$. It follows that

$$S = End_R(M) = \bigoplus End_R(N_i^{p_i}) = \bigoplus M_{p_i}(\Delta_i^o) = \bigoplus S_i$$

Note that $\Delta_i^o = End_R(N_i) = End_R(\Delta_i^{m_i})$. For suppose $T \in End_R(\Delta_i^{m_i})$. Then for any $r \in M_{m_i}(\Delta_i)$, we must have $Tr = rT$ and hence T must necessarily be of the form right multiplication by an element of Δ_i. In fact, $Tr = rT$ for all $r \in M_{m_i}(\Delta_i)$ means that $T = xI$ for some $x \in \Delta_i$. Then $Tr = rx$ for any $r \in M_{m_i}(\Delta_i)$ and we deduce that $T(x_1, ..., x_{m_i})^T = (x_1 x, ..., x_{m_i} x)^T$ for all $x_i \in \Delta_i$. We can view an element of $N_i^{p_i}$ as an $m_i \times p_i$ matrix with entries from Δ_i and the block S_i acts on $N_i^{p_i}$ by right multiplication of an $m_i \times p_i$ matrix with a $p_i \times p_i$ matrix with entries from Δ_i. Note that $End_R(M)$ is the centralizer of R acting on M. Note also that we can write $N_i^{p_i}$ as $M_{m_i,p_i}(\Delta_i)$ and we thus get a decomposition of the centralizer $End_R(M)$ of R on M as a direct sum of m_i copies of $\Delta_i^{op_i}$. Thus, the representation space of $S = End_R(M)$ decomposes into a direct sum of m_i copies of $\Delta_i^{op_i}$.

Representation theory of the Lie algebra $sl(2, \mathbb{C})$ has been discussed. First a basis $\{e, f, h\}$ for the algebra has been constructed satisfying the commutation rules $[e, f] = h, [h, e] = 2e, [h, f] = 2f$. For a representation space V, a weight vector v has been defined as being an eigenvalue for h with the corresponding eigenvalue as the associated weight. It has been shown that the operators e, f respectively lower and raise the weight by two. This fact has been used to construct a convenient basis for an irreducible representation of $sl(2, \mathbb{C})$. The action of the three operators e, f, h on the basis vectors has been obtained. It has then been shown that the irreducible representations can be obtained as symmetric tensor products of a two dimensional vector space. The notion of even and odd representations has been given.

Tensor algebra, symmetric algebra, exterior algebra and universal enveloping algebra of a Lie algebra. Let A be a skew symmetric matrix of size $2k \times 2k$ and let $e_i, i = 1, 2, ..., 2k$ be the standard basis in $\mathbb{R}^{2k}$. Consider the vector

$$\omega(A) = \sum_{i,j} a_{ij} e_i \wedge e_j$$

Let us compute $\wedge^k \omega(A)$. This is denoted by $Pf(A)$ and is called the Pfaffian of A. Suppose B is any matrix of size $2k \times 2k$. Then

$$\omega(B^t AB) = \sum_{i,j,k,m} b_{ri} a_{rs} b_{sj} e_i \wedge e_j = \sum_{r,s} a_{rs} (\sum_i b_{ri} e_i) \wedge (\sum_j b_{sj} e_j)$$

$$= \sum_{r,s} a_{rs} (Be_i) \wedge (Be_j) = (B \wedge B)\omega(A)$$

Suppose B is any matrix of size $n \times n$. We want to compute $\wedge^n B$. This is computed using its defining action on a vector of the form $u_1 \wedge ... \wedge u_n$ where $u_1, ..., u_n \in \mathbb{C}^n$. We have

$$(\wedge^n B)(u_1 \wedge ... \wedge u_n) = Bu_1 \wedge ... \wedge Bu_n$$

$$= \sum_{i_1, j_1, ..., i_n, j_n} b(j_1, i_1) u_1(i_1) ... b(j_n, i_n) u_n(i_n) e_{j_1} \wedge ... \wedge e_{j_n}$$

$$= \sum_{\sigma, \rho \in S_n} b(\sigma 1, \rho 1) ... b(\sigma n, \rho n) u_1(\rho 1) ... u_n(\rho n) e_{\sigma 1} \wedge ... \wedge e_{\sigma n}$$

$$= \sum sgn(\rho)(detB) u_1(\rho 1) ... u_n(\rho n) e_1 \wedge ... \wedge e_n$$

$$= det(B)det((u_i(j))) = det(B)u_1 \wedge ... \wedge u_n$$

This means that

$$\wedge^n B = det(B).I$$

It follows that

$$Pf(B^tAB) = \wedge^k\omega(B^tAB) = \wedge^{2k}B \wedge \wedge^k\omega(A) = det(B) \wedge^k (A) = det(B)P_f(A)$$

Suppose J is a $2k \times 2k$ matrix obtained by taking the direct sum of k copies of the 2×2 matrix $\begin{pmatrix} 0 & 1 \\ -1 & 0 \end{pmatrix}$.
Given any skew symmetric matrix A, there exists a matrix B such that $A = B^tJB$ and then

$$Pf(A) = Pf(B^tJB) = det(B)Pf(J)$$

where

$$Pf(J) = \wedge^k \sum_i (e_{2i} \wedge e_{2i+1}) = n!I$$

On the other hand,

$$det(A) = det(B)^2$$

so that

$$Pf(A)^2 = det(A)$$

The author

78.(Riener Lenz) Group theory in image processing: The author in this book gives a picturesque description of the applications of group theory to image processing problems. He introduces the Haar measure on the scaling and rotation group. He also introduces the left and right invariant Haar measures on the group $G = \mathbb{R}_+ \times \mathbb{R}$ whose action on $\mathbb{R}$ is defined by $(a, b)x = ax + b$. Note that the composition law for elements in G is given by $(a, b).(a', b')x = (a, b)(a'x+b') = a(a'x+b')+b = aa'x+ab'+b = (aa', ab'+b)x$ so that

$$(a, b)(a', b') = (aa', ab' + b)$$

The author gives a general method for obtaining the Haar measure on a Lie group parametrized by real coordinates. The method is based on Jacobian computations. The author then discusses the notion of a homomorphism between groups and then the idea of a representation. Fourier analysis using representation theory on the group $SO(2)$ is explained and then this is used as the starting point for doing Fourier analysis on the translation, scaling and rotation groups. The notion of an irreducible representation is explained in considerable detail. Representations of the rotation group and the associated spherical harmonics are discussed in detail. Representations of the unimodular group two dimensions are also discussed. While discussing the irreducible representations of the rotation group, the author introduces the reader to the infinitesimal method, ie, Lie algebras. The generators of the Lie algebra of the rotation group are identified with the angular momentum operators that appear in quantum mechanics and the ladder operators are constructed. The spherical harmonics are constructed as eigenfunctions of the squared angular momentum operator. While discussing the unimodular group, the author considers $SL(2, \mathbb{C})$ as acting on $\mathbb{C}^2$. This action induces a representation of the same group on functions of

two complex variables in accord with the formula $f(z_1, z_2) \to f(g^{-1}(z_1, z_2))$. As was originally done by Gelfand, one introduces an invariant inner product on functions of two complex variables and proceeds to derive the principal and supplementary series of representations.

Chapter 8

Control Theory

Control theory applied to electromagnetic and quantum mechanics

1. Optimal tracking of mean trajectory of a quantum mechanical system described by the Heisenberg equations of motion. Hamiltonian: $H = H_0 + V(t)$, H_0 is the unperturbed energy operator, $V(t)$ is the perturbing energy. H_0 is time independent while $V(t)$ is time dependent. $X(t)$ is an observable. Its dynamics is described by the Heisenberg equation

$$X'(t) = i[H_0 + V(t), X(t)] = i((H_0 + V(t))X(t) - X(t)(H_0 + V(t))) = i[H_0, X(t)] + i[V(t), X(t)]$$

This can be regarded as a bilinear control system with $V(t), t \geq 0$ as the input and $X(t), t \geq 0$ as the output. The problem is to design the potential $V(t)$, so that given a state density matrix ρ of the system, $Tr(\rho.X(t))$ tracks a given function $f(t)$ subject to the condition that the input energy is a minimum. In other words using a Lagrange multiplier matrix $\Lambda(t)$, the functional to be extremized is given by

$$\mathcal{L}(X, V, \Lambda) = \int_{t_1}^{t_2} Tr(Q(t)V(t)^2)dt + \int_{t_1}^{t_2} (Tr(\rho.X(t)) - f(t))^2 dt$$

$$+ \int_{t_1}^{t_2} Tr(\Lambda(t)(X'(t) - i[H_0, X(t)] - i[V(t), X(t)]))dt$$

We find that

$$\delta L = 2\int Tr(Q(t)V(t)\delta V(t))dt + 2\int Tr(\rho.\delta X(t))(Tr(\rho.X(t)) - f(t))dt$$

$$+ \int Tr(\delta\Lambda(t)(X'(t) - i[H_0, X(t)] - i[V(t), X(t)])dt$$

$$+ \int Tr(\Lambda(t)(\delta X'(t) - i[H_0, \delta X(t)] - i[V(t), \delta X(t)]))dt$$

$$-i\int Tr(\Lambda(t)[\delta V(t), X(t)])dt$$

After integrating by parts, we equate the coefficient of $\delta X(t)$ to zero to get

$$2\rho(Tr(\rho X(t)) - f(t)) - \Lambda'(t) - i.(\Lambda(t)H_0 - H_0\Lambda(t)) - i(\Lambda(t)V(t) - V(t)\Lambda(t)) = 0$$

This can be regarded as the costate equation. Equating to zero the coefficient of $\delta\Lambda(t)$ gives the state equation

$$X'(t) = i[H_0, X(t)] + i[V(t), X(t)]$$

Equating to zero the coefficient of $\delta V(t)$ gives

$$2Q(t)V(t) - i(X(t)\Lambda(t) - \Lambda(t)X(t)) = 0$$

These equations must be solved.

2. Optimal control of the Schrodinger wave equation: Let

$$H_0 = -\frac{h^2}{2m}\nabla^2 + V(\mathbf{r})$$

where $V(\mathbf{r})$ is a known function of position Let $V_1(\mathbf{r})$ also be a known function of position. The input to this quantum system is a scalar function $u(t)$ of time so that the total Hamiltonian operator becomes

$$H(t) = H_0 + u(t)V_1(\mathbf{r})$$

The wave function $\psi(t, \mathbf{r})$ satisfies the time dependent Schrodinger equation

$$ih\partial_t\psi(t, \mathbf{r}) = (H_0 + u(t)V_1)\psi(t, \mathbf{r}), t \geq 0$$

Let $F(\mathbf{r})$ be an observable of the given quantum system. Its average value evolves with time as

$$< F > (t) = \int F(\mathbf{r})|\psi(t, \mathbf{r})|^2 d^3r$$

The aim is to select the input function $u(t)$ so that given a function $f(t)$ of time, $< F > (t)$ tracks $f(t)$ by minimizing the integral of error squared

$$\int_{t_1}^{t_2} (< F > (t) - f(t))^2 dt$$

Since input energy is constrained, we must minimize the sum of the above and $\int_{t_1}^{t_2} u^2(t)dt$. The total functional to be minimized is therefore given by

$$\mathcal{L}(\psi, \bar{\psi}, u, \lambda, \bar{\lambda}) = \int u^2(t)dt + \int (\int \psi(t, r)\bar{\psi}(t, r)F(r)d^3r - f(t))^2 dt$$

$$+ \int \lambda(t, r)(-\frac{h^2}{2m}\nabla^2\psi(t, r) + V(r) + u(t)V_1(r)\psi(t, r) - i\partial_t\psi(t, r))dtd^3r$$

$$+ \int \bar{\lambda}(t, r)(-\frac{h^2}{2m}\nabla^2\bar{\psi}(t, r) + V(r) + u(t)V_1(r) + i\partial_t\bar{\psi}(t, r))dtd^3r$$

Setting the variation of $\mathcal{L}$ to zero with respect to $u, \psi, \bar{\psi}, \lambda, \bar{\lambda}$ gives us the desired state, costate and input differential equations.

3.Quantum optimal control for density matrices: H_0 is the Hamiltonian of a quantum mechanical system.

Prerequisites

1.Wave function: For a system consisting of N particles in space of three dimensions, the wave function is a map $\psi : \mathbb{R}^{3N} \to \mathbb{C}$ satisfying the normalization condition

$$\int_{\mathbb{R}^{3N}} |\psi(\mathbf{x})|^2 d^N x = 1$$

If B is a Borel subset of $\mathbb{R}^{3N}$, then $\mu(B) = \int_B |\psi(\mathbf{x})|^2 d^N x$ is the probability that the N particles will be found in the region B. For example, if $B = B_1 \times ... \times B_N$ with $B_i \in \mathbb{R}^3$, then $\mu(B) = \mu(B_1 \times ... \times B_N)$ is the probability that the ith particle will be found to be inside the Borel set B_i for each $i = 1, 2, ..., N$.

2.Density matrix: Let $\psi_i(\mathbf{r}), i = 1, 2, ..., N$ be a set of orthonormal wave functions in the Hilbert space $L^2(\mathbb{R}^3)$ and let $p_i, i = 1, 2, ..., N$ be non-negative real numbers that add up to unity. Define

$$\rho(r, r') = \sum_{i=1}^{N} p_i \psi_i(r) \psi_i^*(r')$$

Then, ρ is a density operator of a quantum system. It is corresponds to a mixed state in which there is probability p_i of the system being in the state ψ_i. If X is an observable on the system, then it can be represented by a linear operator having kernel $X(r, r')$ and the average value of X in the state ρ equals

$$Tr(\rho.X) = \sum_{i=1}^{N} p_i < \psi_i|X|\psi_i >= \sum_{i=1}^{N} \int_{\mathbb{R}^6} \psi_i^*(r) X(r, r') \psi_i(r') d^3r d^3r'$$

Suppose $\psi_i(t), iu = 1, 2, ..., N$ are wave functions of the system evolving with time in accord with the Hamiltonian H. Thus,

$$\psi_i'(t) = -iH\psi_i(t)$$

Then it is clear that if we set

$$\rho(t) = \sum_i p_i |\psi_i(t) >< \psi_i(t)|$$

with $p_i's$ as constants, then

$$i\rho'(t) = H\rho(t) - \rho(t)H = [H, \rho(t)]$$

Setting $H = H_0 + f(t)V$, where H_0, V are Hermitian operators and $f(t), t \geq 0$ is a scalar valued function of time, the problem is to design the input function $f(t)$, so that the average values of observables $X_i, i = 1, 2, ..., n$ given by $Tr(\rho(t)X_i), i = 1, 2, ..., n$ track given measured functions $x_i(t), i = 1, 2, ..., n$ respectively, ie, such that

$$\sum_{i=1}^{n} \int_{t_i}^{t_f} (Tr(\rho(t)X_i) - x_i(t))^2 dt$$

is minimized.

3. Observables

4. De-Broglie equation $p = h/\lambda$

5. Schrodinger wave equation

6. Heisenberg matrix mechanics

7.Average value of an observable in a state

4.Optimal Control of electromagnetic systems:Design the current density field so that the electric and magnetic fields in space track a given pattern of variations with minimum expenditure: Suppose $\mathbf{J}(t,\mathbf{r})$ is the current density field, $\mathbf{E}(t,\mathbf{r}), \mathbf{H}(t,\mathbf{r})$ are respectively the electric and magnetic fields. The input energy spent over the duration $[t_1, t_2]$ is given by

$$\int_{t_1}^{t_2} \int_{\mathbb{R}^3} (\mathbf{J}, \mathbf{E}) d^3 r dt$$

The relevant Maxwell equations are

$$curl\mathbf{E} = -\mu \frac{\partial \mathbf{H}}{\partial t}, curl\mathbf{H} = \mathbf{J} + \epsilon . \frac{\partial \mathbf{E}}{\partial t}$$

The Maxwell equations are the constraints to be incorporated into the functional to be extermized by the use of Lagrange multiplier fields $\mathbf{M}(t,\mathbf{r}), \mathbf{N}(t,\mathbf{r})$. The functional to be extremized is thus given by

$$S[\mathbf{J}, \mathbf{E}, \mathbf{H}, \mathbf{M}, \mathbf{N}]$$

$$= \int \mathbf{J}.\mathbf{E} d^4x + \int (\mathbf{M}, \nabla \times \mathbf{E} + \mu \frac{\partial \mathbf{H}}{\partial t}) d^4x$$

$$+ \int (\mathbf{N}, curl\mathbf{H} - \mathbf{J} - \epsilon \frac{\partial \mathbf{E}}{\partial t}) d^4x$$

We may in addition wish to design the input source $\mathbf{J}$ so that the em field tracks a given field pattern. Let this pattern be specified by the functions $\mathbf{E}_0(t,\mathbf{r}), \mathbf{H}_0(t,\mathbf{r})$. Then the functional to be minimized reads

$$S[\mathbf{J}, \mathbf{E}, \mathbf{H}, \mathbf{M}, \mathbf{N}]$$

$$= \int \mathbf{J}.\mathbf{E} d^4x + \int \| \mathbf{E} - \mathbf{E}_0 \|^2 d^4x + \int \| \mathbf{H} - \mathbf{H}_0 \|^2 d^4x + \int (\mathbf{M}, \nabla \times \mathbf{E} + \mu \frac{\partial \mathbf{H}}{\partial t}) d^4x$$

$$+ \int (\mathbf{N}, curl\mathbf{H} - \mathbf{J} - \epsilon \frac{\partial \mathbf{E}}{\partial t}) d^4x$$

Setting the variation of S to zero gives

$$\int (E.\delta J + J.\delta E) d^4x + 2 \int (\delta E, E - E_0) d^4x + 2 \int (\delta H, H - H_0) d^4x +$$

$$\int (\delta M, \nabla \times E + \mu H_{,t}) d^4x + \int (M, \nabla \times \delta E + \mu \delta H_{,t}) d^4x$$

$$+\int(\delta N, curlH - J - \epsilon E_{,t})d^4x + \int(N, curl\delta H - \delta J - \epsilon\delta E_{,t})d^4x$$

We integrate by parts so that

$$\int(M, curl\delta E)d^4x = \int(\nabla, \delta E \times M)d^4x - \int(\delta E, \nabla \times M)d^4x$$

Likewise,

$$\int(N, \delta E_{,t})d^4x = -\int(N_{,t}, \delta E)d^4x + \int(N, \delta E)_{,t}d^4x$$

and the second term drops out by virtue of vanishing boundary conditions.

The first term on the right drops out in view of Gauss' divergence theorem combined with vanishing boundary conditions. Likewise,

$$\int(N, curl\delta H)d^4x = -\int(\delta H, curlN)d^4x$$

$$\int(M, H_{,t})d^4x = -\int(M_{,t}, H)d^4x$$

Thus setting to zero the coefficient of δE gives

$$J + 2(E - E_0) - \nabla \times M + \epsilon N_{,t} = 0$$

This is the first costate equation. Setting to zero the coefficient of δH likewise gives another costate equation

$$2(H - H_0) - curlN - \mu M_{,t} = 0$$

Setting to zero the coefficient of δJ gives

$$E - N = 0$$

Thus, we have in effect two three dimensional equations for E, H and two three dimensional equations for M, J and the solution to this set of equations yields the desired optimal control.

5.Optimal control of the field produced in a wave guide. Assume that the waveguide has a rectangular boundary and that its cross section is orthogonal to the z-direction. Assume that the guide is excited by fields $E_z(t, x, y, 0)$ and $H_z = 0$ at $z = 0$. In other words, we are studying the propagation of TM modes (Transverse magnetic modes). The boundary condition $E_z = 0$ at $x = 0, x = a$ and at $y = 0, y = b$ must be met. Thus, we can write

$$E_z(t, x, y, 0) = \sum_{n,m=1}^{\infty} A_{nm}(t)sin(n\pi x/a)sin(m\pi y/b)$$

The input functions $A_{nm}(t)$ must be determined. We write

$$A_{nm}(t) = Re\int_{-\infty}^{\infty} F_{nm}(\omega)exp(i\omega t).d\omega$$

so that F_{nm} is related to the Fourier transform of the real function A_{nm}. Corresponding to the ωth Fourier component, we get the solution

$$E_{z,n,m}(t,x,y,z) = \sum_{n,m} Re(F_{nm}(\omega)exp(i\omega t - \gamma_{nm}(\omega)z))sin(n\pi x/a)sin(m\pi y/b)$$

where

$$-(n\pi/a)^2 - (m\pi/b)^2 + \gamma_{nm}(\omega)^2 + \omega^2 = 0$$

so that

$$\gamma_{nm}(\omega) = j\sqrt{\omega^2 - h_{nm}^2}, h_{nm} = \sqrt{(n\pi/a)^2 + (m\pi/b)^2}$$

The total field inside the guide can thus be expressed as

$$E_z(t,x,y,z) = \sum_{n,m} Re\{F_{nm}(\omega)exp(i\omega t - \gamma_{nm}(\omega)z)\}sin(n\pi x/a)sin(m\pi y/b)$$

From this expression we calculate E_x, E_y, H_x, H_y as linear functionals of the functions $F_{nm}(\omega)$ and from that expression, we evaluate the Poynting vector at $z = 0$ and its integral across the cross sectional area of the guide at $z = 0$. This input power is to be minimized subject to the constraint that the electric field patter E_z in the guide is a good approximation to a given function, ie, a specified field pattern. Assume that the frequency ω is fixed so that $\{F_{nm}\}$ become constants. Then, the input power (integral of the Poynting vector across the cross section at $z = 0$) can be expressed as a quadratic function of $\{F_{nm}\}$ and a minimization subject to the constraints that E_z tracks a given pattern in space ant time can be obtained. We have from the Maxwell equations with $H_z = 0$ and $\gamma = \gamma_{nm}$,

$$E_{z,y} + \gamma.E_y = -i\omega\mu H_x, -\gamma E_x - E_{z,x} = -i\omega\mu H_y,$$

$$\gamma H_y = i\omega\epsilon E_x, -\gamma H_x = i\omega\epsilon E_y$$

so that

$$E_{z,y} + (i\gamma^2/\omega\epsilon)H_x = -i\omega\mu H_x$$

or

$$H_x = [i\omega\epsilon/(\gamma^2 + \omega^2\mu\epsilon)]E_{z,y} = G_1(\omega)E_{z,y}$$

This is in phasor notation. Transforming to real field notation, we get

$$H_x(t,x,y,z) = \sum_{n,m} Re\{G_1(\omega)F_{nm}.exp(i\omega t - \gamma_{nm}z)\}(m\pi/b).sin(n\pi x/a)cos(m\pi y/b)$$

where

$$G_1(\omega) = \frac{i\omega\epsilon}{\gamma^2 + \omega^2\mu\epsilon}$$

We have

$$E_y = (i\gamma/\omega\epsilon)H_x = -(\gamma/(\gamma^2 + \omega^2\mu\epsilon))E_{z,y} = G_2(\omega)E_{z,y}$$

so that

$$E_y = \sum_{n,m} Re\{G_2(\omega).exp(i\omega t - \gamma_{nm}z)\}(m\pi/b)sin(n\pi x/a)cos(m\pi y/b)$$

Likewise,

$$\gamma E_x + E_{z,x} = i\omega\mu H_y = -(\omega^2\mu\epsilon/\gamma)E_x$$

so that

$$E_x = -\gamma E_{z,x}/(\gamma^2 + \omega^2\mu\epsilon) = G_2(\omega)E_{z,x}$$

or

$$E_x = \sum_{n,m} Re(G_2(\omega)F_{nm}.exp(i\omega t - \gamma_{nm}z))(n\pi/a)cos(n\pi x/a)sin(m\pi y/b)$$

$$H_y = (i\omega\epsilon/\gamma)E_x = -(i\omega\epsilon/(\gamma^2 + \omega^2\mu\epsilon))E_{z,x} = -G_1(\omega)E_{z,x}$$

or

$$H_y = -\sum_{n,m} Re\{G_1(\omega).F_{nm}.exp(i\omega t - \gamma_{nm}z)\}(n\pi/a)cos(n\pi x/a)sin(m\pi y/b)$$

At $z = 0$, we have with $<>$ denoting time average,

$$< E_x H_y >= \frac{1}{2}\sum_{n,m,p,q} Re(G_2G_1^*F_{nm}F_{pq}^*)(n\pi/a)(p\pi/a)cos(n\pi x/a)sin(m\pi y/b)cos(p\pi x/a)sin(q\pi y/b)$$

so that

$$\int_S < E_x H_y > dxdy = \frac{\pi^2 b}{4a}\sum_{n,m} Re(G_2G_1^*|F_{nm}|^2)$$

Likewise,

$$\int_S < E_y H_x > dxdy$$

can be computed and the input power into the system expressed as a quadratic function of F_{nm}.

Another approach to this problem is the dynamic method. $A_{nm}(t, z)$ are functions to be determined so that the input power is a minimum and the electric field tracks a given pattern inside the guide. The field is given by

$$E_z(t, x, y, z) = \sum_{n,m} A_{nm}(t, z)sin(n\pi x/a)sin(m\pi y/b)$$

Plugging this into the Helmholtz equation gives us

$$[\frac{\partial^2}{\partial t^2} - c^2\frac{\partial^2}{\partial z^2} + h_{nm}^2]A_{nm}(t, z) = 0$$

Suppose that we are not interested in the input power but are only interested in the fact that the generated field E_z tracks a given field pattern E_0. Then we must minimize $\int(E_z - E_0)^2 dxdydzdt$ subject to the above constraints. This will give us partial differential equations for A_{nm}. It is not clear at one how to proceed from here. A problem of greater interest is to supply a current density over a given subvolume of the guide say in the region $z = 0$ to $z = d$ and to control this current density so that the field in the guide tracks a given pattern. More generally, let V_1, V_2 be two volumes with $V_1 \subset V_2$. Current density $J(r), r \in V_1$ is supplied and the generated fields in V_2 must approximate a given pattern over a given duration of time. The problem is to formulate this in the language of optimal control and derive appropriate integro-differential equations for the input current density.

Optimal control of the field inside a cavity resonator: Assume for simplicity to start with that the field inside a cuboid resonator is given by a single scalar function $V(t,x,y,z)$ that vanishes at the boundary. Then, the field can be expanded as

$$V(t,x,y,z) = \sum_{n,m,k} V[n,m,k,t]sin(n\pi x/a)sin(m\pi y/b)sin(k\pi z/c)$$

Here, a,b,c are the lengths of the three sides of the resonator. Let $f(t,x,y,z)$ be the driving source function. Then f can also be expanded as a half wave Fourier sine series:

$$f(t,x,y,z) = \sum_{n,m,k} f[n,m,k,t]sin(n\pi x/a)sin(m\pi y/b)sin(k\pi z/c)$$

V satisfies the Helmholtz equation with source f, ie,

$$(\nabla^2 - \frac{1}{v^2}\frac{\partial^2}{\partial t^2})V = f$$

This gives

$$\frac{d^2V[n,m,k,t]}{dt^2} + h[n,m,k]^2 V[n,m,k,t] = f[n,m,k,t]$$

where

$$h[n,m,k] = \pi v\sqrt{(n/a)^2 + (m/b)^2 + (k/c)^2}$$

We want the field inside the guide to track a given function $\psi(t,x,y,z)$. We can expand ψ to as a Fourier series,

$$\psi(t,x,y,z) = \sum_{n,m,k} \psi[n,m,k,t]sin(n\pi x/a)sin(m\pi y/b)sin(k\pi z/c)$$

and then the condition of tracking means that

$$\sum_{n,m,k}\int_0^T (V[n,m,k,t] - \psi[n,m,k,t])^2 dt$$

is a minimum. The problem decouples into optimization for each mode separately, ie, it reduces to the problem of minimizing $\int_0^T(\psi(t) - V(t))^2 dt$ with the constraint

$$V''(t) + \omega^2 V(t) = f(t)$$

The function $\{f(t)\}$ must be determined so that the integral of error square is a minimum for a given function $\psi(t)$. Note that the state equations can be brought to first order form. The state is $[V(t), W(t)]$ which satisfies

$$V'(t) = W(t), W'(t) = -\omega^2 V(t) + f(t)$$

and after introducing Lagrange multipliers $\mu(t), \nu(t)$, the action functional to be minimized is given by

$$S[V,W,f,\mu,\nu] = \int (V(t)-\psi(t))^2 dt + \int \mu(t)(V'(t)-W(t))dt + \int \nu(t)(W'(t)+\omega^2 V(t)-f(t))dt$$

More generally, the cost criterion may be dictated by a non-diagonal matrix in which case, coupling between the modes is involved. The functional to be minimized then has the form

$$S[V,W,f,\mu,\nu] = \sum_{n,m,k,p,q,r} \int (V[n,m,k,t]-\psi[n,m,k,t])A[n,m,k|p,q,r](V[p,q,r,t]-\psi[p,q,r,t])dt$$

$$+\sum_{n,m,k} \int \mu[n,m,k,t](V'[n,m,k,t]-W[n,m,k,t])dt$$

$$+\sum_{n,m,k} \int \nu[n,m,k,t](W'[n,m,k,t]+h[n,m,k]^2 V[n,m,k,t]-f[n,m,k,t])dt$$

Optimal control applied to antenna problems: Consider a linear wire antenna say of length L carrying a current $I(t)$ along its length. The far field magnetic vector potential produced by this wire is given by

$$A(t,r,\theta,\phi) = \hat{z} \int_0^L \frac{\mu}{4\pi r} I(t-r/c+\xi.cos\theta)d\xi$$

Let R denote the wire resistance. Then the total Ohmic heating loss in the wire due to this current over the duration $[0,T]$ is given by $W = R\int_0^T I^2(t)dt$. The electromagnetic field can be computed from the expression of A and the control problem is to design the input $\{I(t)\}$ so that the field tracks a given field over the same duration and over a given region on a sphere of radius r. Note that to compute the electric field, we must use the fact that charge accumulates at the ends of the wire. Equivalently, the Lorentz gauge condition can be used. The charge at the point $z=0$ of the wire is given by

$$-Q(t) = -\int_0^t I(s)ds$$

and the charge at the end $z=L$ of the wire is given by

$$+Q(t) = \int_0^t I(s)ds$$

The far field electric potential is given by

$$V(t,r,\theta,\phi) = \frac{Q(t-r/c+L.cos(\theta)/c)}{4\pi\epsilon r} - \frac{Q(t-r/c-Lcos(\theta)/c)}{4\pi\epsilon r}$$

and the electric field is given by

$$E(t,r,\theta,\phi) = -\nabla V - \frac{\partial A}{\partial t}$$

5.Problem on Optimal control for Hamiltonian dynamics: Assume that there are just two phase variables, one position q and the other momentum p. They satisfy the Hamilton equations of motion $q'(t) = H_p(q(t), p(t), u(t)), p'(t) = -H_q(q(t), p(t), u(t))$ where $u(.)$ is a control input. The cost function over the interval $[0, T]$ is given by

$$C[q, p, u] = \int_0^T F(q(t), p(t), u(t))dt$$

This functional is to be minimized subject to the constraints of the equations of motion. We also impose the final time point constraint $q(T) = \alpha, p(T) = \beta$ where α, β are given real numbers.

6.**Optimal control**

Synopsis for the Ph.D thesis of Dr.A.N.Jha: Optimal control applied to electromagnetic field problems. The standard mathematical methods of optimal control theory involve the following steps:First we proved a model for the dynamic evolution of the state of the system with time in the form of coupled first order nonlinear differential equations with an input process. Then we define a cost functional of the state and input process over the duration of the process. The cost functional to be minimized can correspond to one of several situations like trajectory tracking in which the cost functional measures the mean square deviation of the state process from a desired process or to the minimum fuel problem in which the cost functional is only a functional of the input process. It may also be used to solve minimum time problems in which the functional is the duration of the process. The cost functional is to be minimized subject to the constraints like the equations of motion or to the final state condition in which we desire the final state to satisfy certain constraints. The final state condition may for example, be used to correspond to placing a satellite at a given point in space at the end of the duration and impart the satellite a definite final velocity so that the satellite goes into orbit. The constraints involving the equations of motion in the form of first order ordinary differential equations and terminal conditions are incorporated into the cost functional using Lagrange multipliers. The minimization is then carried out using the standard techniques of variational calculus first developed by Euler and Lagrange. The resulting optimal differential equations are known as the Hamilton-Jacobi equation applied to optimal control. Other methods for optimizing the cost functional have been developed by Richard Bellman. In this thesis, we develop analogous methods for dynamics described by partial differential equations. The specific example considered in this thesis is Maxwell's equations of electromagnetism and its quantized version. We first consider radiation enclosed inside a box described by the two Maxwell equations $\nabla \times \mathbf{E} = -\mu\frac{\partial \mathbf{H}}{\partial t}$ and $\nabla \times \mathbf{H} = \epsilon\frac{\partial \mathbf{E}}{\partial t} + \mathbf{J}$ with $\mathbf{J}$ the required input. Optimal control methods are then developed to determine the excitation current density $\mathbf{J}$ that guarantees that the electromagnetic field $\mathbf{E}(t, \mathbf{x}), \mathbf{H}(t, \mathbf{x})$ follows a desired pattern. The differential equations for the current density input $\mathbf{J}$ are now partial differential equations and these will be solved using numerical methods using MATLAB. The other problem to which optimal control methods will be applied is the generation of an appropriate pattern of propagating electromagnetic waves inside a waveguide. The waveguide is assumed to be rectangular and excited by an appropriate pattern of electric and magnetic fields at one end of it across its cross section. Then, in terms of these initial conditions, we can write down the complete solution for the TE and TM waves propagating inside the guide. These fields are then matched to desired fields using the least squares criterion and optimal equations for the initial excitation are obtained which are solved. The same idea can be applied to the generation of a desired pattern of electromagnetic fields by exciting an antenna of a given shape appropriately. The optimal equations for a waveguide or for an antenna are simple algebraic equations infinite in number though and by truncation the problem can be reduced to solving a set of linear algebraic equations. Having applied optimal control methods to electromagnetic problems, we then turn over to applying the same to problems of quantum electrodynamics. Basically, the electromagnetic field is quantized along the following lines (Weinberg, Quantum field theory): We start with the standard Lagrangian $-\frac{1}{2}F_{\mu\nu}F^{\mu\nu}$ of the electromagnetic field and then by adding a gauge term to it of the form $\partial_\mu A^\mu$, we construct the action for the field. The action is used to then compute the propagator $\Delta_{\mu\nu}(x-y) =< T(A_\mu(x)A_\nu(y)) >$ for the free electromagnetic field. This propagator is then used in conjunction with the interaction $\int J^\mu(x)A_\mu(x)d^4x$ between the current density and the electromagnetic four potential to obtain the scattering amplitudes for photons by an external current source. These scattering amplitudes are then matched with predefined amplitudes to determine the structure of the current density $J^\mu(x)$. The method of computation of the scattering amplitudes is based on the Feynman path integral approach to quantum electrodynamics and can be found in the excellent book by Wienberg on Quantum field theory. The other approach to the quantization of the electromagnetic field is the canonical operator theoretic one in which one first pens down the complete Lagrangian for the electromagnetic field interacting with an external current density. The Lagrangian consists of two terms one the field Lagrangian and two the interaction Lagrangian. The Lagrangian density is used to compute the Hamiltonian density by the canonical Legendre transformation. The canonical position and momentum operators introduced in the process correspond to the components of the vector potential and their time derivatives. Canonical commutation rules are introduced between these observables. Then the field dynamics can be described using the Heisenberg picture or the interaction picture. The current density $J^\mu(x)$ are c-numbers and hence commute with all the observables. We then choose a state for the electromagnetic field and determine the expected values of the fields in this state in terms of the current density. The current density is then determined so that the averaged fields are close in distance to given fields in the least squares sense. This optimization is very hard to carry out and intensive computational tools are required for this purpose. The path integral approach is simpler. Here we determine the mean and correlations of the vector potential fields by evaluating the path integrals

$$\frac{\int A_{\mu_1}(x_1)...A_{\mu_r}(x_r)exp(\frac{2\pi i}{h}\int[-\frac{1}{4}F_{\mu\nu}(x)F^{\mu\nu}(x) - J^\mu(x)A_\mu(x)]d^4x)DA_\mu}{\int exp(\frac{2\pi i}{h}\int[-\frac{1}{4}F_{\mu\nu}(x)F^{\mu\nu}(x) - J^\mu(x)A_\mu(x)]d^4x)DA_\mu}$$

as functionals of J^μ and carrying out an appropriate matching of the moments. It is easy to see that the quantity in the exponential inside the path integral is a quadratic polynomial in the vector potential components and the evaluation of the path integral amounts to evaluating the moments of a multivariate normal distribution having imaginary covariances. This evaluation is conveniently carried out in the momentum domain using the Parseval relation. There is another problem involving optimization in which we matter in the form of electrons and positrons is coupled to an external electromagnetic field over which we have control. We wish to control the external field so that scattering amplitudes of electrons and positrons take appropriate forms. For this we need the quantized version of the Dirac equation. The Lagrangian density for the free Dirac field is given by $\bar{\psi}(x)(i\gamma^\mu\partial_\mu + m)\psi(x)$ where $\bar{\psi}(x) = \psi(x)^*\gamma^0$. The Lagrangian density for matter coupled to the external electromagnetic field is given by $\bar{\psi}(x)(\gamma^\mu(i\partial_\mu + eA_\mu(x)) + m)\psi(x)$. This can be split into the sum of the free field part and the interacting part:

$$\mathcal{L} = \bar{\psi}(x)(i\gamma^\mu\partial_\mu + m)\psi(x) + e\bar{\psi}(x)\gamma^\mu A_\mu(x)\psi(x)$$

The canonical position variables are the fields $\psi(x), \psi^*(x)$. The canonical momenta are

$$\Pi(x) = \frac{\delta \mathcal{L}}{\delta \psi_{,0}} = i\bar{\psi}(x)\gamma^0$$

The Hamiltonian density is therefore given by

$$\Pi(x)\psi(x) - \mathcal{L} = \bar{\psi}(x)i\gamma^r\partial_r\psi(x) - m\bar{\psi}(x)\psi(x) - e\bar{\psi}(x)\gamma^\mu A_\mu(x)\psi(x)$$

The third term is responsible for the scattering processes. An example of such an optimization is the following. We let $|p_1, \sigma_1, p_2, \sigma_2>$ denote the initial state of the matter field in which there is one electron having momentum p_1 spin component σ_1 and another electron having momentum p_2 and spin component σ_2. We let $|p_1', \sigma_1', p_2', \sigma_2'>$ denote the final state of the matter field in which there is an electron having momentum p_1' and spin σ_1' and another electron having momentum p_2' and spin σ_2'. The desired scattering amplitude upto second order in the external field is given by

$$(-ie)^2 \int d^4x d^4y < p_1', \sigma_1', p_2', \sigma_2'|T(\bar{\psi}(x)\gamma^\mu\psi(x).\bar{\psi}(y)\gamma^\nu\psi(y)A_\mu(x)A_\nu(y)|p_1, \sigma_1, p_2, \sigma_2 >$$

where T is the time ordering operator. This can be evaluated using the expansion of the free Dirac field in terms of annihilation operators for the electron field and creation operators for the positron field:

$$\psi_l(x) = \int [u_l(p,\sigma)a(p,\sigma)exp(ip.x) + v_l(p,\sigma)b^*(p,\sigma)exp(-ip.x)]d^3p$$

where the four vector p falls on the mass shell. The following anticommutation rules are used in the evaluation

$$\{a(p,\sigma), a^*(p',\sigma')\} = \delta^3(p-p')\delta_{\sigma,\sigma'}$$
$$\{b(p,\sigma), b^*(p',\sigma')\} = \delta^3(p-p')\delta_{\sigma,\sigma'}$$

The scattering matrix can also be represented by a Feynman diagram. The final form of the scattering amplitude upto second order in the external field has the form

$$\int d^4x d^4y S^{\mu\nu}[p_1', \sigma_1', p_2', \sigma_2', p_1, \sigma_1, p_2, \sigma_2, x, y]A_\mu(x)A_\nu(y)$$

where S is a known function. The problem is to determine the fields $A_\mu(x)$ so that this amplitude is the best approximation to a given amplitude function $\Phi[p_1', \sigma_1', p_2', \sigma_2', p_1, \sigma_1, p_2, \sigma_2]$. More generally, one can compute higher order corrections to the scattering amplitude based on perturbation theory. This leads to an expansion of the form

$$\eta[A_\mu, p_1, \sigma_1, p_2, \sigma_2, p_1', \sigma_1', p_2', \sigma_2']$$
$$= \sum_{r=1}^{\infty} S_r^{\mu_1,\ldots,\mu_r}[p_1', \sigma_1', p_2', \sigma_2', p_1, \sigma_1, p_2, \sigma_2, x_1, \ldots, x_r]A_{\mu_1}(x_1)\ldots A_{\mu_r}(x_r)d^4x_1\ldots d^4x_r$$

and the external field is determined by minimizing

$$\sum_\sigma \int |\eta[A_\mu, p, \sigma] - \Phi[p\, sigma]|^2 d^3p_1 d^3p_2 d^3p_1' d^3p_2'$$

with respect to A_μ using a variational principle. To our present knowledge, we do not know of any such work in the literature.

Continuation of Ph.D thesis problem of A.N.Jha: Optimal control in the equations of general relativity. The aim of this class of problems is to design and control the source of the gravitational field, namely matter density and its flux ie velocity field so that the gravitational field produced by it follows a definite pattern. This is analogous to controlling the current distribution in an antenna so that the electromagnetic radiation field produced by it follows a definite pattern in space-time. The application is to the generation of gravitational waves that would enable us to simulate in the laboratory the gravitational waves produced by a pulsar in the sky. Consider a pulsar in the sky that rotates and produces gravitational waves that travel till the earth's surface which are detected by a gravitational wave detector. We measure the pattern of the gravitational waves and attempt to simulate the matter distribution and velocity in the laboratory that will produce a pattern of gravitational waves similar to the one measured due to the pulsar rotation. This is an optimal control problem.

We need first of all, to separate the time derivative part of the Einstein tensor $G_{\mu\nu}$. First start with separating out the time derivative part of the Ricci tensor $R_{\mu\nu}$.

$$R_{\mu\nu} = \Gamma^\alpha_{\mu\alpha,\nu} - \Gamma^\alpha_{\mu\nu,\alpha} - \Gamma^\alpha_{\mu\nu}\Gamma^\beta_{\alpha\beta} + \Gamma^\alpha_{\mu\beta}\Gamma^\beta_{\nu\alpha}$$

$$\Gamma^\alpha_{\mu\alpha} = g^{\alpha\beta}\Gamma_{\beta\mu\alpha}$$

Consider first the term R_{00}. It is given by

$$R_{00} = \Gamma^\alpha_{0\alpha,0} - \Gamma^\alpha_{00,\alpha} - \Gamma^\alpha_{00}\Gamma^\beta_{\alpha\beta} + \Gamma^\alpha_{0\beta}\Gamma^\beta_{0\alpha}$$

$$\Gamma^\alpha_{0\alpha} = \frac{1}{2}g^{\alpha\beta}(g_{\beta 0,\alpha} + g_{\beta\alpha,0} - g_{\alpha 0,\beta})$$

$$= \frac{1}{2}g^{\alpha\beta}g_{\alpha\beta,0}$$

$$\Gamma^\alpha_{0\alpha,0} = \frac{1}{2}g^{\alpha\beta}g_{\alpha\beta,00} + \frac{1}{2}g^{\alpha\beta}_{,0}g_{\alpha\beta,0}$$

$$= \frac{1}{2}g^{\alpha\beta}g_{\alpha\beta,00} - \frac{1}{2}g^{\mu\alpha}g^{\nu\beta}g_{\mu\nu,0}g_{\alpha\beta,0}$$

This equation shows that the only partial derivatives of the metric tensor that appear in the first term of R_{00} are the partial derivatives with respect to time. We shall in what follows, use the Roman indices i, j, k, m etc to assume values $1, 2, 3$, ie, spatial indices and the greek letters $\mu, \nu, \alpha, \beta, \rho, \sigma$ etc. to assume all the four values $0, 1, 2, 3$, ie, space-time indices. Then,

$$\Gamma^\alpha_{00} = \frac{1}{2}g^{\alpha\beta}\Gamma_{\beta 00} = \frac{1}{2}g^{\alpha 0}\Gamma_{000} + \frac{1}{2}g^{\alpha i}\Gamma_{i00}$$

$$= \frac{1}{4}g^{\alpha 0}g_{00,0} + \frac{1}{4}g^{\alpha i}(2g_{i0,0} - g_{00,i})$$

$$\Gamma^\alpha_{00,\alpha} = \Gamma^0_{00,0} + \Gamma^i_{00,i}$$

$$= \frac{1}{4}(g^{00}g_{00,0})_{,0} + \frac{1}{4}(2(g^{0i}g_{0i,0})_{,0} - (g^{0i}g_{00,i})_{,0})$$

$$+\frac{1}{4}(g^{i0}g_{00,0})_{,i} + \frac{1}{4}((2g^{ij}g_{i0,0})_{,j} - (g^{ij}g_{00,i})_{,j})$$

$$= \frac{1}{4}g^{00}g_{00,00} - \frac{1}{4}g^{0\alpha}g^{0\beta}g_{\alpha\beta,0}g_{00,0} + \frac{1}{2}g^{0i}g_{0i,00}$$

$$-\frac{1}{2}g^{0\alpha}g^{i\beta}g_{\alpha\beta,0}g_{0i,0} - \frac{1}{4}g^{0i}g_{00,i0} + \frac{1}{4}g^{0\alpha}g^{i\beta}g_{\alpha\beta,0}g_{00,i}$$

$$+\frac{1}{4}g^{0i}g_{00,i0} - \frac{1}{4}g^{i\alpha}g^{0\beta}g_{\alpha\beta,i}g_{00,0}$$

$$+\frac{1}{2}g^{ij}g_{i0,j0} - \frac{1}{2}g^{i\alpha}g^{j\beta}g_{\alpha\beta,j}g_{i0,0} - \frac{1}{4}g^{ij}g_{00,ij} + \frac{1}{4}g^{i\alpha}g^{j\beta}g_{\alpha\beta,j}g_{00,i}$$

We now focus on the third and fourth terms in R_{00}, ie, the nonlinear terms. The third term is

$$\Gamma^{\alpha}_{0\beta}\Gamma^{\beta}_{0\alpha}$$

$$= \Gamma^{02}_{00} + \Gamma^{0}_{0i}\Gamma^{i}_{00} + \Gamma^{i}_{00}\Gamma^{0}_{0i} + \Gamma^{i}_{0j}\Gamma^{j}_{0i}$$

$$\Gamma^{0}_{00} = g^{0\alpha}\Gamma_{\alpha 00} = \frac{1}{2}g^{0\alpha}(2g_{\alpha 0,0} - g_{00,\alpha})$$

$$\Gamma^{0}_{0i} = g^{00}\Gamma_{00i} + g^{0j}\Gamma_{j0i}$$

$$= \frac{1}{2}g^{00}g_{00,i} + \frac{1}{2}g^{0j}(g_{j0,i} + g_{ji,0} - g_{i0,j})$$

$$\Gamma^{i}_{00} = g^{i0}\Gamma_{000} + g^{ij}\Gamma_{j00}$$

$$= \frac{1}{2}g^{i0}g_{00,0} + \frac{1}{2}g^{ij}(2g_{j0,0} - g_{00,j})$$

$$\Gamma^{i}_{0j} = g^{i0}\Gamma_{00j} + g^{ik}\Gamma_{k0j}$$

$$= \frac{1}{2}g^{i0}g_{00,j} + \frac{1}{2}g^{ik}(g_{k0,j} + g_{kj,0} - g_{j0,k})$$

In this manner, each of the components of $R_{\mu\nu}$ can be decomposed into components containing $g_{\alpha\beta,00}, g_{\alpha\beta,0}\; g_{\alpha\beta,i0}, g_{\alpha\beta,ij}$. The energy-momentum tensor of the matter distribution is $T_{\mu\nu}$ and the Einstein field equation reads

$$R_{\mu\nu} - \frac{1}{2}Rg_{\mu\nu} = KT_{\mu\nu}$$

Contracting the indices gives us

$$-R = KT$$

where

$$T = g^{\mu\nu}T_{\mu\nu}$$

so that the Einstein field equations can be expressed as

$$R_{\mu\nu} + \frac{K}{2} T g_{\mu\nu} = K T_{\mu\nu}$$

or

$$R_{\mu\nu} = K(T_{\mu\nu} - \frac{1}{2} T g_{\mu\nu}) = K S_{\mu\nu}$$

where

$$S_{\mu\nu} = T_{\mu\nu} - \frac{1}{2} T g_{\mu\nu}$$

is the modified energy-momentum tensor of the matter distribution. For the electromagnetic field,

$$T_{\mu\nu} = g^{\alpha\beta} F_{\mu\alpha} F_{\nu\beta} - \frac{1}{4} F_{\alpha\beta} F^{\alpha\beta} g_{\mu\nu}$$

For just matter distribution having density ρ and pressure p, we have

$$T_{\mu\nu} = (p + \rho) v_\mu v_\nu - p g_{\mu\nu}$$

Finally, the Einstein Field equations can be cast in the form

$$g_{\alpha\beta,00}(x) = H_{\alpha\beta}(g_{\mu\nu,0}(x), g_{\mu\nu,i}(x), g_{\mu\nu,i0}(x), g_{\mu\nu,ij}(x)) + S_{\alpha\beta}(x)$$

The problem is to determine the control function $S_{\alpha\beta}(x)$ so that the field pattern $g_{\alpha\beta}(x)$ is closest to a given pattern. Note that the field equations can be cast as first order differential equations with respect to time:

$$g_{\alpha\beta,0} = p_{\alpha\beta},$$

$$p_{\alpha\beta,0} = H_{\alpha\beta}(p_{\mu\nu}, g_{\mu\nu,i}, p_{\mu\nu,i}, g_{\mu\nu,ij}) + S_{\mu\nu}$$

The state vector here has twenty independent components $g_{\alpha\beta}, p_{\alpha\beta}, 0 \leq \alpha \leq \beta \leq 3$. Let $f_{\alpha\beta}$ be the desired pattern for $g_{\alpha\beta}$. The optimal control problem can then be cast as: Minimize

$$\sum_{\alpha\beta} \int_D (f_{\alpha\beta} - g_{\alpha\beta})^2 d^4x$$

subject to the above constraining differential equations.

8.Optimal control applied to the quantized gravitational field. First to quantize the gravitational field in the canonical fashion, we must express R in terms of $g_{\mu\nu}$ and its partial derivatives with respect to the time and spatial variables. The final expression for R must show explicitly the components $g_{\mu\nu,0}, g_{\mu\nu,i}$. The canonical position variable are then to be taken as $g_{\mu\nu}$ and the canonical momenta as $\pi_{\mu\nu} = \frac{\partial L}{\partial g_{\mu\nu,0}}$. The Lagrangian density L can be taken as

$$L = g^{\mu\nu} \sqrt{-g} (\Gamma^\alpha_{\mu\nu} \Gamma^\beta_{\alpha\beta} - \Gamma^\alpha_{\mu\beta} \Gamma^\beta_{\nu\alpha})$$

It is known (see for example Dirac, General Relativity) that the Einstein field equations for vacuum can be derived from the variational principle $\delta \int L d^4x = 0$.

Computation of the Ricci tensor for rotating blackholes by brute force methods. The metric has the form

$$d\tau^2 = a(r,\theta)dt^2 - b(r,\theta)dr^2 - c(r,\theta)(d\phi - \omega(r,\theta)dt)^2 - f(r,\theta)d\theta^2$$

This is for a static black-hole. The rotation takes place around the $\theta = 0$ axis. The form of the metric displays azimuthal symmetry, ie, symmetry with respect to the $\theta = 0$ axis. We can write this metric as

$$d\tau^2 = adt^2 - bdr^2 - c(d\phi^2 + \omega^2 dt^2 - 2\omega d\phi dt) - fd\theta^2$$

$$= (a - c\omega^2)dt^2 - bdr^2 - cd\phi^2 + 2c\omega d\phi dt - fd\theta^2$$

We set

$$x^0 = t, x^1 = r, x^2 = \theta, x^3 = \phi$$

and then the metric can be expressed as

$$d\tau^2 = g_{\mu\nu}dx^\mu dx^\nu$$

$$g_{00} = a - c\omega^2, g_{11} = -b, g_{22} = -f, g_{33} = -c, g_{03} = g_{30} = c\omega$$

This can be expressed in matrix notation as

$$(g_{\mu\nu}) =$$

$$\begin{pmatrix} g_{00} & 0 & 0 & g_{03} \\ 0 & g_{11} & 0 & 0 \\ 0 & 0 & g_{22} & 0 \\ g_{03} & 0 & 0 & g_{33} \end{pmatrix}$$

Let us compute the inverse of this matrix $g^{\mu\nu}$. For doing that, we shall solve the linear system of equations $g_{\mu\nu}x^\nu = y_\mu$. The solution will then be given by $x^\mu = g^{\mu\nu}y_\nu$. Observe that the above linear equations when expanded are

$$g_{00}x^0 + g_{03}x^3 = y_0, g_{11}x^1 = y_1, g_{22}x^2 = y_2, g_{03}x^0 + g_{33}x^3 = y_3$$

We get

$$x^1 = y_1/g_{11}, x^2 = y_2/g_{22},$$

$$x^3 = (y_0 - g_{00}x^0)/g_{03}, g_{03}x^0 + g_{33}(y_0 - g_{00}x^0)/g_{03} = y_3$$

giving

$$(g_{03}^2 - g_{00}g_{33})x^0/g_{03} = y_3 - g_{33}y_0/g_{03}$$

so that

$$x^0 = g^{00}y_0 + g^{03}y_3$$

where

$$g^{00} = -g_{33}/(g_{03}^2 - g_{00}g_{33})$$

and

$$g^{03} = g_{03}/(g_{03}^2 - g_{00}g_{33})$$

Further,

$$x^3 = y_0/g_{03} - g_{00}(g^{00}y_0 + g^{03}y_3)/g_{03}$$

so that

$$g^{33} = -g_{00}g^{03}/g_{03} = -g_{00}/(g_{03}^2 - g_{00}g_{33})$$

Also

$$g^{11} = 1/g_{11}, g^{22} = 1/g_{22}$$

Written out explicitly, we have

$$\begin{aligned} g_{03}^2 - g_{00}g_{33} &= c^2\omega^2 + c(a - c\omega^2) \\ &= c^2\omega^2 + ca - c^2\omega^2 = ca \end{aligned}$$

so that

$$g^{00} = c/ca = 1/a, g^{33} = (c\omega^2 - a)/ca = \omega^2/a - 1/c$$

Note that we can present the contravariant metric tensor as

$$(g^{\mu\nu}) = \begin{pmatrix} g^{00} & 0 & 0 & g^{03} \\ 0 & g^{11} & 0 & 0 \\ 0 & 0 & g^{22} & 0 \\ g^{03} & 0 & 0 & g^{33} \end{pmatrix}$$

Thus, the non-vanishing Christoffel symbols of the first kind are

$$\Gamma_{001} = \frac{1}{2}g_{00,1} = \frac{1}{2}(a_{,r} - (c\omega^2)_{,r})$$

$$\Gamma_{002} = \frac{1}{2}g_{00,2} = \frac{1}{2}(a - c\omega^2)_{,\theta}$$

$$\Gamma_{031} = \frac{1}{2}g_{03,1} = \frac{1}{2}(c\omega)_{,r}$$

$$\Gamma_{032} = \frac{1}{2}g_{03,2} = \frac{1}{2}(c\omega)_{,\theta}$$

$$\Gamma_{111} = \frac{1}{2}g_{11,1} = -\frac{1}{2}b_{,r}$$

$$\Gamma_{112} = \frac{1}{2}g_{11,2} = -\frac{1}{2}b_{,\theta}$$

$$\Gamma_{222} = \frac{1}{2}g_{22,2} = -\frac{1}{2}f_{,\theta}$$

$$\Gamma_{332} = \frac{1}{2}g_{33,2} = -\frac{1}{2}c_{,\theta}$$

$$\Gamma_{331} = \frac{1}{2}g_{33,1} = -\frac{1}{2}c_{,r}$$

$$\Gamma_{133} = -\frac{1}{2}g_{33,1} = \frac{1}{2}c_{,r}$$

$$\Gamma_{313} = \Gamma_{331}$$

$$\Gamma_{233} = \frac{1}{2}(2g_{23,3} - g_{33,2}) = \frac{1}{2}c_{,\theta}$$

$$\Gamma_{100} = -\frac{1}{2}g_{00,1} = -\frac{1}{2}(a - c\omega^2)_{,r}$$

$$\Gamma_{200} = -\frac{1}{2}g_{00,2} = -\frac{1}{2}(a - c\omega^2)_{,\theta}$$

We now evaluate the various components of the Ricci tensor:

$$R_{00} = \Gamma^{\alpha}_{0\alpha,0} - \Gamma^{\alpha}_{00,\alpha} - \Gamma^{\alpha}_{00}\Gamma^{\beta}_{\alpha\beta} + \Gamma^{\alpha}_{0\beta}\Gamma^{\beta}_{0\alpha}$$

The first term $\Gamma^{\alpha}_{0\alpha,0} = 0$ since all the metric tensor coefficients are time independent. We evaluate the second term.

$$\Gamma^{\alpha}_{00,\alpha} = \Gamma^{1}_{00,1} + \Gamma^{2}_{00,2}$$

$$\Gamma^{1}_{00} = g^{11}\Gamma_{100} = -(g^{11}/2)g_{00,1}$$

$$\Gamma^{2}_{00} = -(g^{22}/2)g_{00,2}$$

so that

$$\Gamma^{\alpha}_{00,\alpha} = -\frac{1}{2}((g^{11}g_{00,1})_{,1} + (g^{22}g_{00,2})_{,2})$$

Note that for the above metric corresponding to a rotating blackhole,

$$\Gamma^{\alpha}_{00}\Gamma^{\beta}_{\alpha\beta}$$

$$= \Gamma^{1}_{00}\Gamma^{\beta}_{1\beta} + \Gamma^{2}_{00}\Gamma^{\beta}_{2\beta}$$

$$\Gamma^{1}_{00}\Gamma^{\beta}_{1\beta} = \Gamma^{1}_{00}(\Gamma^{0}_{10} + \Gamma^{1}_{11} + \Gamma^{2}_{12} + \Gamma^{3}_{13})$$

$$\Gamma^{0}_{10} = g^{00}\Gamma_{010} = \frac{1}{2}g^{00}(g_{01,0} + g_{00,1} - g_{01,0}) = \frac{1}{2}g^{00}g_{00,1}$$

$$\Gamma^{1}_{11} = \frac{1}{2}g^{11}g_{11,1}$$

$$\Gamma^{2}_{12} = \frac{1}{2}g^{22}(g_{21,2} + g_{22,1} - g_{21,2}) = \frac{1}{2}g^{22}g_{22,1}$$

$$\Gamma^{3}_{13} = \frac{1}{2}g^{33}(g_{31,3} + g_{33,1} - g_{31,3}) = \frac{1}{2}g^{33}g_{33,1}$$

giving

$$\Gamma^{1}_{00}\Gamma^{\beta}_{1\beta} = -\frac{1}{4}g^{11}g_{00,1}(g^{00}g_{00,1} + g^{11}g_{11,1} + g^{22}g_{22,1} + g^{33}g_{33,1})$$

This expression can be evaluated in terms of a, b, c, ω, f.

$$\Gamma^{\alpha}_{0\beta}\Gamma^{\beta}_{0\alpha}$$

$$= \Gamma^1_{00}\Gamma^0_{01} + \Gamma^2_{00}\Gamma^0_{02} + \Gamma^0_{01}\Gamma^1_{00} + \Gamma$$

Maxwell's equations in the Newman-Penrose formalism: The null tetrad basis is given by $\mathbf{l} = (l^i), \mathbf{n} = (n^i)$, $\mathbf{m} = (m^i)$ and $\bar{\mathbf{m}} = (\bar{m}^i)$. They satisfy the equations

$$\mathbf{l.l} = \mathbf{n.n} = \mathbf{m.m} = \mathbf{m.\bar{m}} = 0,$$

$$\mathbf{l.n} = 1 = \mathbf{n.l}, \mathbf{l.m} = \mathbf{l.\bar{m}} = \mathbf{n.m} = \mathbf{n.\bar{m}} = 0,$$

$$\mathbf{m.\bar{m}} = \mathbf{\bar{m}.m} = -1$$

Note that for any tetrad basis (e^i_a), we have $e_a.e_b = \eta_{ab}$, or in terms of components,

$$g_{ij}e^i_a e^j_b = \eta_{ab}$$

This is the same as saying that

$$\eta^{ab}e^i_a e^j_b = g^{ij}$$

where $(\eta^{ab}) = (\eta_{ab})^{-1}$. The numbers η_{ab} are constant scalars. Here $l^i = e^i_1, n^i = e^i_2, m^i = e^i_3, \bar{m}^i = e^i_4$. Let $F_{ab}, a, b = 1, 2, 3, 4$ denote the Maxwell scalars relative to the Newman-Penrose null tetrad and let F_{ij} denote the Maxwell tensor relative to the given coordinate basis. Define the scalars

$$\phi_0 = F_{13} = F_{ij}l^i m^j, \phi_1 = \frac{1}{2}(F_{12} + F_{43}) = \frac{1}{2}F_{ij}(l^i n^j + \bar{m}^i m^j),$$

$$\phi_2 = F_{42} = F_{ij}\bar{m}^i n^j$$

The Maxwell equations are

$$F_{[ij:k]} = 0, g^{jk}F_{ij:k} = 0$$

or in the Newman-Penrose formalism,

$$F_{[an|m]} = 0, \eta^{nm}F_{an|m} = 0$$

We have

$$A = \phi_{1|1} - \phi_{0|4} = \frac{1}{2}(F_{12|1} + F_{43|1}) - F_{13|4}$$

Now, the Maxwell equations give

$$F_{43|1} - F_{13|4} = -F_{34|1} - F_{13|4} = F_{41|3}$$

Thus,

$$A = \frac{1}{2}(F_{12|1} - F_{13|4} + F_{41|3})$$

on the other hand, the Maxwell equations give

$$0 = \eta^{mn}F_{1a|b} = F_{12|1} - F_{13|4} - F_{14|3}$$

and hence we get

$$A = 0$$

This can be taken as one of the Maxwell equations. Consider now the quantity

$$B = \phi_{2|1} - \phi_{1|4}$$

$$= F_{42|1} - \frac{1}{2}(F_{12|4} + F_{43|4}) = \frac{1}{2}(2F_{42|1} - F_{12|4} - F_{43|4})$$

Now,

$$F_{42|1} - F_{12|4} = F_{42|1} + F_{21|4} = -F_{14|2}$$

so that

$$B = \frac{1}{2}(-F_{14|2} - F_{43|4}) = \frac{1}{2}(F_{41|2} - F_{43|4})$$

On the other hand,

$$0 = \eta^{mn} F_{4m|n} = F_{41|2} - F_{43|4}$$

We thus get

$$B = 0$$

and this can be taken as the other Maxwell equation. Next,

$$C = \phi_{1|3} - \phi_{0|2} = \frac{1}{2}(F_{12|3} + F_{43|3}) - F_{13|2}$$

$$= \frac{1}{2}(F_{12|3} + F_{43|3} - 2F_{13|2})$$

Now,

$$F_{12|3} - F_{13|2} = F_{12|3} + F_{31|2} = -F_{23|1}$$

so that

$$C = \frac{1}{2}(F_{43|3} - F_{23|1} - F_{13|2}) = \frac{1}{2}(-F_{34|3} + F_{32|1} + F_{31|2})$$

$$= \frac{1}{2}\eta^{nm} F_{3n|m} = 0$$

Thus, $C = 0$ can also be taken as one of the Maxwell equations. Next,

$$D = \phi_{2|3} - \phi_{1|2} = F_{42|3} - \frac{1}{2}(F_{12|2} + F_{43|2})$$

$$= \frac{1}{2}(2F_{42|3} - F_{12|2} - F_{43|2})$$

Now,

$$F_{42|3} - F_{43|2} = F_{42|3} + F_{34|2} = -F_{23|4}$$

so that

$$D = \frac{1}{2}(F_{42|3} - F_{23|4} - F_{12|2}) = \frac{1}{2}(F_{21|2} - F_{24|3} - F_{23|4}) = \frac{1}{2}\eta^{nm} F_{2n|m} = 0$$

Thus, $D = 0$ is also a Maxwell equation. We compute

$$\phi_{0|4} = F_{13|4}$$

Note that

$$F_{ab,c} = e^i_c F_{ab,i} = e^i_c (F_{km} e^k_a e^m_b)_{:i}$$
$$= F_{km:i} e^k_a e^m_b e^i_c + F_{km} e^k_{a:i} e^i_c e^m_b + F_{km} e^i_c e^k_a e^m_{b:i}$$
$$= F_{ab|c} + F_{km} e^k_{a:i} e^i_c e^m_b + F_{km} e^i_c e^m_{b:i} e^k_a$$

Now,

$$e_{fk} e^k_{a:i} e^i_c = \gamma_{fac}$$

so that

$$e^k_{a:i} e^i_c = \gamma_{fac} e^{fk}$$

This yields

$$F_{km} e^k_{a:i} e^i_c e^m_b = F_{km} \gamma_{fac} e^{fk} e^m_b = \eta^{fd} F_{db} \gamma_{fac}$$

and likewise for the other term.

Newman-Penrose null tetrad for the Schwarzchild metric:

$$ds^2 = \frac{\Delta}{r^2} dt^2 - \frac{r^2}{\Delta} dr^2 - r^2 d\theta^2 - r^2 sin^2(\theta) d\phi^2$$

where $\Delta = r^2 - 2Mr$. For radial geodesics, we set $d\theta = d\phi = 0$ giving

$$L = ds/d\lambda = [\frac{\Delta}{r^2} t^2_{,\lambda} - \frac{r^2}{\Delta} r^2_{,\lambda}]^{1/2}$$

The Euler-Lagrange equation for t gives

$$(\Delta/r^2) dt/ds = E$$

where E is a constant. Assuming the geodesic to be null, it follows that E is an infinite constant. The null condition gives the relationship between r and t:

$$\frac{\Delta}{r^2} dt^2 = \frac{r^2}{\Delta} dr^2$$

or

$$dr/dt = \Delta/r^2$$

The Newman-Penrose null vector l^i associated with this null radial geodesic is given by

$$(l^i) = (r^2/\Delta, 1, 0, 0)$$

In fact we have for the null radial geodesic, $dr = \Delta dt/r^2$ or $dt = (r^2/\Delta) dr$ so that

$$(dt, dr, d\theta, d\phi) = dr(r^2/\Delta, 1, 0, 0)$$

The other null radial geodesic is given by the differential equation

$$dr/dt = -\Delta/r^2$$

and the corresponding Newman-Penrose null vector (n^i) is given by

$$(n^i) = \frac{1}{2r^2}(r^2, -\Delta, 0, 0)$$

The Newman-Penrose null vector (m^i) is given by

$$(m^i) = \frac{1}{r\sqrt{2}}(0, 0, 1, i.cosec(\theta))$$

so that the final Newman-Penrose null vector is

$$(\bar{m}^i) = \frac{1}{r\sqrt{2}}(0, 0, 1, -i.cosec(\theta))$$

We can check the orthogonality conditions

$$l.l = g_{ij}l^i l^j = g_{00}(r^2/\Delta)^2 + g_{11} = (\Delta/r^2)(r^2/\Delta)^2 - r^2/\Delta = 0$$

$$l.n = g_{ij}l^i n^j = g_{00}(r^2/\Delta)(1/2) + g_{11}(-\Delta/2r^2)$$

$$= (\Delta/r^2)(r^2/2\Delta) + (r^2/\Delta)(\Delta/2r^2) = 1/2 + 1/2 = 1$$

$$n.n = g_{00}n_0^2 + g_{11}n_1^2 = (\Delta/r^2)(1/2)^2 - (r^2/\Delta)(-\Delta/2r^2)^2 = \Delta/4r^2 - \Delta/4r^2 = 0$$

$$l.m = l.\bar{m} = n.m = n.\bar{m} = 0$$

since the nonzero components of the pair of vectors occurring in any of the above scalar products are non-overlapping. Finally,

$$m.\bar{m} = \frac{1}{2r^2}(g_{22} + g_{33}cosec^2(\theta)) = \frac{1}{2r^2}(r^2 + r^2 sin^2(\theta)cosec^2(\theta)) = 1$$

We find using

$$l_i = g_{ij}l^j, n_i = g_{ij}n^j, m_i = g_{ij}m^j, \bar{m}_i = g_{ij}\bar{m}^j$$

that

$$(l_i) = (1, -r^2/\Delta, 0, 0),$$

$$(n_i) = (g_{00}n^0, g_{11}n^1, 0, 0)$$

$$g_{00}n^0 = (\Delta/2r^2)$$

$$g_{11}n^1 = (-r^2/\Delta)(-\Delta/2r^2) = 1/2$$

Thus,

$$(n_i) = (\Delta/2r^2, 1/2, 0, 0) = \frac{1}{2r^2}(\Delta, r^2, 0, 0)$$

$$(m_i) = (g_{00}m^0, g_{11}m^1, g_{22}m^2, g_{33}m^3) = (0, 0, -r^2(1/r\sqrt{2}), -r^2 sin^2(\theta)(icsc(\theta)/r\sqrt{2}))$$

$$= (0, 0, -r/\sqrt{2}, -ir.sin(\theta)/\sqrt{2}) = \frac{1}{r\sqrt{2}}(0, 0, -r^2, -ir^2.sin(\theta))$$

Note that

$$l^i = e_1^i, n^i = e_2^i, m^i = e_3^i, \bar{m}^i = e_4^i$$

We calculate the spin coefficients:

$$\lambda_{abc} = e_{bi,j}(e_a^i e_c^j - e_a^j e_c^i)$$

$$\lambda_{122} = e_{2i,j}(e_1^i e_2^j - e_1^j e_2^i)$$

$$e_{2i,j}e_1^i e_2^j = n_{i,j}l^i n^j = n_{0,1}l^0 n^1 = (1/2)(1 - 2M/r)_{,r}(r^2/\Delta)(-\Delta/2r^2)$$
$$= (1/2)(2M/r^2)(-1/2) = -M/2r^2$$
$$e_{2i,j}e_1^j e_2^i = n_{i,j}l^j n^i = n_{0,1}n^0 l^1 = (\Delta/2r^2)_{,r}(1/2)(1) = M/2r^2$$

Subtracting, we get

$$\lambda_{122} = -M/2r^2 - M/2r^2 = -M/r^2$$

$$\lambda_{243} = e_{4i,j}(e_2^i e_3^j - e_2^j e_3^i)$$

Now,

$$e_{4i,j}e_2^i e_3^j = \bar{m}_{i,j}n^i m^j = \bar{m}_{2,1}n^2 m^1 + \bar{m}_{3,1}n^3 m^1 + \bar{m}_{3,2}n^3 m^2 = 0$$
$$e_{4i,j}e_2^j e_3^i = \bar{m}_{i,j}n^j m^i = \bar{m}_{2,1}n^1 m^2 = (-r/\sqrt{2})_{,r}(-\Delta/2r^2)(1/r\sqrt{2})$$

9.Quantum optimal control: The dynamics of the state is given by a Schrodinger equation with Hamiltonian dependent upon a control input $u(t)$. The equation of motion of the state can be expressed as

$$\psi'(t) = -iH(u(t))\psi(t)$$

The problem is to design the control input $u(t), 0 \le t \le T$ so that $\psi(T)$ is closest in distance to a given state vector ψ_f, ie, so that $\| \psi(t) - \psi_f \|$ is a minimum. Let $u(t)$ assume values in a set $\mathcal{U}$, the control space at any given time t. Let $\mathcal{L}$ denote the Lie algebra generated by the operators $iH(u), u \in \mathcal{U}$. Then any element of the corresponding Lie group $exp(\mathcal{L})$ can be expressed as a limit of the unitary operators of the form $exp(L_1)...exp(L_n)$ with $L_1, ..., L_n \in \mathcal{L}$, ie, For a given initial state ψ_i, the possible final states are all of the form $U\psi_i$ where $U \in exp(\mathcal{L})$. If the initial state ρ_i is mixed, then the possible final states are all of the form $U\rho U^*$ with $U \in exp(\mathcal{L})$. The quantum dynamical system is controllable iff given any initial state ψ_i and any final state ψ_f, there exists a $T > 0$ and an input $u(t), 0 < t < T$ such that $\psi_f = U(T)\psi_i$ where

$$U'(t) = -iH(u(t))U(t), t \ge 0, U(0) = I$$

10.Optimal control in scattering theory problems. The unperturbed Hamiltonian is H_0 and the perturbed Hamiltonian is $H = H_0 + V$. Let $|\psi>$ be an unperturbed eigenstate with energy E. Then, after the perturbation V is switched on, the corresponding perturbed eigenstate is $|\psi+>$ which satisfies the Lippman Schwinger equation:

$$|\psi+>=|\psi>+(E-H_0+i\epsilon)^{-1}V|\psi+>$$

We write $|E,a>$ for $|\psi>$ and $|Ea+>$ for $|\psi+>$. Assume that the unperturbed state is a superposition $\int\phi(E)|Ea>dE$. It evolves with time under the free Hamiltonian as $\phi(t)=\int exp(-iEt)\phi(E)|Ea>dE$. The corresponding perturbed state is $\int\phi(E)|Ea+>dE$ and this evolves in time as $\phi+(t)=\int exp(-iEt)\phi(E)|Ea+>dE$. We have

$$lim_{t\to-\infty}(\phi+(t)-\phi(t))=0$$

This fact has been proved in Weinberg's book, "The quantum theory of fields". Note that

$$<E'a'|\psi+>=<E'a'|\psi>+\frac{<E'a'|V|\psi+>}{E-E'+i\epsilon}$$

It is not hard to see that $lim_{t\to-\infty}(\phi+(t)-\phi(t))=0$. The details can be found in the book by Arno Bohm, "Quantum mechanics, foundations and applications". Write $V(t)=exp(-iH_0t), U(t)=exp(-iHt)$ so that

$$|\phi(t)>=V(t)|\phi(0)>,|\phi+(t)>=U(t)|\phi+(0)>$$

and hence we have

$$|\phi+(0)>=lim_{t\to-\infty}U(t)^*V(t)|\phi(0)>=\Omega+|\phi(0)>$$

where

$$\Omega+=lim_{t\to-\infty}U(t)^*V(t)$$

This is the Moller scattering operator. $\Omega+$ is defined on an appropriate domain. Likewise, we define

$$|\phi-(0)>=\Omega-|\phi(0)>,\Omega-=lim_{t\to\infty}U(t)^*V(t)$$

The scattering matrix is defined by

$$S=\Omega-^*\Omega_+$$

S is an isometry. We shall not prove this fact here (Refer to the book, Scattering theory by Amrein, Jauch and Sinha). The problem is to design the potential V so that the scattering matrix S is closest to a given matrix. One way to do this is to take a real T sufficiently large so that $\Omega+$ can be approximated by $exp(-iHT)exp(iH_0T)$ and $\Omega-$ by $exp(iHT)exp(-iH_0T)$. Then S is approximated by $exp(iH_0T)exp(-2iHT)exp(iH_0T)$. The problem is that given an isometry S and T, choose V so that

$$\| S-exp(iH_0T)exp(-2iHT)exp(iH_0T)\|$$

is a minimum. Note that the potential V that can be controlled appears as an additive term in H. The corresponding problem in matrix theory is the following: We are given a unitary matrix S and a Hermitian matrix H_0 both of size $n\times n$. The problem is to determine a Hermitian matrix V also of size $n\times n$ so that the quantity

$$F(V)=\| S-exp(iH_0T)exp(-2i(H_0+V)T)exp(iH_0T)\|$$

is a minimum. We can try to solve this problem by using the Campbell-Baker-Hausdorff formula in which $exp(2i(H_0+V)T)$ is expressed in terms of $exp(iH_0 2T).exp(iV2T)$ and commutators between H_0 and V.

11.Optimal control problems for motion in a gravitational field. Assume to start with that the metric of space-time is Newtonian with a gravitational potential $U(t,x,y,z)$ that can be controlled over a region of space-time. The problem is to determine the potential so that motion in this field tracks a given trajectory. The equations of motion in this metric are obtained from the Lagrangian

$$\mathcal{L}(x,y,z,x',y',z',t) = [(1+2U(x,y,z,t)/c^2) - \frac{1}{c^2}(x'^2+y'^2+z'^2)]^{1/2}$$

The equations of motion are then obtained from the Euler-Lagrange equations

$$\frac{d}{dt}\frac{\partial \mathcal{L}}{\partial(x',y',z')} = \frac{\partial \mathcal{L}}{\partial(x,y,z)}$$

To be able to formulate the optimal control problem in more elementary terms, we assume that the shape of the potential U is known but the magnitude can be controlled by us, ie,

$$U(x,y,z,t) = \phi(t)\Phi(x,y,z)$$

where Φ is a known function of spatial location but $\phi(t)$ can be controlled by us. The problem is to design the function $\phi(t)$ that would guarantee optimal tracking of a given trajectory $(X(t),Y(t),Z(t))$. We shall, in what follows, derive the approximate optimal control equations by making approximations based on the inequality $|U| << c^2$.

$$\frac{\partial \mathcal{L}}{\partial(x',y',z')} = -(m(x',y',z')/c^2\mathcal{L})$$

Now,

$$\mathcal{L}^{-1} \approx 1 - U/c^2 + v^2/2c^2$$

and

$$\frac{\partial \mathcal{L}}{\partial(x,y,z)} = \nabla U/c^2\mathcal{L}$$

so that the approximate equations of motion read

$$-\frac{d}{dt}(m(x',y',z'))(1-U/c^2+v^2/2c^2)) = (\nabla U)(1-U/c^2+v^2/2c^2)$$

or

$$m\mathbf{r}''$$

Chapter 9

Dynamical Systems

Dynamical Systems on C^∞-manifolds

1. Let $\mathcal{M}$ be a C^∞-manifold. This means that $\mathcal{M}$ is a topological space with a collection of local charts called an atlas :$\{(U_\alpha, \phi_\alpha) : \alpha \in \Gamma\}$. By chart, we mean the following: Each U_α is a non-empty open subset of $\mathcal{M}, \mathcal{M} = \bigcup_{\alpha\in\Gamma} U_\alpha$, $\phi_\alpha : U_\alpha \to \phi_\alpha(U) \subset \mathbb{R}^n$ is a homeomorphism where $\phi_\alpha(U_\alpha)$ is open in $\mathbb{R}^n$. n is a fixed positive integer independent of the index $\alpha \in \Gamma$. Further, suppose that for some $\alpha \neq \beta, \alpha, \beta \in \Gamma$, we have that $\phi_\alpha(U_\alpha) \cap \phi_\beta(U_\beta) \neq \phi$. Then, we require that

$$\phi_\beta o \phi_\alpha^{-1} : \phi_\alpha(U_\alpha \cap U_\beta) \to \phi_\beta(U_\alpha \cap U_\beta)$$

be C^∞. Then the manifold $\mathcal{M}$ is said to be a C^∞-manifold. We now introduce the notion of tangent space at a point in $\mathcal{M}$. Let $x \in \mathcal{M}$. A curve in $\mathcal{M}$ passing through x is defined as a map $c : I \to \mathcal{M}$ where I is an interval in $\mathbb{R}$ containing 0 and $c(0) = x$. Let c be such a curve. Let (U, ϕ) be a local chart at x, ie, U is an open subset of $\mathcal{M}$ containing x and $\phi : U \to \phi(U) \subset \mathbb{R}^n$ is a homeomorphism with $\phi(U)$ being open in $\mathbb{R}^n$. If $c : I \to \mathcal{M}$ is continuous with respect to the topology I inherits from $\mathbb{R}$ and the given topology on $\mathcal{M}$, we say that c is a continuous curve. If $t \to \phi(c(t))$ is differentiable at some $t \in I$, then we say that the curve c is differentiable at t. Assume that c is differentiable at $t = 0$. Then, consider the class of all curves that pass through x. If c_1, c_2 are any two such curves, we say that c_1, c_2 are equivalent if

$$\frac{d}{dt}\phi(c_1(t))|_{t=0} = \frac{d}{dt}\phi(c_2(t))|_{t=0}$$

we then say that c_1 and c_2 are equivalent. This relation is an equivalence relation. Further, this relation is independent of the chart (U, ϕ) chosen. For if (V, ψ) is another chart at $t = 0$, then

$$\frac{d}{dt}\psi(c(t))|_{t=0} = \frac{d}{dt}(\phi o \psi^{-1})(\psi(c(t)))|_{t=0} = T(\phi o \psi^{-1})_{\psi(x)} \frac{d}{dt}\psi(c(t)))|_{t=0}$$

which shows that

$$\frac{d}{dt}\psi(c_1(t))|_{t=0} = \frac{d}{dt}\psi(c_2(t))|_{t=0}$$

implies

$$\frac{d}{dt}\psi(c_1(t))|_{t=0} = \frac{d}{dt}\psi(c_2(t))|_{t=0}$$

Note that the definition of differentiability of a curve at a given point is independent of the local chart. Indeed, suppose $(U,\phi),(V,\psi)$ are two local charts at x and $c: I \to \mathcal{M}$ is a curve with $c(0)=x$ so that $0 \in I$. Then, $\psi o \phi^{-1}$ is differentiable at $\phi(x)$ and hence

$$\psi(c(t)) = \psi o \phi^{-1}(\phi(c(t))$$

will be differentiable at $t=0$ iff $\phi(c(t))$ is differentiable at $t=0$. In fact, in this case, we have the identity

$$\frac{d}{dt}\psi(c(t))|_{t=0} = T(\psi o \phi^{-1})_x \frac{d}{dt}\phi(c(t))|_{t=0}$$

and conversely,

$$\frac{d}{dt}\phi(c(t))|_{t=0} = T(\phi o \psi^{-1})_x \frac{d}{dt}\psi(c(t))|_{t=0}$$

(See the book "Foundations of Mechanics", by Ralph Abraham and J.E.Marsden).

2.Problem :Discuss the process of linearization of the dynamics for a C^1 vector field defined on a C^∞-manifold. Consider the special case when linearization is carried around a fixed point of the vector field.

Hint: In terms of coordinates, we have

$$\frac{dx_i(t)}{dt} = X_i(x_1(t),...,x_n(t)), i = 1,2,...,n$$

Let $x(t)$ be a given trajectory and $x(t)+\delta x(t)$ the perturbed trajectory. Then,

$$\frac{d\delta x_i(t)}{dt} = \sum_j \frac{\partial X_i(x(t))}{\partial x_j}\delta x_j(t)$$

Define the matrix valued function of time

$$\mathbf{A}(t) = ((\frac{\partial X_i(x(t))}{\partial x_j}))_{1\le i,j\le n}$$

Then the dynamics can be expressed as

$$\frac{d\delta\mathbf{x}(t)}{dt} = \mathbf{A}(t)\delta\mathbf{x}(t)$$

In particular, if $x(t)=x_0$ is a fixed point of the dynamical system in the sense that

$$X(x_0) = 0$$

then the matrix $\mathbf{A}(t)$ becomes independent of time:

$$\mathbf{A} = ((\frac{\partial X_i(x_0)}{\partial x_j}))_{1\le i,j\le n}$$

3.Suppose X is a vector field on $\mathbb{R}^n$ such that for all $x, y \in \mathbb{R}^n$, we have

$$|X(x) - X(y)| \leq K.|x - y|$$

where K is a positive constant. Let $x_0(t) = x_0$ for all t and define the process $x_n(t), n = 1, 2, ...$ recursively by the iteration

$$x_{n+1}(t) = x_0 + \int_0^t X(x_n(s))ds$$

Then, we have

$$|x_{n+1}(t) - x_n(t)| \leq \int_0^t |X(x_n(s)) - X(x_{n-1}(s)|ds$$

$$\leq K \int_0^t |x_n(s) - x_{n-1}(s)|ds, n \geq 1$$

We can write this identity as

$$h_{n+1}(t) \leq K \int_0^t h_n(s)ds \leq K^2 \int_{0<u<s<t} h_{n-1}(u)duds = ... = K^{n+1}|x_0| \int_{0<u_1<...<u_n<t} du_1du_2...du_ndu_{n+1}$$

so that

$$h_n(t) \leq K^n t^n |x_0|/n!$$

from which we get for $m > n$,

$$|x_m(t) - x_n(t)| =\leq_{r=n+1}^{m} h_r(t) \to 0, n, m \to \infty$$

so that $lim_{n \to \infty} x_n(t)$ exists and is differentiable (Uniform convergence over compact intervals). The limit $x(t) = lim x_n(t)$ evidently satisfies the integral equation

$$x(t) = x_0 + \int_0^t X(x(s))ds$$

4.Consider a vector field X on $\mathbb{R}^n$. Denote its flow by $F(x, t)$. Thus, $(x, t) \to F(x, t)$ is a mapping from $\mathbb{R}^n \times \mathbb{R} \to \mathbb{R}^n$ such that

$$\frac{dF(x,t)}{dt} = X(F(x,t)), F(x,0) = x$$

Define a new coordinate system on $\mathbb{R}^n$ via the equation

$$x = \phi(y), \phi^{-1}(x_1, ..., x_n) = F(x_1, ..., x_{n-1}, 0, x_n)$$

It is clear that ϕ^{-1} is a local diffeomorphism (in a nhood of $x = 0$) provided that $X_n(\mathbf{0}) \neq 0$. We can ensure this condition by applying an appropriate diffeomorphism to $\mathbb{R}^n$. We want to see what the flow of the vector field looks like in the new coordinate system. The flow in the new coordinate system can be computed easily. We have

$$F(\phi^{-1}(x), t) = F(F(x_1, ..., x_{n-1}, 0, x_n), t) = F(x_1, ..., x_{n-1}, 0, x_n + t) = \phi^{-1}(x_1, ..., x_{n-1}, x_n + t)$$

so that

$$\phi o F(\phi^{-1}(x), t) = (x_1, ..., x_{n-1}, x_n + t)$$

from which, it follows that the flow in the new coordinate system corresponds to simply a uniform translation of the last coordinate in time.

(Abraham and Marsden, Foundations of Mechanics).

5.Define a differentiable manifold and hence define the notion of a vector field on a differentiable manifold. What do you understand by dynamics on a differentiable manifold corresponding to a vector field X ?

6.What do you understand by a derivation on a manifold. Let X be a vector field on a manifold M and let $\phi : M \to M$ be a diffeomorphism. Then $\phi_* X$ is a vector field on the same manifold defined by its action on functions,

$$\phi_* X(f)(x) = X(fo\phi)(\phi^{-1}(x))$$

which in terms of components, reads

$$\phi_* X = T\phi o X o \phi^{-1}$$

This is the same as

$$(\phi_* X)^i(x) = \sum_j \frac{\partial \phi^i(z)}{\partial z^j}(\phi^{-1}x)X^j(\phi^{-1}(x))$$

Derivation: Let M be a differentiable manifold and let (U, ϕ) be a local chart at $x \in M$. Let $f : U \to \mathbb{R}$ be a differentiable function. $\mathcal{F}(U)$ is the class of all differentiable functions defined on U. This means that $f : U \to \mathbb{R}$ is an element of $\mathcal{F}(U)$ iff $fo\phi^{-1}$ is differentiable at each $\phi(x), x \in U$. Let θ be a derivation on $\mathcal{F}(U)$. Then we claim that

$$\theta(f) = \sum_i \frac{\partial fo\phi^{-1}(y)}{\partial y^i}\theta(\phi^i)(x), y = \phi(x)$$

Indeed, let $u \in U$. Then for $u, y \in \phi(U)$, we have

$$fo\phi^{-1}(y) - fo\phi^{-1}(u) = \sum_j (y^j - u^j)\frac{\partial fo\phi^{-1}(z)}{\partial z^j}|_{z=u+t(y-u)}$$

for some $t(y) \in [0, 1]$. u here is fixed say $u = \phi(x_0), y = \phi(x)$ with $x_0, x \in U$. Note that $t(u) = 0$. We can write this equation as

$$f(x) - f(x_0) = \sum_j (\phi^j(x) - \phi^j(x_0))\frac{\partial fo\phi^{-1}(z)}{\partial z^j}|_{z=\phi(x_0)+t(\phi(x)-\phi(x_0))}$$

Let θ be a derivation. We have by definition of derivation that if $f(x) = g(x)h(x)$, then

$$\theta(f.g)(x) = f(x)\theta(g)(x) + g(x)\theta(f)(x)$$

so that

$$\theta(f)(x_0) = \sum_j \theta(\phi^j)(x_0)\frac{\partial fo\phi^{-1}(z)}{\partial z^j}(\phi(x_0))$$

or equivalently

$$\theta(f) = \sum_j \theta(\phi^j)\frac{\partial fo\phi^{-1}}{\partial z^j}|_{z=\phi(x_0)}$$

Since $\theta(\phi^j)$ is a smooth function of x, it follows that the local form of any derivation is given by

$$\theta_x = \sum_j a_j(z)\frac{\partial}{\partial z^j}, z = \phi(x)$$

Remark:Lagrange's theorem can be derived from Rolle's theorem. It states that if f is continuous on $[a,b]$, differentiable on (a,b) and $f(a) = f(b) = 0$, then for some $c \in (a,b)$, we have $f'(c) = 0$. Lagrange's theorem states that if f is continuous on $[a,b]$ and differentiable on (a,b), then for some $c \in (a,b)$,

$$f(b) - f(a) = f'(c)$$

7.Problem: Let M be a differentiable symplectic manifold and let $H(x)$ be a Hamiltonian on this manifold. Let $f_1, ..., f_k$ be independent integrals of the motion, ie $\{f_i, H\} = 0, i = 1, 2, ..., k$, or equivalently, if X_H is the vector field corresponding to the Hamiltonian H, then $X_H(f_i) = 0, i = 1, 2, ..., k$. Consider the level surfaces

$$\Sigma_{\mathbf{c}} = \{x \in M : f_i(x) = c_i, i = 1, 2, ..., k\}$$

or equivalently, with

$$\mathbf{F}(x) = (f_1(x), ..., f_k(x))$$

and

$$\mathbf{c} = (c_1, ..., c_k),$$

we have

$$\Sigma_{\mathbf{c}} = \{x \in M : \mathbf{F}(x) = \mathbf{c}\} = \mathbf{F}^{-1}(\mathbf{c})$$

We want to prove the existence of a unique invariant volume $\mu_{\mathbf{c}}$ on $\Sigma_{\mathbf{c}}$. The surface $\Sigma_{\mathbf{c}}$ is a manifold of codimension k. Let Ω be the invariant volume measure on M. We can then write

$$\Omega = df_1 \wedge ... \wedge df_k \wedge d\eta$$

We have $L_{X_H}\Omega = 0$ (Liouville's theorem) and since $L_{X_H}f_i = 0$, it follows that

$$0 = df_1 \wedge ... \wedge df_k \wedge L_{X_H}\eta$$

and hence

$$L_{X_H}\eta = \sum_{i=1}^{k} df_i \wedge \rho_i$$

Suppose η' is another form such that

$$\Omega = df_1 \wedge ... \wedge df_k \wedge \eta'$$

Then we will have

$$0 = df_1 \wedge ... \wedge df_k \wedge (\eta - \eta')$$

whence

$$\eta - \eta' = \sum_{i=1}^{k} df_i \wedge \sigma_i$$

Let $i : \Sigma_c \to M$ denote the inclusion map. Then i^* (the pullback of i) maps forms on M to forms on Σ_c. WE have since f_i is constant on Σ_c that

$$i^*\eta - i^*\eta' = i^*(\sum_{i=1}^{k} df_i \wedge \sigma_i = 0$$

and hence we can define a volume form on Σ_c via the equation

$$\mu_c = i^*\eta$$

This volume is uniquely defined as was just proved. We have further,

$$L_{X_H}\mu_c = L_{X_H}i^*\eta = i^*L_{X_H}\eta = 0$$

since

$$L_{X_H}\eta = \sum_{i=1}^{k} df_i \wedge \rho_i$$

and df_i is zero on Σ_c. The construction of the invariant volume measure on Σ_c is now complete.

8.Problem: Clarify the notions of push forward of a vector field and pull back of a one form.

Ans: Let M be a differentiable manifold and $\phi : M \to M$ a smooth map. Let X be a vector field on M. Then $\phi_* X$ is the vector field on M having components $T\phi oXo\phi^{-1}$. In terms of differential operators, we have

$$(\phi_* X)(f)(x) = X(fo\phi)(\phi^{-1}(x))$$

We thus have

$$(\phi_* X)(f)(\phi(x)) = X(fo\phi)(x)$$

or

$$\phi^* L_{\phi_* X} = L_X \phi^*$$

or equivalently,

$$L_{\phi_* X} = \phi_* L_X \phi^*$$

Note that for functions, ϕ^*, ϕ_* are defined as

$$\phi^* f(x) = f(\phi(x)), \phi_* f(x) = f(\phi^{-1}(x))$$

9.State and prove Darboux theorem on exact two differential forms. Cast the equations of Hamiltonian mechanics in the language of symplectic geometry.

10. Action angle variables formulation (Kompaneyetc, p.111): First we discuss canonical transformations: Let (q,p) be the original coordinates and momenta and (Q,P) the new coordinates and momenta. The two are connected by a canonical transformation. Let $H(q,p,t)$ be the original Hamiltonian and $K(Q,P,t)$ the new Hamiltonian. Then, the original Lagrangian is given by $p_a q'_a - H$ and the new Lagrangian by $P_a Q'_a - K$, summation over the repeated variable a being assumed. The two Lagrangians must differ by a total time derivative for them to correspond to the same equations of motion. Thus, there must exist a function V which we may to start with, assume to be a function of (q,Q,t) so that

$$p_a q'_a - H = P_a Q'_a - K + \frac{dV}{dt} = P_a Q'_a - K + \frac{\partial V}{\partial q_a} q'_a + \frac{\partial V}{\partial Q'_a} Q'_a + \frac{\partial V}{\partial t}$$

This gives

$$p_a = \frac{\partial V}{\partial q_a}, P_a = -\frac{\partial V}{\partial Q_a}, K = H + \frac{\partial V}{\partial t}$$

Define V' by

$$V'(q,P,t) = V(q,Q,t) + P_a Q_a$$

so that

$$p_a q'_a - H = P_a Q'_a - K + \frac{\partial V'}{\partial t} - P_a Q'_a - P'_a Q_a + \frac{\partial V'}{\partial q_a} q'_a + \frac{\partial V'}{\partial P_a} P'_a$$

giving

$$p_a = \frac{\partial V'}{\partial q'_a},$$

$$Q_a = \frac{\partial V'}{\partial} \partial P_a,$$

$$K = H + \frac{\partial V'}{\partial t}$$

Consider a transformation function $V(q,Q,t)$ chosen so that the new Hamiltonian is identically zero. Then, we get from the previous set of transformation equations

$$H(q, \frac{\partial V}{\partial q}, t) + \frac{\partial V}{\partial t} = 0$$

Once we solve this partial differential equation for the function V, we obtain the complete solution to the equations of motion. Indeed, since P_a, Q_a are constants in view of the new Hamiltonian being zero, we can solve the system

$$P_a = -\frac{\partial V(q,Q,t)}{\partial Q_a}, a = 1, 2, ..., N$$

for the variables $q_a(t)$ in terms of the constants P, Q and then get the solution to $p_a(t)$ from

$$p_a(t) = \frac{\partial V(q(t), Q, t)}{\partial q_a}$$

Example:The Kepler problem. The Lagrangian is given by

$$L(r, r', \theta, \theta', \phi, \phi') = \frac{m}{2}(r'^2 + r^2\theta'^2 + r^2 sin^2(\theta)\phi'^2) - U(r)$$

The generalized momenta are

$$p_r = \frac{\partial L}{\partial r'} = mr', p_\theta = \frac{\partial L}{\partial \theta'} = mr^2\theta',$$

$$p_\phi = \frac{\partial L}{\partial \phi'} = mr^2 sin^2(\theta)\phi'$$

so that the Hamiltonian is given by

$$H(r, \theta, \phi, p_r, p_\theta, p_\phi)$$

$$= (p_r^2/2m + p_\theta^2/2mr^2 + p_\phi^2/2mr^2 sin^2(\theta)) + U(r)$$

The Hamilton-Jacobi equation for the transformation function S is thus given by

$$(S_{,r})^2/2m + (S_{,\theta})^2/2mr^2 + (S_{,\phi})^2/2mr^2 sin^2(\theta)) + U(r) + S_{,t} = 0$$

Define the shortened action $S_0(r, \theta)$ by the equation

$$S = S_0 - Et + p_\phi \phi$$

Plugging this into the HJ equation gives the following pde for S_0:

$$(S_{0,r})^2/2m + (S_{0,\theta})^2/2mr^2 + p_\phi^2/2mr^2 sin^2(\theta)) + U(r) = E$$

Try a solution of the form

$$S_0(r, \theta) = A(r) + B(\theta)$$

which gives on substitution,

$$A_{,r}^2/2m + B_{,\theta}^2/2mr^2 + p_\phi^2/2mr^2 sin^2(\theta)) + U(r) = E$$

or

$$r^2 E - r^2 A'^2/2m - -r^2 U(r) = B'^2/2m + p_\phi^2/2m.sin^2(\theta)$$

where $A' = A'(r)$ and $B' = B'(\theta)$. The left side of this equation is a function of only r while the right side is a function of only θ. It follows that both sides must equal a constant, say $M^2/2m$. This gives

$$r^2 E - r^2 A'^2/2m - r^2 U(r) = M^2/2m,$$

$$B'^2 + p_\phi^2/sin^2(\theta) = M^2$$

This gives

$$A(r) = \int_0^r [2m(E - U(r) - M^2/2mr^2)]^{1/2} dr$$

$$B(\theta) = \int_0^\theta [M^2 - p_\phi^2/sin^2(\theta)]^{1/2} d\theta$$

Thus, the complete solution to the Kepler problem for any central potential $U(r)$ is given by the generating function (action)

$$S = -Et + p_\phi \phi + \int_0^r [2m(E - U(r) - M^2/2mr^2)]^{1/2} dr + B(\theta) + \int_0^\theta [M^2 - p_\phi^2/sin^2(\theta)]^{1/2} d\theta$$

We take as the constants $\{Q_a\}$ E, p_ϕ, M and the derived constants $\{P_a\}$ are obtained by taking the negative derivative of S with respect to these constants. Thus, denoting P_1 by t_0, we have

$$t_0 = P_1 = -\frac{\partial S}{\partial E} = t - \int m(2m(E - U - M^2/2mr^2))^{-1/2} dr$$

or

$$t - t_0 = \int m(2m(E - U - M^2/2mr^2))^{-1/2}$$

This equation determines the radial coordinate r as a function of time. Next, denoting the constant P_2 by $-\phi_0$, we have

$$-\phi_0 = P_2 = -\frac{\partial S}{\partial p_\phi} = -\phi + \int (p_\phi d\theta/sin^2(\theta))(M^2 - p_\phi^2/sin^2(\theta))^{-1/2} d\theta$$

or

$$\phi - \phi_0 = \int (p_\phi d\theta/sin^2(\theta))(M^2 - p_\phi^2/sin^2(\theta))^{-1/2} d\theta$$

This equation gives ϕ as a function of θ. The solution will be complete if we know θ as a function of r and this is achieved by

$$-P_3 = \alpha = \frac{\partial S}{\partial M} = \int M(M^2 - p_\phi^2/sin^2(\theta))^{-1/2} d\theta$$

$$-\int r^{-2} M(2m(E - U(r) - M^2/2mr^2))^{-1/2} dr$$

11. Let f be a differentiable function on a manifold and X a vector field. We have $df(X) = X(f) = i_X df$ where i_X is the contraction operator. The Lie derivative $L_X f = X(f)$, so we have $i_X df = L_X f$. Now let θ be a one form. We have that $d\theta$ is a two form and $(i_X d\theta)(Y) = d\theta(X, Y)$. On the other hand, $(d i_X \theta)(Y) = Y(i_X \theta) = Y(\theta(X))$. We have

$$(L_Y \theta)(X) = Y(\theta(X)) - \theta(L_Y X)$$

In terms of local coordinates and components, we have

$$(L_Y\theta)_i X^i = Y^j(\theta_k X^k)_{,j} - \theta_i[Y,X]^i$$

$$= Y^j\theta_{k,j}X^k + Y^j\theta_k X^k_{,j} - \theta_i(Y^k X^i_{,k} - X^k Y^i_{,k})$$

$$= (Y^j\theta_{k,j} + \theta_i Y^i_{,k})X^k$$

so that

$$(L_Y\theta)_k = Y^j\theta_{k,j} + \theta_j Y^j_{,k}$$

Also

$$(i_X d\theta)_i Y^i = \theta_{i,j}(X^j Y^i - X^i Y^j)$$

so that

$$(i_X d\theta)_i = \theta_{i,j}X^j - \theta_{j,i}X^j = (\theta_{i,j} - \theta_{j,i})X^j$$

Thus,

$$(L_X\theta)_k - (i_X d\theta)_k = X^j\theta_{k,j} + \theta_j X^j_{,k} - (\theta_{k,j} - \theta_{j,k})X^j$$

$$= \theta_j X^j_{,k} + \theta_{j,k}X^j = (\theta_j X^j)_{,k} = (di_X\theta)_k$$

This gives

$$L_X\theta = i_X d\theta + di_X\theta$$

This equation in fact holds for any form in place of θ.

12. Define a tensor field on a differentiable manifold. Explain the pullback and pushforward of tensor fields.

Hint: First consider a finite dimensional vector space $\mathbf{E}$ and its dual $\mathbf{E}^*$. t is said to be a tensor of class (r,s), if t is a multilinear map from $E^{*r} \times E^s$. Thus, for $\alpha^j, j = 1,2,...,r \in E^*$ and $v_j \in E, j = 1,2,...,s$, we have $t(\alpha^1,...,\alpha^r,v_j,...,v_s)$ a scalar such that t is linear in each of its arguments α_i, v_j. Let $\phi : E \to E$ be an isomorphism. Then for $\alpha \in E^*$, $\phi^*\alpha \in E^*$ is defined by

$$(\phi^*\alpha)(u) = \alpha(\phi u)$$

which in standard notation means that $\phi^* = \phi^T$, the transpose of ϕ. Note that if ϕ, ψ are two isomorphisms, then

$$(\psi\phi)^*\alpha(u) = \alpha(\psi\phi u) = (\psi^*\alpha)(\phi u) = \phi^*(\psi^*\alpha)(u)$$

so that

$$(\psi\phi)^* = \phi^*\psi^*$$

Now for $t \in T(r,s)$ (tensor of type (r,s)), we set

$$(\phi_* t)(\alpha^1,...,\alpha^r,v_1,...,v_s) = t(\phi^*\alpha_1,...,\phi^*\alpha_r,\phi^{-1}v_1,...,\phi^{-1}v_r)$$

We then have

$$(\psi\phi)_* t(\alpha^1,...,\alpha^r,v_1,...,v_s) = t((\psi\phi)^*\alpha_1,...,(\psi\phi)^*\alpha_r,(\psi\phi)^{-1}v_1,...,(\psi\phi)^{-1}v_r)$$

$$= t(\phi^*\psi^*\alpha^1, ..., \phi^*\psi^*\alpha^r, \phi^{-1}\psi^{-1}v_1, ..., \phi^{-1}\psi^{-1}v_r)$$
$$= (\phi_* t)(\psi^*\alpha^1, ..., \psi^*\alpha^r, \psi^{-1}v_1, ..., \psi^{-1}v_r)$$
$$= \psi_*\phi_* t(\alpha^1, ..., \alpha^r, v_1, ..., v_r)$$

or equivalently,

$$(\psi\phi)_* t = \psi_*\phi_* t$$

So far we have been dealing with vector spaces. The map ϕ_* acting on tensors is called the pushforward map. To generalize this to differentiable manifolds, we consider a manifold M and $T(r,s)(M)$ to be the collection of all tensor fields on M of type (r,s). Thus for any $x \in M$, $t_x : TM_x^{*r} \times TM_x^s \to \mathbb{R}$ is a multilinear map. Let $\phi : M \to M$ be a diffeomorphism. Then

$$(\phi_* t)_{\phi(x)}(\alpha^1_{\phi(x)}, ..., \alpha^r_{\phi(x)}, v_{1\phi(x)}, ..., v_{s\phi(x)})$$

$$= t_x((\phi^*\alpha^1)_x, ..., (\phi^*\alpha^r)_x, (\phi^* v_1)_x, ..., (\phi_s^* v_s)_x)$$

where

$$(\phi^*\alpha)_x = \phi^*\alpha_{\phi(x)}$$

is defined for any $v \in TM_x$ via the equation

$$(\phi^*\alpha)_x(v) = \alpha_{\phi(x)}(\phi_* v) = \alpha_{\phi(x)}(T\phi_x(v))$$

and likewise,

$$(\phi^* v)_x = (T\phi)^{-1}_{\phi(x)}(v_{\phi(x)})$$

where α is a one form on M while v is a vector field on M.

13.Let $H(x)$ be a Hamiltonian function on a symplectic manifold (M,ω). Define the Hamiltonian vector field X_H by the equation

$$\omega(X_H, Y) = dH(Y)$$

or more precisely,

$$\omega_x(X_H(x), Y(x)) = dH(x)(Y(x))$$

Let $F_t(x)$ be the flow generated by the Hamiltonian vector field X_H. We want to show that F_t is a symplectic map, ie, $F_t^*\omega = \omega$. To this end, we observe that

$$(F_t^*\omega)_x(X,Y) = \omega_{F_t(x)}((TF_t)_x(X(x)), (TF_t)_x(Y(x)))$$

From this how can the conclusion be reached ? We've got to use the condition that ω is closed, ie, $d\omega = 0$. In terms of coordinates, we have

$$\omega_x = \omega_{ij}(x)dx^i \wedge dx^j$$

so that the condition $d\omega = 0$ gives

$$\omega_{ij,k}(x)dx^k \wedge dx^i \wedge dx^j$$

The computation leading to the desired result is long and tedious. Instead we provide a coordinate free proof.

$$\frac{d}{dt}F_t^* = F_t^* L_{X_H}$$

gives

$$\frac{d}{dt}F_t^*\omega = F_t^* L_{X_H}\omega = F_t^*(d(i_{X_H}\omega) + i_{X_H}d\omega)$$

$$= F_t^*(ddH + i_{X_H}d\omega) = 0$$

where we have used the definition of X_H, namely

$$i_{X_H}\omega = dH$$

Let α be a one form and let X be a vector field with flow F_t. We compute $\frac{d}{dt}F_t^*\alpha$. We have by definition of pullback, for any $v \in TM_x$,

$$(F_t^*\alpha)_x(v) = \alpha_{F_t(x)}((TF_t)_x(v))$$

or in terms of components,

$$(F_t^*\alpha)_i(x) = \alpha_j(F_t(x))\frac{\partial F_t^j(x)}{\partial x^i}$$

so that

$$\frac{d}{dt}(F_t^*\alpha)^i(x) = \alpha_{j,k}(F_t(x))X^k(F_t(x))F_{t,i}^j(x) + \alpha_j(F_t(x))\frac{\partial}{\partial x^i}X^j(F_t(x))$$

$$= \alpha_{j,k}(y)X^k(y)F_{t,i}^j(x) + \alpha_j(y)X_{,m}^j(y)F_{t,i}^m(x)$$

where

$$y = F_t(x)$$

This can be expressed as

$$\frac{d}{dt}(F_t^*\alpha)_i(x) = [\alpha_{j,k}(y)X^k(y)F_{t,i}^j(x) + \alpha_k(y)X_{,j}^k(y)]F_{t,i}^j(x)$$

$$= (L_X\alpha)_j(y)F_{t,i}^j(x)$$

This is the same as

$$\frac{d}{dt}(F_t^*\alpha) = F_t^* L_X\alpha$$

proving the claim. Let us now compute

$$\frac{d}{dt}F_t^* Z$$

where Z is a vector field. Let α be a one form. Then $\alpha(Z)$ is a smooth function. We have

$$(F_t^*\alpha)_x(Z_x) = \alpha_{F_t(x)}((TF_t)_x Z_x)$$

or equivalently,

$$(F_t^*\alpha)(Z)(x) = \alpha(F_{t*}Z)(F_t(x))$$

On differentiating both sides with respect to t and using the above formula for $\frac{d}{dt}F_t^*\alpha$, we get

$$(F_t^* L_X\alpha)(Z)(x) = \alpha(\frac{d}{dt}F_{t*}Z)(F_t(x)) + X(\alpha(F_{t*}(Z))(F_t(x))$$

or equivalently,

$$\alpha(\frac{d}{dt}F_{t*}Z)(x) = (F_t^* L_X\alpha)(Z)(F_{-t}(x)) - X(\alpha(F_{t*}Z))(x)$$

From this, we conclude that

$$\alpha(\frac{d}{dt}F_{t*}Z)(x) = (F_t^* L_X\alpha)(Z)(F_{-t}(x)) - (L_X\alpha)(F_{t*}Z)(x) - \alpha(L_X F_{t*}Z)(x)$$

$$= -\alpha(L_X F_{t*}Z)(x)$$

giving finally,

$$\frac{d}{dt}F_{t*}Z = -L_X F_{t*}Z$$

14.Notion of second order equation: Let M be a manifold and for each $v \in M$, let E_v be a vector space of dimension n. Then $F = \bigcup_{v\in M}\{v\} \times E_v$ is a vector bundle. A function $f : E \to \mathbb{R}$ is defined by $f(v, x)$ for each $v \in M$ and $x \in E_v$. E_v is called the fiber over v. An example of a vector bundle is the tangent bundle on the manifold. In this case, we have $E_v = TM_v$, the tangent space over v. Suppose E, F are two vector bundles. A map $f : E \to F$ is a vector bundle mapping if it has the form

$$f(v, x) = (w, T(v)x)$$

where w is a base point of F and $T(v) \in \mathcal{L}(E_v, F_w)$.

15.Fiber derivative:Let E, F be two fiber bundles and $f : E \to F$ a map. Assume that M is the base manifold of E and N the base manifold of F. Fix a $v \in M$ and consider the map $x \to f(v, x) \in F$. We can write $f(v, x) = (w, \phi(v, x))$. Define $f_v(x) = f(v, x)$. Then $f_v : E_v \to F_{f(v)}$ where $w = f(v)$. The differential of $f_v(x)$ is given by $Tf_v(x) : E_v \to F_{v(v)}$ where the tangent spaces of E_v and $F_{f(v)}$ have been identified with themselves. This is also called the fiber derivative of f and is denoted by Ff. Thus, $Ff : E \to L(E, F)$.

Let E, F be two fiber bundles and $f : E \to F$ a fiber bundle map. Thus $(v, x) \to (\phi(v), \psi(v, x))$ where $\phi(v)$ is an element of the base point of F and $\psi(v, x) \in F_{\phi(v)}$ for $x \in E_v$. The fiber derivative of f, Ff is the differential of the function $x \to \psi(v, x)$ with v fixed. It is a linear mapping from TE_v into $TF_{\phi(v)}$ or equivalently, identifying E_v with TE_v and $F_{\phi(v)}$ with $TF_{\phi(v)}$, it follows that $Ff : E \to L(E_v, F_{\phi(v)})$.

Example of fiber derivative: Let M be a differentiable manifold and TM its tangent bundle. A Lagrangian function $L : TM \to \mathbb{R}$ is given. We write a point in TM as (q, q') so that $q \in M$ and $q' \in TM_q$. Then $FL : TM \to L(TM, \mathbb{R}) = T^*M$. Its components are $\frac{\partial L(q,q')}{\partial q'_i}$. In mechanics, these components are identified with the canonical momenta of the system. Thus, the canonical momentum is an element of the cotangent bundle.

16.Notion of second order equation: Let M be a manifold and TM is tangent bundle. Let X be a vector field on TM. We denote a point in TM by (u, e) so that $u \in M$ and $e \in TM_u$. Then, an element of TTM has the form (u, e, e_1, e_2) where $(e_1, e_2) \in T(TM)_{(u,e)}$. We have $X(u, e) = (e_1, e_2)$ for some (e_1, e_2). We can thus write the vector field as $(u, e, X_1(u, e), X_2(u, e))$. We say that the vector field X is a second order equation if $X_1(u, e) = e$. Suppose that this is the case. Let $c(t)$ be an integral curve for this vector field. We write $c(t) = (c_1(t), c_2(t))$. Then we have

$$\frac{dc_1(t)}{dt} = X_1(c_1(t), c_2(t)) = c_2(t), \frac{dc_2(t)}{dt} = X_2(c_1(t), c_2(t))$$

This gives

$$\frac{d^2c_1(t)}{dt^2} = X_2(c_1(t), c_2(t))$$

These equations can be compressed into a single equation

$$\frac{d^2c_1}{dt^2} = X_2(c_1, \frac{dc_1}{dt})$$

17.Let Q be a differentiable manifold and TQ the tangent manifold. A point in TQ has the form (q, v) where $q \in Q$ and $v \in TQ_q$. The Lagrangian is a smooth function $L : TQ \to \mathbb{R}$. In terms of components, we write $L(q, v)$ or $L(q, q')$ so that $v, q' \in TQ_q$. The fiber derivative of L is given by $FL : TQ \to T^*Q$. In terms of components we write $p_i = \frac{\partial L(q,q')}{\partial q'_i}$ and then $p \in T^*_qQ$ is defined by

$$p(v) = \sum_i p_i v_i = \sum_i \frac{\partial L}{\partial q'_i} v_i$$

We write $\theta_0 = \sum_i p_i dq_i$. Write $M = TQ$ and then $TM = TTQ = \chi(M)$ is the set of all vector fields on M. θ_0 is a vector field on M. Let $\alpha_q \in TQ^*_q$ or more precisely, $(q, \alpha_q) \in TQ^*_q$. Let $w_q \in T_{(q,\alpha_q)}T^*Q$. We should visualize w_q as $(q, \alpha_q, \delta q, \delta\alpha_q)$ where (q, α_q) is the base point in TT^*Q and $(\delta q, \delta\alpha_q) \int T_{(q,\alpha_q)}T^*Q$. $\tau_Q : TQ \to Q$ is the map $\tau_Q(q, v) = q$. $T\tau_Q : TTQ \to TQ$ is the mapping

$$T\tau_Q(q, v, \delta q, \delta v) = (q, \delta q)$$

or simply δq the base point q being understood. We define $\theta_0 \in \chi(M)$ so that

$$\begin{aligned}\theta_0(q, \alpha_q)(w_q) &= \theta_0(q, \alpha_q)(q, \alpha_q, \delta q, \delta\alpha_q)\\ &= \alpha_q(\delta q)\end{aligned}$$

or using our shorthand notation,

$$\theta_0(\alpha_q)(w_q) = \alpha_q(T\tau_Q w_q)$$

Since $\alpha_q = (p_i)$ in coordinates, it follows that

$$\theta_0(\alpha_q)(w_q) = \sum_i p_i \delta q_i$$

Writing the point of TQ as

$$\delta q = \sum_i \delta q_i \frac{\partial}{\partial q_i}$$

it follows that

$$\theta_0(\alpha_q) = \sum_i p_i dq_i$$

Note that we are here to interpret dq_i as a linear functional on TT^*Q rather than on TQ. Specifically,

$$dq_i(q, \alpha_q, \delta q, \delta\alpha_q) = \delta q_i$$

The differential of $-\theta_0$ is the canonical symplectic form on $M = T^*Q$. Consider the fiber derivative $FL : TQ \to T^*Q$. Then $(FL)^* : T^*T^*Q \to T^*TQ$. The canonical symplectic form ω_0 on T^*Q is given by

$$\omega_0 = -d\theta_0 = \sum_i dq_i \wedge dp_i$$

Write

$$(FL)^*\theta_0 = \theta_L$$

Thus, writing $\omega_L = -d\theta_L$, we have

$$\omega_L = (FL)^*\omega_0$$

In terms of coordinates,

$$-\omega_L = \sum_i dq_i \wedge d(\frac{\partial L}{\partial q'_i})$$

ω_L is a symplectic form on T^*Q. It can be expanded to give

$$-\omega_L = \sum_{i,j} \frac{\partial^2 L}{\partial q_j \partial q'_i} dq_i \wedge dq_j$$

$$+\sum_{i,j} \frac{\partial^2 L}{\partial q'_j \partial q'_i} dq_i \wedge dq'_j$$

Define the action as

$$A = \sum_i p_i q'_i = FL(v)(v)$$

where $v = q'$ is the velocity and $FL(v)$ is the Fiber derivative of L at v, ie,

$$FL(v)(v) = \sum_i \frac{\partial L}{\partial q'_i} q'_i$$

Define the energy

$$E(v) = A - L = \sum_i \frac{\partial L}{\partial q'_i} q'_i - L(q, q')$$

Let X_E be the vector field corresponding to the symplectic form ω_L generated by E, ie,

$$dE(Y) = i_{X_E}\omega_L$$

or equivalently,

$$dE(Y) = \omega_L(X_E, Y)$$

Suppose X_E is a second order equation. Then we want to deduce the validity of the Euler-Lagrange equations of motion for an integral curve of X_E. Indeed, in this case we have in terms of components,

$$X_E(q,v) = (q, v, X_{E1}(q,v), X_{E2}(q,v))$$

Since X_E is a second order equation, we have $X_{E1}(q,v) = v$. The condition for $(q(t), v(t))$ to be an integral curve for this field is that

$$q' = v = X_{E1}(q,v), v' = X_{E2}(q,v)$$

Let us now write down the condition for X_E to be a second order equation, ie, $v = X_{E1}(q,v)$. We have with $p_i = \frac{\partial L(q,v)}{\partial v_i}$,

$$\omega_L = \sum_{i,j} L_{,q_j q_i'} dq_i \wedge dq_j + L_{,q_j' q_i'} dq_i \wedge dq_j'$$

so that,

$$\omega_L(X_E, Y) = L_{,q_j q_i'}(X_{E1i}Y_{1j} - X_{E1j}Y_{1i}) + L_{,q_j' q_i'}(X_{E1i}Y_{2j} - Y_{1i}X_{E2j})$$

On the other hand,

$$dE(Y) = E_{,q_i}Y_{1i} + E_{,q_i'}Y_{2i}$$

equating these two expressions for arbitrary Y gives us

$$E_{,q_i} = L_{,q_i q_j'}X_{E1j} - L_{,q_j q_i'}X_{E1j} - L_{,q_j' q_i'}X_{E2j},$$

$$E_{,q_i'} = L_{,q_j' q_i'}X_{E1j}$$

Now writing down the condition for X_E to be a second order equation, replacing q_i' by v_i, we get

$$E_{,q_i} = L_{,q_i v_j}v_j - L_{,q_j v_i}v_j - L_{,v_j v_i}X_{E2j}$$

$$E_{,v_i} = L_{,v_j v_i}v_j$$

Since

$$E = A - L = L_{,v_i}v_i - L$$

we have

$$E_{,q_i} = L_{,q_i v_j}v_j - L_{,q_i}, E_{,v_i} = L_{,v_i v_j}v_j + L_{,v_i} - L_{,v_i} = L_{,v_i v_j}v_j$$

Thus we get on substituting this into the above relations,

$$L_{,q_i} = L_{,q_j v_i}v_j + L_{,v_j v_i}X_{E2j}$$

$$L_{,v_i v_j}v_j = L_{,v_i v_j}v_j$$

The second equation is redundant. The condition for (q, v) to be an integral curve of X_E is that $q' = v, v' = X_{E2}$. Combining this with the first, we get

$$L_{,q_i} = L_{,q_j q_i'} q_j' + L_{,q_j' q_i'} q_j''$$

or equivalently,

$$\frac{d}{dt} L_{,q_i'} = L_{,q_i}$$

which is the Euler-Lagrange equation of motion.

18.Dynamical Systems:

Equivariant momentum maps: Let (P, ω) be a symplectic manifold and G a group acting on P, via the map $\Phi : G \times P \to P$. This means that $(g, x) \to \Phi(g, x)$ satisfies

$$\Phi(g_2, \Phi(g_1, x)) = \Phi(g_2 g_1, x), g_1, g_2 \in G, x \in P$$

Let $\xi \in \}$, the Lie algebra of G. Then, $g(t) = exp(t\xi), t \in \mathbb{R}$ is a one parameter subgroup of G. We define

$$\xi_P(x) = \frac{d}{dt} \Phi(exp(t\xi), x)|_{t=0}$$

Then ξ_P is a vector field on P. Suppose $J : P \to \}^*$ is a function such that

$$d\hat{J}(\xi) = i_{\xi_P} \omega$$

where $\hat{J}(\xi)(x) = J(x)(\xi)$. Then we say that J is an equivariant momentum map on P. Note that if $X_{\hat{J}(\xi)}$ is the Hamiltonian vector field corresponding to the function $\hat{J}(\xi)$, then we have by definition

$$i_{X_{\hat{J}(\xi)}} \omega = d\hat{J}(\xi)$$

This means that we can write the condition for $\hat{J}$ to be an equivariant momentum map as

$$i_{X_{\hat{J}(\xi)}} \omega = i_{\xi_P} \omega$$

or equivalently,

$$X_{\hat{J}(\xi)} = \xi_P$$

A useful formula while dealing with equivariant momentum maps is that if $g \in G$ and $\xi \in \}$, then

$$(Ad_{g^{-1}} \xi)_P = \Phi_g^* \xi_P$$

To prove this, we use the result that

$$exp(t.Ad_{g^{-1}} \xi) = g^{-1}.exp(t\xi).g$$

It follows that

$$(Ad_{g^{-1}} \xi)_P(x) = \frac{d}{dt} \Phi(exp(t.Ad_{g^{-1}} \xi), x)|_{t-0}$$

But observe that

$$\Phi(exp(t.Ad_{g^{-1}}\xi), x) = \Phi(g^{-1}.exp(t\xi).g, x)$$
$$= \Phi(g^{-1}, \Phi(exp(t\xi), \Phi(g, x)))$$

Differentiating with respect to t at $t = 0$, we get

$$(Ad_{g^{-1}}\xi)_P(x) = T\Phi_{g^{-1}}\xi_P(\Phi(g, x)) = (\Phi_g^*\xi_P)(x)$$

Note that Φ_g^* is the pullback map. In other words, we have established that

$$(Ad_{g^{-1}}\xi)_P = \Phi_g^*\xi_P$$

Suppose $H : P \rightarrow \mathbb{R}$ is invariant under the group action, ie,

$$H(\Phi_g(x)) = H(x), g \in G, x \in P$$

The notation we are using here is $\Phi_g(x) = \Phi(g, x)$. Then we claim that $\hat{J}(\xi)$ is a constant of the motion generated by the Hamiltonian H. To see this, we choose any $\xi \in \}^*$ and consider the invariance condition

$$H(\Phi(exp(t\xi), x)) = H(x)$$

Differentiating both sides with respect to t at $t = 0$ gives us

$$dH(x)(\xi_P(x)) = 0$$

or using equivariance of J,

$$dH(X_{\hat{J}(\xi)}) = 0$$

This is the same as

$$\{H, \hat{J}(\xi)\} = 0$$

where $\{.,.\}$ is the Poisson bracket. Note that in terms of coordinates, we have the following: Let f, g be smooth functions on P and let X_f, X_g be the corresponding Hamiltonian vector fields generated by f and g respectively. In other words,

$$df = i_{X_f}\omega, dg = i_{X_g}\omega$$

The Poisson bracket between f, g is defined by

$$\{f, g\} = df(X_g) = (i_{X_f}\omega)(X_g) = \omega(X_f, X_g)$$

Note that the definition implies

$$\{g, f\} = -\{f, g\}$$

We have thus proved that $\hat{J}(\xi)$ is a constant of the motion generated by the Hamiltonian vector field H. Q is a manifold and $P = T^*Q$ is the cotangent manifold. Let θ be the canonical one form on T^*Q, ie,

$$\theta = \sum p_i dq^i$$

Its differential is the symplectic form on T^*Q, ie,

$$-d\theta = \sum_i dq^i \wedge dp_i$$

Suppose G acts on Q via the map Φ_g. Then this action induces an action on P, namely, Φ_g^*. This map preserves θ. In fact, let $F : Q \to Q$ be any smooth map. Write $Q = F(q)$. This is an abbreviation for $Q_i = F_i(q_1, ..., q_n), i = 1, 2, ..., n$. The map induced by F on the cotangent manifold is given in terms of coordinates by

$$p \to P$$

which is an abbreviation for

$$(p_1, ..., p_n) \to (P_1, ..., P_n)$$

We have

$$\sum p_i dq^i = \sum_{i,j} p_i \frac{\partial dq_i}{\partial Q_j} dQ_j = \sum P_j dQ^j$$

where

$$P_j = \sum_i p_i \frac{\partial q_i}{\partial Q_j}$$

In other words, any map of coordinates preserves the canonical one form in the cotangent manifold. We define for $\xi \subset \}$,

$$\hat{J}(\xi) = i_{\xi_P}\theta$$

Then we claim that J is an equivariant momentum map. To see this, we compute

$$d\hat{J}(\xi) = di_{\xi_P}\theta = L_{\xi_P}\theta - i_{\xi_P}d\theta$$

where L_X denotes the Lie derivative along the vector field X. Now the invariance condition $\Phi_g^*\theta = \theta$ gives on replacing g by $exp(t\xi)$ and differentiating with respect to t at $t = 0$, $L_{\xi_P}\theta = 0$. Note that $\Phi_g^*\theta = \theta$ means that for any vector field X on Q, we have

$$\theta(gx)(T\Phi_g X(x)) = \theta(x)(X(x))$$

and replacing g by $exp(t\xi)$ and differentiating with respect to t at $t = 0$ gives us $L_{\xi_P}\theta = 0$. It follows that

$$d\hat{J}(\xi) = i_{\xi_P}\omega$$

which is the condition for J to be a momentum map. We want to prove equivariance of J, ie,

$$J(g.x)(\xi) = J(x)(ad_{g^{-1}}\xi)$$

or equivalently,

$$Jo\Phi_g(x) = ad^*_{g^{-1}}J(x)$$

This is equivalent to proving that

$$(i_{\xi_P}\theta)(\Phi_g(x)) = J(x)(ad_{g^{-1}}\xi) = (i_{(ad_{g^{-1}}\xi)_P}\theta)(x)$$

or equivalently,

$$(i_{\xi_P}\theta)(\Phi_g(x)) = i_{\Phi_g^*\xi_P}\theta(x)$$

Now since

$$\Phi_g^*\theta = \theta,$$

we have

$$i_{\Phi_g^*\xi_P}\theta(x) = i_{\Phi_g^*\xi_P}\Phi_g^*\theta(x)$$

$$= (\Phi_g^*\theta)(\Phi_g^*\xi_P)(x)$$

$$= \theta_{gx}(T\Phi_g(\Phi_g^*\xi_P)_x) = \theta_{gx}(T\Phi_g T\Phi_g^{-1}\xi_{Pgx}) = \theta_{gx}(\xi_{gx}) = (i_{\xi_P}\theta)(\Phi_g(x))$$

This proves the claim of equivariance made above.

19. Let Q be a manifold and $P = T^*Q$. Assume that a group G acts on Q via $(g,x) \to \Phi(g,x) = \Phi_g(x)$. This group action induces an action on $P = T^*Q$ that preserves the canonical one form $\theta = \sum p_i dq^i$. Let $\tau_Q^* : T^*Q \to Q$ be defined by $(q, \alpha_q) \to q$. The action Φ_g of G on Q induces a mapping Φ_g^{T*} of T^*Q onto itself. This is called the lifted action. It is defined by the equation

$$\Phi_g^{T*}\alpha = T\Phi_{g^{-1}}^*\alpha$$

More precisely, this relation can be expressed as

$$(\Phi_g^{T*}\alpha)_q(X_q) = \alpha_{g^{-1}q}(T\Phi_{g^{-1}}X_q)$$

We claim that

$$\tau_Q^* o \Phi_g^{T*} = \Phi_g o \tau_Q^*$$

Φ_g^{T*} maps $T^*Q_{g^{-1}q}$ onto T^*Q_q. It follows that Φ_g^{T*} maps T^*Q_q onto T^*Q_{gq}. Note that for $\alpha_q \in T^*Q_q$ and $X \in TQ_{gq}$, we have

$$\Phi_g^{T*}(\alpha_q)(X) = \alpha_q(T\Phi_{g^{-1}}X)$$

In other words,

$$\Phi_g^{T*}\alpha_q = \alpha_q o T\Phi_{g^{-1}} \in T^*Q_{gq}$$

It follows that

$$\tau_Q^* o \Phi_g^{T*}(\alpha_q) = gq,$$

Also

$$\tau_Q^*(\alpha_q) = q, \Phi_g o \tau_Q(\alpha_q) = g.q$$

and this completes the proof of the above equation. Note that Φ_g^{T*} satisfies the homomorphism property of group action, ie, if $g, h \in G$, then $\Phi_{gh}^{T*} = \Phi_g^{T*}\Phi_h^{T*}$. This follows from

$$\Phi_{gh}^{T*}\alpha_q = \alpha_q o T\Phi_{(gh)^{-1}} = \alpha_q o T\Phi_{h^{-1}g^{-1}}$$

$$= \alpha_q o T\Phi_{h^{-1}} o T\Phi_{g^{-1}} = \Phi_g^{T*}\alpha_q o T\Phi_{h^{-1}}$$

$$= \Phi_g^{T*}\Phi_h^{T*}\alpha_q$$

This gives

$$\Phi_{gh}^{T*} = \Phi_g^{T*}\Phi_h^{T*}, g, h \in G$$

Consider once again the equation

$$\tau_Q^*\Phi_g^{T*} = \Phi_g o \tau_Q^*$$

Set $g = exp(t\xi)$ where $\xi \in \}$, the Lie algebra of G. We then have on differentiating this expression with respect to t at $t = 0$,

$$T\tau_Q^*\xi_P = \xi_Q o \tau_Q^*$$

It follows that

$$i_{\xi_P}\theta(\alpha_q) = \alpha_q(T\tau_Q^* o \xi_P(\alpha_q))$$
$$= \alpha_q(\xi_Q o \tau_Q^*(\alpha_q)) = \alpha_q(\xi_Q(q))$$

This shows that if we define

$$J(\alpha_q)(\xi) = \alpha_q(\xi_Q(q))$$

then J is an equivariant momentum map on P. What does this map look like in terms of coordinates ?

20. Let Q be a manifold and TQ its tangent bundle, T^*Q its cotangent bundle. We define a one form θ on TT^*Q as follows. Thus, $\theta \in T^*T^*Q$. For $\alpha_q \in T^*Q_q$, we have $\theta(\alpha_q)$ is a linear functional on $T_{\alpha_q}T^*Q$. It is defined as follows. Let $w_q \in T_{\alpha_q}T^*Q$. Then,

$$\theta(\alpha_q)(w_q) = \alpha_q(T\tau_Q^* w_q)$$

Let us see what θ looks like in terms of coordinates. Let $q \in Q$ and U a nhood of q in Q. Let $\phi : U \to V \subset \mathbb{R}^n$ be a local chart around q. We claim that θ in coordinates is $\sum_i p_i dq^i$.

21. Vector bundle: Let S be a set. Let E be a finite dimensional vector space and U and open subset of E. We call $U \times E$ a local vector bundle. Suppose $U \times E$ is a local vector bundle and $V \times F$ is another local vector bundle and suppose $\phi : U \times E \to V \times F$ is a map of the form

$$\phi(u, e) = (\phi_1(u), \phi_2(u)e)$$

where $\phi_1 : U \to V$ is a smooth map and $\phi_2(u)$ is for each $u \in U$ an element of $\mathcal{L}(E, F)$.

22. Mechanics:

Some concepts from differential geometry

Let $\gamma(t)$ be a smooth curve on a Riemannian manifold M. Let $\tau_{t,s} : TM_{\gamma(s)} \to TM_{\gamma(t)}$ be the linear transformation of parallel translation along the curve from $\gamma(s)$ to $\gamma(t)$. Then for $v(s) \in TM_{\gamma(s)}$, we set $v(t) \in TM_{\gamma(t)}$. We know from the general theory of parallel tranlsates (See the book "Foundations of Mechanics", by Ralph Abraham and J.E.Marsden) that in a local coordinate chart,

$$\frac{dv^i(t)}{dt} + \Gamma_{jk}^i(\gamma(t))v^j(t)\gamma^{k'}(t) = 0$$

Recall how one defines the parallel translate via the connection: If X, Y are two vector fields, then $\nabla_X(Y)$ is another vector field. It satisfies (a) $\nabla_{fX}(Y) = f\nabla_X(Y)$ and (b) $\nabla_X(fY) = X(f)Y + f\nabla_X Y$. In terms of a local chart x^i, we set

$$\nabla_{\partial_i}\partial_j = \Gamma^k_{ij}\partial_k$$

Suppose $\gamma(t)$ is a curve. $\gamma'(t)$ is the tangent vector to the curve at t. We take a vector field $X(x)$ on the manifold and define $DX(\gamma(t))/dt$, the covariant derivative of X along the curve $\gamma(t)$. It is defined by $\nabla_{\gamma'(t)}X(\gamma(t))$. In terms of local coordinates, $\gamma'(t) = \gamma^{i'}(t)\partial_i$ and

$$\begin{aligned}\nabla_{\gamma'(t)}X(\gamma(t)) &= \nabla_{\gamma^{i'}(t)\partial_i}(X^j(\gamma(t))\partial_j)\\ &= \gamma^{i'}(t)X^j_{,i}(\gamma(t))\partial_j + X^j(\gamma(t))\gamma^{i'}(t)\nabla_{\partial_i}\partial_j\\ &= [\nabla_{\gamma'(t)}X(\gamma(t))]^j\partial_j\end{aligned}$$

where

$$\begin{aligned}&[\nabla_{\gamma'(t)}X(\gamma(t))]^k =\\ &= \gamma^{i'}(t)X^k_{,i}(\gamma(t)) + \Gamma^k_{ij}(\gamma(t))\gamma^{i'}(t)X^j(\gamma(t))\\ &= \frac{dX^k(\gamma(t))}{dt} + \Gamma^k_{ij}(\gamma(t))\frac{d\gamma^i(t)}{dt}X^j(\gamma(t)) = 0\end{aligned}$$

If $D\gamma'(t)/Dt = 0$, we say that γ is a geodesic curve. The condition for γ to be a geodesic curve is therefore

$$\frac{d^2\gamma^k(t)}{dt^2} + \Gamma^k_{ij}(\gamma(t))\frac{d\gamma^i(t)}{dt}\frac{d\gamma^j(t)}{dt} = 0$$

Let γ be a curve and $\tau_{s,t}$ the parallel translate operator from s to t along γ. Then for $v(s) \in TM_{\gamma(s)}$, we set $v(t) = \tau_{s,t}(v(s)) \in TM_{\gamma(t)}$. By definition of parallel translation, it satisfies the differential equations

$$\frac{dv^i(t)}{dt} + \Gamma^i_{jk}(\gamma(t))v^j(t)\gamma^{k'}(t) = 0$$

It is easy to see that $\tau^{-1}_{s,t} = \tau_{t,s}$. We Let $v(t) = \tau_{0,t}(v(0))$ and then consider

$$\begin{aligned}\frac{d}{ds}\tau_{s,t}(v(s))|_{s=t} &= \frac{d}{dt}\tau^{-1}_{t,s}(v(s))|_{s=t}\\ &= -\frac{d}{ds}\tau_{t,s}(v(t))|_{s=t} + \frac{dv(t)}{dt}\end{aligned}$$

In terms of components,

$$\begin{aligned}\frac{d}{ds}\tau_{s,t}(v(s))^i|_{s=t} &= \frac{dv^i(t)}{dt} + \Gamma^i_{jk}(\gamma(t))\gamma^{j'}(t)v^k(t)\\ &= \frac{Dv^i(t)}{dt}\end{aligned}$$

the covariant derivative of v along the curve γ.

23.Killing vector field:Let X be a vector field on a Riemannian manifold M having metric g. Note that the notation $\chi(M)$ stands for the space of all smooth vector fields on M. The map $(X,Y) \to g(X,Y)$ from $\chi(M) \times \chi(M) \to \mathcal{F}(M)$ is bilinear and symmetric, ie,

$$g(c_1X_1 + c_2X_2, Y) = c_1g(X_1, Y) + c_2g(X_2, Y), g(Y, X) = g(X, Y), X, X_1, X_2, Y \in \chi(M)$$

A vector field X on M is called a Killing vector field if $L_X g = 0$, ie the Lie derivative of g along X vanishes. This means that for all $Y, Z \in \chi(M)$, we must have

$$X(g(Y,Z)) - g([X,Y],Z) - g(Y,[X,Z]) = 0$$

in terms of local coordinates,

$$g(Y,Z) = g_{ij}Y^iZ^j$$

and

$$X(g(Y,Z)) = X^k(g_{ij}Y^iZ^j)_{,k} = g_{ij,k}X^kY^iZ^j + g_{ij}X^kY^i_{,k}Z^j + g_{ij}X^kY^iZ^j_{,k}$$

$$[X,Y]^i = X^jY^i_{,j} - Y^jX^i_{,j}$$

so that

$$g([X,Y],Z) = g_{ij}(X^kY^i_{,k} - Y^kX^i_{,k})Z^j$$

$$g(Y,[X,Z]) = g_{ij}Y^i[X,Z]^j = g_{ij}Y^i(X^kZ^j_{,k} - Z^kX^j_{,k})$$

Substituting these expressions into the above equation gives us the condition for X to be killing:

$$g_{ij,k}X^kY^iZ^j + g_{ij}X^i_{,k}Y^kZ^j + g_{ij}X^j_{,k}Y^iZ^k = 0$$

This can be rewritten as

$$g_{ij,k}X^kY^iZ^j + g_{jk}X^k_{,i}Y^iZ^j + g_{ik}X^k_{,j}Y^iZ^j = 0$$

or since Y, Z are arbitrary fields, the condition for X to be killing reduces to

$$g_{ij,k}X^k + g_{jk}X^k_{,i} + g_{ik}X^k_{,j} = 0$$

Now observe that

$$X_{j,i} = (g_{jk}X^k)_{,i} = g_{jk}X^k_{,i} + g_{jk,i}X^k$$

so that the condition for X to be a killing field is the same as

$$X_{j,i} + X_{i,j} + (g_{ij,k} - g_{jk,i} - g_{ik,j})X^k = 0$$

$$X_{j:i} = X_{j,i} - \Gamma^k_{ij}X_k = X_{j,i} - \Gamma_{kij}X^k$$

Thus,

$$X_{j:i} + X_{i:j} = X_{j,i} + X_{i,j} - (\Gamma_{kij} + \Gamma_{kji})X^k$$

$$= X_{j,i} + X_{i,j} - (g_{ki,j} + g_{kj,i} - g_{ij,k})X^k = 0$$

Thus the condition for X to be a killing field is that

$$X_{i:j} + X_{j:i} = 0$$

24.Push forward and pull back:Let M be a differentiable manifold and X a vector field on M. Let $\phi : M \to M$ be a diffeomorphism. Then $\phi_* X$, the pushforward of X by ϕ is the vector field defined by

$$(\phi_* X)(x) = (T\phi)X(\phi^{-1}(x))$$

Equivalently,

$$\phi_* X = T\phi o X o \phi^{-1}$$

The pullback $\phi^* X$ of X by ϕ is the vector field defined by

$$(\phi^* X)(x) = T\phi^{-1} o X o \phi(x)$$

or

$$(\phi^* X)(x) = T\phi^{-1} o X o \phi$$

Suppose Y is another vector field on M and F_t the flow generated by Y. Then

$$(F_{t*} X)(x) = TF_t o X o F_{-t}(x)$$

In terms of components,

$$(F_{t*} X)^i(x) = \sum_j \frac{\partial F_t^i(y)}{\partial y^j} X^j(y)$$

where

$$y = F_{-t}(x)$$

It follows that

$$\frac{d}{dt}(F_{t*} X)^i(x)|_{t=0} = \sum_j \frac{\partial Y^i(x)}{\partial x^j} X^j(y) - \sum_{j,k} \delta_j^i Y^k(x) \frac{\partial X^j(x)}{\partial x^k}$$

$$= \sum_j X^j Y^i_{,j} - Y^j X^i_{,j} = [X, Y] = L_X Y$$

Likewise, we find that

$$\frac{d}{dt} F_t^* X|_{t=0} = L_Y X = [Y, X]$$

Now observe that

$$F_{t*} F_{s*} = F_{(t+s)*}, t, s \in \mathbb{R}$$

This can be seen by explicitly writing down the components:

$$F_{s*} X = TF_s o X o F_s^{-1}$$

$$F_{t*} F_{s*} X = TF_t o TF_s o X o F_s^{-1} o F_t^{-1} = T(F_t o F_s) o X o (F_t o F_s)^{-1} = TF_{t+s} o X o F_{t+s}^{-1} = F_{(t+s)*} X$$

Likewise,

$$F_{t+s}^* = (F_t o F_s)^* = F_s^* F_t^*$$

Moreover F_t^* and F_{t*} are both linear operators. For example, if U, V are two vector fields on M and c a scalar, then

$$F_t^*(c.X+Y) = TF_t^{-1}(c.X+Y)oF_t = cTF_t^{-1}XoF_t + TF_t^{-1}YoF_t = cF_t^*X + F_t^*Y$$

Thus,

$$\frac{d}{dt}F_t^*X = \frac{d}{ds}F_s^*F_t^*X|_{s=0} = L_Y F_t^*X$$

Also

$$\frac{d}{dt}F_t^*X = \frac{d}{ds}F_t^*F_s^*X|_{s=0} = F_t^*\frac{d}{ds}F_s^*X|_{s=0} = F_t^*L_YX$$

In other words, we have proved that when operators act on vector fields, then

$$\frac{d}{dt}F_t^* = F_t^*L_Y = L_YF_t^*$$

Let us now evaluate the rate of change of the flow generated by Y on a one form, ie, if $\alpha \in TM^*$, so that for each $x \in M$, we have $\alpha_x \in TM_x^*$, then we want to evaluate

$$\frac{d}{dt}F_t^*\alpha$$

Note that

$$(F_t^*\alpha)(x)(X) = \alpha(F_t(x))(TF_t(X(x))$$

From this, it follows that

$$(F_{t+s}^*\alpha)(x)(X) = \alpha(F_{t+s}(x))(TF_{t+s}(X(x)))$$

$$= \alpha(F_toF_s(x))(TF_tTF_s(X(x)))$$

$$= (F_t^*\alpha)(F_s(x))(TF_s(X(x))) = (F_s^*F_t^*\alpha)(x)(X(x))$$

In other words,

$$F_{t+s}^*\alpha = F_s^*F_t^*\alpha = F_t^*F_s^*\alpha$$

It follows from the equation

$$\frac{d}{dt}F_t^*\alpha|_{t=0} = L_Y\alpha$$

that

$$\frac{d}{dt}F_t^*\alpha = F_t^*L_Y\alpha = L_YF_t^*\alpha$$

Now let ω be a two form on M. We shall compute $\frac{d}{dt}F_t^*\omega$. Suppose for simplicity, we assume that

$$\omega(X,Y) = \alpha(X)\beta(Y)$$

where α, β are one forms and X, Y arbitrary vector fields on M. Then,

$$(F_t^*\omega)(x)(X,Z) = \omega(F_t(x))(TF_t(X(x)), TF_t(Z(x)))$$

$$= (F_t^*\alpha)(x)(X(x))(F_t^*\beta)(x)(Z(x))$$

These equations can be expressed as

$$\omega = \alpha \otimes \beta$$

$$F_t^* \omega = F_t^* \alpha \otimes F_t^* \beta$$

and hence

$$\frac{d}{dt} F_t^* \omega = F_t^* L_Y \alpha \otimes F_t^* \beta + F_t^* \alpha \otimes F_t^* L_Y \beta$$
$$= F_t^* L_Y (\alpha \otimes \beta)$$

proving formula for two forms. The same argument works for r forms for all $r \geq 1$. Suppose now that t is a tensor field. For simplicity, to start with, we shall assume that t is a $(1,1)$ tensor field. Then the action of t at the point x on a pair (X, α) consisting of a vector field X and a one form α is given by the scalar $t(x)(X(x), \alpha(x))$ abbreviated as $t(X, \alpha)(x)$. We have

$$(F_t^* t)(x)(X, \alpha) = t(F_t(x))(TF_t(X(x)), (F_{t*}\alpha)(F_t(x)))$$
$$= t(F_t(x))((F_{t*}X)(F_t(x)), (F_{t*}\alpha)(F_t(x)))$$

This is the pullback of t by the flow generated by Y. It follows that

$$(F_t^* t)(F_{-t}(x))(X, \alpha) = t(x)((F_{t*}X)(x), (F_{t*}\alpha)(x))$$

We consider the identity

$$(F_t^* t)(x)(X, \alpha) =$$
$$t(F_t(x))((F_{t*}X)(F_t(x)), (F_{t*}\alpha)(F_t(x)))$$

Differentiate both sides with respect to t at $t = 0$ to get

$$\frac{d}{dt}(F_t^* t)(x)(X, \alpha)|_{t=0}$$

$$= t(x)(-(L_Y X)(x), \alpha(x)) + t(x)(X(x), -(L_Y \alpha)(x)) + L_Y(t(x)(X(x), \alpha(x)))$$
$$= L_Y(t(X, \alpha)) - t(L_Y X, \alpha) - t(X, L_Y \alpha) = (L_Y t)(X, \alpha)(x)$$

Note that $L_Y(t(X, \alpha)) = Y(t(X, \alpha))$. We have proved the fundamental formula

$$\frac{d}{dt} F_t^* t|_{t=0} = L_Y t$$

From this, and the equation

$$F_{t+s}^* t = F_t^* F_s^* t = F_s^* F_t^* t$$

it follows that

$$\frac{d}{dt} F_t^* t = F_t^* L_Y t = L_Y F_t^* t$$

25. **Mechanics**

1.Let G be a Lie group and $\}$ its Lie algebra. Let TG denote the tangent bundle of G. Note that TG_e is $\}$ by left invariance. Define the map $\lambda : TG \to G \times \}$ so that $\lambda(v) = (g, TL_g^{-1}v)$ for $v \in TG_g$. Also define the map $\rho : TG \to G \times \}$ so that $\rho(v) = (g, TR_g^{-1}v)$. $\lambda(v)$ describes the vector v in body coordinates and $\rho(v)$ the same in space coordinates. G can be taken as $SO(3)$, the configuration space of a rotating body. Let $x(t)$ be a curve in G. Its rate of change with time is $dx(t)/dt \in TG_{x(t)}$. The vector $x(t)^{-1}dx(t)/dt$ which should be read as $TL_{x(t)}^{-1}dx(t)/dt$ is an element of TG_e and should be interpreted as the rate of change of $x(t)$ in body coordinates. Suppose the body coordinates of the motion are (g, ξ). g can be interpreted as describing the configuration of the spinning top and ξ its rate of change in body coordinates. Then, the space coordinates of the motion are

$$\rho\lambda^{-1}(g, \xi) = \rho(TL_g\xi) = (g, TR_g^{-1}TL_g\xi) = (g, Ad_g\xi)$$

Suppose that the rigid body moves so that its configuration traces out a curve $x(t)$ in G. Let $v_0(t) \in TG_{x(t)}$. Then $v_0(t)$ in body coordinates is given by $\xi(t) = TL_{x(t)}^{-1}v_0(t)$. The same in space coordinates is given by $\tilde{\xi}(t) = Ad_{x(t)}\xi(t)$. Its rate of change can be computed as follows. First if η is a fixed vector, we compute $\frac{d}{dt}Ad_{x(t)}\eta$. We perform the computations for a matrix Lie group and generalize the formulas to an arbitrary Lie group. Note that $Ad_{x(t)}\eta = x(t)\eta x(t)^{-1}$ and therefore,

$$\begin{aligned}\frac{d}{dt}Ad_{x(t)}\eta &= (dx/dt)\eta x^{-1} - x\eta x^{-1}(dx/dt)x^{-1} \\ &= x.x^{-1}(dx/dt)\eta x^{-1} - x\eta x^{-1}(dx/dt)x^{-1} \\ &= x[x^{-1}dx/dt, \eta]x^{-1} = Ad_{x(t)}[TL_{x(t)}^{-1}(dx/dt), \eta]\end{aligned}$$

An alternate expression is

$$\begin{aligned}\frac{d}{dt}Ad_{x(t)}\eta &= (dx/dt)x^{-1}x\eta x^{-1} - x\eta x^{-1}(dx/dt)x^{-1} \\ &= [TR_x^{-1}(dx/dt), Ad_x\eta]\end{aligned}$$

so that if we define

$$\eta(t) = Ad_{x(t)}\eta$$

then we have

$$\frac{d\eta(t)}{dt} = [TR_{x(t)}^{-1}(dx(t)/dt), \eta(t)]$$

Note that $TR_{x(t)}^{-1}(dx(t)/dt) = (dx(t)/dt)_{space}$ is the rate of change of x in space coordinates. we can write the above formula as

$$\frac{d\eta(t)}{dt} = [(dx(t)/dt)_{space}, \eta(t)]$$

Now consider the rate of change of $Ad_{x(t)}\xi(t) = [\xi(t)]_{space}$. Note that $\xi(t)$ specifies the motion in body coordinates. Its rate of change is given by

$$\begin{aligned}\frac{d[\xi(t)]_{space}}{dt} &= Ad_{x(t)}(d\xi(t)/dt) + [(dx(t)/dt)_{space}, [\xi(t)]_{space}] \\ &= [d\xi(t)/dt]_{space} + [(dx(t)/dt)_{space}, [\xi(t)]_{space}]\end{aligned}$$

Let us compute the operation of left translation on the tangent manifold in body and space coordinates. In body coordinates, left translation reads

$$\lambda o T L_g o \lambda^{-1}(h,\xi) = \lambda o T L_g \dot{o}(T L_h \xi) = \lambda o T L_{gh} \xi = (gh, \xi)$$

so left translates in body coordinates does not alter the velocity vector. On the other hand in space coordinates,

$$\begin{aligned} \rho o T L_g o \rho^{-1}(h,\xi) &= \rho o T L_g o T R_h \xi = (gh, T R_{gh}^{-1} T L_g T R_h \xi) \\ &= (gh, T R_g^{1} T L_g \xi) = (gh, Ad_g \xi) \end{aligned}$$

In space coordinates, left translation acts in accordance with the adjoint action. Let us now look at the cotangent formulation: G as before is a Lie group and $\}$ is its Lie algebra. $\bar{\lambda} : T^*G \to G \times \}^*$ is a map that transforms the motion of the cotangent in T^*G to body coordinates while $\bar{\rho} : T^*G \to G \times \}^*$ transforms the motion in the cotangent space to space coordinates. Then, for $\alpha \in T^*G_g$, we have

$$\bar{\lambda}(\alpha) = (g, T L_g^* \alpha) = (g, \alpha o T L_g)$$

$$\bar{\rho}(\alpha) = (g, T R_g^* \alpha) = (g, \alpha o T R_g)$$

Note that $\alpha o T L_g$ and $\alpha o T R_g$ are elements of $\}^*$ or equivalently, linear functionals on TG_e. Let us look at how left translations act in body and space coordinates. In body coordinates, left translations look as follows:

$$\begin{aligned} \bar{\lambda} o T^* L_g^{-1} o \bar{\lambda}^{-1}(h,\alpha) &= \bar{\lambda} o T^* L_g^{-1} T^* L_h^{-1}(\alpha) \\ &= \bar{\lambda} o \alpha o T L_h^{-1} T L_g^{-1} = \bar{\lambda} o \alpha o T L_{gh}^{-1} \\ &= (gh, \alpha) \end{aligned}$$

Thus left translations do not alter the element of $\}^*$, they alter only the base point in G. Consider now left translations in space coordinates:

$$\bar{\rho} o T^* L_g^{-1} o \bar{\rho}^{-1}(h,\alpha) =$$

$$\begin{aligned} &= \bar{\rho} o T^* L_g^{-1} T^* R_h^{-1} \alpha = \bar{\rho} o \alpha T R_h^{-1} T L_g^{-1} = (gh, \alpha o T R_h^{-1} T L_g^{-1} T R_{gh}) \\ &= (gh, \alpha o Ad_{g^{-1}}) = (gh, Ad^*_{g^{-1}} \alpha) \end{aligned}$$

Digression:Let Q be a differentiable manifold and T^*Q the cotangent manifold. Let $\tau_Q^* : T^*Q \to Q$ be defined as the canonical projection, ie, $(q, \alpha_q) \in T^*Q$ implies $\tau_Q^*(q, \alpha_q) = q$. Let a group G act on Q in accord with $(g,q) \to \Phi_g(q) = \Phi(g,q)$. Let Φ^{T*} be the lifted action of G on T^*Q, so that

$$\Phi_g^{T*} \alpha_q = (gq, \alpha_q o T \Phi_g^{-1})$$

Note that while $\alpha_q \in T^*Q_q$, we have $\alpha_q o T \Phi_g^{-1} \in T^*Q_{gq}$. This formula is sometimes written as

$$(\Phi_g^{T*} \alpha)_{gq} = (gq, \alpha_q o T \Phi_g^{-1})$$

We have

$$\tau_Q^* \Phi_g^{T*}(\alpha_q) = gq,$$

on the one hand and on the other,

$$\Phi_g o \tau_Q^*(\alpha_q) = gq$$

This means that

$$\tau_Q^* o \Phi_g^{T*} = \Phi_g o \tau_Q^*$$

Let $\xi \in \}^*$ and we define the vector field ξ_Q on Q so that

$$\xi_Q(q) = \frac{d}{dt}\Phi(exp(t\xi), q)|_{t=0} = \frac{d}{dt}\Phi_{exp(t\xi)}(q)|_{t=0}$$

Likewise, we define the vector field $\xi_{T^*Q}(\alpha_q)$ on T^*Q so that

$$\xi_{T^*Q}(\alpha_q) = \frac{d}{dt}\Phi^{T*}_{exp(t\xi)}(\alpha_q)|_{t=0}$$

Then we obtain from the identity

$$\tau_Q^* o \Phi_g^{T*}(\alpha_q) = \Phi_g o \tau_Q^*(\alpha_q)$$

on setting $g = exp(t\xi)$ and differentiating with respect to t at $t = 0$,

$$T\tau_Q^* \xi_{T^*Q} = \xi_Q o \tau_Q^*$$

Let θ be the canonical one form on T^*Q, ie, $\theta \in T^*T^*Q$. We recall the definition

$$\theta(\alpha_q)(w_q) = \alpha_q(T\tau_Q^* w_q)$$

Let $P = T^*Q$. We have proved that $J(\xi)(\alpha_q) = i_{\xi_P}\theta(\alpha_q)$ is an equivariant momentum mapping on P. We write

$$J(\xi)(\alpha_q) = \alpha_q(T\tau_Q^*\xi_P) = \alpha_q(\xi_Q o \tau_Q^*)$$

This shows that an equivariant momentum mapping for the action of G on T^*Q induced by the action of G on Q is given by the formula

$$J(\xi)(\alpha_q) = \alpha_q(\xi_Q(q))$$

When the Hamiltonian on T^*Q is invariant under G-action, then we get corresponding to this momentum map the classical form of Noether's conservation theorem. To see this explicitly, we consider a Hamiltonian $H(q,p)$ specified as a function of the coordinates q and the momenta p. The group G acts on the variable q and induces an action on the momentum p:

$$q_i \to \Phi_i(g, q), i = 1, 2, ..., n$$

$$p_i \to \sum_j p_j \Phi_{j,i}(g^{-1}, q)$$

where

$$\Phi_{j,i}(g, q) = \frac{\partial \Phi_j(g, q)}{\partial q_i}$$

Suppose that H is invariant under G-action, ie,

$$H(\{q_i\},\{p_i\}) = H(\{\Phi_i(g,q)\},\{\sum_j \Phi_{j,i}(g^{-1},q)p_j\}), g \in G$$

For infinitesimal g, we can write

$$\Phi_i(g,q) = \Phi_i(\epsilon\xi,q) = q_i + \epsilon\xi_i(q) + O(\epsilon^2)$$

Then

$$\Phi_{j,i}(g^{-1},q) = \delta_{ij} - \epsilon\xi_{j,i}(q) + O(\epsilon^2)$$

Thus, under the infinitesimal transformation,

$$q_i \to q_i + \epsilon\xi_i(q),$$

$$p_i \to p_i - \epsilon\sum_j p_j\xi_{j,i}(q)$$

and invariance of H under infinitesimal G-action gives

$$\sum_i H_{,q_i}\xi_i(q) - \sum_{i,j} H_{,p_i}\xi_{j,i}(q)p_j = 0$$

If q,p vary along a trajectory of the Hamiltonian system, then $p_i' = -H_{,q_i}, q_i' = H_{,p_i}$ and the above becomes

$$\sum_i p_i'\xi_i(q) + \sum_{i,j} q_i'\xi_{j,i}p_j = 0$$

or

$$\frac{d}{dt}(p_i\xi_i(q)) = 0$$

so that the Noether current $\sum_i p_i\xi_i(q)$ is conserved along the trajectory.

26.Abraham and Marsden, Foundations of mechanics: The crucial ideas in this book are focussed around differential forms applied to the formulation of the fundamental equations of classical mechanics. The calculus of forms is used to formulate the Hamilton and Lagrange equations of motion in the language of symplectic geometry. The starting point for these investigations is the notion of a tangent manifold. $\mathcal{M}$ is a differentiable manifold given to us and for each point $x \in \mathcal{M}$, one can introduce the tangent space $T\mathcal{M}_x$ at x. This is a vector space each of whose elements is the equivalence class of all curves passing through x in $\mathcal{M}$ Equivalently, given a chart (U,ϕ) at x, consider any curve $t \to c(t)$ passing through x at $t = 0$. This curve has a image curve in $\mathbb{R}^n$, namely $\phi(c(t))$. All curve $c(.)$ for which $\frac{d}{dt}\phi(c(t))|_{t=0}$ is the same form an equivalence class. The vector $\frac{d}{dt}\phi(c(t))|_{t=0} \in \mathbb{R}^n$ is the image of the tangent vector at x defined by the equivalence class of curves at x containing $c(.)$. The next stage is the introduction of a vector field X on a manifold. It is a mapping $X : M \to TM$ so that $(x, X(x)) \in TM_x$ for all x. Suppose that in local coordinates, $X(x)$ has components $X^i(x), i = 1,2,...,n$. Then the integral curve/flow generated by this vector field is given by solving the sequence of coupled ordinary differential equations

$$dx^i/dt = X^i(x), i = 1,2,...,n, x = (x^i)$$

Denote the solution corresponding to an initial condition x_0 by $F_t(x_0)$. Then, $F_t, t \in \mathbb{R}$ is a one parameter group of diffeomorphisms of $\mathbb{R}^n$. Various properties of flows generated by vector fields have been discussed in this book. The next notion introduced in this book is that of a form. Consider for each $x \in M$ a skewsymmetric multilinear mapping $\Omega_x : TM_x^k \to \mathbb{R}$. Then Ω is called a k-form on the manifold. Suppose (x^i) is a local coordinate chart. Any vector field X having components $X^i(x)$ can be regarded as a linear first order differential operator $\sum_i X^i(x)\partial_i$. Suppose $f : M \to \mathbb{R}$ is a smooth function. Then if $c(t)$ is a curve on the manifold, such that in local coordinates $c(0) = x, c'(0) = v$, then regarding v as a local differential operator $D_v = v^i\partial_i$, it follows that

$$\frac{df(c(t))}{dt}|_{t=0} = \sum_i v^i\partial_i f(x)$$

which is the reason why vectors on the manifold are represented as linear differential operators.

27. Poincare variables, Delaunay variables and Poincare variables. The equation of the ellipse is given by

$$r = \frac{l}{1 + e.cos(\phi)}, r^2 d\phi/dt = \beta$$

where l is the latus rectum of the ellipse and e is the eccentricity of the orbit. β is the angular momentum. The origin here is one of the focii. The nearest distance of the orbit from the origin is

$$\frac{l}{1+e}$$

and the farthest distance is

$$\frac{l}{1-e}$$

It follows that the semimajor axis is given by

$$a = \frac{1}{2}(l/(1+e) + l/(1-e)) = \frac{l}{1-e^2}$$

the x, y coordinates of the orbit are respectively given in parametric form by

$$X(\phi) = r.cos(\phi) = \frac{l.cos(\phi)}{1 + e.cos(\phi)}, Y(\phi) = r.sin(\phi) = \frac{l.sin(\phi)}{1 + e.cos(\phi)}$$

We also have the angular momentum equation

$$r^2 d\phi/dt = L$$

where L is the total angular momentum. L is a constant of the motion. The first aim is to calculate explicitly an expression for L in terms of the semimajor axis a and the orbit eccentricity e. To do so, we start with the energy equation

$$\frac{1}{2}(r'^2 + r^2\phi'^2) - GM/r = E$$

and $r^2\phi' = L$. Then, from these two equations

$$\frac{1}{2}((dr/d\phi)^2 + r^2) - GMr^3/L^2 - Er^4/L^2 = 0$$

Letting $u = 1/r$ gives us $r' = -u'/u^2$ so that

$$\frac{1}{2}(u'^2 + u^2) - GMu/L^2 = E/L^2$$

Here, prime denotes derivative with respect to ϕ. Differentiating both sides gives us

$$u'' + u = GM/L^2$$

the solution to which is

$$u(\phi) = GM/L^2 + A.cos(\phi)$$

To determine the constant A, we compute

$$u' = -A.sin(\phi)$$

Plugging this into the first order differential equation satisfied by u gives us

$$\frac{1}{2}((GM/L^2 + A.cos(\phi))^2 + A^2.sin^2(\phi)) - (GM/L^2)(GM/L^2 + A.cos(\phi)) = E/L^2$$

from which it follows that

$$(A^2 + (GM/L^2)^2)/2 - (GM/L^2)^2 = E/L^2$$

or

$$A^2 - (GM/L^2)^2 = 2E/L^2$$

or

$$A = \sqrt{(GM/L^2)^2 + 2E/L^2}$$

The eccentricity of the orbit is now

$$e = AL^2/GM = \sqrt{1 + 2EL^2/(GM)^2}$$

The semimajor axis is

$$a = \frac{1}{2}\left(\frac{L^2}{GM(1+e)} + \frac{L^2}{GM(1-e)}\right) = \frac{L^2}{GM(1-e^2)}$$

so that

$$L^2/GM = a(1-e^2)$$

or

$$L = \sqrt{GMa(1-e^2)}$$

We also have from the above,

$$1 + 2EL^2/(GM)^2 = e^2 = 1 - L^2/GMa$$

giving

$$E = -GM/2a$$

It is standard to take $GM = 1$ and then we get $L = \sqrt{a(1-e^2)}$ and $E = -1/2a$.

Problem:Assume that the initial position and velocity $x(0), y(0), x'(0), y'(0)$ of the orbiting particle in the two body problem are specified. Calculate from these the quantities (a) the inclination of the perihelion with respect to the x axis, (b) the angle between the radius vector of the particle at time zero with respect to the origin which is one of the focii relative to the perihelion, (c) the eccentricity and (d) the semimajor axis of the elliptical orbit.

Solution: Let $r(t) = \sqrt{x(t)^2 + y(t)^2}$ and $\phi(t) = tan^{-1}(y(t)/x(t))$. Then,

$$r(0) = \sqrt{x(0)^2 + y(0)^2}, \phi(0) = tan^{-1}(y(0)/x(0)),$$

$$r'(0) = (x(0)x'(0) + y(0)y'(0))/\sqrt{x(0)^2 + y(0)^2},$$

$$\phi'(0) = (x(0)y'(0) - y(0)x'(0))/(x(0)^2 + y(0)^2)$$

We are assuming $GM = 1$ and then we have

$$r^2(0)\phi'(0) = \sqrt{a(1-e^2)}, \frac{1}{2}(r'^2(0) + r(0)^2\phi'^2(0)) - 1/r(0) = E = -1/2a$$

These two algebraic equations can be solved to give a, e as functions of $r(0), \phi(0)$. The equation of the orbit is

$$r = L^2/(1 + e.cos(\phi - \alpha))$$

where α is a constant equal to the inclination of the perihilion with respect to the x axis. We know $L = r(0)^2\phi'(0)$ and e is also known in terms of L, E which are in terms known in terms of $r(0), \phi(0), r'(0), \phi'(0)$. Thus,

$$r(0) = L^2/(1 + e.cos(\phi(0) - \alpha))$$

can be solved to get α in terms of $r(0), \phi(0), \phi'(0)$. $\phi(0) - \alpha$ is then the angle between the radius vector at time zero relative to the perihelion.

28. Flow of a vector field on a differentiable manifold: Let $\mathcal{M}$ be a C^∞-manifold and $X(x)$ a vector field on this manifold. In terms of a local chart, $X(x) = (X^i(x), i = 1, 2, ..., n)$ where $dimTM_x = n$ $\forall x \in \mathcal{M}$.

29. Mechanics:Critical points in the restricted three body problem. There are two centres of force both moving around the origin in circles. The first is at $(\mu.cos(t), \mu.sin(t))$ and the second at $-((1-\mu).cos(t), -(1-\mu).sin(t))$. The Lagrangian of the system is

$$\mathcal{L}(q_1, q_2, q_1', q_2', t) = \frac{1}{2}(q_1'^2 + q_2'^2) - V(t, q_1, q_2)$$

where

$$V(t, q_1, q_2) = -((q_1 - \mu.cos(t))^2 + (q_2 - \mu.sin(t))^2)^{-1/2} - ((q_1 + (1-\mu).cos(t))^2 + (q_2 + (1-\mu).sin(t))^2)^{-1/2}$$

This Lagrangian depends explicitly on time. The corresponding Hamiltonian will also depend explicitly on time. However, by a canonical transformation, the Lagrangian and Hamiltonian can be made time independent. The general theory of canonical transformations can be developed along the following lines. Let q, p be the original coordinates and momenta and let the corresponding Hamiltonian be $H(q, p, t)$. The corresponding Lagrangian is given by $pq' - H$. The new set of coordinates and momenta are given by Q, P and the corresponding Hamiltonian is $K(Q, P, t)$. The corresponding Lagrangian is $PQ' - K$ and since the two Lagrangians correspond to equivalent equations of motion and hence must differ from a total time derivative. We write

$$PQ' - K = pq' - H + dS/dt$$

Writing $S = S(q, Q, t)$, we get

$$PQ' - K = pq' - H + S_{,q}q' + S_{,Q}Q' + S_{,t}$$

from which we infer that

$$S_{,q} = -p, S_{,Q} = P, K = H - S_{,t}$$

S is the generating function of the canonical transformation. Suppose that we are given a one form $pdq - Hdt$ and the generating function S transforms this one form to $PdQ - Kdt$. Then,

$$PdQ - Kdt = pdq - Hdt + dS$$

This is the same as the above equation if we note that

$$dS = S_{,q}dq + S_{,Q}dQ + S_{,t}dt$$

By taking the differential of the above, we get

$$dP \wedge dQ + dt \wedge dK = dp \wedge dq + dt \wedge dH$$

or equivalently,

$$dP \wedge dQ - dp \wedge dq = dt \wedge d(H - K)$$

Writing $S_{,t} = F$, we have since $H - K = F$ that

$$dP \wedge dQ - dp \wedge dq = dt \wedge dF$$

which is an equivalent condition for a canonical transformation.

30.More elaboration on canonical transformations. Let (P, ω) be a symplectic manifold and let $\pi : \mathbb{R} \times P \to P$ be the canonical projection on the second coordinate, ie, $\pi(t, x) = x$ for all $x \in P$. Let $\tilde{\omega} = \pi^*\omega$. Then $(\mathbb{R} \times P, \tilde{\omega})$ is a contact manifold. Let $(\mathbb{R} \times P_i, \tilde{\omega}_i), i = 1, 2$ be two contact manifolds and let $F : \mathbb{R} \times P_1 \to \mathbb{R} \times P_2$ be a diffeomorphism such that $F^*t = t$ where $t : \mathbb{R} \times P_i \to \mathbb{R}, i = 1, 2$ is the canonical projection on the first coordinate. The condition $F^*t = t$ means that F has the form $F(\lambda, x) = (\lambda, F_0(\lambda, x))$. Secondly we assume that there is a function K_F such that

$$F^*\tilde{\omega}_2 = \tilde{\omega}_1 + dK_F \wedge dt$$

Then, F is called a canonical transformation. Canonical transformations on the same contact manifold $\mathbb{R} \times P$ form a group. This follows from the following argument: Let F, G be canonical transformations. We must then show that FoG^{-1} It is clear that

$$(FoG^{-1})^*t = to(FoG^{-1}) = (toF)0G^{-1} = toG^{-1} = t$$

since F, G both preserve time and since G preserves time and is a diffeomorphism, G^{-1} also preserves time. Further,

$$(FoG^{-1})^*\tilde{\omega} = (G^{-1})^*F^*\tilde{\omega}$$

$$F^*\tilde{\omega} = \tilde{\omega} + dK_F \wedge dt$$

$$G^*\tilde{\omega} = \tilde{\omega} + dK_G \wedge dt$$

so that

$$\begin{aligned}(G^{-1})^*\tilde{\omega} &= \tilde{\omega} - dK_G oG^{-1} \wedge dtoG^{-1} \\ &= \tilde{\omega} - dK_G oG^{-1} \wedge dt\end{aligned}$$

giving

$$(FoG^{-1})^*\tilde{\omega} = \tilde{\omega} + d(K_F oG^{-1} - K_G oG^{-1}) \wedge dt$$

proving that FoG^{-1} is a canonical transformation with

$$K_{FoG^{-1}} = (K_F - K_G)oG^{-1}$$

Let $(P_i, \omega_i), i = 1, 2$ be two symplectic manifolds and let $\pi_i : \mathbb{R} \times P_i \to P_i$ be the canonical projection on the second coordinate. Consider the contact manifold $(\mathbb{R} \times P_i, \tilde{\omega}_i), i = 1, 2$ where $\tilde{\omega}_i = \pi_i^*\omega_i$. Suppose that there exists a one form α on $\mathbb{R} \times P_1$ such that a mapping $F : \mathbb{R} \times P_1 \to \mathbb{R} \times P_2$ satisfies

$$F^*\tilde{\omega}_2 = \tilde{\omega}_1 + \alpha \wedge dt$$

Then $F_t : ¶_1 \to P_2$ is symplectic. Define $j_t : P \to \mathbb{R} \times P$ so that $j_t(x) = (t, x)$. Also π_2 denotes the canonical projection on the second coordinate. We have $F(t, x) = (t, F_t(x))$ so that $\pi_2 oFoj_t = F_t$. Then, Now,

$$\begin{aligned}F_t^*\omega_2 &= (\pi_2 oFoj_t)^*\omega_2 = j_t^*F^*\pi_2^*\omega_2 \\ &= j_t^*F^*\tilde{\omega}_2 = j_t^*(\tilde{\omega}_1 + \alpha \wedge dt) \\ &= j_t^*\pi_2^*\omega_1 + j_t^*\alpha \wedge d(toj_t) \\ &= (\pi_2 oj_2)^*\omega_1 + j_t^*\alpha \wedge d(toj_t)\end{aligned}$$

Now since t in the function j_t is fixed, it follows that toj_t is a constant :$toj_t(x) = t(t, x) = t$ and hence $d(toj_t) = 0$. Also $\pi_2 oj_2(x) = \pi_2(t, x) = x$, ie, $\pi_2 oj_t$ is the identity and hence we get

$$F_t^*\omega_2 = \omega_1$$

proving that F_t is symplectic. Conversely, suppose F_t is symplectic. We then have to show that there exists a one form α on $\mathbb{R} \times P_1$ such that

$$F^*\tilde{\omega}_2 = \tilde{\omega}_1 + \alpha \wedge dt$$

Let

$$\beta = F^*\tilde{\omega}_2 - \tilde{\omega}_1$$

Then,

$$j_t^*\beta = j_t^* F^* \pi_2^* \omega_2 - j_t^* \pi_2^* \omega_1$$

$$= (\pi_2 o F o j_t)^* \omega_2 - (\pi_2 o j_t)^* \omega_1$$

$$= F_t^* \omega_2 - \omega_1 = 0$$

Now write

$$\beta = \gamma + \alpha \wedge dt$$

where γ does not involve dt. We claim that $\pi_2^* j_t^* \beta = \gamma$ and hence $\gamma = 0$. Indeed, note that $j_t o \pi_2(s, x) = j_t(x) = (t, x)$ and hence the tth (first) component of $j_t o \pi_2$ is a constant. It follows that $t o j_t o \pi_2 = t$ a constant and hence $\pi_2^* j_t^* dt = d(t o j_t o \pi_2) = 0$ implying that $\pi_2^* j_t^* (\alpha \wedge dt) = 0$. Also it is clear that $(j_t o \pi_2)^* \gamma = \gamma$ since γ does not involve dt and $\pi_2 o j_t o \pi_2(s, x) = (t, x)$ so that the Jacobian matrix of the second component of $j_t o \pi_2$ with respect to the second component is the identity.

31. Klingenberg, Differential geometry:

Differential geometry fundamentals:Notion of a curve in $\mathbb{R}^n$, notion of a surface in $\mathbb{R}^n$, tangent to a curve and tangent space on a surface at a point. A curve in $\mathbb{R}^3$ is a map $t \to (x(t), y(t), z(t)) = \mathbf{r}(t)$. its tangent vector at the point $\mathbf{r}(t)$ is given by

$$\mathbf{v}(t) = \mathbf{r}'(t) = (x'(t), y'(t), z'(t))$$

The length of the tangent vector can be normalized to equal unity by defining the distance parameter s so that

$$s'(t) = \sqrt{x'^2(t) + y'^2(t) + z'^2(t)}$$

or equivalently,

$$ds = \sqrt{dx^2 + dy^2 + dz^2}$$

The normalized tangent vector is given by

$$\mathbf{t} = d\mathbf{r}/ds = (s'(t))^{-1}(d\mathbf{r}/dt)$$

Note that we have

$$\mathbf{t}.\mathbf{t} = 1$$

since

$$d\mathbf{r}.d\mathbf{r} = ds^2$$

We have on differentiating the equation

$$(d\mathbf{r}/ds).(d\mathbf{r}/ds) = 1$$

that

$$(d^2\mathbf{r}/ds^2).(d\mathbf{r}/ds) = 0$$

implying that if we set

$$\mathbf{n} = d^2\mathbf{r}/ds^2$$

then

$$\mathbf{n.t} = 0$$

$\mathbf{n}$ is called the principal normal. Note that $\mathbf{n}$ does not have unit length in general. In fact, the length of $\mathbf{n}$ is the local curvature K of the curve. We write this as

$$d\mathbf{t}/ds = K\mathbf{n}$$

We shall now obtain a formula for K when the curve is specified as a function of an arbitrary parameter t. We have

$$d\mathbf{r}/ds = \mathbf{r}'/|\mathbf{r}'|$$

where prime denotes derivative with respect to t. Thus,

$$d^2\mathbf{r}/ds^2 = [|\mathbf{r}'|\mathbf{r}'' - \mathbf{r}'(\mathbf{r}'', \mathbf{r}')/|\mathbf{r}'|]/|\mathbf{r}'|^3$$

$$= [|\mathbf{r}'|^2\mathbf{r}'' - \mathbf{r}'(\mathbf{r}', \mathbf{r}'')]/|\mathbf{r}'|^4$$

$$= \mathbf{r}' \times (\mathbf{r}'' \times \mathbf{r}')/|\mathbf{r}'|^4$$

and taking magnitudes on both sides, we get

$$K = |\mathbf{r}' \times \mathbf{r}''|/|\mathbf{r}'|^3$$

Problem: Suppose that the curve is in the plane, ie specified as $y = f(x)$. Find out the expression for the curvature K.

Solution: $\mathbf{r} = x\hat{x} + f(x)\hat{y}$, so that

$$\mathbf{r}' = \hat{x} + f'(x)\hat{y}, \mathbf{r}'' = f''(x)\hat{y}$$

and

$$\mathbf{r}' \times \mathbf{r}'' = f''(x)\hat{z}$$

and thus,

$$|\mathbf{r}' \times \mathbf{r}''| = |f''(x)|$$

Also

$$|\mathbf{r}'| = \sqrt{1 + f'^2(x)}$$

giving

$$K = \frac{|f''|}{(1 + f'^2)^{3/2}}$$

The curvature has the following physical interpretation: $\mathbf{t}$ is the unit tangent vector to the curve at s and $\mathbf{t} + d\mathbf{t}$ is the unit tangent vector to the curve at $s + ds$. It follows that if $d\theta$ is the angle between these two neighbouring tangent vectors, then

$$d\theta = sin(d\theta) = |\mathbf{t} \times (\mathbf{t} + d\mathbf{t})| = |\mathbf{t} \times d\mathbf{t}| = |d\mathbf{t}|$$

and hence the change in angle per unit length is given by

$$d\theta/ds = |d\mathbf{t}/ds| = K$$

Thus, the curvature of the curve is simply the change in angle per unit length of the curve.

32. Problem: Show that if a surface has the parametric equation $\mathbf{r} = \mathbf{r}(u, v)$, then the area element on this surface is given by $dS(u, v) = \|\mathbf{r}_{,u} \times \mathbf{r}_{,v}|dudv$.

33. Problem: A particle moves along a curve in $\mathbb{R}^3$ specified in cylindrical coordinates as $(r(t), \phi(t), z(t))$. Determine the distance travelled by the particle in the duration $[0, T]$.

Answer:

$$s[0,T] = \int_0^T ds(t) = \int_0^T (r'^2(t) + r^2(t)\phi'^2(t) + z'^2(t))^{1/2} dt$$

34. Problem: A particle moves along a curve in $\mathbb{R}^3$ specified in spherical polar coordinates as $(r(t), \theta(t), \phi(t))$. Determine the distance travelled by the particle in the duration $[0, T]$.

Answer:

$$s[0,T] = \int_0^T (r'^2(t) + r^2(t)\theta'^2(t) + r^2(t)sin^2(\theta(t))\phi'^2(t))^{1/2} dt$$

35. Problem: Determine the formula for the unit normal to a two dimensional surface immersed in $\mathbb{R}^3$ specified as $f(x, y, z) = 0$ at the point (x, y, z).

Answer:

$$\mathbf{n} = \nabla f/|\nabla f| = (f_x\hat{x} + f_y\hat{y} + f_z\hat{z})/\sqrt{f_x^2 + f_y^2 + f_z^2}$$

Chapter 10

Quantum Stochastic Calculus

1.The triplet $(\mathcal{H}, \mathcal{P}(\mathcal{H}), \rho)$ of a Hilbert space, the lattice of projections and a density operator defines a quantum probability space in contrast to the triplet $(\Omega, \mathcal{F}, P)$ with Ω the sample space, $\mathcal{F}$ the σ algebra of events and P a probability measure on $\mathcal{F}$. Compare and contrast.

2.Quantum stochastic integral, definition. First give the definition of an adapted quantum process. $\langle_0$ is the initial Hilbert space and $\mathcal{H}$ another Hilbert space. $\langle_0 \otimes \Gamma_s(\mathcal{H})$ is the Fock space tensored with the initial Hilbert space. For $f \in \langle$ and $u \in \mathcal{H}$, we have $fe(u) = f \otimes e(u) \in \langle \otimes \Gamma_s(\mathcal{H})$. Let $X_t, t \geq 0$ is a family of operators acting in the space $\langle \otimes \Gamma_s(\mathcal{H})$. Let $\xi(.)$ be a spectral measure on $\mathbb{R}_+$ in the Hilbert space $\mathcal{H}$ without any atoms, ie, $\xi(\{t\}) = 0$ for all $t \in \mathbb{R}_+$. Define $\mathcal{H}_{t]} = \xi[0,t]\mathcal{H}$, $\mathcal{H}_{[t} = \xi[t,\infty)\mathcal{H}$ and $\mathcal{H}_{[s,t]} = \xi[s,t]\mathcal{H}$ for $s \leq t$. We have the following upto isomorphism for all $0 = t_0 < t_1 < ... < t_n$:

$$\mathcal{H} = \bigoplus_{i=0}^{n} \mathcal{H}_{[t_i,t_{i+1}]})$$

where $[t_0, t_1]$ is to be interpreted as $[0, t_1]$ so that $\mathcal{H}_{[t_0,t_1]} = \mathcal{H}_{t_1]}$ and $[t_n, t_{n+1}]$ is to be interpreted as $[t_n, \infty)$ so that $\mathcal{H}_{t_n,t_{n+1}]} = \mathcal{H}_{[t_n}$. Thus,

$$\mathcal{H} = \mathcal{H}_{t_1]} \oplus \mathcal{H}_{[t_1,t_2]} \oplus ... \oplus \mathcal{H}_{[t_{n-1},t_n]} \oplus \mathcal{H}_{[t_n}$$

It follows that

$$\Gamma_s(\mathcal{H}) = \bigotimes_{i=0}^{n} \Gamma_s(\mathcal{H}_{[t_i,t_{i+1}]}$$

upto isomorphism.

Remark: Let $\mathcal{H} = \mathcal{H}_1 \oplus \mathcal{H}_2$. Then $\Gamma_s(\mathcal{H})$ is isomorphic to $\Gamma_s(\mathcal{H}_1) \otimes \Gamma_s(\mathcal{H}_2)$. Indeed, let $u \in \mathcal{H}$. Then $u = u_1 + u_2$ where $u_i \in \mathcal{H}_i, i = 1,2$ uniquely. Then define $Ue(u) = e(u_1) \otimes e(u_2)$. This map preserves inner products and since $\{e(u), u \in \mathcal{H}\}$, $\{e(u) : u \in \mathcal{H}_i\}, i = 1,2$ are respectively total in $\mathcal{H}$ and $\mathcal{H}_i, i = 1,2$ respectively, it follows that U extends to a unique unitary isomorphism between the Hilbert spaces $\Gamma_s(\mathcal{H})$ and $\Gamma_s(\mathcal{H}_1) \otimes \Gamma_s(\mathcal{H}_2)$. (Let $P_i : \mathcal{H} \to \mathcal{H}_i, i = 1,2$) be the canonical orthogonal projections. Then we have for $u, v \in \mathcal{H}$,

$$< e(u), e(v) >= exp(< u, v >) = exp(< P_1u, P_1v > + < P_2u, P_2v >)$$

$$= exp(< P_1u, P_1v >).exp(< P_2u, P_2v >) =< e(P_1u), e(P_1v) >< e(P_2u), e(P_2v) >$$

$$=< e(P_1u) \otimes e(P_2u), e(P_1v) \otimes e(P_2v) >$$

showing that $e(u)$ under the isomorphism is identified with $e(P_1u) \otimes e(P_2u)$. $X_t, t \geq 0$ is an adapted process if for all $f \in \langle, u \in \mathcal{H}$, we have the following factorization

$$X_t fe(u) = (X_t fe(u_{t]})).e(u_{[t})$$

By this we mean that if $g \in \langle$ and $v \in \mathcal{H}$ are arbitrary, then

$$< ge(v), X_t fe(u) >=< ge(v_{t]}), X_t fe(u_{t]}) >< e(v_{[t}), e(u_{[t}) >$$

3.K.R.Parthasarathy, An introduction to quantum stochastic calculus: The author starts with obtaining a quantum version of the expectation value of a random variable on a probability space. Specifically, if X is a random variable that assumes values $X(i), i = 1, 2, ..., n$ with probabilities $p(i), i = 1, 2, ..., n$, then $\mathbb{E}X = \sum_i p(i)X(i)$ can be expressed as $Tr(\rho X)$ where $\rho = diag[p(1), ..., p(n)]$ and $X = diag[X(1), ..., X(n)]$ and it can also be expressed as $< \psi|X|\psi >$ where $\psi = row(\sqrt{p}_k exp(i\theta_k), k = 1, 2, ..., n)$. The author then introduces the Hilbert space $L^2(\mu, \langle)$ where $(S, \mathcal{F}, \mu)$ is a measure space and $\langle$ a Hilbert space. This space consists of all measurable maps $f : S \to \langle$ for which $\int \| f \|^2 d\mu < |infty$. Direct sum of a countable family of Hilbert spaces $\mathcal{H}_n, n = 1, 2, ...$ is introduced next. Any element in this space is an infinite sequence $(u_n, n = 1, 2, ...)$ with $u_n \in \mathcal{H}_n$ and $\sum_n \| u_n \|^2 < \infty$. The notion of a bounded linear operator in a Hilbert space is discussed next. This leads one to the notion of operator norm, ie a norm defined on the space of bounded linear operators induced by the norm on the Hilbert space. If $\mathcal{H}$ is a Hilbert space and $T : \mathcal{H} \to \mathcal{H}$ a linear operator, then T is said to be a bounded linear operator if $sup_{\|u\| \leq 1, u \in \mathcal{H}} \| Tu \| < \infty$. In this case, we denote this quantity by $\| T \|$. This is the operator norm. Any linear operator satisfying this finiteness condition is called a bounded linear operator and the class of all bounded linear operators is denoted by $\mathcal{B}(\mathcal{H})$. The class of observables in a Hilbert space $\mathcal{H}$, namely $O(\mathcal{H})$ is the set of bounded Hermitian operators in the space, ie, $T \in O(\mathcal{H})$ iff $T \in \mathcal{B}(\mathcal{H})$ and $T^* = T$. The set of orthogonal projections in a Hilbert space $\mathcal{H}$, namely $\mathcal{P}(\mathcal{H})$ forms a lattice. If $X, Y \in O(\mathcal{H})$, then $i(XY - YX), XY + YX \in O(\mathcal{H})$. If $E_\alpha, \alpha \in \Gamma$ is a family in $\mathcal{P}(\mathcal{H})$, then $\bigcup E_\alpha$ is the projection onto the smallest subspace of $\mathcal{H}$ containing $R(E_\alpha), \alpha \in \Gamma$ and $\bigcap E_\alpha$ is the projection onto $\bigcap R(E_\alpha)$. De-Morgan's rules are stated for these operations. For example, if $E_1, E_2 \in \mathcal{P}(\mathcal{H})$, then $(E_1 \cup E_2)^\perp = E_1^\perp \cap E_2^\perp$. This follows from the fact that if $u \in (R(E_1) \cup R(E_2))^\perp$, then $E_1u = E_2u = 0$ and hence $u \in R(E_1)^\perp \cap R(E_2)^\perp$ and conversely, if $u \in R(E_i)^\perp, i = 1, 2$, then $E_iu = 0, i = 1, 2$ so that $u \in (R(E_1) \cup R(E_2))^\perp$. Likewise, the De-Morgan rule $(E_1 \cap E_2)^\perp = E_1^\perp \cup E_2^\perp$ has been stated. If $u \in E_1^\perp \cup E_2^\perp$, then $u = E_1^\perp u_1 + E_2^\perp u_2$ so that if $v = E_1v_1 = E_2v_2$, then $< u, v >= 0$. This combined with a dimension argument then yields the desired identity. The notions of pure and mixed states have been introduced here and the idea that a mixed state of rank k can be expressed as a convex combination of k pure states has been stated. The distributive part of De-Morgan's rules fail to be valid. Properties of the rank one operator $|u >< v|$ have been stated. A thorough discussion of finite dimensional quantum probability spaces has been made. The notion of a state ρ as a positive operator with unit trace has been given and $tr(\rho E), E \in \mathcal{P}(\mathcal{H})$ has been interpreted as the probability of the event E in the state ρ. If ρ is a state and X an observable in a finite dimensional Hilbert space, then using the spectral theorem, the event that X assumes a given value and the probability of this event in the state ρ has been given. This is part of the axiomatic development of the subject of quantum probability. Using, this, the notion of expectation of an observable in a given state for finite dimensional quantum probability spaces has been given. Using the spectral theorem, the quantity $exp(itX), t \in \mathbb{R}$ can be computed and this when combined with the notion of quantum expectation yields the formula for the quantum characteristic function. This formula has been given. If X is an observable and ρ a state and U a unitary operator, then the identity $tr(\rho.U.X.U^*) = tr(U^*.\rho.U.X)$ has been stated. The UXU^* can be interpreted as the evolution of the observable X after a certain duration and $U^*\rho U$ as the corresponding evolution of the state ρ after the same duration. This leads to the Heisenberg-Schrodinger duality: In the Heisenberg picture, the state is fixed and observables evolve while in the Schrodinger picture, the observables are fixed and the states evolve. The expected values of observables yield the same result in both the pictures. A neat proof the Hahn-Hellinger theorem is presented, namely that if ξ is a spectral measure on a measurable space $(\Omega, \mathcal{F})$ with values in the projection lattice $\mathcal{P}(\mathcal{H})$ of a Hilbert space $\mathcal{H}$, then ξ is equivalent to multiplication by indicators in the Hilbert space

$$\bigoplus_{k=1}^{\infty} \bigoplus_{m=1}^{k} L^2(\Omega, \mathcal{F}, \mu_k)$$

where $\mu_k, k = 1, 2, ...$ are mutually singular measures on $(\Omega, \mathcal{F})$. Spectral integration has been discussed in great detail. Suppose $(\Omega, \mathcal{F})$ is a measurable space and ξ is a spectral measure on this space with values in the projection lattice of a Hilbert space $\mathcal{H}$. Let $f \in L^2(\Omega, \mathcal{F}, \mu_u)$, ie,

$$\int_\Omega |f(\omega)|^2 d\mu_u(\omega) < \infty$$

where

$$\mu_u(E) =< u, \xi(E)u >, E \in \mathcal{F}$$

The set of all $u \in \mathcal{H}$ for which $\int |f(\omega)|^2 d\mu_u(\omega) < \infty$ is denoted by D_f. In other words, D_f is the set of all $u \in \mathcal{H}$ for which $f \in L^2(\Omega, \mathcal{F}, \mu)$. Here, f is any complex valued measurable function on $(\Omega, \mathcal{F})$. It is easy to see that D_f is dense in $\mathcal{H}$. Indeed, let $u \in \mathcal{H}$ be arbitrary. Then define for $n = 1, 2, ...,$

$$u_n = \xi(E_n)u, E_n = \{\omega \in \Omega : |f(\omega)| \le n\}$$

Then, $u_n \to u$ as $n \to \infty$ and further $u_n \in D_f$ for

$$\int |f|^2 d\mu_{u_n} \le n^2 < u, \xi(E_n)u >< \infty$$

It is now easy to define $(\int f(\omega)\xi(d\omega))u$ for all $u \in D_f$. The trick is to note that if $u \in D_f$, then $f \in L^2(\Omega, \mathcal{F}, \mu)$ and hence a sequence f_n of bounded measurable functions on $(\Omega, \mathcal{F})$ can be constructed so that

$$\int |f - f_n|^2 d\mu_u \to 0, n \to \infty$$

Since f_n is bounded, $\int f_n(\omega)\xi(d\omega)$ is defined and we set

$$v_n = (\int f_n(\omega)\xi(d\omega))u$$

Then it follows that

$$\| v_n - v_m \|^2 = \int |f_n - f_m|^2 d\mu_u \to 0, n, m \to \infty$$

and hence, there exists a $v \in \mathcal{H}$ such that $v_n \to v$. We set

$$(\int f(\omega)\xi(d\omega))u = v$$

and prove that this definition of $\int f d\xi$ is independent of the approximating sequence f_n. This shows how the spectral integral can be constructed for any measurable f so that it is defined on a dense domain in $\mathcal{H}$. The spectral integral will be defined on the whole of $\mathcal{H}$ if f is a bounded measurable function.

Symmetric and antisymmetric tensor products of Hilbert spaces has been dealt with in great detail. Let $\mathcal{H}$ be a Hilbert space with countable dimension, ie, $\mathcal{H}$ is a separable Hilbert space. Then using the theory of stochastic processes, the author has explained how to define $\mathcal{H} \otimes \mathcal{H}$. Specifically, for $u_1, u_2, v_1, v_2 \in \mathcal{H}$, we define

$$K((u_1, u_2), (v_1, v_2)) = < u_1, v_1 >< u_2, v_2 >$$

Since the map $(u, v) \to < u, v >$ is positive definite(

$$\sum_{i,j=1}^{n} \bar{c}_i c_j < u_i, u_j > = \| \sum_i c_i u_i \|^2 \geq 0$$

it follows from an elementary result in matrix theory that K is a positive definite kernel. Hence we can define using Kolmogorov's existence theorem on stochastic processes, a stochastic process $(u, v) \to \xi(u, v)$ on $\mathcal{H} \times \mathcal{H}$ such that

$$\mathbb{E}[\xi(u_1, u_2)\xi(v_1, v_2)] = < u_1, v_1 >< u_2, v_2 >$$

Then, it can be shown using this that the map $(u, v) \to \xi(u, v)$ is bilinear and we define the closure of the span of the random variables $\xi(u, v), u, v \in \mathcal{H}$ as the tensor product $\mathcal{H} \otimes \mathcal{H}$. The author has also published a monograph in springer lecture notes series dealing with this construction of the tensor product of Hilbert spaces. The main idea of the construction of quantum stochastic processes in this book is the Weyl representation which is defined as follows. First let $\mathcal{H}^{\otimes_s n}$ denote the n fold symmetric tensor product of the Hilbert space $\mathcal{H}$ with itself. This means that any vector in this space can be expressed as a finite linear combination of vectors of the form

$$\sum_{\sigma \in S^n} u_{\sigma 1} \otimes ... \otimes u_{\sigma n}$$

where S^n is the symmetric group of degree n, ie, any element $\sigma \in S^n$ is a permutation of n elements. KRP describes a formula for computing the orthogonal projection of $\mathcal{H}^{\otimes n}$ onto $\mathcal{H}^{\otimes_s n}$. This formula is given by

$$E = \frac{1}{n!} \sum_{\sigma \in S^n} U_\sigma$$

where

$$U_\sigma(u_1 \otimes ... \otimes u_n) = u_{\sigma^{-1}1} \otimes ... \otimes u_{\sigma^{-1}n}$$

We need to show first that E is an orthogonal projection operator, ie, $E^2 = E^* = E$. It is then clear that since the range of E is the whole of $\mathcal{H}^{\otimes_s n}$, that E is the orthogonal projection on this space. Note that U_σ is a unitary operator since since

$$< U_\sigma(u_1 \otimes ... \otimes u_n), U_\sigma(v_1 \otimes ... \otimes v_n) >$$

$$=< u_{\sigma^{-1}1} \otimes ... \otimes u_{\sigma^{-1}n}, v_{\sigma^{-1}1} \otimes ... \otimes v_{\sigma^{-1}n} >$$

$$= \Pi_{i=1}^n < u_{\sigma^{-1}i}, v_{\sigma^{-1}i} >= \Pi_{i=1}^n < u_i, v_i >$$

$$=< u_1 \otimes ... \otimes u_n, v_1 \otimes ... \otimes v_n >$$

Also

$$U_\sigma^{-1} = U_{\sigma^{-1}}$$

as follows from the defining formula for U_σ. Likewise, let $\mathcal{H}^{\otimes_a n}$ denote the antisymmetric tensor product of the Hilbert space $\mathcal{H}$ with itself n times. Thus, any vector in this space is a finite linear combination of vectors in $\mathcal{H}^{\otimes n}$ of the form

$$\sum_{\sigma \in S^n} sgn(\sigma) u_{\sigma 1} \otimes ... \otimes u_{\sigma n}$$

We define $F : \mathcal{H}^{\otimes n} \to \mathcal{H}^{\otimes_a n}$ so that

$$F - \frac{1}{n!} \sum_{\sigma \in S^n} sgn(\sigma) U_\sigma$$

Then, it is easily shown that F is the orthogonal projection operator of $\mathcal{H}^{\otimes n}$ onto $\mathcal{H}^{\otimes_a n}$. To see this, we first note that F is onto from the very definition of the antisymmetric tensor product. Secondly,

$$F^* = \frac{1}{n!} \sum sgn(\sigma) U_\sigma^* = \frac{1}{n!} \sum sgn(\sigma) U_{\sigma^{-1}} = \frac{1}{n!} \sum sgn(\sigma) U_\sigma$$

where the last equality is a consequence of the fact that $sgn(\sigma^{-1}) = sgn(\sigma)$. Finally,

$$F^2 = \frac{1}{n!^2} \sum_{\sigma,\rho} sgn(\sigma) sgn(\rho) U_\sigma U_\rho$$

$$= \frac{1}{n!^2} \sum sgn(sigma\rho) U_{\sigma\rho} = \frac{1}{n!} \sum_\tau sgn(\tau) U_\tau = F$$

Note that $EF = 0$. The symmetric and antisymmetric tensor products of Hilbert spaces are at the two extremes. A natural question to ask is how to construct intermediate spaces in a natural fashion, ie a family of mutually orthogonal subspaces of $\mathcal{H}^{\otimes n}$ such that the symmetric and antisymmetric tensor products appear in this family and such that the direct sum of this family is the whole of $\mathcal{H}^{\otimes n}$. The answer is provided by an exercise in KRP's book: Let π be any irreducible unitary representation of S^n with character χ. Thus, $\chi(\sigma) = Tr(\pi(\sigma))$. Now define

$$E_\chi = \frac{d_\chi}{n!} \sum_{\sigma \in S^n} \chi(\sigma) U_\sigma$$

Then E_χ is an orthogonal projection. Indeed,

$$E_\chi^* = \frac{d_\chi}{n!}\sum \bar{\chi}(\sigma)U_\sigma^*$$

But,

$$\bar{\chi}(\sigma) = \sum_i \bar{\pi}_{ii}(\sigma) = \sum_i \pi_{ii}(\sigma^{-1}) = \chi(\sigma^{-1})$$

showing that

$$E_\chi^* = \frac{d_\chi}{n!}\sum \chi(\sigma^{-1})U_{\sigma^{-1}} = \frac{d_\chi}{n!}\sum \chi(\sigma)U_\sigma = E_\chi$$

Further,

$$\begin{aligned} E_\chi^2 &= \frac{d_\chi^2}{n!^2}\sum_{\sigma,\rho} \chi(\sigma)\chi(\rho)U_{\sigma\rho} \\ &= \frac{d_\chi^2}{n!^2}\sum_{\sigma,\tau} \chi(\sigma)\chi(\sigma^{-1}\tau)U_\tau \end{aligned}$$

Now observe that

$$\begin{aligned} &\frac{1}{n!}\sum_\sigma \chi(\sigma)\chi(\sigma^{-1}\tau) \\ &= \frac{1}{n!}\sum_{\sigma,i,j} \pi_{ii}(\sigma)\pi_{jj}(\sigma^{-1}\tau) \\ &= \frac{1}{n!}\sum_{\sigma,i,j,k} \pi_{ii}(\sigma)\pi_{jk}(\sigma^{-1})\pi_{kj}(\tau) \\ &= \frac{1}{n!}\sum_{\sigma,i,j,k} \pi_{ii}(\sigma)\bar{\pi}_{kj}(\sigma)\pi_{kj}(\tau) \\ &= d_\chi^{-1}\sum_{i,j,k} \delta_{ik}\delta_{ij}\pi_{kj}(\tau) = d_\chi^{-1}\chi(\tau) \end{aligned}$$

It follows that

$$E_\chi^2 = \frac{d_\chi}{n!}\sum \chi(\tau)U_\tau = E_\chi$$

completing the proof that E_χ is an orthogonal projection. Suppose π, π' are two inequivalent irreducible unitary representations of S^n with characters χ, χ' respectively. Then, we claim that $E_\chi E_{\chi'} = 0$, ie, the subspaces $\mathcal{R}(E_\chi)$ and $\mathcal{R}(E_{\chi'})$ are mutually orthogonal.

Chapter 11

Statistical Mechanics

1. Statistical mechanics:Maxwell-Boltzmann, Fermi-Dirac and Bose-Einstein statistics. Let $H(q,p,N)$ be the Hamiltonian of an N particle system in which the particle positions and momenta are $(q,p) = (q_1,...,q_{3N},p_1,...,p_{3N})$. Consider a subsystem of this large system consisting of N_1 particles. The Hamiltonian of this subsystem is given by $H_1(q_1,p_1,N_1)$ and the Hamiltonian of the bath, ie, the system in which the smaller subsystem is immersed is $H_2(q_2,p_2,N_2)$ where $N_2 = N - N_1$. The probability law of the phase variables of the small subsystem is given by

$$\rho(q_1,p_1,N_1) = \frac{N!}{N_1!N_2!}\frac{exp(-\beta.H_1(q_1,p_1,N_1)).\int exp(-\beta.H_2(q_2,p_2,N_2))dq_2dp_2}{\int exp(-\beta.H(q,p,N))dqdp}$$

The particles are assumed to be indistinguishable and the factor $N!/N_1!N_2!$ arises from the fact that in these many ways, one can select N_1 particles out of the N particles of the large system and put them into the smaller subsystem. The canonical partition function is for the N particle system is defined by

$$Q_N(V,T) = \frac{1}{N!}\int exp(-\beta.H(q,p,N))dqdp = exp(-\beta.A(N,V,T))$$

where $A(N,V,T)$ is the Helmholtz free energy. The Helmholtz free energy of the bath is given by

$$A(N_2,V_2,T) = \frac{1}{N_2!}\int exp(-\beta.H(q_2,p_2,N_2))dq_2dp_2 = A(N-N_1,V-V_1,T)$$

where V_1 is the volume occupied by the system and V_2 the volume occupied by the bath. How do we justify that A as defined above is the Helmholtz free energy ? We calculate the average energy of the canonical ensemble (In the canonical ensemble, the number of particles is fixed while in the grand canonical ensemble, the number of particles is variable). We have the following expression for the average energy:

$$U = \int H.exp(-\beta H)dqdp/\int exp(-\beta H)dqdp = -\frac{d}{d\beta}log\int exp(-\beta H)dqdp$$

so that if we define

$$A = -\beta^{-1}log\int exp(-\beta.H)dqdp$$

or equivalently,

$$exp(-\beta.A) = \int exp(-\beta.H)dqdp$$

or

$$\int exp(\beta.(A(N,V,T) - H(q,p,N)))dqdp = 1$$

then we have

$$U = \frac{d}{d\beta}(\beta.A)$$

For the canonical ensemble, the entropy is given by

$$S = -k\int flogf dqdp, f = exp(-\beta H)/\int exp(-\beta H)dqdp = exp(\beta(A-H))$$

so that

$$S = -k\beta(A-U) = -(A-U)/T$$

or

$$F = U - TS = A$$

is the Helmholtz free energy. Note that $\beta = 1/kT$. Note that using the first law of thermodynamics, we have

$$dA = dU - TdS - SdT = -pdV - SdT$$

so that

$$p = -\frac{\partial A}{\partial V}, S = -\frac{\partial A}{\partial T}$$

The latter expression can be used to check our formula for S. We have

$$A = -\frac{1}{\beta}log\int exp(-\beta.H)dqdp$$

so that

$$\frac{\partial A}{\partial T} = -k\beta^2\frac{\partial A}{\partial \beta}$$

$$= -k\beta^2((1/\beta^2).log\int exp(-\beta.H)dqdp + \beta^{-1}\int H.exp(-\beta.H)dqdp/\int exp(-\beta.H)dqdp)$$

$$= -k.log\int exp(-\beta.H)dqdp - k\beta.\int H.exp(-\beta.H)dqdp/\int exp(-\beta.H)dqdp)$$

so that

$$-\frac{\partial A}{\partial T} = k.log\int exp(-\beta H)dqdp + k\beta U$$

on the other hand,

$$-k\int f.log(f)dqdp = -k\int(exp(-\beta H)/Z)log(exp(-\beta H)/Z)dqdp = k\beta U + k.log(Z)$$

where $Z = \int exp(-\beta H)dqdp$. This confirms our expression for the entropy. Now observe that if our subsystem consisting of N_1 particles enclosed in the volume V_1 is small in comparison with the original system, we have the approximation

$$A(N_2, V_2, T) = A(N - N_1, V - V_1, T) \approx A(N, V, T) - N_1 \frac{\partial A(N, V, T)}{\partial N} - V_1 \frac{\partial A(N, V, T)}{\partial V}$$

$$= A(N, V, T) - \mu(N, V, T).N_1 + p(N, V, T)V_1$$

where $\mu = \frac{\partial A}{\partial N}$ is the chemical potential and p the pressure. Thus, with this approximation,

$$\rho(q_1, p_1, N_1) = \frac{1}{N_1!} exp(-\beta.H_1(q_1, p_1, N_1)).exp(-\beta(A(N_2, V_2, T) - A(N, V, T)))$$

$$= \frac{1}{N_1!}.exp(-\beta.H_1(q_1, p_1, N_1)).exp(\beta(\mu.N_1 - pV_1))$$

The fugacity is defined as

$$z = exp(\beta\mu)$$

and then we have

$$\rho(q_1, p_1, N_1) = \frac{1}{N_1!}.exp(-\beta.H_1)z^{N_1} exp(-\beta pV_1)$$

Note that we have defined

$$Q_N(V, T) = \frac{1}{N_1!} \int exp(-\beta H(q, p, N))dqdp$$

and hence we get

$$1 = \sum_{N_1} \int \rho(q_1, p_1, N_1)dq_1 dp_1 = \sum Q_{N_1}(V_1, T)z^{N_1} exp(-\beta pV_1)$$

or equivalently defining the function

$$Z(z, V, T) = \sum_N Q_N(V, T)z^N$$

we get

$$exp(\beta pV) = Z(z, V, T)$$

or

$$pV/kT = log(Z(z, V, T))$$

This is the equation of state of the gas.

Example:Fermi gas. Let n_p denote the number of particles having momentum p. The energy of any one such particle is ϵ_p. The total number of particles in the system is given by

$$N = \sum_p n_p$$

and the total energy of the system of particles is given by

$$E = \sum_p \epsilon_p n_p$$

The canonical partition function of the system of particles is given by

$$Q_N = \sum_{\sum_p n_p = N} exp(-\beta . \sum_p \epsilon_p n_p)$$

and the grand canonical partition function is given by

$$Z(V,T) = \sum_N Q_N(V,T) z^N = \sum_{\{n_p\}} \Pi_p (exp(-\beta \epsilon_p))^{n_p} z^{n_p}$$

Since the gas is assumed to be a Fermi gas, we have the restriction $n_p \in \{0,1\}$ and hence we get

$$Z(V,T) = \Pi_p (1 + z.exp(-\beta \epsilon_p))$$

The equation of state of the gas is then given by

$$pV/kT = logZ = \sum_p log(1 + z.exp(-\beta \epsilon_p))$$

In order to make computations such as the average number of particles in each momentum state, we need to introduce an additional set of parameters $\{g_p\}$ which will be set equal to unity at the end of the calculations. With this modification, the canonical partition function of the Fermi gas is given by

$$Q_N(V,T,z,\{g_i\}) = \sum_{\sum n_p = N, n_p = 0,1} \Pi_p (g_p . z.exp(-\beta \epsilon_p))^{n_p}$$

$$= \sum_{\sum n_p = N, n_p = 0,1} (\Pi g_p^{n_p} z^{n_p}) exp(-\beta . \sum n_p \epsilon_p)$$

from which we deduce the formula for the grand canonical partition function:

$$Z(z,V,T,\{g_p\}) = \sum_N z^N Q_N$$

$$= \Pi_p (1 + g_p . z.exp(-\epsilon_p))$$

The average number of particles having momentum p is given by

$$< n_p >= g_p \frac{\partial}{\partial g_p} log(Z(z,V,T,\{g_p\})|_{g_p=1} = z.exp(-\beta \epsilon_p)/(1 + z.exp(-\beta \epsilon_p))$$

or

$$< n_p >= \frac{1}{z^{-1}.exp(\beta \epsilon_p) + 1}$$

2.Statistical mechanics: All particles in the gas are assumed to have the same mass. $f(r,v,t)d^3rd^3v$ is the number of particles at time t that fall inside the spatial volume element d^3r and have velocity in the range d^3v. A particle having velocity v collides with a particle having velocity v_1 and after the collision, the first particle has velocity v' and the second has velocity v'_1. If Before the collision, the first particle moves relative to the second with a velocity $v-v_1$. Assume that the closest distance between the two particles prior to the collision is s. Then after the collision, the first particle gets scattered into a solid angle $d\Omega$ which corresponds to the polar angles (θ,ϕ) relative to the direction of the second particle. Thus $d\Omega = sin(\theta)d\theta.d\phi$. The number of particles incident per unit area per unit time having velocity in the range d^3v relative to the second particle is $f(r,v,t)d^3v|v-v_1|$ and the number of particles incident per unit time having impact parameter in the range ds is given by $2\pi sds|v-v_1|f(r,v,t)d^3v$. Writing $\sigma(\Omega) = sds/sin(\theta)d\theta$, the cross section of scattering (Note that $\int_0^{2\pi}\sigma(\Omega)sin(\theta)d\phi = 2\pi sds/sin(\theta)d\theta$, it follows that the number of collsions per unit time of particles having velocity in the range d^3v with a particle having velocity v_1 is given by

$$d^3v\int|v-v_1|\sigma(\Omega)f(r,v,t)d^3vd\Omega = d^3v\int 2\pi s|v-v_1|f(r,v,t)ds$$

It follows that the total number of collisions per unit time per unit volume at the spatial location r is given by

$$\int|v-v_1|\sigma(\Omega)f(r,v,t)f(r,v_1,t)d^3vd^3v_1d\Omega$$

$f(r,v,t)d^3v$ is the number of particles per unit volume having velocities in the range d^3v at time t. We wish to determine its rate of increase due to collisions. The number of collisions per unit volume of such particles is given by

$$d^3v\int f(r,v,t)f(r,v_1,t)\sigma(v',v'_1,v,v_1)|v-v_1|d^3v_1d\Omega$$

where $v'-v'_1$ has orientation $\Omega = (\theta,\phi)$ relative to the incident relative direction $v-v_1$. This term contributes to a decrease in the number of particles inside the phase volume d^3rd^3v. Likewise, consider collisions of particles having velocities v',v'_1 such that after the collision the former particle has velocity v. Such collisions contribute to an increase in the number of particles in the phase volume d^3rd^3v. This is given by

$$\int d^3v'f(r,v',t)f(r,v'_1,t)\sigma(v,v_1,v',v'_1)|v'-v'_1|d^3v'_1d\Omega$$

This integral is evaluated keeping v fixed. Noting that $\sigma(v',v'_1,v,v_1) = \sigma(v,v_1,v',v'_1)$ and $d^3v'd^3v'_1 = d^3vd^3v_1$, this integral can also be written as

$$d^3v\int f(r,v',t)f(r,v'_1,t)\sigma(v',v'_1,v,v_1)d^3v_1d\Omega$$

Remark: For a fixed Ω, v',v'_1 are functions of v,v_1,Ω. Thus we can write

$$v' = v'(v,v_1,\Omega),\ v'_1 = v'_1(v,v_1,\Omega)$$

and we have the identity $d^3v'd^3v'_1 = d^3vd^3v_1$ keeping Ω fixed. It follows that

$$(\frac{\partial f(r,v,t)}{\partial t})_{coll} = \int\sigma(\Omega)|v-v_1|(f(r,v',t)f(r,v'_1,t) - f(r,v,t)f(r,v_1,t))d\Omega d^3v_1$$

is the rate of increase in the number of particles in unit phase volume due to collisions. Let us take an example of collision in the $x-y$ plane. The relevant equations are

$$v+v_1=v'+v_1', v^2+v_1^2={v'}^2+{v_1'}^2$$

Consider the collision in the centre of mass frame. The velocity of the centre of mass is

$$V=(v+v_1)/2$$

The velocities of the two particles relative to the centre of mass are respectively

$$u=v-V, u_1=v_1-V$$

prior to the collision and after the collision

$$u'=v'-V, u_1'=v_1'-V$$

We thus have

$$u+u_1=0=u'+u_1'$$

or equivalently,

$$u_1=-u, u_1'=-u'$$

These equations are consequences of the momentum conservation equation. The energy conservation equation gives us

$$(V+u)^2+(V+u_1)^2=(V+u')^2+(V+u_1')^2$$

so that

$$u^2+u_1^2={u'}^2+{u_1'}^2$$

or

$$u^2={u'}^2$$

Thus, we can write

$$u'=|u|(cos(\theta), sin(\theta))$$

Assume that θ is known. We have

$$v_x'=V_x+u_x=\frac{1}{2}(v_x+v_{1x})+|v-V|cos\theta=\frac{1}{2}(v_x+v_{1x})+\frac{1}{2}|v-v_1|cos(\theta)$$

and likewise,

$$v_y'=\frac{1}{2}(v_y+v_{1y})+\frac{1}{2}|v-v_1|sin(\theta)$$

$$v_{1x}'=V_x+u_{1x}=V_x-u_x=\frac{1}{2}(v_x+v_{1x})-\frac{1}{2}|v-v_1|cos(\theta)$$

$$v_{1y}'=\frac{1}{2}(v_y+v_{1y})-\frac{1}{2}|v-v_1|sin(\theta)$$

We can verify by explicit computation that the Jacobian determinant

$$\frac{\partial(v_x', v_y', v_{1x}', v_{1y}')}{\partial(v_x, v_y, v_{1x}, v_{1y})}=1$$

Boltzmann kinetic transport equation: The energy E is specified as a function of the momentum or equivalently, as a function of the wave vector $\mathbf{k}$. We write $E = E(\mathbf{k})$. The velocity of the particle is given by $\mathbf{v} = \nabla_{\mathbf{k}} E(\mathbf{k})/h$ where h is Planck's constant divided by 2π. Let $f_0(E)$ denote the equilibrium Fermi-Dirac distribution. Thus, the probability of a particle occupying a state having energy E equals

$$f_0(E) = (exp((E - F)/kT) + 1)^{-1}$$

The number of states per unit volume the momentum range d^3k is given by (taking h=1) d^3k. It follows that the number of particles (Fermi particles) per unit volume in phase space (ie, number of particles per unit volume in momentum space per unit volume in position space) equals

$$f_0(E(\mathbf{k}) = (exp((E(\mathbf{k}) - F)/kT) + 1)^{-1}$$

After the perturbation to the Fermi system by an electromagnetic field $\mathbf{E}, \mathbf{B}$ has been applied let the particle distribution be given by $f(\mathbf{r}, \mathbf{k}, t)$, ie, the number of particles in the phase volume d^3rd^3k is $f(\mathbf{r}, \mathbf{k}, t)d^3rd^3k$. Let $w(k', k)d^3k'$ denote the probablity per unit time of a particle having momentum k to get scattered after a collision into the phase volume d^3k' centered at k'. Then the rate of decrease in the number of particles per unit volume in phase space is given by

$$\int w(k', k)f(r, k, t)(1 - f(r, k', t))d^3k'$$

and the rate of increase in the number of particles per unit volume in phase space is given by

$$\int w(k, k')f(r, k', t)(1 - f(r, k, t))d^3k'$$

The factor $1 - f$ arises due to the Pauli exclusion principle, ie, scattering intot the momentum volume element d^3k takes place iff the state is empty and $1 - f$ is the probability of it being empty. The total rate of increase in the number of particles inside a unit volume in phase space due to collisions is therefore given by

$$(\frac{\partial f}{\partial t})|_{coll} = \int [w(k, k')f(r, k', t)(1 - f(r, k, t)) - w(k', k)f(r, k, t)(1 - f(r, k', t))]d^3k'$$

If we assume that the transition probability per unit time $w(k, k')$ is a symmetric kernel, then the above reduces to

$$\int w(k', k)(f(r, k', t) - f(r, k, t))d^3k'$$

and the Boltzmann kinetic transport equation reads

$$\frac{\partial f(r, k, t)}{\partial t} + (\mathbf{v}, \nabla_k)f(r, k, t) + (F, \nabla_k f(r, k, t)) = \int w(k', k)(f(r, k', t) - f(r, k, t))d^3k'$$

We can approximate the rate $(\partial f/\partial t)_{coll}$ by $(f_0(r, k, t) - f(r, k, t))/\tau(k)$ where $\tau(k)$ equals the relaxation time. This time is given by

$$\tau(k)^{-1} = \int w(k', k)(f(r, k', t) - f(r, k, t))d^3k'/(f_0(r, k, t) - f(r, k, t))$$

3.Translation operator for a crystal: Consider a three dimensional crystal with basis vectors $\mathbf{a}_i, i = 1,2,3$ for a unit cell. The nuclei of the crystal are therefore located at the points $n_1\mathbf{a}_1 + n_2\mathbf{a}_2 + n_3\mathbf{a}_3, n_i \in \mathbb{Z}, i = 1,2,3$. The vectors $\mathbf{a}_i, i = 1,2,3$ are assumed to be linearly independent so that they form a parallelopiped having volume $(\mathbf{a}_1, \mathbf{a}_2 \times \mathbf{a}_3)$. The reciprocal lattice basis vectors are given by $\mathbf{b}_i, i = 1,2,3$ where $(\mathbf{b}_i, \mathbf{a}_j) = \delta_{ij}, i,j = 1,2,3$. It is easy to construct these vectors:

$$\mathbf{b}_1 = (\mathbf{a}_2 \times \mathbf{a}_3)/V, \mathbf{b}_2 = (\mathbf{a}_3 \times \mathbf{a}_1)/V, \mathbf{b}_3 = (\mathbf{a}_1 \times \mathbf{a}_2)/V,$$

where

$$V = (\mathbf{a}_1, \mathbf{a}_2 \times \mathbf{a}_3)$$

is the volume of the unit cell in the original lattice. The potential energy generated by the positively charged nuclei of the lattice is a periodic function with periods being the lattice basis vectors. If $U(\mathbf{r})$ is the potential energy, then

$$U(\mathbf{r} + n_1\mathbf{a}_1 + n_2\mathbf{a}_2 + n_3\mathbf{a}_3) = U(\mathbf{r}), n_1, n_2, n_2 \in \mathbb{Z}, \mathbf{r} \in \mathbb{R}^3$$

We can therefore expand $U(\mathbf{r})$ in a Fourier series:

$$U(\mathbf{r}) = \sum c(n_1, n_2, n_3) exp(i2\pi(n_1\mathbf{b}_1 + n_2\mathbf{b}_2 + n_3\mathbf{b}_3, \mathbf{r}))$$

It is clear that any periodic function U with periods $\mathbf{a}_1, \mathbf{a}_2, \mathbf{a}_3$ has this expansion. To check periodicity, we note that

$$(n_1\mathbf{b}_1 + n_2\mathbf{b}_2 + n_3\mathbf{b}_3, \mathbf{a}_i) = n_i$$

since

$$(\mathbf{b}_j, \mathbf{a}_i) = \delta_{ij}$$

Thus the above Fourier series is justified. The computation of the Fourier series coefficients is achieved is to be carried out. We leave that part as an exercise for the reader. The Schrodinger equation for an electron moving in the crystal is given by

$$(-\frac{h^2}{2m}\nabla^2 + U(\mathbf{r}))\psi(\mathbf{r}) = E\psi(\mathbf{r})$$

$\psi(\mathbf{r})$ has the general form

$$\psi_{\mathbf{k}}(\mathbf{r}) = exp(i\mathbf{k}.\mathbf{r})\phi_{\mathbf{k}}(\mathbf{r})$$

where $\phi_{\mathbf{k}}$ is periodic with periods $\mathbf{a}_1, \mathbf{a}_2, \mathbf{a}_3$. We see that $\psi_{\mathbf{k}}$ as defined here is an eigenfunction of the translation operators

$$T(n_1, n_2, n_3) = exp(-i(n_1\mathbf{a}_1 + n_2\mathbf{a}_2 + n_3\mathbf{a}_3, \nabla))$$

or using our shorthand notation,

$$T(\mathbf{n}) = exp(-i(\mathbf{n}, \nabla)$$

where

$$\mathbf{n} = n_1\mathbf{a}_1 + n_2\mathbf{a}_2 + n_3\mathbf{a}_3$$

We see that

$$T(\mathbf{n})\psi(\mathbf{r}) = \psi(\mathbf{r}-\mathbf{n})$$

Thus,

$$T(\mathbf{n})U(\mathbf{r}) = U(\mathbf{r}-\mathbf{n}) = U(\mathbf{r})$$

Also for any function $\psi(\mathbf{r})$, we have

$$\nabla T(\mathbf{n})\psi(\mathbf{r}) = \nabla f(\mathbf{r}-\mathbf{n}) = T(\mathbf{n})\nabla f(\mathbf{r})$$

ie

$$[T(\mathbf{n}), \nabla] = \mathbf{0}$$

or

$$[T(\mathbf{n}), \mathbf{P}] = \mathbf{0}$$

Thus

$$[T(\mathbf{n}), H] = 0$$

and hence the translation operator and the energy operator can be jointly diagonalized. The eigenfunctions are precisely the Bloch wave functions $\psi_{\mathbf{k}}(\mathbf{r})$ defined earlier. We have

$$T(\mathbf{n})\psi_{\mathbf{k}}(\mathbf{r}) = exp(i\mathbf{k}.(\mathbf{r}-\mathbf{n}))\phi_{\mathbf{k}}(\mathbf{r}-\mathbf{n}) = exp(-i\mathbf{k}.\mathbf{n})\psi_{\mathbf{k}}(\mathbf{r})$$

Thus the eigenvalues of $T(\mathbf{n})$ are $exp(-i\mathbf{k}.\mathbf{n})$ where $\mathbf{k}$ is an arbitrary real vector. The periodic functions $\phi_{\mathbf{k}}(\mathbf{r})$ are to be selected so that $\psi_{\mathbf{k}}(\mathbf{r})$ become eigenfunctions of the energy operator. We have

$$\nabla\psi_{\mathbf{k}} = i\mathbf{k}\psi_{\mathbf{k}} + exp(i\mathbf{k}.\mathbf{r})\nabla\phi_{\mathbf{k}}$$

$$\nabla^2\psi_{\mathbf{k}} = -k^2\psi_{\mathbf{k}} + exp(i\mathbf{k}.\mathbf{r})\nabla^2\phi_{\mathbf{k}} + +2iexp(i\mathbf{k}.\mathbf{r})\mathbf{k}.\nabla)\phi_{\mathbf{k}}$$

Plugging this into the Schrodinger equation gives us

$$-\frac{h^2}{2m}(-k^2\phi_{\mathbf{k}} + \nabla^2\phi_{\mathbf{k}} + 2i\mathbf{k}.\nabla\phi_{\mathbf{k}}) + U\phi_{\mathbf{k}} = E\phi_{\mathbf{k}}$$

or

$$\nabla^2\phi_{\mathbf{k}} + 2i\mathbf{k}.\nabla\phi_{\mathbf{k}} + ((2m/h^2)(E-U) - k^2)\phi_{\mathbf{k}} = 0$$

Assume that

$$\phi_{\mathbf{k}} = \sum d(\mathbf{n})exp(i2\pi(n_1\mathbf{b}_1 + n_2\mathbf{b}_2 + n_3\mathbf{b}_3, \mathbf{r}))$$

This Fourier series expansion is justified since $\phi_{\mathbf{k}}$ is periodic with periods $\mathbf{a}_i, i = 1, 2, 3$. We further have

$$U(\mathbf{r})\phi_{\mathbf{k}}(\mathbf{r}) = \sum_{n_1,n_2,n_3} exp(i2\pi(n_1\mathbf{b}_1+n_2\mathbf{b}_2+n_3\mathbf{b}_3,\mathbf{r})) \sum_{m_1,m_2,m_3} c(m_1,m_2,m_3)d(n_1-m_1,n_2-m_2,n_3-m_3)$$

Thus the Schrodinger equation for the Bloch wave function becomes an algebraic equation for its Fourier series coefficients:

$$-(n_1\mathbf{b}_1 + n_2\mathbf{b}_2 + n_3\mathbf{b}_3)^2 d(n_1,n_2,n_3) - 4\pi(\mathbf{k}, n_1\mathbf{b}_1 + n_2\mathbf{b}_2 + n_3\mathbf{b}_3)d(n_1,n_2,n_3)$$
$$+(2mE/h^2 - k^2)d(n_1,n_2,n_3) - (2m/h^2)$$
$$\sum_{m_1,m_2,m_3} c(m_1,m_2,m_3)d(n_1-m_1,n_2-m_2,n_3-m_3) = 0, n_1,n_2,n_3 \in \mathbb{Z}$$

If we truncate the Fourier series, then we get an algebraic eigenvalue problem with the eigenvalues E being the energy levels of the electron in the crystal.

Chapter 12

Antenna Theory:Analysis and Synthesis

1.Radiation pattern. The Maxwell equations with magnetic current density **M** and magnetic charge density ρ_m for sinusoidal excitation included are

$$\nabla.E = \rho/\epsilon_0, \nabla\times E = -j\omega\mu H + M, \nabla\times H = J + j\omega\epsilon E, \nabla.H = \rho_m$$

The solution can be written as the sum of two parts; one part produced by (ρ, J) and another part by (ρ_m, M). Denote the first part by E_1, H_1 and the second part by E_2, H_2. Then we have

$$\nabla.E_1 = \rho/\epsilon, \nabla\times E_1 = -j\omega\mu H_1, \nabla\times H_1 = J + j\omega\epsilon E_1, \nabla.H_1 = 0$$

$$\nabla.E_2 = 0, \nabla\times E_2 = -j\omega\mu H_2 - M, \nabla\times H_2 = j\omega\epsilon E_2, \nabla.H_2 = \rho_m/\mu$$

We can thus write

$$H_1 = \nabla\times A/\mu, E_1 = -\nabla A - j\omega\Phi,$$

$$E_2 = -\nabla\times F/\epsilon, H_2 = -\nabla\Phi_m - j\omega F$$

A is the magnetic vector potential and F the electric vector potential. Φ is the electric scalar potential and Φ_m is the magnetic scalar potential. By applying the Lorentz gauge condition, we find that

$$\nabla^2 A - \mu\epsilon A = -\mu J, \nabla^2\Phi - \mu\epsilon\Phi = -\rho/\epsilon,$$

$$\nabla^2 F - \mu\epsilon F = -\epsilon M, \nabla^2\Phi_m - \mu\epsilon\Phi_m = -\rho_m/\mu$$

Note the duality between the fields (E_1, H_1) produced by electric charges and electric current densities and the fields (E_2, H_2) produced by magnetic charges and magnetic current densities: If we make the changes

$$E_1 \to H_2, H_1 \to -E_2, \epsilon \to \mu, \mu \to \epsilon, J \to M, \rho \to \rho_m$$

then the equations for (E_1, H_1) go over to the equations for (E_2, H_2). Note that under this transformation, $\Phi \to \Phi_m$ and $A \to -F$. The complete solution to the field equations is given by

$$E = E_1 + E_2, H = H_1 + H_2$$

or equivalently,

$$E = -\nabla A - j\omega\Phi - \nabla \times F/\epsilon,$$

$$H = \nabla \times A/\mu - \nabla\Phi_m - j\omega F$$

The solutions for the vector and scalar potentials both electric and magnetic can be penned down (Constantine Balanis, Antenna Theory)

$$A(r) = \frac{\mu}{4\pi}\int J(r')exp(-jk|r-r'|)d^3r'/|r-r'|$$

$$F(r) = \frac{\epsilon}{4\pi}\int M(r')exp(-jk|r-r'|)d^3r'/|r-r'|$$

The far field patterns are

$$A(\mathbf{r}) = \frac{\mu.exp(-jkr)}{4\pi r}\int J(\mathbf{r}')exp(jk\mathbf{n}.\mathbf{r}')d^3\mathbf{r}'$$

$$F(\mathbf{r}) = \frac{\epsilon exp(-jkr)}{4\pi r}\int M(\mathbf{r}')exp(jk\mathbf{n}.\mathbf{r}')d^3\mathbf{r}'$$

where

$$\mathbf{n} = \mathbf{r}/r, r = |\mathbf{r}|$$

Define the vectors

$$\mathbf{N} = \int J(\mathbf{r}')exp(jk\mathbf{n}.\mathbf{r}')d^3\mathbf{r}'$$

$$\mathbf{L} = \in M(\mathbf{r}')exp(jk\mathbf{n}.\mathbf{r}')d^3r'$$

It follows that in spherical polar coordinates,

$$N_r = \int [J_x(r',\theta',\phi')cos(\phi)sin(\theta) + J_y(r',\theta',\phi')sin(\phi)sin(\theta) + J_z(r',\theta',\phi')cos(\theta')]$$

$$\times exp(jkr'(cos(\phi-\phi')sin(\theta)sin(\theta') + cos(\theta)cos(\theta')))r'^2 sin(\theta')dr'd\theta'd\phi'$$

$$N_\theta = \int [J_x(r',\theta',\phi')cos(\phi)cos(\theta) + J_y(r',\theta',\phi')sin(\phi)cos(\theta) - J_z(r',\theta',\phi')sin(\theta)]$$

$$\times exp(jkr'(cos(\phi-\phi')sin(\theta)sin(\theta') + cos(\theta)cos(\theta')))r'^2 sin(\theta')dr'd\theta'd\phi'$$

Radiation by a circular loop: First let us calculate the components of the vector potential produced by a current distribution where the current distribution and the vector potential are specified respectively in cylindrical and spherical polar coordinates. The current density is given by

$$\mathbf{J} = J_\rho(\rho',\phi',z')\hat{\rho} + J_\phi(\rho',\phi',z')\hat{\phi} + J_z(\rho',\phi',z')\hat{z}$$

We denote $J_\rho(\rho',\phi',z'), J_\phi(\rho',\phi',z'), J_z(\rho',\phi',z')$ respectively by $J'_\rho, J_{\phi'}, J'_z$. Then using

$$\hat{\rho} = \hat{x}.cos(\phi') + \hat{y}.sin(\phi'), \hat{\phi} = -\hat{x}.sin(\phi') + \hat{y}.cos(\phi'),$$

we deduce that

$$\mathbf{J} = J'_\rho(\hat{x}.cos(\phi') + \hat{y}.sin(\phi')) + J'_\phi(-\hat{x}.sin(\phi') + \hat{y}.cos(\phi')) + J'_z\hat{z}$$

and hence

$$A_x = (\mu/4\pi)\int(J'_\rho.cos(\phi') - J'_\phi.sin(\phi'))exp(-jkR)d^3r'/R$$

$$A_y == (\mu/4\pi)\int(J'_\rho sin(\phi') + J'_\phi cos(\phi'))exp(-jkR)d^3r'/R$$

$$A_z = (\mu/4\pi)\int J'_z exp(-jkR)d^3r'/R$$

where

$$d^3r' = r'^2 sin(\theta')dr'd\theta'd\phi'$$

and

$$R = [(r.cos(\phi)sin(\theta) - \rho' cos(\phi'))^2 + (r.sin(\phi)sin(\theta) - \rho' sin(\phi'))^2 + (r.cos(\theta) - z')^2]^{1/2}$$
$$= (r^2 + \rho'^2 + z'^2 - 2r\rho' sin(\theta)cos(\phi - \phi'))^{1/2}$$

We compute

$$A_r = A_x cos(\phi)sin(\theta) + A_y sin(\phi)sin(\theta) + A_z cos(\theta)$$

Using

$$(J'_\rho cos(\phi') - J'_\phi sin(\phi'))cos(\phi)sin(\theta) + (J'_\rho sin(\phi') + J'_\phi cos(\phi'))sin(\phi)sin(\theta)$$
$$+J'_z cos(\theta)$$
$$= J'_\rho sin(\theta)cos(\phi - \phi') + J'_\phi sin(\theta)sin(\phi - \phi') + J'_z cos(\theta)$$

we get

$$A_r(r, \theta, \phi)$$
$$= (\mu/4\pi)\int[J'_\rho sin(\theta)cos(\phi - \phi') + J'_\phi sin(\theta)sin(\phi - \phi') + J'_z cos(\theta)]exp(-jkR)d^3r'/R$$

Likewise,

$$A_\theta = A_x.cos(\phi)cos(\theta) + A_y.sin(\phi)cos(\theta) - A_z sin(\theta)$$

and using

$$(J'_\rho cos(\phi') - J'_\phi sin(\phi'))cos(\phi)cos(\theta) + (J'_\rho sin(\phi') + J'_\phi cos(\phi'))sin(\phi)cos(\theta) - J'_z sin(\theta)$$
$$= J'_\rho cos(\theta)cos(\phi - \phi') + J'_\phi sin(\phi - \phi')cos(\theta) - J'_z sin(\theta)$$

we get

$$A_\theta(r, \theta, \phi) = (\mu/4\pi)\int(J'_\rho cos(\theta)cos(\phi - \phi') + J'_\phi sin(\phi - \phi')cos(\theta) - J'_z sin(\theta))exp(-jkR)d^3r'/R$$

Finally,

$$A_\phi = -A_x.sin(\phi) + A_y.cos(\phi)$$

and using

$$-(J'_\rho cos(\phi') - J'_\phi sin(\phi'))sin(\phi) + (J'_\rho sin(\phi') + J'_\phi cos(\phi'))cos(\phi)$$
$$= J'_\phi cos(\phi - \phi')$$

we get

$$A_\phi(r,\theta,\phi) = (\mu/4\pi)\int J'_\phi cos(\phi-\phi')exp(-jkR)d^3r'/R$$

For the special case of current flowing along the circular loop, we have

$$\mathbf{J} = J_\phi\hat{\phi}$$

We leave it as an exercise to simplify the above expressions and derive the expansions for the vector potential in powers of the radius a of the loop. For the loop we have the following exact expression (Balanis, Antenna Theory)

$$A_r(r,\theta,\phi) = (\mu I.a.sin(\theta)/4\pi)\int_0^{2\pi} sin(\phi')(exp(-jkR)/R)d\phi'$$

where

$$R = (r^2 + a^2 - 2ar.sin(\theta)cos(\phi'))^{1/2}$$

It is easy to see that the integral evaluates to zero, ie,

$$A_r = 0$$

Likewise,

$$A_\theta(r,\theta,\phi) = (\mu.I.a.cos(\theta)/4\pi)\int sin(\phi')exp(-jkR)d\phi'/R$$

It again follows that this evaluates to zero.

Remark: The dependence of R on ϕ' has the form $f(cos(\phi'))$ and

$$-\int_0^{2\pi} f(cos(\phi'))sin(\phi')d\phi' = \int_1^1 f(x)dx = 0$$

Finally,

$$A_\phi(r,\theta,\phi) = (\mu Ia/4\pi)\int_0^{2\pi} cos(\phi')exp(-jkR)d\phi'/R$$

The integral here to be evaluated(approximately) is

$$\int_0^{2\pi} cos(u)exp(-jk\sqrt{r^2+a^2-2ar.sin(\theta).cos(u)})du/\sqrt{r^2+a^2-2ar.sin(\theta)cos(u)}$$

Define

$$F(a) = exp(-jk(r^2+a^2-2ar.sin(\theta)cos(u)))/\sqrt{r^2+a^2-2ar.sin(\theta)cos(u)}$$

We have

$$(r^2+a^2-2a\xi)^{1/2} = r(1+a^2/r^2-2\xi a/r^2)^{1/2} = r(1+a^2/2r^2-\xi a/r^2+((1/2)(-1/2)/2)(2\xi a/r^2)^2)+O(a^3)$$

Infinitesimal dipole placed above the ground plane. Compute the electromagnetic field pattern due to this. Assume that the dipole is placed vertically at a height h above the ground. First assume that the ground plane is perfectly conducting. Let the origin of the coordinates be at a point on the ground directly below the dipole. Let (r, θ, ϕ) be the polar coordinates measured relative to axes with origin on the ground as just mentioned. Let r_1 denote the distance of the point where the field is to be calculated from the dipole and r_2 the same distance from the image dipole. Assuming that $r >> h$, we have approximately $r_1 = r - h.cos(\theta)$ and $r_2 = r + h.cos(\theta)$. The reflection coefficient is $R_h = -1$ since the plane is perfectly conducting and hence the electric field pattern is given by

$$E_\theta \approx -\frac{j\omega\mu.exp(-jkr)}{4\pi r}(exp(jkhcos(\theta)) - exp(-jkhcos(\theta)))sin(\theta)$$

$$= \frac{\omega\mu.exp(-jkr)}{2\pi r}sin(\theta)sin(kh.cos(\theta))$$

When however the ground plane is not perfectly conducting, we model the earth below the ground plane as a dielectric having permittivity ϵ and then determine the reflection coefficient for parallel polarization and determine the field thereof. Note that the reflection coefficient will be a function of the angle θ. The procedure for calculating the reflection coefficient is as follows. For specified θ we determine the angle of incidence of the wave coming from the vertical dipole so that on reflection from the ground plane, it meets the given point. Call this angle of incidence as θ_i. Note that by Snell's law, the angle of reflection is $\theta_r = \theta_i$. It is clear from the geometry that the electric field is in a plane parallel to the plane of incidence. So using conventional electromagnetic theory with appropriate boundary conditions on the tangential components of the electric field and normal components of the magnetic field, we can determine the reflection coefficient $R_h(\theta_i)$. Note that θ_i is a function of h and θ. The approximate expression for the electric field at the given point is then given by

$$-\frac{j\omega\mu.exp(-jkr).sin(\theta)}{4\pi r}(exp(jkh.cos(\theta)) + R_h(\theta_i).exp(-jkh.cos(\theta)))$$

Radiation pattern for an aperture: Let the aperture be defined by the region $S = \{0 \le x \le a, 0 \le y \le b\}$. The magnetic field in the incident wave in space is given by $\mathbf{H}(x, y, z)$. The surface electric current density at the aperture is then given by $J_S(x, y) = \hat{z} \times \mathbf{H}(x, y, 0)$. This equals

$$J_S(x, y) = -H_y(x, y, 0)\hat{x} + H_x(x, y, 0)\hat{y}$$

and the far field magnetic vector potential is then given by

$$\mathbf{A}(x, y, z) = \frac{\mu.exp(-ikr)}{4\pi.r}\int_S J_S(x', y')exp(jk(xx' + yy')/r)dx'dy', r = \sqrt{x^2 + y^2 + z^2}$$

or in terms of polar coordinates,

$$\mathbf{A}(r, \theta, \phi) = \frac{\mu.exp(-ikr)}{4\pi r}\int_S J_S(x', y')exp(jk(x'cos(\phi) + y'sin(\phi))sin(\theta)))dx'dy'$$

Antenna array theory:Amplitude pattern produced by complex current excitation $I_m exp(jm\beta)$ for a linear array with uniform spacing:

$$E(\theta) = \sum_{m=0}^{N-1} I_m exp(jm(kd.cos(\theta) + \beta))$$

The different sensors are assumed to be isotropic sources of electromagnetic waves. For a uniformly excited array, $I_m = 1$ and we get

$$E(\theta) = \frac{exp(jN(kd.cos(\theta) + \beta)) - 1}{exp(j(kd.cos(\theta) + \beta)) - 1}$$

$$= exp(jk(N-1)d.cos(\theta)/2).\frac{sin(N((kd.cos(\theta)) + \beta)/2)}{sin((kd.cos(\theta) + \beta)/2)}$$

and the power pattern is given by

$$P(\theta) = |E(\theta)|^2 = \frac{sin^2(N(kd.cos(\theta) + \beta)/2)}{sin^2((kd.cos(\theta) + \beta)/2)}$$

Note that θ is the angle between the array direction and the point where the electric field is being computed. There are two main classes of arrays, the broadside array and the endfire array. Taking $\beta = 0$ (relative phase between the current of neighboring sensors) gives us

$$P(\theta) = sin^2(Nkd.cos(\theta)/2)/sin^2(kd.cos(\theta)/2)$$

we find that $P(\pi/2) = N^2$ is the maximum. Thus the array power pattern has a maximum at $\theta = \pi/2$, ie, perpendicular to the array. This is the broadside array. For general β, we find that

$$P(0) = sin^2(N(kd + \beta)/2)/sin^2((kd + \beta)/2)$$

and this is a maximum when $kd + \beta = 0$, or when $\beta = -2\pi d/\lambda$. In other words, when this condition prevails, the antenna power pattern has a maximum along the direction of the array. This is the endfire array.

Approximations: Assuming that $kd << 1$, ie, $d << \lambda$, we can approximate the power pattern for the broadside array by

$$P(\theta) \approx sin^2(Nkd.cos(\theta)/2)/(kd.cos(\theta)/2)^2 = sin^2(Z)/Z^2$$

where

$$Z = kd.cos(\theta)/2$$

The total power radiated is then given by the integral

$$P_{tot} = \int_0^{\pi} P(\theta).2\pi.sin(\theta)d\theta = \int_0^{\pi} sin^2(NZ)2\pi.sin(\theta)d\theta)/Z^2$$

Now observe that

$$sin(\theta)d\theta = -2dZ/kd$$

and $Z = kd/2$ when $\theta = 0$ while $Z = -kd/2$ when $\theta = \pi$, so that

$$P_{tot} = \frac{2}{kd}\int_{-kd/2}^{kd/2} sin^2(NZ)dZ/Z^2$$

The exact formula will read

$$P_{tot} = \frac{2}{kd}\int_{-kd/2}^{kd/2} sin^2(NZ).dZ/sin^2(Z)$$

The evaluation of this integral in closed form is impossible.

Antenna pattern modified by a weak gravitational field:

$$g_{00} = 1 + 2U, g_{rr} = -1, r = 1,2,3$$

$$g^{00} = 1 - 2U, g^{rr} = -1, r = 1,2,3$$

$$A^0 = V, A^1 = A_x, A^2 = A_y, A^3 = A_z$$

$$A_0 = g_{00}V = (1+2U)V, A_1 = g_{11}A_x = -A_x, A_2 = -A_y, A_3 = -A_z$$

$$F_{0r} = A_{r,0} - A_{0,r} = -A^r_{,0} - ((1+2U)V)_{,r} = E_r - 2(UV)_{,r}$$

$$F_{12} = A_{2,1} - A_{1,2} = -B_3, F_{23} = A_{3,2} - A_{2,3} = -B_1, F_{31} = -B_2$$

$$F^{0r} = g^{00}g^{rr}F_{0r} = -(1-2U)F_{0r} = -(1-2U)(E_r - 2(UV)_{,r}) = -E_r + 2UE_r + 2(UV)_{,r}$$

with neglect of $O(U^2)$ terms.

$$F^{12} = g^{11}g^{22}F_{12} = -B_3, F^{23} = -B_1, F^{31} = -B_2$$

Consider the Maxwell equations

$$(F^{\mu\nu}\sqrt{-g})_{,\nu} = \mu_0 J^\mu$$

We have for $\mu = 0$,

$$(F^{0r}(1+U))_{,r} = \mu_0 J^0$$

Now,

$$F^{0r}(1+U) = (1+U)(-E_r + 2UE_r + 2(UV)_{,r}) = -E_r + UE_r + 2(UV)_{,r}$$

and Gauss, law, ie, the $\mu = 0$ equation becomes

$$divE - div(UE) - 2\nabla^2(UV) = -\rho/\epsilon_0$$

Note that

$$E = -\nabla V - A_{,t}$$

For $\mu = 1$, we get

$$(F^{12}\sqrt{-g})_{,2} + (F^{13}\sqrt{-g})_{,3} + (F^{10}\sqrt{-g})_{,0} = \mu_0 J^1$$

Now,

$$F^{12}\sqrt{-g} = -B_3(1+U), F^{13}\sqrt{-g} = B_2(1+U), F^{10}\sqrt{-g} = (E_1-2UE_1-2(UV)_{,1})(1+U) = E_1-UE_1-2(UV)_{,1}$$

so that

$$-(B_3(1+U))_{,2} + (B_2(1+U))_{,3} + (E_{1,t} - (UE_1)_{,t} - 2(UV)_{,1t}) = -\mu_0 J^1$$

or

$$B_{3,2} - B_{2,3} + (UB_3)_{,2} - (UB_2)_{,3} = \mu_0 J^1 + E_{1,t} - (UE_1)_{,t} - 2(UV)_{,1t}$$

with similar equations for the other two components. These equations can be combined into a single vector equation:

$$\nabla \times B + \nabla \times (UB) = \mu_0 \mathbf{J} + \mathbf{E}_{,t} - (U\mathbf{E})_{,t} - 2\nabla(UV)_{,t}$$

Let us now formulate the Lorentz gauge condition:

$$(A^\mu \sqrt{-g})_{,\mu} = 0$$

This expands to give

$$(V(1+U))_{,t} + div\mathbf{A} = 0$$

or

$$div\mathbf{A} + V_{,t} + (UV)_{,t} = 0$$

The fundamental Maxwell equations for electromagnetic fields in a gravitational field are therefore given by

$$divE - div(UE) - 2\nabla^2(UV) = \rho/\epsilon_0,$$

$$\nabla \times B + \nabla \times (UB) = \mu_0 J + E_{,t} - (UE)_{,t} - 2\nabla(UV)_{,t}$$

along with the above gauge condition. Substituting for E in terms of the potential V, A, we get

$$-\nabla^2 V - \nabla A_{,t} - 2\nabla^2(UV) + div(U\nabla V) + div(UA_{,t}) = \rho/\epsilon_0$$

Using the gauge condition, we get

$$-\nabla^2 V + V_{,tt} + (UV)_{,tt} - 2\nabla^2(UV) + div(U\nabla V) + div(UA_{,t}) = \rho/\epsilon_0$$

This is the modified wave equation for the electrostatic potential V. Substituting for E, B in the other Maxwell equation in terms of A gives us

$$\nabla \times \nabla \times A + \nabla \times (U\nabla \times A) = \mu_0 J - \nabla V_{,t} - A_{,tt} - (UE)_{,t} - 2\nabla(UV)_{,t}$$

giving

$$\nabla^2 A - A_{,tt} - \nabla(divA) - \nabla V_{,t} - (UE)_{,t} - 2\nabla(UV)_{,t} = -\mu_0 J$$

or using the gauge condition,

$$\nabla^2 A - A_{,tt} + \nabla(UV)_{,t} - (UE)_{,t} - 2\nabla(UV)_{,t} = -\mu_0 J$$

This is the modified wave equation for the magnetic vector potential. These equations may be used to derive first order corrections to the antenna pattern in the presence of a weak gravitational field.

Consider a static gravitational field $U(x,y,z)$. For sinusoidal variations of the current with frequency ω, the equations for the phasors V, A are

$$\nabla^2 V + \omega^2 V + \omega^2 UV + 2\nabla^2(UV) - div(U\nabla V) - j\omega UA = -\rho/\epsilon_0$$

$$\nabla^2 A + \omega^2 A + j\omega\nabla(UV) - j\omega UE - 2j\omega\nabla(UV) = -\mu_0 J$$

The potentials when $U = 0$(absence of the gravitational field) are given by

$$V_0(\mathbf{r}) = \frac{1}{4\pi\epsilon_0}\int \rho(\mathbf{r}')exp(-jk|\mathbf{r}-\mathbf{r}'|)d^3r'/|\mathbf{r}-\mathbf{r}'|$$

$$A_0(\mathbf{r}) = \frac{\mu_0}{4\pi}\int J(\mathbf{r}')exp(-jk|\mathbf{r}-\mathbf{r}'|)d^3r'/|\mathbf{r}-\mathbf{r}'|$$

Let V_1, A_1 be the corrections to these potentials produced by gravitation. Then from the above equations, we get

$$\nabla^2 V_1 + \omega^2 V_1 = \psi$$

where

$$\psi = -\omega^2 UV_0 - 2\nabla^2(UV_0) + div(U\nabla V_0) + j\omega UA_0$$

Likewise,

$$\nabla^2 A_1 + \omega^2 A_1 = -j\omega\nabla(UV_0) + j\omega UE_0 + 2j\omega\nabla(UV_0) = \nu$$

say. The solutions are

$$V_1(\mathbf{r}) = -\int \psi(r')exp(-jk|r-r'|)d^3r'/4\pi|r-r'|$$

$$A_1(r) = -\int \nu(r')exp(-jk|r-r'|)d^3r'/4\pi|r-r'|$$

Antenna theory continued: (Problem from Balanis) Consider an infinitesimal dipole located at the origin and directed along the $\hat{x}$ axis. The far field vector potential is given by

$$A(r,\theta,\phi) = \frac{\mu I.dl}{4\pi r}exp(-jkr)\hat{x}$$

The scalar potential can be easily computed from this using the Lorentz gauge condition:

$$divA = -j\omega\mu\epsilon V$$

We have

$$divA = \frac{\mu I}{4\pi}\hat{x}.\hat{r}\frac{d}{dr}(exp(-jkr)/r)$$

$$= \frac{\mu I}{4\pi}cos(\phi)sin(\theta)(-jk)exp(-jkr)/r$$

on retaining only $O(1/r)$ terms. It follows that

$$V = (k/\omega\mu\epsilon)(\mu I/4\pi)cos(\phi)sin(\theta)exp(-jkr)/r$$

and

$$\nabla V = (-jk^2/\omega\mu\epsilon)(\mu Idl/4\pi)cos(\phi)sin(\theta)exp(-jkr)/r = (-j\omega\mu Idl/4\pi)cos(\phi)sin(\theta)exp(-jkr)/r$$

where only $O(1/r)$ terms have been retained. We have made use of the identity

$$\omega^2\mu\epsilon = k^2$$

Also,

$$-j\omega A_r = (-j\omega\mu Idl/4\pi r)exp(-jkr)cos(\phi)sin(\theta)$$

We see that

$$-j\omega A_r\hat{r} - \nabla V = 0$$

and hence the far electric field has nonzero components E_θ, E_ϕ only, not E_r. We calculate

$$E_\theta = -j\omega A_\theta = -j\omega\hat{\theta}.\hat{x}(\mu Idl/4\pi r)exp(-jkr)$$

Now,

$$\hat{\theta}.\hat{x} = cos(\phi)cos(\theta)$$

giving

$$E_\theta = (-j\omega\mu Idl/4\pi r)exp(-jkr)cos(\phi)cos(\theta)$$

and likewise

$$E_\phi = (j\omega\mu Idl/4\pi r)exp(-jkr)sin(\phi)$$

The expressions for the magnetic field can be derived immediately from this. We compute the polarization of the electric field in the $\phi = 0$ and $\phi = \pi/2$ planes. When $\phi = 0$, we get $E_\phi = 0$ and hence the field has linear polarization along the $\hat{\theta}$ direction. When $\phi = \pi/2$, $E_\theta = 0$ and hence the field once again has linear polarization but now along the $\hat{\phi}$ direction. A similar calculation can be performed when the dipole is oriented along the $\hat{y}$ direction.

Pocklington's integral equation:

Step 1:Consider a cylindrical sheet of length L having radius a. Let $J_s(z)\hat{z}$ be the surface current density on this sheet. The vector potential at (x, y, z) is then given by the formula

$$A_z(x, y, z) = \frac{\mu.a}{4\pi}\int_{-L/2}^{L/2} J_s(z')\int_0^{2\pi} exp(-jkR)d\phi' dz'/R$$

where

$$R = \sqrt{(x-x')^2 + (y-y')^2 + (z-z')^2}$$

where

$$x = \rho.cos(\phi), y = \rho.sin(\phi), x' = a.cos(\phi'), y' = a.sin(\phi')$$

We find

$$(x-x')^2 + (y-y')^2 = \rho^2 + a^2 - 2a\rho.cos(\phi - \phi')$$

so that

$$R^2 = \rho^2 + a^2 - 2a\rho.cos(\phi - \phi') + (z - z')^2$$

When the point (x, y, z) falls on the antenna surface, we can set

$$\rho = a$$

and get

$$R^2 = 4a^2 sin^2((\phi - \phi')/2) + (z - z')^2$$

giving

$$A_z(\rho = a) = \frac{\mu.a}{4pi}\int_{-L/2}^{L/2} J_s(z')G(z,z')dz'$$

where

$$G(z,z') = \int_0^{2\pi} exp(-jkR)d\phi'/R$$

Suppose now that we have an electromagnetic field E_i incident on the antenna. This incident field will excite a surface current $J_s(z')$ in the antenna and the excited current will in turn produce a scattered field E_s whose vector potential is given by the above formula. We can compute E_s on the antenna surface. It is obtained as

$$dA_z/dz = -j\omega\mu\epsilon V$$

so that

$$V = (j/\omega\mu\epsilon)dA_z/dz$$

giving

$$E_{sz} = -dV/dz - j\omega A_z = -(j/\omega\mu\epsilon)d^2A_z/dz^2 - j\omega A_z$$

$$= -(j/\omega\mu\epsilon)(d^2A_z/dz^2 + k^2A_z)$$

The boundary conditions require that the sum of the tangential components of the incident and scattered electric field be zero on the antenna surface, ie,

$$E_{iz}(\rho = a) = (j/\omega\mu\epsilon)(d^2A_z/dz^2 + k^2A_z)(\rho = a)$$

This furnishes us with an integral equation for the function $J_s(z)$ and it is called Pocklington's integral equation.

Finite wire antenna: $I(z) = I_m sin(k(l/2 - |z|)$. We compute the far field magnetic vector potential:

$$A_z(r,\theta) = \frac{\mu I_m.exp(-jkr)}{4\pi r}\int_{-L/2}^{L/2} sin(k(L/2 - |z'|))exp(jkz'.cos(\theta))dz'$$

The integral is split into an integral from $-L/2$ to 0 and another from 0 to $L/2$. We then get for the first integral

$$\int_{-L/2}^{0} sin(k(L/2 + z'))exp(jkz'cos(\theta))dz'$$

$$= (2j)^{-1}[exp(jkL/2)\int_{-L/2}^{0} exp(jkz'(1+cos(\theta))dz' - exp(-jkL/2)\int_{-L/2}^{0} exp(jkz'(cos(\theta)-1))dz']$$

$$= \frac{1}{2j}(exp(jkL/2)(jk(1+cos(\theta))^{-1}(1-exp(-jkL(1+cos(\theta))/2)) - exp(-jkL/2)(jk(cos(\theta)-1))^{-1}$$

$$(1-exp(-jkL(cos(\theta)-1)/2)))$$

with a similar expression for the other integral. We can rewrite the above expression as

$$[(exp(jkL/2)/jk(1+cos(\theta))exp(-jkL(1+cos(\theta))/4)sin(kL(1+cos(\theta))/4)-(exp(-jkL/2)/jk(cos(\theta)-1))$$

$$exp(-jkL(cos(\theta)-1)/4)sin(kL(cos(\theta)-1)/4)]$$

Chapter 13

Quantum Field Theory

Reference:Steven Weinberg, The Quantum Theory of Fields, Cambridge University Press.

1.$\psi_{p,\sigma}$ is the state of a particle having four momentum p and spin σ. Λ is a Lorentz transformation of space-time. $U(\Lambda)$ is the representation of the Lorentz group in the space of one particle wave functions. Let $k = (k^\mu)$ be a fixed four momentum vector and let $L(p) = (L^\mu_\nu(p))$ be a Lorentz transformation that takes (k^μ) to (p^μ), ie, $p = L(p)k$, or equivalently,

$$p^\mu = L^\mu_\nu(p)k^\nu$$

We then define the wave function of the particle at four momentum p and spin σ to be

$$\psi_{p,\sigma} = N(p)U(L(p))\psi_{k,\sigma}$$

where $N(p)$ is a normalization constant. Then

$$U(\Lambda)\psi_{p,\sigma} = N(p)U(\Lambda)U(L(p))\psi_{k,\sigma}$$

$$= N(p).U(L^{-1}(\Lambda p))U(L(\Lambda p)\Lambda L(p))\psi_{k,\sigma}$$

Define

$$W(\Lambda, p) = L^{-1}(\Lambda p)\Lambda L(p)$$

$W(\Lambda, p)$ acts on k to take it to p followed by Λp and back to k. So $W(\Lambda, p)$ belongs to the little group that fixes k. It follows that we can write

$$U(W(\Lambda, p))\psi_{k,\sigma} = \sum_{\sigma'} D_{\sigma',\sigma}(W(\Lambda, p))\psi_{k,\sigma'}$$

and hence the Lorentz transformation law of one particle states is given by

$$U(\Lambda)\psi_{p,\sigma} = N(p).\sum_{\sigma'} D_{\sigma',\sigma}(W(\Lambda, p))U(L(\Lambda p))\psi_{k,\sigma'}$$

$$= (N(p)/N(\Lambda p))\sum_{\sigma'} D_{\sigma',\sigma}(W(\Lambda, p))\psi_{\Lambda p,\sigma'}$$

How does one determine the normalization constant $N(p)$?. We consider the scalar product

$$(\psi_{p,\sigma},\psi_{p',\sigma'})=\delta^3(p-p')\delta_{\sigma,\sigma'}$$

It follows that

$$\begin{aligned}
&\delta^3(p-p')\delta_{\sigma,\sigma'}\\
&=(\psi_{p,\sigma},N(p')U(L(p'))\psi_{k,\sigma'})\\
&=N(p')(U(L^{-1}(p'))(\psi_{p,\sigma},\psi_{k,\sigma'})\\
&=N(p')((N(p)/N(k'))(\sum_{\sigma''}D_{\sigma'',\sigma}(W(L^{-1}(p'),p))\psi_{k',\sigma''},\psi_{k,\sigma'})\\
&=(N(p')\bar{N}(p)/\bar{N}(k'))\sum_{\sigma''}\bar{D}_{\sigma'',\sigma}(W(L^{-1}(p'),p)\delta_{\sigma'',\sigma'}\delta^3(k'-k)\\
&=(N(p')\bar{N}(p)/\bar{N}(k'))\bar{D}_{\sigma',\sigma}(W(L^{-1}(p'),p))\delta^3(k'-k)
\end{aligned}$$

where

$$k'=L^{-1}(p')p$$

When $k'=k$, it follows that $p'=p$ and since $N(k)=1$ and $W(L^{-1}(p),p)=I$, we deduce that

$$(\psi_{p,\sigma},\psi_{p',\sigma'})=|N(p)|^2\delta_{\sigma',\sigma}\delta^3(k'-k)$$

We want this to equal $\delta^3(p-p')\delta_{\sigma,\sigma'}$. In other words, we want that

$$|N(p)|^2\delta^3(k)=\delta^3(p)$$

To obtain this we shall use the identity

$$p^0\delta^3(p)=k^0\delta^3(k)$$

where $p=(p^0,\mathbf{p})$ is on the mass shell, ie, $p^0=\sqrt{m^2+\mathbf{p}^2}$. For let $\Lambda=L(p)$ be orthochronous. Let $f(p)$ be a Lorentz scalar. Consider the Lorentz invariant integral $I=\int f(p)\theta(p^0)\delta(p^2-m^2)d^4p$ where $p^2=p^{02}-\mathbf{p}^2$. Note that since the mass shell is defined by the equation $p^2-m^2=0$, it follows that the above integral can be reduced to one over the mass shell. First observe that

$$p^2-m^2=p^{02}-\mathbf{p}^2-m^2$$

so that

$$\delta(p^2-m^2)=\frac{1}{2\sqrt{\mathbf{p}^2+m^2}}(\delta(p^0-\sqrt{\mathbf{p}^2+m^2})+\delta(p^0+\sqrt{\mathbf{p}^2+m^2}))$$

It follows that the above invariant integral can be expressed as

$$I=\int f(\sqrt{m^2+\mathbf{p}^2},\mathbf{p})d^3p/2\sqrt{m^2+\mathbf{p}^2}$$

and hence, it follows that the three volume element

$$d^3p/\sqrt{m^2+\mathbf{p}^2}$$

is a Lorentz invariant. We define $F(\mathbf{p}) = f(\sqrt{m^2+\mathbf{p}^2}, \mathbf{p})$, so that F is a Lorentz invariant defined on the mass shell (note that the mass shell is a Lorentz invariant), and observe that

$$F(\mathbf{p}') = \int F(\mathbf{p})\sqrt{m^2+\mathbf{p}^2}\delta(\mathbf{p}-\mathbf{p}')d^3p/\sqrt{m^2+\mathbf{p}^2}$$

Let Λ be a Lorentz transformation. Applying Λ to both sides of the above equation, we deduce using Lorentz invariance of $d^3p/\sqrt{m^2+\mathbf{p}^2}$ that $\sqrt{m^2+\mathbf{p}^2}\delta(\mathbf{p}-\mathbf{p}')$ is a Lorentz invariant. Setting $p^0 = \sqrt{m^2+\mathbf{p}^2}$, it follows that

$$p^0\delta(\mathbf{p}-\mathbf{p}') = k^0\delta(\mathbf{k}-\mathbf{k}')$$

It follows that $|N(p)|^2$ is proportional to $1/p^0$. We can thus write $N(p) = (p_0)^{-1/2}$ and we deduce the fundamental transformation formula

$$U(\Lambda)\psi_{p,\sigma} = [(\Lambda p)^0/p^0]^{1/2}\sum_{\sigma'} D_{\sigma',\sigma}(W(\Lambda,p))\psi_{\Lambda p,\sigma'}$$

Λ is an element of the orthochronous Lorentz group (Lorentz transformations having unit determinant that preserve the forward light cone. (Λ, a) where Λ is a Lorentz transformation and a a space-time translation, is an element of the Poincare group. Its action on a four vector ξ is defined by

$$(\Lambda, a)\xi = \Lambda\xi + a$$

Let $(\Lambda, a) \to U(\Lambda, a)$ be a unitary representation of the Poincare group acting on the momentum wave functions having definite spin. We have

$$U(I, a)\psi_{p,\sigma} = exp(-ip.a)\psi_{p,\sigma}$$

since $U(I, a)$ is a translation by the four vector a. It follows that

$$U(\Lambda, a)\psi_{p,\sigma} = U(I, a)U(\Lambda, 0)\psi_{p,\sigma}$$

$$= U(I, a)[(\Lambda p)^0/p^0]^{1/2}\sum_{\sigma'} D_{\sigma',\sigma}(W(\Lambda,p))\psi_{\Lambda p,\sigma'}$$

$$= [(\Lambda p)^0/p^0]^{1/2}exp(-i\Lambda p.a)\sum_{\sigma'} D_{\sigma',\sigma}(W(\Lambda,p))\psi_{\Lambda p,\sigma'}$$

This equation is the fundamental transformation law of the momentum-spin state wave functions under a Lorentz transformation followed by a space-time translation.

2.Lorentz transformation law of the creation and annihilation fields: Let Φ_0 be the vacuum state and $\Phi_{p,\sigma}$ the state in which there is a single particle having four momentum p and spin component σ. Then if $a(p, \sigma)$ is the operator that annihilates a particle of momentum p and spin component σ and $a^*(p, \sigma)$ the operator that creates a particle of momentum p and spin σ, we have

$$\Phi_{p,\sigma} = a^*(p, \sigma)\Phi_0$$

Then

$$U(\Lambda, a)a^*(p, \sigma)\Phi_0 = U(\Lambda, a)a^*(p, \sigma)U^{-1}(\Lambda, a)\Phi_0$$

$$= U(\Lambda, a)\Phi_{p,\sigma} = [(\Lambda p)^0/p^0]^{1/2} exp(-i\Lambda p.a) \sum_{\sigma'} D_{\sigma',\sigma}(W(\Lambda, p))\Phi_{\Lambda p,\sigma'}$$

$$= [(\Lambda p)^0/p^0]^{1/2} exp(-i\Lambda p.a) \sum_{\sigma'} D_{\sigma',\sigma}(W(\Lambda, p)) a^*(\Lambda p, \sigma')\Phi_0$$

Note that p and Λp fall on the mass shell ie, $p^{02} - \mathbf{p}^2 = m^2$ and $(\Lambda p)^{02} - (\Lambda \mathbf{p})^2 = m^2$. We thus obtain

$$U(\Lambda, a)a^*(p, \sigma)U^{-1}(\Lambda, a) = [(\Lambda p)^0/p^0]^{1/2} exp(-i\Lambda p.a) \sum_{\sigma'} D_{\sigma',\sigma}(W(\Lambda, p)) a^*(\Lambda p, \sigma')$$

Noting that

$$U^{-1}(\Lambda, a) = U^*(\Lambda, a), D^*(W(\Lambda, p)) = D^{-1}(W(\Lambda, p))$$

we get on taking adjoints,

$$U(\Lambda, a)a(p, \sigma)U^{-1}(\Lambda, a) = [(\Lambda p)^0/p^0]^{1/2} exp(i\Lambda p.a) \sum_{\sigma'} D_{\sigma,\sigma'}(W^{-1}(\Lambda, p)) a(\Lambda p, \sigma')$$

These Poincare transformation laws of the creation and annihilation operators play a fundamental role in the applications to quantum field theory. Consider the Lorentz transformation law of the fields $\psi_l^+(x)$ defined by the equation

$$\psi_l^+(x) = \int u_l(x, p, \sigma)a(p, \sigma)d^3p$$

where it is assumed that there is only one species of the particle. The field adjoint is given by

$$\psi_l^-(x) = \int u_l^*(x, p, \sigma)a^*(p, \sigma)d^3p$$

Its transformation law is given by

$$U(\Lambda, a)\psi_l^+(x)U^{-1}(\Lambda, a) = \sum_{l'} D_{l,l'}(\Lambda^{-1})\psi_{l'}^+(\Lambda x + a)$$

$$= \sum_{l',\sigma} D_{l,l'}(\Lambda^{-1}) \int a(p, \sigma)u_{l'}(\Lambda x + a, p, \sigma)d^3p$$

$$= \sum_{l',\sigma} D_{l,l'}(\Lambda^{-1}) \int a(\Lambda p, \sigma)u_{l'}(\Lambda x + a, \Lambda p, \sigma)d^3(\Lambda p)$$

On the other hand, using the transformation law of the annihilation operators, we have

$$U(\Lambda, a)\psi_l^+(x)U^{-1}(\Lambda, a) = \sum_{\sigma} \int u_l(x, p, \sigma)U(\Lambda, a)a(p, \sigma)U^{-1}(\Lambda, \sigma)d^3p$$

$$= \sum_{\sigma} \int [(\Lambda p)^0/p^0]^{1/2} exp(i\Lambda p.a) \sum_{\sigma'} D_{\sigma',\sigma}(W^{-1}(\Lambda, p)) a(\Lambda p, \sigma')u_l(x, p, \sigma)d^3p$$

and since d^3p/p^0 is a Lorentz invariant, we can replace d^3p by $d^3(\Lambda p)p^0/(\Lambda p)^0$, so that the above expression equals

$$\sum_{\sigma,\sigma'}\int (p^0/(\Lambda p)^0)^{1/2} exp(i\Lambda p.a)D_{\sigma',\sigma}(W^{-1}(\Lambda,p))a(\Lambda p,\sigma')u_l(x,p,\sigma)d^3(\Lambda p)$$

Equating coefficients of $d^3(\Lambda p)a(\Lambda p,\sigma)$, we get

$$\sum_{\sigma'}[(\Lambda p)^0/p^0]^{1/2} exp(i\Lambda p.a)D_{\sigma,\sigma'}(W^{-1}(\Lambda,p))u_l(x,p,\sigma')$$

$$=\sum_{l'} D_{l,l'}(\Lambda^{-1})u_{l'}(\Lambda x+a,\Lambda p,\sigma)$$

This is the fundamental equation for analyzing the quantum theory of fields produced by particles having arbitrary mass and spin.

Note:Let$\psi_l(x)$ be components of a quantized field. (Λ,a) denotes an element of the Poincare group corresponding to a Lorentz transformation Λ and a space-time translation a. Its Lorentz transformation law is given by

$$U(\Lambda,a)\psi_l(x)U^{-1}(\Lambda,a)=\sum_{l'} D_{l,l'}(\Lambda^{-1})\psi_{l'}(\Lambda.x+a)$$

and likewise, for its adjoint,

$$U(\Lambda,a)^{-1}\psi_l^*(x)U^{-1}(\Lambda,a)=\sum_{l'} D_{l,l'}(\Lambda^{-1})\psi_{l'}^*(\Lambda.x+a)$$

Note that it follows from the above with $a=0$ that

$$U(\Lambda_2\Lambda_1)\psi_l(x)U(\Lambda_2\Lambda_1)^{-1}=U(\Lambda_2)U(\Lambda_1)\psi_l(x)U(\Lambda_1)^{-1}U(\Lambda_2)^{-1}$$

$$\sum_{l'} D_{l,l'}(\Lambda_1^{-1})U(\Lambda_2)\psi_{l'}(\Lambda_1 x)U(\Lambda_2^{-1})$$

$$=\sum_{l',l''} D_{l,l'}(\Lambda_1^{-1})D_{l',l''}(\Lambda_2^{-1})\psi_{l''}(\Lambda_2\Lambda_1 x)$$

$$=\sum_{l''} D_{l,l''}((\Lambda_2\Lambda_1)^{-1})\psi_{l''}(\Lambda_2\Lambda_1 x)$$

which is a consistent equation.

3.Problem:

$$\psi_l^+(x)=\sum_{\sigma}\int u_l(x,p,\sigma)a(p,\sigma,n)d^3p$$

$\psi_l^+(x)$ are the annihilation fields. $a(p,\sigma,n)$ annihilates a particle of type n having momentum p and spin σ. Also,

$$\psi_l^-(x)=\sum_{\sigma}\int v_l(x,p,\sigma)a^*(p,\sigma,n^c)d^3p$$

$a^*(p,\sigma,n^c)$ creates a particle of type n^c (the antiparticle of n) having momentum p and spin σ. Using the Poincare transformation formulae for ψ_l^+, ψ_l^-, derive the corresponding transformation laws for u_l, v_l. Ref:See Weinberg's book, The quantum theory of fields.

4.Schrodinger and Heisenberg dynamics in the interaction picture: $H(t) = H_0 + V(t)$. The state vector $|\psi(t)>$ satisfies Schrodinger's equation

$$\frac{d}{dt}|\psi(t)>= -iH(t)|\psi(t)>$$

Define

$$|\phi(t)>= exp(itH_0)|\psi(t)>$$

Then

$$\frac{d}{dt}|\phi(t)>= -iexp(itH_0)V(t)|\psi(t)>= -i\bar{V}(t)|\phi(t)>$$

where

$$\bar{V}(t) = exp(itH_0)V(t)exp(-itH_0)$$

The evolution of $|\phi(t)>$ describes the evolution of states in the interaction picture. Likewise, we wish to determine the evolution of observables in the interaction picture. Let X be an observable at time $t = 0$. Define

$$X(t) = U(t)X(0)U(t)^*$$

where

$$U'(t) = -iH(t)U(t)$$

$X(t)$ is the evolution of the observable X in the Heisenberg picture. We have

$$X'(t) = i[H(t), X(t)] = i[H_0, X(t)] + i[V(t), X(t)]$$

Define

$$\bar{X}(t) = exp(-itH_0)X(t)exp(itH_0)$$

Then

$$\bar{X}'(t) = iexp(-itH_0)[V(t), X(t)]exp(itH_0) == i[\tilde{V}(t), \bar{X}(t)]$$

where

$$\tilde{V}(t) = exp(-itH_0)V(t)exp(itH_0)$$

Remark: Suppose we define $\bar{X}(t) = exp(itH_0)X(0)exp(-itH_0)$. Then

$$X(t) = U(t)exp(-itH_0)\bar{X}(t)exp(itH_0)U^*(t)$$

5.Feynman rules for pairing: The Hamiltonian density of the field in the interaction picture is given by $\mathcal{H}(x)$. The total Hamiltonian is thus given by

$$H = \int \mathcal{H}(x)d^3x$$

$\mathcal{H}$ must behave like a scalar under Poincare transformations. This means that if $U(\Lambda, a)$ is the representative of the Poincare transformation (Λ, a), then

$$U(\Lambda, a)\mathcal{H}(x)U^{-1}(\Lambda, a) = \mathcal{H}(\Lambda x + a)$$

Let ψ_l^+ be an annihilation field and ψ_l^- the corresponding creation field. This means that if antiparticles exist then

$$\psi_l^+(x) = \int a(p,\sigma,n)u_l(p,\sigma,n)exp(ip.x)d^3p,$$

$$\psi_l^-(x) = \int a^*(p,\sigma,n^c)v_l(p,\sigma,n)exp(-ip.x)d^3p$$

so that the total field is given by

$$\psi_l(x) = \psi_l^+(x) + \psi_l^-(x)$$

Assume that the Hamiltonian density is given by

$$\mathcal{H}(x) = \sum_{l_1,...,l_k,m_1,...,m_r} g(l_1,...,l_k,m_1,...,m_r)\psi_{l_1}^-(x)...\psi_{l_k}^-(x)\psi_{m_1}^+(x)...\psi_{m_r}^+(x)$$

Then the coefficients $g(l_1,...,l_k,m_1,...,m_r)$ must satisfy certain conditions for the Hamiltonian to be Lorentz covariant. This condition is obtained by noting that

$$U(\Lambda, a)\psi_l^\pm(x)U^{-1}(\Lambda, a) = \sum_{l'} D_{l,l'}(\Lambda^{-1})\psi_{l'}^\pm(\Lambda x + a)$$

It follows that for the Hamiltonian to be Lorentz covariant, we must have

$$\sum g(l_1,...,l_k,m_1,...,m_r)D_{l_1,l_1'}(\Lambda^{-1})...D_{l_k,l_k'}(\Lambda^{-1})D_{m_1,m_1'}(\Lambda^{-1})...D_{m_r,m_r'}(\Lambda^{-1})$$
$$\psi_{l_1'}^-(\Lambda x + a)...\psi_{m_r'}^+(\Lambda x + a)$$
$$= \sum g(l_1',...,l_k',m_1',...,m_r')\psi_{l_1'}^-(\Lambda x + a)...\psi_{m_r'}(\Lambda x + a)$$

implying that

$$g(l_1',...,l_k',m_1',...,m_r') =$$
$$\sum_{l_1,...,m_r} g(l_1,...,l_k,m_1,...,m_r)D_{l_1,l_1'}(\Lambda^{-1})...D_{m_r,m_r'}(\Lambda^{-1})$$

In other words, the coefficients $g(l_1,...,m_r)$ must transform like a multiindex tensor quantity.

Ref:Weinberg, The quantum theory of fields, Vol.I

6.Propagator computations: Assume first that antiparticles exist. Then

$$\psi_l(x) = \sum_\sigma \int [a(p,\sigma,n)u_l(p,\sigma)exp(ip.x) + a^*(p,\sigma,n^c)v_l(p,\sigma)exp(-ip.x)]d^3p$$

p assumes values on the mass shell, ie,

$$p^0 = \pm\sqrt{m^2 + \mathbf{p}^2}$$

Then

$$[\psi_l(x), \psi_m^*(y)] = \sum_\sigma \int [u_l(p,\sigma)u_m^*(p,\sigma)exp(ip.(x-y)) - v_l(p,\sigma)v_m^*(p,\sigma)exp(-ip.(x-y))]d^3p$$

Consider the matrix element

$$< vac|T(\psi_l(x)\psi_m^*(y))|vac >$$

For $x^0 > y^0$, this is the same as

$$< vac|\psi_l^+(x)\psi_m^{+*}(y)|vac >= \int \sum_\sigma u_l(p,\sigma)u_m^*(p,\sigma)exp(ip.(x-y))d^3p$$

7.Canonical field theory (problems): Let $\mathcal{L}(\psi_l, \partial_\mu\psi_l)$ be a Lagrangian density. From the variational principle

$$\delta \int \mathcal{L}d^4x = 0$$

show that the resulting field equations are given by

$$\partial_\mu \frac{\partial \mathcal{L}}{\partial \partial_\mu \psi_l} - \frac{\partial \mathcal{L}}{\partial \psi_l} = 0$$

Define the canonical momentum densities

$$\Pi_l = \frac{\partial \mathcal{L}}{\partial \partial_0 \psi_l}$$

Construct the Hamiltonian density

$$\mathcal{H}(\psi_l, \nabla\psi_l, \Pi_l) = \sum_l \Pi_l \partial_0 \psi_l - \mathcal{L}$$

The Hamiltonian of the field is given by

$$H = \int \mathcal{H}d^3x$$

We compute the variation in two different ways: On the one hand,

$$\delta H = \int [(\delta\mathcal{H}/\delta\psi_l)\delta\psi_l + (\delta\mathcal{H}/\delta\Pi_l)\delta\Pi_l]d^3x$$

where

$$\delta\mathcal{H}/\delta\psi_l = \frac{\partial\mathcal{H}}{\partial\psi_l} - \nabla.\frac{\partial\mathcal{H}}{\partial\nabla\psi_l}$$

is the variational derivative. On the other hand,

$$\delta H = \int \sum_l [\Pi_l \delta\partial_0\psi_l + \partial_0\psi_l \delta\Pi_l - \frac{\delta\mathcal{L}}{\delta\psi_l}\delta\psi_l - \frac{\delta\mathcal{L}}{\delta\partial_0\psi_l}\delta\partial_0\psi_l]d^3x$$

$$= \int \sum_l [(\delta \Pi_l)\partial_0 \psi_l - (\delta \mathcal{L}/\delta \psi_l)\delta \psi_l] d^3x$$

It follows that

$$\partial_0 \psi_l = \delta \mathcal{H}/\delta \Pi_l,$$

$$-\delta \mathcal{L}/\delta \psi_l = \delta \mathcal{H}/\delta \psi_l$$

Using the Euler-Lagrange field equations and the definition of Π_l, this becomes

$$\partial_0 \psi_l = \delta \mathcal{H}/\delta \Pi_l,$$

$$\partial_0 \psi_l = -\delta \mathcal{H}/\delta \Pi_l$$

These are precisely the Hamilton equations of motion for the field.

8.Radiative corrections: Consider first the free Dirac field. It Lagrangian density is given by $\bar{\psi}(x)(i\gamma^\mu \partial_\mu + m)\psi(x)$ We write

$$\psi(x) = \sum [a(p,\sigma)u(p,\sigma)exp(ip.x) + b^*(p,\sigma)v(p,\sigma)]exp(-ip.x)d^3p$$

where ψ, u, v have each four components. The electron propagator is given by

$$S_{lm}(x-y) =< vac|T\{\psi_l(x)\psi_m^*(y)\}|vac >$$

Assume $x^0 > y^0$. Then

$$S_{lm}(x-y) =< vac|\psi_l(x)\psi_m^*(y)|vac >= \int \sum_\sigma u_l(p,\sigma)u_m^*(p,\sigma)exp(ip.(x-y))d^3p$$

Assume now that $y^0 > x^0$. Then,

$$S_{lm}(x-y) =< vac|\psi_m^*(y)\psi_l(x)|vac >= \int \sum_\sigma v_m^*(p,\sigma)v_l(p,\sigma)exp(ip.(y-x))d^3p$$

The evaluation of these sums using Lorentz transformation considerations (Weinberg, Quantum theory of fields) shows that

$$\sum_\sigma u_l(p,\sigma)u_m^*(p,\sigma)\gamma^0 = (i\gamma.p + m)$$

with a similar expression for the $v's$. Using these, the electron propagator can be calculated. It is given by

$$((S_{lm}(x))) = \frac{(i\gamma.p + m)\gamma^0}{p^2 - m^2}$$

where

$$p^2 = p^{02} - \mathbf{p}^2$$

The calculation is based on the following integral. Define

$$\Delta_F(x) = \int exp(ip.x)d^4p/(p^2 - m^2)$$

The p^0 integral is done as follows. First note that $1/(p^2 - m^2) = 1/(p^{02} - (\mathbf{p}^2 + m^2))$ has poles at $p^0 = \pm\sqrt{\mathbf{p}^2 + m^2}$. Let $x^0 > 0$. Then the p^0 integral can be extended over the entire semicircle in the $Imp^0 > 0$ region. The integral over the semicircle vanishes since $ip^0 x^0 \to -\infty$ as $Imp^0 \to \infty$. The contour encloses the single pole at $\sqrt{\mathbf{p}^2 + m^2}$.

$$\Delta_F(x) = \int [exp(ip.x)/2p^0] d^3p$$

where the integral is over the mass shell, ie, $p^0 = \sqrt{\mathbf{p}^2 + m^2}$. For $x^0 < 0$, the integral is extended over the semicircle in the lower half p^0 plane. This gives

$$\Delta_F(x) = \int exp(i(-\sqrt{\mathbf{p}^2 + m^2}x^0 - \mathbf{p}.\mathbf{x}))d^3p/\sqrt{\mathbf{p}^2 + m^2}$$

$$= \int exp(-ip.x)d^3p/2p^0$$

Thus, we find that for all x,

$$\Delta_F(x) = \theta(x^0)\int exp(ip.x)d^3p/2p^0 + \theta(-x^0)\int exp(-ip.x)d^3p/2p^0$$

This is the Feynman propagator. The Dirac propagator can be expressed as

$$S(x) = (i\gamma.\partial + m)\Delta_F(x)$$

Consider now the interaction between matter and radiation. Both the fields are assumed to be quantized. The interaction Hamiltonian density is given by $e\bar{\psi}(x)\gamma^\mu\psi(x)A_\mu(x)$ where $\bar{\psi}(x) = \psi^*(x)\gamma^0$. The approximate electron propagator is thus given by

$$\tilde{S}_{lm}(x-y) = < vac|\psi_l(x)\bar{\psi}_m(y)|vac >$$

$$+ \int < vac|\psi_l(x)\bar{\psi}_m(y)\bar{\psi}(u)\gamma^\mu\psi(u)\bar{\psi}(v)\gamma^\nu\psi(v)A_\mu(u)A_\nu(v)|vac > d^4u d^4v$$

The first term on the right is the bare electron propagator $S(x-y)$. The second term can be computed as follows. Observe that

$$\int \psi_l(x)\bar{\psi}_m(y)\bar{\psi}_a(u)(\gamma^\mu)_{ab}\psi_b(u)\bar{\psi}_c(v)(\gamma^\nu)_{cd}\psi_d(v)A_\mu(u)A_\nu(v)d^4u d^4v$$

The pairing can be done as follows (Time ordering operators are to be introduced in all the pairs).

$$< \psi_l(x)\bar{\psi}_a(u) > (\gamma^\mu)_{ab} < \psi_b(u)\bar{\psi}_c(v) > (\gamma^\nu)_{cd} < \psi_d(v)\bar{\psi}_m(y) >< A_\mu(u)A_\nu(v) >$$

This gives

$$\int S_{la}(x-u)(\gamma^\mu)_{ab}S_{bc}(u-v)(\gamma^\nu)_{cd}S_{dm}(v-y)\Delta_{\mu\nu}(u-v)d^4u d^4v$$

In terms of Fourier transforms, this is

$$\int S_{la}(p_1)exp(ip_1.(x-u)S_{bc}(p_2)exp(ip_2.(u-v))S_{dm}(p_3)exp(ip_3.(v-y))$$

$$\Delta_{\mu\nu}(p_4)exp(ip_4.(u-v))d^4ud^4vd^4p_1d^4p_2d^4p_3d^4p_4(\gamma^\mu)_{ab}(\gamma^\nu)_{cd}$$

The exponential factors can be arranged as

$$exp(i(p_1.x-p_3.y))exp(i(-p_1+p_2+p_4).u)exp(i(-p_2+p_3-p_4).v))$$

The integrals over u and v result into

$$exp(i(p_1.x-p_3.y))\delta(-p_1+p_2+p_4)\delta(-p_2+p_3-p_4)$$

Thus, the above becomes after integrating over p_1 and p_3

$$\int exp(i(p_2+p_4).(x-y))S_{la}(p_2+p_4)S_{bc}(p_2)S_{dm}(p_2+p_4)\Delta_{\mu\nu}(p_4)(\gamma^\mu)_{ab}(\gamma^\nu)_{cd}d^4p_2d^4p_4$$

which is the same as

$$\int exp(ip.(x-y))S_{la}(p)(\gamma^\mu)_{ab}S_{bc}(p_2)(\gamma^\nu)_{cd}S_{dm}(p)\Delta_{\mu\nu}(p-p_2)d^4pd^4p_2$$

This is same as the $(lm)^{th}$ entry of the matrix

$$\int exp(ip.(x-y))S(p)\gamma^\mu S(p_2)\gamma^\nu S(p)\Delta_{\mu\nu}(p-p_2)d^4pd^4p_2$$

Extracting out the coefficient of $exp(ip.(x-y))$ gives us the Fourier transform of the second order correction to the electron propagator

$$\tilde{S}(p) = S(p).[\gamma^\mu S(p_2)\gamma^\nu \Delta_{\mu\nu}(p-p_2)d^4p_2].S(p)$$

Or the equivalently,

$$\tilde{S}(p) = S(p)\Pi^*(p)S(p)$$

where

$$\Pi^*(p) = \int \Delta_{\mu\nu}(k)\gamma^\mu S(p-k)\gamma^\nu d^4k$$

The idea can be generalized to obtain higher order corrections to the electron propagator. The complete series for the exact propagator reads

$$\tilde{S}(p) = S(p) + S(p)\Pi^*(p)S(p) + S(p)\Pi^*(p)S(p)\Pi^*(p)S(p) + ... = S(p)(I-\Pi^*(p)S(p))^{-1}$$

Noting that

$$S(p) = (i\gamma.p+m)^{-1}$$

this gives

$$\tilde{S}(p) = (i\gamma.p+m)^{-1}(I-\Pi^*(p)(i\gamma.p+m)^{-1})^{-1} = (i\gamma.p+m-\Pi^*(p))^{-1}$$

Note that the photon propagator

$$\Delta_{\mu\nu}(k) = \eta_{\mu\nu}/k^2$$

9.Quantization of the electromagnetic field:

Field tensor:

$$F_{\mu\nu} = A_{\nu,\mu} - A_{\mu,\nu}$$

Lagrangian density

$$\mathcal{L} = -\frac{1}{4}F_{\mu\nu}F^{\mu\nu} + J^{\mu}A_{\mu}$$

Canonical positions:

$$A_{\mu}, \mu = 0, 1, 2, 3$$

Canonical momenta:

$$\Pi^{\mu} = \frac{\partial \mathcal{L}}{\partial A_{\mu,0}}$$

Now the terms involving $A_{\mu,0}$ in the Lagrangian density are

$$-\frac{1}{4}F^{\mu 0}(A_{0,\mu} - A_{\mu,0})$$

so that after attaching a factor of two (due to that coming from $F^{\mu\nu}$, we get

$$\Pi^{\mu} = \frac{1}{2}F^{\mu 0}$$

Note that

$$\Pi^{0} = 0$$

and this is a primary constraint. Now note that the pure field part of the Lagrangian density contains the term

$$-\frac{1}{4}F^{r0}(A_{0,r} - A_{r,0}) = -\frac{1}{4}F^{r0}F_{r0}$$

It follows that

$$\frac{\partial \mathcal{L}}{\partial F_{r0}} = -\frac{\partial \mathcal{L}}{\partial A_{r,0}} = -\Pi^{r} = \frac{\partial \mathcal{L}}{\partial A_{0,r}}$$

and hence we have from the field equations,

$$\partial_r \Pi^r = -\frac{\partial \mathcal{L}}{\partial A_0} = -J^0$$

or equivalently,

$$\partial_r \Pi^r + J^0 = 0$$

This is not an equation of motion since it does not involve time derivatives. Hence this equation must be regarded as a constraint equation. It this latter constraint equation is satisfied at one time, then it is satisfied at all times. This follows from

$$\partial_r \frac{\partial \mathcal{L}}{\partial F_{ri}} = \partial_r \frac{\partial \mathcal{L}}{\partial A_{i,r}}$$

$$= -\partial_0 \frac{\partial \mathcal{L}}{\partial A_{i,0}} + J^r$$

so that

$$\partial_0(\partial_i \frac{\partial \mathcal{L}}{\partial F_{i0}} - J^0) =$$

$$-\partial_i \partial_0 \frac{\partial \mathcal{L}}{\partial A_{i,0}} - \partial_0 J^0$$

$$= -\partial_0 J^0 - \partial_r J^r + \partial_i \partial_r \frac{\partial \mathcal{L}}{\partial F_{ri}} = 0$$

if we assume that the current is conserved. It follows that the second constraint equation can be regarded as an initial condition.

Chapter 14

Statistical Signal Processing Problems in Music

1. Raag Mohanam. The swaras occurring here upward are s, r, g, p, d, S where small letters stand for the lower order swaras and big S stands for large S. While going downward, the swaras occurring are S, d, p, g, r, s. Each swaram corresponds to a definite frequency. The required frequency is generated on a given instrument in a given fashion. For example, while playing on the Veena, the frequency is generated by changing the length of the vibrating string. We denote the frequency of a given swaram x by $\omega(x)$. Note that ω is a function of the shruti chosen as well as of the swaram chosen. Let θ be the shruti chosen. Then the swaram x has the frequency $\omega(x, \theta)$. Going up in the line of swaras is called aarohanam while coming down is called avarohanam. Consider for example, the well known Geetam Vara Veena:

Vara Veena, Mridu Paani, Vana Rugalo, Chanu Raani, Sura Chita Pampara Memi, Sura nuta Kalyani.

Nirupam Supa Guna Lola, Nirutajayaa paga Sheela, Varadaa priya, Ranga Naayaki, Vaanchita pala, Daayaki,

Sara sija jana janani, Jaya Jaya Jaya.

In swarams, this Geetam reads

ga,ga pa, pa, da pa, sa, sa, ri sa, dada, pa, dapa, gaga, ri, ga, ga, dapa, ga, pa, ga, ga, ri, sa. ga, ga, ga, ga, ri ga pa ga, papa. ga,ga, dapa, daa, da pa, saa, saa, da ga ri, ri sa, sa, da sa, da, da da, pa ga, pa da sa da pa, da pa ga ga ri sa, sa ga gaa gaa ga ri paa ga ri, sa ri sa ga ri sa.

When a swaram x is played and a pause of duration τ is given, the resulting signal can be represented as a truncated sinusoid over a duration τ, say $[t_0, t_0 + \tau]$. The signal can be represented therefore as $sin(\omega(x)(t - t_0))(u(t - t_0) - u(t - t_0 - \tau))$. For example, let $\omega(ga)$ be the frequency associated with the swaram ga. We play ga, ga. The duration of each swaram is τ which means that the signal corresponding to this doublet is given by

$$x(t) = sin(\omega(ga)t)(u(t) - u(t - \tau)) + sin(\omega(ga)(t - \tau))(u(t - \tau) - u(t - 2\tau))$$

This is more conveniently represented by introducing the pulse

$$p(t) = u(t) - u(t - \tau)$$

giving

$$x(t) = sin(\omega(ga)t)p(t) + sin(\omega(ga)(t - \tau))p(t - \tau)$$

The doublet ga, ga is followed by pa, which is pa of duration τ followed by a pause also of duration τ. The pa lasts from the time 2τ to 3τ and the pause of duration τ extends the continuation of pa from the time 3τ to the time 4τ. Thus the $pa,.$ (pa followed by a pause) can be represented by the signal

$$y(t) = sin(\omega(pa)(t - 2\tau))(u(t - 2\tau) - u(t - 4\tau)) = sin(\omega(pa)(t - 2\tau))(p(t - 2\tau) + p(t - 3\tau))$$

The entire sequence $ga, ga, pa, ., pa, .,$ may therefore be represented by the signal

$$z(t) = x(t) + y(t) + y(t - 2\tau)$$

In general, suppose we are given a sequence of swarams $x_1 x_2 ... x_m$ without any pause. Then, the resulting corresponding signal has the form

$$s(t) = \sum_{p=1}^{m} sin(\omega(x_p)(t - (p-1)\tau))p(t - (p-1)\tau)$$

Such a sequence can be Fourier analyzed as follows. Let $P(\omega)$ denote the Fourier transform of $p(t)$ and $X(\omega)$ that of a signal $x(t)$. Then, the Fourier transform of the signal $x(t - a)p(t - b)$ is given by

$$\mathcal{F}\{x(t-a)p(t-b)\} = \frac{1}{2\pi}[exp(-i\omega a)X(\omega)] * [exp(-i\omega b)P(\omega)]$$

$$= \frac{1}{2\pi}\int_{-\infty}^{\infty} exp(-i(\omega - u)a)X(\omega - u)exp(-iub)P(u)du$$

$$= \frac{1}{2\pi}exp(-i\omega a)\int_{-\infty}^{\infty} exp(iu(a - b))X(\omega - u)P(u)du$$

The Fourier transform of $sin(\omega_0 t)p(t)$ is given by

$$\mathcal{F}\{sin(\omega_0 t)p(t)\} = -j\pi(\delta(\omega - \omega_0) - \delta(\omega + \omega_0)) * P(\omega)$$

$$= -j\pi(P(\omega - \omega_0) - P(\omega + \omega_0))$$

It follows that the Fourier transform of the signal $s(t)$ is given by

$$S(\omega) = \mathcal{F}\{s(t)\} = -(j/2)\sum_{p=1}^{m} exp(-j\omega(p-1)\tau)(P(\omega - \omega(x_p)) - P(\omega + \omega(x_p)))$$

Suppose we play a signal $sin(\omega(x)(t - a))$ over the duration $t \in [a, b]$. Then the corresponding signal is given by $sin(\omega(x)(t - a))(u(t - a) - u(t - b)) = z(t - a)$ where

$$z(t) = sin(\omega(x)t)(u(t) - u(t - b + a))$$

The Fourier transform of $z(t)$ is given by

$$Z(\omega) = (-j/2)(\delta(\omega - \omega(x)) - \delta(\omega + \omega(x))) * P(\omega)$$

$$= (-j/2)(P(\omega - \omega(x)) - P(\omega + \omega(x)))$$

where

$$P(\omega) = \mathcal{F}\{u(t) - u(t - \Delta))\} = \int_0^\Delta exp(-j\omega t)dt = j^{-1}(1 - exp(-j\omega\Delta)), \Delta = b - a$$

or

$$P(\omega) = 2.exp(-j\omega\Delta/2).sin(\omega\Delta/2)$$

Problem:Calculate the Fourier transform of the first line of the Geetam Vara Veena. Hint: The signal representing the first line

$$ga, ga, pa, .pa, ., da, pa, sa, .sa, .$$

is

$$sin(\omega(ga)t)p(t) + sin(\omega(ga)t)p(t - \tau) + sin(\omega(pa)(t - 2\tau))(p(t - 2\tau) + p(t - 3\tau))$$

$$+sin(\omega(pa)(t - 4\tau))(p(t - 4\tau) + p(t - 5\tau)) + ...$$

We next wish to investigate the transformation of swaram sequences to saahityam. The swaram sequence $ga, ga, pa., pa.$, transforms to the sahityam sequence *varaveena*,. The frequency, ie, pitch of "vara" is still $omega(ga)$ but the amplitude gets modulated a little bit. Thus we can write this signal as

$$vara = a(1)sin(\omega(ga)t)p(t) + a(2)sin(\omega(ga)(t - \tau))p(t - \tau)$$

This signal can be also be represented as an amplitude modulated waveform:

$$(a(1)p(t) + a(2)p(t - \tau)cos(\omega(ga)\tau))sin(\omega(ga)t) - a(2)p(t - \tau)sin(\omega(ga)\tau)cos(\omega(ga)t)$$

In general, the sahityam associated with a swaram x will have the amplitude modulated form, ie, $\xi(t).cos(\omega(x)t + \phi)$ and its Fourier analysis can be carried out easily. The same analysis can be carried out for Sanskrit slokas. For example consider the following sequence of slokas occurring in the Visnu Sahasranamam.

Shuklambara dara Visnum, Shashi Varnam Chaturbhujam, Prasanna vadanam dhyaayet, sarva vigno pa shaantaye.

Yasya dviradha vaktradhyaa, paarishadyaa, parashshatam, vignam nignanti satatam vishwaksenam tamaashraye.

Vyaasam, vashishta naptaaram shakte pautrama kalmasham, paraasharaatmajam vande, shukataatam tapo nidhim.

When this sequence of sahityaams is played on an instrument like the flute, for each sloka, the same swaram sequence is adopted. So it is impossible to distinguish the slokas from the instrumental pattern. The swaram sequence is generated by a certain short time frequency pattern and the associated saahityam sequence by an amplitude modulated version of the corresponding sinuosoid. Alternately, the given saahityam sequence corresponding to a swaram sequence can be generated more accurately by adopting a combination of amplitude and frequency modulations. Thus, a swaram x is associated with a frequency $\omega(x)$ and the corresponding saahityam over the duration $[0, \tau]$ can be generated as

$$\xi(t) = a(t)p(t).cos(\omega(x)t + \eta(t))$$

where $a(t), \eta(t)$ are slowly varying functions of time.

In the transformation of a sequence of sahitya in the Visnu Sahasranamam into a swaram sequence, we once again use the reverse of amplitude and frequency modulation. For example, the phrase $ShuklambaradaramVisnum$ when transformed into swarams becomes $dada, paga, sada, dapa.$ The swarams $dada$ correspond to the sahitya sequence *shukla* and paga, sada, dapa translates into $mbaradaram\ Visnum.$ Let da correspond to the frequency $\omega(da)$. Then $dada$ is represented by the signal

$$cos(\omega(da)t)p(t) + cos(\omega(da)(t - \tau))p(t - \tau)$$

To arrive at the sahitya shukla, we should amplitude and frequency modulate this signal. In other words, the above waveform is replaced with

$$a(t)cos(\omega(da)t + \phi(t))p(t) + b(t)cos(\omega(da)(t - \tau) + \psi(t))p(t - \tau)$$

where $a(t), b(t), \phi(t)$ and $\psi(t)$ are now slowly time varying functions. The Fourier transforms can be computed using the standard methods. Suppose $exp(j\phi(t))$ has the Fourier transform $\eta(\omega)$. Then the Fourier transform of $m(t).exp(j\phi(t))$ is given by $\frac{1}{2\pi}M(\omega) * \eta(\omega)$. The FM signal $cos(\omega(x)t + \phi(t))$ can be expressed as the sum of two signals $x(t) = \frac{1}{2}exp(j(\omega(x)t + \phi(t)))$ and $y(t) = \frac{1}{2}exp(-j(\omega(x)t + \phi(t)))$. Note that $y(t) = \bar{x}(t)$ so that $Y(\omega) = \bar{X}(-\omega)$. Now $X(\omega) = \frac{1}{2}\eta(\omega - \omega(x))$ and the Fourier transformation problem is thus completely solved.

Chapter 15

Miscellaneous Study Project Problems

1. Numerical integration: $f:[a,b]\to\mathbb{R}$ is given. Approximate the Riemann integral $\int_a^b f(x)dx$ by a discrete sum. A better approximation can be obtained using truncated Taylor expansions. This involves setting $\Delta=(b-a)/N$ and for $x\in[k\Delta,(k+1)\Delta)$, we write

$$f(x)\approx f(k\Delta)+f'(k\Delta)(x-k\Delta)+...+f^{(n)}(k\Delta)(x-k\Delta)^n/n!$$

and approximate the derivatives of f by finite differences. Then,

$$\int_a^b f dx\approx\sum_{k=0}^{N-1}f(k\Delta)\Delta+f'(k\Delta)\Delta^2/2!+...+f^{(n)}(k\Delta)\Delta^{n+1}/(n+1)!$$

2. Differential equation description of control systems: Let $x(t)$ be the input and $y(t)$ the output. Then x,y are related by a differential equation of the form

$$D^n y=f(D^{n-1}y,D^{n-2}y,...,Dy,y,u(t))$$

This system can be represented using a block diagram as follows. Draw a line from a point and call the point $D^n y(t)$. Pass $D^n y$ through a sequence of n integrators and draw outputs $D^{n-i}y(t),i=1,2,...,n$. Draw the input point $x(t)$ and take wires from $x,D^{n-i}y,i=1,2,...,n$ and feed these wires as inputs to a functional block denoted by f. Take the output of this block and join it to the point y, thus completing the block diagram.

3. Finite group, definition: A finite set $G=\{g_1,...,g_n\}$ with a mapping $\phi:G\times G\to G$ satisfying the associativity property, ie, $\phi(g_1,\phi(g_2,g_3))=\phi(\phi(g_1,g_2),g_3)$, a unique element $e\in G$ such that $\phi(g,e)=g$ for all $g\in G$ and for each $g\in G$, a unique element $g^{-1}\in G$ with the property $\phi(g,g^{-1})=e$. Example: $G=\{0,1,...,n-1\}$ with $e=0$ and $\phi(x,y)=x+y(modn)$.

4. Example of non-commutative finite groups:The most typical example is the permutation group S_n of n elements denoted by $G=\{\sigma:\sigma\in S_n\}$ where $\sigma:\{1,2,...,n\}\to\{1,2,...,n\}$ is a bijection.

5. Lie derivative: X, Y are vector fields on a differentiable manifold. Then, $L_X Y = [X, Y]$ is the Lie derivative of Y along X. Let $\phi_t(x)$ be the flow induced by X. Then

$$L_X Y = lim_{t \to 0}(\phi_t^* Y - Y)/t$$

Here, if $T : M \to M$ is a smooth map, then $T^* X(x)$ has components

$$\xi^i(x) = \sum_j \frac{\partial T^i}{\partial x^j}(T^{-1}x) X^j(T^{-1}x)$$

Thus,

$$\xi^i = (\sum_j \frac{\partial T^i}{\partial x^j} X^j) o T^{-1}$$

6.Prove that symplectic maps form a group

7. Convolution of measures on a group: μ is one measure on the Borel subsets of a metric group G and ν is another measure on the same. Let E be a Borel subset of G. We assume that μ and ν are probability measures corresponding to independent G valued random variables X, Y. Let $Z = XY$. Then, for any Borel subset E of G, we have

$$Pr(Z \in E) = Pr(XY \in E) = \int Pr(Y \in x^{-1}E)\mu(dx) = \int_G \nu(x^{-1}E)\mu(dx)$$

8. Assume that the metric tensor on a Riemannian manifold is parameterized by a random parameter θ. Determine the statistics of the geodesic trajectory. Assume to begin with that the metric can be expressed as an affine linear function of the parameter:

$$g_{\mu\nu}(x, \theta) = g_{\mu\nu}^{(0)}(x) + \theta . g_{\mu\nu}^{(1)}(x)$$

where θ has pdf $f(\theta)$. To do this problem, express the trajectory as a power series in the parameter θ and derive recursive differential equations for the successive approximations.

9. Definition of a group: A set G with a binary operation $\phi : G \times G \to G$ that respects associativity, ie, $\phi(g_3, \phi(g_2, g_1)) = \phi(\phi(g_3, g_2), g_1)$ for all $g_1, g_2, g_3 \in G$. Further, there exists a unique element $e \in G$ called the identity such that $\phi(g, e) = \phi(e, g) = g$ for all $g \in G$. An example of a group is obtained by taking a set S, finite or infinite and considering the collection of all bijective transformations $T : S \to S$. The identity transformation e on this set is defined by $e(g) = g, g \in G$. Given two bijective transformations $T_i, i = 1, 2$ on this set, the composition $T_2 o T_1$ is also bijective. Note that its inverse is given by $T_1^{-1} o T_2^{-1}$.

10. Groups of transformations on a set. Let G be a Lie group acting smoothly on a differentiable manifold M. Let $\}$ be the Lie algebra of G. Then the infinitesimal action of $\}$ on M is defined by

$$x \to x + \delta t \frac{d}{dt} exp(t\xi).x|_{t=0}$$

where $\xi \in \}$. The vector field on M induced by $\xi \in \}$ is given by

$$\xi_M(x) = \frac{d}{dt} exp(t\xi).x|_{t=0}$$

Let (M,ω) be a symplectic manifold and let

11. Define a representation of a group. Let G be a group and let X be a vector space. Let $GL(X)$ be the general linear group of X, ie, the algebra of all non-singular linear transformations acting on X. A map $g \to T(g)$ from G into $GL(X)$ that is a homomorphism in the sense that $T(g_1g_2) = T(g_1)T(g_2), g_1, g_2 \in G$ is called a representation of G in X. In terms of matrix elements, we write $T(g) = ((t_{ij}(g)))$ and the definition of representation boils down to

$$t_{ik}(g_1g_2) = \sum_m t_{im}(g_1)t_{mk}(g_2), i, k = 1, 2, ..., n$$

where we are assuming that $dimX = n < \infty$.

12. Semidirect product: G is a group, H, K are subgroups of G such that K is normal in G. Further, every element $g \in G$ can be uniquely expressed as $g = hk$ with $h \in H, k \in K$. Then we have for $g_i = h_ik_i, i = 1, 2, h_i \in H, k_i \in K$,

$$g_1g_2 = h_1k_1h_2k_2 = h_1h_2h_2^{-1}k_1h_2k_2 = h_1h_2.k_1'k_2, k_1' = h_2^{-1}k_1h_2 \in K$$

13. Multiresolution analysis of $L^2(\mathbb{R})$: A family of subspace $\{V_j : j \in \mathbb{Z}\}$ such that $V_j \subset V_{j+1}$ and $f(x) \in V_j$ iff $f(2x) \in V_{j+1}, \bigcap V_j = \{0\}$ and $Cl\{\bigcup V_j\} = L^2(\mathbb{R})$. Further, there exists a $\phi \in V_0$ such that $\phi(x-k), k \in \mathbb{Z}$ is a basis for V_0. Note that suppose we choose an $f(x) \in V_j$. Then $f(x) \in V_{j+1}$ and hence $f(x/2) \in V_j$. In general, $f(x/2^m) \in V_j$ for all $m = 1, 2, \ldots$. This formula suggest that V_j consists functions along with their dilated versions with the dilation factor being two.

14. Read about the following topics in wavelet theory.

[1] Scaling sequence,

[2] Construction of the wavelet basis from the scaling sequence

[3] Approximation of signals using wavelet bases

[4] Construction of the Haar wavelet

[5] Construction of the Daubechies $D4, D6$ wavelets.

15. Flow of a compressible fluid in a plane. Solution by Fourier methods. The functions are $\rho(t,x,y), v_x(t,x,y), v_y(t,x,y), p(t,x,y)$. The Navier -Stokes equations are

$$\rho(v_{x,t} + v_xv_{x,x} + v_yv_{x,y}) = \nu(v_{x,xx} + v_{x,yy}) - p_{,x},$$

$$\rho(v_{y,t} + v_xv_{y,x} + v_yv_{y,y}) = \nu(v_{y,xx} + v_{y,yy}) - p_{,y},$$

The mass conservation equation is

$$(\rho v_x)_{,x} + (\rho v_y)_{,y} + \rho_{,t} = 0$$

and finally, the equation of state of the fluid is

$$p = f(\rho)$$

where f is a known function. To solve this numerically, we discretize x, y, t space so that $t = n\tau, x = k\delta_x, y = m\delta_y$ and set $\rho(n\tau, k\delta_x, m\delta_y) = \rho[n,k,m]$ and likewise for v_x, v_y. The iteration proceeds as follows.

Chapter 16

BE Projects

1. Numerical simulation of the Boltzmann kinetic transport equation. Under a given potential $U(\mathbf{r}, t)$, the probability density $f(\mathbf{r}, \mathbf{v}, t)$ of the position-velocity vector of a particle (or equivalently, particle number density in phase space) obeys the Boltzmann equation

$$\frac{\partial f}{\partial t} + (\mathbf{v}, \nabla f) + (\mathbf{F}, \nabla_v f) = (\frac{\partial f}{\partial t})_{coll}$$

where $\mathbf{F} = -\nabla U$ and where the term on the right is the Boltzmann collision integral. The aim of the project will be to simulate this evolution and to calculate averaged quantities in a gas like average number density field, average velocity field and momentum tensor field using the distribution function f.

2. Study of fluid dynamical problems using finite difference scheme: The project will involve simulating the motion of a fluid in a finite region of two and three dimensional space using the finite difference scheme applied to the Navier-Stokes equation. The problems will include (a) flow past a cylinder, (b) flow past a sphere, (c) flow through a pipe, (d) gravity waves, (e) flow inside a square region. The flow past a cylinder and sphere is treated using the method of velocity potential which satisfies Laplace's equation $\nabla^2\Phi = 0$. The velocity potential of the two dimensional problem is expressed as

$$\Phi(\rho, \phi) = \sum_{n \geq 0} \rho^{-n}(A_n cos(n\phi) + B_n sin(n\phi))$$

and appropriate boundary conditions are applied to this. Likewise the flow past a sphere is described by a velocity potential having azimuthal symmetry $\Phi(r, \theta)$ and the solution is expressed in terms of the spherical harmonics. The boundary conditions for both the problems involve constant velocity at ∞. These are exact solutions and a comparison with the solution using the finite difference scheme will be dealt with in this project. The finite difference scheme is especially applicable to problems involving unsteady fluid flow where the initial velocity field is given and its evolution in time is to be determined.

3. Numerical study of the motion of charged particles in a static magnetic field. The vector potential $\mathbf{A}$ here is assumed to depend on the spatial coordinates but is independent of the time coordinate. The

Lagrangian for a charged particle e moving in this field (assuming non-relativistic velocities) is given by

$$\mathcal{L}(\mathbf{r},\mathbf{v}) = m|\mathbf{v}|^2/2 + e(\mathbf{v},\mathbf{A}(\mathbf{r}))$$

If there are several charged particles, then modifications to this equation must be made for apart from the external magnetic field, there is a magnetic field produced by the motion of each charged particle and there is also an electric field produced by each charged particle. The magnetic vector potential at the point $\mathbf{r}_i$ where the ith particle is located will be given by the formula

$$\mathbf{A}(\mathbf{r_i}) = A_{ext}(\mathbf{r}_i) + \sum_{j:j\neq i} e_j\mathbf{v}_j/|\mathbf{r}_j - \mathbf{r}_i|$$

and the electrostatic potential at the location of the ith charged particle is given by

$$\Phi(\mathbf{r}_i) = \Phi_{ext}(\mathbf{r}_i) + \sum_{j:j\neq i} e_j/|\mathbf{r}_j - \mathbf{r}_i|$$

The total Lagrangian for the system of particles is then given by

$$\mathcal{L}(\mathbf{r}_i,\mathbf{v}_i, i = 1,2,...,N)$$

$$= \sum_i m_i|\mathbf{v}_i|^2/2 - \sum_i e_i(\Phi_{ext}(\mathbf{r}_i) - (\mathbf{v}_i,\mathbf{A}_{ext}(\mathbf{r}_i)))$$

$$-\frac{1}{2}\sum_{i\neq j} e_ie_j/|\mathbf{r}_i - \mathbf{r}_j| + \frac{1}{2}\sum_{i\neq j} e_ie_j(\mathbf{v}_i,\mathbf{v}_j)/|\mathbf{r}_i - \mathbf{r}_j|$$

4. Rules for combining angular momenta in quantum mechanics. If L_1 is the angular momentum vector operator for the first particle and L_2 that for the second particle, then $L_1 + L_2$ is the total angular momentum vector of the two particles. L_1^2 assumes values $l_1(l_1+1)$ while L_2^2 assumes values $l_2(l_2+1)$ where $l_i = 0,1,2,...$. In the subspace in which L_1^2 assumes the value $l_1(l_1+1)$ and L_2^2 the value $l_2(l_2+1)$, it can be shown that $(L_1+L_2)^2$ has eigenvalues $l(l+1)$ where $l = |l_1-l_2|, |l_1-l_2|+1, ..., l_1+l_2-1, l_1+l_2$. The argument for this may be found in the book by Kompaneyetc. The result of adding angular momenta can also be verified via MATLAB simulations as follows. First using MATLAB, generate the spherical harmonic functions $Y_{l_1,m_1}(\theta,\phi), Y_{l_2,,2}(\theta,\phi)$ with $|m_1| \leq l_1, |m_2| \leq l_2$. Then construct the differential operators for $L_{ix}, L_{iy}, L_{iz}, i = 1,2$ as first order partial differential operators in the variables $(\theta_i,\phi_i), i = 1,2$ respectively. Compute the partial differential operator

$$L^2 = (L_{1x}+L_{2x})^2 + (L_{1y}+L_{2y})^2 + (L_{1z}+L_{2z})^2$$

as a second order partial differential operator in the variables $(\theta_1,\phi_1,\theta_2,\phi_2)$ and then determine its matrix elements

$$A[m_1,m_2|m_1'm_2'] = \int Y_{l_1,m_1}(\theta_1,\phi_1)Y_{l_2,m_2}(\theta_2,\phi_2)L^2(Y_{l_1,m_1'}(\theta_1,\phi_1)Y_{l_2,m_2'}(\theta_2,\phi_2))dS(\theta_1,\phi_1)dS(\theta_2,\phi_2)$$

and compute the eigenvalues of the resulting $(2l_1+1)(2l_2+1)\times(2l_1+1)(2l_2+1)$ matrix. An alternate method for computing the eigenvalues of L^2 is to use the Ladder operators. These operators can be used to determine the matrix elements of L_{1x}, L_{1y}, L_{1z} relative to a basis $|m_1>, m_1 = -l_1, -l_1+1, ..., l_1-1, l_1$ and likewise the matrix elements of L_{2x}, L_{2y}, L_{2z} relative to a basis $|m_2>, m_2 = -l_2, -l_2+1, ..., l_2-1, l_2$ and then we can compute

$$L^2 = (L_{1x}\otimes I + I\otimes L_{2x})^2 + (L_{1y}\otimes I + I\otimes L_{2y})^2 + (L_{1z}\otimes I + I\otimes L_{2z})^2$$

relative to the basis

$$|m_1, m_2> = |m_1> \otimes |m_2>, |m_1| \le l_1, |m_2| \le l_2$$

5. A numerical method for computing the spherical harmonic functions: Let r denote the radial coordinate. $1/r$ satisfies the Laplace equation $\nabla^2(1/r) = 0$. Take l constant vectors $a_1, ..., a_l$ and form the function

$$F(r,\theta,\phi) = \Pi_{j=1}^{l}(a_j, \nabla)r^{-1}$$

Consider now

$$\nabla r^{-1} = \hat{r}/r^2$$

so that

$$(a_1, \nabla)r^{-1} = f_1(\theta,\phi)/r^2$$

In general, we get

$$F(r,\theta,\phi) = f_l(\theta,\phi)/r^{l+1}$$

and hence we get from the radial form of the Laplacian:

$$\nabla^2 = \frac{1}{r}\frac{\partial^2}{\partial r^2}r - \frac{1}{r^2}L^2$$

the equation

$$L^2 f_l(\theta,\phi) = l(l+1)f_l(\theta,\phi)$$

which means that f_l is an element of the subspace $\{Y_{lm} : |m| \le l\}$. By varying the $a_j's$, we can obtain different bases for this subspace.

6.Quasi-Classical approximation for quantum systems: We plug into the Schrodinger equation

$$\frac{-h^2}{2m}\psi''(x,t) + V(x,t)\psi(x,t) = ih\psi_t(x,t)$$

where prime denotes partial derivative with respect to x while a subscript t denotes partial derivative with respect to time. We substitute

$$\psi(x,t) = A(x,t)exp(iS(x,t)/h)$$

getting

$$\psi_t = (A_t + iAS_t/h)exp(iS/h), \psi' = (A' + iAS'/h)exp(iS/h),$$

$$\psi'' = (A'' + iA'S'/h + iAS''/h - AS'^2/h^2)exp(iS/h)$$

Plugging these expressions into the Schrodinger equation gives us

$$(-h^2/2m)(A'' + iA'S'/h + iAS''/h - AS'^2/h^2) + VA = ih(A_t + iAS_t/h)$$

Equating coefficients of equal powers of h gives a sequence of differential equations for the functions A, S. Details can be found in the book "Quantum Mechanics" by Landau and Lifshitz.

7.Problem:Explain why the current through an inductor and the voltage across a capacitor should be continuous functions of time. Hint:The energy stored in the magnetic field in an inductor is proportional to the square of the instantaneous current while the energy stored in the electrostatic field of a capacitor is proportional to the instantaneous voltage across it. If the inductor current or capacitor voltage is discontinuous, then the instantaneous energy stored will be discontinuous which means that the power in the corresponding fields will be infinity and hence the element will burn out.

8.Maupertuis' principle of least action: $L(q(t), q'(t))$ is a given Lagrangian function. The generalized momentum is $p_i(t) = \frac{\partial L}{\partial q'_i}$. The energy along the path is given

$$E(q(t), q'(t)) = \sum_i p_i(t)q'_i(t) - L(q(t), q'(t))$$

We consider a time change $t \to \tau(t)$ that keeps the coordinates of the endpoints fixed, ie, $q(\tau(a)) = q_a$ and $q(\tau(b)) = q_b$ and that also keeps the energy fixed, ie,

$$E(q(\tau(t)), q'(\tau(t)) = e$$

where e is a constant. The Maupertius' action integral along the path is given by

$$S = \sum_i \int_{\tau(a)}^{\tau(b)} p_i(t).q'_i(t)dt = \int_{\tau(a)}^{\tau(b)} (L(q(t), q'(t)) + e)dt$$

We write down the condition for this action to be stationary with respect to small changes in the function τ about the identity that keep the energy fixed and also the coordinates of the endpoints fixed. Let $\alpha(t)$ denote the variation in the function $\tau(t)$ and $v(t)$ the variation in the function $q(t)$. The condition $q(\tau(a)) = q_a$ gives on variation

$$v(\tau(a)) + q'(\tau(a))\alpha(a) = 0$$

and likewise,

$$v(\tau(b)) + q'(\tau(b))\alpha(b) = 0$$

Also the variation in $q'(t)$ is $v'(t)$, so that Then $\delta S = 0$ at $\tau = id, q = q_0$ with q_0 as the classical path obtained using the Euler-Lagrange formalism gives

$$(L(q(\tau(b)), q'(\tau(b)) + e)\alpha(b) - (L(q(\tau(a)) + e)\alpha(a)$$

$$+ \int_{\tau(a)}^{\tau(b)} \frac{\partial L(q(t), q'(t))}{\partial q(t)} v(t) + \frac{\partial L(q(t), q'(t)}{\partial q'}$$

9. **Group theory**: Let X, Y be elements of the Lie algebra of a Lie group G and let $f : G \to \mathbb{R}$ be a smooth function. Define

$$F(u,v) = f(x.exp(uX + vY)), x \in G, u, v \in \mathbb{R}$$

Then we have

$$\frac{d^n}{dt^n} F(tu, tv)|_{t=0} = (uX + vY)^n f(x)$$

It follows from Taylor's theorem that

$$F(u,v) = \sum_{n=0}^{\infty} ((uX + vY)^n/n!) f(x)$$

On other other hand,

$$F(u,v) = \sum_{r,s=0}^{\infty} \frac{u^r v^s}{r!s!} \frac{\partial^{r+s}}{\partial u^r \partial v^s} F(0,0)$$

from which we conclude that $\frac{\partial^{r+s}}{\partial u^r \partial v^s} F(0,0)$ equals the coefficient of $u^r v^s$ in the expansion of $(r!s!/(r+s)!)(uX + vY)^{r+s}$. For example, we get

$$(uX + vY)^3 = u^3 X^3 + v^3 Y^3 + u^2 v(X^2 Y + XYX + YX^2) + uv^2(XY^2 + YXY + Y^2 X)$$

so that

$$\frac{1}{3}(X^2 Y + XYX + YX^2) f(x) = \frac{\partial^3}{\partial u^2 \partial v} f(x.exp(uX + vY))|_{u=v=0}$$

Differentiable manifolds and Lie groups

10.Let M, N be C^∞-manifolds of dimension m, n respectively. Let $\pi : M \to N$ be a C^∞-map. Let $x \in M$ and $y = \pi(x) \in N$. Suppose $(d\pi)_x$ is injective. Then $m \le n$ and in local coordinates, π looks like $(t_1, ..., t_m) \to (t_1, ..., t_m, 0, ..., 0)$. Read about the proof of this.

11.**Studies in Optimal control** (Synopsis submitted by A.N.Jha to the Faculty of Technology)

Abstract: The basic issue in most of the studies in optimal control is the following. We are given a dynamical system specified by a set of coupled linear/nonlinear first order differential equations. The dynamical variables constitute the state vector. The dynamical equations involve a set of input processes which are to be determined over a finite duration of time such that a cost functional of the state and input process over that duration is minimized. Constraints are of course the dynamical equations for the state and possibly some final point conditions like the fact that a certain function of the state variables forming the output of the system attains a specified set of values at the final time. A typical example of this is the problem of throwing a satellite in space via a rocket within a given time duration so that the fuel spent by the rocket over the duration is minimized and such that at the final time point, the rocket has reached a specified position in space and has a specified velocity so that the satellite thrown by it into space has acquired a given initial velocity that will enable it to orbit around the earth. Another example is that of minimum time optimal control problems. Here, the the problem is to treat the final time point as unknown and to minimize it subject to the constraints on the state and input imposed by the dynamical equations and final point condition. In this thesis, we propose to extend the scope of optimal control theory to include dynamical systems described by partial differential equations and apply the theory specifically to the problems of controlling the electromagnetic field inside a waveguide or a cavity resonator. A field is a quantity specified as a function of four space-time coordinates.

12.**Quantum mechanics**

1.Define the creation and annihilation operator fields on the antisymmetric Fock spaces. Show via an explicit calculation that these operators are adjoints of each other.

2.Suppose G is a group and H an open subgroup of G. If $g_1, g_2 \in G$ then either $g_1 H = g_2 H$ or else $g_1 H \cap g_2 H = \phi$. We can thus write

$$G = \bigcup_{g \in I} gH$$

where $I \subset G$ is such that for any two distinct elements g_1, g_2 in I, $g_1 H \cap g_2 H = \phi$. We may assume that $e \in I$. It follows that

$$H = G - \bigcup_{g \in I, g \neq e} gH$$

and hence H is the complement of a union of open sets. H is therefore closed. Thus, every open subgroup of a topological group is also closed.

13.**Mechanics**

1.State and prove Darboux's theorem (every non-degenerate skew symmetric from on a differentiable manifold whose exterior derivative vanishes, can be locally expressed in the from $\sum_{i=1}^{n} dx^i \wedge dx^{i+n}$, ie, we can find a local system of coordinates (x^i) satisfying this equation. The manifold must of course, be assumed to be even dimensional.

14.**Probability theory**

1.Let Ω be a set and $\mathcal{F}$ a field of subsets of Ω. Let $P : \mathcal{F} \to [0,1]$ be a probability measure on $\mathcal{F}$. The aim of the extension theorem is to show that if $\sigma(\mathcal{F})$ is the σ field generated by $\mathcal{F}$, then there is a unique probability measure Q on $\sigma(\mathcal{F})$ whose restriction to $\mathcal{F}$ equals P. This theorem was first proved by Constantin Caratheodory and is called the Caratheodory extension theorem. The steps involved are as follows: For any $A \subset \Omega$, define $P^*(A)$ to be the infimum of $\sum_n P(E_n)$ where $E_n \in \mathcal{F}, n = 1, 2, ...$ are such that $A \subset \bigcup_n E_n$. Then we show that P^* is monotone and countably subadditive. Then we define $\mathcal{M}$ to be the set of all $A \subset \Omega$ such that for every $E \subset \Omega$, we have

$$P^*(EA) + P^*(EA^c) = P^*(E)$$

In view of the fact that P^* is subadditive, it follows that $\mathcal{M}$ is the collection of all $A \subset \Omega$ for which $P^*(EA) + P^*(EA^c) \leq P^*(E)$ for all $E \subset \Omega$. We then show that $\mathcal{M}$ is a field by proving that it is closed under complementation (trivial to see) and finite intersections. We then prove that P^* is finitely additive on $\mathcal{M}$. Then we prove that $\mathcal{M}$ is closed under countable disjoint unions and is therefore a σ-field. Finally, we prove that P^* is countably additive on $\mathcal{M}$. Since $\mathcal{M}$ is a σ-field containing $\mathcal{F}$, it follows that $\mathcal{M}$ also contains $\sigma(\mathcal{F})$ and the restriction of P^* to $\sigma(\mathcal{F})$ is the desired extension of P. The proofs may be found in the book "Probability and Measure", by P.Billingsley and also in the book "An introduction to probability and measure" by K.R.Parthasarathy.

15.(Electronics) Draw the small signal equivalent circuit of an *npn* transistor; Determine the h-parameters of this model. By taking sample values of the constants $I_{CO}, I_{EO}, \alpha_C, \alpha_E, V_T$, determine the approximate range of the values of $h_{11} = h_{ie}, h_{12} = h_{re}, h_{21} = h_{fe}, h_{22} = h_{oe}$.

16. (Signal theory)

1.Define a signal in most general terms and describes the various kinds of classification.

2.Let $\phi_n(t), n = 1, 2, ...$ be an orthonormal basis for $L^2[0, T]$ and let $\lambda_n, n = 1, 2, ...$ be a sequence of non-negative real numbers such that $\sum \lambda_n < \infty$. Define $R(u, v) = \sum_n \lambda_n \phi_n(u)\phi_n(v)$ and $Q(u, v) = \sum_n \lambda_n^{-1} \phi_n(u)\phi_n(v)$. Show that R, Q are inverse kernels of each other, ie,

$$\int_0^T R(u, v).dv.Q(v, w) = \int_0^T Q(u, v).dv.R(v, w) = \delta(u - w)$$

Hint use orthonormality of $\phi_n' s$ and

$$\sum_n \phi_n(u)\phi_n(v) = \delta(u - v)$$

3.Let V be the vector space of all polynomials with real coefficients having degree $\leq n$. Show that $dimV = n + 1$. Let $D : V \to V$ be defined by

$$Df(t) = f'(t)$$

Show that D is a well defined linear operator in V and determine the matrix of D in the basis $\{1, t, t^2, ..., t^n\}$.

17.Problems in quantum field theory

Commutation relations for the Poincare group.

Infinitesimal Lorentz transformations.

Derivation of the formula for the Feynman propagator and Dirac propagator.

Derivation of the Dirac equation from symmetry considerations.

Clifford algebra anticommutation rules.

18.Group representation theory problems

1.Define the general linear group $GL(V)$ where V is a finite dimensional vector space. Define $SL(V)$ to be the set of all $g \in GL(V)$ for which $det(g) = 1$. Show that $SL(V)$ is a subgroup of $GL(V)$. Let V

be a finite dimensional vector space and let $\otimes$ be a tensor product. Consider for $g \in GL(V)$, the element $g^{\otimes n} \in GL(V)^{\otimes n}$. Show that $g \to g^{\otimes n}$ is a representation of $GL(V)$. Note that

$$g^{\otimes n} v^{\otimes n} = (gv)^{\otimes n}$$

The representation space for the representation $g \to g^{\otimes n}$ acts in the symmetric tensor space $V^{\otimes_s n}$. Any symmetric tensor can be expressed as linear combinations of tensors of the form $v^{\otimes n}$. For example,

$$v_1 \otimes v_2 + v_2 \otimes v_1 = (v_1 + v_2) \otimes (v_1 + v_2) - v_1 \otimes v_1 - v_2 \otimes v_2$$

Note how permutations act on the tensor space. Let n be a positive integer and $v_1, ..., v_n \in V$. Let σ be a permutation of $(1, 2, ..., n)$. Then,

$$\sigma(v_1 \otimes ... \otimes v_n) = v_{\sigma^{-1}1} \otimes ... \otimes v_{\sigma^{-1}n}$$

so that if τ is another permutation, then

$$\tau\sigma(v_1 \otimes ... \otimes v_n) = \tau(v_{\sigma^{-1}1} \otimes ... \otimes v_{\sigma^{-1}n})$$

$$= v_{\sigma^{-1}\tau^{-1}1} \otimes ... \otimes v_{\sigma^{-1}\tau^{-1}n}$$

$$= v_{(\tau\sigma)^{-1}1} \otimes ... \otimes v_{(\tau\sigma)^{-1}n}$$

as it should be for an action.

19. (Probability theory)

1.Define a Markov chain and show that the time reversed version of a Markov chain is also a Markov chain.

2.Give a rigorous definition of a transition probability measure. Let $(X_i, \mathcal{B}_i), i = 1, 2$ be two measurable spaces and consider a map $\lambda : X_1 \times \mathcal{B}_2 \to [0, 1]$ so that for every $B \in \mathcal{B}_2$, the map $x \to \lambda(x, B)$ is $\mathcal{B}_1$ -measurable and for every $x \in X_1$, the map $B \to \lambda(x, B)$ is a probability measure. Then $\lambda(.,.)$ is called a transition probability measure.

20. MATLAB Problems:

1.Mass is distributed over a rectangular region such that if (i, j) is the location of the pixel having coordinates (i, j), then $m(i, j)$ is the mass inside this pixel. The coordinates of the centre of mass of the region are given by

$$x = \frac{\sum_{i,j=1}^{n} i.m(i, j)}{\sum_{i,j=1}^{n} m(i, j)}, y = \frac{\sum_{i,j=1}^{n} j.m(i, j)}{\sum_{i,j=1}^{n} m(i, j)}$$

Write a MATLAB program to calculate this.

2.Two projectiles are thrown from a point O with initial velocities v_1, v_2 respectively making angles θ_1, θ_2 respectively with the horizontal. After time t, the position coordinates of the two projectiles are respectively given by

$$(x_1(t), y_1(t)) = (v_1 cos(\theta_1)t, v_1 sin(\theta_1)t - gt^2/2), (x_2(t), y_2(t)) = (v_2 cos(\theta_2)t, v_2 sin(\theta_2)t - gt^2/2)$$

The distance between two projectiles at time t is given by

$$d(t) = \sqrt{(x_1(t) - x_2(t))^2 + (y_1(t) - y_2(t))^2}$$

Compute the times T_1, T_2 at which the respective projectiles hit the ground. Also compute the distance $d(t)$ between the two projectiles at times $0, T/N, 2T/N, ..., (N-1)T/N, T$ where $N = 1000$ and $T = min(T_1, T_2)$.

3.Simulate the motion of particles in a radially symmetric metric

Dr. Sudipta's Ph.D thesis: Estimating transistor parameters from Volterra representation of amplifier circuits. The first part of this work focuses on estimating the parameters of a second order ordinary differential equation with no input, ie, an autonomous system. Such an equation can describe either an LCR circuit or a spring-mass system with damping. One way to estimate the parameters of such a model is to use a least squares approach. For example, if the differential equation is $y''(t) + ay'(t) + by(t) = 0$ and noisy measurements on $\{y(t)\}$ are made over the time interval $[0, T]$, then a, b may be estimated by minimizing

$$\int_0^T (y''(t) + ay'(t) + by(t))^2 dt$$

In practice, the data is digital, so that if measurements are taken with a sampling interval of Δ, then a, b may be estimated by minimizing

21. Problem in stochastic processes: Consider the one dimensional sde

$$dx(t) = f(x(t))dt + g(x(t))dB(t)$$

Determine the mean and variance propagation equations here. Let $< x(t) >$ be the mean of $x(t)$. Then we have

$$d < x(t) > = < f(x(t)) > dt$$

and

$$f(x(t)) \approx f(< x(t) >) + f'(< x(t) >)\delta x(t) + \frac{1}{2} f''(< x(t) >)(\delta x(t))^2$$

where

$$\delta x(t) = x(t) - < x(t) >$$

It follows that with

$$P(t) = < (\delta x(t))^2 > = Var(x(t)),$$

$$d < x(t) > /dt \approx f(< x(t) > + \frac{1}{2} f''(< x(t) >)P(t)$$

22.Some additional project problems

1.Summary of research reports coming out from the DSP laboratory: Currently planning to guide the construction of antenna arrays for signal reception: If we take a uniformly spaced antenna array consisting of isotropic sources with spacing d, then the amplitude pattern function can be expressed as

$$\sum_{n=0}^{N-1} exp(jn\psi), \psi = kd.cos(\theta) + \beta$$

where $k = 2\pi/\lambda$ is the wave number, d is the interelement spacing and β is the phase difference between the currents in two successive elements. This sums up to

$$F(\theta) = (exp(jN\psi) - 1)/(exp(j\psi) - 1 = exp(j(N-1)\psi/2)sin(N\psi/2)/sin(\psi/2)$$

By adjusting the phase β appropriate directional properties of the antenna can be obtained. Special cases are (a) the broad side array and (b) the end fire array. (Ref:Balanis, Antenna Theory). The directivity can be computed for the variety of cases, namely the ratio of the maximum amplitude to the root mean square amplitude:

$$D = |F_{max}|/\sqrt{\int |F(\psi))|^2 2\pi sin(\theta)d\theta}$$

This array can be utilized to receive signals from radio-astronomial sources in the sky. Multiple correlation interferometry can be done using such arrays.

2.Special lectures:Lectured to M.Tech students at the Ambedkar Institute of Technology on topics in Signal Theory. The lectures focussed on advanced techniques in deterministic and statistical signal analysis. For deterministic signal analysis, the Fourier transform as a unitary operator in a Hilbert space along with the spectral representation for unitary operators was introduced. Spectral integration was covered in depth starting with the finite dimensional case. Spectral theorem for normal linear operators for the finite dimensional case was also discussed. The orthogonal projection operator onto a subspace of a Hilbert space was introduced in the form of the best approximation theorem. The Gram-Schmidt orthonormalization process was then discussed. A little bit of time series analysis using linear models $AR, MA, ARMA$ was introduced. Basic detection theory was covered. The Bayes criterion was introduced leading to the maximum likelihood and maximum aposteriori techniques for signal detection and parameter estimation. The Neyman-Pearson criterion for signal detection was introduced after that, based on the fact that for a fixed probability of false alarm, the probability of missing the target is minimized. The Karhunen-Loeve expansion for L^2-stochastic processes was discussed. After that, a brief outline of quantum stochastic processes was presented as developed by R.L.Hudson and K.R.Parthasarathy. For doing so, in the lectures we introduced the Boson Fock space and exponential vectors and followed it up with a discussion of the Weyl operator acting on the Boson Fock space. The From the Weyl operator, two kinds of one parameter unitary subgroups were derived, the former leading to the creation and annihilation processes and the latter to the conservation processes. Using these processes, it was shown how the Ito differential rule of stochastic calculus could be generalized to a quantum Ito formula and then using this formula, the notion of quantum stochastic differential equation was introduced. This remarkable developement due to Hudson and K.R.Parthasarathy could be used to construct noisy versions of the Schrodinger and Heisenberg equations in quantum mechanics.

3.Particle moving along a random trajectory, statistics of the radiation field. If the particle trajectory is given by $\mathbf{R}(t)$, then then upto a first order approximation, the retarded time is given by

$$t' = t - |\mathbf{r} - \mathbf{R}(t)|/c$$

and the vector potential produced by this charge is given approximately by

$$\mathbf{A}(t,\mathbf{r}) = \frac{\mu Q\mathbf{V}(t')}{4\pi(|\mathbf{r}-\mathbf{R}(t)| - (\mathbf{V}(t'),\mathbf{r}-\mathbf{R}(t'))/|\mathbf{r}-\mathbf{R}(t')|)}$$

When $\mathbf{R}(t)$ is a diffusion process, it is a problem to calculate the correlation functions of the vector potential as well as those of the electric field.

4.Correlations in the wave function of a quantum mechanical particle. Assume that the particle is moving in a potential field $U(x,\theta)$ where θ is a random parameter. Solve Schrodinger's wave equation and determine the correlations in the wave function. To start with, assume that $U(x,\theta) = U_0(x) + \theta.U_1(x)$ where θ is small and determine the expansion for the wave function $\psi(x,\theta)$ in powers of the parameter θ. Hint:

$$\psi''(x,\theta) + \frac{2m}{h^2}(E - U_0(x) - \theta.U_1(x))\psi(x,\theta) = 0$$

We start with this equation and apply perturbation theory to this in the form

$$\psi(x,\theta) = \sum_{m=0}^{\infty} \theta^m \psi_m(x)$$

Equate coefficients of each power of θ to zero and derive differential equations for the successive corrections $\psi_m, m = 0,1,2,....$

5.Schwarzchild metric in cartesian coordinates:

$$d\tau^2 = (1-2m/r)dt^2 - ((1-2m/r)^{-1} - 1)dr^2 + dr^2 + r^2 d\theta^2 + r^2 sin^2(\theta)d\phi^2$$

$$(1-2m/r)dt^2 = (1-2m/\sqrt{x^2+y^2+z^2})dt^2,$$

$$dr = (xdx + ydy + zdz)/r,$$

$$dr^2 = (x^2dx^2 + y^2dy^2 + z^2dz^2 + 2xydxdy + 2yzdydz + 2xzdxdz)/r^2$$

The final problem is to obtain explicit expressions for $\{g_{tt}, g_{xx}, g_{yy}, g_{zz}, g_{+xy}, g_{yz}, g_{xz}\}$.

6.Problem: Define the Lie algebra of a Lie group and the Lie group generated by a Lie algebra. Give examples.

7.Quantum stochastic calculus(quiz): Define the creation, annihilation and conservation operator fields and derive the commutation relations between these fields. Start from the definition of the Weyl operator.

8.Supersymmetry: Define a graded vector space $V = V_0 \oplus V_1$ is a vector space V_0 are the even elements and V_1 are the odd elements. This is a graded vector space graded by $\mathbb{Z}/2\mathbb{Z}$. If V has a product such that

$$xy = (-1)^{p(x)p(y)} yx$$

for homogeneous elements, then A is said to be a graded commutative algebra. The elements of $V_i, i = 0, 1$ are called the homogeneous elements.

9.Clifford Algebra: Let V be a vector space and Q a quadratic form on V. Let $\tau(V)$ denote the tensor algebra on V and let J denote the two sided ideal of $\tau(V)$ generated by the elements $x \otimes x - Q(x)$ (or equivalently by $x \otimes y + y \otimes x - B(x, y)$ where B is the symmetric bilinear form associated with Q). Then $C = \tau(V)/J$ is the Clifford algebra on V defined by the quadratic form Q. This algebra is a graded algebra since the tensor algebra τ is graded. Suppose $x, y \in V$. Then let $\bar{x} = x + J, \bar{y} = y + J$. We have $\bar{x}\bar{y} + \bar{y}\bar{x} = B(x, y) + J = B(x, y)\bar{1}$ where $\bar{1} = 1 + J$. For simplicity, we write xy in place of $x \otimes y$. Note that

$$\tau = \tau_0 \oplus \tau_1$$

where τ_0 is the vector space spanned by elements of τ of the form $x_1 \otimes ... \otimes x_n$ with n even and τ_1 the same with n odd. Suppose $V = V_1 \oplus V_2$ with V_1, V_2 orthogonal with respect to the symmetric bilinear form $B(.,.)$, ie, $B(x, y) = 0$ for all $x \in V_1, y \in V_2$. Then for $x \in V_1, y \in V_2$, we have $\bar{x}\bar{y} + \bar{y}\bar{x} = B(x, y) = 0$. Suppose Let C_1 be the Clifford algebra on V_1 and C_2 the Clifford algebra on V_2. Then what it the Clifford algebra on $V = V_1 \oplus V_2$? Choose $x_i \in V_i, i = 1, 2$. Then $x_1 + x_2 \in V$ and using the same notation x for $\bar{x}$, it follows that

$$(x_1 + x_2)^2 = x_1^2 + x_2^2 + x_1x_2 + x_2x_1 = Q(x_1) + Q(x_2)$$

Consider now another element $y = y_1 + y_2 \in V$ with $y_1 \in V_1, y_2 \in V_2$. Then

$$xy = (x_1 + x_2)(y_1 + y_2) = x_1y_1 + x_2y_2 + x_1y_2 + x_2y_1$$

Now,

$$x_1y_1 \in C_1, x_2y_2 \in C_2, x_1y_2 \in C_1 \otimes C_2.x_2y_1 = -y_1x_2 \in C_1 \otimes C_2$$

so it follows that

$$xy \in C_1 \otimes C_2$$

if we identify C_1, C_2 as the subspaces $1 \otimes C_i, i = 1, 2$ of $C_1 \otimes C_2$. Define a map

$$\Phi : C_1 \otimes C_2 \to C$$

so that

$$\Phi(a \otimes b) = ab, a \in C_1, b \in C_2$$

and extend Φ by linearity. More precisely, if $a \in \tau(V_1), b \in \tau(V_2)$ and J_i is the two sided ideal in τ_i generated by elements of the form $x^2 - Q(x), x \in V_i$, then

$$\Phi((a + J_1) \otimes (b + J_2)) = a \otimes b + J_1 \otimes J_2$$

This map, after extension, becomes an isomorphism from $C_1 \otimes C_2$ onto C. The onto property is clear. The one one property can be proved as follows by means of an example.

Chapter 17

Objective Type Questions and Interview Problems

1. Suppose two diodes having voltage current relations $i = f(v)$ are connected in series. What is the voltage current relation for the resulting circuit ?

2. Write down the Ebers-Moll equations for a transistor giving the collector and emitter currents in terms of the base emitter and base collector voltage. Assume that the transistor is pnp.

3. For quantum mechanical tunneling through a rectangular potential barrier of width $[0, a]$ and height V_0, the Schrodinger equation in the three different regions are given by

4. The autocorrelation of white noise equals

5. The autocorrelation function of the Wiener process equals

6. The interarrival times of a Poisson process are distributed in accordance with what law ?

7. The sum of n identically distributed exponential random variables has what distribution ?

8. According to Gauss' theorem, if $\mathcal{V}$ is a volume bounded by the closed surface S and $\mathbf{F}$ is a vector field in space, then $\int_{\mathcal{V}} div\mathbf{F}d^3x$ equals

9. According to Green's theorem, if $\mathcal{V}$ is a volume bounded by the closed surface S and f, g are scalar fields, then $\int_{\mathcal{V}}(f\nabla^2 g - g\nabla^2 f)d^3x$ equals

10. If $\nabla \times \mathbf{f} = \mathbf{0}$, then $\mathbf{f}$ must be of the form

(a) $\nabla.\mathbf{g}$ (b) $\nabla \times \mathbf{g}$, (c) $\nabla\phi$, (d) None of the above.

11. If ψ is the wave function of a quantum mechanical particle, then the probability density of the particle is given by and the current density by

(a) $\psi^2, \psi\nabla\psi^*$

(b) $|\psi|^2, iK(\psi^*\nabla\psi - \psi\nabla\psi^*)$

(c) $|\psi|^4, \nabla|\psi|^2$

(d) None of the above.

12. If $G(s)$ is the feed forward transfer function and $H(s)$ the negative feedback transfer function, then the overall transfer function of the closed loop system is given by

(a) $G(s)H(s)$

(b) $G(s)/(1+H(s))$

(c) $\frac{G(s)}{1+G(s)H(s)}$

(d) None of the above.

13. The spin of a quantum mechanical particle is an operator defined by

(a) A vector with scalar components

(b) A vector with 3×3 matrix components

(c) A vector with 2×2 matrix components

(d) None of the above.

14. If $m(t)$ is the modulating wave and $c(t)$ the carrier, then the modulated wave for AM has the from

(a) $m(t)c(t)$

(b) $(1+\mu.m(t))c(t)$

(c) $(1+\mu.c(t))m(t)$

(d) None of the above

15. The Fourier transform of the product $x(t)y(t)$ equals

(a) $X(j\omega).Y(j\omega)$

(b) $\int_{-\infty}^{\infty} X(j(\omega-u))Y(ju)du$

(c) $\frac{1}{2\pi}\int_{-\infty}^{\infty} X(j(\omega-u))Y(ju)du$

(d) None of the above.

16. Let V be a vector space and assume that $V = V_1 \oplus V_2$ where V_1, V_2 are subspaces. Then let B_1 be an ordered basis for V_1 and V_2 an ordered basis for V_2. Let $T : V \to V$ be a linear transformation. Then T relative to the ordered basis $B = B_1 \cup B_2$ has the matrix

$$[T]_B = \begin{pmatrix} A_{11} & A_{12} \\ A_{21} & A_{22} \end{pmatrix}$$

where the sizes of $A_{11}, A_{12}, A_{21}, A_{22}$ are respectively

(a) $dimV_1 \times dimV_1$, $dimV_1 \times dimV_2$, $dimV_2 \times dimV_1$, $dimV_2 \times dimV_2$

17. Let V be a vector space over the field $\mathbb{C}$ and let V^* be the set of all linear maps $f : V \to \mathbb{C}$. Then if $dimV = n$, we have $dimV^* =$

(a) $n-1$, (b) $n+1$, (c) 0, (d) n

18. Let V be a vector space of dimension n. For each $x \in V$ and $f \in V^*$, define the the linear operator $T(f,x)$ on V so that $T(f,x)(y) = f(y)x$. More then, if f_1, f_2 are nonzero in V^* and x_1, x_2 are linearly independent in V, then $T = T(f_1, x_1) + T(f_2, x_2)$ has rank

(a) two (b) one (c) three, (d) none of the above.

19. Let ρ denote the charge density in space and j the current density. Define $j^0 = \rho$. Then, j^μ is a

(a) scalar, (b) vector, (c) four vector, (d) none of the above.

20. Let ρ be the charge density and $j^r, r = 1,2,3$ the cartesian components of the current density. Define $j^0 = \rho$. Let $A^0 = \Phi$ be the electrostatic potential, $A^r, r = 1,2,3$ the magnetic vector potential, then $j^\mu A_\mu = \rho\Phi - \sum_{r=1}^3 j^r A^r$ is

(a) invariant under Lorentz transformations, (b) transforms like a three vector, (c) transforms like a four vector, (d) transforms like the time component of a four vector.

21. The first term in the relativistic correction to the kinetic energy of a particle having rest mass m and three momentum p is proportional to

(a) p^2, (b) p^3, (c) p^4, (d) none of the above.

22. The magnetic field produced by a particle having charge Q and moving with velocity $\mathbf{v}$ is given by

(a) $Q\mathbf{v}/r$, (b) $Q\mathbf{v} \times \mathbf{r}/r^3$, (c) $Q\mathbf{v} \times (\mathbf{v} \times \mathbf{r})/r^3$, (d) none of the above.

23. Let V be a vector space and let $\otimes$ be a tensor product of V. Let $\wedge$ be the wedge product on V. Then $\wedge$ satisfies

(a) $x \wedge y = -y \wedge x$, (b) $x \wedge y = y \wedge x$, (c) $x \wedge y = 0$, (d) none of the above.

24. Let ω be a p antilinear form on a vector space V. Then if σ is any permutation of $1, 2, ..., n$, it follows that

(a) $\omega(v_{\sigma 1}, ..., v_{\sigma n}) = sgn(\sigma)\omega(v_1, ..., v_n)$

(b) $\omega(v_1, v_1, v_2, ..., v_{n-1}) \neq 0$

(c) $\omega(v_1, v_2, ..., v_n) = \omega(v_2, v_1, ..., v_n)$

25. If $H(r, r')$ is the kernel of the Hamiltonian operator of a quantum mechanical system and $u(r)$ is the wave function of the particle, then the average energy of the particle is given by

(a) $\int H(r,r)|u(r)|^2 d^3r$

(b) $\int \bar{u}(r)H(r,r')u(r')d^3rd^3r'$

(c) $\int \bar{H}(r,r')\bar{u}(r)u(r')d^3rd^3r'$

(d) none of the above.

26. The interaction part of the Lagrangian density of an electromagnetic field interacting with a current is given by

(a) ρV where ρ is the charge density and V the electrostatic potential

(b) $J.A$ where J is the current density and A the magnetic vector potential

(c) $\rho V - J.A$

(d) $-\rho V + J.A$

27. A thin straight wire carrying phasor current $I(z)$ at frequency ω_0 and extending along the z axis from $z = -L/2$ to $z = +L/2$ radiates out electromagnetic fields in space. The exact magnetic vector potential phasor is given by

(a) $\mathbf{A}(\mathbf{r}) = \frac{\mu_0}{4\pi}\int_{-L/2}^{L/2} I(z')exp(-j\omega_0|\mathbf{r} - z'\hat{z}|/c)dz'/|\mathbf{r} - z'\hat{z}|$

(b) $\mathbf{A}(\mathbf{r}) = \frac{\mu_0 exp(-jkr)}{4\pi r}\int_{-L/2}^{L/2} I(z')exp(j\omega_0 rz'cos(\theta))dz'$

(c) $\mathbf{A}(\mathbf{r}) = \frac{exp(-jkr)}{4\pi\epsilon_0 r}\int_{-L/2}^{L/2} I(z')exp(j\omega_0 rz'cos(\theta)/c)dz'$

28. For a plane electromagnetic wave with electric field $\mathbf{E} = \mathbf{E}_0 exp(i\omega t - ik.r)$ and magnetic field $\mathbf{H} = \mathbf{H}_0 exp(i\omega t - i.k.r)$, the Maxwell equations are

(a) $k \times E = \omega\mu H, k \times H = -\omega\epsilon E$

(b) $k \times E = -\omega\mu H, k \times H = \omega\epsilon E$

(c) $k \times E = \omega\epsilon H, k \times H = -\omega\mu E$

(d) None of the above.

29. Let M be a manifold and let F be a finite dimensional vector space. Then $E = M \times F$ is a vector bundle. A field defined on the space M with values in F is simply a section of the vector bundle E. Specifically, if ϕ is such a field, then for each $x \in M$, $\phi(x) \in F$. A Lagrangian density is a function $L(\phi, \partial_\mu\phi)$ where ϕ is a smooth field. The Lagrangian field equations can be formulated from the action principle $\delta\int_M L(\phi, \partial_\mu\phi)d^mx = 0$. This variational principle leads to the field equations

(a) $\partial_\mu\frac{\partial L}{\partial\partial_\mu\phi} = \frac{\partial L}{\partial\phi}$

(b) $\partial_\mu\frac{\partial L}{\partial\phi} = \frac{\partial L}{\partial\partial_\mu\phi}$

(c) eqn (b) with the sign changed on one side of the equation

(d) None of the above.

30. The action principle $\delta\int(\partial^\mu\phi\partial_\mu\phi - m^2\phi^2)d^4x = 0$ gives rise to the Klein-Gordon equation which has the structure

(a) $\partial^\mu\partial_\mu\phi + m^2\phi = 0$

(b) $\nabla^2\phi = m^2\phi$

(c) $(\partial_t\phi)^2 = m^2\phi$

(d) none of the above.

31. The action functional $\int(\partial^\mu\phi\partial_\mu\phi - m^2\phi^2)d^4x$ when expressed in the momentum domain for the purpose of computing path integrals with the four dimensional Fourier transform of the field given by $\phi(x) = \int\hat{\phi}(p)exp(ip.x)d^4p, p.x = p^\mu x_\mu$ has the form

(a) $\int(p^2 - m^2)|\hat{\phi}(p)|^2 d^4p, p^2 = p_\mu p^\mu$

(b) $\int(p^2 + m^2)|\hat{\phi}(p)|^2 d^4p$

(c) $-\int(p^2 + m^2)|\hat{\phi}(p)|^2 d^4p$

32. Let $F(x, y, z, t)$ be a physical quantity in a fluid defined as a function of position and time. Let v_x, v_y, v_z be the velocity field components of the fluid as a function of position and time. Then the material derivative of F with respect to time, denoted by DF/Dt is given by

(a) $\frac{\partial F}{\partial t}$

(b) $\frac{\partial F}{\partial t} + v_x\frac{\partial F}{\partial x}$

(c) $\frac{\partial F}{\partial t} - \mathbf{v}.\nabla F$

(d) $\frac{\partial F}{\partial t} + \mathbf{v}.\nabla F$

33. Suppose the permittivity $\epsilon(\mathbf{E})$ of a medium is a function of the electric field $\mathbf{E}$. Then the displacement current is given by

(a) $\epsilon\frac{\partial \mathbf{E}}{\partial t}$

(b) $\frac{\partial(\epsilon\mathbf{E})}{\partial t}$

(c) $\frac{\partial\epsilon}{\partial t}\mathbf{E}$

(d) none of the above.

34. In the hysterisis curve of magnetic field versus magnetic field intensity, B specified as a function of H is

(a) an ordinary memoryless function

(b) a function having memory

(c) a function having memory and also depending on the future values

(d) a discontinuous function

35. Let E be the energy of a particle undergoing quantum mechanical tunneling through a rectangular barrier of height V_0. Assuming that $E < V_0$, the form of the wave function within the barrier is given by

(a) $C_1.exp(kx) + C_2exp(-kx)$, k real

(b) $C_1.exp(ikx) + C_2.exp(-ikx)$, k imaginary

(c) $C.exp(kx)$

(d) none of the above.

36. Let $L(q, q')$ be a Lagrangian with q as the generalized coordinate and q' as the generalized velocity. Then, the canonical momentum is given by

(a) $p = \frac{\partial L}{\partial q'}$

(b) $p = \frac{\partial L}{\partial q}$

(c) $p = \frac{\partial^2 L}{\partial q \partial q'}$

(d) none of the above.

37. If $\alpha = \alpha_i dx^i$ is a one form on a differentiable manifold, then its exterior derivative $d\alpha$ is the two form given by

(a) $\frac{\partial \alpha_i}{\partial x^j} dx^i \wedge dx^j$

(b) $\frac{\partial \alpha_i}{\partial x^j} dx^j \wedge dx^i$

(c) none of the above.

38. Let $A_z \hat{z}$ be the vector potential field phasor for a sinusoidally varying electromagnetic field. Then the scalar potential field in the Lorentz gauge is given by

(a) $\Phi = \frac{j}{\omega\epsilon\mu} \frac{\partial A_z}{\partial z}$

(b)

39. Let $A_z \hat{z}$ be the vector potential field for a sinusoidally varying electromagnetic field. Then the z component of the electric field is given by

(a) $E_z = (-j/\omega\epsilon\mu)(\frac{\partial^2 A_z}{\partial z^2} + k^2 A_z)$

40. Suppose we are given N thin straight wire antennas of lengths $L_n, n = 1, 2, ..., N$. The nth antenna is stretched over the region $(x_n, y_n, z), |z| \leq L_n/2$ and carries a current $I_n(z)$. Then the total z component of the electric field is given by

ans:

$$\sum_{n=1}^{N} \frac{\mu}{4\pi} \int_{-L_n/2}^{L_n/2} I_n(\xi)\mathcal{L}exp(-jk((x - x_n)^2 + (y - y_n)^2 + (z - \xi)^2)^{1/2})d\xi/((x - x_n)^2 + (y - y_n)^2 + (z - \xi)^2)^{1/2})$$

where the linear differential operator $\mathcal{L}$ is given by

$$\mathcal{L} = \frac{j}{\omega\epsilon\mu}(\frac{\partial^2}{\partial z^2} + k^2)$$

41. Pocklington's integral equation in antenna theory is based on the principle that

ans:The tangential component of the total electric field on the surface of a perfectly conducting antenna is identically zero.

42. If $L(q, q')$ is the Lagrangian of a particle and $S[q] = \int_{t_1}^{t_2} L(q(t), q'(t))dt$ is the action, then the variation in the action produced by a variation $\delta q(t)$ in the coordinates is given by

$$\delta S = \int_{t_1}^{t_2} (\frac{\partial L}{\partial q} - \frac{d}{dt}\frac{\partial L}{\partial q'})\delta q dt + \frac{\partial L}{\partial q'}\delta q|_1^2$$

43. The symplectic form $\sum_i dp_i \wedge dq_i$ on the cotangent manifold can be expressed as the exterior derivative of a one form. This one form is given by

ans: $\sum_i p_i dq^i$

44. Noether's theorem states that

ans:If the Lagrangian is invariant under a group of transformations, then a current is conserved.

45. If ϕ is a field and $L(\phi)$ is a Lagrangian function for the field, then we can view $L(\phi)$ as an n form on the manifold where n is the dimension of the manifold. Then the variational equation for L can be expressed in the language of differential forms as

$$\delta L = DL + d\gamma$$

where δL and DL are linear in the field variations and γ is also linear in the field variation but is an $n-1$ form on the manifold. Thus, for a given field variation, δL and DL are densities and so is $d\gamma$.

46. Suppose L is a Lagrangian density and ξ is a vector field on the manifold. Suppose α_ξ is an $n-1$ form such that

$$i_\xi \delta L = d\alpha_\xi$$

47. Consider the covariant derivative $D_\mu = \partial_\mu + iA_\mu$ where A_μ are real functions of space-time. Let ψ be the wave function of the quantum mechanical particle moving in the electromagnetic field determined by the four potential A_μ. If the Lagrangian density is $D^\mu\psi(D_\mu\psi)^*$ then derive the wave equation $D^\mu D_\mu\psi = 0$ from the action principle and explain the fact that when ψ undergoes the transformation $\psi \rightarrow exp(i\alpha)\psi$ and simultaneously, A_μ undergoes the gauge transformation $A_\mu \rightarrow A_\mu + \partial_\mu\alpha$, the Lagrangian is invariant. What are the physical implications of this fact ?

48. Let $\mathbf{J}(\mathbf{r})$ be the phasor current density. Then the far field vector potential is given by

ans:

$$\mathbf{A}(\mathbf{r}) = \frac{\mu.exp(-jkr)}{4\pi r}\int \mathbf{J}(\mathbf{r}')exp(jk\hat{r}.\mathbf{r}')d^3r'$$

49. For the far field magnetic vector potential determined by problem 48, calculate the far field electric and magnetic fields.

50. The Lie bracket between two vector fields $X = \sum X^i\partial_i$ and $Y = \sum_i Y^i\partial_i$ on a differentiable manifold is given in terms of components as

ans:

$$[X, Y]^i = \sum_j (X^j\partial_i Y^j - Y^j\partial_j X^i)$$

51. Let M, N be manifolds and $u : M \to N$ a map. Then the pullback induced by u of a function $f : N \to \mathbb{R}$ is the function $g : M \to \mathbb{R}$ defined by the equation

ans: $g = fou$

52. Let M, N be smooth manifolds and $u : M \to N$ a smooth map. Then the pushforward of a tangent vector $v \in TM_x$ to $TM_{u(x)}$ is given in terms of coordinates by

ans:

$$v^i = \sum_j \frac{\partial u^i(x)}{\partial x^j} v^j$$

53. With M, N, u as in problem 52, the pullback of a one form $\alpha \in TN^*_{u(x)}$ to TM^*_x is given by the formula

$$\beta(v) = \alpha(T_x u(v)), v \in TM_x$$

where $T_x u$ is the differential of u at x. In terms of components,

$$\beta_j v^j = \alpha_i \frac{\partial u^i}{\partial x^j} v^j$$

or

$$\beta_j = \alpha_i \frac{\partial u^i}{\partial x^j}$$

53. If T, S are linear operators in a finite dimensional vector space that commute with each other, and both are diagonable, then what can be said about them ?

ans: Both T, S will be simultaneously diagonable, ie, there is a basis B for the vector space such that $[T]_B$ and $[S]_B$ are both diagonal. Equivalently, if the dimension of the vector space V over the field $\mathbb{F}$ is n, then there exists an isomorphism $U : V \to \mathbb{F}^n$ such that both UTU^{-1} and USU^{-1} are diagonal.

54. The spherical wave solution to the wave equation in three dimensions

$$c^2 \nabla^2 \psi(t, r) - \frac{\partial^2 \psi(t, r)}{\partial t^2} = 0$$

has the form

ans:

$$\psi(t, r) = f(t - r/c)/r$$

55. The eigenfunctions of the angular part of the Laplacian operator in three dimensions are known as and denoted by

ans:known as spherical harmonics and denoted by $Y_{lm}(\theta, \phi), l = 0, 1, 2, ..., m = -l, -l+1, ..., l-1, l$.

56. Let $x \to \pi(x)$ be a representation of a group G in a finite dimensional vector space V. Then for any function $f : G \to \mathbb{C}$, define the linear operator $T_f = \int_G f(x)\pi(x)dx$ in V, where dx is the left invariant Haar measure on G. Then $T_f T_h$ is given by

ans: T_{f*h} where $f * h$ is the convolution of f and h.

57. The expression for the gradient operator in cylindrical coordinates is given by

ans:

$$\nabla = \hat{\rho}.\frac{\partial}{\partial \rho} + \frac{\hat{\phi}}{\rho}\frac{\partial}{\partial \phi} + \hat{z}\frac{\partial}{\partial z}$$

58. Suppose q_1, q_2, q_3 is an orthogonal curvilinear coordinate system. Define the Lame coefficients for this system as

$$H_i = |\frac{\partial \mathbf{r}}{\partial q_i}|, i = 1, 2, 3$$

Let $\mathbf{e}_i$ denote the unit vector along the direction of q_i. Then the gradient operator in the curvilinear system is given by

ans:

$$\nabla = \sum_i (\mathbf{e}_i/H_i)\frac{\partial}{\partial q_i}$$

59. For the cylindrical coordinate system, the Lame coefficients are given by

ans:

$$H_\rho = 1, H_\phi = \rho, H_z = 1$$

60. Suppose $a(x), a^*(x), x \in \mathbb{R}^3$ are operators in a Hilbert space corresponding respectively to the annihilation and creation fields of bosons. Then, their commutation relations are given by

ans:

$$[a(x), a^*(y)] = \delta(x - y)$$

Suppose $f : \mathbb{R}^3 \to \mathbb{C}$ is a function and we define

$$a(f) = \int f(x)a(x)d^3x, a^*(f) = \int \bar{f}(x)a^*(x)d^3x$$

Then the commutation relations for these fields is given by ans:

$$[a(f), a^*(g)] = \int f(x)\bar{g}(y)\delta(x - y)d^3xd^3y = \int f(x)\bar{g}(x)d^3x =< f, g >$$

61. In the quantum harmonic oscillator corresponding to a natural frequency of ω and a mass m, the Heisenberg equations of motion are

ans:

$$dq/dt = i[H, q] = p/m, dp/dt = i[H, p] = -m\omega^2 q$$

62. The differential equations governing the electric field of a plane electromagnetic wave of frequency ω in free space are

ans:

$$div\mathbf{E} = 0, \nabla^2\mathbf{E} + \omega^2\mathbf{E}/c^2 = \mathbf{0}$$

63. The differential equations governing the electric field in a plane electromagnetic wave of wave number **k** and frequency ω translate to algebraic equations that are given by

ans:

$$\mathbf{k.E} = 0, |\mathbf{k}|^2 = \omega^2/c^2$$

64. What do you understand by a Lie group of transformations acting on a differentiable manifold? Each element of the Lie algebra induces a vector field on the manifold. Explain this notion.

65. What do you understand by a super algebra ? What do the commutations in a super algebra look like for even and odd elements ? A super algebra is a super vector space with a product such that the product of even elements is even, the product of odd elements is even and the product of an even and an odd element is odd.

66. What is a commutative superalgebra ?

ans: $A = A_0 \oplus A_1$ where A_0 are formed from the even elements, A_1 are formed from the odd elements and if $p(a)$ denotes the parity of a, ie, $p(a_0) = 0, p(a_1) = 1$, then

$$ab = (-1)^{p(a)p(b)}ba$$

67. What is a morphism on a superalgebra $A(p|q)$?

ans:First note that $A(p|q)$ is a super commutative algebra whose underlying vector space has a basis of $p+q$ elements with the first p elements $e_1, ..., e_p$ being even and the last q elements $f_1, ..., f_q$ being odd. A point in this vector space has the form $\sum_{i=1}^{p} t_i e_i + \sum_{i=1}^{q} \theta_i f_i$. Here, (t_i) are the even coordinates and (θ_i) are the odd coordinates. The superalgebra $A(p|q)$ is the space spanned by elements of the form $t_{i_1}...t_{i_r}\theta_{j_1}...\theta_{j_s}$ where $t_i's$ commute with all the $t's$ and $\theta's$ while $\theta's$ anticommute amongst themselves. For example

$$t_1\theta_1 t_2\theta_2 = t_1 t_2 \theta_1 \theta_2 = -t_1 t_2 \theta_2 \theta_1$$

Consider a matrix of size $p+q \times p+q$ expressed in block form

$$T = \begin{pmatrix} T_{11} & T_{12} \\ T_{21} & T_{22} \end{pmatrix}$$

where T_{11} is a $p \times p$ block consisting of even elements. T_{12} is a $p \times q$ matrix consisting of odd elements. T_{21} is a $q \times p$ matrix consisting of odd elements and finally T_{22} is a $q \times q$ matrix consisting of even elments. Then T can operate on a super vector to give a super vector bearing in mind the even-odd rule, namely, the product of two even elements is even, the product of two odd elements is even and the product of an even and an odd element is odd.

68. Describe the interaction of an atom with a radiation field using the Dirac equation. What is the interaction Hamiltonian ?

Hint: The Dirac Hamiltonian is given by

$$H = (\alpha, p + eA) + \beta m - eA_0$$

So the interaction part is given by the Hamiltonian

$$H_I = e(\alpha, A) - eA_0$$

and all transition probabilities between stationary states of the atom must be calculated using this Hamiltonian.

69. What do you understand by a connection on a Riemannian manifold M?.

Ans: Let $\chi(M)$ denote the space of vector fields on M. Then a connection is a map $\nabla : \chi(M) \times \chi(M) \to \chi_M$ defined by $(X,Y) \to \nabla_X Y$ so that if $\mathcal{F}(M)$ denotes the space of all smooth functions on M, then

$$\nabla_{fX}(Y) = f\nabla_X Y$$

and

$$\nabla_X(fY) = X(f)Y + f\nabla_X Y$$

Another way to look at a connection is (Deligne et.al.: Quantum fields and strings) that if $\Omega^1(M)$ denotes the space of all one forms on M, then the connection is a map $\nabla : \chi(M) \to \Omega^1(M) \otimes \chi(M)$ so that

$$\nabla(fv) = df \otimes v + f\nabla v$$

Note that $v \in \chi(M)$. If $X \in \chi(M)$, then for $\alpha \in \Omega^1(M)$ and $v \in \chi(M)$, we define

$$< X, \alpha \otimes v >=< X, \alpha > v$$

where $< X, \alpha >= \alpha(X)$. It follows that

$$\nabla_X(fv) =< X, \nabla(fv) >=< X, df \otimes v + f\nabla v >= df(X)v + f\nabla_X v$$

$$= X(f)v + f\nabla_X v$$

and this coincides with our earlier definition. We can extend the domain of ∇ to the $\Omega(M)\otimes\chi(M)$ where $\Omega(M)$ is the space of all forms on M, ie, $\Omega(M) = \bigcup_{k\geq 0} \Omega^k(M)$. If $\alpha \in \Omega(M)$ and $v \in \chi(M)$, then we set

$$\nabla(\alpha.v) = d\alpha.v + (-1)^k \alpha\nabla v$$

where $\alpha \in \Omega^k(M)$. Note that $\Omega^0(M) = \mathcal{F}(M)$. The curvature F of a connection ∇ is defined by

$$\nabla^2 x = Fx, x \in \chi(M)$$

We wish to show that

$$F(X,Y) = [\nabla_X, \nabla_Y] - \nabla_{[X,Y]} = \nabla_X\nabla_Y - \nabla_Y\nabla_X - \nabla_{[X,Y]}$$

Let

$$\nabla x = \alpha \otimes y = \alpha.y$$

where we are using an abbreviated notation for the tensor product. Then

$$\nabla_X x = \alpha(X)y$$

$$\nabla\nabla_X x = \nabla(\alpha(X)y) = d\alpha(X).y + \alpha(X)\nabla y$$

so that

$$\nabla_Y\nabla_X x = d\alpha(X)(Y).y + \alpha(X)\nabla_Y y$$

Interchanging X and Y gives us

$$\nabla_X\nabla_Y x = d\alpha(Y)(X)y + \alpha(Y)\nabla_X y$$

Also

$$\nabla_{[X,Y]}x = \alpha([X,Y])y$$

Thus,

$$([\nabla_X, \nabla_Y] - \nabla_{[X,Y]})x$$
$$= d\alpha(Y)(X)y - d\alpha(X)(Y)y + \alpha(Y)\nabla_X y - \alpha(X)\nabla_Y y - \alpha([X,Y])y$$

On the other hand,

$$\nabla x = \alpha \otimes y = \alpha.y,$$
$$\nabla^2 x = d\alpha.y - \alpha\nabla y$$

so that

$$\nabla^2 y(X,Y) = d\alpha(X,Y).y - \alpha(X)\nabla_Y y + \alpha(Y)\nabla_X y$$

In terms of components,

$$\alpha(X) = \alpha_i X^i$$
$$d\alpha(X,Y) = \alpha_{i,j}(Y^j X^i - Y^i X^j)$$
$$d\alpha(Y)(X) = (\alpha_j Y^j)_{,i}X^i = \alpha_{j,i}Y^j X^i + \alpha_j X^i Y^j_{,i}$$
$$d\alpha(X)(Y) = (\alpha_j X^j)_{,i}Y^i = \alpha_{j,i}X^j Y^i + \alpha_j Y^i X^j_{,i}$$

and so,

$$d\alpha(Y)(X) - d\alpha(X)(Y) - \alpha([X,Y]) = \alpha_{j,i}(Y^j X^i - X^j Y^i) = d\alpha(X,Y)$$

which proves the claim.

71. Consider a Riemannian manifold M with metric tensor components in terms of local coordinates $(g_{ij}(x))$. For a trajectory $t \to X^i(t)$ on this manifold, the action functional is defined by $S[X] = \frac{1}{2}\int_{t_1}^{t_2} g_{ij}(X(t))X^{i'}(t)X^{j'}(t)dt$. Write down the equations of motion.

ans: The variation of the action is given by

$$\delta S[X] = \frac{1}{2}\int g_{ij,k}(X)X^{i'}X^{j'}\delta X^k dt + \int g_{ij}(X)X^{i'}\delta X^{j'}dt$$

Integrating the second term by parts and setting the coefficient of δX^k to zero gives the equations of motion as

$$-(g_{ik}X^{i'})' + \frac{1}{2}g_{ij,k}X^{i'}X^{j'} = 0$$

which can be expressed as

$$g_{ik}X^{i''} + g_{ik,j}X^{i'}X^{j'} - \frac{1}{2}g_{ij,k}X^{i'}X^{j'} = 0$$

or

$$g_{ik}X^{i''} + \frac{1}{2}(g_{ik,j} + g_{jk,i} - g_{ij,k})X^{i'}X^{j'} = 0$$

or

$$g_{ik}X^{i''} + \Gamma_{kij}X^{i'}X^{j} = 0$$

or equivalently,

$$DX^{i'}/Dt = 0$$

where D/Dt is the covariant derivative along the curve.

70. Write down the Schwarzchild metric in cartesian coordinates.

Ans: The Schwarzchild metric in spherical polars is given by

$$d\tau^2 = \alpha(r)dt^2 - \alpha(r)^{-1}dr^2 - r^2(d\theta^2 + sin^2(\theta)d\phi^2)$$

where

$$r = \sqrt{x^2 + y^2 + z^2}$$

We can write

$$d\tau^2 = \alpha(r)dt^2 - (\alpha(r)^{-1} - 1)dr^2 - (dx^2 + dy^2 + dz^2)$$

In these formulae,

$$\alpha(r) = 1 - 2GM/rc^2$$

Note that we've used the formula

$$dx^2 + dy^2 + dz^2 = dr^2 + r^2d\theta^2 + r^2sin^2(\theta)d\phi^2$$

Now note that

$$dr = (xdx + ydy + zdz)/r$$

so that

$$dr^2 = r^{-2}(x^2dx^2 + y^2dy^2 + z^2dz^2 + 2xydxdy + 2yzdydz + 2zxdzdx)$$

and putting all these together, we arrive at the Schwarzchild metric in cartesian form.

71.Show that the Schwarzchild metric in cartesian form can be put as

$$d\tau^2 = g_{tt}dt^2 + g_{xx}dx^2 + g_{yy}dy^2 + g_{zz}dz^2 + 2g_{xy}dxdy + 2g_{yz}dydz + 2g_{xz}dxdz$$

where $g_{tt}, g_{xx}, g_{yy}, g_{zz}, g_{xy}, g_{yz}, g_{xz}$ are all functions of (x, y, z) alone, not involving t.

72. Calculate the inverse of $(g_{\alpha\beta})$, the Schwarzchild metric in cartesian coordinates and show that its nonzero components are $g^{tt}, g^{xx}, g^{yy}, g^{zz}, g^{xy} = g^{yx}, g^{yz} = g^{zy}, g^{xz} = g^{zx}$ with all these being functions of x, y, z alone, ie, not involving time. Also calculate $g = det((g_{\alpha\beta}))$

73. Let $A^0 = V, A^1 = A_x, A^2 = A_y, A^3 = A_z$ denote the electromagnetic contravariant four potential. The components of the covariant four potential are given by

$$A_0 = g_{tt}V, A_1 = g_{xx}A_x + g_{xy}A_y + g_{xz}A_z,$$

$$A_2 = g_{xy}A_x + g_{yy}A_y + g_{yz}A_z, A_3 = g_{xz}A_x + g_{yz}A_y + g_{zz}A_z$$

Calculate the components of the covariant antisymmetric field tensor.

Ans:

$$F_{01} = F_{tx} = A_{1,0} - A_{0,1} = A_{1,t} - A_{0,x} = (g_{xx}A_x + g_{xy}A_y + g_{xz}A_z)_{,t} - (g_{tt}V)_{,x}$$

$$= j\omega(g_{xx}A_x + g_{xy}A_y + g_{xz}A_z) - g_{tt,x}V - g_{tt}V_{,x}$$

where sinusoidal dependence with frequency ω has been assumed. Likewise,

$$F_{02} = j\omega(g_{xy}A_x + g_{yy}A_y + g_{yz}A_z) - g_{tt,y}V - g_{tt}V_{,y}$$

and

$$F_{03} = j\omega(g_{xz}A_x + g_{yz}A_y + g_{zz}A_z) - g_{tt,z}V - g_{tt}V_{,z}$$

$$F_{12} = A_{2,1} - A_{1,2} = (g_{xy}A_x + g_{yy}A_y + g_{yz}A_z)_{,x} - (g_{xx}A_x + g_{xy}A_y + g_{xz}A_z)_{,y}$$

$$= (g_{xy,x} - g_{xx,y})A_x + (g_{yy,x} - g_{xy,y})A_y + (g_{yz,x} - g_{xz,y})A_z +$$

$$g_{xy}A_{x,x} + g_{yy}A_{y,x} + g_{yz}A_{z,x} - g_{xx}A_{x,y} - g_{xy}A_{y,y} - g_{xz}A_{z,y}$$

and likewise,

$$F_{23} =$$

74. Yang-Mills fields in the classical theory: G is a matrix Lie group $\}$ is the Lie algebra of G. $A = A_\mu(x)dx^\mu$ is a $\}$ valued one form on $\mathbb{R}^4$. Thus, for each $x \in \mathbb{R}^4$, $A_\mu(x)$ is a matrix. Define the connection ∇ so that

$$\nabla = \partial + A$$

or in terms of components,

$$\nabla_\mu = \partial_\mu + A_\mu$$

Let ψ be a vector wave function on $\mathbb{R}^4$ having $dimG$ components. G therefore acts on such wave functions. A gauge transformation of a wave function ψ takes this wave function to $g(x)\psi(x)$ where $g : \mathbb{R}^4 \to G$ is a space-time dependent group element. Under this transformation, suppose that the gauge fields A_μ transform to $A_\mu(x) + \xi_\mu(x)$. We require that under this gauge transformation, $\nabla_\mu\psi$ transforms to $g\nabla_\mu\psi$. Thus, we must have

$$g(\partial_\mu + A_\mu)\psi = (\partial_\mu + A_\mu + \xi_\mu)g\psi$$

Note that g depends on x. It follows that

$$g\partial_\mu + gA_\mu = \partial_\mu g + g\partial_\mu + A_\mu g + \xi_\mu g$$

and hence

$$gA_\mu = \partial_\mu g + (A_\mu + \xi_\mu)g$$

or

$$A_\mu + \xi_\mu = gA_\mu g^{-1} - (\partial_\mu g)g^{-1}$$

In other words, under the gauge transformation $g(x)$, the gauge fields undergo the transformation

$$A_\mu \to gA_\mu g^{-1} - (\partial_\mu g)g^{-1}$$

Define the field F so that

$$F = 2dA + A \wedge A$$

where

$$A \wedge A = [A_\mu, A_\nu]dx^\mu \wedge dx^\nu$$

Note that

$$[\nabla_\mu, \nabla_\nu] = [\partial_\mu + A_\mu, \partial_\nu + A_\nu]$$

$$= A_{\nu,\mu} - A_{\mu,\nu} + [A_\mu, A_\nu] = F_{\mu\nu}$$

so that

$$\nabla \wedge \nabla = [\nabla_\mu, \nabla_\nu]dx^\mu \wedge dx^\nu$$

$$= (A_{\nu,\mu} - A_{\mu,\nu})dx^\mu \wedge dx^\nu + A \wedge A$$

$$= 2dA + A \wedge A$$

we have

$$F = F_{\mu\nu}dx^\mu \wedge dx^\nu$$

Then, let us derive the field equations from the action principle.

$$F_{\mu\nu}F^{\mu\nu} = g^{\mu\alpha}g^{\nu\beta}(A_{\nu,\mu} - A_{\mu,\nu} + [A_\mu, A_\nu])(A_{\beta,\alpha} - A_{\alpha,\beta} + [A_\alpha, A_\beta])$$

$$= g^{\mu\alpha}g^{\nu\beta}(A_{\nu,\mu} - A_{\mu,\nu} + [A_\mu, A_\nu])F_{\alpha\beta}$$

and the variation of its trace is therefore given by

$$\delta Tr(F_{\mu\nu}F^{\mu\nu}) = 2g^{\mu\alpha}g^{\nu\beta}Tr(\delta A_{\nu,\mu} - \delta A_{\mu,\nu} + ((\delta A_\mu)A_\nu + A_\mu\delta A_\nu - (\delta A_\nu)A_\mu - A_\nu\delta A_\mu)F_{\alpha\beta})$$

On getting rid of total differentials (ie, divergences), this becomes

$$2g^{\mu\alpha}g^{\nu\beta}Tr(-\delta A_\nu.F_{\alpha\beta,\mu} + \delta A_\mu.F_{\alpha\beta,\nu} + \delta A_\mu A_\nu F_{\alpha\beta} + \delta A_\nu F_{\alpha\beta}A_\mu$$

$$-\delta A_\nu A_\mu F_{\alpha\beta} - \delta A_\mu F_{\alpha\beta}A_\nu)$$

Setting the coefficient of δA_μ to zero results in the field equations

$$-g^{\nu\alpha}g^{\mu\beta}F_{\alpha\beta,\nu} + g^{\mu\alpha}g^{\nu\beta}F_{\alpha\beta,\nu} + g^{\mu\alpha}g^{\nu\beta}[A_\nu, F_{\alpha\beta}]$$

$$+g^{\nu\alpha}g^{\mu\beta}[F_{\alpha\beta}, A_\nu] = 0$$

This simplifies, using the antisymmetry of $F^{\mu\nu}$ and $F_{\mu\nu}$ to

$$F^{\mu\nu}_{,\nu} + [A_\nu, F^{\mu\nu}] = 0$$

These are the nonlinear field equations in the Yang-Mills theory.

75. Suppose V is a vector space and $f : V \to \mathbb{C}$ a smooth function. For $Y \in V$, define the operator $\partial(Y)$ so that

$$\partial(Y)f(X) = \frac{d}{dt}f(X+tY)|_{t=0}$$

Then show that $\partial(Y)\partial(Z) = \partial(Z)\partial(Y)$ for all $Y, Z \in V$.

ans:

$$\partial(Z)\partial(Y)f(X) = \frac{\partial^2}{\partial t \partial s}f(X+tY+sZ)|_{t=s=0}$$
$$= \partial(Y)\partial(Z)f(X)$$

ref:HarishChandra, collected papers vol.II, edited by V.S.Varadarajan.

76. The context is that of the previous problem. Let $B : V \times V \to \mathbb{C}$ be a symmetric, non-degenerate bilinear form. For $X \in V$, let f_X denote the linear functional

$$f_X(Y) = B(X,Y)$$

ie,

$$f_X = B(X,.)$$

Let a Lie group G act on V smoothly. For any function $f \in C^\infty(V)$, define the function $f^x \in C^\infty(V)$ so that

$$f^x(X) = f(x^{-1}X)$$

For $X \in V$, let $\partial(X)^x = \partial(xX)$. Then we have

$$\partial(X)^x f(Y) = \frac{d}{dt}f(Y+txX)|_{t=0}$$

$$= \frac{d}{dt}f(x(x^{-1}Y+tX))|_{t=0} = \frac{d}{dt}f^{x^{-1}}(x^{-1}Y+tX)|_{t=0}$$
$$= (\partial(X)f^{x^{-1}})^x$$

More formally, we can write this as follows. Define the operator τ_x acting on functions defined on V so that $\tau_x f(X) = f(x^{-1}X)$. Then,

$$\partial(X)^x = \tau_x \partial(X)\tau_x^{-1}$$

Since $\tau_{xy} = \tau_x\tau_y$, it is immediate from this formula that

$$\partial(X)^{xy} = \tau_{xy}\partial(X)\tau_{xy}^{-1} = \tau_x\tau_y\partial(X)\tau_y^{-1}\tau_x^{-1}$$

$$= \tau_x \partial(X)^y \tau_x^{-1} = (\partial(X)^x)^y$$

77. Express the wave operator

$$\square = \frac{\partial^2}{\partial x^2} + \frac{\partial^2}{\partial y^2} + \frac{\partial^2}{\partial z^2} - \frac{\partial^2}{\partial t^2}$$

in four dimensional notation.

Ans: $(x^\mu) = (x, y, z, t)$,

$$(\partial_\mu) = (\frac{\partial}{\partial x}, \frac{\partial}{\partial y}, \frac{\partial}{\partial z}, \frac{\partial}{\partial t})$$

and

$$(x_\mu) = (x, y, z, -t)$$

$$(\partial^\mu) = (\frac{\partial}{\partial x}, \frac{\partial}{\partial y}, \frac{\partial}{\partial z}, -\frac{\partial}{\partial t})$$

Note that the metric tensor is given by

$$(g_{\mu\nu}) = (1, 1, 1, -1)$$

so that

$$x_\mu = g_{\mu\nu} x^\nu, \partial_\mu = g_{\mu\nu} \partial^\nu$$

78. Write down the Feynman path integral expression for a particle moving in one dimension in a potential $V(q)$ in discretized form between the times $t = 0$ and $t = T$. Assume that $q(0) = a, q(T) = b$.

ans:The action functional is given by

$$S[q] = \int_0^T (\frac{mq'^2(t)}{2} - V(q(t)))dt$$

which in discretized form reads

$$S[q] \approx \sum_{n=0}^{N-1} (\frac{m}{2\Delta}(q[n+1] - q[n])^2 - V(q[n])\Delta)$$

where

$$N\Delta = T, q[n] = q(n\Delta)$$

It follows that the path integral is approximated by the n-variate integral

$$\int exp(\frac{2\pi i}{h} \sum_{n=0}^{N-1} (\frac{m}{2\Delta}(q[n+1] - q[n])^2 - V(q[n])\Delta))dq[1]dq[2]...dq[N-1]$$

where $q[0] = a, q[N] = b$

79.Derive the quadratic (Gaussian) approximation to the Feynman path integral for a particle moving in a potential $V(q)$.

ans: The Lagrangian has the form

$$\mathcal{L}(q, q') = mq'^2/2 - V(q)$$

Let $q_0(t)$ denote the classical path and $\delta q(t)$ the variation of the path about the classical path. Then, we have upto second order

$$\delta S[q] = \int_0^T [m\delta q'^2(t)/2 - V''(q_0(t))\delta q(t)^2/2]dt$$

so that the discretized action with $\xi(t) = \delta q(t)$ is given upto quadratic terms by the expression

$$\delta S[q_0, \xi] \approx \sum_{n=0}^{N-1}(m(\xi[n+1] - \xi[n])^2/2\Delta - A[n]\xi[n]^2\Delta/2)$$

with

$$A[n] = V''(q_0[n])$$

The significant contribution to the path integral comes in the vicinity of the classical path since away from the classical path, the action is no longer stationary and hence the phase in the path integrand varies rapidly producing cancellations. Thus the path integral is approximately given by a Gaussian integral with imaginary covariance:

$$\int_{\mathbb{R}^{N-1}} exp(\frac{2\pi i}{h}\sum_{n=0}^{N-1}(m(\xi[n+1] - \xi[n])^2/2\Delta - A[n]\xi[n]^2\Delta/2)\Pi_{n=1}^{N-1}d\xi[n]$$

80. Consider a time varying linear system defined by the differential equation

$$\frac{d\mathbf{X}(t)}{dt} = \mathbf{A}(t)\mathbf{X}(t) + \mathbf{Z}(t)$$

Here, $\mathbf{X}(t)$ is an $n \times 1$ vector for each $t \in \mathbb{R}$ and for each $t \in \mathbb{R}$, $\mathbf{A}(t)$ is an $n \times n$ matrix. $\mathbf{Z}(t)$ is an $n \times 1$ vector. If $\mathbf{A}(t) = \mathbf{A}$ is a constant matrix, then the state transition matrix for this system is simply $\Phi(t) = exp(t\mathbf{A})$ and the solution can be expressed as

$$\mathbf{X}(t) = \int_0^t \Phi(t-s)\mathbf{Z}(s)ds$$

assuming zero initial condition. In the general case when $\mathbf{A}(t)$ is a non-constant matrix, the state transition matrix $\Phi(t, s)$ satisfies the differential equation

$$\frac{\partial \Phi(t, s)}{\partial t} = \mathbf{A}(t)\Phi(t, s), t \geq s$$

with the initial condition

$$\Phi(s,s) = \mathbf{I}$$

By repeated integration we get a series expansion for Φ:

$$\Phi(t,s) = \mathbf{I} + \sum_{n=1}^{\infty} \int_{s<t_n<t_{n-1}<...<t_2<t_1<t} \mathbf{A}(t_1)\mathbf{A}(t_2)...\mathbf{A}(t_n)dt_n dt_{n-1}...dt_2 dt_1$$

Note that if

$$M = sup_{u\in[s,t]} \parallel \mathbf{A}(u) \parallel$$

then

$$\parallel \Phi(t,s) \parallel \leq exp((t-s)M)$$

The solution to the state equations can now be expressed as

$$\mathbf{X}(t) = \Phi(t,0)\mathbf{X}(0) + \int_s^t \Phi(t,u)\mathbf{Z}(u)du$$

81. We are given a quantum Hamiltonian

$$H = -\frac{1}{2}\frac{d^2}{dx^2} + V(x)$$

and we wish to express the partition function $Z(\beta) = Tr(exp(-\beta H))$ as a path integral. Let $f(x)$ be a function and consider the function

$$u(\beta, x) = exp(\beta(0.5D^2 - V))f(x), D = d/dx$$

Then

$$u_{,\beta} = (0.5D^2 - V)u$$

This is the heat equation with a potential term. Let $W(t)$ be the standard Brownian motion process and consider

$$v(t,x) = \mathbb{E}[f(x+W(t))exp(\int_0^t V(x+W(s))ds]$$

we have

$$v(0,x) = f(x)$$

Here $W(0) = 0$. Suppose we drop the constraint $W(0) = 0$. Then, we have

$$v(t,x) = \mathbb{E}(f(W(t)).exp(\int_0^t V(W(s))ds)|W(0) = x)$$

consider

$$v(s,t,x) = \mathbb{E}(f(W(t)).exp(\int_s^t V(W(u))du)|W(s) = x)$$

Then,

$$v(s,t,x) = (1 + h.V(x))\mathbb{E}\mathbb{E}(f(W(t)).exp(\int_{s+h}^{t} V(W(u))du)|W(s+h))|W(s) = x)$$
$$= (1 + h.V(x))\mathbb{E}(v(s+h,t,W(s+h))|W(s) = x)$$

where $h \to 0$ and $o(h)$ terms have been neglected. Hence application of Ito's rule gives

$$\frac{\partial v(s,t,x)}{\partial s} + 0.5\frac{\partial^2 v(s,t,x)}{\partial x^2} + V(x)v(s,t,x) = 0$$

It is obvious that v depends on s and t through the difference $t-s$. Hence writing $v(s,t,x) = v(t-s,x)$, it follows that

$$\frac{\partial v(t,x)}{\partial t} = 0.5\frac{\partial^2 v(t,x)}{\partial x^2} + V(x)v(t,x)$$

which proves that

$$v(t,x) = exp(t(0.5D^2 + V))f(x)$$

Let $f_n(x)$ be an orthonormal basis for $L^2(\mathbb{R})$. Then we have

$$v_n(t,x) = exp(t(0.5D^2 + V))f_n(x) = \mathbb{E}[f_n(x + W(t))exp(\int_0^t V(x + W(s))ds)]$$

hence

$$\sum_n < f_n, v_n(t,.) >= \sum_n < f_n, exp(t(0.5D^2 + V))f_n >$$

$$= \sum_n \int_{-\infty}^{\infty} dx\mathbb{E}[f_n(x)f_n(x + W(t))exp(\int_0^t V(x + W(s))ds)]$$

$$= \mathbb{E}[\delta(W(t))exp(\int_0^t V(x + W(s))ds)]$$

This can be expressed as a limit which yields a path integral formula for the partition function. Indeed the approximation to the above is

$$\int \Pi_{i=0}^{n-1}(2\pi\delta)^{-1/2}exp(-(W_{i+1} - W_i)^2/2\delta)exp(\sum_{i=0}^{n-1} V(x + W_i)\delta)dW_1...dW_n$$

where

$$W_0 = 0$$

82. Explain the notion of energy momentum tensor with respect to path integrals.

Suppose that the action functional is given by $S[\gamma, x] = \int_0^T \gamma_{ij}(x(t))x^{i'}(t)x^{j'}(t)dt$ where γ is the metric and x the particle trajectory. Then the variation with respect to γ gives the energy momentum tensor T_{ij}. Thus,

$$T^{ij} = v^i v^j$$

83. Consider the operator $L = -d^2/dx^2$ operating on $C^2[0, L]$ with vanishing boundary conditions. Determine the eigenfunctions and eigenvalues of this operator and hence evaluate the partition function $Tr(exp(-\beta L))$.

ans:

$$L\psi = \lambda\psi$$

gives

$$\psi'' = -\lambda\psi$$

the solutions to which for vanishing boundary conditions are

$$\psi(x) = A.sin(n\pi x/L), \lambda = (n\pi/L)^2$$

It follows that the partition function is

$$Z(\beta) = \sum_{n=1}^{\infty} exp(-\beta n^2\pi^2/L^2)$$

84. Consider a metric $\gamma_{ij}(x)$ on a p dimensional Riemannian manifold. The Lagrangian of a particle is given by

$$L =< v, v >_\gamma /2 = \gamma_{ij}v^i v^j/2 = \gamma_{ij}(x)x^{i'}x^{j'}/2$$

Compute the Hamiltonian function for this and write down the corresponding Schrodinger equation.

ans: The canonical momenta are given by

$$p_i = \frac{\partial L}{\partial x^{i'}} = \gamma_{ij}x^{j'}$$

solving this gives the generalized velocities as

$$x^{i'} = \gamma^{ij}p_j$$

and hence the Hamiltonian is given by

$$H(x, p) = p_i x^{i'} - L = \gamma^{ij}(x)p_i p_j/2$$

The symmetrized Hamiltonian operator is given by

$$H(x, p) = \frac{1}{2}p_i\gamma_{ij}(x)p_j$$

where now the $p_i's$ are the operators $-i\partial/\partial x^i$.

85.What do you understand by the interaction representation in quantum mechanics.

Ans:Suppose that the Hamiltonian of the system can be written as $H(t) = H_0 + V(t)$. Then the Schrodinger equation for the wave function is

$$i\psi'(t) = (H_0 + V(t))\psi(t)$$

Define the modified state vector

$$\phi(t) = exp(itH_0)\psi(t)$$

or equivalently,

$$\psi(t) = exp(-itH_0)\phi(t)$$

Then from the Schrodinger equation,

$$i\phi'(t) = V_I(t)\phi(t)$$

where

$$V_I(t) = exp(itH_0)V(t)exp(-itH_0)$$

is the interaction in the new representation. Suppose $V(.)$ does not vary with time. Then $V_I(t) = exp(itH_0)V.exp(-itH_0)$ The evolution operator in the interaction representation $U_I(t,s)$ satisfies $U_I(s,s) = I$ and

$$i\frac{\partial U_I(t,s)}{\partial t} = V_I(t)U_I(t,s)$$

We wish to solve this. Note that

$$\psi(t) = exp(-itH)\psi(0) = exp(-itH)\phi(0)$$

so that

$$\phi(t) = exp(itH_0)\psi(t) = exp(itH_0)exp(-itH)\phi(0)$$

It follows from this that

$$U_I(t,0) = exp(itH_0)exp(-itH)$$

More generally for $t > s$, we have

$$\psi(t) = exp(-i(t-s)H)\psi(s) = exp(-i(t-s)H)exp(-isH_0)\phi(s)$$

so that

$$\phi(t) = exp(itH_0)\psi(s) = exp(itH_0)exp(-i(t-s)H)exp(-isH_0)\phi(s)$$

giving

$$U_I(t,s) = exp(itH_0)exp(-i(t-s)H)exp(-isH_0)$$

or equivalently,

$$U_I(t,s) = exp(itH_0)exp(-itH)exp(isH)exp(-isH_0)$$

for the evolution operator in the interaction representation. We can check directly that $U_I(t,s)$ satisfies the Schrodinger equation in the interaction representation. We get on differentiation,

$$\frac{\partial U_I(t,s)}{\partial t} = exp(itH_0)i(H_0 - H)exp(-i(t-s)H)exp(-isH_0)$$

$$= -i.exp(itH_0)V.exp(isH)exp(-isH_0) = -i.exp(itH_0)V.exp(-i(t-s)H)exp(-isH_0)$$
$$= -iV_I(t)U_I(t,s)$$

proving the claim.

86. **Derive the formula for the classical action of a harmonic oscillator between two specified endpoints (t_a, q_a) and (t_b, q_b).**

ans: Equation of motion:

$$q''(t) = -\omega^2 q(t)$$

solution:

$$q(t) = A.cos(\omega t) + B.sin(\omega t)$$
$$q' = -A\omega.sin(\omega t) + B\omega.cos(\omega t)$$
$$\int_{t_a}^{t_b} q'^2(t)dt$$

The constants A, B are determined from the equations

$$q_a = q(t_a) = A.cos(\omega t_a) + B.sin(\omega t_a)$$

$$q_b = q(t_b) = A.cos(\omega t_b) + B.sin(\omega t_b)$$

in matrix notation,

$$\begin{pmatrix} q_a \\ q_b \end{pmatrix}$$

$$= \begin{pmatrix} cos(\omega t_a) & sin(\omega t_a) \\ cos(\omega t_b) & sin(\omega t_b) \end{pmatrix} \begin{pmatrix} A \\ B \end{pmatrix}$$

so that

$$A = (q_a sin(\omega t_b) - q_b sin(\omega t_a))/sin(\omega(t_b - t_a))$$

and likewise,

$$B = (q_a cos(\omega t_b) - q_b cos(\omega t_a))/sin(\omega(t_a - t_b))$$

The action integral is

$$S[q] = \int_{t_a}^{t_b} (mq'^2(t)/2 - m\omega^2 q^2(t)/2)dt$$

We calculate

$$q'(t) = -A\omega.sin(\omega t) + B\omega.cos(\omega t)$$
$$q'^2(t) = A^2\omega^2 sin^2(\omega t) + B^2\omega^2 cos^2(\omega t) - 2AB\omega^2 sin(\omega t)cos(\omega t)$$
$$= A^2\omega^2(1 - cos(2\omega t))/2 + B^2\omega^2(1 + cos(2\omega t))/2 - AB\omega^2 sin(2\omega t)$$

It follows that

$$\int_{t_a}^{t_b} q'^2(t)dt = (A^2\omega^2/2)(t_b - t_a - 0.5sin(2\omega t_b) + 0.5sin(2\omega t_a))/2 + B^2\omega^2(t_b - t_a + 0.5sin(2\omega t_b) - 0.5sin(2\omega t_a))$$

$$+AB\omega(cos(2\omega t_b) - cos(2\omega t_a))/2$$

Likewise, we can evaluate

$$\int_{t_a}^{t_b} q^2(t)dt == \int_{t_a}^{t_b} (A.cos(\omega t) + B.sin(\omega t))^2 dt$$

$$= A^2 \int_{t_a}^{t_b} cos^2(\omega t)dt + B^2 \int_{t_a}^{t_b} sin^2(\omega t)dt + AB \int_{t_a}^{t_b} sin(2\omega t)dt$$

$$= (A^2/2) \int_{t_a}^{t_b} (1 + cos(2\omega t))dt + (B^2/2) \int_{t_a}^{t_b} (1 - cos(2\omega t))dt + AB \int_{t_a}^{t_b} sin(2\omega t)dt$$

$$= (A^2/2)(t_b - t_a + (sin(2\omega t_b) - sin(2\omega t_a))/2) + (B^2/2)(t_b - t_a - (sin(2\omega t_b) - sin(2\omega t_a))/2)$$

$$-AB(cos(2\omega t_b) - cos(2\omega t_a))$$

Substituting the values of A, B in terms of t_a, q_a, t_b, q_b will give us the desired formula for the action in terms of these endpoint time and spatial coordinates.

87. Explain the Fourier representation of the action integral for the harmonic oscillator:

We let $q_0(t)$ denote the classical trajectory and $q(t) = q_0(t) + \delta q(t)$ the perturbed trajectory. At the end points we have $\delta q(t) = 0$. We assume that motion takes place over the interval $t \in [0, T]$. The action after subtracting out the classical action.

88. Determine the fundamental equation for electromagnetic wave propagation in a nonlinear medium, ie, when the permittivity and permeability are non-linear memoryless functions of the electric and magnetic field vectors.

89.Determine the fundamental wave propagation equation in an anisotropic but homogeneous medium, ie, when the permittivity and permeability are constant 3×3 matrices.

ans:The relevant Maxwell equations are written down with the substitutions

$$\mathbf{B} = \mu.\mathbf{H}, \mathbf{D} = \epsilon.\mathbf{E}$$

where μ, ϵ are 3×3 constant matrices. This means that

$$B_x = \mu_{xx}H_x + \mu_{xy}H_y + \mu_{xz}H_z,$$

$$B_y = \mu_{yx}H_x + \mu_{yy}H_y + \mu_{yz}H_z,$$

$$B_z = \mu_{zx}H_x + \mu_{zy}H_y + \mu_{zz}H_z$$

and

$$D_x = \epsilon_{xx}E_x + \epsilon_{xy}E_y + \epsilon_{xz}E_z,$$

$$D_y = \epsilon_{yx}E_x + \epsilon_{yy}E_y + \epsilon_{yz}E_z,$$

$$D_z = \epsilon_{zx}E_x + \epsilon_{zy}E_y + \epsilon_{zz}E_z$$

For sinusoidal time dependence, it is instructive to write down these equations.

90. Write down the expression of the partition function of a quantum system having Hamiltonian operator H at temperature $T = 1/k\beta$ and hence derive the formulae for the average energy and mean square energy fluctuations in the quantum canonical ensemble.

ans:

Partition function:

$$Z(\beta) = Tr(exp(-\beta H))$$

Average energy:

$$< U >= Tr(H.exp(-\beta H))/Z(\beta) = -\frac{d}{d\beta}log(Z(\beta))$$

Mean square energy fluctuations:

$$\sigma_U^2 =< (U- < U >)^2 >=< U^2 > - < U >^2$$

$$< U^2 >= Tr(H^2 exp(-\beta H))/Z(\beta) = Z''(\beta)/Z(\beta)$$

Hence,

$$\sigma_U^2 = Z''(\beta)/Z(\beta) - (Z'(\beta)/Z(\beta))^2 = (ZZ'' - Z'^2)/Z^2 = (\log Z)''(\beta)$$

90. What is a maximal compact subgroup of a semisimple Lie group ? What is a conjugation on a Lie algebra ? What is a compact real form. $\}$ is a Lie algebra of a semisimple Lie group G.

Let $\}_0$ be a real form of $\}$ and let η be the conjugation relative to $\}_0$. Let $\sqcap$ be a compact real form of $\}$ and let η be the conjugation relative to $\sqcap$. Define $\approx_0 = \}_0 \cap \sqcap$ and $\surd_0 = \}_0 \cap i\sqcap$. Then $\approx_0$ is a Lie subalgebra of $\}_0$ and if K denotes the analytic subgroup of $\approx_0$, then K is a compact subgroup of $\}_0$. We have the direct sum decomposition $\}_0 = \approx_0 \oplus \surd_0$ and let $\theta_1 = \eta o \theta$. Let $\approx$ be the complexification of $\approx_0$ and $\surd$ the complexification of $\surd_0$. Then, $\} = \approx \oplus \surd$. Let $\theta_1 = \eta o \theta$. Then for $X \in \approx, Y \in \surd$, we have $\theta_1(X+Y) = X - Y$. We have

$$\approx_0 + i\surd_0 = \}_0 \cap \sqcap + i\}_0 \cap \sqcap = (\}_0 + i\}_0) \cap \sqcap$$

$$= \} \cap \sqcap = \sqcap$$

We also note that

$$[\approx_0, \approx_0] \subset \approx_0$$

since $\approx_0$ is a Lie algebra. Further,

$$[\approx_0, \surd_0] = [\}_0 \cap \sqcap, \}_0 \cap i\sqcap] \subset i[\sqcap, \sqcap] \cap \}_0$$

$$= \subset i\sqcap \cap \}_0 = \surd_0$$

and finally,

$$[\surd_0, \surd_0] = [i\sqcap \cap \}_0, i\sqcap \cap \}_0]$$

$$\subset [\sqcap, \sqcap] \cap \}_0 \subset \sqcap \cap \}_0 = \approx_0$$

91. What do you understand by a group of transformations that preserve a bilinear form on a vector space.

ans:Let V be the vector space and $f : V \times V \to \mathbb{C}$ a non-degenerate bilinear form. Let G denote the set of all $T \in \mathcal{L}(V)$ such that $f(Tx, Ty) = f(x, y)\ \forall x, y \in V$. Then, G is a group. Indeed, suppose $T_1, T_2 \in G$. Then, $f(T_2T_1x, T_2T_1y) = f(T_1x, T_1y) = f(x, y)\ \forall x, y \in V$ so G is closed under composition. To prove that T^{-1} exists for all $T \in G$, we note that suppose $Tx = 0$. Then, $0 = f(Tx, Ty) = f(x, y)$ for all $y \in V$ implying that $x = 0$ since V is non-degenerate.

92. If L is a Lorentz transformation corresponding to a rotation R around the z axis by an angle θ followed by a boost in the x direction with a velocity v, then write down the matrix of L.

ans:

$$L = BR$$

where

$$R = \begin{pmatrix} 1 & 0 & 0 & 0 \\ 0 & cos(\theta) & -sin(\theta) & 0 \\ 0 & sin(\theta) & cos(\theta) & 0 \\ 0 & 0 & 0 & 1 \end{pmatrix}$$

93. Write down the DSP implementation of the Heat equation and solve it using z-transforms.

ans:

$$u_t(t, x) = D.u_{xx}(t, x)$$

is the original heat equation. Here, D is the diffusion coefficient. After discretizing in time with a step size of τ and in space with a step size of Δ, we get

$$u[n+1, k] - u[n, k] = (D\tau/\Delta^2)(u[n, k+1] - u[n, k] + u[n, k-1])$$

This difference equation can be solved by taking z transforms with respect to the variable k.

94.Suppose $\phi(t, x, y, z)$ is the Klein-Gordon field inside the unit cube with periodic boundary conditions. Develop a Fourier series expansion for this field.

ans: Set

$$\phi(t, x, y, z) = \sum_{n,m,p \in \mathbb{Z}} c(n, m, p, t).exp(2\pi i(nx + my + pz))$$

Plugging this into the differential equation

$$\nabla^2\phi - \partial_t^2\phi - M^2\phi = 0$$

gives us

$$(2\pi)^2(n^2+m^2+p^2)c(n,m,p,t)+\partial_t^2 c(n,m,p,t)+M^2 c(n,m,p,t)=0$$

so that

$$c(n,m,p,t)=A(n,m,p).exp(iE(n,m,p)t)+B(n,m,p)exp(-iE(n,m,p)t)$$

where

$$E(n,m,p)=(4\pi^2(n^2+m^2+p^2)+M^2)^{1/2}$$

95.Explain how using Fourier series, you can solve the quantum mechanical problem in a periodic potential.

Hint: The potential $U(x,y,z)$ satisfies the periodic conditions $U(x+nL,u+mL,z+pL)=U(x,y,z)$ for all $n,m,p\in\mathbb{Z}$. So, it can be expanded in a Fourier series as

$$U(x,y,z)=\sum c(nmp)exp(2\pi i(nx+my+pz)/L)$$

The wave function is a Bloch wave function having the form

$$\psi(x,y,z)=\phi(x,y,z).exp(2\pi i(ax+by+cz)/NL)$$

where NL is the size of the box. We thus have $\psi(x+NL,y,z)=\psi(x,y,z)$ and likewise for translates of y,z by NL. Here, a,b,c are integers. We plug this expression into the Schrodinger wave equation and determine algebraic eigenequations for the Fourier series coefficients of ϕ. Here, we are assuming that ϕ is periodic in all its three arguments with period L.

96.Explain the DSP implementation of Lagrangian field theory for the Klein-Gordon field.

ans: The actual Lagrangian density is given by

$$L(\phi,\partial_\mu\phi)=\frac{1}{2}\partial_\mu\phi\partial^\mu\phi-m^2\phi^2/2$$

The action functional $\in Ld^4x$ must be discretized in space-time and its optimization must be carried out. We can assume the integral to be carried out over a cube of sidelength L using a discrete sum over pixels. The details we shall leave out here.

97. Derive the propagator for the Klein-Gordon field using Feynman's functional integral formalism:

$$<\Omega|T\{\phi(x)\phi(y)\}|\Omega>=$$

$$\frac{\int\phi(x)\phi(y)exp(i\int\mathcal{L}(\phi)d^4x)\mathcal{D}\phi}{\int exp(i\int\mathcal{L}(\phi)d^4x)\mathcal{D}\phi}$$

where

$$\mathcal{L}(\phi)=\frac{1}{2}\partial_\mu\phi\partial^\mu\phi-m^2\phi^2/2$$

is the Klein-Gordon Lagrangian. The functional integrals on the numerator and denominator may be evaluated in the frequency domain using Fourier transforms. These integrals are equivalent to evaluating Gaussian integrals for the total area under the Gaussian curve and the covariance of Gaussian random variables.

98. Consider a Lagrangian $\mathcal{L}(\phi)$. Let $J(.)$ be a source field and consider the functional integral

$$Z[J] = \int \mathcal{D}\phi.exp(i\int(\mathcal{L}(\phi) + J\phi)d^4x)$$

This is like the partition function in classical statistical mechanics. Show that the functional derivatives of Z with respect to the source at zero source are

$$(-i)^n \frac{\delta Z[J]}{\delta J(x_1)\delta J(x_2)...\delta J(x_n)}|_{J=0} = \int \mathcal{D}\phi.\phi(x_1)...\phi(x_n).exp(i\int \mathcal{L}(\phi)d^4x)$$

Show that this is the same as

$$Z[0] < \Omega|T\{\phi(x_1)...\phi(x_n)\}|\Omega >$$

where T is the time ordering operator.

99. Write down the Fourier resolution of the Klein-Gordon field in terms of the creation and annihilation operator fields and set up the commutation relations. From that show that the Fourier transform of the Feynman propagator has the form

$$\int < \Omega|T\{\phi(x)\phi(0)\}|\Omega > exp(ip.x)d^4x = \frac{1}{p^2 - m^2 + i\epsilon}$$

Derive this propagator using the functional/path integral approach.

Hint for the path integral approach:

$$S[\phi] = \int \mathcal{L}(\phi)d^4x = \frac{1}{2}\int((\partial_\mu\phi)(\partial^\mu\phi) - m^2\phi^2)d^4x$$

$$= -\frac{1}{2}\int \phi(\partial^2 + m^2)\phi d^4x$$

after integrating by parts. It follows from the theory of the Gaussian integral that

$$\int \mathcal{D}\phi.\phi(x)\phi(y)exp(iS[\phi])/\int \mathcal{D}\phi.exp(iS[\phi])$$

$$= (\partial^2 + m^2)^{-1}$$

which in the Fourier domain gives the correct Feynman formula $(p^2 - m^2)^{-1}$ where $p^2 = p_\mu p^\mu$.

100. Evaluate the functional integral $\int exp(i\int(\mathcal{L}(\phi)+J\phi)d^4x)\mathcal{D}\phi$ for the Klein-Gordon Lagrangian and form the functional derivatives with respect to J to arrive at the expression for the Klein-Gordon propagator.

101. Explain the DSP implementation of fluid dynamics by discretizing the mass and momentum conservation equations.

102. Symmetry breaking.

$$\mathcal{L} = \frac{1}{2}(\partial_\mu\phi\partial^\mu\phi - m^2\phi^2) - \lambda\phi^4/24$$

This Lagrangian is invariant under the transformation $\phi \to c\phi$ where c is a complex number of unit magnitude. Let $\mu^2 = -m^2$ and consider the minimum of the potential part

$$V(\phi) = -\mu^2\phi^2/2 + \lambda\phi^4/24$$

The minimizing equation $V'(\phi_0) = 0$ gives $-\mu^2\phi_0 + \lambda\phi_0^3/6 = 0$ so that

$$\phi_0 = \mu\sqrt{6/\lambda}$$

This is the ground state of the system. Define the field $\sigma = \phi - \phi_0$. Then express $\mathcal{L}$ in terms of σ and show that the above symmetry is broken, ie, the Lagrangian is no longer invariant under the transformation $\sigma \to c\sigma$ where c is a complex number of unit magnitude.

103. Generalize the symmetry breaking principle for multivariate fields with Lagrangian

$$\mathcal{L}(\phi) = \frac{1}{2}\sum_i(\partial^\mu\phi_i\partial_\mu\phi_i - m^2\phi_i^2) - \lambda(\sum_i\phi_i^2)^2$$

This Lagrangian is invariant under the rotation group $O(N)$ of $\mathbb{R}^N$. Define the ground state corresponding to a minimum of the potential and then show that the shifted Lagrangian is invariant only under the subgroup $O(N-1)$ of $O(N)$.

104. Goldstone Bosons: Suppose that the Lagrangian of a field ϕ can be expressed as

$$\mathcal{L}(\phi) = A_1 - V(\phi)$$

where A_1 contains terms involving the partial derivatives of ϕ. Suppose that V attains its mininum at some ϕ_0, ie, $\frac{\partial V}{\partial\phi^a}|_{\phi=\phi_0} = 0$. Expand $V(\phi)$ about this minimum giving

$$V(\phi) = V(\phi_0) + \frac{1}{2}\sum_{a,b} m(a,b)(\phi^a - \phi_0^a)(\phi^b - \phi_0^b)$$

where

$$m_{ab} = \frac{\partial V}{\partial\phi^a\partial\phi^b}(\phi_0)$$

Assume that the Lagrangian is invariant under the gauge transformation $\phi^a \to \phi^a + \Delta^a(\phi)$. Then, the potential V is also invariant under this, ie,

$$V(\phi + \Delta(\phi)) = V(\phi)$$

which gives

$$\Delta^a(\phi)\frac{\partial V(\phi)}{\partial\phi^a} = 0$$

Differentating both sides with respect to ϕ^b and set $\phi = \phi_0$. We then get

$$\Delta^a(\phi_0)m(a,b) = 0$$

Suppose that the ground state ϕ_0 is not invariant under this gauge transformation (ie, the ground state is not symmetric), then it follows from the above equation that the mass matrix $((m(a,b)))$ has zero as an eigenvalue and hence corresponds to the existence of massless particles. In other words, symmetry breaking of the ground state leads to the presence of massless particles called Goldstone Bosons.

105. The photon propagator computed using Feynman's path integral. A^μ is the contravariant electromagnetic four potential and A_μ is the covariant electromagnetic four potential. The antisymmetric field tensor is given by

$$F_{\mu\nu} = \partial_\mu A_\nu - \partial_\nu A_\mu$$

The Lagrangian density is given by

$$L(A) = -\frac{1}{4}F_{\mu\nu}F^{\mu\nu}$$

and the corresponding action is the integral of this over the whole of space-time:

$$S[A] = \int L(A)d^4x = -\frac{1}{4}\int_{\mathbb{R}^4} F_{\mu\nu}F^{\mu\nu}d^4x$$

We have

$$F_{\mu\nu}F^{\mu\nu} = (A_{\nu,\mu} - A_{\mu,\nu})(A^{\nu,\mu} - A^{\mu,\nu})$$

$$= 2A^{\nu,\mu}(A_{\nu,\mu} - A_{\mu,\nu}) = 2A^\nu(\partial^\mu\partial_\nu A_\mu - \partial^2 A_\nu)$$

on neglecting a perfect divergence. This is permissible since the action remains invariant under the addition of a perfect divergence. Thus,

$$F_{\mu\nu}F^{\mu\nu} = 2A_\nu(\partial^\mu\partial^\nu - \partial^2 g^{\mu\nu})A_\mu$$

$$= 2A_\mu(\partial^\mu\partial^\nu - g^{\mu\nu}\partial^2)A_\nu$$

Thus

$$S[A] = \frac{1}{2}\int A_\mu(g^{\mu\nu}\partial^2 - \partial^\mu\partial^\nu)A_\nu d^4x$$

Suppose we evaluate the photon propagator as

$$D_{\mu\nu}(x-y) = <\Omega|T(A_\mu(x)A_\nu(y)|\Omega> = \int \Pi_\alpha DA_\alpha . A_\mu(x)A_\nu(y)exp(iS[A])/\int \Pi_\alpha DA_\alpha . exp(iS[A])$$

Then from the general theory of the Gaussian integral, this will evaluate to

$$D_{\mu\nu}(x-y) = i(g^{\mu\nu}\partial^2 - \partial^\mu\partial^\nu)^{-1}\delta^4(x-y)$$

or equivalently, $D_{\mu\nu}(x)$ will satisfy the differential equation

$$(g^{\mu\nu}\partial^2 - \partial^\mu\partial^\nu)D_{\nu\rho}(x) = i\delta^\mu_\rho\delta^4(x)$$

In the Fourier domain, this equation reads

$$(-g^{\mu\nu}k^2 + k^\mu k^\nu)D_{\nu\rho}(k) = i\delta^\mu_\rho$$

This equation has no solution since the matrix $(g^{\mu\nu}k^2 - k^\mu k^\nu)$ is singular. In fact

$$(g^{\mu\nu}k^2 - k^\mu k^\nu)k_\nu = k^2 k^\mu - k^\mu k^2 = 0$$

Hence, one has to use a different scheme to obtain a well defined propagator. The method is based on exploitation of gauge invariance of the Lagrangian. Define $G(A) = \partial^\mu A_\mu(x) - \omega(x)$ where $\omega(x)$ is some function of space-time. Consider the Gauge transformation

$$A^\alpha_\mu = A_\mu + \partial_\mu \alpha$$

The Lagrangian and hence action is invariant under this transformation. Thus, $S[A^\alpha] = S[A]$. We have

$$\int \delta(G(A^\alpha))det(\delta G(A^\alpha)/\delta\alpha)D\alpha = \int \delta(G(A^\alpha))DA^\alpha = 1$$

Further,

$$G(A^\alpha) = \partial^\mu A_\mu + \partial^2\alpha - \omega$$

so that

$$det(\delta G(A^\alpha)/\delta\alpha) = det(\partial^2)$$

Hence, we can write the path integral

$$\int A_\mu(x)A_\nu(y)exp(iS[A])DA =$$

$$\int A^\alpha_\mu(x)A^\alpha_\nu(y)exp(iS[A^\alpha])DA$$

since $DA = DA^\alpha$ (A^α is obtained by shifting A). We can write the above as

$$det(\partial^2)\int A_\mu(x)A_\nu(y)exp(iS[A])\delta(G(A^\alpha))D\alpha DA$$

$$= det(\partial^2)\int A_\mu(x)A_\nu(y)exp(iS[A^\alpha])DA.D\alpha\delta(\partial^\mu A^\alpha_\mu - \omega)exp(-i\int \omega^2 d^4x/2\xi)D\omega$$

after neglecting a normalization factor. This is equivalent to

$$\int A^\alpha_\mu(x)A^\alpha_\nu(y)exp((i/2)\int A^\alpha_\mu(g^{\mu\nu}\partial^2 - \partial^\mu\partial^\nu)A^\alpha_\nu d^4x)exp(-i\frac{1}{2\xi}\int (\partial^\mu A^\alpha_\mu)^2 d^4x)DA^\alpha$$

since $DA = DA^\alpha$. This is a functional Gaussian integral for the covariance and is easily seen to evaluate to

$$i[(g^{\mu\nu}\partial^2 - \partial^\mu\partial^\nu) + \partial^\mu\partial^\nu/\xi]^{-1}$$
$$= i(g^{\mu\nu}\partial^2 - (1-1/\xi)\partial^\mu\partial^\nu)^{-1}$$

106. DSP implementation of the wave propagation equations inside a waveguide. The z-dependence is $exp(-\gamma z)$ and the waveguide equations are

$$E_{z,y} + \gamma E_y = -j\omega\mu H_x, \gamma E_x + E_{z,x} = -j\omega\mu H_y,$$

$$E_{y,x} - E_{x,y} = -j\omega\mu H_z,$$

$$H_{z,y} + \gamma H_y = j\omega\epsilon E_x, \gamma H_x + H_{z,x} = j\omega\epsilon E_y,$$

$$H_{y,x} - H_{x,y} = j\omega\epsilon E_z$$

These equations must be discretized in the spatial variables and a discrete form of the Helmholtz equation must be derived therefrom. The discretization in the spatial variables yields the equations

$$\Delta^{-1}(E_z[n, m+1, k] - E_z[n, m, k]) + \gamma E_y[n, m, k]) = -j\omega\mu H_x[n, m, k]$$

$$\gamma E_x[n, m, k] + \Delta^{-1}(E_z[n+1, m, k] - E_z[n, m, k]) = -j\omega\mu H_y[n, m, k]$$

$$\Delta^{-1}(E_y[n+1, m, k] - E_y[n, m, k] - E_x[n, m+1, k] + E_x[n, m, k]) = -j\omega\mu H_z[n, m, k]$$

with similar equations for H.

107.Determine the Lie algebra commutation relations for the Lorentz group.

Hint: An inifinitesimal Lorentz transformation has the form

$$x^{\mu'} = x^\mu + \epsilon^\mu_\nu x^\nu$$

where ϵ^μ_ν are numbers of the first order of smallness. We set $\epsilon_{\mu\nu} = g_{\mu\alpha}\epsilon^\alpha_\nu$ Then the condition for Lorentz invariance can be written as

$$(x^\mu + \epsilon^\mu_\nu x^\nu)(x_\mu + \epsilon_{\mu\rho}x^\rho) = x^\mu x_\mu$$

Thus,

$$\epsilon^\mu_\nu x^\nu x_\mu + \epsilon_{\mu\rho}x^\mu x^\rho = 0$$

or

$$\epsilon_{\mu\nu}x^\mu x^\nu = 0$$

which is satisfied for all x iff $((\epsilon_{\mu\nu}))$ is an antisymmetric matrix, ie, iff $\epsilon_{\mu\nu} = -\epsilon_{\nu\mu}$. Any antisymmetric matrix $((\epsilon_{\mu\nu}))$ can be expressed as a linear combination of the antisymmetric matrices $J^{\mu\nu}$ where

$$(J^{\mu\nu})_{\alpha\beta} = \delta^\mu_\alpha\delta^\nu_\beta - \delta^\mu_\beta\delta^\nu_\alpha$$

We can write down the commutation relations for this Lie algebra:

$$(J^{\mu\nu}J^{\rho\sigma})_{\alpha\beta} =$$

$$(J^{\mu\nu})_{\alpha\gamma}(J^{\rho\sigma})^\gamma_\beta$$

Now observe that

$$(J^{\rho\sigma})^\gamma_\beta = g^{\rho\gamma}\delta^\sigma_\beta - g^{\sigma\gamma}\delta^\rho_\beta$$

so that

$$(J^{\mu\nu}J^{\rho\sigma})_{\alpha\beta} =$$

$$(\delta^{\mu}_{\alpha}\delta^{\nu}_{\gamma} - \delta^{\mu}_{\gamma}\delta^{\nu}_{\alpha})(g^{\rho\gamma}\delta^{\sigma}_{\beta} - g^{\sigma\gamma}\delta^{\rho}_{\beta})$$

$$= g^{\rho\nu}\delta^{\mu}_{\alpha}\delta^{\sigma}_{\beta} - g^{\rho\mu}\delta^{\nu}_{\alpha}\delta^{\sigma}_{\beta}$$

$$+g^{\mu\sigma}\delta^{\nu}_{\alpha}\delta^{\rho}_{\beta} - g^{\nu\sigma}\delta^{\mu}_{\alpha}\delta^{\rho}_{\beta}$$

Interchanging the pairs (μ,ν) and (ρ,σ) gives us

$$(J^{\rho\sigma}J^{\mu\nu})_{\alpha\beta} = g^{\mu\sigma}\delta^{\rho}_{\alpha}\delta^{\nu}_{\beta} - g^{\mu\rho}\delta^{\sigma}_{\alpha}\delta^{\nu}_{\beta}$$

$$+g^{\rho\nu}\delta^{\sigma}_{\alpha}\delta^{\mu}_{\beta} - g^{\nu\sigma}\delta^{\rho}_{\alpha}\delta^{\mu}_{\beta}$$

108. Explain the Rayleigh-Ritz variational method for determining the ground state energy of a quantum Hamiltonian.

Hint: Let H denote the Hamiltonian of the quantum system and let $|u_n>, n = 0, 1, 2, ...$ be the energy eigenstates of H with eigenvalues $E_n, n = 0, 1, 2, ...$ respectively arranged in increasing order, ie, $E_0 < E_1 < ... < E_n < E_{n+1} < ....$ Then let $|u>$ be any normalized state so that $|u>$ can be expanded as $|u>= \sum_n c_n|u_n>$ where $\sum_n |c_n|^2 = 1$. Then

$$<u|H|u>= \sum_n E_n|c_n|^2 \geq E_0 \sum |c_n|^2 = E_0$$

with equality iff $u = u_0$. This suggests the following scheme for determining the ground state wave function and energy eigenvalue. Let u_θ be any normalized wave function parameterized by θ, a set of unknown parameters. Then choose θ so that

$$<u_\theta|H|u_\theta>= \int u_\theta(x)^* H u_\theta(x)dx$$

is a minimum. The minimum value of this function is then a good approximation to the ground state wave function and the corresponding $|u_\theta>$ is a good approximation to the ground state wave function.

109. Write down the complete Lagrangian for photons interacting with electrons and positrons.

ans: Let A_μ denote the electromagnetic four potential and ψ the Dirac wave function of the electron. Then the total Lagrangian density for the Maxwell-Dirac field is given by

$$L(A,\psi) = -\frac{1}{4}F_{\mu\nu}F^{\mu\nu} + i\bar{\psi}\gamma^\mu D_\mu\psi$$

where

$$F_{\mu\nu} = A_{\nu,\mu} - A_{\mu,\nu}$$

and

$$D_\mu = \partial_\mu - ieA_\mu$$

We can express this as a sum of three terms, (a) the Maxwell term for photons, (b) the Dirac term for electrons and (c) the interaction term between photons and electrons:

$$L_{Maxwell} = -\frac{1}{4}F_{\mu\nu}F^{\mu\nu},$$

$$L_{Dirac} = i\bar{\psi}\gamma^{\mu}\partial_{\mu}\psi,$$

$$L_{int} = e\bar{\psi}\gamma^{\mu}\psi.A_{\mu}$$

Note that

$$\bar{\psi} = \psi^{*}\gamma^{0}$$

where ψ^* is the conjugate transpose of ψ.

110. Explain how you would draw the Feynman diagrams for the scattering of Yukawa particles and Fermions.

Hint: The Yukawa field is defined by a single scalar field $\phi(x)$ and the Fermionic field by a Dirac four component wave function $\psi(x)$. The Lagrangian density for the Yukawa field is

$$\mathcal{L}_{Yukawa} = \frac{1}{2}\partial_{\mu}\phi\partial^{\mu}\phi - m\phi^{2}/2$$

The Lagrangian density for the Fermionic field is

$$\mathcal{L}_{Dirac} = i\bar{\psi}\gamma^{\mu}\partial_{\mu}\psi$$

and finally the Lagrangian for the interaction of the Yukawa field with the Dirac field is

$$\mathcal{L}_{int} = K\bar{\psi}\psi\phi$$

In the interaction representation, the relevant matrix elements for the scattering of a Fermion having momentum p and a Yukawa Boson having momentum k into a Fermion having momentum p' and a Yukawa boson having momentum k' is given by

$$< p'k'|exp(-iK\int\bar{\psi}\psi\phi d^{4}x)|pk >$$

and the first nonzero contribution to this matrix element comes from the second order term, ie,

$$(-iK)^{2} < p'k'|\int\bar{\psi}(x)\psi(x)\phi(x)\bar{\psi}(y)\psi(y)\phi(y)d^{4}xd^{4}y|pk >$$

Appropriate pairing can be performed here. We omit the details which can be found in Peskin and Schroeder.

111. What is a Cartan subalgebra of a semisimple Lie group. What do you understand by solvability of a Lie algebra and what is the radical of a Lie algebra. What is a regular element of a Lie algebra and

what is the Cartan algebra associated with a regular element. Prove that in a compact semisimple Lie algebra, any two Cartan subalgebras are conjugate to each other.

112. Given a Lie algebra }, what do you understand by the symmetric algebra $S(\})$. The symmetric algebra is a quotient of the tensor algebra. Clarify this point. What is the universal enveloping algebra of a Lie algebra. The symmetric algebra is isomorphic as a vector space with the universal enveloping algebra via the λ-map. Clarify this point.

113. Suppose } is a Lie algebra and $U(\})$ is the universal enveloping algebra of }. Then let $X_1, ..., X_n$ be a basis for }. Show that $X_1^{m_1} X_2^{m_2} ... X_k^{m_k}, m_1, ..., m_k \geq 0$ is a base for $U(\})$. If X, Y are two elements of } such that $[X, Y] = aX + bY$, then express YX in terms of XY, X, Y in $U(\})$. (ans: $YX = XY - [X, Y] = XY - aX - bY$.

114. Suppose G is a Lie group and } is its Lie algebra. Let $f : \} \to \mathbb{C}$ be a smooth function. Consider the function $F : G \times \} \to \mathbb{C}$ defined by the equation

$$F(g, X) = f(Ad(g)(X)) = f(gXg^{-1})$$

Let $Y, Z \in \}$. Then obtain a formula for $F(g; Y, X : Z)$ defined by the equation

$$F(g; Y, X; Z) = \frac{\partial^2}{\partial t \partial s} F(g.exp(tY), X + sZ)|_{t=s=0}$$

More generally, let p be any polynomial on }. Assuming } to be semisimple, the Cartan Killing form $B(.,.)$ on } is non-singular by duality with respect to this form, there is a unique element of $S(\})$, the symmetric algebra on } associated with p. Denote this element by the same symbol p. Let $\partial(p)$ denote the differential operator corresponding to $\partial(p)$ in the sense that if $p = X_1 X_2 ... X_k$ with $X_i \in \}$, then

$$f(X, \partial(X_1 ... X_k)) = \frac{\partial^k}{\partial t_1 ... \partial t_k} f(X + t_1 X_1 + ... + t_k X_k)|_{t_1 = ... = t_k = 0}$$

Then obtain a formula for

$$F(g; b, X, \partial(p))$$

where b is an element of $U(\})$. Do this by taking $b = Z_1 ... Z_r$ with $Z_1, ..., Z_r \in \}$.

Hint:

$$F(g.exp(tY), X + sZ) = f(g.exp(tY)(X + sZ).exp(-tY)g^{-1})$$

Differentiating this expression with respect to s at $s = 0$ gives us

$$\partial(g.exp(tY)Zexp(-tY)g^{-1})f(g.exp(tY)Xexp(-tY)g^{-1})$$

Differentiating this expression with respect to t at $t = 0$ then gives us

$$\partial(g[Y, Z]g^{-1})f(gXg^{-1}) + \partial(g[Y, X]g^{-1})\partial(gZg^{-1})f(gXg^{-1})$$

Thus,

$$F(g;Y,X;Z) = f(gXg^{-1}, g\partial(q)g^{-1})$$

where

$$q = [Y,Z] + [Y,X]Z = (ad(Y) + L_[Y,X])Z$$

In general, extending this argument, we have

$$F(g;Y_1...Y_k,X;\partial(p)) = f(gXg^{-1}, g\partial(q)g^{-1})$$

where

$$q = (ad(Y_1) + L_{[Y_1,X]})...(ad(Y_k) + L_{[Y_k,X]})p$$

To prove this we use induction. Assume that the result is true with k replaced by $k-1$. This means that

$$F(g,Y_2...Y_k,X;\partial(p)) = f(gXg^{-1}, g\partial(q')g^{-1}) = f^g(X,\partial(q'))$$
$$= \partial(gq'g^{-1})f(gXg^{-1})$$

Here

$$f^g(X) = f(gXg^{-1})$$

Then,

$$F(g;Y_1...Y_k,X;\partial(p)) = \frac{d}{dt}\partial(g.exp(tY_1)q'exp(-tY_1)g^{-1})f(g.exp(tY_1)X.exp(-tY_1)g^{-1})|_{t=0}$$

$$= \partial(g[Y_1,q']g^{-1})f(gXg^{-1}) + \partial([Y_1,X]q')f(gXg^{-1})$$
$$= \partial(gqg^{-1})f(gXg^{-1})$$

where

$$q' = (ad(Y_2) + L_{[Y_2,X]})...(ad(Y_k) + L_{[Y_k,X]})p$$

and

$$q = [Y_1,q'] + [Y_1,X]q' = (ad(Y_1) + [Y_1,X])q' = (ad(Y_1) + L_[Y_1,X])...(ad(Y_k) + L_{[Y_k,X]})p$$

Reference:HarishChandra, Differential operators on a Lie algebra. The mapping that assigns the element $q \in S(\})$ to each $(b,p) \in U(\}) \times S(\})$ plays a fundamental role in applications involving determination of the characters of a Lie algebra. We define $q = \Gamma(b,p)$ so that the map $\Gamma : U(\}) \times S(\}) \to S(\})$ is onto.

115. What do you understand by frequency independent antennas.

ans: The characteristic frequency at which an antenna operates is inversely proportional to its linear dimension. Suppose the surface of the antenna is defined by the equation $r = f(\theta,\phi)$. A rotation about the z axis should produce a proportional scaling in r, ie all dimensions can be obtained by performing a rotation. This means that for each real number C, there should be a unique real number $K(C)$ such that $K(C)r = f(\theta,\phi + C)$. Solving this equation gives us $f(\theta,\phi) = g(\theta).exp(a\phi)$ where a is some constant. Thus frequency independent antennas must be spirals.

116. Explain without supplying the details how you would perform an analysis of a planar spiral antenna.

ans:Assume that the plane of the spiral is $\theta = \pi/2$ and that the two bounding curves of this spiral are respectively $\rho = K_1.exp(a\phi)$ and $\rho = K_2.exp(a\phi)$. The region of the spiral is then $0 \le \phi \le 2m\pi$ and $K_1.exp(a\phi) \le \rho \le K_2.exp(a\phi)$. Assume that the surface current density on the spiral is given by

$$\mathbf{J} = J_\rho(\rho,\phi)\hat{\rho} + J_\phi(\rho,\phi)\hat{\phi}$$

In terms of the cartesian unit vectors, we have

$$\mathbf{J} = J_\rho(\hat{x}.cos(\phi) + \hat{y}.sin(\phi)) + J_\phi(-\hat{x}.sin(\phi) + \hat{y}.cos(\phi))$$

$$= J_x(\rho,\phi)\hat{x} + J_y(\rho,\phi)\hat{y}$$

where

$$J_x = J_\rho.cos(\phi) - J_\phi sin(\phi),$$
$$J_y = J_\rho.sin(\phi) + J_\phi cos(\phi)$$

The magnetic vector potential produced by this surface current density is given by

$$\mathbf{A}(\rho,\phi,z) = \frac{\mu}{4\pi}\int_0^{2m\pi}\int_{K_1.exp(a\phi)}^{K_2.exp(a\phi)} (\hat{x}.J_x(\rho',\phi') + \hat{y}.J_\phi(\rho,\phi))$$

$$\frac{exp(-jK\sqrt{\rho^2+\rho'^2-2\rho\rho'cos(\phi-\phi')+(z-z')^2})}{\sqrt{\rho^2+\rho'^2-2\rho\rho'.cos(\phi-\phi')+(z-z')^2}}\rho' d\rho' d\phi'$$

Using this vector potential, the scalar potential is calculated and then x, y components of the electric field are determined. These are set to zero at the planar surface pixels except at the feed point $\phi = 0, K_1 \le \rho \le K_2$. The current density at the feed point is given by $J_\rho(\rho, 0) = 0, J_\phi(\rho, 0) = I_0/(K_2 - K_1)$. Thus just as we analyze a Yagi array, we can analyze the planar spiral antenna using Pocklington's theory.

117. In what way does Hallen's integral equation differ from Pocklington's integral equation. Where can it be used more successfully in comparison with Pocklington's equation.

118. Given an electric field on an aperture, how would you compute the equivalent surface magnetic current density.

119. Explain renormalization theory with respect to the Yukawa theory of interaction of Yukawa Bosons and Fermions.

Hint: The Lagrangian density of the Fermion and Yukawa Boson field is given by the sum of three terms. First, the Dirac Lagrangian density, second, the Klein-Gordon Lagrangian density and three the interaction Lagrangian density:

$$\mathcal{L} = i\bar{\psi}\gamma^\mu\partial_\mu\psi + \frac{1}{2}(\partial_\mu\phi\partial^\mu\phi - m\phi^2) + g\bar{\psi}\psi\phi$$

Note that in quantum electrodynamics, the last term is replaced by $\bar{\psi}\gamma^\mu\psi A_\mu$. For renormalization, we set

$$\phi = Z\phi_r$$

where Z is a parameter and ϕ_r is the renormalized Yukawa field. We also renormalize the mass by setting $m = m_0 + \delta m$. Then, the Lagrangian density is

$$\bar{\psi}\gamma^\mu \partial_\mu \psi + \frac{1}{2}(Z^2 \partial_\mu \phi_r \partial^\mu \phi_r - (m_0 + \delta m) Z^2 \phi_r^2) + gZ\bar{\psi}\psi\phi_r$$

We define

$$Z - 1 = \delta_Z$$

so that the renormalized Hamiltonian density is given by

$$\mathcal{H} = \bar{\psi}\gamma^\mu \partial_\mu \psi + \frac{1}{2}(\partial_\mu \phi_r \partial^\mu \phi_r - m_0 \phi_r^2) + g\bar{\psi}\psi\phi_r$$

$$+\delta Z \partial_\mu \phi_r \partial^\mu \phi_r + \delta m \phi_r^2 + g\delta Z \bar{\psi}\psi\phi_r$$

where we have neglected terms of second order of smallness, ie involving $(\delta Z)^2$. We are now in a position to evaluate the propagator and counter terms for the renormalized Lagrangian. In the absence of the counter terms and the interaction, the propagator of the Yukawa field is the same as that of the Klein-Gordon operator. It can be evaluated using the path integral formalism as

$$D_\phi(x - y) = \int \phi(x)\phi(y) exp(\frac{i}{2} \int (\partial_\mu \phi \partial^\mu \phi - m^2 \phi^2) d^4x) D\phi$$

$$= (\partial_\mu \partial^\mu + m^2)^{-1} \delta(x - y)$$

or in the momentum domain via a Fourier transform as

$$\hat{D}_\phi(p) = (p^2 - m^2)^{-1}$$

where

$$p^2 = p_\mu p^\mu = p^{02} - \mathbf{p}^2 - m^2$$

When there is an interaction, we can compute the second order contribution to the scattering amplitude of a Yukawa boson. This is given by

$$< p_2| \int \bar{\psi}(x)\psi(x)\bar{\psi}(y)\psi(y)\phi(x)\phi(y) d^4x d^4y |p_1 >$$

and it can be expressed as

$$\bar{u}(p_2) u(p_1) \int S_{ba}(y - x) S_{ab}(x - y) exp(i(p_1.x - p_2.y)) d^4x d^4y$$

where $S(x - y)$ is the Fermionic Dirac propagator:

$$S(x - y) =< 0|T(\psi(x)\bar{\psi}(y))|0 >$$

or in terms of matrix elements,

$$S_{ab}(x - y) =< 0|T(\psi_a(x)\bar{\psi}_b(y))|0 >$$

This can be evaluated as

$$\bar{u}(p_2)u(p_1)\int S_{ba}(-\xi)S_{ab}(\xi)exp(i(p_1.(y+\xi)-p_2.y))d^4yd^4\xi$$

$$=\bar{u}(p_1)u(p_1)\delta(p_1-p_2)\int S_{ba}(-\xi)S_{ab}(\xi)exp(ip_1.\xi)d^4\xi$$

$$=|u(p_1)|^2\delta(p_1-p_2)\int \hat{T}r(\hat{S}(k+p_1)S(k))d^4k$$

120. Explain how by taking into account the gauge invariance of the Maxwell theory, you would develop a path integral formalism to quantize it.

Hint: Let A_μ be the covariant electromagnetic four potential and let $F_{\mu\nu}$ be the antisymmetric field tensor corresponding to this four potential, ie,

$$F_{\mu\nu}=A_{\nu,\mu}-A_{\mu,nu}$$

Let $\bar{\psi}$ denote the matter field second quantized wave function corresponding to the Dirac equation. Then the total Lagrangian of the matter and field is given by

$$\mathcal{L}(A_\mu, A_{\nu,\mu}, \psi)=-\frac{1}{4}F_{\mu\nu}F^{\mu\nu}+i\bar{\psi}\gamma^\mu\partial_\mu\psi-e\bar{\psi}\gamma^\mu\psi.A_\mu$$

The path integral of this functional must be formed and Feynman diagrams obtained for the Fourier domain exact propagator.

121. Explain the analogy between the Poisson bracket in classical mechanics and the Lie bracket in quantum mechanics.

ans: In classical mechanics, the state of the system at any given time is specicified by means of n generalized coordinates and n generalized momenta $(q_1,...,q_n,p_1,...,p_n)$. The dynamics of the system is specified by means of an Hamiltonian function $H(q_1,...,q_n,p_1,...,p_n)$. The dynamics is the Hamilton equations of motion:

$$q_i' = h_{,p_i}, p_i' = -H_{,q_i}$$

An observable of the system is a differentiable function of the generalized coordinates, the generalized momenta and time, ie, $F:\mathbb{R}^{2n+1}\to\mathbb{R}$ or equivalently as

$$(q_1,...,q_n,p_1,...,p_n,t)\to F(q_1,...,q_n,p_1,...,p_n,t)$$

Its rate of change in time is given by

$$\frac{dF}{dt}=\sum_i(F_{,q_i}q_i'+F_{,p_i}p_i')+F_{,t}$$

$$=\sum_i(F_{,q_i}H_{,p_i}-F_{,p_i}H_{,q_i})+F_{,t}=\{F,H\}+F_{,t}$$

where $\{.,.\}$ is the canonical Poisson bracket. The Poisson bracket satisfies the properties:

$$\{u,v\}=-\{v,u\},\{uv,w\}=\{u,w\}v+u\{v,w\},\{\{u,v\},w\}+\{v,w\},u\}+\{\{w,u\},v\}=0$$

The Lie bracket between two operators u, v is defined by

$$[u,v]=uv-vu$$

It also satisfies the above three properties:

$$[v,u]=vu-uv=-(uv-vu)=-[u,v],$$

$$[uv,w]=uvw-wuv=u(vw-wv)+(uw-wu)v=u[v,w]+[u,w]v$$

$$[[u,v],w]+[[v,w],u]+[[w,u],v]=$$

$$(uv-vu)w-w(uv-vu)+(vw-wv)u-u(vw-wv)+(wu-uw)v-v(wu-uw)$$

$$=uvw-vuw-wuv+wvu+vwu-wvu-uvw+uwv+wuv-uwv-vwu+vuw=0$$

The Lie bracket can be derived from properties of the Poisson bracket if we assume noncommutativity. Thus we have for any four observables u, v, w, x, and a bracket $[.,.]$ assumed to satisfy the above three properties of the Poisson bracket,

$$[uv,wx]=[uv,w]x+w[uv,x]=u[v,w]x+[u,w]vx+wu[v,x]+w[u,x]v$$

on the one hand and on the other,

$$[uv,wx]=u[v,wx]+[u,wx]v=u[v,w]x+uw[v,x]+[u,w]xv+w[u,x]v$$

Equating these two expressions gives us

$$[u,w]vx+wu[v,x]=uw[v,x]+[u,w]xv$$

so that

$$[u,w](vx-xv)=(uw-wu)[v,x]$$

from which we conclude that $[u,w]$ is a constant times $uw-wu$.

122. What do you understand by a generalized measurement in the quantum theory.

ans: Let $\mathcal{H}$ denote the Hilbert space of the quantum system and let ρ be a state, ie, a density operator in $\mathcal{H}$. This means that $(u,\rho u)\geq 0$ for all $u\in\mathcal{H}$ and $Tr(\rho)=1$, ie, if $(e_n : n=1,2,...)$ is an orthonormal basis for $\mathcal{H}$, then

$$\sum_n (e_n,\rho e_n)=1$$

Let $M_m, m=1,2,...p$ be a set of p bounded operators in $\mathcal{H}$ satisfying the condition

$$\sum_m M_m M_m^* = I$$

We call this set a generalized measurement. We say that that outcomes of this measurement are the p real numbers $1, 2, ..., p$. If this measurement is made when the state of the system is ρ, the probability of obtaining the outcome m is given by $p(m) = (M_m, \rho M_m) = Tr(\rho M_m M_m^*)$. In this case, the state of the system collapses to the state

$$\rho_m = M_m \rho M_m^* / p(m)$$

If the measurement is made but no outcome is observed, then the state of the system collapses after the measurement to the state

$$\rho_M = \sum_{m=1}^{p} p(m)\rho_m = \sum_{m=1}^{p} M_m \rho M_m^*$$

As an example, suppose $e_n, n = 1, 2, ...$ is an orthonormal basis for $\mathcal{H}$ and set $M_m = |e_m >< e_m|, m = 1, 2, ...$ Then we obviously have $\sum_m M_m = I$. If the measurement is taken the probability of the outcome m is given by $p(m) = Tr(\rho < e_m|\rho|e_m >$ which is the expected value of ρ in the pure state $|e_m >$. If the outcome m occurs, the system collapses to the state

$$\rho_m = (M_m \rho M_m)/p(m) = |e_m >< e_m|$$

or equivalently, to the pure state $|e_m >$.

123. Let $\mathcal{H}$ be the Hilbert space of pure states of a quantum system having Hamiltonian operator H. Let G be a group of unitary transformations in $\mathcal{H}$ that leave the Hamiltonian H invariant, ie, $g.H.g^{-1} = H, \forall g \in G$. Then show that each energy eigenvalue is degenerate and the restriction of G to the corresponding eigenstate defines a representation of G.

124. What do you understand by the term "phase centre" of an antenna.

ans:That point in space relative to which the antenna pattern has radial symmetry.

125. Where is the phase centre of a parabolic reflector located ?

ans:at the focus of the antenna.

126. Show that the expressions for the electric and magnetic fields in terms of the scalar and vector potentials are invariant under gauge transformations of the potentials, ie, under the transformations

$$A \rightarrow A - \nabla f, V \rightarrow V + \partial_t f$$

with f any smooth function of the space-time coordinates.

127. Prove that if B is any positive definite $n \times n$ real matrix, then

$$\int_{\mathbb{R}^n} exp(-\frac{1}{2}\mathbf{x}^T \mathbf{B} \mathbf{x}) d^n x = (2\pi)^{n/2} (detB)^{-1/2}$$

Using this formula, evaluate the Feynman path integral

$$\int exp(-\frac{1}{2}\int D(x,y)\phi(x)\phi(y)dxdy).exp(\int g(x)\phi(x)dx)\Pi_x d\phi(x)$$

where D is a positive definite kernel and g an arbitrary function.

128. Prove that if δ denotes the delta function on $\mathbb{R}^n$ and $\mathbf{B}$ is a positive definite matrix, then for any smooth function $f : \mathbb{R}^n \to \mathbb{R}$, we have

$$\int f(\mathbf{x})\delta(\mathbf{Bx})d^n x = (det\mathbf{B})^{-1} f(\mathbf{0})$$

Hence conclude that

$$\delta(\mathbf{Bx}) = (det\mathbf{B})^{-1}\delta(\mathbf{x})$$

129. Consider the linear autoregressive time series

$$x(n) + \sum_{k=1}^{p} a(k)x(n-k) = w(n)$$

where $w(.)$ is Gaussian white noise. Assume that this system is causal and define the signal autocorrelation function by $r(m,n) = \mathbb{E}(x(n)x(m))$. Prove that (a) $\mathbb{E}(w(n)x(n-k)) = 0, k > 0$ and secondly

$$r(n, n-m) + \sum_{k=1}^{p} a(k)r(n-k, n-m) = 0, m = 1, 2, ...$$

(c) Show that if w in addition to being white is stationary with variance σ^2, then the power spectral density of x can be expressed as

$$S_{xx}(\omega) = \frac{\sigma^2}{|1 + \sum_{k=1}^{p} a(k)exp(-j\omega k)|^2}$$

130. Let $w(n)$ be a non-Gaussian white noise sequence. Show that if this process is stationary, then there is a constant γ_p for each $p > 0$ such that

$$\mathbb{E}(w(n)w(n+m_1)...w(n+m_p)) = \gamma.\delta(m_1)...\delta(m_p)$$

Hence show that for a causal time series of the form

$$x(n) + \sum_{k=1}^{p} a(k)x(n-k) = w(n)$$

we have for $m_1, ..., m_r \geq 0$,

$$\mathbb{E}(x(n)x(n-m_1)...x(n-m_r)) + \sum_{k=1}^{p} a(k)\mathbb{E}(x(n-k)x(n-m_1)...x(n-m_r)) = \gamma.\delta(m_1)...\delta(m_r)$$

Show that if x is stationary, then this equation can be expressed as

$$r(m_1, ..., m_r) + \sum_{k=1}^{p} a(k)r(m_1-k, m_2-k, ..., m_r-k) = \gamma.\delta(m_1)...\delta(m_r), m_1, ..., m_r \geq 0$$

131. Consider a second order Volterra system with input-output relation given by

$$y(n) = \sum_{k=0}^{p} h(k)x(n-k) + \sum_{k,m=0}^{p} g(k,m)x(n-k)x(n-m) + w(n)$$

Obtain the matrix equations for determining the first and second order filter coefficients $\{h(k)\}$ and $\{g(k,m)\}$ by minimizing the sum of error squares $\sum_{n=0}^{N-1} w(n)^2$. The algorithm should tell us the estimated filter coefficients in terms of the time averaged joint moments of input-output data.

132. For a second order Volterra system, formulate the least mean square algorithm for estimating the filter coefficients from input output data in an adaptive fashion.

Hint:

$$e(n) = y(n) - \sum_{k=0}^{p} h(k,n)x(n-k) + \sum_{k,m=0}^{p} g(k,m,n)x(n-k)x(n-m)$$

with the algorithm being

$$h(k,n+1) = h(k,n) - \mu\frac{\partial}{\partial h(k,n)}e(n)^2,$$

$$g(k,m,n+1) = g(k,m,n) - \mu\frac{\partial}{\partial g(k,m,n)}e(n)^2$$

133. Formulate the recursive least squares algorithm for estimating the filter coefficient vector for an FIR system recursively in time. To do this, you must make use of the matrix inversion lemma.

134. By discretizing the relationship between charge density and potential as per Coulomb's law, show that this relationship can be cast as a multidimensional FIR filter.

ans:

$$V(x,y,z) = \frac{1}{4\pi\epsilon}\int \frac{\rho(x',y',z')}{((x-x')^2+(y-y')^2+(z-z')^2)^{1/2}}dx'dy'dz'$$

In other words, defining the Green's function

$$G(u,v,w) = \frac{1}{4\pi\epsilon(u^2+v^2+w^2)^{1/2}}$$

the potential is expressible as a convolution between the Green's function and the charge density:

$$V(x,y,z) = \int G(x-x',y-y',z-z')\rho(x',y',z')dx'dy'dz'$$

On discretizing this equation, V becomes a three dimensional discrete convolution between the discretized Green's function and the discretized charge density. The parameter ϵ is an unknown parameter of the system and can be estimated using a least squares or an LMS algorithm.

135. Show that in the presence of an anisotropic dielectric, Poisson's equation for the electrostatic potential gets modified to

$$(\nabla, \epsilon.\nabla V) = -\rho$$

Determine the Green's function for this using Fourier domain techniques.

136. Suppose $\phi : \mathbb{R}^4 \to \mathbb{C}^2$ is a field with Lagrangian density given by

$$\mathcal{L}(\phi, \partial_\mu \phi) = \frac{1}{2}\partial_\mu \phi^* \partial^\mu \phi - m^2 \phi^* \phi$$

Show that this Lagrangian is invariant under the group $SU(2)$. What is the Hamiltonian density corresponding to this Lagrangian density ?

Hint:The $SU(2)$ transformation of the field is given by $\phi \to g.\phi$ where $g \in SU(2)$. Then, assuming that g is a constant matrix, we have $\partial_\mu \phi \to g\partial_\mu \phi$ and $\phi^* \to \phi^* g^* = \phi^* g^{-1}$ while $\partial_\mu \phi^* \to \partial_\mu \phi^* g^{-1}$.

137. Explain the concept of symmetry breaking by introducing a shift relative to the vacuum expectation of the fields. Show how symmetry breaking leads to the production of massless Goldstone bosons.

138. Suppose $\}$ is a semisimple Lie algebra over the complex field $\mathbb{C}$. For $X, Y \in \}$, define the complex number $B(X, Y) = sp(ad(X).ad(Y))$ where sp denotes spur or trace. Show that $B : \} \times \} \to \mathbb{C}$ is a non-degenerate bilinear form.

ref:Take a look at Varadarajan's book, Lie groups, Lie algebras and their representations in which he proves this result using Chevalley's theory of replicas.

139. Let $\}$ be a semisimple Lie algebra and let B denote the Cartan-Killing form on $\}$, ie $B(X, Y) = sp(ad(X).ad(Y))$. Show that B is invariant under the adjoint mapping, ie,

$$B(Ad(g)X, Ad(g)Y) = B(X, Y), g \in G$$

where G is the Lie group corresponding to $\}$. Suppose $\langle$ is a Cartan subalgebra of $\}$ and consider the corresponding rootspace decomposition

$$\} = \langle \oplus \bigoplus_{\alpha \in P} (\}_\alpha \oplus \}_{-\alpha})$$

where P is the set of positive roots. Then prove the following statements:(1) $B(.,.)$ is non-singular on $\langle$, (2) $B(X, [Y, Z]) = B([X, Y], Z)\ \forall X, Y, Z \in \}$, (3) If $H, H' \in \langle$ and $X \in \}_\alpha$ for some root α, ie, $\alpha \in P \cup (-P)$, then

$$\alpha(H')B(H, X) = B(H, [H', X]) = B([H, H'], X) = 0$$

implying that if H' is selected so that $\alpha(H') \neq 0$, then $B(H, X) = 0$. This proves that $\langle \perp \}_\alpha$ for all roots α. Show that if $X \in \}_\alpha$ and $Y \in \}_\beta$, where α, β are any two roots, positive or negative, then for $H \in \langle$, we have since

$$[H, [X, Y]] + [X, [Y, H]] + [Y, [H, X]] = 0$$

the relation

$$[H, [X, Y]] = (\alpha(H) + \beta(H))[X, Y]$$

showing that $[X,Y] \in \}_{\alpha+\beta}$. In particular, if α is a positive root and β is a negative root, then, $[X,Y]=0$ unless $\beta=-\alpha$, in which case, $[X,Y] \in \langle$.

140. Let $\}$ be a semisimple Lie algebra and $\langle$ a Cartan subalgebra. Let $B(.,.)$ denote the Cartan-Killing form on $\}$ and consider the polynomial $\omega(X)=B(X,X)$. We denote by the same symbol ω, the element of the symmetric algebra $S(\})$ corresponding to ω. Let α be a positive root and let H_α denote the unique element of $\langle$ such that $B(H_\alpha,H)=\alpha(H)$ for all $H \in \langle$. Let the same symbol $\bar{\omega}$ denote the element of $S(\langle)$ corresponding to the polynomial $\bar{\omega}$ on $\langle$. Then obtain an explicit formula for $\bar{\omega}$ in terms of $\{H_\alpha\}$. We have

$$\bar{\omega}(H)=\omega(H)=B(H,H)=sp(ad(H)^2)=2\sum_{\alpha\in P}\alpha(H)^2$$

since $ad(H)$ is semisimple and has zero eigenvalues on $\langle$ and eigenvalue $\alpha(H)$ on the one dimensional subspace $\}_\alpha$ for any root α. Now,

$$2\sum_{\alpha\in P}\alpha(H)^2=2\sum_{\alpha\in P}B(H,H_\alpha)^2$$

It follows from this expression that

$$\bar{\omega}=2\sum_{\alpha\in P}H_\alpha^2$$

and if we arrange things so that $H_{-\alpha}=-H_\alpha$, then

$$\bar{\omega}=\sum_\alpha H_\alpha^2$$

the sum being over all roots α. Let $X_\alpha \in \}_\alpha$ be selected so that $[X_\alpha,X_{-\alpha}]=H_\alpha$ for all $\alpha \in P$. Then for any $X \in \}$, we have the unique decomposition

$$X=H+\sum_{\alpha\in P}(c_\alpha X_\alpha+c_{-\alpha}X_{-\alpha})$$

for some complex constants $\{c_\alpha\}$. We then have

$$\omega(X)=\omega(H)+2\sum_{\alpha\in P}c_\alpha c_{-\alpha}$$

Note that we also have

$$B(X_\alpha,X_{-\alpha})=1$$

In fact, we can normalize the $X'_\alpha s$ so that this equation is satisfied and then we would get

$$[X_\alpha,X_{-\alpha}]=H_\alpha$$

for

$$B([X_\alpha,X_{-\alpha}],H)=B(X_\alpha,[X_{-\alpha},H])=\alpha(H).B(X_\alpha,X_{-\alpha})=\alpha(H)=B(H_\alpha,H)$$

Suppose that we extend the domain of $\bar{\omega}$ so that it is zero on the root spaces. Then we have from the above

$$\omega(X) = \bar{\omega}(X) + \sum_{\alpha \in P} B(X, X_\alpha).B(X, X_{-\alpha}), X \in \}$$

The element in $S(\langle\rangle)$ that corresponds to $\bar{\omega}$ on $\}$ as defined here is precisely $\bar{\omega}$ since $B(H_\alpha, X) = 0$ if X belongs to any one of the root spaces. It follows then on inversion of the above identity that as a relation between elements of $S(\})$, we have

$$\omega = \bar{\omega} + 2\sum_{\alpha \in P} X_\alpha.X_{-\alpha}$$

141. Obtain an expression for the integral that is used in obtaining the far field electric and magnetic field pattern of a helical antenna carrying uniform current at frequency ω.

142.A Microstrip antenna of rectangular shape and height h carries transverse magnetic modes inside its cavity. Assuming that the side walls of the antenna cavity are perfect magnetic conductors and the top and bottom surfaces are perfect electric conductors, determine the far field electromagnetic field pattern corresponding to the TM_{mnp} mode.

Hint: For the transverse magnetic field mode, the fields inside the cavity are determined by a single component A_x of the magnetic vector potential, where x is the direction of the height of the microstrip antenna. We have $0 \leq x \leq h$. The length and width of the antenna are respectively L along the y axis and W along the z axis. The top and bottom surfaces of the antenna are perfect conductors so that the tangential components of the electric field vanish at these points and on the other hand, the side walls of the antenna are perfect magnetic conductors, so that the tangential components of the magnetic field vanish on these surfaces. There are in effect, just two radiating slots and two nonradiating slots defined respectively by opposite side walls.

143. Explain the transmission line model of a microstrip antenna and the notion of fringing.

Hint: Consider the two radiating slots separated by a distance L. There is an electric field that exists between the top and bottom portions of each radiating slot. It is therefore possible to regard the combination of this pair of radiating slots as a transmission line of length L and apply the method of impedance transformation to such a line. Fringing is the effect caused by sudden truncation of the rectangular microstrip patch at its sidewalls within the substrate. It causes the electric field lines to acquire curvature rather than being straight and hence the effective length of the antenna is slightly more than its physical length.

144.Explain how you would perform the field analysis of a corner reflector.

Hint: Assume that the angle of the corner reflector is $\pi/2$ and that the surface of the reflector are perfectly conducting. Assume that the source is placed on the plane that bisects the two walls of the corner reflector. Then we can replace the reflector by four image sources, two of which have opposite polarity compared with the original source and one of which has the same polarity. The effective pattern is then obtained by applying the basic principles of array processing to this system of four sources.

Reference:Balanis, Antenna theory.

145. Explain what do you mean by effective dielectric constant of a microstrip antenna.

146. Explain how you would perform the field analysis of a parabolic reflector.

Hint: Assuming that the incident field emerges from the focus, we have to compute the reflected fields. We compute the value of the incident field at the point where the incident ray hits the parabolic reflecting surface. An appropriate polarization of the incident wave is assumed, for example, we can assume the electric field vector of the incident field to fall in the plane formed by the incident ray and the normal to the reflector. The polarization of the reflected ray is determined from the image condition, ie, the condition that the tangential component of the incident and reflected magnetic fields at the parabolic surface are identical. Hence, one can compute the reflected fields on the focal plane, ie, the plane passing through the focus and perpendicular to the axis of the parabolic reflector. Then using standard aperture theory, one can compute the reflected fields everywhere in space. This involves forming the equivalent electric and magnetic surface current densities on the focal plane and then performing the appropriate integrations.

147. Circular array with uniform excitation: Isotropic elements are located at the points $R.cos(2k\pi/N)\hat{x} + R.sin(2k\pi/N)\hat{y}$ with $k = 0, 1, ..., N-1$. The far field ampliude pattern neglecting polarization is given by

$$A(\theta, \phi) = \sum_{k=0}^{N-1} I(k)exp(jK.R.cos(\phi - 2k\pi/N)sin(\theta))$$

To sum this up, we note that

$$exp(jx.sin(\phi) = \sum_{n\in\mathbb{Z}} c(n).exp(jn\phi)$$

where

$$c(n) = J_n(x)$$

with J_n being the Bessel function of order n. It follows that

$$exp(jx.cos(\phi)) = \sum j^n J_n(x).exp(jn\phi)$$

and hence,

$$A(\theta, \phi) = \sum_{k,m} I(k)J_m(KR.sin(\theta)exp(jm(\phi - 2k\pi/N))$$

Defining the function

$$F(\theta, \phi) = \sum J_m(KR.sin(\theta)).exp(jm\phi)$$

we can express this amplitude pattern as

$$A(\theta, \phi) = \sum_{k=0}^{N-1} I(k)F(\theta, \phi - 2k\pi/N)$$

148. Limit of a uniform circular array when the number of elements increases to ∞. The far field amplitude pattern is then given by

$$A(\theta) = \int_0^{2\pi} exp(jK.sin(\theta)cos(\phi')d\phi'$$

This pattern has azimuthal symmetry, ie, it is independent of ϕ.

149. Explain in words how you would determine the mutual impedance between two antennas.

ans: Assume two antennas with appropriate geometry are located in space. Apply an excitation current I_0 at some point of the first antenna and assume that this current produces a current distribution $I_1(s)$ along the length of the first antenna and a current distribution $I_2(s)$ along the length of the second antenna. Here, we are assuming the two antennae to be thin wires and s is a parameter that varies along the length of the wires. For example, s can be the length parameter. We compute the two distributions $I_1(s)$ and $I_2(s)$ using Pocklington's integral equation. Having done so, we can also compute the tangential component of the electric field on the second antenna. Denote this by $E_{21}(s)$. The subscript 21 corresponds to the fact that this field is produced along the length of the second antenna due to the excitation of the first antenna. The power dissipated in the second antenna is then the line integral

$$W_2 = \int_{\Gamma_2} E_{21}(s)I_2(s)ds$$

where Γ_2 is the curve of the second antenna. Assume that the feeder point of the second antenna is $s = 0$. Then the mutual impedance is computed as

$$Z_{21} = \frac{W_2}{I_0 I_2(0)}$$

150. Give a closed form integral expression for the far field amplitude pattern produced by a circular disc of radius R in the $x-y$ plane carrying a surface current $\mathbf{J}_s = J_s\hat{\phi} = J_s(-sin(\phi)\hat{x} + cos(\phi)\hat{y})$.

ans:

$$\mathbf{A}(r,\theta,\phi) = \frac{\mu.exp(-jkr)}{4\pi.r}\int_0^R\int_0^{2\pi} J_s(-sin(\phi')\hat{x} + cos(\phi')\hat{y})exp(jk\rho'.sin(\theta).cos(\phi'-\phi))\rho' d\rho' d\phi'$$

It follows that

$$A_\phi(r,\theta,\phi) = -sin(\phi)A_x + cos(\phi)A_y$$

$$= \frac{\mu.J_s.exp(-jkr)}{4\pi.r}\int cos(\phi').exp(jk\rho'.cos(\phi')sin(\theta))\rho' d\rho' d\phi'$$

This formula can be expressed in terms of the Bessel functions.

$$J_0(x) = (2\pi)^{-1}\int_0^{2\pi} exp(jx.cos(\phi))d\phi$$

giving

$$J_0'(x) = (2\pi)^{-1}j\int_0^{2\pi} cos(\phi).exp(jx.cos(\phi))d\phi$$

so that

$$A_\phi(r,\theta,\phi) = -\frac{j\mu.J_s.exp(-jkr)}{2r}\int_0^R J_0'(k\rho'.sin(\theta))d\rho'$$

149. Let $\}$ be a Lie algebra and G the corresponding Lie group. Then, determine a closed form formula for the quantity

$$\frac{\partial^2}{\partial t \partial s} f(g.exp(tX)(Y+sZ)exp(-tX)g^{-1})|_{s=t=0}$$

where $f : \} \to \mathbb{C}$ is a smooth mapping.

ans:Differentiating with respect to s at $s = 0$ gives

$$\partial(g.exp(tX)Zexp(-tX)g^{-1})f(g.exp(tX)Y.exp(-tX)g^{-1})$$

Differentiating this with respect to t at $t = 0$ then gives us

$$\partial(g[X,Z]g^{-1})f(gYg^{-1}) + \partial(g[X,Y]g^{-1})\partial(gZg^{-1})f(gYg^{-1})$$

Thus, defining

$$q = [X,Z] + [X,Y]Z = (L_{[X,Y]} + ad(X))Z$$

an element of the symmetric algebra $S(\})$, it follows that setting $f(g : X) = f(gXg^{-1})$, we have

$$f(g; X_1...X_n : Y, \partial(p)) = f(gXg^{-1}, g\partial(q)g^{-1})$$

where $p \in S(\})$ and

$$q = (L_{[X_1,Y]} + ad(X_1))(L_{[X_2,Y]} + ad(X_2)...(L_{[X_n,Y]} + ad(X_n))p$$

We set

$$q = \Gamma_Y(X_1...X_n \times p)$$

Then, for each $Y \in \}$, we have the map $\Gamma_Y : \mathcal{B} \times S(\}) \to S(\})$ where $\mathcal{B}$ is the universal enveloping algebra generated by $\}$. The map Γ_Y is clearly onto if we define for scalars c, $\Gamma(c,p) = cp$, ie, $\Gamma(1,p) = p$. Let now $\}$ be semisimple and $\langle$ a Cartan subalgebra. Let $\int$ be the direct sum of all the root spaces of $\}$ relative to $\langle$. Then, we have the direct sum decomposition $\} = \langle + \int$. Let $S(\int)$ denote the symmetric algebra generated by $\int$ and let λ denote the symmetrizer map. λ maps the symmetric algebra isomorphically (as a vector space) onto the universal enveloping algebra. Let $H \in \langle$ be such that $\alpha(H) \neq 0$ for all roots α. Let $Z_1, ..., Z_r \in \int$. The action of adH on $\int$ is nonsingular since $ad(H)X = \alpha(H)X$ for $X \in \}_\alpha$ and $\alpha(H) \neq 0$ if α is a root. Hence, there exist elements $Y_1, ..., Y_r \in \int$ such that $adH(Y_i) = -Z_i, i = 1, 2, ..., r$. It then follows that for $Z'_i \in \langle, i = 1, 2, ..., k$, we have

$$\Gamma_H(Y_1...Y_r, Z'_1...Z'_k)) = (-L_{[H,Y_1]} + ad(Y_1))...(-L_{[H,Y_r]} + ad(Y_r))(Z'_1...Z'_k)$$

$$= (Z_1 + ad(Y_1))...(Z_r + ad(Y_r))Z'_1...Z'_k = Z_1...Z_rZ'_1...Z'_k$$

modulo $S_{r+k-1}(\})$. Now $Z_1...Z_r$ and $\lambda(Z_1...Z_r)$ are equal modulo $\lambda(S_{r-1}(\}))$, ie, modulo the space of elements in $\mathcal{B}$ having order $\leq r-1$. It follows therefore that

$$\Gamma_H(\lambda(Y_1...Y_r), Z'_1...Z'_k) = Z_1...Z_rZ'_1...Z'_k$$

modulo $\lambda(S_{r-1}(\}))$ and by the Poincare-Birkhoff Witt theorem, it follows that Γ_H maps $\lambda(S(\int)) \times S(\langle)$ onto $\mathcal{B}$. From a dimensional argument, it is then easily concluded that this mapping is one-one too.

Reference: Harish Chandra, Differential Operators on a Lie Algebra.

150. Suppose $L(q,q')$ is the Lagrangian function of a mechanical system and let $\bar{L}(q,q')$ be another Lagrangian such that for some function $f(q,q')$, we have

$$L(q,q') - \bar{L}(q,q') = df(q,q')/dt$$

Then what is the relationship between the Hamiltonians H and $\bar{H}$ corresponding to these two Lagrangians? State with a proof why the two Lagrangians L and $\bar{L}$ yield the same equations of motion.

151. Volterra system arising in the motion of a simple pendulum:

$$\theta''(t) = -g.sin(\theta(t)) + f(t)$$

The approximation to this system is

$$\theta''(t) = -g\theta(t) + g\theta^3(t)/6 + f(t)$$

Using perturbation theory we write

$$\theta(t) = \theta^{(0)}(t) + \theta^{(1)}(t)$$

so that

$$\theta^{(0)''}(t) = -g.\theta^{(0)}(t) + f(t),$$

$$\theta^{(1)''}(t) = -g.\theta^{(1)}(t) + g\theta^{(0)3}(t)$$

This has solutions using Laplace transforms

$$\theta^{(0)}(t) = g^{-1/2}\int_0^t sin((t-u)\sqrt{g})f(u)du$$

and

$$\theta^{(1)}(t) = \sqrt{g}\int_0^t sin((t-u)\sqrt{g})\theta^{(0)3}(u)du$$

The expression for $\theta^{(1)}(t)$ can be expressed in the form

$$\theta^{(1)}(t) = \int_0^t\int_0^t\int_0^t h(t-u,t-v,t-w)f(u)f(v)f(w)dudvdw$$

Problem:Determine the kernel h in terms of generalized functions, ie, δ functions.

152. What do you understand by a partial order on a set. Give examples.

ans: A relation $\geq$ between elements of a set E for which $a \geq a$, $a \geq b$ and $b \geq a$ imply $a = b$ and finally, $a \geq b$ and $b \geq c$ together imply $a = b$. For example the usual $\geq$ on $\mathbb{R}$. Another example is let E be the collection of all subsets of a set Ω and let $A \geq B$ iff $B \subset A, A, B \in E$. Then we get a partial order on E. Note that in the previous example, all elements are comparable, ie $a \geq b$ or $b \geq a$ holds for all $a, b \in \mathbb{R}$. In this example of subsets of a set, however, there can exist elements that are not comparable.

153. Define the adjoint representation of a Lie group on its Lie algebra starting from the notion of tangent spaces on a differentiable manifold and the differential of a smooth map.

154. If $\mathcal{M}$ and $\mathcal{N}$ are differentiable manifolds and $f : \mathcal{M} \to \mathcal{N}$ a smooth map, then we define its differential at $x \in \mathcal{M}$ as a linear map $df(x) : T\mathcal{M}_x \to T\mathcal{N}_{f(x)}$. This is defined so that if $v \in T\mathcal{M}_x$ and g is defined in a nhood of $f(x)$, then $df(x)(v)(g) = v(gof)$. Using this definition, show that relative to local bases for $\mathcal{M}$ and $\mathcal{N}$, $df(x)$ is a Jacobian matrix.

155. Suppose we second quantize both the photon radiation and the electron-positron matter fields. The Dirac current operator is then $J^\mu = \bar{\psi}\gamma^\mu\psi$. Let A_μ be the quantized electromagnetic four potential. The interaction energy between currents and radiation is then $\int J^\mu A_\mu d^3x$. Express this in terms of the creation and annihilation operators of the photon field in momentum space as well as the creation and annihilation operators of the electron and positron fields in momentum space.

hint: The Dirac field ψ, after second quantization, can be expressed as a linear superposition of the annihilation operators of the electrons and the creation operators of the positrons:

$$\psi(x) = \int (2p^0)^{-1/2}(u_s(p)a_s(p)exp(-ip.x) + v_s(p)b_s(p)^*exp(ip.x))d^3p$$

where

$$p = (p^0, \mathbf{p}), p^0 = \sqrt{m^2 + \mathbf{p}^2}$$

$u_s(p)$ and $v_s(p)$ satisfy the Dirac equation

$$(\gamma.p - m)u_s(p) = 0, (\gamma.p + m)v_s(p) = 0, s = 1, 2$$

s specifies the polarization state of the electrons and positrons. $\gamma = (\gamma^\mu)$ are the 4×4 Dirac matrices and $\gamma.p = \gamma^\mu p_\mu$. In the standard representation,

$$\gamma^0 = \begin{pmatrix} 0 & I \\ I & 0 \end{pmatrix}$$

and

$$\gamma^r = \begin{pmatrix} 0 & \sigma_r \\ -\sigma_r & 0 \end{pmatrix}$$

where $\sigma_r, r = 1, 2, 3$ are the Pauli matrices. Note that

$$\gamma^\mu\gamma^\nu + \gamma^\nu\gamma^\mu = 2g^{\mu\nu}$$

Let u be a solution to the Dirac equation in the momentum domain, ie,

$$(\gamma.p - m)u = 0$$

Set

$$u = \begin{pmatrix} \phi \\ \chi \end{pmatrix}$$

where ϕ and χ are two component spinors. Then,

$$\gamma.p = \gamma^0 p^0 - \gamma^r p^r = \begin{pmatrix} 0 & p^0 + (\sigma, \mathbf{p}) \\ p^0 - (\sigma, \mathbf{p}) & 0 \end{pmatrix}$$

so that the Dirac equation reads

$$(p^0 + (\sigma, \mathbf{p}))\chi = m\phi, (p^0 - (\sigma, \mathbf{p}))\phi = m\chi$$

There is another representation of the Dirac γ matrices. The $\gamma^r, r = 1, 2, 3$ are defined as before but

$$\gamma^0 = \begin{pmatrix} I & 0 \\ 0 & -I \end{pmatrix}$$

The same anticommutation relations are satisfied in this representation. For this representation, we find that

$$(p^0 - m)\phi = (\sigma, \mathbf{p})\chi, (\sigma, \mathbf{p})phi = (p^0 + m)\chi$$

In this representation, we can write

$$\psi = \begin{pmatrix} (p^0 + m)w \\ (\sigma, \mathbf{p})w \end{pmatrix}$$

where w is any 2×1 vector. $a_s(p)$ annihilates an electron of four momentum p and polarization state s while $b_s(p)$ annihilates a positron of four momentum p and polarization state s. $b_s(p)^*$ creates a positron of four momentum p and polarization state s. Likewise, we have for the conjugate field

$$\bar{\psi}(x) = \int (2p^0)^{-1/2}(\bar{u}_s(p)a_s(p)^* exp(ip.x) + \bar{v}_s(p)b_s(p)exp(-ip.x))d^3p$$

156. Let G be a group and H a subgroup of G. Define the centralizer of H in G. Show that the centralizer is a subgroup.

ans: The centralizer of H in G is the set

$$\mathcal{C}(H) = \{g \in G : gh = hg \forall h \in H\}$$

If $g_1, g_2 \in G$, then $g_2 g_1 h = g_2 h g_1 = h g_2 g_1$ implying that $g_2 g_1 \in G$.

157. What do you understand by the conjugacy class of an element in a group.

If G is a group and $x \in G$, then the conjugacy class of x is the set

$$N[x] = \{gxg^{-1} : g \in G\}$$

$N[x]$ is the set of all elements of G that are conjugate to x in G.

158: Let G be a group and let $x \in G$. Define

$$C(x) = \{g \in G : gx = xg\}$$

Show that $C(x)$ is a subgroup of G. Define $T : G/C(x) \to N(x)$ so that $T(gC(x)) = gxg^{-1}$. Then show that if T is a well defined map. $g_2C(x) = g_1C(x)$, then $T(g_2C(x)) = T(g_1C(x))$. Hence show that T is a bijection and hence conclude that $\mu(N(x)) = \mu(G)/\mu(C(x))$.

158. Let $\bar{\psi}$, as before, denote the second quantized electron-positron field and $A_\mu(x)$ the second quantized photon field. Determine second order corrections to the transition probability amplitude for electron-electron scattering.

Hint: The matrix element

$$< p_1'p_2'|T(\int A_\mu(x)A_\nu(y)\bar{\psi}(x)\gamma^\mu\psi(x)\bar{\psi}(y)\gamma^\nu\psi(y)d^4xd^4y)|p_1p_2 >$$

must be evaluated. This factors as

$$\int < p_1'p_2'| \int \bar{\psi}(x)\gamma^\mu\psi(x)\bar{\psi}(y)\gamma^\nu\psi(y)dxdy.D_{\mu\nu}(x-y)|p_1p_2 >$$

where $D_{\mu\nu}$ is the photon propagator, ie,

$$D_{\mu\nu}(x-y) =< 0|T(A_\mu(x)A_\nu(y))|0 >$$

To evaluate the photon propagator, we must expand $A_\mu(x)$ in terms of the annihilation and creation operators of the photon field in momentum space and make use of the commutation relations

$$[a_s(p), a_s^*(p')] = \delta_{ss'}\delta^3(p-p')$$

etc. To evaluate the matrix element

$$< p_1'p_2'|\bar{\psi}(x)\gamma^\mu\psi(x)\bar{\psi}(y)\gamma^\nu\psi(y)|p_1p_2 >$$

we must write $|p_1p_2 >= a(p_1)^*a(p_2)^*|0 >$ and $< p_1'p_2'| =< 0|a(p_1)a(p_2)$ and use the anticommutation rules to contract the ψ terms with the creation and annihilation operators. The anticommutation rules are

$$\{a_s(p_1), a_m(p_2)^*\} = \delta_{sm}\delta^3(p_1-p_2)$$

$$\{a_s(p_1), a_m(p_2)\} = \{a_s(p_1)^*, a_m(p_2)^*\} = 0$$

where $\{X, Y\}$ is the anticommutator of X with Y, ie, $\{X, Y\} = XY + YX$.

159(a). Let $\}$ be a semisimple Lie algebra and let $X_1, ..., X_n$ be a basis for this Lie algebra. Define $p(X_1, ..., X_n) = \sum c(m_1, ..., m_n)X_1^{m_1}...X_n^{m_n}$ an element of the symmetric algebra $S(\})$. Determine $\partial(X_1^{m_1}...X_n^{m_n})p(0)$.

ans: $m_1!...m_n!c(m_1, ..., m_n)$.

159(b).Derive the partial differential equations governing the stream function in two dimensional incompressible fluid flow problems.

ans:

$$v = v_x(t,x,y)\hat{x} + v_y(t,x,y)\hat{y}$$

The incompressibility condition reads

$$v_{x,x} + v_{y,y} = 0$$

which means the existence of a function $\psi(t,x,y)$ such that

$$v_x = \psi_{,y}, v_y = -\psi_{,x}$$

The vorticity is

$$\Omega = curlv = \hat{z}(v_{y,x} - v_{x,y}) = -\Delta\psi$$

where $\Delta = \partial^2/\partial x^2 + \partial^2/\partial y^2$ is the Laplacian. The Navier-Stokes equation is

$$\rho(v_{,t} + \Omega \times v) = -\rho\nabla\Phi - \nabla p$$

and taking the curl this gives

$$\Omega_{,t} + \nabla \times (\Omega \times v) = 0$$

Now,

$$\Omega \times v = \Omega_z \hat{z} \times (v_x\hat{x} + v_y\hat{y}) = \Omega_z(v_x\hat{y} - v_y\hat{x})$$

Thus,

$$\begin{aligned}\nabla \times (\Omega \times v) &= \hat{z}((\Omega_z v_x)_{,x} + (\Omega_z v_y)_{,y}) \\ &= \hat{z}(-((\Delta\psi)\psi_{,y})_{,x} + ((\Delta\psi)\psi_{,x})_{,y}) \\ &= \hat{z}((\Delta\psi_{,y})\psi_{,x} - \Delta\psi_{,x}\psi_{,y})\end{aligned}$$

It follows that ψ satisfies the pde

$$-\Delta\psi_{,t} - (\Delta\psi_{,y})\psi_{,x} - \Delta\psi_{,x}\psi_{,y} = 0$$

This is for a inviscid fluid. If the fluid has viscosity, then the right side of the Navier-Stokes equation acquires a term $\nu\Delta v$ which on taking curl becomes $\nu\Delta\Omega = -\nu\Delta\Delta\psi$. We then get

$$-\Delta\psi_{,t} - (\Delta\psi_{,y})\psi_{,x} - (\Delta\psi_{,x})\psi_{,y} + \nu\Delta\Delta\psi = 0$$

159(c). Derive the recursion for the Delta modulator.

Hint: Let $m(t)$ be the message signal sampled at the rate of Δ^{-1} and let $\hat{m}[k]$ denote the Delta modulated signal. Then $\hat{m}[k+1] = \hat{m}[k] + \delta$ if $\hat{m}[k] < m(k\Delta)$ and $\hat{m}[k+1] = \hat{m}[k] - \delta$ if $\hat{m}[k] > m(k\Delta)$. Thus the recursion can be cast as

$$\hat{m}[k+1] = \hat{m}[k] + \delta.sgn(m(k\Delta) - \hat{m}[k])$$

159(d). Let G be a complex Lie group with Lie algebra $\}$. Let $\}_0$ be a real Lie algebra such that $\}$ equals the complexification of $\}_0$. Let $\sqcap$ be a compact real form of $\}$ and let $\approx_0 = \sqcap \cap \}_0$ and $\surd_0 = i\sqcap \cap \}_0$. Then show that $\}_0$ is the direct sum of $\approx_0$ and $\surd_0$. Show that $\approx_0 + i\surd_0 = \sqcap$. Let $\approx$ be the complexification of $\sqcup_0$ and $\surd$ the complexification of $\surd_0$. Show that $\}$ is the direct sum of $\approx$ and $\surd$. For example, let $\}$ be $sl(n,\mathbb{C})$, $\}_0 = sl(n,\mathbb{R})$ and let $\sqcap = so(n,\mathbb{R})$. Then, $\approx_0 = so(n,\mathbb{R})$. $\approx_0$ is precisely the set of skewsymmetric real matrices and $\surd_0$ is precisely the set of purely imaginary Hermitian matrices. What are $\approx$ and $\surd$ here. Consider the general case now. Let $\dashv_{p0}$ be a maximal Abelian subspace of $\surd_0$. On complexification, we get $\dashv_p$ as a maximal Abelian subspace of $\surd$. $\dashv_{p0}$ can be extended to get a Cartan subalgebra $\dashv_0$ of $\}_0$. Let $\dashv_{t0} = \dashv_0 \cap \}_0$. Then $\dashv_0$ is the direct sum of $\dashv_{p0}$ and $\dashv_{t0}$. As usual we denote by $\dashv_t$, the complexification of $\dashv_{t0}$. Then $\dashv$ is the direct sum of $\dashv_p$ and $\dashv_t$. Let Q be a system of positive roots of $\}$ relative to $\dashv$, ie, if $\beta \in Q$, we have a nonzero $X_\beta \in \}$ such that for all $\Gamma \in \dashv$, $[\Gamma, X_\beta] = \beta(\Gamma)X_\beta$. Here, β is a linear functional on $\dashv$. Let $\updownarrow$ denote the centralizer of $\dashv_p$ in $\approx$. Let Q_+ denote the set of positive roots, ie, elements of Q that do not vanish identically on $\dashv_p$. Let $q = \sum_{\beta\in Q_+} \mathbb{C}X_\beta$ and $q- = \sum_{\beta\in Q_+} \mathbb{C}X_{-\beta}$. Note that X_β is an eigenvector of each element of $\dashv_p$ corresponding to a nonzero eigenvalue for at least on element in $\dashv_p$. The elements of $\updownarrow$ all have zero eigenvalue for every element in $\dashv_p$. We have the direct sum decomposition

$$\} = q - + \updownarrow + \dashv_p + q$$

Note that $\dashv_t \subset \updownarrow$ since $\dashv = \dashv_t + \dashv_p$ is an Abelian algebra. Let θ denote the standard conjugation on $\}$ relative to the compact real form $\sqcap$. Let B be the Cartan-Killing form on $\}$. For $X, Y \in \}$, define $< X, Y >= -B(X, \theta(Y))$. Then $< .,. >$ is an inner product on $\}$. Indeed, if $X \in \sqcup_0$, then X is skew symmetric real and hence skew Hermitian in the special case considered above. Then, adX has all purely imaginary eigenvalues and hence $-Tr(adX)^2$ is positive. In the general case, the adjoint representation of a compact group in its Lie algebra can be transformed into a unitary representation relative to an appropriately chosen inner product on the Lie algebra. It follows that the eigenvalues of each element in this representation will fall on the torus and hence the eigenvalues of the differential of this adjoint representation will be purely imaginary. But $\frac{d}{dt}Ad(exp(tX))|_{t=0} = ad(X)$. It follows that adX will have purely imaginary eigenvalues if X varies in the Lie algebra of a compact group. It follows that for $X \in \approx_0$,

$$-B(X, \theta(X)) = -B(X, X) \geq 0$$

and

$$-B(iX, \theta(iX)) = B(iX, iX) = -B(X, X) \geq 0$$

proving that $-B(X, \theta(X)) \geq 0$ for $X \in \approx$. On the other hand, for $Y \in \surd_0$, we have $iY \in \sqcap$ and hence

$$0 \leq -B(iY, \theta(iY)) = -B(iY, iY) = B(Y, Y) = -B(Y, \theta(Y))$$

Thus,

$$B(Y, \theta(Y)) \leq 0, Y \in \surd$$

Note how the conjugation works: $\theta(X) = X$ for $X \in \approx_0$ and $\theta(Y) = -Y$ for $Y \in \sqrt{\ }_0$. Thus, $\theta(iX) = -iX$ for $X \in \approx_0$ and $\theta(iY) = iY$ for $Y \in \sqrt{\ }_0$. It follows that $\theta(X+iY) = X+iY$ for $X \in \approx_0$ and $Y \in \sqrt{\ }_0$, ie, $\theta(Z) = Z$ for $Z \in \sqcap$ as it should be. Suppose $X \in \approx$ and $Y \in \sqrt{\ }$. Then write $X = X_1+iX_2, Y = Y_1+iY_2$ where $X_1, X_2 \in \approx_0$ and $Y_1, Y_2 \in \sqrt{\ }_0$. Then

$$X + Y = X_1 + Y_1 + i(X_2 + Y_2)$$

$$\theta(X + Y) = X_1 - Y_1 - i(X_2 - Y_2)$$

Then since $B(U, V) = 0$ if $U \in \sqcup_0$ and $V \in \sqrt{\ }_0$, it follows that

$$B(X + Y, \theta(X + Y)) = B(X_1 + iX_2, X_1 - iX_2) + B(Y_1 + iY_2, iY_2 - Y_1)$$

$$= B(X_1, X_1) + B(X_2, X_2) - B(Y_1, Y_1) - B(Y_2, Y_2) \geq 0$$

160. Show that the solution to the vector differential equation

$$\mathbf{x}'(t) = \mathbf{V}(t)\mathbf{x}(t)$$

with initial condition $\mathbf{x}(0) = \mathbf{a}$ is given by

$$\mathbf{x}(t) = \mathbf{a} + (\sum_{n=1}^{\infty} \int_{0<t_n<...<t_1<t} \mathbf{V}(t_1)...\mathbf{V}(t_n)dt_n dt_{n-1}...dt_2 dt_1)\mathbf{a}$$

161. Explain what you mean by integrating factor for a first order differential equation.

hint: Consider the differential equation

$$x'(t) = a(t)x(t) + r(t)$$

where x is the function to be determined and $a(t)$ and $r(t)$ are known functions. Define $\phi(t) = exp(-\int_0^t a(s)ds)$. Then, the above equation can be expressed as

$$\frac{d}{dt}(x(t).exp(-\int_0^t a(s)ds)) = r(t).exp(-\int_0^t a(s)ds)$$

and hence x can be expressed in terms of an integral of a known function and its initial value.

162. State the general form of the heat equation in one spatial dimension and explain how you would solve it using the separation of variables technique.

163. Determine the Green's function for the Poisson equation in three dimensions.

$$\Delta G(xyz|x'y'z') = \delta(x-x')\delta(y-y')\delta(z-z')$$

164. What do you understand by a reductive Lie algebra. Show that for a reductive Lie algebra $\}$, we have that $\}$ is the direct sum of its centre and its commutator.

ans: Let $\}$ be a Lie algebra and $\updownarrow$ a Lie subalgebra of $\}$. For $X \in \updownarrow$ and $Y \in \}$, define $\sigma(X)(Y) = [X,Y] = ad(X)Y$, ie, $\sigma(X) = adX$. Then σ is a representation of $\updownarrow$ in $\}$. If the representation σ is semisimple, then $\updownarrow$ is said to be reductive in $\}$. In this case we can write

$$\} = \bigoplus_{i=1}^{n} \}_i$$

where $\}_i$ are simple ideals in $\}$. $\}_i$ is irreducible under σ for every i. It follows that either σ vanishes on $\}_i$ in which case, $\}_i$ is in the centre of $\updownarrow$, or else, $\}_i$ provides us with the substratum of a nonzero irreducible reprsentation of $\updownarrow$. Suppose $\}$ is reductive in itself. Then, we say that $\}$ is a reductive Lie algebra. In this case, in the above decomposition, for each i, either $\}_i$ is in the centre of $\}$ or else, $ad(X)g_i$ as X varies over $\}$ is the whole of $\}_i$. Now $[\}_i, \}_j] = 0$ if $i \neq j$ since $\}_i$ is an ideal in $\}$ and $\}$ is the direct sum decomposition of the $\}_i's$. It follows that if $\}_i, i \in J$ is the set of all the $\}_i's$ which are not contained in the centre of $\}$, then

$$[\}, \}] = \sum_{i \in J} [\}_i, \}_i] = \sum_{i \in J} [\}, \}_i] = \sum_{i \in J} \}_i$$

Hence, we have the direct sum decomposition

$$\} = \rfloor \oplus [\}, \}]$$

where $\rfloor$ is the centre of $\}$.

165.State the Iwasawa decomposition for Lie groups and Lie algebras.

166.We have seen the following. Suppose $\}$ is a semisimple Lie algebra with Cartan subalgebra $\langle$ and let $\int$ denote the direct sum of all the root spaces of $\}$ relative to $\langle$. Let $S(\int)$ denote the symmetric algebra over $\int$ and $\mathcal{S} = \lambda(S(\int))$ is then a subalgebra of $\mathcal{B}$, the universal enveloping algebra of $\}$. We have then seen that $\Gamma : \mathcal{S} \times S(\langle) \to S(\})$ is a bijection satisfying

$$f(x; b : H; \partial(p)) = f(x : H; \partial(\Gamma(b \times p)))$$

where $f(xH) = f(Ad(x)H)$. Let $p \in S(\})$. Then it follows that we can find elements $s_i \in S'$, $h_i \in S(\langle)$ and $\beta \in S(\langle)$ such that

$$p = \Gamma(1 \times \beta + \sum_i s_i \times h_i)$$

Here S' consists of elements of S having degree at least one. Then we get

$$f(x : H; \partial(p)) = f(x : H; \partial(\beta)) + \sum f(x : H; \partial(\Gamma(s_i \times h_i)))$$

$$= f(x : H; \partial(\beta)) + \sum f(x; s_i : H, \partial(h_i))$$

We set $\beta = \beta(p)$ and $\partial(\beta(p)) = \delta'(p)$. The mapping $p \to \delta'(p)$ is well defined. Here, p is an element of $S(\})$ and $\delta'(p)$ is an element of $\partial(S(\langle))$. It can be shown easily that the map $p \to \delta'(p)$ is a homomorphism. Ideally, $\delta'(p)$ depends on H_0 and hence we should write $\delta'_{H_0}(p)$ in place of $\delta'(p)$.

With the usual notation, we set $\omega(X) = sp(adX)^2 = B(X, X)$. Then we have seen that the root vectors X_α and $X_{-\alpha}$ with α ranging over the positive roots P may be determined so that $[X_\alpha, X_{-\alpha}] = H_\alpha$ and $B(X_\alpha, X_{-\alpha}) = 1$. We choose $X \in \}$ and set

$$X = H + \sum_{\alpha \in P} c_\alpha X_\alpha + c_{-\alpha} X_{-\alpha}$$

so that

$$B(X, X_\alpha) = c_{-\alpha}, B(X, X_{-\alpha}) = c_\alpha$$

Also $B(H, X_\alpha) = 0$. Thus,

$$B(X, X) = B(H, H) + 2 \sum_{\alpha \in P} c_\alpha c_{-\alpha}$$

giving

$$\omega = \bar{\omega} + 2 \sum_{\alpha \in P} X_\alpha X_{-\alpha}$$

in the symmetric algebra. Note that the polynomial ω can be viewed as an element of the symmetric algebra $S(\})$ and the restriction of ω to $\langle$ is $\bar{\omega}$ which can be viewed an element of the symmetric algebra $S(\langle)$. For $H_0 \in \langle'_0$, (ie, $H_0 \in \langle_0$ and $\alpha(H_0) \neq 0$ for all $\alpha \in P$), we have

$$\Gamma_{H_0}(X_\alpha . X_{-\alpha} \times 1) = (L_{[X_\alpha, H_0]} + ad(X_\alpha))(L_{[X_{-\alpha}, H_0]} + ad(X_{-\alpha}))1$$

$$= (L_{[X_\alpha, H_0]} + ad(X_\alpha))[X_{-\alpha}, H_0] = (-\alpha(H_0) L_{X_\alpha} + ad(X_\alpha))(-\alpha(H_0) X_{-\alpha})$$

$$= \alpha(H_0)^2 X_\alpha X_{-\alpha} - \alpha(H_0)[X_\alpha, X_{-\alpha}]$$

$$= \alpha(H_0)^2 X_\alpha X_{-\alpha} - \alpha(H_0) H_\alpha$$

Now to go further and use the isomorphism property of Γ on $\lambda(S(\int)) \times S(\langle)$, we must express the element $X_\alpha . X_{-\alpha}$ in $\mathcal{B}$ in terms of $\lambda(S(\int))$. This is achieved by noting that in $\mathcal{B}$, we have

$$X_\alpha X_{-\alpha} = \frac{1}{2}(X_\alpha . X_{-\alpha} + X_{-\alpha} . X_\alpha) + \frac{1}{2}[X_\alpha, X_{-\alpha}]$$

$$= \frac{1}{2}(X_\alpha X_{-\alpha} + X_{-\alpha} X_\alpha) + H_\alpha/2$$

and clearly the first term here is in $\lambda(S(\mathfrak{f}))$. It follows that

$$\Gamma_{H_0}(\frac{1}{2}(X_\alpha X_{-\alpha}+X_{-\alpha}X_\alpha)+(H_\alpha/2)\times 1)$$
$$=\alpha(H_0)^2 X_\alpha X_{-\alpha}-\alpha(H_0)H_\alpha$$

Now observe that since $ad(H_\alpha)1=0$ and $[H_\alpha,H_0]=0$, it follows that

$$\Gamma_{H_0}(H_\alpha\times 1)=0$$

and hence we get from the above,

$$\Gamma_{H_0}(\alpha(H_0)^{-2}(X_\alpha X_{-\alpha}+X_{-\alpha}X_\alpha)\times 1+1\times\bar{\omega}+2\alpha(H_0)^{-1}(1\times H_\alpha))$$
$$=\bar{\omega}+2\sum_{\alpha\in P}X_\alpha X_{-\alpha}=\omega$$

and hence it follows that

$$\delta'(\partial(\omega))=\partial(\bar{\omega})+2\sum_{\alpha\in P}\alpha(H_0)^{-1}\partial(H_\alpha)$$

Choose a basis $H_1,...,H_l$ for $\mathfrak{h}$ such that $B(H_i,H_j)=\delta_{ij}$. Then we shall evaluate $\bar{\omega}$ in terms of this basis. For $H\in\mathfrak{h}$,

$$\bar{\omega}(H)=sp(adH)^2=B(H,H)=\sum_i B(H,H_i)^2$$

In other words,

$$\bar{\omega}=\sum H_i^2$$

and hence

$$\{\partial(\bar{\omega}),\alpha\}=\sum\{\partial(H_i)^2,\alpha\}$$
$$=\sum\partial(H_i)\{\partial(H_i),\alpha\}+\{\partial(H_i),\alpha\}\partial(H_i)$$
$$=2\sum\alpha(H_i)\partial(H_i)=2\partial(\sum\alpha(H_i)H_i)$$

Now,

$$\sum\alpha(H_i)H_i=\sum B(H_\alpha,H_i)H_i=H_\alpha$$

It follows therefore that

$$\{\partial(\bar{\omega}),\alpha\}=2\partial(H_\alpha)$$

Let $\alpha_1,...,\alpha_r$ denote the positive roots and define the homogeneous polynomial of degree r

$$\pi=\Pi_{j=1}^r\alpha_j$$

167. Suppose V is a finite dimensional real vector space and $B : V \times V \to \mathbb{R}$ is a bilinear form on V. For $f : V \to \mathbb{C}$ with f rapidly decreasing, define the Fourier transform of f using the formula

$$Ff(X) = \int_V f(Y)exp(i.B(X,Y))dY$$

where dY is the Lebesgue measure on V relative to some ordered basis. What is the Fourier transform of foT^{-1}, where T is a non-singular linear transformation on V ?

Ans:

$$F(foT^{-1})(X) = \int f(T^{-1}Y)exp(iB(X,Y))dY = \int f(Y)exp(iB(X,TY))|detT|dY$$

Suppose B is T-invariant,ie, $B(TX,TY) = B(X,Y)$ for all $X, Y \in V$. Then, it follows from the above that

$$F(foT^{-1})(X) = |detT|.F(f)(T^{-1}X)$$

On the other hand, suppose $B(TX,TY) = \lambda(T)B(X,Y)$ where $\lambda(T)$ is a real scalar. Then, $B(X,TY) = B(TT^{-1}X,TY) = \lambda(T)B(T^{-1}X,Y)$ whence

$$F(foT^{-1})(X) = |detT|F(f)(\lambda(T)T^{-1}X)$$

168. Solve the generalized wave equation with source in n dimensions:

$$\sum_{\alpha,\beta=1}^{n} a_{\alpha,\beta}\frac{\partial^2 f(\mathbf{x},t)}{\partial x_\alpha \partial x_\beta} - \frac{\partial^2 f(\mathbf{x},t)}{\partial t^2} = g(\mathbf{x},t)$$

Hint: Use the Fourier transform in n dimensions.

169. Prove Green's identity in n dimensions:If $u, v : \mathbb{R}^n \to \mathbb{R}$ are twice differentiable, then

$$div(u\nabla v - v\nabla u) = u\nabla^2 v - v\nabla^2 u$$

implying that if S is an $n-1$ dimensional surface bounding a volume V, and having a unit normal n_i, then

$$\int_S n_i(uv_{,i} - vu_{,i})dS = \int_V (uv_{,ii} - vu_{,ii})dV$$

170. State the method of quasiclassical approximation in quantum mechanics.

ans: The wave function is set equal to $\psi = A.exp(iS/h)$ where A, S are functions of position and time. We plug this into the time dependent Schrodinger wave equation and expand S in powers of h, thus

$$S = S_0 + hS_1 + h^2 S_2 + ...$$

Compute first the Laplacian of ψ:

$$\nabla\psi = (\nabla A + iA\nabla S/h)exp(iS/h)$$

$$\nabla^2\psi = (\nabla^2 A + iA\nabla^2 S/h - A|\nabla S|^2/h^2 + 2i(\nabla A, \nabla S)/h)exp(iS/h)$$

171. Suppose $F(x)$ is a function to be minimized with respect to x by solving the equation $F'(x) = 0$. Justify the gradient algorithm that an approximate solution can be arrived at via the recursion

$$x[k+1] = x[k] - \mu.F'(x[k])$$

where μ is some positive constant. Suppose x^* is the true minimum. Then we have the approximation

$$F'(x[k]) \approx F''(x^*)(x[k] - x^*)$$

so that the above recursion approximately becomes

$$x[k+1] = x[k] - \mu F''(x^*)(x[k] - x^*)$$

Use this equation to determine approximate bounds on μ for convergence of the gradient algorithm.

172. Define the free particle Fock space associated with a Hilbert space $\mathcal{H}$.

ans: First, a tensor product on $\mathcal{H}$ must be introduced. This can be done via the GNS principle as is done in the book "An introduction to quantum stochastic calculus" by K.R.Parthasarathy. After doing so, we define

$$\Gamma(\mathcal{H}) = \mathbb{C} \oplus \bigoplus_{n=1}^{\infty} \mathcal{H}^{\otimes n}$$

Note how the inner product on $\mathcal{H}$ induces the inner product on $\Gamma(\mathcal{H})$: If $u_{n,1}, ..., u_{n,n}$ are in $\mathcal{H}$, and so are $v_{n,1}, ..., v_{n,n}$, and $u_0, v_0 \in \mathbb{C}$, then

$$< u_0 + \sum_{n=1}^{\infty} u_{n,1} \otimes ... \otimes u_{n,n}, v_0 + \sum_{n=1}^{\infty} v_{n,1} \otimes ... \otimes v_{n,n} >$$

$$= u_0\bar{v}_0 + \sum_{n=1}^{\infty} \Pi_{i=1}^{n} < u_{n,i}, v_{n,i} >$$

172. Explain the construction of the quantum Ito integral on the Boson Fock space associated with a Hilbert space.

hint:Taking the method from K.R.Parthasarathy's book, "an introduction to quantum stochastic calculus", we define the annihilation and conservation processes on the boson Fock space with respect to a martingale. Specifically, given a Hilbert space $\mathcal{H}$ and a spectral family $\xi(B)$ on the Borel subsets of the real line and taking values in the lattice $P(\mathcal{H})$, we consider a process $m(t), t \geq 0$ with values in $\mathcal{H}$ for which $\xi[0,s]m(t) = m(s), t \geq s$, ie, m is a ξ-martingale. Then we define let $e(u), u \in \mathcal{H}$ denote the exponential vectors in $\Gamma_s(\mathcal{H})$ and define for $U \in \mathcal{U}(\mathcal{H})$ and $v \in \mathcal{H}$, the Weyl operator $W(u,U)$ so that

$$W(u,U)e(v) = exp(- \parallel u \parallel^2 /2- < u, Uv >)e(Uv+u)$$

We then observe that W is a unitary operator:

$$< W(u,U)e(v), W(u,U)e(w) >= exp(-|u|^2 - < u, Uw > - < Uv, u >)exp(< Uv+u, Uw+u >) = 1$$

We have made use of the equation

$$< e(u), e(v) >= exp(< u, v >)$$

We then introduce the one parameter subgroups of $\sqcap(\Gamma_s(\mathcal{H}))$ $t \to W(tu, I)$ where $t \in \mathbb{R}$ and define the momentum field $p(u)$ as the generator of this one parameter group:

$$W(tu, I) = exp(-itp(u))$$

We can calculate the action of $p(u)$ on the exponential vectors:

$$W(tu, I)e(v) = exp(-t^2|u|^2/2 - t < u, v >)e(v+tu)$$

so that

$$< e(w), W(tu, I)e(v) >= exp(-t^2|u|^2/2 - t < u, v >)exp(< w, v+tu >)$$

giving

$$-i < e(w), p(u)e(v) >= \frac{d}{dt} < e(w), W(tu, I)e(v) > |_{t=0}$$

$$= (< w, u > - < u, v >) < e(w), e(v) >$$

We also get

$$-i < e(w), p(iu)e(v) >= (< w, iu > - < iu, v >) < e(w), e(v) >= (i < w, u > +i < u, v >) < e(w), e(v) >$$

so that

$$< e(w), (p(u) + ip(iu))e(v) >= (i < w, u > -i < u, v > -i < w, u > -i < u, v >) < e(w), e(v) >$$

$$= -2i < u, v >< e(w), e(v) >$$

$$< e(w), (p(u) - ip(iu))e(v) >= (i < w, u > -i < u, v > +i < w, u > +i < u, v >) < e(w), e(v) >$$

$$= 2i < w, u >< e(w), e(v) >$$

Define the operator

$$a(u) = -(p(u) + ip(iu))/2i$$

Note that $p(u)$ is Hermitian and so,

$$a(u)^* = (p(u) - ip(iu)^*)/2i$$

whence from the above equations,

$$<e(w),a(u)e(v)>=<u,v><e(w),e(v)>$$

$$<e(w),a(u)^*e(v)>=<w,u><e(w),e(v)>$$

as expected. The totality of the exponential vectors in Boson Fock space implies that

$$a(u)e(v)=<u,v>e(v)$$

We have

$$\frac{d}{dt}<e(w),e(v+tu)>|_{t=0}=\frac{d}{dt}exp(<w,v+tu>)|_{t=0}=<w,u><e(w),e(v)>$$

and hence, we get the nice formula

$$a(u)^*e(v)=\frac{d}{dt}e(v+tu)|_{t=0}$$

Now observe that

$$e(tu)=1\oplus\bigoplus_{n=1}^{\infty}t^nu^n/\sqrt{n!}$$

where $u^n=u\otimes...\otimes u$ n times. It follows therefore that

$$\frac{d^n}{dt^n}e(tu)|_{t=0}=\sqrt{n!}u^n$$

and hence

$$\sqrt{n!}a(v)u^n=\frac{d^n}{dt^n}a(v)e(tu)|_{t=0}=\frac{d^n}{dt^n}t<v,u>e(tu)|_{t=0}$$

$$=n!/\sqrt{(n-1)!}<v,u>u^{n-1}$$

so that

$$a(v)u^n=\sqrt{n}<v,u>u^{n-1}$$

Thus, as its name suggests, the annihilation operator $a(v)$ transforms an n-particle state to an $n-1$-particle state. On the other hand, let us compute $a^*(v)u^{\otimes n}$. We have

$$\sqrt{n!}a^*(v)u^n=\frac{d}{ds}\frac{d^n}{dt^n}e(tu+sv)|_{t=s=0}$$

$$=\sqrt{n!}\sum_{i=0}^{n}u^ivu^{n-i}$$

so that

$$a^*(v)u^n=\sum_{i=0}^{n}u^ivu^{n-i}$$

173.Show that if G is a compact Lie group, then any two Cartan subalgebras of the Lie algebra } of G are conjugate.

hint:Take a look at Varadarajan's book, "Lie groups, Lie algebras and their representations".

174.Show by means of an example, that in a semisimple Lie algebra that is not compact, two Cartan subalgebras need not be conjugate.

hint:Take a look at $Sl(2,\mathbb{R})$. It has two Cartan subgroups, the first one consisting of matrices of the form

$$u_\theta = \begin{pmatrix} cos(\theta) & -sin(\theta) \\ sin(\theta) & cos(\theta) \end{pmatrix}$$

with θ taking values in $[0, 2\pi)$, the second consists of all matrices of the form

$$m_t = \begin{pmatrix} e^t & 0 \\ 0 & e^{-t} \end{pmatrix}$$

with t taking real values. It can easily be shown that these two are maximal Abelian subgroups and not conjugate to each other. Let us prove maximal Abelian property of the second subgroup. Suppose

$$A = \begin{pmatrix} a & b \\ c & d \end{pmatrix}$$

is an element of $SL(2,\mathbb{R})$ and that $Am_t = m_t A$ for all $t \in \mathbb{R}$. Then,

$$ae^t = e^t a, be^{-t} = e^t b, ce^t = e^{-t} c, de^t = e^t d$$

for all $t \in \mathbb{R}$, whence $b = c = 0$, or A is diagonal and hence of the form m_s for some $s \in \mathbb{R}$. We can also prove that the subgroup $\{u_\theta : \theta \in [0, 2\pi)\}$ is maximal Abelian in $SL(2,\mathbb{R})$. For suppose A is as above and commutes with all the u_θ. Choose two complex unit norm vectors x, y that are eigenvectors of all the u_θ. Writing $T = [u, v]$, we find that T is a unitary matrix and

$$Tu_\theta T^* = diag[exp(i\theta), exp(-i\theta)]$$

Then writing

$$B = TAT^*$$

we have that

$$B.diag[exp(i\theta), exp(-i\theta)] = diag[exp(i\theta), exp(-i\theta)]B, \theta \in \mathbb{R}$$

and hence by the same argument as for the previous case, B must be diagonal, ie, $B = diag[\alpha, \beta]$ where $\alpha\beta = 1$. It then follows from the form of T that for $A = T^*BT$ to be a real matrix, we must necessarily have α, β to be complex conjugates of each other and hence, both of them fall on the unit circle. This proves the statement.

175. What do you understand by a Turing machine. Give an example showing how a turing machine can perform elementary computations.

176. Define a classical probability space and a quantum probability space. In what respects to these two spaces differ from each other.

177. Give an example of Schrodinger's equation in the presence of white noise and derive statistical properties of its solution.

178. If $B(t), t \geq 0$ is Brownian motion, then evaluate the characteristic function of the random variable $X(f) = \int_0^T f(t)dB(t)$, where f is a deterministic function.

179. Explain the polar decomposition and singular value decomposition for a rectangular complex matrix.

180. What is the action of the quantum gate

$$T_\alpha = \begin{pmatrix} 1 & 0 \\ 0 & exp(i\alpha) \end{pmatrix}$$

on the base states $|0>, |1>$ defined in the standard way.

181. The CNOT gate in quantum computation is defined so that $|0, x> \rightarrow |0, x>$ and $|1, x> \rightarrow |1, \bar{x}>$ where $x = 0, 1$ and $\bar{x}$ is the complement of x. Show that the CNOT gate can be expressed as

$$|x, y> \rightarrow |x, x \oplus y>$$

182. If $(X_i, \mathcal{F}_i, P_i), i = 1, 2, ..., n$ are probability spaces, then explain how you would construct the product probability space.

183. If μ, ν are finite measures on spaces $(\Omega_1, \mathcal{F}_1)$ and $(\Omega_2, \mathcal{F}_2)$ respectively then for $E \in \mathcal{F}_1 \otimes \mathcal{F}_2$, define $E_x = \{y \in \Omega_2 : (x, y) \in E\}$ for $x \in \Omega_1$. Then, show that

$$\lambda(E) = \int_{\Omega_1} \nu(E_x) d\mu(x)$$

defines a measure on the product space.

184. Let $x(t)$ be a Gaussian random process input to an LTI system having impulse response $h(t)$. The output of the system is then given by $y(t) = \int_{-\infty}^{\infty} h(\tau)x(t-\tau)d\tau$. Prove that $y(.)$ is also a Gaussian random process.

185. A random process $x(t)$ defined on a probability space is said to be a Gaussian random process if given any finite subset $\{t_1, ..., t_r\}$ of $\mathbb{R}$ and real numbers $c_1, ..., c_r$, the random variable $Z = \sum_{i=1}^{r} c_i x(t_i)$ is Gaussian. Show that if the process is Gaussian and almost surely continuous, then for any continuous function $h(t), 0 \leq t \leq T$, the random variable $\int_0^T h(t)x(t)dt$ is also a Gaussian random variable.

Hint:Compute the characteristic function of $\int_0^T h(t)x(t)dt$.

186. Prove that if X is a random variable and $\phi_X(t) = \mathbb{E}exp(itX)$ is its characteristic function, then ϕ_X is positive definite, ie, for arbitrary complex numbers $c_1, ..., c_n$ and real numbers $t_1, ..., t_n$, the inequality

$$\sum_{i,j=1}^{n} \bar{c}_i c_j \phi_X(t_i - t_j) \geq 0$$

holds.

186. If $B(t)$ is standard Brownian motion, and $f(t)$ and $g(t)$ are two nonrandom functions square integrable over the interval $[0,T]$, then

$$\mathbb{E}[(\int_0^T f(t)dB(t))(\int_0^T g(t)dB(t))] = \int_0^T f(t)g(t)dt$$

Do this problem by considering the Ito approximants to the stochastic integrals $\int_0^T f(t)dB(t)$ and $\int_0^T g(t)dB(t)$.

187. Let $B(t)$ be standard Brownian motion and let A be a square matrix. Let X be another square matrix. Then determine a differential equation satisfied by $\mathbb{E}[exp(B(t)A)Xexp(-B(t)A)]$. Solve this differential equation.

ans:

Let

$$ad(A)(X) = [A, X] = AX - XA$$

Then,

$$exp(B(t)ad(A))X = exp(B(t)A)Xexp(-B(t)A)$$

and we also have with

$$L(t) = exp(B(t)ad(A))X,$$

$$dL(t) = ad(A)(L(t))dB(t) + \frac{1}{2}(adA)^2(L(t))dt$$

so that

$$d\mathbb{E}L(t) = \frac{1}{2}(adA)^2\mathbb{E}L(t)dt$$

so that

$$\mathbb{E}L(t) = exp(\frac{t}{2}(adA)^2)X$$

Note that

$$(adA)^2Y = [A, [A, Y]] = A^2Y + YA^2 - 2AYA$$

188. Suppose $X(t), t \geq 0$ is a Markov process with transition probability density $P(t, x, A) = Pr(X(t+s) \in A | X(s) = x)$ for all $t \geq s \geq 0$. Then show that for $0 \leq t_1 \leq ... \leq t_n$, we have

$$P(X(t_i) \in A_i, i = 1, 2, ..., n) = \int_{A_1 \times ... \times A_n} \mu_{t_1}(dx_1)P(t_2 - t_1, x_1, dx_2)P(t_3 - t_2, x_2, dx_3)...$$

$$P(t_n - t_{n-1}, x_{n-1}, dx_n)$$

where

$$\mu_t(E) = Pr(X(t) \in E)$$

189. Outline the basic steps involved in obtaining an extension of a finite measure from a field of subsets of a set to the σ-field generated by the field.

190. Suppose U, V are stationary Gaussian random fields defined on $\mathbb{R}^3$. Determine a formula for the mean and correlation function of the random field $U\nabla^2 V$.

hint: Let

$$\mathbb{E}(U(x)U(y)) = R_U(x-y), \mathbb{E}(V(x)V(y)) = R_V(x-y), \mathbb{E}(U(x)V(y)) = R_{UV}(x-y)$$

Then

$$\mathbb{E}[U(x)\nabla^2 V(y)] = \nabla_y^2 \mathbb{E}(U(x)V(y)) = \nabla_y^2 R_{UV}(x-y) = \nabla_x^2 R_{UV}(x-y)$$

It follows that

$$\mathbb{E}[U(x)\nabla^2 V(x)] = \nabla^2 R_{UV}(0)$$

and further using a basis property of Gaussian random vectors,

$$\mathbb{E}[U(x)\nabla^2 V(x)U(y)\nabla^2 V(y)] = \mathbb{E}[U(x)U(y)]\mathbb{E}[\nabla^2 V(x)\nabla^2 V(y)]$$

$$+\mathbb{E}[U(x)\nabla^2 V(y)]\mathbb{E}[U(y)\nabla^2 V(x)] + \mathbb{E}[U(x)\nabla^2 V(x)]\mathbb{E}[U(y)\nabla^2 V(y)]$$

$$= R_U(x-y)(\nabla^2)^2 R_V(0) + \nabla^2 R_{UV}(c-y)\nabla^2 R_{UV}(y-x) + (\nabla^2 R_{UV}(0))^2$$

191. Suppose $\sigma(t,x)$ satisfies the inequality

$$|\sigma(t,x) - \sigma(t,y)| \le K|x-y|$$

Let $B(.)$ be standard Brownian motion and define a process $\xi_n(t)$ recursively via the Ito stochastic integral equation

$$\xi_n(t) = x + \int_0^t \sigma(s, \xi_{n-1}(s))dB(s)$$

where

$$\xi_0(t) = x$$

Show that

$$\mathbb{E}|\xi_{n+1}(t) - \xi_n(t)|^2 \le K^2 \int_0^t \mathbb{E}|\xi_n(s) - \xi_{n-1}(s)|^2 ds$$

Show that

$$max_{0<s<t}|\xi_{n+1}(s) - \xi_n(s)|^2 = max_{0<s<t}|\int_0^s (\sigma(u, \xi_n(u)) - \sigma(u, \xi_{n-1}(u))dB(u)|^2$$

and hence deduce using Doob's maximal inequality for martingales that if

$$\Delta_n(t) = \mathbb{E}max_{0<s<t}|\xi_n(s) - \xi_{n-1}(s)|^2$$

then

$$\Delta_{n+1}(t) \leq K'\mathbb{E}|\int_0^t (\sigma(u,\xi_n(u)) - \sigma(u,\xi_{n-1}(u))dB(u)|^2$$

$$\leq K''\mathbb{E}\int_0^t |\xi_n(u) - \xi_{n-1}(u)|^2 du \leq K'' \int_0^t \Delta_n(u)du$$

Deduce that

$$\Delta_n(t) \to 0, n \to \infty$$

Use this result to prove the existence and uniqueness result for solutions to stochastic differential equations.

191:Let G be a compact group and π a representation of G in a Hilbert space $\mathcal{H}$. Let dg denote the Haar measure on G and define the operator $P = \int_G \pi(g)dg$. Show that $P\pi(k) = \pi(k)P = P$ for all $k \in G$. Show that if π is a unitary representation then in addition P is an orthogonal projection. What is the subspace of $\mathcal{H}$ on which P projects orthogonally ?

192. Statistical analysis of music

1.Raagam Bihalari:Determine the mean number and mean square number of each of the swarams in each line of the song Raara venu Gopaala, rajita sadguna jayasreeraa.

saari gapa da sa nida paadapa magarisarisanida saa. saari gaapa maaga paada risa nida paamagaari saari ga ga ga, ga riga paa pa paa paa paa paadapa saasa saa saa, gari saani nidaa paa, pada pa maa ga gaa ri, gapa maga risa nida sa.

ans: The frequency table per line of the song is

sa: $4, 3, 4, 1, 2$.

ri: $3, 4, 1, 2, 1$

ga: $2, 6, 2, 2, 1$

pa: $3, 3, 7, 3, 1$

The average number of *sa* is

$$(4 + 3 + 4 + 1 + 2)/5 = 14/5 = \mu$$

Ther mean square number of $sa's$ per line is given by

$$\sigma^2 = ((4 - \mu)^2 + (3 - \mu)^2 + (4 - \mu)^2 + (1 - \mu)^2 + (2 - \mu)^2)/5$$

The mean and variance for the other swarams can likewise be computed.

2. In the previous problem, associated a sinusoid of a definite frequency to each swaram. For example, the frequency associated with *sa* is denoted by $\omega(sa)$, that associated with ri is denoted by $\omega(ri)$ etc. Assume that the duration of each sinusoid is T. Then write down the expression for the time domain signal corresponding to the entire song and determine its Fourier transform.

3. In the previous problem show that the signal has the form

$$x(t) = \sum_{n=1}^{N} x_{\omega(n)}(t - nT)$$

where $x_\omega(t)$ is the signal associated with the frequency ω.

4. Raagam Mohanam, song Ninnu Kori. The swaram string for the first part is given by ga, ga, rii, sa sa ri ri ga ga ri ri, sa ri ga ri sa ri sa da, sa ri ga paa ga ri sa ri, ga paa ga ga ri sa ri ga, ri ri sa da, sa ri ga ri ga paa, ga pa da pa da sa da da pa ga, da pa ga ri, ga, ga, paa, ga ga pa pa da da paa, da saa da da pa ga, da pa da ga da pa ga ri sa ri, ga ga paa paa, da pa da sa da sa ri sa ga ri sa, da sa ri sa da pa da sa daa, pa ga ri sa ri,...

Explain how as per the preceding problems, one can represent this song by a signal and carry out spectral analysis of this signal.

ans:We assign a truncated sinusoid having a frequency dependent on the swaram to each swaram. Let this signal corresponding to the swaram ξ be given by $x(t, \xi)$ and let the duration of this signal be T. Then the total signal associated with the entire song has the form

$$z(t) = \sum_{n=0}^{N-1} x(t - nT, \xi[n])$$

where for each n, $\xi[n]$ is one of the swarams $sa, ri, ga, ma, pa, da, nisa$. The Fourier transform of this signal is then given by

$$Z(\omega) = \int_{-\infty}^{\infty} z(t) exp(-j\omega t)dt = \sum_{n=0}^{N-1} exp(-j\omega nT) X(\omega, \xi[n])$$

where

$$X(\omega, \xi) = \int_{-\infty}^{\infty} x(t, \xi) exp(-j\omega t)dt$$

is the Fourier transform of the truncated sinusoid associated with the swaram ξ. We have

$$x(t, \xi) = (u(t) - u(t - T)) cos(\omega[\xi]t)$$

and hence,

$$X(\omega, \xi) = \int_0^T cos(\omega[\xi]t) exp(-j\omega t)dt = 0.5 \int_0^T (exp(j(\omega[\xi] - \omega)t) + exp(-j(\omega[\xi] + \omega)t))dt$$

$$= 0.5\left(\frac{exp(j(\omega[\xi] - \omega)T) - 1}{\omega[\xi] - \omega} + \frac{1 - exp(-j(\omega[\xi] + \omega)T)}{\omega[\xi] + \omega}\right)$$

Note that the total energy of the signal corresponding to the entire song is given by

$$\int_0^{\infty} z^2(t)dt = \sum_n \int_0^T x^2(t, \xi[n])dt = \frac{1}{2\pi} \sum_n \int_{-\infty}^{\infty} |X(\omega, \xi[n])|^2 d\omega$$

Now we compute the autocorrelation function of the above signal. We have

$$R_{zz}(\tau) = \int_{-\infty}^{\infty} z(t)z(t+\tau)dt = \sum_{n,m=0}^{N-1} \int_0^{\infty} x(t-nT, \xi[n])x(t+\tau-mT, \xi[m])dt$$

$$= \frac{1}{2\pi}\int_{-\infty}^{\infty} |Z(\omega)|^2 d\omega$$

$$= \frac{1}{2\pi}\sum_{n,m}\int_{-\infty}^{\infty} X(\omega, \xi[n])\bar{X}(\omega, \xi[m])exp(j\omega(\tau+(m-n)T))d\omega$$

5. Write down the time domain signal associated with the swaram string

sa sa ri ri ga, sa ri ga, sa sa ri ri ga ga ma ma,

ri ri ga ga ma, ri ga ma, ri ri ga ga ma ma pa pa,

ga ga ma ma pa, ga ma pa, ga ga ma ma pa pa da da,

ma ma pa pa da, ma pa da, ma ma pa pa da da ni ni,

pa pa da da ni pa da ni pa pa da da ni ni sa sa,

Associate a definite frequency to each swaram and compute the short time Fourier transform of the resulting signal. Also if $x(t)$ denotes the resulting signal, then evaluate the signal energy $\int_0^{NT} x^2(t)dt$. If instead of associating a definite frequency with each swaram, we associate a finite duration damped sinusoid with each swaram, then what will its Fourier transform look like ?

193. A Lie algebra } is generated by three elements H, X, Y that satisfy the commutation relations

$$[H, X] = 2X, [H, Y] = -2Y, [X, Y] = H$$

Calculate the Casimir polynomial

$$\omega(t, x, y) = -Tr(ad(tH + xX + yY)^2)$$

hint: This computation can be carried out in the basis $\{H, X, Y\}$ as follows:

$$ad(tH + xX + yY)(H) = x[X, H] + y[Y, H] = -2xX + 2yY$$

$$ad(tH + xX + yY)^2(H) = [tH + xX + yY, -2xX + 2yY]$$
$$= -2tx[H, X] + 2ty[H, Y] + 2xy[X, Y] - 2xy[Y, X]$$
$$= -4tx.X - 4ty.Y + 2xy.H$$

$$ad(tH + xX + yY)(X) = [tH + xX + yY, X] = t[H, X] + y[Y, X] = 2t.X - y.H$$

$$ad(tH + xX + yY)^2(X) = [tH + xX + yY, 2tX - yH]$$

$$= 2t^2[H, X] - xy[X, H] + 2ty[Y, X] - y^2[Y, H]$$

$$= 2t^2(2X) - xy(-2X) + 2ty(-H) - 2y^2Y = 4t^2X + 2xyX - 2tyH - 2y^2Y$$

Finally,

$$ad(tH + xX + yY)(Y) = [tH + xX + yY, Y] = t[H, Y] + x[X, Y] = -2tY + xH$$

$$ad(tH + xX + yY)^2(Y) = [tH + xX + yY, -2tY + xH]$$

$$= -2t^2[H, Y] - 2tx[X, Y] + x^2[X, H] + xy[Y, H] = 4t^2Y - 2txH + 4txX - 2xyY$$

Use these expressions to evaluate $\omega(t, x, y)$. The coefficient of H in $ad(tH + xX + yY)^2(H)$ is given by $2xy$. The coefficient of X in $ad(tH + xX + yY)^2(X)$ is given by $4t^2 + 2xy$ and finally the coefficient of Y in $ad(tH + xX + yY)^2(Y)$ is given by $4t^2 - 2xy$. Adding these three gives us

$$\omega(t, x, y) = 8t^2 + 2xy$$

Check the steps once again to ensure no calculation error has been made.

194. Consider a potential in one dimension $V(x) = V_1$ for $0 < x < a$ and $V(x) = V_2$ for $a < x < l$. Assume that V is periodic over the entire real line with a period of l. Then show that if $\psi(x)$ is a solution to the stationary state Schrodinger equation corresponding to the energy eigenvalue E, then $\psi(x + l) = \rho\psi(x)$ where ρ is a complex number of unit magnitude. Note that the condition of ρ being bounded follows from $\psi(x+nl) = \rho^n\psi(x)$ for all integers n and hence $|\rho| = 1$ if ψ is bounded. By solving Schrodinger's equation in the three regions (i) $0 < x < a$, (ii) $a < x < l$ and (iii) $l < x < l + a$ and matching the boundary conditions on ψ and its derivative, obtain the transcendental equation for the energy level E.

195.Suppose G is a semisimple Lie group and let $\}$ denote its Lie algebra. Let $\langle$ be a Cartan subalgebra of $\}$ and let $\approx$ and $\surd$ be as earlier (see the papers of HarishChandra). Thus, $\approx$ is the Lie algebra of a maximal compact subgroup K of G. Define $\langle_p = \langle \cap \surd$. Consider the root space decomposition corresponding to $\langle$. In this setup, state the Iwasawa decomposition of G as KAN where A is Abelian and N is Nilpotent.

196. Let $(\Omega, \mathcal{F}, P)$ be a probability space and let $\mathcal{F}_t, t \geq 0$ be a filtration on this space. Let $\theta(t), t \geq 0$ be a martingale on this space and let τ be a stop time with respect to this filtration. Then show that $\theta(t \wedge \tau), t \geq 0$ is also a martingale with respect to this filtration. Do this by considering the discrete time case first. Necessary conditions on θ like integrability may be assumed.

197. What do you understand by outer measure. Write down the basic inequalities associated with the outer measure.

198. Let $\}$ be a Lie algebra and let $\mathcal{V}$ be any subset of $\}$. Show that the centralizer of $\mathcal{V}$ in $\}$ is a Lie subalgebra of $\}$.

Hint: $[a, v] = 0, [b, v] = 0$ implies $[[a, b], v] = 0$ by the Jacobi identity.

199. Let A, B be two linear operators in a Hilbert space such that $[A, B] = c$ where c is a scalar. Show that $[A, B^n] = cnB^{n-1}$ and hence $[A, f(B)] = cf'(B)$

200. Find the eigenfunctions of the operator d/dx if no restriction is made on the nature of the eigenfunction except that it be differentiable.

201. Find the eigenfunctions of the operator d^2/dx^2 on the positive real line if the assumption of square integrability is made.

202. Write down Leibniz rule for $D^n(\phi\psi)$ along with a derivation using an induction argument.

203. Consider the Hermite polynomials $H_n(x), n = 0, 1, 2, \ldots$ defined via the generating function

$$exp(tx - t^2/2) = \sum_{n=0}^{\infty} t^n H_n(x)/n!$$

Prove the orthogonality of these polynomials with respect to the normal distribution.

204. If X is a normed linear space, then a linear function $f : X \to \mathbb{C}$ is said to be continuous if for every sequence $\{x_n\}$ in X such that $\| x_n - x \| \to 0$, one has $f(x_n) \to f(x)$. The linear function f is said to be bounded if $sup(|f(x)| : x \in X, \| x \| \leq 1\} < \infty$. Show that continuity and boundedness for linear functions are equivalent concepts.

205. Let $X(t), t \geq 0$ be martingale with respect to a filtration $\mathcal{F}_t, t \geq 0$. Show that if ϕ is a convex function on $\mathbb{R}$, then $\phi(X(t)), t \geq 0$ is a submartingale.

ans: for $t > s$,

$$\mathbb{E}(\phi(X(t))|\mathcal{F}_s) \geq \phi(\mathbb{E}X(t)|\mathcal{F}_s) = \phi(X(s))$$

206. Let $X(t)$ be a submartingale with respect to a filtration $\mathcal{F}_t$. Show that if ϕ is a convex non-decreasing function on $\mathbb{R}$, then $\phi(X(t))$ is a submartingale.

ans: Just use the same argument as above combined with the submartingale property and the non-decreasing property of ϕ.

207. Let $X(t)$ be a martingale. Then prove that $|X(t)|$ is a submartingale.

208. Let $X(t)$ be a diffusion process with drift coefficient $\mu(t, x)$ and diffusion coefficient $\sigma(t, x)$. Define the partial differential operator

$$L_t = \sum_i \mu_i \frac{\partial}{\partial x_i} + \frac{1}{2}\sum_{i,j} b_{ij} \frac{\partial^2}{\partial x_i \partial x_j}$$

where

$$b = \sigma\sigma^T$$

Then let f be a well behaved function and define the process

$$M(t) = f(X(t)) - \int_0^t (L_s f)(X(s))ds$$

show that M is a martingale.

Hint:Use the Ito differential rule to conclude that

$$dM(t) = \sum_{i,j} f_{,i}(X(t))\sigma_{ij}(t, X(t))dB_j(t)$$

where $B(.)$ is the Brownian motion that drives the process X. Estimating the parameters of transistors using a Volterra model combined with wavelets.

209. Let V be a vector space and $B : V \times V \to \mathbb{C}$ a non-degenerate bilinear functional. For $X \in V$, define the linear functional $f_X = B(X, .)$. Let $Y \in V$ and define $g(Y) = exp(f_X(Y))$. Show that if $Z \in V$, then

$$g(Y; Z) = \frac{d}{dt} g(Y + tZ)|_{t=0} = f_X(Z)g(Y)$$

ans:

$$g(Y) = exp(B(X, Y))$$

$$g(Y + tZ) = exp(B(X, Y + tZ)) = exp(B(X, Y))exp(tB(X, Z)) = g(Y)exp(tB(X, Z))$$

so that

$$\frac{d}{dt} g(Y + tZ)|_{t=0} = B(X, Z)g(Y)$$

More generally if p is any element of the symmetric algebra, then

$$g(Y, \partial(p)) = p(X)g(Y)$$

For example, let $p = Z^2$. Then $\partial(p) = \partial(Z)^2$ and

$$g(Y, \partial(p)) = g(Y, \partial(Z)^2) = \frac{\partial^2}{\partial t \partial s} g(Y + tZ + sZ)|_{t=s=0} = \frac{\partial^2}{\partial t^2} g(Y + tZ)|_{t=0}$$

$$= \frac{\partial^2}{\partial t^2} g(Y)exp(tB(X, Z))|_{t=0} = B(X, Z)^2 g(Y) = p(X)g(Y)$$

Note that we are identifying the element $Z \in V$ with the linear functional $B(., Z)$ and hence Z^p is identified with the linear polynomial $B(., Z)^2$.

210. Consider a second order Volterra system defined by the input output equation

$$y(n) = \sum_{k=0}^{p} h(k, \theta)x(n - k) + \sum_{k,m=0}^{p} g(k, m, \theta)x(n - k)x(n - m) + \epsilon(n)$$

where θ is an unknown parameter to be estimated. We can assume that this equation corresponds to fitting a Volterra system to input output data with $\epsilon(.)$ being the error involved in this model fitting. The problem is to design a gradient algoritym for estimating the parameter θ. Display the relevant equations in the gradient algorithm.

hint

$$E(\theta) = \sum_{n=p}^{N}(y(n) - \sum_{k=0}^{p} h(k,\theta)x(n-k) - \sum_{k,m=0}^{p} g(k,m,\theta)x(n-k)x(n-m))^2$$

and the gradient algorithm reads

$$\theta[n+1] = \theta[n] - \mu.\nabla_\theta E(\theta[n])$$

211. Let $\}$ be a semisimple Lie algebra and $\langle$ a Cartan subalgebra of $\}$. Let $\int$ be the direct sum of all the root spaces of $\}$ relative to this Cartan subalgebra. Then we have the well known direct sum decomposition $\} = \int \oplus \langle$. Let $\mathcal{S}$ be the subalgebra of the universal enveloping algebra $\mathcal{B}$ of $\}$ generated by $\int$. Then as in an earlier problem, we have a bijection

$$\Gamma_X : \mathcal{S} \times S(\langle) \to \mathcal{B}$$

for all $X \in \langle'$ where $\langle'$ is the set of $H \in \langle$ for which $\pi(H) \neq 0$, where π is the product of the positive roots. This bijection satisfies

$$f(x; b : X; p) = f(xX, \partial(q)^x), q = \Gamma_X(b \times p), b \in \mathcal{S}, p \in S(\langle)$$

Let now p be an invariant polynomial on $\}$. Then by the above, we can find elements $\beta_H(p) \in S(\langle)$ and $b_i \in \mathcal{S}\}, p_i \in S(\langle), i = 1, 2, .., n$, such that

$$p = \Gamma_H(1 \times \beta_H(p)) + \sum_{i=1}^{n} \Gamma_H(b_i \times p_i), H \in \langle'$$

and hence

$$f(x : H; \beta_H(p)) + \sum_{i=1}^{n} f(x; b_i : H; \partial(p_i)) = f(xH, \partial(p))$$

Here, $f(x : X) = f(xX)$ and $xX = Ad(x)X$. Let $X \in \langle$ It is easy to see that

$$\int_G f(x : b_i : H, \partial(p_i))dx = 0$$

since $b_i \in \mathcal{S}\}$. Hence,

$$\int_G f(x : H; \beta_H(p))dx = \int_G f(xH, \partial(p))dx$$

For precise details, see HarishChandra's paper, "Differential operators on a Lie algebra".

212. Navier-Stokes equation for the flow of a fluid in the region between two concentric cylinders.

We use cylindrical coordinates. The radius of the inner cylinder is a and that of the outer cylinder b. The common axis of the two cylinders coincides with the z axis. The velocity field has the form

$$\mathbf{v} = v_\rho(\rho)\hat{\rho} + v_\phi(\rho)\hat{\phi}$$

where

$$\rho = \sqrt{x^2+y^2}, \phi = tan^{-1}(y/x)$$

The unit vectors are

$$\hat{\rho} = \hat{x}.cos(\phi) + \hat{y}.sin(\phi), \hat{\phi} = -\hat{x}.sin(\phi) + \hat{y}.cos(\phi)$$

We then compute $\mathbf{v}.\nabla\mathbf{v}$.

$$\mathbf{v}_{,\rho} = v'_\rho(\rho)\hat{\rho} + v'_\phi(\rho)\hat{\phi},$$

$$\mathbf{v}_{,\phi} = v_\rho\hat{\phi} - v_\phi\hat{\rho}$$

We also need $\nabla^2\mathbf{v}$ for the viscous term.

213. Notations for a semisimple Lie algebra $\}$ as usual. $\}_0$ is a real form of $\}$ and $\langle_0$ is a Cartan subalgebra of $\}_0$. $\sqcap$ is a compact real form of $\}$ and $\approx_0 = \sqcap\cap\}_0$ and $\sqrt{}_\rho = i\sqcap\cap\}_0$. θ denotes conjugation with respect to $\sqcap$ and ζ is the conjugation with respect to $\}_0$. $\tilde{\theta} = \theta o\zeta$. We obviously have $\theta(X) = X$ if $X \in \approx_0$ and $\theta(X) = -X$ if $X \in \sqrt{}_\rho$. Let P denote the set of positive roots and Q the set of positive roots α for which α^2 is real on $\langle_0$. We assume that $\theta(\langle_0) = \langle_0$ and then it follows that

$$\langle_0 = \langle_{\approx_0} + \langle_{\sqrt{}_\rho}$$

where

$$\langle_{\approx_0} = \langle_0 \cap \sqcup_0, \langle_{\sqrt{}_\rho} = \langle_0 \cap \sqrt{}_\rho$$

Indeed, let $X \in \langle_0$. Then, since $\langle_0 \subset \}_0$ and $\}_0 = \approx_0 + \sqrt{}_\rho$, it follows that

$$X = X_1 + X_2, X_1 \in \approx_0, X_2 \in \sqrt{}_\rho$$

Applying θ to both sides then gives us

$$\theta(X) = X_1 - X_2$$

and hence

$$X_1 = \frac{1}{2}(X + \theta(X)) \in \langle_0, X_2 = \frac{1}{2}(X - \theta(X)) \in \langle_0$$

since

$$\theta(X) \in \langle_0$$

This proves that

$$X_1 \in \langle_{\approx_0}, X_2 \in \langle_{\sqrt{}_\rho}$$

Suppose α is a root relative to $\langle_0$. Then if $H \in \langle_{\approx_0}$, $\alpha(H)$ is pure imaginary because if $H \in \langle_{\approx_0}$, then $exp(ad(H)) \in K$ where K is the compact group corrresponding to $\approx_0$. Now, K acts on $\}_0$ via the adjoint representation and since K is compact, we can introduce an inner product on $\}_0$ such that $X \to Ad(k)X, k \in K$ is a unitary representation of K in $\}_0$. It follows that for each $k \in K$, $Ad(k)$ has all its eigenvalues on the unit circle and taking $k = exp(ad(H))$, it follows that $ad(H)$ has all pure imaginary eigenvalues. In other words, $\alpha(H)$ is pure imaginary for every $H \in \langle_{\approx_0}$. Suppose $H \in \langle_{\iota 0}$. Then $H \in \langle_0$ and $iH \in \sqcap$. Thus, $exp(i.ad(H)) \in U$ where U is the compact group corresponding to $\sqcap$. Now, once again, by the same argument as above, since U is compact, $exp(iad(H))$ has eigenvalues all falling on the unit circle and hence $ad(H)$ has all real eigenvalues. In other words, $\alpha(H)$ is real. Thus, α is real on $\langle_{\iota 0}$. We have thus concluded that for every $\alpha \in P$, α is real on $i\langle_{\approx_0} + \langle_{\sqrt{\,}\rho}$. Suppose that $\alpha \in Q$. Then α must be zero either on $\langle_{\approx_0}$ or on $\langle_{\sqrt{\,}\rho}$. This can be seen via the following argument. Suppose $X_1 \in \langle_{\approx_0}$ and $X_2 \in \langle_{\sqrt{\,}\rho}$ and $\alpha(X_1), \alpha(X_2) \neq 0$. Then, $X = X_1 + X_2 \in \langle_0$ and $\alpha(X_1)$ is nonzero pure imaginary while $\alpha(X_2)$ is nonzero pure real. It follows that $\alpha(X) = \alpha(X_1) + \alpha(X_2)$ is complex with both real and imaginary parts nonzero. Thus, α^2 is not real on $\langle_0$, a contradiction. Conversely, if α is zero either on $\langle_{\approx_0}$ or on $\langle_{\sqrt{\,}\rho}$, then α^2 is real on $\langle_0$ since in this case, for every $H \in \langle_0$, $\alpha(H)$ is either real or pure imaginary. Let P_+ denote the set of $\alpha \in P$ which do not vanish identically on $\langle_{p0}$. Suppose $\alpha \in P_+$. We have for $H \in \langle$

$$[H, X_\alpha] = \alpha(H)X_\alpha$$

Let $H \in \langle_{p0}$. Then $\theta(H) = -H$ and hence, we get from the above on applying the operator θ to both sides,

$$[-H, \theta(X_\alpha)] = \alpha(H)\theta(X_\alpha)$$

It follows that $\theta(X_\alpha) = C.X_{-\alpha}$ for some constant C.

213. Write down the equation for the velocity field of an irrotational fluid having specified compressibility.

ans: Consider a fluid that is irrotational but has a definite compressibility s. The equations for the velocity field are then

$$\nabla . v = s, \nabla \times v = 0$$

The second equation has solution

$$v = -\nabla\phi$$

where ϕ is a scalar function called the velocity potential. Substituting this into the first equation gives us

$$\nabla^2\phi = -s$$

the solution to which is

$$\phi(x) = \frac{1}{4\pi}\int \frac{s(x')}{|x - x'|} d^3x'$$

Suppose the compressibility field s has the form

$$s(x) = \delta(x - x_0)$$

Then

$$\phi(x) = \frac{s(x_0)}{4\pi|x - x_0|}$$

Suppose more generally that

$$s(x) = \sum_{i=1}^{n} s_i \delta(x - x_i)$$

Then,

$$\phi(x) = \sum_{i=1}^{n} \frac{s_i}{|x - x_i|}$$

Suppose

$$s(x) = s_0(\delta(x - d/2) - \delta(x + d/2))$$

where

$$d \to 0, s_0 \to \infty, s_0 d = p$$

with p finite. Then,

$$\phi(x) = \frac{s_0}{4\pi}(|x - d/2|^{-1} - |x + d/2|^{-1}) \to \frac{p.x}{4\pi |x|^3}$$

$v(x)$ must be calculated by forming the gradient of $\phi(x)$. This problem is analogous to electrostatics. s corresponds to the charge density while v corresponds to the electric field.

214. Give the physical interpretation of $divv$ where v is the velocity field of a fluid.

ans: Consider a closed surface S bounding a volume V of the fluid. A fluid particle on the surface has a velocity v and its normal component is $v.n$. In time δt, this element moves a distance $v.n\delta t$ and an infinitesimal area dS on the surface about this point, moves the same distance in which process, the volume increases by $v.ndS\delta t$. The total increase in the volume bounded by the surface in time δt is therefore $(\int_S v.ndS)\delta t$, which by Gauss' theorem is the same as $\delta t \int divvdV$. Thus, $divv$ gives the volume increase per unit volume per unit time. If $divv \neq 0$ at some point, then depending upon whether $divv$ is positive or negative, the fluid flows outward or inward from that point thereby increasing or decreasing the volume of a fluid element at that point.

215. Suppose a fluid is incompressible but has a specified vorticity. Then explain how you would compute the velocity field.

ans:

$$\nabla.v = 0, \nabla \times v = \omega$$

The first implies

$$v = \nabla \times b$$

for some vector field b. We can add an arbitrary gradient to b without affecting v and this gradient can be so chosen that $divb = 0$. Then we get from the second equation

$$\omega = \nabla \times \nabla \times b = \nabla(\nabla.b) - \nabla^2 b = -\nabla^2 b$$

the solution to which is

$$b(x) = \int \frac{\omega(x')}{4\pi |x - x'|} d^3x'$$

We thus get

$$v(x) = \nabla \times b(x) = \int \frac{\omega(x') \times (x - x')}{4\pi|x - x'|^3} d^3x'$$

This problem is analogous to magnetostatics. b corresponds to the magnetic vector potential while v corresponds to the magnetic field.

216. Suppose $\mathcal{B}$ is an associative algebra and $\pi : \mathcal{B} \to L(V)$ is an irreducible representation in a vector space V. Let

$$\mathcal{M} = Ker\pi = \{x \in \mathcal{B} : \pi(x) = 0\}$$

Then we claim that $\mathcal{M}$ is a maximal two sided ideal of $\mathcal{B}$. For let $\mathcal{M}'$ be a two sided ideal of $\mathcal{B}$ containing $\mathcal{M}$.

217. Consider coordinates $x^i, i = 1, 2, ..., p$ on a p-dimensional differentiable manifold. Consider another set of coordinates $\bar{x}^i$. What is the transformation law of $dx^i \to d\bar{x}^i$. Use this to derive the transformation law of a contravariant vector field.

218. Consider coordinates x^i on a p-dimensional differentiable manifold. What is the transformation law of $\frac{\partial}{\partial x^i}$. Use this formula to derive the transformation law of a covariant vector field.

219. By considering the tensor product between vector fields, derive the transformation law of a tensor field having p contravariant and q covariant indices.

220.Write down the differential equation for the geodesic curve on the surface of a sphere. By integrating this equation, show that the geodesics are great circles.

221.A particle enters a uniform magnetic field in space. Show that its trajectory is a circle if the initial velocity when it enters the field is perpendicular to the magnetic field.

222. Write down the differential equation for a particle in a uniform magnetic field when space-time is curved with a definite metric. Analyze this differential equation using perturbation theory.

223. Write down the Navier-Stokes equation for a charged fluid in an electromagnetic field. Using this equation, deduce the relationship between the vorticity and the magnetic field.

224. A uniform magnetic field exists between two parallel plates of infinite extent. A charged particle enters the region between the two plates. Derive an expression for the point at which the particle hits the second plate if the point on the first plate at which the particle enters is specified.

225.State Faraday's law of electromagnetic induction. By considering a rectangular loop with a sliding metallic wire and a uniform magnetic field prevalent inside the loop area, explain using Faraday's law of induction how current is induced in the loop.

226.What do you understand by motional emf ? Give an example.

227. Let $x[n], y[n], n \in \mathbb{Z}$ be two discrete time signals. Prove that if the DFT of x is X and that of y is Y, ie,

$$X(\omega) = \sum_n x[n]exp(-j\omega n), Y(\omega) = \sum_n y[n]exp(-j\omega n)$$

then the DFT of $x[n]y[n]$ is given by

$$\sum_n x[n]y[n]exp(-j\omega n) = \frac{1}{2\pi}\int_0^{2\pi} X(\omega-u)Y(u)du$$

Prove also the convolution theorem ie the DFT of $x[n]*y[n]$ defined by

$$x[n]*y[n] = \sum_k x[k]y[n-k]$$

is given by $X(\omega).Y(\omega)$.

228. Let G be a locally compact group and let H be a subgroup of G. Let μ be a left invariant measure on G. Then for $f : G/H \to \mathbb{C}$ define

$$I(f) = \int_G f(xH)d\mu(x)$$

Show that if $f^y(U) = f(yU), U \in G/H$, then

$$I(f^y) = I(f), y \in G$$

ie, I defines an invariant measure on G/H.

229. Let G be a semisimple group and H a Cartan subgroup of G. Let $\}$ and $\langle$ denote the corresponding Lie subalgebras. Let P denote the set of positive roots. Thus the set of all roots is $\Delta = P \cup -P$. Consider the map $\Phi(xH, h) = xhx^{-1}$ where $x \in G, h \in H$. This map is well defined since H is Abelian. Let $\}_\alpha$ denote the root space corresponding to the root α. We then have the direct sum decomposition

$$\} = \langle + \sum_{\alpha\in\Delta} \}_\alpha$$

Fix $h \in H$ and compute the differential of the function $x \to xhx^{-1}$ at $x = e$ in the direction X_α where $X_\alpha \in \}_\alpha$. hint: To do so, we note that for $H \in \langle$,

$$Z(t,s) = x.exp(tX_\alpha).exp(sH).exp(-tX_\alpha)x^{-1}$$

$$= x.exp(s.Ad(exp(tX_\alpha))(H))x^{-1}$$

Now,

$$ad(X_\alpha)(H) = -\alpha(H)X_\alpha$$

so that

$$Ad(exp(tX_\alpha))(H) = exp(t.ad(X_\alpha))(H) = H - t\alpha(H)X_\alpha$$

This gives

$$Z(t,s) = x.exp(sH - st\alpha(H)X_\alpha)x^{-1}$$

$$= exp(sH^x - st\alpha(H)X_\alpha^x)$$

where $U^x = Ad(x)U$ for $x \in G$ and $U \in \}$. The Jacobian determinant can be obtained from this general formula.

230. Define the notion of a measurable function. What do you understand by the statement: Single points in a measurable space are measurable.

231. Define the notions Lie algebra, Tensor algebra, Universal enveloping algebra, Symmetric algebra, Symmetrizer mapping.

232. Let $x(t_1, ..., t_k)$ be a k-dimensional signal. Its Fourier transform is defined by

$$X(\omega_1, ..., \omega_k) = \int_{\mathbb{R}^k} x(t_1, ..., t_k) exp(-j \sum_{i=1}^{k} w_i t_i) dt_1 ... dt_k$$

Prove the Parseval identity for this Fourier transform:

$$(2\pi)^{-k} \int_{\mathbb{R}^k} |X(\omega_1, ..., \omega_k)|^2 d\omega_1 ... d\omega_k = \int_{\mathbb{R}^k} |x(t_1, ..., t_k)|^2 dt_1 ... dt_k$$

Assuming that x is square integrable over $\mathbb{R}^k$.

233. Suppose V is the vector space of all $n \times n$ complex matrices. Define for $X, Y \in V$

$$< X, Y >= Tr(X^* Y)$$

Prove that $< .,. >$ is an inner product on V. In particular, establish the identity

$$|Tr(X^* Y)|^2 \leq Tr(X^* X).Tr(Y^* Y)$$

where star denotes complex conjugate.

234. Let $\mathcal{H}$ be a Hilbert space and let $\{e_i : i = 1, 2, ...\}$ be an orthonormal basis for this space. Let A be a linear operator in $\mathcal{H}$ with the property

$$\sum_i < e_i | A^* A | e_i >< \infty$$

Then we say that A is a Hilbert Schmidt operator in $\mathcal{H}$. Prove that the Hilbert Schmidt property does not depend on the orthonormal basis chosen.

Hint: Let $\{f_i : i = 1, 2, ...\}$ be another orthonormal basis. Then

$$\sum_i < f_i | A^* A | f_i >= \sum_{i,j,k} < f_i | e_j >< e_j | A * A | e_k >< e_k | f_i >$$

and

$$\sum_i < e_k | f_i >< f_i | e_j >=< e_k | e_j >= \delta_{kj}$$

235. In the context of the previous problem, show that if A, B are Hilbert-Schmidt operators, then

$$|\sum_i < e_i|A^*B|e_i > |^2 \leq \sum_i < e_i|A^*A|e_i > \sum_i < e_i|B^*B|e_i >$$

Hence deduce that if we define for a Hilbert Schmidt operator A its norm as

$$\| A \| = (\sum_i < e_i|A^*A|e_i >)^{1/2}$$

then $\| . \|$ satisfies the triangle inequality.

236. (Arthi's Ph.D thesis) Method of moments applied to a cavity resonator in the Schwarzchild space-time.

The metric is first expressed in the cartesian system. It has the form

$$ds^2 = g_{tt}dt^2 + g_{xx}dx^2 + g_{yy}dy^2 + g_{zz}dz^2 + 2g_{xy}dxdy + 2g_{yz}dydz + 2g_{xz}dxdz$$

where $g_{tt}, g_{xx}, g_{yy}, g_{zz}, g_{xy}, g_{yz}, g_{xz}$ are all functions of x, y, z and are independent of t. The Maxwell equations in such a metric have the form

$$(F^{\mu\nu}\sqrt{-g})_{,\nu} = 0$$

and the gauge condition reads

$$(A^\mu\sqrt{-g})_{,\mu} = 0$$

This gauge condition can be expressed as

$$A^\mu_{,\mu}\sqrt{-g} + A^\mu(\sqrt{-g})_{,\mu} = 0$$

We have

$$F_{\mu\nu} = A_{\nu,\mu} - A_{\mu,\nu}$$

so that

$$F^{\mu\nu} = g^{\mu\alpha}g^{\nu\beta}F_{\alpha\beta}$$
$$= g^{\mu\alpha}g^{\nu\beta}(A_{\beta,\alpha} - A_{\alpha,\beta})$$
$$F^{\mu\nu}\sqrt{-g} = g^{\mu\alpha}g^{\nu\beta}\sqrt{-g}(A_{\beta,\alpha} - A_{\alpha,\beta})$$

so that

$$(F^{\mu\nu}\sqrt{-g})_{,\nu} = (g^{\mu\alpha}g^{\nu\beta}\sqrt{-g})_{,\nu}(A_{\beta,\alpha} - A_{\alpha,\beta})$$
$$+g^{\mu\alpha}g^{\nu\beta}\sqrt{-g}(A_{\beta,\alpha} - A_{\alpha,\beta})_{,\nu}$$

This can be brought to the form

$$F^{\mu\nu}\sqrt{-g})_{,\nu} = c(\mu,\nu,\alpha,\beta)A_{\nu,\alpha\beta} + d(\mu,\nu,\alpha)A_{\nu,\alpha} + f(\mu,\alpha)A_\alpha = 0$$

We write the full expanded form of this system of equations and replace partial derivative with respect to time by multiplication by $j\omega$. In this way, we end up with a system of linear partial differential equations in the variables x, y, z alone. We then expand the potentials as

$$A_\mu = \sum_{n,m,k=1}^{N} c(n,m,k)sin(n\pi x/a)sin(m\pi y/b)sin(k\pi z/c)$$

where a, b, c are the lengths of the sides of the cuboidal cavity. Then take the inner product with the same functions to end up with a generalized eigenvalue equation for the frequencies of oscillation ω. To be more specific, we write the above equation as

$$c(\mu,\nu,x,x)A_{\nu,xx} + c(\mu,\nu,y,y)A_{\nu,yy} + c(\mu,\nu,z,z)A_{\nu,zz}+$$

$$2c(\mu,\nu,x,y)A_{\nu,xy} + 2c(\mu,\nu,y,z)A_{\nu,yz} + 2c(\mu,\nu,x,z)A_{\nu,xz}$$

$$-2\omega^2 c(\mu,\nu,t,t)A_{\nu,tt} + d(\mu,\nu,x)A_{\nu,x} + d(\mu,\nu,y)A_{\nu,y} + d(\mu,\nu,z)A_{\nu,z}$$

$$+2i\omega c(\mu,\nu,t,x)A_{\nu,x} + 2i\omega c(\mu,\nu,t,y)A_{\nu,y} + 2i\omega c(\mu,\nu,t,z)A_{\nu,z}$$

$$+i\omega.d(\mu,\nu,t)A_\nu + f(\mu,\nu)A_\nu = 0$$

We then plug in the expressions

$$A_{\mu,x} = \sum c(n,m,k)(n\pi/a).cos(n\pi x/a)sin(m\pi y/b)sin(k\pi z/c)$$

$$A_{\mu,y} = \sum c(n,m,k)(m\pi/b)sin(n\pi x/a)cos(m\pi y/b)sin(k\pi z/c)$$

$$A_{\mu,xx} = -\sum c(n,m,k)(n\pi/a)^2 sin(n\pi x/a)sin(m\pi y/b)sin(k\pi z/c)$$

$$A_{\mu,yy} = -\sum c(n,m,k)(m\pi/b)^2 sin(n\pi x/a)sin(m\pi y/b)sin(k\pi z/c)$$

$$A_{\mu,zz} = -\sum c(n,m,k)(k\pi/c)^2 sin(n\pi x/a)sin(m\pi y/b)sin(k\pi z/c)$$

$$A_{\mu,xy} = -\sum c(n,m,k)(n\pi/a)(m\pi/b)cos(n\pi x/a)cos(m\pi y/b)sin(k\pi z/c)$$

$$A_{\mu,yz} = -\sum c(n,m,k)(n\pi/a)(k\pi/c)sin(n\pi x/a)cos(m\pi y/b)cos(k\pi z/c)$$

$$A_{\mu,xz} = -\sum c(n,m,k)(n\pi/a)(k\pi/c)cos(n\pi x/a)sin(m\pi y/b)cos(k\pi z/c)$$

We assume that the range of the summation in all these expressions is $1 \le n, m, k \le N$. Plugging these expressions into the system of partial differential equations gives us an equation of the form

237.Explain the Zeeman effect in the context of splitting of the energy levels of an atom in a magnetic field.

238.Explain the Rayleigh-Ritz method for determining the energy levels of a Hamiltonian based on the variational principle.

239. Define the creation, annihilation and conservation processes in the quantum stochastic calculus of Hudson and Parthasarathy.

240. Let G be a group and a differentiable manifold such that the map $(g, h) \rightarrow gh^{-1}$ is differentiable. Then, we say that G is a Lie group. Clarify this statement with an elementary example.

241. Let V be a finite dimensional vector space and let $\mathcal{L}(V)$ be the family of all linear operators in V. Prove that $\mathcal{L}(V)$ is a vector space of dimension $(dimV)^2$.

241. Let V be a finite dimensional vector space over the field $\mathbb{F}$ and let $M(V, n)$ be the family of all multilinear maps $f : V^n \rightarrow \mathbb{F}$. Show that $M(V, n)$ is a vector space of dimension $(dimV)^n$.

242. Let V be a finite dimensional vector space over the field $\mathbb{F}$ and let $B = \{e_1, ..., e_n\}$ be a basis for V. Define for $f, g \in V^*$, $f \otimes g : V^2 \rightarrow \mathbb{F}$ so that $f \otimes g(v, w) = f(*v)g(w)$ where $v, w \in V$. Then show that a basis for $M(V, 2)$ is given by $\{e^i \otimes e^j : 1 \leq i, j \leq n\}$ where $\{e^1, ..., e^n\}$ is the dual basis of B. More generally show that if for $f_1, ..., f_k \in V^*$, $f_1 \otimes ... \otimes f_k \in M(V, k)$ is defined so that

$$f_1 \otimes ... \otimes f_k(v_1, ..., v_k) = \Pi_{i=1}^k f_i(v_i), v_1, ..., v_k \in V$$

then a basis for $M(V, k)$ is given by $\{e^{i_1} \otimes ... \otimes e^{i_k} : 1 \leq i_1, ..., i_k \leq n\}$.

243. Let V be a finite dimensional vector space over a field $\mathbb{F}$ and let $T \in \mathcal{L}(V)$. Let $B = \{e_1, ..., e_n\}$ be a basis for V. Define $[T]_B = ((t_{ij}))$ where $T(e_i) = \sum_{j=1}^n t_{ji}e_j, i = 1, 2, ..., n$. Show that if also $S \in \mathcal{L}(V)$, then $[TS]_B = [T]_B.[S]_B$. For $x \in V$, define $[x]_B = (x_i)$ where $x = \sum_i x_i e_i$. Then show that $[Tx]_B = [T]_B[x]_B$. Deduce the preceding relationship from this one.

244. Let V be a finite dimensional vector space over a field $\mathbb{F}$ and let $T \in \mathcal{L}(V)$. Suppose B, B' are two bases for V, then show that there exists a non-singular matrix U such that for all $T \in \mathcal{L}(V)$, one has $[T]_{B'} = U.[T]_B.U^{-1}$. What is this relationship between $[T]_{B'}$ and $[T]_B$ called ?

245. Let V be a finite dimensional vector space over a field $\mathbb{F}$. For $T \in \mathcal{L}(V)$ and B a basis for V, define $f_T(t) = det(t.I - [T]_B)$. Show that f_T is a polynomial independent of B and that $f_T(T) = 0$. What is f_T called ?

ans: The characteristic polynomial of T.

246. Let V be a finite dimensional vector space and let V^* be its dual. Let G be a group and let $\pi : G \rightarrow GL(V)$ be a representation of G in V. Here, $GL(V)$ is the group of all non-singular linear transformations on V. For $f \in V^*$ and $g \in G$, define $(\rho(g)f)(v) = f(\pi(g^{-1})v), v \in V$. Then show that ρ is a representation of G in V^*.

247. Explain how using left invariant vector fields, you can construct the left invariant Haar measure on a Lie group G. Your argument should first start with the construction of the Haar measure using left invariant one forms on the Lie group.

248. Suppose G is a Lie group and $\mathcal{M}$ is a differentiable manifold on which G acts differentiably. Using this, explain how you would define an infinitesimal transformation of an element ξ of the Lie algebra $\}$ of G on $\mathcal{M}$.

hint: If $X \in \}$ and $p \in \mathcal{M}$, then for $f : \mathcal{M} \rightarrow \mathbb{C}$, consider $\frac{d}{dt}f(exp(tX).p)|_{t=0}$. Denote this by $X_p f(x)$.

249. Define the Ito integral for adapted processes with respect to the Brownian motion process. Clarify the relationship between the Ito and Stratanovich integrals.

Hint for the second part: Suppose

$$dX = f(X)dt + g(X)dB$$

Then

$$h(X + dX/2)dB = h(X)dB + \frac{1}{2}h'(X)g(X)dt$$

250. Assume that a nonlinear two port network (for example, a transistor) has the voltage current relations given by

$$i_1 = f_1(v_1, v_2), i_2 = f_2(v_1, v_2)$$

Let (V_1, V_2) denote the dc operating point of the network and let $\delta v_1, \delta v_2$ be the infinitesimal variations in the voltages about the operating point and $\delta i_1, \delta i_2$ the corresponding infinitesimal variations in the currents. Determine the infinitesimal Y-parameters of this network about the operating point.

251. Describe the Born-Oppenheimer method for determining approximately computing the energy levels of a system of heavy nuclei and light electrons.

hint: Denote by $\mathbf{R}, \mathbf{P}$ the positions and momenta of the heavy nuclei and by $\mathbf{r}, \mathbf{p}$ those of the light electrons. Then the Hamiltonian of the whole system has the form

$$H = \mathbf{P}^2/2M + \mathbf{p}^2/2m + V_1(\mathbf{r}, \mathbf{R}) + V_2(\mathbf{R})$$

Assume that the wave function has the form

$$\psi(\mathbf{r}, \mathbf{R}) = \phi_{\mathbf{R}}(\mathbf{r})\Phi(\mathbf{R})$$

where $\phi_{\mathbf{R}}(\mathbf{r})$ is the wave function of the electrons for fixed nuclei. Let $E_e(\mathbf{R})$ denote the corresponding electronic eigenvalue. Then,

$$(\mathbf{p}^2/2m + V_1(\mathbf{r}, \mathbf{R}))\phi_{\mathbf{R}}(\mathbf{r}) = E_e(\mathbf{R})\phi_{\mathbf{R}}(\mathbf{r})$$

Use this equation to derive the differential equation satisfied by the nuceli wave function $\Phi(\mathbf{R})$. Use the following facts:

$$\nabla_R(\phi_{\mathbf{R}}(\mathbf{r})\Phi(\mathbf{R})) = (\nabla_R\phi_R(r))\Phi(R) + \phi_R(r)\nabla_R\Phi(R)$$

$$\nabla_R^2(\phi\Phi) = \phi\nabla_R^2\Phi + \Phi\nabla_R^2\phi + 2(\nabla_R\phi, \nabla_R\Phi)$$

Another way to perform the calculation is to calculate the average value of H in the state $\phi_R(r)\Phi(R)$ and then apply the variational principle. Note that

$$< P^2 >= - < \nabla_R^2 >$$

and

$$< \nabla_R^2 >= \int (\phi_R^*(r)\Phi^*(R)\nabla_R^2(\phi_R(r)\Phi(R)))dRdr$$

$$= \int \Phi^*(R)\nabla_R^2\Phi(R)dR + 2\int \phi_R^*(r)\Phi^*(R)(\nabla_R\phi_R(r), \nabla_R\Phi(R))drdR$$

$$+ \int |\Phi(R)|^2\phi_R^*(r)\nabla_R^2\phi_R(r)dRdr$$

Define

$$A(R) = \int \phi_R^*(r)\nabla_R\phi_R(r)dr$$

Then

$$(\nabla_R + A(R))^2\Phi(R) = \nabla_R^2\Phi(R) + A(R)^2\Phi(R) + 2(A(R), \nabla_R\Phi(R)) + (\nabla_R, A(R))\Phi(R)$$

and

$$\int \Phi^*(R)(\nabla_R + A(R))^2\Phi(R)dR = \int \Phi^*(R)\nabla_R^2\Phi(R)dR + \int \Phi(R)^*A(R)^2\Phi(R)dR$$

$$+2\int \Phi^*(R)(A(R), \nabla_R\Phi(R))dR + \int \Phi^*(R)(\nabla_R, A(R))\Phi(R)dR$$

These expressions can be simplified further and hence $< \nabla_R^2 >$ can be expressed in terms of the latter integral.

252. What do you understand by an analytic function of a complex variable in a domain of the complex plane ? Give an example.

ans:The function should be differentiable at each point of the domain.

253. When do you say that a function of a complex variable in a domain is differentiable at a given point ?

ans:When the derivative is independent of the direction along which the point converges, ie, the limit of $(f(z) - f(z_0))/(z - z_0)$ as $z \to z_0$ is independent of the direction along which z approaches z_0.

254. Is the function $z \to |z|$ differentiable at $z = z_0$?

255. Let X be a normed linear space and $x(n) \in X, n = 0, 1, 2, ...$ For $z \in \mathbb{C}$ define $X(z) = \sum_{n=0}^{\infty} x(n)z^n$. Show that $X(z)$ is a well defined analytic function in the region $|z| \leq r$ when $\sum_n \| x(n) \| r^n < \infty$.

256. Suppose $f(z)$ is an analytic function of the complex variable z in a domain D. Write

$$f(x + iy) = u(x, y) + iv(x, y), u(x, y), v(x, y) \in \mathbb{R}$$

Then establish the Cauchy-Riemann equations

$$\frac{\partial u}{\partial x} = \frac{\partial v}{\partial y},$$

$$\frac{\partial u}{\partial y} = -\frac{\partial v}{\partial x}$$

Hint: Differentiate the equation $f(x+iy) = u(x,y) + iv(x,y)$ with respect to x and y and equate real and imaginary parts of the equation that results after eliminating f from these two.

257. Suppose X is a Banach space and $x(n), n = 1, 2, ...$ are elements of X. Then show that $\sum x(n)$ is well defined as the limit of partial sums provided that $\sum \| x(n) \| < \infty$.

258. Let G be a group and let H, K be two subgroups of G such that H is normal in G, ie $gHg^{-1} = H$ for all $g \in G$. Assume further that for each $x \in G$ there exist unique elements $h \in H, k \in K$ such that $x = hk$. Then show that if $x_1 = h_1k_1, x_2 = h_2k_2$, then $x_1x_2 = h_1k_1h_2k_2 = h_1k_1h_2k_1^{-1}k_1k_2 = h_1\tau_{k_1}(h_2)k_1k_2$ where $k \to \tau_k$ is a homomorphism from K into the group of all automorphisms of H. What is the product associated with such a situation called in the Group theory literature ?

ans:Semidirect product.

259. Define a Hausdorff topological space and give an example.

ans: A topological space X is called a Hausdorff topological space if given any two distinct points $x, y \in X$, there exist disjoint open subsets U, V of X such that $x \in U, y \in V$. An example is $\mathbb{R}$ with the usual topology.

260. Let f be an analytic function of a complex variable in a given domain D. Prove using the Cauchy-Riemann equations, the real and imaginary parts of f are harmonic functions.

261. Let $x(n), n = 0, 1, 2, ...$ be a sequence in a Banach space X such that for some $r > 0$, $\sum_n \| x(n) \| r^n < \infty$. Let f be a bounded linear functional on X. For $z \in \mathbb{C}, |z| < r$ show that if $X(z) = \sum_n x(n)z^n$, then $f(X(z)) = \sum_n f(x(n))z^n$ and that this function is analytic for $|z| < r$. Show in particular that

$$|f(X(z))| \leq \| f \| \sum_n \| x(n) \| r^n, |z| \leq r$$

(Ref:See the papers of HarishChandra on representations of semisimple Lie groups).

262.Suppose $F(z) = \sum_{n \geq 0} x(n)z^n$ is analytic for $|z| \leq r$ and suppose $G(\xi) = \sum_{n \geq 0} y(n)\xi^n$ is analytic for $|\xi| < s$. Then what can you say about the analyticity of the function $F(G(\xi))$.

hint: Choose $\delta \in (0, s)$ such that $|G(\xi)| < r$ when $|\xi| < \delta$. This can be achieved by choosing δ small enough so that $\sum_n |y(n)|\delta^n < r$. Then $F(G(\xi))$ will be analytic for $|\xi| < \delta$.

263. Generalize the result of the preceding problem to power series in several complex variables.

hint: Suppose

$$F(z_1, ..., z_n) = \sum_{r_1, ..., r_n \geq 0} x(r_1, ..., r_n) z_1^{r_1} ... z_n^{r_n}$$

is absolutely convergent for $|z_i| < r, i = 1, 2, ..., n$ and

$$G_i(\xi_1, ..., \xi_k) = \sum_{q_1, ..., q_k \geq 0} y(q_1, ..., q_k) \xi_1^{q_1} ... \xi_k^{q_k}$$

is absolutely convergent for $|\xi_i| < s, i = 1, 2, ..., k$. Then choose $\delta < s$ so that $|G_i(\xi_1, ..., \xi_k)| < r$ whenever $|\xi_i| < \delta$. Then $F(G_1(\xi_1, ..., \xi_k), ..., G_n(\xi_1, ..., \xi_k))$ is analytic for $|\xi_i| < \delta, i = 1, 2, ..., k$.

264. Suppose $\}$ is a semisimple Lie algebra and $($ a Cartan subalgebra. Let P denote the set of positive roots so that the root-space decomposition is given by

$$\} = (+ \sum_{\alpha \in P} (\}_\alpha + \}_{-\alpha})$$

Then if $\alpha, \beta \in \mathcal{P}$, show that $[\}_\alpha, \}_\beta] \subset \}_{\alpha+\beta}$. Deduce that $\sum_{\alpha \in P} \}_\alpha$ is a nilpotent Lie algebra and so is $\sum_{\alpha \in P} \}_{-\alpha}$.

265. Let A be a nilpotent matrix. Then show that $exp(A)$ is a polynomial in A.

266. Let A, B be nilpotent matrices. Then is $A + B$ also nilpotent in general ?

267. Let A, B, X be matrices of the same size and define $L_A(X) = AX, R_A(X) = XA, L_B(X) = BX, R_B(X) = XB$. Then prove that $L_A R_B = R_B L_A$ and

$$(L_A - R_A)(X) = AX - XA = Ad(A)(X)$$

so that

$$Ad(A)^m(X) = \sum_{k=0}^{m} (-1)^k \binom{m}{k} A^{m-k} X A^k$$

Deduce that if A is nilpotent, then so is $ad(A)$.

268. Suppose that the inductance in a circuit is current dependent, say $L(i)$. Then show that the voltage across this element can be expressed as

$$v = L'(i) i'(t) i(t) + L(i) i'(t)$$

Suppose that this element is connected in series with a resistor R to a voltage source $v_i(t)$. Then write down the differential equation for the current in the circuit and assuming that $L(i) = L_0 + \epsilon . L_1(i)$, obtain an integral expression for i upto $O(\epsilon)$ terms.

269. In a coupled coil, derive using basic physics the equations for the fluxes $\Phi_i, i = 1, 2$ through the coils in terms of the currents $i_k, k = 1, 2$ through them:

$$\Phi_1 = L_{11} i_1 + L_{12} i_2, \Phi_2 = L_{21} i_1 + L_{22} i_2$$

Justify using the Maxwell equations that $L_{12} = L_{21}$.

270. Suppose $H = p^2/2m + V(q)$ for a quantum mechanical system. Show that $i[H, q] = p/m, i[H, p] = -V'(q)$.

271. Let $\}$ be a semisimple Lie algebra. Then every representation ρ of $\}$ is semisimple, ie, decomposable into a direct sum of irreducibles, true or false ?

272. Let $\}$ be a semisimple Lie algebra and $($ a Cartan subalgebra of $\}$. Let $\}$ be the complexification of $\}_0$ and let $($ be the complexification of $(_0$. Let $\approx_0$ and $\imath_0$ be as usual. Let α be any root. Then $\alpha(H)$ is pure imaginary for all $H \in \eqsim_0$, true or false.

ans:True for consider the adjoint representation of $\approx_0$ in $\}$. Since $\approx_0$ is compact, we can put an inner product on $\}$ so that $ad(X)$ is skew-Hermitian for every $X \in \approx_0$. This in turn implies that $ad(X)$

has pure imaginary eigenvalues. Taking $X = H \in \langle_0$ gives us the result that $ad(H)$ has pure imaginary eigenvalues and hence $\alpha(H)$ is pure imaginary for all $H \in \langle_0$.

273. Let $\}$ be a semisimple Lie algebra and let $\}$ be the complexification of $\}_0$. Let $\sqcap$ be a compact real form of $\}$ and let $\tilde{\theta}$ be the conjugation with respect to $\sqcap$, ie for $X, Y \in \sqcap$, we have $\tilde{(X+iY)} = X - iY$. Let η be the conjugation with respect to $\}_0$. Put $\approx_0 = \sqcap \cap \}_0$ and $\imath_0 = i\sqcap \cap \}_0$. Then let $\approx$ be the complexification of $\approx_0$ and $\imath$ the complexification of $\imath_0$, ie,

$$\approx = \approx_0 + i\approx_0, \imath = \imath_0 + i\imath_0$$

Then $\theta = \theta o \eta$ is a conjugation that satisfies

$$\theta(X+Y) = X - Y, X \in \approx, Y \in \imath$$

true or false ?

ans: Let $X = X_1 + iX_2, Y = Y_1 + iY_2$ where $X_1, X_2 \in \approx_0, Y_1, Y_2 \in \imath_0$. Then

$$X + Y = X_1 + Y_1 + i(X_2 + Y_2)$$

so that

$$\eta(X+Y) = X_1 + Y_1 - i(X_2 + Y_2)$$

and

$$\tilde{\theta} o \eta(X+Y) = \tilde{(X_1 + Y_1 - i(X_2 + Y_2))} = X_1 - Y_1 + iX_2 - iX_2 = X - Y$$

so the statement is true.

274. The solution to the diffusion equation

$$N_{,t}(t,x) = DN_{,xx}(t,x)$$

in one dimension is given by

ans:

$$N(t,x) = \frac{1}{\sqrt{4\pi Dt}} \int_{-\infty}^{\infty} exp(-(x-u)^2/4Dt)N(0,u)du$$

275. If A is any square matrix and $B = SAS^{-1}$ where S is a nonsingular matrix, then for any polynomial f, $f(B)$ equals

ans:

$$f(B) = Sf(A)S^{-1}$$

276. The magnetic vector potential produced by an infinitesimal current element $I.dl$ is given according to the Biot-Savart law by

ans:

$$B = \frac{\mu I}{4\pi r^3} dl \times r$$

277. Wnat happens in the preceding problem if I is time dependent ?

hint:Make use of the retarded potentials.

278. Write down the equation of a helix and calculate the distant magnetic field produced by a sinusoidal current through this helix. Note that the integrals will have to be evaluated numerically.

279. Calculate from first principles the radiation resistance of a $\lambda/2$ length dipole with centre fed sinusoidal current distribution.

hint:

The current distribution has the form $I(z) = I_0 cos(kz)$ with $|z| \leq \lambda/4$. Then the far field magnetic vector potential has the form

$$A_z(r,\theta) = \frac{\mu I_0}{4\pi r} exp(-jkr) \int_{-\lambda/4}^{\lambda/4} cos(kz') exp(jkz' cos(\theta)) dz'$$

The electric and magnetic fields can be obtained by differentiating this expression. Then the Poynting vector is calculated and its surface integral over a large spherical surface surrounding the antenna is calculated and set equal to $I_0^2 R/2$ with R being the radiation resistance.

280.State and prove the reciprocity theorem for electromagnetic fields.

hint: Consider two current sources J_1 and J_2 and let (E_1, H_1) and (E_2, H_2) denote the electric and magnetic fields produced by these two sources respectively. Write down the Maxwell equations and using vector calculus, compute $\nabla.(E_1 \times H_2)$ and $\nabla.(E_2 \times H_1)$ in terms of these two sources. Then apply Gauss' theorem and obtain a formula for the surface integral of $E_1 \times H_2 - E_2 \times H_1$ over a surface of large radius.

281. Explain how you would calculate the mutual impedance between two antennas. Assume that the antennas are dipoles.

hint:Let $I_i(z), i = 1, 2$ be respectively the currents through the two dipoles. Let $E_{21}(z)$ denote the electric field produce in the second antenna due to the current in the first antenna. Then, the voltage drop across the second antenna due to the current in the first is given by

$$V_2 = I_2(0)^{-1} \int_2 E_{21}(z) I_2(z) dz$$

and hence the mutual impedance is given by

$$Z_{21} = V_2/I_1(0) = \frac{1}{I_1(0)I_2(0)} \int_2 E_{21}(z) I_2(z) dz$$

282. Suppose a medium has refractive index $\mu(x, y, z)$. Then calculate the path of a light ray in this medium.

hint: The velocity of light in the medium at the point (x, y, z) is given by $c = c_0/\mu(x, y, z)$. Then the total time taken for a light ray to propagate along a path $x(t), y(t), z(t)$ is given by

$$\int (\mu(x, y, z)/c_0)(dx^2 + dy^2 + dz^2)^{1/2}$$

Now according to Fermat's principle, this time taken is a minimum. Using the Euler-Lagrange variational calculus, we can now obtain a differential equation for the light ray.

283. Prove that if $\}$ is a semisimple Lie algebra and $\langle$ is a Cartan subalgebra then the root space decomposition holds:

$$\} = \langle + \sum_{\alpha \in P} \}_\alpha + \}_{-\alpha}$$

where P is the set of positive roots relative to an order.

284. What do you understand by a partial order on a set ?

285. Let $\}_0$ be a real Lie algebra and $\}$ its complexification. Assume that $\}$ is semisimple. Let $\sqcap$ be a compact real form of $\}$ and define $\approx_0 = \sqcap \cap \}_0$ and $\surd_0 = i\sqcap \cap \}_0$. Show that $[\approx_0, \approx_0], \subset \approx_0$, $[\approx_0, \surd_0] \subset \surd_0$ and $[\surd_0, \surd_0] \subset \approx_0$.

286. Let $\}_0$ be a real Lie algebra and $\}$ its complexification. Let $\sqcap$ be a compact real form of $\}$ and let $\approx_0 = \sqcap \cap \}_0$ and $\surd_0 = i\sqcap \cap \}_0$. Let $\langle_0$ be a Cartan subalgebra of $\}_0$ such that $\langle_0 \subset \approx_0$. Then if $\langle$ is the complexification of $\langle_0$, $\langle$ is a Cartan subalgebra of $\}$. Let α be a root of $\}$ relative to $\langle$ and let X_α be the corresponding root vector. We have

$$[H, X_\alpha] = \alpha(H)X_\alpha, H \in \langle$$

If $H \in \langle_0$, it follows that $\alpha(H)$ is pure imaginary and hence if θ is the conjugation relative to $\approx_0 + i\surd_0$, we have for $H \in \approx_0$, $\theta(H) = H$

$$[H, \theta(X_\alpha)] = -\alpha(H)\theta(X_\alpha)$$

while if $H \in \rangle\langle_0$, we have $\theta(H) = -H$ and since $\alpha(H)$ is real in this case, we once again get

$$[H, \theta(X_\alpha)] = -\alpha(H)\theta(X_\alpha)$$

It follows that for all $H \in \langle$, we have

$$[H, \theta(X_\alpha)] = -\alpha(H)X_\alpha$$

and this implies that

$$\theta(\}_\alpha) = \}_{-\alpha}$$

Note that if θ were the conjugation relative to $\}_0$, the same result would have ensued.

287. State and illustrate Babinet's principle on complementary screens.

ans:Let S_1 denote a screen in space and S_2 a complementary screen so that $S_1 \cup S_2$ is an infinite plane screen extending in all directions. Let sources of electromagnetic fields be located on one side of this plane. Suppose with screen S_1 present but S_2 absent, the electromagnetic field on the other side of the plane is (E_1, H_1) and with the screen S_2 present but S_1 absent, the field is (E_1, H_2). Then the sum of these fields $(E_1 + E_2, H_1 + H_2)$ equals the field if no screen were present.

ref:Kraus, Antennas.

287. Consider a bob of mass m_1 attached to a string of length L_1 whose other end is tied to a nail on the wall. Another bob of mass m_2 is connected to another string of length L_2 whose other end is connected to the bob having mass m_1. This system therefore forms a double pendulum and oscillations take place in a plane. Let θ_1 denote the angle between the first string and the vertical and θ_2 that between the second string and the vertical. Write down the Lagrangian of this system and by applying the Legendre transformation, derive the Hamiltonian.

hint: In terms of cartesian coordinates, the position of the first bob is

$$(x_1, y_1) = (L_1.sin(\theta_1), L_1.cos(\theta_1))$$

and that of the second bob is

$$(x_2, y_2) = (L_1sin(\theta_1) + L_2sin(\theta_2), L_1cos(\theta_1) + L_2cos(\theta_2))$$

The kinetic energy of the system is then given by

$$T = \frac{m_1}{2}(x_1'^2 + y_1'^2) + \frac{m_2}{2}(x_2'^2 + y_2'^2)$$

Express, using the chain rule for differentials, T in terms of $\theta_1, \theta_2, \theta_1', \theta_2'$ and then note that the potential energy of this system is given by

$$V = -m_1gL_1.cos(\theta_1) - m_2g(L_1.cos(\theta_1) + L_2.cos(\theta_2))$$

The Lagrangian of the system is

$$\mathcal{L}(\theta_1, \theta_2, \theta_1', \theta_2') = T(\theta_1, \theta_2, \theta_1', \theta_2') - V(\theta_1, \theta_2)$$

The canonial momenta are

$$p_i = \frac{\partial \mathcal{L}}{\partial \theta_i'}, i = 1, 2$$

These equations can be solved to obtain $\theta_i', i = 1, 2$ in terms of $\theta_i, p_i, i = 1, 2$. We do so and apply the Legendre transformation to get the Hamiltonian as

$$H = p_1\theta_1' + p_2\theta_2' - \mathcal{L}$$

The Hamiltonian equations of motion are then given by

$$\theta_i' = H_{,p_i}, p_i' = -H_{,\theta_i}, i = 1, 2$$

These can be discretized using the truncated Taylor expansion. The general approach is as follows. Consider a dynamical system with differential equations

$$x_i' = f_i(x_1, ..., x_N), i = 1, 2, ..., N$$

Then using a second order truncated Taylor expansion, we have

$$x_i(t+\tau) \approx x_i(t) + \tau.x_i'(t) + \frac{\tau^2}{2}x_i''(t)$$

and

$$x_i' = f_i(x), x_i'' = \sum_j f_{i,j}x_j' = \sum_j f_{i,j}f_j$$

so that the discretized equations are

$$x_i[n+1] = x_i[n] + \Delta.f_i(x[n]) + \frac{\Delta^2}{2}\sum_j f_{i,j}(x[n])f_j(x[n]), i = 1, 2, ..., N$$

288. Introduction to Tarun's thesis: The problems of this thesis are woven around the fabric of stochastic differential equations driven by Brownian motion. In the mathematics literature, there exists an enormous number of papers and books dealing with the Ito stochastic calculus. In summary, this calculus is focussed around the heuristic formula $(dB(t))^2 = dt$ where $B(.)$ is the Brownian motion process. This formula implies that the length of the Brownian trajectory between two time points t_1 and t_2 is infinite, ie $\int_{t_1}^{t_2} |dB(t)| = \infty$. This formula also implies that the Brownian trajectories are nowhere differentiable. In the physical world, these properties are nearly satisfied by many kinds of disturbances. It follows that many kinds of motion that occur in the physical world that are subject to random disturbances from various directions, can be modeled using the Ito stochastic differential equation formalism. A part of this thesis deals with modeling the motion of aircraft and satellites using stochastic differential equations and then by taking measurements on the position of the satellite at discrete time instants, to obtained filtered estimates of the trajectory. In brief, the model for the motion has the form

$$d\mathbf{x}(t) = \mathbf{f}(t, \mathbf{x}(t))dt + \mathbf{b}(t, \mathbf{x}(t))d\mathbf{B}(t)$$

where $\mathbf{x}(t)$, the state vector, has components comprising position, velocity, and angular velocity. The disturbing forces in this motion are contained in the second term $\mathbf{b}(t, \mathbf{x}(t))d\mathbf{B}(t)$ and these forces cause the satellite or aircraft to execute fluctuations about a mean trajectory. The object is to obtain estimates of the state vector $\mathbf{x}(t)$. These estimates are based upon measurements of the trajectory taken at discrete time instants $t_k, k = 1, 2, ...$ Such measurements can be modeled using the general scheme

$$\mathbf{z}_k = \mathbf{h}(t_k, \mathbf{x}(t_k)) + \mathbf{v}_k$$

where $\mathbf{h}$ is a known function and $\mathbf{v}_k$ are independent Gaussian random vectors. Ideally, it is possible to construct a conditional probability density of the state at time t given the observations upto time t and then maximize this conditional probability density to obtain the optimal maximum likelihood estimate of the state at time t based on past observations. However this scheme results in a highly nonlinear set of equations for the estimate and implementation is computationally very expensive. Alternate suboptimal methods exist for recursively computing the estimate and one of the goals of this thesis is precisely applying these suboptimal techniques to the trajectory estimation problem.

289. Discuss the estimation of transistor parameters using Volterra representation combined with wavelets.

(Dr.Sudipta's thesis)

Introduction:The Ebers Moll model provides an accurate characterization of a transistor. The model involves parameters $I_{C0}, I_{E0}, \alpha_C, \alpha_E$ and V_T. In this paper, we assume that all these parameters except V_T are unknown. We consider an amplifier circuit in which the transistor is one of the components and set up a system of complete differential equations, ie, the Kirchchoff current and voltage equations for the currents and nodal voltages in the circuit. The Eber's Moll model is used for the transistor herein. The Ebers-Moll model expresses the collector and emitter currents I_C and I_E as nonlinear functions of V_{BC} and V_{BE}. We express these nonlinear relationships as the sum of a strong linear relation and a weak nonlinear relation. This is achieved by expanding the exponential $exp(V/V_T)$ appearing in the Ebers-Moll model in powers of V/V_T and retaining only the linear component in the strong linear part. We then incorporate a parameter in the weak nonlinear part and expand the currents and voltages in the circuit in powers of the parameter retaining only terms upto linear order in the parameter as is done in standard perturbation theory for quantum mechanical systems. The zeroth order components of the currents and voltages are can be obtained by convolving the input voltage with impulse response functions while the first order components can be calculated by convolving nonlinear functions of the zeroth order components with impulse response functions. In this way, we obtain a second order Volterra representation of the transistor amplifier. The nonlinear functions mentioned here are obtained from the Ebers-Moll model. The transistor parameter V_T, the cutin voltage, appears as an unknown parameter in the first and second order Volterra kernels herein. Using this approximate input-output equation for the transistor, we can design algorithms for estimating the parameter V_T. One way is to discretize the Volterra input output equation giving a discrete second order Volterra relation and then apply the least mean square or the least squares algorithm to the equation. However the discretized version of the Volterra system can be inaccurate leading to serious errors. For example, if there is a spike in the continuous time Volterra kernels at some time point, this spike may get ignored in the discretized system. A more accurate way to estimate the parameters is not to discretize the Volterra system but rather represent the Volterra system in an appropriate orthognormal basis and then truncate the result. Such a scheme, requires three steps, one to represent the continuous time Volterra kernels as linear combinations of the orthonormal basis functions and then plug this into the given Volterra equation, two to take the inner product on both sided of the resulting equation with the orthonormal basis and three to truncate the resulting relationship. This procedure results in a linear relation between the Volterra kernel coefficients in the orthonormal basis and the coefficients of the output signal in the orthonormal basis. One particularly nice choice of the orthonormal basis is the wavelet basis because an appropriate truncation of this basis can be chosen to preserve all the time localization and frequency of variation properties of the Volterra kernels. In other words, if we expand the first and second order Volterra kernels in the wavelet basis and retain only a finite number of terms, we can still obtain a very good approximation to the kernels in terms of time localization and frequency resolution properties unlike what we would get on performing a direct discretization. Having obtained a discrete representation of the Volterra equation using the wavelet basis, we can apply the least squares method to estimate V_T. This objective has been achieved in this paper.

Problem formulation: Consider the single transistor amplifier shown in fig. The KCL and KVL equations for the currents and voltages are as follows. The nonlinear functions f and g in the Ebers-Moll model of the transistor used in this system of equations are decomposed as follows into a strong linear part and a weak nonlinear part. We then express the voltages as the sum of zeroth order component and a first order component where the zeroth order component is obtained as the solution to the linearized system of equations, ie, with the nonlinear component neglected and the first order component is generated by the quadratic component of the nonlinearity. The linear differential equations for the zeroth order component are solved as follows to yield these signals as a convolution of the input voltage with impulse responses. The first order part of the equation yields the following solutions. Combining the zeroth and first order parts, we obtain the following second order Volterra model for our transistor amplifier. It is clear from our expressions of the first and second order Volterra kernels that the transistor parameter V_T enters into these kernels and this suggest that we can develop algorithms for estimating this parameter from input output data. Discretizing the Volterra equation gives us the a discrete time second order Volterra equation relating the input and the output. These are as follows. In order to develop wavelet based algorithms for estimating the parameter V_T, we expand the first and second order Volterra kernels in terms of the wavelet basis and then take the wavelet transform of the output. The result is the following discrete second order nonlinear system. Based on this system, we shall in the next section use the gradient search algorithm for estimating the parameter V_T.

290. Problems in antenna theory and wave propagation:

[1] A rectangular aperture of dimensions $a \times b$ in the xy plane carries an electric field phasor

$$\mathbf{E}(x,y) = E_{x0} sin(\pi y/b)\hat{x} + E_{y0} sin(\pi x/a)\hat{y}$$

Determine the equivalent magnetic current density and the far field electric field pattern.

[2]. What do you understand by the following terms :(a) maximum effective area, (b) directivity, (c) polarization loss, (d) radiation resistance, (e) power density.

[3] A thin wire loop of circular shape and radius R lies in the xy plane. It carries a current I_0 at frequency ω. Assume that this loop is distorted slightly to an elliptic shape with major axis $R+\delta R$ and minor axis $R-\delta R$. Determine the far field vector potential upto order $(\delta R)^2$.

[4] (a) Design a Chebyshev array with spacing $d = \lambda/4$ and five elements such that the mainlobe to sidelobe ratio is 10.

[5] Obtain using separation of variables the general solution to the two dimensional wave equation in plane polar coordinates:

$$\frac{1}{\rho}\frac{\partial}{\partial\rho}(\rho.\frac{\partial\psi}{\partial\rho}) + \frac{1}{\rho^2}\frac{\partial^2\psi}{\partial\rho^2}$$

$$-\frac{1}{c^2}\frac{\partial^2\psi}{\partial t^2} = 0$$

[6] Prove that the extra phase factor introduced in a horn antenna with curvature along the x direction and radius of curvature R is given for $|x| << R$ by $\delta_x = Kx^2/2R$. For the TE_{10} mode use this to derive the far field electric and magnetic field. Assume that the waveguide that feeds the horn has dimensions a, b resepectively along the x and y axes.

[7] Two finite length thin wires of equal length L are placed parallely to each other with directions pointing along the z axis. The spacing between the two wires is d. One of the wires is fed with a current phasor I_0 at its centre. Assuming currents $I_1(z)$ and $I_2(z)$ on the wires, (a) write down the expression for the total electric field in space, (b) write down the conditions that the z components of the electric fields vanish on the two wires at all points except at the centre of the feeder element, (c) Assuming appropriate Fourier series expansions for $I_1(z)$ and $I_2(z)$, derive a complete set of linear equations for the Fourier coefficients of the currents. Note that this is a two element Yagi array.

[8] (a) State and prove the pattern multiplication theorem for an antenna array. (b) A thin dipole pointing along the direction (θ_0, ϕ_0) is located at a height h above the ground plane assumed to be of infinite extent and conducting. By using the method of images, derive an expression for the far field electric field pattern.

References

1.Ralph Abraham and Jerrold Marsden, Foundations of Mechanics.

2.S.Chandrasekhar, The mathematical theory of blackholes.

3.Steven Weinberg, Gravitation and Cosmology.

4.Steven Weinberg, The Quantum Theory of Fields, Vol.I

5.K.R.Parthasarathy, An introduction to probability and measure.

6.K.R.Parthasarathy, An introduction to Quantum Stochastic Calculus.

7.K.R.Parthasarathy, Quantum Computation

8.K.R.Parthasarathy, Coding theorems of classical and quantum information theory.

9.V.S.Varadarajan, Lie groups, Lie algebras and their representations.

10.C.Procesi, Lie groups, an approach through invariants.

11.Revuz and Yor, Continuous martingales and Brownian motion.

12.V.I.Arnold, Mathematical methods of classical mechanics.

13.H.Goldstein, Classical mechanics.

14.L.Schiff, Quantum mechanics.

15.Arno Bohm, Quantum mechanics, foundations and applications.

16.Landau and Lifshitz, The classical theory of fields.

17.J.D.Jackson, Classical electrodynamics.

18.P.Billingsley, Probability and measure.

19.D.W.Stroock and S.R.S.Varadhan, Multidimensional diffusion processes.

20.Erwyn Kreyszig, Introductory Functional analysis.

21.Landau and Lifshitz, Fluid dynamics.

22.G.K.Bachelor, Fluid mechanics.

23.Nielsen and Chuang, Quantum computation and Quantum Information

24.Kerson Huang, Statistical mechanics.

25.K.R.Parthasarathy, Mathematical foundations of quantum mechanics.

26.Feller Vols I and II, An introduction to probability theory and its applications.

27.Proakis, Nikias, Rader and Ling, Advanced digital signal processing.

28.H.L.Van Trees, Detection, Estimation and Modulation Theory.

29.S.J.Orfanidis, Optimum Signal Processing.

30.C.R.Rao, Linear statistical inference and its applications.

31.Sternberg, Group theory and physics.

32.H.Weyl, Space, time and matter

33.H.Weyl, Group theory and quantum mechanics.

34.A.Papoulis, Probability, random variables and stochastic processes.

35.Balanis, Antenna theory.

36.Feynman lectures on physics, vols.II and III

37.Oppenheim and Schaffer, Digital signal processing.

38.Hoffman and Kunze, Linear algebra.

39.P.A.M.Dirac, Principles of quantum mechanics

40.P.A.M.Dirac, General relativity

41.E.Wigner, Group theory and its application to the quantum mechanics of atomic spectra.

42.Karatzas and Shreve, Brownian motion and stochastic calculus.

43.Jazwinsky, Stochastic processes and filtering theory.

44.Sage and Melsa, Estimation theory with applications to communications and control.

45.A good book on optimal control theory.

46.Landau and Lifshitz, Quantum mechanics.

47.Nannapenni Narayana Rao, Engineering electromagnetics

48.Jordan and Balmain, Electromagnetic waves and radiating systems.

49.W.Rudin, Functional analysis.

50.Landau and Lifshitz, Quantum Electrodynamics.

51.Zelobenko, Compact Lie groups and their representations.

52.Barry Simon, Representations of compact groups.

53.Reiner Lenz, Group theory in image processing, Springer lecture notes in computer science.

54.D.F.Lawden, An introduction to tensor calculus, relativity and cosmology.

55. V.A.Fock, Quantum Mechanics

56.V.S.Varadarajan, Harmonic analysis on semisimple Lie groups.

57.Millman and Halkias, Integrated Electronics

58. Ph.D theses of Sharma, Vipin, Tarun and Arathi.

59. Collected papers of HarishChandra edited by V.S.Varadarajan Springer.

Index